The UK Pesticide Guide 1999

Editor – R. Whitehead BA, MSc

CABI *Publishing* (a division of CAB International) is one of the world's foremost publishers of databases, books and journals in agriculture, and applied life sciences. It has a worldwide reputation for producing high quality, value-added information, drawing on its links with the scientific community.

From its Headquarters in Wallingford, UK, CABI *Publishing* runs a worldwide operation, distributing books, journals and electronic products to customers in over 150 countries, and selling its products through a network of international agents. For further information, please contact CABI *Publishing*, CAB International, Wallingford, Oxon OX10 8DE, UK.

The British Crop Protection Council (BCPC), is a self-supporting limited company with charity status which was formed in 1968 to promote the knowledge and understanding of the science and practice of crop protection. The corporate members include government departments and research councils; advisory services; associations concerned with the farming industry, agrochemical manufacturers, agricultural engineering, agricultural contracting and distribution services; universities; scientific societies; organizations concerned with the environment; and some experienced independent members. Further details available from BCPC, 49 Downing Street, Farnham GU9 7PH.

ISBN 0 85199 316 8

Cover design by David Pratt
Printed and bound in the UK at the University Press, Cambridge

CONTENTS

EDITOR'S NOTE

Over the 12 years of its existence, this *Guide* has grown in size and scope. In part this reflects the steady increase in the numbers of active ingredients and products available for British farmers and growers. We try to provide a complete current listing of pesticides and adjuvants on the UK market by inviting manufacturers and suppliers to notify, every year, additions, deletions and amendments to their product entries. Most do so, and each new edition contains several hundred changes from its predecessor. Nevertheless we are conscious that a few products may fail to get included, and readers are asked to notify the Editor of any such omissions.

The *Guide* is now used by people across the whole spectrum of agriculture, horticulture, forestry and amenity use, as well as those with commercial or teaching interests, all of whom have different needs. In response to an increasing number of requests to make the *Guide* available in electronic format, this twelfth printed edition is matched by simultaneous publication of the first edition of *The Electronic UK Pesticide Guide* on CD-ROM. We anticipate that the CD version will be updated, enhanced and published annually, to coincide with each new printed edition.

The two versions have much information in common, but they differ in two important respects. We aim to keep this printed volume to a manageable size, which inevitably limits the amount of information we can include. The CD version is not constrained in this way and it contains much more information about the active ingredients and the products as well as more extensive background information. Secondly, electronic technology allows rapid searching and retrieval of information.

For many people the printed book will continue to be the most convenient and accessible format. For those needing more information, or wanting to carry out structured searches of the data, the CD-ROM version is ideal. A lot of people will find it advantageous to have both.

R. Whitehead
Editor

DISCLAIMER

Every effort has been made to ensure that the information in this book is complete and correct at the time of going to press but the Editor and the publishers do not accept liability for any error or omission in the content, or for any loss, damage or other accident arising from the use of products listed herein. Omission of a product does not necessarily mean that it is not approved and available for use.

The information has been selected from official sources and from suppliers' labels and product manuals of pesticides approved under the Control of Pesticides Regulations 1986 and available for pest, disease and weed control applications in the United Kingdom.

It is essential to follow the instructions on the approved label before handling, storing or using any crop-protection product. Approved 'off-label' uses are made entirely at the risk of the user.

The contents of this publication are based on information received up to October 1998.

CHANGES SINCE 1998 EDITION

Pesticides and adjuvants added or deleted since the 1998 edition are listed below. Every effort has been made to ensure the accuracy of this list, but late notifications to the Editor are not included.

Products Added

Products new to this edition are shown here. In addition, products that were listed in the previous edition whose MAFF Approval number *and* supplier have changed are included. Where only the MAFF number has changed, the product is not listed.

PRODUCT	MAFF NO.	SUPPLIER
Actirob B	ADJ0278	Stefes
Activus	09174	Cyanamid
Admix-P	ADJ0301	Newman
Affinity	08639	DuPont
Afrisect 10	09114	Stefes
Agate	08826	Bayer
Agrys	08382	Novartis
Alfacron 10 WP	08587	Novartis A H
Alfadex	08591	Novartis A H
Ally Express	08640	DuPont
Alpha Terbalin 35 SC	04792	Makhteshim
Alto 100 SL	08350	Novartis
Alto Combi	08465	Novartis
Alto Elite	08467	Novartis
Alto Major	08468	Novartis
Amazon TP	08384	Novartis
Amistar Pro	08871	Zeneca
Angle	08385	Novartis
Aplan 240 EC	08355	Novartis
Aplan SL	08351	Novartis
Apres	08881	Dow
Apron Combi FS	08386	Novartis
Apron Elite	08770	Novartis
Ashlade Solace	08087	Ashlade
Atlas Cropgard	09123	Atlas
Atum WDG	07778	RP Agric.
Aura	08388	Novartis
Axit GR	08892	Fargro
Barclay Banshee XL	09201	Barclay
Barclay Cleanup	09179	Barclay
Barclay Goldpost	08964	Barclay
Barclay Holdup 600	08794	Barclay
Barclay Holdup 640	08795	Barclay
Barclay Rebel II	08897	Barclay
BASF 3C Chlormequat 720	06514	BASF
Beret Gold	08390	Novartis
Betanal Flo	08898	AgrEvo
Bolero	08392	Novartis
Boscor	08682	Novartis
Bravo 500	09059	Zeneca
Bravo 720	09104	Zeneca
Bravocarb	09105	Zeneca
Broadsword	09140	United Phosphorus
Cadence	08796	Novartis
Capture	06881	RP Agric.
Casoron G	09022	Uniroyal
Casoron G	09023	Nomix-Chipman
Casoron G4	09215	Uniroyal
CDA Vanquish	08577	RP Amenity
Cirsium	08729	Mirfield
Clarosan	08396	Novartis
Cleanrun 2	09271	Scotts
Clenecorn II	08880	FCC
Clinic	08579	Nufarm
Clovotox	05354	RP Amenity
Cogito	08397	Novartis
Contest	09024	Cyanamid
Corniche	08947	AgrEvo
Cyclade	08958	BASF
Daconil Turf	09265	Scotts
Darmycel Agarifume Smoke Generator	07904	Sylvan
Decade	08402	Novartis
Devrinol	06653	United Phosphorus
Dimilin Flo	08769	Uniroyal
Divora	08960	Novartis
Dosaflo	08473	Novartis
Dynamec	08701	Novartis BCM
Enforcer	09288	Scotts

PRODUCT	MAFF NO.	SUPPLIER
Erysto	08697	Dow
Evade	08071	Nomix-Chipman
Evict	08731	Bayer
Favour 600 SC	08405	Novartis
Field Marshal	08956	United Phosphorus
Flagon 400 EC	08875	Makhteshim
Folio 575 SC	08406	Novartis
Fongarid	08407	Novartis
Foundation	08475	Novartis
Fubol 58 WP	08534	Novartis
Fubol 75 WP	08409	Novartis
Fubol Gold	08812	Novartis
Garnet	06391	Bayer
Gesagard	08410	Novartis
Gesaprim	08411	Novartis
Gesatop	08412	Novartis
Gladio	08413	Novartis
Glint	08414	Novartis
Goltix 90	08654	Bayer
Goltix Flowable	08986	Bayer
Graphic	08415	Novartis
Grenadier	08136	RP Agric.
Guide	ADJ0208	Newman
Harlequin 500 SC	08416	Novartis
Hawk	08417	Novartis
Hispor 45 WP	08418	Novartis
Hyban-P	09129	Agrichem
Hygrass-P	09130	Agrichem
Hyprone-P	09125	Agrichem
Hysward-P	09052	Agrichem
Ingot	08321	RP Agric.
Intrepid	09266	Scotts
ISK 375	09103	Zeneca
Jester	08681	Novartis
Judge	08163	RP Agric.
Jupital DG	09181	Zeneca
Katamaran	09049	BASF
Keytin	08894	Chiltern
Klerat	06869	Scotts
Klerat Wax Blocks	06827	Scotts
Landgold Amidosulfuron	09021	Landgold
Landgold Clodinafop	09017	Landgold
Landgold Fenpropidin 750	08973	Landgold
Landgold PQF 100	08976	Landgold
Landgold Rimsulfuron	08959	Landgold
Landgold Strobilurin KF	09196	Landgold
Landgold TFS 50	08941	Landgold
Landmark	08889	BASF
Lentagran WP	08478	Novartis
Lexus 50 DF	09026	DuPont
Lexus XPE WSB	08542	DuPont
Lo-Gran 20 WG	08421	Novartis
Lonpar	08686	Dow
Lucifer	08129	RP Agric.
Luxan Chlorotoluron 500 Flowable	09165	Luxan
Luxan Dichlobenil Granules	09250	Luxan
Lyrex	ADJ0177	Batsons
Mallard	08662	Novartis
Mantis	08423	Novartis
Mantle	08424	Novartis
Mantra	08886	BASF
Maraud	09274	Scotts
MCC 25 EC	09115	Chiltern
Midstream	09267	Scotts
Minuet EW	08820	PBI
Mistral	08425	Novartis
Moddus	08801	Novartis
Moot	08479	Novartis
MSS Diuron 500 FL	08171	Mirfield
MSS Limpet	ADJ0283	Mirfield
MSS Linuron 500	08893	Mirfield
MSS Mircarb	08788	Mirfield
MSS Thor	08817	Mirfield
Nemathorin 10G	08915	Zeneca
Neporex 2SG	08589	Novartis A H
Nimrod-T	09268	Scotts
Nortron Flo	08154	AgrEvo
Novagib	08954	Fine
Nufarm Nu-Shot	09139	Nufarm
Nuvan 500 EC	08590	Novartis A H
Octolan	08480	Novartis
Opogard	08427	Novartis
Optimol	09061	PBI
Osprey 58 WP	08428	Novartis
Pasturol Plus	08581	FCC
Patriot EC	08990	AgrEvo
Phase	ADJ0279	Newman
Pilot D	08041	AgrEvo
Platform S	08638	DuPont
Plover	08429	Novartis
Prebane	08432	Novartis
Prima	ADJ0310	Newman
Prophet	08433	Novartis
Quickstep	08557	Novartis
Ranger	ADJ0296	Newman
Ratak Wax Blocks	06829	Scotts
Raxil Secur	08966	Bayer
Recoil	08483	Novartis
Renovator 2	09272	Scotts
Renown	09058	Stefes
Ridomil mbc 60 WP	08437	Novartis
Ridomil Plus	08353	Novartis

PRODUCT	MAFF NO.	SUPPLIER
Ripost Pepite	08485	Novartis
Rival	09220	Monsanto
Sambarin 312.5 SC	08439	Novartis
Sapecron 240 EC	08440	Novartis
Satellite	08969	Cyanamid
Scotts Octave	09275	Scotts
Semeron	08441	Novartis
Sheen	08442	Novartis
Shirlan Programme	08761	Zeneca
Shortcut	09254	Scotts
Sibutol Secur	09131	Bayer
Skirmish	08444	Novartis
SL 567A	08811	Novartis
Slaymor	08592	Novartis A H
Slaymore Bait Bags	08593	Novartis A H
Sovereign	08533	Novartis
Sparkle 45 WP	08450	Novartis
Speedway 2	09273	Scotts
Standon Alpha-C10	08823	Standon
Standon Carfentrazone MM	09159	Standon
Standon Chlorothalonil 500	08597	Standon
Standon CIPC 300 HN	09187	Standon
Standon Cycloxydim	08830	Standon
Standon Dichlobenil 6G	08874	Standon
Standon Diflufenican-IPU	09175	Standon
Standon Epoxifen	08972	Standon
Standon Etridiazole 35	08778	Standon
Standon Fluazinam 500	08670	Standon
Standon Flupyrsulfuron MM	09098	Standon
Standon Imazaquin 5C	08813	Standon
Standon IPU	08671	Standon
Standon Kresoxim FM	08922	Standon
Standon Pencycuron DP	08774	Standon
Standon Propaquizafop	09120	Standon
Standon Quinoxyfen 500	08924	Standon

PRODUCT	MAFF NO.	SUPPLIER
Standon Spiroxamin 500	08916	Standon
Stefes 880	08914	Stefes
Stefes Abet	ADJ0259	Stefes
Stefes Competitor	08753	Stefes
Stefes Eros	ADJ0257	Stefes
Stefes Leyclene	08173	Stefes
Stefes Oberon	ADJ0292	Stefes
Stefes Tandem	08906	Stefes
Stroby WG	08653	BASF
Stronghold	09134	BASF
Stryper	08310	Uniroyal
Supertox 30	09102	RP Amenity
Swipe P	08452	Novartis
Sydex	06412	Vitax
Teal	08453	Novartis
Teldor	08955	Bayer
Tern	08660	Novartis
Thiovit	08493	Novartis
Tigress Ultra	08946	AgrEvo
Tilt	08456	Novartis
Tolkan Turbo	06795	RP Agric.
Tomahawk	09249	Makhteshim
Tomcat 2	08257	Antec
Tomcat 2 Blox	08261	Antec
Topas	08458	Novartis
Topas C 50 WP	08459	Novartis
Topik	08461	Novartis
Touchdown LA	09270	Scotts
Triumph	08740	AgrEvo
Twincarb	08777	Vitax
Unison	08318	PBI
Unix	08764	Novartis
UPL Grassland Herbicide	08934	United Phosphorus
Xanadu	09228	Monsanto
Zapper	08605	RP Amenity
ZP Rodent Pellets	07814	Antec
Zulu	08464	Novartis

Products Deleted

The appearance of a product name in the following list may not necessarily mean that it is no longer available. In cases of doubt refer to the supplier.

PRODUCT	MAFF NO.	SUPPLIER
Accrue	08335	Bayer
Ambush C	06632	Zeneca
Antec Durakil	05147	Antec
Ashlade Solace	06472	Ashlade
Asulox	05235	RP Amenity
Atlas Red	03091	Hortichem
Avadex BW	00173	Monsanto
Bactospeine WP	02913	Koppert
Barclay Canter	07530	Barclay
Barclay Carbosect	05512	Barclay
Barclay Champion	06903	Barclay
Barclay Cluster	08792	Barclay
Barclay Cypersect XL	06509	Barclay
Barclay Dodex	09055	Barclay
Barclay Dryfast XL	ADJ0121	Barclay
Barclay Karaoke	08185	Barclay
Barclay Manzeb 455	07990	Barclay
Barclay Manzeb 80	05944	Barclay
Barclay Manzeb Flow	05872	Barclay
Barclay Mecrop	07578	Barclay
Barclay Rebel	08608	Barclay
Barclay Winner	06986	Barclay
Basagran	00188	BASF
Basamid	00192	BASF
Baytan Flowable	06635	Zeneca
Benocap	07646	Dow
Betanal E	07248	AgrEvo
Betanal Montage	07250	AgrEvo
Betanal Tandem	07254	AgrEvo
BH CMPP/2,4-D	05393	RP Amenity
BH MCPA 75	05395	RP Amenity
Blade	H4839	Chemsearch
Bravocarb	05119	ISK Biosciences
Casoron G	08065	Zeneca
Casoron G-SR	06856	Zeneca Prof.
Centex	00456	Chemsearch
Certan	00465	Steele & Brodie
Cheetah R	07308	AgrEvo
Cirrus	06367	Quadrangle
Clarion	06749	Zeneca
Clenecorn	00542	FCC
Compatibility Agent 2	ADJ3530	Ciba Agric.
Contest	07001	Cyanamid
Curalan CL	07174	BASF
Cyper 10	03242	Quadrangle
Cytro-Lane	00626	Cyanamid
Dynamec	06804	MSD AGVET
Dyrene	06982	Bayer
Enforcer	07079	Zeneca Prof.
Ferrax	06662	Zeneca
Fettel	06399	Zeneca
Flarepath	05502	Tech Nova
Folio	08547	Novartis
FS Thiram 15% Dust	07428	Ford Smith
Gainer	08692	Bayer
Galben M	05904	Isagro
Gastratox 6G Slug Pellets	04066	Truchem
Gramonol Five	06673	Zeneca
Green-Up Weedfree Spot Weedkiller for Lawns	06321	Vitax
GS 800	ADJ0150	Techsol
Halophen RE 49	04636	McMillan
Hero	07821	Zeneca
Holdfast D	06858	Zeneca Prof.
Hyban	01084	Agrichem
Hygrass	01090	Agrichem
Hymec	01091	Agrichem
Hyprone	01093	Agrichem
Hysward	01096	Agrichem
Hy-Vic	06247	Agrichem
I T Propyzamide WP	08271	I T Agro
Jupital	05554	ISK Biosciences
Karathane Liquid	01126	Rohm & Haas
Keetak	06950	BASF
Knave	02534	Hortichem
Landgold Thifensulfuron	08523	Landgold
Landgold TM 75	08376	Landgold
Manex	05731	Chiltern
Match	07186	Cyanamid
Mission	07264	AgrEvo
Monarch	05160	Cyanamid
Morestan	01376	Hortichem
MSS Chlorothalonil	08366	Mirfield
MSS Linuron 50	07862	Mirfield
Myriad	07584	RP Agric.
New Squadron	05981	Quadrangle
Nortron	07266	AgrEvo
Nuvanol N 500 SC	H4951	Ciba Agric.
Optimol	06688	Zeneca

PRODUCT	MAFF NO.	SUPPLIER
Opus Plus	07363	BASF
Parable	06692	Zeneca
Paramos	H4524	Chemsearch
Pasturol	01545	FCC
Pentac Aquaflow	05741	Dow
Pilot	07268	AgrEvo
Quad Mini Slug Pellets	01670	Quadrangle
Raxil S	07484	Zeneca
Ripcord	07014	Cyanamid
Rizolex Nova	07434	AgrEvo
Roundup A	08375	Monsanto
Sage	08334	Bayer
Scythe LC	05877	Cyanamid
Sibutol	07487	Zeneca
Siege II	H5251	Chemsearch
Source	07427	Chiltern
Spectron	07284	AgrEvo
Speedway	06861	Zeneca Prof.
Sportak 45 HF	07287	AgrEvo
Sportak Focus EW	08002	AgrEvo
Sportak Sierra HF	07290	AgrEvo
Stampede	06327	Zeneca
Standon Chlorothalonil 50	05922	Standon
Strate	07430	RP Agric.
Sydex	06412	Vitax
Sypex	04650	BASF
Thiodan 35 EC	07182	Promark
Tigress	07337	AgrEvo
Topas D 275 SC	05855	Ciba Agric.
Touchdown LA	06444	Zeneca Prof.
Tracker	03847	PBI
Tripart Ludorum 700	03999	Tripart
Turbair Permethrin	02246	Fargro
Turbair Resmethrin Extra	02247	Fargro
Turbair Systemic Insecticide	02250	Fargro
Unicrop 6% Mini Slug Pellets	02275	Unicrop
Vitaflo Extra	07048	Uniroyal
Vitax Lawn Sand	04352	Vitax
Vitax Weed 'N' Feed Extra	06506	Vitax
Whip X	07339	AgrEvo
Wildcat	07340	AgrEvo

SECTION 1
INTRODUCTORY INFORMATION

ABOUT THE UK PESTICIDE GUIDE

Purpose

The primary aim of this book is to provide a practical guide to what pesticides, plant growth regulators and adjuvants the farmer or grower can realistically and legally obtain in the UK, and to indicate the purposes for which they may be used. It is designed to help in the identification of products appropriate for a particular problem, and a Crop/Pest guide is included to facilitate this. In addition to uses recommended on product labels, details are provided of those uses which do not appear on labels but which have been granted off-label approval (p. 15). Such uses are not endorsed by the manufacturer, and are undertaken entirely at the risk of the user.

As well as identifying the products available, the book provides guidance on how to use them safely and effectively, but without giving details of doses, volumes, spray schedules or approved tank mixtures. Other sections provide essential background information on a whole range of pesticide-related issues including legislation, codes of practice, poisons and treatment of poisoning, products for use in special situations, and weed and crop growth stage keys.

While we have tried to cover all other important factors, this book does not provide a full statement of product recommendations. **Before using any pesticide product it is essential that the user should read the label carefully and comply strictly with the instructions it contains.**

Scope

Individual profiles are provided of some 350 individual active ingredients, and a further 180 profiles of mixtures containing one or more of these active ingredients. Each profile has a list of available approved products for use in agriculture, horticulture (including amenity horticulture), forestry and areas near water, and the supplier from whom each may be obtained. Within these fields all types of pesticide covered by the Control of Pesticides Regulations 1986 are included. This embraces acaricides, algicides, fungicides, herbicides, insecticides, lumbricides, molluscicides, nematicides and rodenticides, together with plant growth regulators used as straw shorteners, sprout inhibitors and for various horticultural purposes. The total number of pesticide products covered in the *Guide* is about 1400.

In addition, the *Guide* includes information on more than 100 authorized adjuvants which, although not themselves pesticides, may be added to pesticides to improve their effectiveness in use.

The *Guide* does **not** cover home garden, domestic, food storage, public health or animal health uses.

Sources of Information

The information for this edition has been drawn from these authoritative sources:

- approved labels and product manuals received from the suppliers of pesticides up to October 1997
- the MAFF/HSE publication *Pesticides 1998*

- entries in *The Pesticides Register*, published monthly by MAFF and HSE, listing new UK approvals (including off-label approvals), up to and including the issue for September 1998
- MAFF lists of approval expiries

Criteria for Inclusion

To be included in the Guide, a product must meet the following conditions:

- it must have MAFF/HSE Approval under the 1986 Regulations
- information on the approved uses must have been provided by the supplier
- the product must be expected to be on the UK market in 1999

Active ingredients which have been banned in the period since 1987 are listed, but no further details are given.

When a company changes its name, whether by merger, take-over or joint venture, it is obliged to re-register its products in the name of the new company, and new MAFF numbers are assigned. Where stocks of the previously registered products remain legally available, both old and new numbers are included in the *Guide* and will remain until approval for the former lapses or stocks are notified as exhausted.

Products that have been withdrawn from the market and whose approval will finally lapse during 1999 are identified in each profile. After the indicated date, sale, storage or use of the product *bearing that approval number becomes illegal.* Where there is a direct replacement product, it is indicated.

HOW TO USE THE GUIDE

The book consists of five main sections:

- Introductory Information
- Adjuvants
- Crop/Pest Guide
- Pesticide Profiles
- Appendices

Introductory Information

This section summarises legislation covering approval, storage, sale and use of pesticides in the UK. Chemicals subject to the Poisons Laws are listed and there is a summary of first aid measures if pesticide poisoning should be suspected. The section also provides lists of products approved for use in or near water, in forestry and as seed treatments. Products specifically approved for aerial application are tabulated.

Adjuvants

Adjuvants are listed in a table ordered alphabetically by product name. For each product, details are shown of the main supplier, the authorisation number and the mode of action (e.g. wetter, sticker, etc.) as shown on the label. A brief statement of the uses of the adjuvant is given. Protective clothing requirements and label precautions are listed using the codes from Appendix 4.

Crop/Pest Guide

This section (p. 75) enables the user to identify which active ingredients are approved for a particular crop/pest combination. The crops are grouped as shown in the Crop Index (p. 77), but indications from the Crop/Pest Guide must always be followed up by reference to the specific entry in the pesticide profiles section. This is because chemicals indicated as having uses in cereals, for example, may only be approved for use on winter wheat and barley, not for other cereals, and the profile entry must be consulted to find out which particular products are approved for which particular crops. Because of differences in the wording of product labels it may sometimes be necessary to refer to broader categories of organism in addition to specific organisms (e.g. annual grasses and annual weeds in addition to annual meadow grass).

Pesticide Profiles

Each active ingredient has a separate numbered profile entry, as does each mixture of active ingredients. The entries are arranged in alphabetical order using the common names approved by the British Standards Institution, and used in *The Pesticide Manual*. Where an active ingredient is available only in mixtures, this is stated. The ingredients of the mixtures are themselves ordered alphabetically and the entries appear at the correct point for the first named ingredient.

Within each profile entry a table lists the products approved and available on the market, in the following style:

Product name	*Main supplier*	*Active ingredient content*	*Formulation type*	*Registration No.**
3 Dursban 4	**DowElanco**	**480 g/l**	EC	**05735**
4 Lorsban T	**Rigby-Taylor**	**480 g/l**	EC	**05970**
5 Spannit	**PBI**	**480 g/l**	EC	**01992**

*Normally refers to registration with MAFF. In a few cases where products registered with the Health and Safety Executive are used in agriculture, HSE numbers are quoted, e.g. H0462

Many of the product names are registered trade-marks but no special indication of this is given. Individual products are indexed under their **entry number** in the Index of Proprietary Names at the back of the book. The main supplier indicates the marketing outlet through which the product may be purchased, and may not be the approval holder. Full addresses and telephone/fax numbers of suppliers are listed in Appendix 1. Some website and e-mail addresses are also shown. For mixtures, the active ingredient contents are given in the same order as they appear in the profile heading. The formulation types are listed in full in the key to abbreviations and acronyms (Appendix 5).

The **Uses** section lists all approved uses (on-label and off-label) notified by suppliers to the Editor by October 1998, giving both the principal target organisms and the recommended crops or situations. Where there is an important condition of use, or the approval is off-label, this is shown in parentheses. Numbers in square brackets refer to the numbered products in the table above. Thus a typical use approved for Dursban 4 and Lorsban T (products 3 and 4) but not for Spannit (product 5) appears as:

Cutworms in LETTUCE (*outdoor only*) [3, 4]

Below the **Uses** paragraph, **Notes** are listed under the following headings. Unless otherwise stated any reference to dose is made in terms of product rather than active ingredient. Where notes refer to particular products, rather than to the entry generally, this is indicated by numbers in square brackets as described above.

Efficacy — Factors important in making the most effective use of the treatment. Certain factors, such as the need to apply chemicals uniformly, the need to spray at appropriate volume rates and the need to prevent settling out of active ingredient in spray tanks, are assumed to be important for all treatments and are not emphasized in individual profiles.

Crop safety/ Restrictions — Factors important in minimizing the risk of crop damage including statutory conditions relating to the maximum permitted number of applications, or the maximum total dose. Recommended timing of treatment in relation to crop growth stage is included, with reference to standardized growth stage keys where possible (GS numbers – see Appendix 3). Where a statutory 'latest time of application' has been established this is given in the section 'latest application/Harvest interval (HI)'.

Special precautions/ Environmental safety

Notes are included in this section where products are subject to the Poisons Law, where Maximum Exposure Limits apply and where the label warns about organophosphorus and/or anticholinesterase compounds. Where the label specifies a Hazard Class this is noted, together with the associated risk phrases. Any other special operator precautions are specified.

Environmental hazards are also noted here, including potential dangers to livestock, game, wildlife, bees and fish. The need to avoid drift onto neighbouring crops and to wash out equipment thoroughly after use are important with all pesticide treatments, but may receive special mention here if of particular significance.

Protective clothing/ Label precautions

The COSHH regulations require that wherever there is a label recommendation for use of protective clothing it should be preceded by the phrase '*Engineering control of operator exposure must be used where reasonably practicable in addition to the following personal protective equipment:*' and followed by '*However, engineering controls may replace personal protective equipment if a COSHH assessment shows that they provide an equal or higher standard of protection.*'

These phrases are not repeated in the profiles but an indication is given of the items of protective clothing which may be required in using the various products. Without a greater degree of uniformity in the wording of labels, it is not possible to show the items required for different operations e.g. handling concentrate, spraying with hand lance, cleaning equipment etc. However, the items which may be needed for one or other of the operations involved in using a product are given as a series of letters on the first line of this section and refer to the items of protective clothing listed in Appendix 4.

Other label precautions not concerned with choice of protective clothing are given on the second line as a list of numbers referring to the key to standard phrases in Appendix 4. **The letters referring to protective clothing and the numbered precautions are given for information only and should not be used for the purpose of making a COSHH assessment without reference to the current product label**.

Withholding period

Any requirements relating to pesticides used on grazing land with potentially harmful effects on livestock are given here.

Latest application/ Harvest interval (HI)

The 'Latest time of application' given in this section is laid down as a statutory condition of use for individual crops. Where quoted in terms of the period which must elapse between the last application and harvesting for human or animal consumption this 'Harvest Interval' is also given.

Approval

Includes notes on approval for aerial or ULV application and 'Off-label' approvals, giving references to the official approval document numbers (OLA numbers), copies of which can be obtained from ADAS or NFU and must be consulted before the treatment is used.

Where a product has been withdrawn by its manufacturer, and its approval will finally expire in 1999, the expiry date is shown here, together with the MAFF number of its direct replacement, if any.

MRL

If MRLs have been set for any/all of the active ingredients in the profile, they are listed here, or else a cross-reference is given to the "parent" profile where they may be found.

DEFINITIONS

The descriptions used in this *Guide* for the crops or situations in which products are approved for use are those used on the approved product labels. These can sometimes be ambiguous or imprecise, especially where simply limiting use to a certain crop is not relevant, leading to misunderstandings among advisors and users. To clarify matters for all concerned, the Pesticides Safety Directorate (PSD) has commenced a periodic issue of definitions drawn up after wide consultation with the industry. Further guidance can be obtained from PSD.

The definitions already published cover non-edible crop, amenity and forestry situations:

Ornamental plant production: All ornamental plants that are grown for sale or are produced for replanting into their final growing position (e.g. *flowers*, *house plants*, *nursery stock*, *bulbs grown in containers or in the ground*).

Managed amenity turf: Frequently mown, intensively managed turf (e.g. *public parks*, *golf courses*, *sports fields*).

Amenity grassland: Areas of semi-natural or planted grassland with minimal management (e.g. *railway and motorway embankments*, *airfields*).

Amenity vegetation: Areas of semi-natural or planted herbaceous plants, trees and shrubs.

Land not intended to bear vegetation: Soil or man-made surfaces where it is intended that no, or minimal, vegetation will be grown for several years (e.g. *pavements*, *tennis courts*, *industrial areas*, *railway ballast*). It does *not* include the land between rows of crops.

Green cover on land temporarily removed from production: Fields covered by natural regeneration or by a planted green cover crop which will not be consumed by humans or livestock, but which will be growing harvested crops in other years (e.g. *green cover on setaside*).

Forest nursery: Areas where young trees are raised outside for subsequent forest planting.

Forest: Groups of trees being grown in their final positions. Covers all woodland grown for whatever objective, including commercial timber production, amenity and recreation, conservation and landscaping, ancient traditional coppice and farm forestry, and trees from natural regeneration, colonisation or coppicing. Also includes restocking of established woodlands and new planting on both improved and unimproved land.

Farm forestry: Groups of trees established on arable land or improved grassland including those planted for short rotation coppicing.

These phrases will be phased into product labels over a period of time, and others will be published in due course.

PESTICIDE LEGISLATION

Anyone who advertises, sells, supplies, stores or uses a pesticide is affected by legislation, including those who use pesticides in their own homes, gardens and allotments. There are numerous statutory controls but the major legal instruments are outlined below.

The Food and Environment Protection Act 1985 (FEPA) and Control of Pesticides Regulations 1986 (COPR)

FEPA introduced statutory powers to control pesticides with the aims of protecting human beings, creatures and plants, safeguarding the environment, ensuring safe, effective and humane methods of controlling pests and making pesticide information available to the public. Control of pesticides is achieved by COPR, which lay down the Approvals required before any pesticide may be sold, stored, supplied, advertised or used, and allow for the general requirements set out in various Consents, which specify the conditions subject to which approval is given. Consent A relates to advertisement, Consent B to sale, supply and storage, Consent C (i) to use and Consent C (ii) to aerial application of pesticides. The conditions of the Consents may be changed from time to time. Details are given in the MAFF/HSE reference book *Pesticides 1998* (revised annually) and are updated in the monthly publication *The Pesticides Register*.

The controls currently in force include the following:

- Only approved products may be sold, supplied, stored, advertised or used
- Only products specifically approved for the purpose may be applied from the air
- A recognised Storeman's Certificate of Competence is required by anyone who stores for sale or supply pesticides approved for agricultural use
- A recognised Certificate of Competence is required by anyone who gives advice when selling or supplying pesticides approved for agricultural use
- Users of pesticides must comply with the Conditions of Approval relating to use
- A recognised Certificate of Competence is required by all contractors and persons born after 31 December 1964 applying pesticides approved for agricultural use (unless working under direct supervision of a certificate holder). A proposal is under consideration that every user of agricultural pesticides will have to hold a Certificate of Competence regardless of age or supervision.
- Only those adjuvants authorised by MAFF may be used (see Section 2)
- Regarding tank-mixes, 'no person shall combine or mix for use two or more pesticides which are anti-cholinesterase compounds unless the approved label of at least one of the pesticide products states that the mixture may be made; and no person shall combine or mix for use two or more pesticides if all the conditions of the approval relating to this use cannot be complied with'

The 'Authorisation' Directive

European Council Directive 91/414/EEC, known as the 'Authorisation' Directive, is intended to harmonise national arrangements for the authorisation of plant protection products within the European Union. It became effective on 25 July 1993. Under the provisions of the Directive, individual Member States will be responsible for authorisation within their own territory of products containing active substances that appear in a list agreed at Community level. This list, to be known as Annex I, will be created over a period of time by review of existing active ingredients and authorisation of new ones. The process of reviewing existing active ingredients is scheduled to be completed by 2003, but progress to date makes it unlikely that this target will be achieved.

Individual Member States are amending their national arrangements and legislation in order to meet the requirements of Directive 91/414/EEC. In the UK this has been achieved by a series of Plant Protection Products Regulations (PPPR), under which, over a period of time, all agricultural and horticultural pesticides will come to be regulated. Meanwhile existing product approvals are being maintained under COPR, and new ones are granted for products containing active ingredients already on the market by 25 July 1993. Products containing new active substances not on the market at this date may be granted provisional approval under PPPR in advance of Annex I listing of their active ingredients.

The Directive also provides for a system of mutual recognition of products registered in other Member States, subject to a number of constraints.

The Control of Substances Hazardous to Health Regulations 1988 (COSHH)

The COSHH regulations, which came into force on 1 October 1989, were made under the Health and Safety at Work Act 1974 and are also important as a means of regulating the use of pesticides. The regulations cover virtually all substances hazardous to health, including those pesticides classed as Very toxic, Toxic, Harmful, Irritant or Corrosive, other chemicals used in farming or industry and substances with occupational exposure limits. They also cover harmful micro-organisms, dusts and any other material, mixture or compound used at work which can harm people's health.

The original Regulations, together with all subsequent amendments, have been consolidated into a single set of regulations: The Control of Substances Hazardous to Health Regulations 1994 (COSHH 1994).

The basic principle underlying the COSHH regulations is that the risks associated with the use of any substance hazardous to health must be assessed before it is used and the appropriate measures taken to control the risk. The emphasis is changed from that pertaining under the Poisonous Substances in Agriculture Regulations 1984 (now repealed), whereby the principal method of ensuring safety was the use of protective clothing, to the prevention or control of exposure to hazardous substances by a combination of measures. In order of preference the measures should be:

a. substitution with a less hazardous chemical or product
b. technical or engineering controls (e.g. the use of closed handling systems etc.)
c. operational controls (e.g. operators located in cabs fitted with air-filtration systems etc.)
d. use of personal protective equipment (PPE), which includes protective clothing

Consideration must be given as to whether it is necessary to use a pesticide at all in a given situation and, if so, the product posing the least risk to humans, animals and the environment must be selected. Where other measures do not provide adequate control of exposure and the use of PPE is necessary, the items stipulated on the product label must be used as a minimum. It is essential that equipment is properly maintained and the correct procedures adopted.

Where necessary, the exposure of workers must be monitored, health checks carried out and employees must be instructed and trained in precautionary techniques. Adequate records of all operations involving pesticide application must be made and retained for at least 3 years.

Certificates of Competence – the roles of BASIS and NPTC

COPR, COSHH and other legislation places certain obligations on those who handle and use pesticides. Minimum standards are laid down for the transport, storage and use of pesticides and the law requires those who act as storekeepers, sellers and advisors to hold recognised Certificates of Competence (see above).

BASIS is an independent Registration Scheme for the pesticide industry, recognised under COPR. It is responsible for organising training courses and examinations to enable such staff to obtain a Certificate of Competence.

In addition, BASIS undertakes annual assessment of pesticide supply stores, enabling distributors, contractors and seedsmen to meet their obligations under the Code of Practice for Suppliers of Pesticides. Further information can be obtained from BASIS (see Appendix 2).

Certain spray operators also require Certificates of Competence (see above). These are issued by the National Proficiency Tests Council (see Appendix 2).

Maximum Residue Levels

A small number of pesticides are liable to leave residues in foodstuffs, even when used correctly. Where residues can occur, statutory limits, known as Maximum Residue Levels (MRLs), have been established. MRLs provide a check that products have been used as directed; they are not safety limits. However, they do take account of consumer safety because they are set at levels that ensure that normal dietary intake of residues presents no risk to health. Wide safety margins are built in and eating food containing residues above the MRL does not automatically imply a risk to health. Nevertheless, it is an offence to put into circulation any produce where the MRL is exceeded.

The UK has set statutory MRLs since 1988. The European Union intends eventually to introduce MRLs for all pesticide/commodity combinations. These are being introduced initially by a series of priority lists but will subsequently be covered by the review programme under Directive 91/414 (see above).

MRLs apply to imported as well as to home-produced foodstuffs. Details of those set for the first list of 87 active ingredients are contained in *The Pesticides (Maximum Residue Levels in Crops, Food and Feeding Stuffs) Regulations 1994*, effective from 26 July 1994. A further 21 active ingredients (12 of which are included in this Guide) are covered by amendments to these Regulations which came into force in 1995 and 1996. These Statutory Instruments are available from The Stationery Office.

The MRLs applicable to the active ingredients included in this Guide are shown in the relevant pesticide profiles (Section 4). However, it is essential to consult the Regulations for full details and definitions, and for information on chemicals not currently marketed in Britain.

MRLs are shown by crop in the pesticide profiles (Section 4). To make the information more manageable, crops have been grouped according to the structure used in the Regulations and reproduced below. Where the same MRL applies to all crops in a group, only the group name is given. Thus *pome fruits 0.5* indicates that a MRL of 0.5 mg/kg has been set for all crops in the pome fruit sub-group (apples, pears, quinces). *Apples, pears 0.5* indicates that either a MRL has not been set for quinces, or that a different level has been set and appears elsewhere in the list. *Fruits 0.5* indicates the level set for the entire fruits group.

Group	**Sub-group**	**Crops/products**
FRUITS	Citrus fruits	Grapefruit, lemons, limes, mandarins (including clementines & similar hybrids), oranges, pomelos, others
	Tree nuts	Almonds, brazil nuts, cashew nuts, chestnuts, coconuts, hazelnuts, macadamia nuts, pecans, pine nuts, pistachios, walnuts, others
	Pome fruits	Apples, pears, quinces, others
	Stone fruits	Apricots, cherries, peaches, nectarines, plums, others
	Berries & small fruit	Grapes, strawberrries, cane fruits (blackberries, loganberries, raspberries, others), bilberries, cranberries, currants, gooseberries, wild berries, others
	Miscellaneous fruit	Avocados, bananas, dates, figs, kiwi fruit, kumquats, litchis, mangoes, olives, passion fruit, pineapples, pomegranates, others
VEGETABLES	Root & tuber vegetables	Beetroot, carrots, celeriac, horseradish, Jerusalem artichokes, parsnips, parsley root, radishes, salsify, sweet potatoes, swedes, turnips, yams, others
	Bulb vegetables	Garlic, onions, shallots, spring onions, others
	Fruiting vegetables	Tomatoes, peppers, aubergines, cucumbers, gherkins, courgettes, melons, squashes, watermelons, sweet corn, others
	Brassica vegetables	Broccoli, cauliflowers, Brussels sprouts, head cabbages, Chinese cabbage, kale, kohlrabi, others

Group	Sub-group	Crops/products
	Leafy vegetables/herbs	Lettuces (cress, lamb's lettuce, lettuce, scarole), spinach, beet leaves, watercress, witloof, herbs (chervil, chives, parsley, celery leaves), others
	Legume vegetables	beans (with pods), beans (without pods), peas (with pods), peas (without pods), others
	Stem vegetables	Asparagus, cardoons, celery, fennel, globe artichokes, leeks, rhubarb, others
	Fungi	wild and cultivated mushrooms
PULSES		Beans, lentils, peas, others
OILSEEDS		Linseed, peanuts, poppy seed, sesame seed, sunflower seed, rape seed, soya bean, mustard seed, cotton seed, others
POTATOES		Early potatoes, ware potatoes
TEA		
HOPS		
CEREALS		Wheat, rye, barley, oats, triticale, maize, rice, others
ANIMAL PRODUCTS		Meat, fat and preparations of meat, milk, dairy produce, eggs

APPROVAL (ON-LABEL AND OFF-LABEL)

Only officially approved pesticides may be marketed and used in the UK. Approvals are normally granted only in relation to individual products and for specified uses. It is an offence to use non-approved products or to use approved products in a manner that does not comply with the statutory conditions of use except where the crop or situation is the subject of an off-label extension of use.

Statutory Conditions of Use

Statutory conditions have been laid down for the use of individual products and may include:

- field of use (e.g. agriculture, horticulture, etc.)
- crop or situations for which treatment is permitted
- maximum individual dose
- maximum number of treatments or the maximum total dose
- maximum area or quantity which may be treated
- latest time of application or harvest interval
- operator protection or training requirements
- environmental protection requirements
- any other specific restrictions relating to particular pesticides.

All products must now display these statutory conditions of use in a 'statutory box' on the label.

Types of Approval

There are three categories of approval that may be granted, under the Control of Pesticides Regulations, to products containing active ingredients that were already on the market by 25 July 1993:

- *Full* (granted for an unstipulated period)
- *Provisional* (granted for a stipulated period)
- *Experimental Permit* (granted for the purposes of testing and developing new products, formulations or uses). Products with only an Experimental Permit may not be advertised or sold and do not appear in this *Guide*.

Products containing new active substances, and those containing older ingredients once they have been listed in Annex I to Directive 91/414 (see previous section), will be granted approval under the Plant Protection Products Regulations. Again, there are three categories, similar to those listed above:

- *Standard approval* (granted for a period not exceeding ten years). Only applicable to products whose active substances are listed in Annex I.
- *Provisional approval* (granted for a period not exceeding three years, but renewable). Applicable where Annex I listing of the active substance is awaited.
- *Approval for research and development* to enable field experiments or tests to be carried out with active substances not otherwise approved.

The official list of approved products, including all the above categories except those for experimental purposes, is the MAFF/HSE Reference Book 500, *Pesticides 199x*. Details of new approvals, amendments to existing approvals and off-label approvals (see below) are published monthly in *The Pesticides Register*.

Withdrawal of Approval

Product approvals may be reviewed, amended, suspended or revoked at any time. Revocation may occur for various reasons, such as commercial withdrawal or failure by the approval holder to meet data requirements. Where there are no safety concerns, a new approval will normally be issued for up to 2 years to allow the using up of stocks by persons other than the approval holder. This is known as the phased revocation procedure. Where safety considerations make it necessary, however, immediate revocation may occur.

Approval of Commodity Substances

Some chemicals have minor uses as pesticides but are predominantly used for non-pesticidal purposes. Approval is granted to such commodity substances for use only. They may not be sold, supplied, stored or advertised as pesticides unless specific approval has been granted. 17 such substances have been approved for certain specified uses as laid down in Annex D of *Pesticides 1998*. Of these, carbon dioxide, formaldehyde, methyl bromide, sodium chloride, sodium hypochlorite, strychnine hydrochloride and sulfuric acid are approved for agricultural or horticultural use and are included in this Guide.

Off-label Extension of Use

Products may legally be used in a manner not covered by the printed label in several ways:

- in accordance with the "Off-label Arrangements" (see below)
- in accordance with a specific off-label approval (SOLA). SOLAs are uses for which approval has been sought by individuals or organisations other than the manufacturers. The Notices of Approval are published by MAFF and are widely available from ADAS or NFU offices. Users of SOLAs must first obtain a copy of the relevant Notice of Approval and comply strictly with the conditions laid down therein
- in tank mixture with other approved pesticides in accordance with Consent C(i) made under FEPA. Full details of Consent C(i) are given in Annex A of *Pesticides 1998* but there are two essential requirements for tank mixes. Firstly, all the condititons of approval of all the components of a mixture must be complied with. Secondly, no person may mix or combine pesticides which are anticholinesterase compounds unless allowed by the label of at least one of the pesticides in the mixture
- in conjunction with authorised adjuvants
- in reduced spray volume under certain conditions ,
- the use of certain herbicides on specified set-aside areas subject to restrictions which differ between Scotland and the rest of the UK
- by mutual recognition of a use fully approved in another Member State of the European Union and authorised by PSD

Although approved, off-label uses are not endorsed by manufacturers and such treatments are made entirely at the risk of the user.

The Off-label Arrangements

Since 1 January 1990, arrangements have been in place allowing many approved products to be used for additional specific minor uses. These were revised to current standards in December 1994 to become *The Revised Long Term Arrangements for Extension of Use 1995*, and are valid until 31 December 1999. The arrangements were revised in August 1997 to provide additional guidance notes to improve clarity.

The arrangements are set out in full at Annex C of *Pesticides 1998*. The following is a summary of the main points and uses.

Specific restrictions for extension of use

Certain restrictions are necessary to ensure that the extension of use does not increase the risk to the operator, the consumer or the environment:

- extensions of use may be made only from label recommendations. Extrapolations may not be made from specific off-label approvals (SOLAs).
- safety precautions and statutory conditions relating to use as specified on the label must be observed. The application method must be as stated on the product label and in accordance with relevant Codes of Practice and COSHH requirements. Where application of a pesticide under these arrangements is made by hand-held equipment, or broadcast air-assisted sprayers particular restrictions apply (consult Annex C of *Pesticides 1998*).
- pesticides must only be used in the same situation as that on the product label i.e. outdoor or protected. Approval for use on tomatoes, cucumbers, lettuces, chrysanthemums and mushrooms include protected crops unless otherwise stated. Use on other protected crops must be specifically permitted on the label.
- off-label use in or near water, or by aerial application, is not permitted under the arrangements.
- rodenticides are not included, nor is use on land not intended for cropping.
- pesticides classed as harmful, dangerous, extremely dangerous or high risk to bees must not be used off-label on any flowering crop. Where use is recommended on-label on flowering crops of peas, cereals or oilseed rape this relates only to those crops.
- the user of a pesticide under these arrangements must ensure that such use does not result in a breach of any statutory MRL.

The arrangements apply as follows:

Non-edible crops and plants

Subject to the specific restrictions set out above, pesticides approved for use on any growing crop may be used on commercial holdings and forest nurseries on: (i) ornamental crops including hardy nursery stock, plants, bulbs, flowers and seed crops where neither the seed nor any part of the plant is to be consumed by humans or animals; (ii) forest nursery crops prior to final planting out.

In addition, pesticides approved for use on any growing *edible* crop (except seed treatments) may be used on non-ornamental crops grown for seed subject to the same consumption restrictions as above (but not including seed crops of potatoes, cereals, oilseeds, peas and beans).

Pesticides (except seed treatments) approved for use on oilseed rape may be used on hemp grown for fibre.

Farm forestry and rotational cropping

Subject to the specific restrictions set out above, *herbicides* approved for use on cereals may be used in the first five years of establishment in farm forestry on land previously under arable cultivation or improved grassland.

In addition, herbicides approved for use on cereals, oilseed rape, sugar beet, potatoes, peas and beans may be used in the first year of regrowth after cutting in coppices established on land previously under arable cultivation or improved grassland.

Nursery fruit crops and hops

Subject to the specific restrictions set out above, pesticides approved for use on any crop for human or animal consumption may be used on hops in specified circumstances and in commercial holdings on nursery fruit trees, nursery vines prior to final planting out, bushes, canes and non-fruiting strawberries provided any fruit harvested within one year is destroyed. Applications must not be made if fruit is present.

Crops used partly or wholly for consumption by humans or livestock

Subject to the specific restrictions set out above, pesticides may be used on commercial holdings on crops in the first column below if they have been approved for use on the crop(s) opposite them in the second column. These extrapolations do not include uses in store, which are subject to separate arrangements (see below).

MINOR USE	CROPS ON WHICH USE IS APPROVED	ADDITIONAL SPECIAL CONDITIONS
Arable crops		
Poppy (grown for oilseed)	Sunflower	
Sesame	Sunflower	
Mustard	Oilseed rape	
Linseed	Oilseed rape	
Evening primrose	Oilseed rape	
Honesty	Oilseed rape	
Linola	Oilseed rape, linseed	
Flax (oilseed and fibre)	Oilseed rape, linseed	
Borage (grown for oilseed)	Oilseed rape	Seed treatments are not permitted
Grass seed crop	Wheat, barley, oats, rye, triticale	Treated crops must not be grazed or cut for fodder. Seed treatments are not permitted
Grass seed crop	Grass for grazing or fodder	
Rye Triticale	Wheat, barley Wheat, barley	Treatments applied before second node detectable stage only
Durum wheat	Wheat	
Lupins	Combining peas, field beans	

MINOR USE	CROPS ON WHICH USE IS APPROVED	ADDITIONAL SPECIAL CONDITIONS
Fruit crops		
Almond Application	Apple, cherry, plum	For herbicides used on the orchard floor ONLY
Chestnut to the	Apple, cherry, plum	
Hazelnut orchard	Apple, cherry, plum	
Walnut floor ONLY	Apple, cherry, plum	
Quince	Apple, pear	
Crab apple	Apple, pear	
Almond Chestnut Hazelnut Walnut	Products approved for use on two of the following: almond, chestnut, hazelnut, walnut	
Nectarine	Peach	
Apricot	Peach	
Blackcurrant	Redcurrant	
Blackberry	Raspberry	
Rubus species (e.g tayberry, loganberry)	Raspberry	
Dewberry	Raspberry	
Redcurrant	Blackcurrant	
Whitecurrant	Blackcurrant, redcurrant	
Bilberry	Blackcurrant, redcurrant	
Cranberry	Blackcurrant, redcurrant	
Vegetable crops		
Parsley root	Carrot, radish	
Fodder beet	Sugar beet	
Mangel	Sugar beet	
Horseradish	Carrot, radish	
Parsnip	Carrot	
Salsify	Carrot, celeriac	
Swede	Turnip	
Turnip	Swede	
Garlic	Bulb onion	
Shallot	Bulb onion	
Aubergine	Tomato	
Squash	Melon	
Pumpkin	Melon	
Marrow	Melon	
Watermelon	Melon	

MINOR USE	CROPS ON WHICH USE IS APPROVED	ADDITIONAL SPECIAL CONDITIONS
Broccoli	Calabrese	
Calabrese	Broccoli	
Roscoff cauliflower	Cauliflower	
Collards	Kale	
Lamb's lettuce, frise, radicchio	Lettuce	
Beet leaves	Spinach	
Cress	Lettuce	
Scarole	Lettuce	
Leaf herbs and edible flowers*	Lettuce, spinach, parsley, sage, mint, tarragon	
Edible podded peas (e.g. mange-tout, sugar snap)	Edible podded beans	
Runner beans	Dwarf French beans	
Rhubarb	Celery	
Cardoon	Celery	
Edible fungi other than mushroom (e.g. oyster mushroom)	Mushroom	

*This extension of use applies to the following leaf herbs and edible flowers: angelica, balm, basil, bay, borage, burnet (salad), caraway, chamomile, chervil, chives, clary, coriander, dill, fennel, fenugreek, feverfew, hyssop, land cress, lovage, marjoram, marigold, mint, nasturtium, nettle, oregano, parsley, rocket, rosemary, rue, sage, savory, sorrel, tarragon, thyme, verbena (lemon), woodruff.

Application in store on crops used partly or wholly for human or animal consumption

Subject to the specific restrictions set out above, pesticides may be used on commercial holdings on the crops listed in the first column below if they have been approved for use in store on the crops listed opposite them in the second column. *Seed treatments are not covered by this arrangement.*

MINOR USE	CROPS ON WHICH USE IS APPROVED
Rye	Wheat
Barley	Wheat
Oats	Wheat
Buckwheat	Wheat
Millet	Wheat

MINOR USE	CROPS ON WHICH USE IS APPROVED
Sorghum	Wheat
Triticale	Wheat
Dried peas	Dried beans
Dried beans	Dried peas
Mustard	Oilseed rape
Sunflower	Oilseed rape
Linola	Oilseed rape
Flax	Oilseed rape
Honesty	Oilseed rape
Poppy (grown for oilseed)	Oilseed rape
Borage (grown for oilseed)	Oilseed rape
Evening primrose	Oilseed rape
Sesame	Oilseed rape
Linseed	Oilseed rape

GUIDANCE ON THE USE OF PESTICIDES

The information given in the *UK Pesticide Guide* provides some of the answers needed to assess health risks, including the hazard classification and the level of operator protection required. However, the Guide cannot provide all the details needed for a complete hazard assessment, which must be based on the product label itself and, where necessary, the Health and Safety Data Sheet and other official literature.

Detailed guidance on how to comply with the Regulations is available from several sources:

1. *Pesticides: Code of practice for the safe use of pesticides on farms and holdings*, 1998. (MAFF Publication PB3528)

The Code of Practice is available from MAFF Publications and has been substantially revised and updated from the previous (1990) edition. It gives comprehensive guidance on how to comply with COPR and the COSHH regulations under the headings User training and certification, Planning and preparation, Working with pesticides, Disposal of pesticide waste and Keeping records. In particular it details what is involved in making a COSHH Assessment. The principal source of information for making such an assessment is the approved product label. In most cases the label provides all the necessary information but in certain circumstances other sources must be consulted and these are listed in the Code of Practice.

2. Other codes of practice:

- *Code of Practice for the Use of Approved Pesticides in Amenity & Industrial Areas* (ISBN 1 871140 12 9)
- *The Safe Use of Pesticides for Non-Agricultural Purposes – Approved Code of Practice* (ISBN 0 11 885673 1)
- *Code of Practice for Suppliers of Pesticides to Agriculture, Horticulture and Forestry* (MAFF Booklet PB3529)
- *Code of Good Agricultural Practice for the Protection of Soil* (MAFF Booklet PB0617)
- *Code of Good Agricultural Practice for the Protection of Water* (MAFF Booklet PB0587)
- *Code of Good Agricultural Practice for the Protection of Air* (MAFF Booklet PB0618)

3. Other guidance and practical advice:

HSE (from HSE Books – see Appendix 2)

- *A step by step guide to COSHH assessment*, HS(G) 97 (ISBN 0 11 886 379 7)
- *Food and Environment Protection Act 1985/Control of Pesticides Regulations 1986. An open learning course* (ISBN 0 11 8857 43 6)
- *COSHH in agriculture*, AS 28
- *COSHH in forestry*, AS 30
- *Occupational Exposure Limits 1995*, Guidance Note EH 40/95 (ISBN 0 7176 0876 X)
- *Biological monitoring of workers exposed to organophosphorus pesticides*, Guidance Note MS 17 (ISBN 0 11 883951 9)

MAFF (from The Stationery Office – see Appendix 2)

- *Pesticides 1998* (ISBN 0 11 243032 5)
- *Pesticides Register* (monthly) (ISSN 0955 7458)

British Agrochemicals Association (see Appendix 2)

- *The plain man's guide to pesticides and the COSHH assessment*
- *Amenity Handbook: A guide to the selection and use of amenity pesticides*
- *Pesticides and Water Quality*
- *Sprayer Cleaning*
- *Container Cleaning*
- *Agrochemical Disposal*
- *Hand Protection*
- *Good Agrochemical Storage*

British Crop Protection Council (see Appendix 2)

- *The Pesticide Manual* (ISBN 1 901396 11 8)
- *Hand-operated Sprayers Handbook* (ISBN 0 948404 30 2)
- *Boom Sprayers Handbook* (ISBN 0948404 50 7)
- *Fruit Sprayers Handbook* (ISBN 0948404 60 4)
- *Using Pesticides: A Complete Guide to Safe, Effective Spraying* (ISBN 0 98404 94 9)

The Environment Agency (see Appendix 2)

- *The Prevention of Pollution of Controlled Waters by Pesticides* (Leaflet PPG9)
- *The Use of Herbicides in or near Water*

PROTECTION OF HONEY BEES

Honey bees are a source of income for their owners and important to farmers and growers as pollinators of their crops. It is irresponsible and unnecessary to use pesticides in such a way that may endanger them. Pesticides vary in their toxicity to bees, but those that present a special hazard are classed as "harmful", "dangerous" or "extremely dangerous" to bees. Products so classified carry a specific warning in the precautions section of the label. These are indicated in this *Guide* in the protective clothing/label precautions section of the pesticide profile by the numbers E12b, E12c, E12d or E12e, depending on which phrase applies. The phrases are stated in full in Appendix 4.

In July 1996 a new classification was introduced that more accurately reflects the actual risk to honeybees when a product is used. Products assessed by the Pesticides Safety Directorate as posing such a risk are classified as "High risk to bees". Such products carry a new warning in the precautions section of the label, indicated in this Guide by the number E12a.

Where use of a product that is hazardous to bees is contemplated, there are simple guidelines to follow, all of which will be stated on the product label. Most important of these are:

- give local beekeepers as much warning of your intention as possible
- avoid spraying crops in flower or where bees are actively foraging
- keep down flowering weeds

The British Beekeepers' Association (see Appendix 2) can give further advice on protecting honey bees from pesticides.

THE CAMPAIGN AGAINST ILLEGAL POISONING OF WILDLIFE

The Campaign Against Illegal Poisoning of Wildlife, aimed at protecting some of Britain's rarest birds of prey and wildlife whilst also safeguarding domestic animals, was launched in March 1991 by the Ministry of Agriculture, Fisheries and Food and the Department of the Environment, Transport and the Regions. The main objective is to deter those who may be considering using pesticides illegally.

The Campaign is supported by a range of organisations associated with animal welfare, nature preservation, field sports and game keeping including the RSPB, English Nature, the British Field Sports Society and the Game Conservancy Trust.

The three objectives are:

- To advise farmers, gamekeepers and other land managers on legal ways of controlling pests;
- To advise the public on how to report illegal poisoning incidents and to respect the need for legal alternatives;
- To investigate incidents and prosecute offenders.

A freephone number (0800 321 600) is available to make it easier for the public to report incidents and numerous leaflets, posters, postcards and stickers have been created to publicise the existence of the Campaign. A video has also been produced to illustrate the many talks, demonstrations and exhibitions presented by ADAS Consulting Ltd, on behalf of MAFF.

The Campaign arose from the results of the MAFF Wildlife Incident Investigation Scheme for the investigation of possible cases of illegal poisoning. Under this scheme, all reported incidents are considered and thoroughly investigated where appropriate. Enforcement action is taken wherever sufficient evidence of an offence can be obtained and numerous prosecutions have been made since the start of the Campaign.

Further information about the Campaign is available from: Mrs Alison Hall, Pesticides Safety Directorate, Room 317, Mallard House, 3 Peasholme Green, York YO1 7PX.

CHEMICALS SUBJECT TO THE POISONS LAW

Certain products in this book are subject to the provisions of the Poisons Act 1972, the Poisons List Order 1982 and the Poisons Rules 1982 (copies of all these are obtainable from The Stationery Office). These Rules include general and specific provisions for the storage and sale and supply of listed non-medicine poisons.

The chemicals approved for use in the UK and included in this book are specified under Parts I and II of the Poisons List as follows:

Part I Poisons (sale restricted to registered retail pharmacists and to registered non-pharmacy businesses provided sales do not take place on retail premises)

aluminium phosphide
chloropicrin
magnesium phosphide
methyl bromide
sodium cyanide
strychnine

Part II Poisons (sale restricted to registered retail pharmacists and listed sellers registered with a local authority)

aldicarb
alphachloralose
carbofuran (a)
chlorfenvinphos (a,b)
demeton-S-methyl
dichlorvos (a, e)
disulfoton (a)
endosulfan
fentin acetate
fentin hydroxide
fonofos (a)
formaldehyde
methomyl (f)
nicotine (c)
oxamyl (a)
paraquat (d)
phorate (a)
sulfuric acid
zinc phosphide

(a) Granular formulations containing up to 12% w/w of this, or a combination of similarly flagged poisons, are exempt
(b) Treatments on seeds are exempt
(c) Formulations containing not more than 7.5% of nicotine are exempt
(d) Pellets containing not more than 5% paraquat ion are exempt
(e) Preparations in aerosol dispensers containing not more than 1% a.i. are exempt
Materials impregnated with dichlorvos for slow release are exempt
(f) Solid substances containing not more than 1% w/w a.i. are exempt

OCCUPATIONAL EXPOSURE LIMITS

A fundamental requirement of the COSHH Regulations is that exposure of employees to substances hazardous to health should be prevented or adequately controlled. Exposure by inhalation is usually the main hazard, and in order to measure the adequacy of control of exposure by this route various substances have been assigned occupational exposure limits.

There are two types of occupational exposure limits defined under COSHH: Occupational Exposure Standards (OES) and Maximum Exposure Limits (MEL). The key difference is that an OES is set at a level at which there is no indication of risk to health; for a MEL a residual risk may exist and the level takes socio-economic factors into account. In practice, MELs have been most often allocated to carcinogens and to other substances for which no threshold of effect can be identified and for which there is no doubt about the seriousness of the effects of exposure.

OESs and MELs are set on the recommendations of the Advisory Committee on Toxic Substances (ACTS). Full details are published by HSE in *EH 40/95 Occupational Exposure Limits 1995* (ISBN 0 7176 0876 X).

As far as pesticides are concerned, OESs and MELs have been set for relatively few active ingredients. This is because pesticide products usually contain other substances in their formulation, including solvents, which may have their own OES/MEL. In practice inhalation of solvent may be at least, or more, important than that of the active ingredient. These factors are taken into account in the approval process of the pesticide product under the Control of Pesticide Regulations. This indicates one of the reasons why a change of pesticide formulation usually necessitates a new approval assessment under COPR.

POISONING BY PESTICIDES – FIRST AID MEASURES

If pesticides are handled in accordance with the required safety precautions, as given on the container label, poisoning should not occur. It is difficult however, to guard completely against the occasional accidental exposure. Thus, if a person handling, or exposed to, pesticides becomes ill, it is a wise precaution to apply first aid measures appropriate to pesticide poisoning even though the cause of illness may eventually prove to have been quite different. An employer has a legal duty to make adequate first aid provision for employees. Regular pesticide users should consider appointing a trained first aider, even if numbers of employees are not large, since there is a specific hazard.

The first essential in a case of suspected poisoning is for the person involved to stop work, to be moved away from any area of possible contamination and for a doctor to be called at once. If no doctor is available the patient should be taken to hospital as quickly as possible. In either event it is most important that the name of the chemical being used should be recorded and preferably the whole product label or leaflet should be shown to the doctor or hospital concerned.

Some pesticides which are unlikely to cause poisoning in normal use are extremely toxic if swallowed accidentally or deliberately. In such cases get the patient to hospital as quickly as possible, with all the information you have.

General Measures

Measures appropriate in all cases of suspected poisoning include:

- Remove any protective or other contaminated clothing (taking care to avoid personal contamination)
- Wash any contaminated areas carefully with water or with soap and water if available
- In cases of eye contamination, flush with plenty of clean water for at least 15 min
- Lay the patient down, keep at rest and under shelter. Cover with one clean blanket or coat etc. Avoid overheating
- Monitor level of consciousness, breathing and pulse-rate
- If consciousness is lost, place the casualty in the recovery position (on his/her side with head down and tongue forward to prevent inhalation of vomit)
- If breathing ceases or weakens commence mouth to mouth resuscitation. Ensure that the mouth is clear of obstructions such as false teeth, that the breathing passages are clear, and that tight clothing around the neck, chest and waist has been loosened. If a poisonous chemical has been swallowed, it is essential that the first aider is protected by the use of a resuscitation device (several types are available on the market)

Specific measures

In case of poisoning with particular chemical groups, the following measures may be taken before transfer to hospital:

Dinitro compounds* (dinoseb, DNOC, dinocap etc.)
Keep the patient at rest and cool.

*Approvals for use of dinoseb, dinoseb acetate. dinoseb amine, dinoterb, and binapacryl were revoked on 22 Jan 1988. Approval for storage of these materials was withdrawn on 30 June 1988. Approvals for supply of DNOC were revoked in Dec 1988 and for its use in Dec 1989.

Organophosphorus and carbamate insecticides
(aldicarb, azinphos-methyl, benfuracarb, carbofuran, chlorfenvinphos, chlorpyrifos, dichlorvos, heptenophos, malathion, methiocarb, mevinphos, propoxur, quinalphos, thiometon, triazophos etc.)
Keep the patient at rest. The patient may suddenly stop breathing so be ready to give artificial respiration.

Organochlorine compounds (aldrin†, dienochlor, endosulfan, gamma-HCH)
If convulsions occur, do not interfere unless the patient is in danger of injury; if so any restraint must be gentle. When convulsions cease, place the casualty on his or her side with head down, tongue forward (recovery position).

Paraquat, diquat
Irrigate skin and eye splashes copiously with water. If any chemical has been swallowed, take to hospital for tests.

Cyanide (including sodium cyanide)
Send for medical aid. Remove casualty to fresh air, if necessary using breathing apparatus and protective clothing. Remove casualty's contaminated clothing. Gently brush solid particles from the skin, making sure you protect your own skin from contamination. Wash the skin and eyes copiously with water. Transfer casualty to nearest accident and emergency hospital by the quickest possible means together with the first aid cyanide antidote, if held on the premises.

Reporting of Pesticide Poisoning

Any cases of poisoning by pesticides must be reported without delay to an HM Agricultural Inspector of the Health and Safety Executive. In addition any cases of poisoning by substances named in schedule 2 of The Reporting of Injuries, Diseases and Dangerous Occurrences Regulations 1985, must also be reported to HM Agricultural Inspectorate (this includes organophosphorus chemicals, mercury and some fumigants).

Cases of pesticide poisoning should also be reported to the manufacturer concerned.

Additional information
General advice on the safe use of pesticides is given in a range of Health and Safety Executive leaflets available from HSE Books (see Appendix 2)

A useful booklet produced by GCPF (Global Crop Protection Federation) titled 'Guidelines for emergency measures in cases of pesticide poisoning' is available from GCPF, Avenue Louise 143, B-1050 Brussels, Belgium.

The major agrochemical companies are able to provide authoritative medical advice about their own pesticide products. Detailed advice is also available to doctors from The National Poisons Information Service, New Cross Hospital, London SE14 5ER (0171 635 9191) and from regional centres (see Appendix 2).

† Approvals for supply, storage and use of aldrin were revoked in May 1989.

AGRICULTURAL PESTICIDES AND WATER

The Food and Environment Protection Act 1985 (FEPA) places a special obligation on users of pesticides to "safeguard the environment and in particular avoid the pollution of water". Under the Water Resources Act 1991 it is an offence to pollute any controlled waters (watercourses or groundwater), either deliberately or accidentally. Protection of controlled waters from pollution is the responsibility of the Environment Agency (EA).

Users of pesticides therefore have a duty to adopt responsible working practices and, unless they are applying herbicides in or near water, to prevent them getting into water. Guidance on how to achieve this is given in the MAFF Code of Good Agricultural Practice for the Protection of Water.

The duty of care covers not only the way in which a pesticide is sprayed, but also its storage, preparation and disposal of surplus, sprayer washings and the container. Products in this *Guide* that are a major hazard to fish, other aquatic life or aquatic higher plants carry one of several specific label precautions in their profile, depending on the assessed hazard level.

Advice on any water pollution problems is available from EA and the British Agrochemicals Association.

No-spray (Buffer) Zones

Surface waters are particularly vulnerable to contamination. One of the best ways of preventing those pesticides that carry the greatest risk to aquatic wildlife from reaching surface waters is to prohibit their application within a boundary adjacent to the water. Such areas are known as no-spray, or buffer, zones. Certain products in this *Guide* are restricted in this way and have a legally binding label precaution to reduce the potential exposure of aquatic organisms and to enable users to manage the risk.

The width of the protected zone is 2 metres for hand-held or knapsack sprayers, 6 metres for tractor-mounted applications, and a variable distance (but often 18 metres) for broadcast air-assisted applications, such as in orchards.

"Surface water" includes lakes, ponds, reservoirs, streams, rivers and watercourses (natural or artificial). The definition of watercourses is currently under review. However, at present it includes temporarily or seasonally dry ditches, which have the potential to carry water at different times of the year.

Buffer zone restrictions do not necessarily apply to all products containing the same active ingredient. Those in formulations that are not likely to contaminate surface water through spray drift do not pose the same risk to aquatic life and are not subject to the restrictions.

A proposal to replace buffer zones by a system of Local Environmental Risk Assessments for Pesticides (LERAPS) is currently under review. The aim is to reduce the impact of regulation on users by allowing flexibility where local conditions make it safe to do so. In the meantime the statement of a buffer zone restriction on a product label is a legally binding constraint on the use of that product.

USE OF HERBICIDES IN OR NEAR WATER

Products in this Guide approved for use in or near water are listed below. Before use of any product in or near water, the appropriate water regulatory body (Environment Agency/Local Rivers Purification Authority or, in Scotland, the local River Purification Board) must be consulted. Guidance and definitions of the situations covered by approved labels are given in the MAFF publication *Guidelines for the Use of Herbicides on Weeds in or near Watercourses and Lakes*. Always read the label before use.

CHEMICAL	PRODUCT	WEEDS CONTROLLED	SAFETY INTERVAL BEFORE IRRIGATION
asulam	Asulox	docks, bracken	nil
2,4-D	Atlas 2,4-D Dormone MSS 2,4-D Amine	water weeds, dicotyledon weeds on banks	3 wk
dalapon + dichlobenil	Fydulan G	weeds in or near water	–
dichlobenil	Casoron G Casoron G-SR Luxan Dichlobenil Granules	aquatic weeds	2 wk
diquat	Midstream Reglone	aquatic weeds	10 d
fosamine-ammonium	Krenite	woody weeds	nil
glyphosate	Alpha Glyphogan Barclay Gallup Amenity Clinic Glyfos Glyfos Pro-Active Glyper Helosate MSS Glyfield Rival Roundup Amenity Roundup Biactive Roundup Biactive Dry Roundup Pro	Reeds, rushes, sedges, waterlilies, grass weeds on banks	nil

CHEMICAL	PRODUCT	WEEDS CONTROLLED	SAFETY INTERVAL BEFORE IRRIGATION
	Roundup Pro Biactive Spasor Spasor Biactive		
maleic hydrazide	Regulox K	suppression of grass growth on banks	3 wk
terbutryn	Clarosan Clarosan 1 FG	aquatic weeds	7 d

USE OF PESTICIDES IN FORESTRY

The table below lists those products, the labels of which refer to approval for use in forestry.

CHEMICAL	PRODUCT	USE
aluminium phosphide	Amos Talunex Phostek Phostoxin	Mole and rabbit control
aluminium ammonium sulfate	Guardsman B Guardsman M Liquid Curb Crop Spray	Reduction of damage by deer, hares, rabbits, birds
ammonium sulfamate	Amcide Root-Out	Rhododendron and other woody weed control
asulam	Asulox	Bracken control
atrazine	Atlas Atrazine Atrazol	Weed control in conifers
azaconazole + imazalil	Nectec Paste	Canker, silver leaf
carbosulfan	Marshal/suSCon	Weevil control
chlorpyrifos	Barclay Clinch II Cyren Dursban 4 Lorsban T	Weevil and beetle control in transplant lines and cut logs
copper hydroxide	Spin Out	Root control in forest nursery containers
cycloxydim	Standon Cycloxydim	Annual grasses
2,4-D	Dicotox Extra	Perennial and woody weed control
	MSS 2,4-D Ester	Perennial and woody weed control
2,4-D + dicamba + triclopyr	Nufarm Nu-Shot	Perennial and woody weed control
dalapon + dichlobenil	Fydulan G	Grass and perennial weed control
dicamba	Tracker	Bracken and perennial weed control

CHEMICAL	PRODUCT	USE
dichlobenil	Casoron G Casoron G4 Prefix D	Annual and perennial weed control in plantations of certain tree species established for at least 2 years
diflubenzuron	Dimilin Flo	Caterpillar control on forest trees
diquat + paraquat	PDQ	Pre-planting weed control, pre-emergence control in nurseries, directed sprays or ring weeding in plantations. Firebreak maintenance
fluazifop-P-butyl	Citadel Corral Fusilade 250 EW	Perennial grass weed control
fosamine-ammonium	Krenite	Woody weed control in forestry land (not conifer plantations)
glufosinate-ammonium	Challenge Harvest	Control of annual and perennial weeds (directed sprays)
glyphosate	Alpha Glyphogan Barbarian Barclay Gallup Barclay Gallup Amenity Barclay Garryowen Clinic Glyfos Glyfos ProActive Glyper Helosate Hilite MSS Glyfield Portman Glyphosate 360 Portman Glyphosate Rival Roundup Amenity Roundup Pro Roundup Pro Biactive Stefes Glyphosate Stefes Kickdown Stirrup	Control of annual, perennial and woody weeds (directed sprays and wiper application). Chemical thinning

CHEMICAL	PRODUCT	USE
isoxaben	Flexidor 125 Gallery 125	Pre-emergence weed control in forestry transplants, second year undercuts, forest and woodland plantings
paraquat	Barclay Total Dextrone X Gramoxone 100	Pre-planting weed control, pre-emergence control in nurseries, directed sprays or ring-weeding in plantations. Firebreak maintenance
propaquizafop	Falcon Landgold PQF 100 Standon Propaquizafop	Control of annual and perennial grass weeds
propyzamide	Headland Judo Headland Redeem Flo Kerb Flo Kerb Granules Kerb Pro Flo Kerb Pro Granules	Grass weed control in forest trees
simazine	Alpha Simazine 50SC Ashlade Simazine 50FL Gesatop Sipcam Simazine Flowable Unicrop Simazine FL	Annual weed control in nursery beds and transplant lines
sulfonated cod liver oil	Scuttle	Reduction of damage by deer, rabbits
triclopyr	Garlon 4 Timbrel	Woody weed control
ziram	Aaprotect	Bird and animal repellent

PESTICIDES USED AS SEED TREATMENTS (including treatments on seed potatoes)

Information on the target pests for these products can be found in the relevant pesticide profile in Section 4.

CHEMICAL	PRODUCT	FORMULATION	CROP(S)
aluminium ammonium sulphate	Guardsman STP Seed Dressing Powder	DP	Field crops, corms, flower bulbs
2-aminobutane	Hortichem 2-aminobutane	VP	Potatoes
bendiocarb	Seedox SC	FS	Maize, sweetcorn
benomyl	Benlate	WP	Field beans, peas
bitertanol + fuberidazole	Sibutol	FS	Oats, rye, triticale, wheat
bitertanol + fuberidazole + imidacloprid	Sibutol Secur	LS	Oats, wheat
carbendazim + tecnazene	Hickstor 6 + MBC	DP	Potatoes
	New Hickstor 6 + MBC	DS	Potatoes
	Hortag Tecnacarb	DS	Potatoes
	New Arena Plus	DS	Potatoes
	Tripart Arena Plus	DS	Potatoes
carboxin + gamma-HCH + thiram	Vitavax RS	FS	Oilseed rape
carboxin + thiram	Anchor	FS	Wheat, barley, oats, rye, triticale
ethirimol + flutriafol + thiabendazole	Ferrax	FS	Barley
fludioxonil	Beret Gold	FS	Wheat, barley
fonofos	Fonofos Seed Treatment	CS	Barley, wheat
fuberidazole + triadimenol	Baytan Flowable	FS	Cereals

CHEMICAL	PRODUCT	FORMULATION	CROP(S)
gamma-HCH	Gammasan 30	FS	Cereals
	Kotol FS	FS	Cereals
gamma-HCH + fenpropimorph + thiram	Lindex-Plus FS Seed Treatment	FS	Oilseed rape
gamma-HCH + thiram	Hydraguard	LS	Cabbage, Brussels sprouts, kale, swedes, turnips, oilseed rape
guazatine	Panoctine	LS	Wheat, barley, oats
guazatine + imazalil	Panoctine Plus	LS	Barley, oats
hymexazol	Tachigaren 70WP	WP	Sugar beet
imazalil	Fungazil 100 SL	LS	Potatoes
	Stryper	LS	Barley
imazalil + pencycuron	Monceren IM	DS	Potatoes
	Monceren IM Flowable	FS	Potatoes
imazalil + thiabendazole	Extratect Flowable	FS	Potatoes
imidacloprid	Gaucho	WS	Sugar beet
imidacloprid + tebuconazole + triazoxide	Raxil Secur	LS	Barley
iprodione	I.T. Iprodione	WP	Brassicas, oilseed rape, potatoes, linseed, mustard, swedes, turnips, ornamentals
	Rovral Liquid FS	FS	Oilseed rape, linseed, potatoes
maneb + zinc oxide	Mazin	WP	Potatoes
metalaxyl + thiabendazole + thiram	Apron Combi FS	FS	Beans, peas

CHEMICAL	PRODUCT	FORMULATION	CROP(S)
pencycuron	Monceren DS Monceren Flowable Standon Pencycuron DP	DS FS DS	Potatoes Potatoes Potatoes
prochloraz	Prelude 20 LF	LS	Linseed, flax
tebuconazole + triazoxide	Raxil S	FS	Barley
tecnazene	Various products, see tecnazene entry	Various	Potatoes
tecnazene + thiabendazole	Hytec Super Storite SS	DS FS	Potatoes Potatoes
tefluthrin	Force ST	LS	Sugar and fodder beet
thiabendazole	Hykeep Storite Clear Liquid Storite Flowable	DS LS FS	Potatoes Potatoes Potatoes
thiabendazole + thiram	Hy-TL	FS	Peas
thiram	Agrichem Flowable Thiram	FS	Beans, peas, maize, vegetables, flowers
tolclofos-methyl	Rizolex Rizolex Flowable	DS FS	Potatoes Potatoes

AERIAL APPLICATION OF PESTICIDES

Only these products specifically approved for aerial application may be so applied and they may only be applied to specific crops or for specified uses. A list of products approved for application from the air was published as Annex B in *Pesticides 1998*. The list given below is taken mainly from this source with updating from issues of *The Pesticides Register* and product labels.

It is emphasized that the list is for guidance only — reference must be made to the product labels for detailed conditions of use which must be complied with. The list does not include those products which have been granted restricted aerial application approval limiting the area which may be treated.

Detailed rules are imposed on aerial application regarding prior notification of the Nature Conservancy, water authorities, bee keepers, Environmental Health Officers, neighbours, hospitals, schools etc. and the conditions under which application may be made. The full conditions are available from MAFF and must be consulted before any aerial application is made.

CHEMICAL	PRODUCT	CROPS/USES
asulam	Asulox	Grassland, forestry, non-crop land (bracken control)
carbendazim	Ashlade Carbendazim FL	Winter wheat, winter and spring barley, oilseed rape
	Carbate Flowable	Wheat, barley, field beans, oilseed rape
	Hinge	Winter wheat, barley
	Tripart Defensor FL	Winter wheat, winter and spring barley, oilseed rape
carbendazim + maneb	Ashlade Mancarb FL	Wheat, barley
	Headland Dual	Wheat, barley
	Tripart Legion	Winter wheat, winter barley
carbendazim + maneb + tridemorph	Cosmic FL	Winter wheat, winter and spring barley
carbendazim + propiconazole	Hispor 45 WP Sparkle 45 WP	Winter wheat, winter barley
chlormequat	Ashlade 460 CCC	Wheat, winter barley, oats, rye

CHEMICAL	PRODUCT	CROPS/USES
	Ashlade 700 CCC	Wheat, winter barley, oats, rye
	Atlas Chlormequat 46	Wheat, oats
	Atlas Chlormequat 700	Wheat, oats
	Atlas Terbine	Wheat, oats
	Atlas Tricol	Wheat, oats
	Barclay Holdup	Wheat
	BASF 3C Chlormequat 720	Wheat, oats, winter barley
	Hyquat 70	Wheat, oats, winter barley
	Mandops Chlormequat 700	Wheat, oats, winter barley
	MSS Mirquat	Wheat, oats
	Quadrangle Chlormequat 700	Wheat, oats
	Sigma PCT	Wheat, oats
	Stefes CCC 700	Wheat, oats, winter barley
	Tripart Brevis	Wheat, oats
	Uplift	Wheat, oats, winter barley
chlormequat + choline chloride	Atlas 5C Chlormequat	Wheat, oats
	Atlas 460:46	Wheat, oats
	Atlas Quintacel	Wheat, oats
	Barclay Take 5	Wheat, oats
	MSS Mircell	Wheat, oats
	New 5C Cycocel	Wheat, oats, triticale, barley, rye
2-chloroethyl phosphonic acid	Barclay Coolmore	Winter barley
	Cerone	Winter barley

CHEMICAL	PRODUCT	CROPS/USES
chlorothalonil	Barclay Corrib 500	Wheat, barley, potatoes, field beans
	Bombardier FL	Potatoes
	Bravo 500	Winter wheat, potatoes, field beans, peas
	ISK 375	Potatoes
	Mainstay	Winter wheat, potatoes, field beans, peas
	Sipcam UK Rover 500	Wheat, potatoes
	Tripart Faber	Wheat, potatoes
chlorotoluron	Tripart Ludorum	Winter wheat, winter barley, durum wheat, triticale
chlorpropham + linuron	Profalon	Daffodils, narcissi, tulips
copper oxychloride	Cuprokylt	Potatoes
cymoxanil + mancozeb	Ashlade Solace	Potatoes
	Standon Cymoxanil Extra	Potatoes
	Systol M	Potatoes
cymoxanil + mancozeb + oxadixyl	Ripost Pepite	Potatoes
	Trustan WDG	Potatoes
diflubenzuron	Dimilin Flo	Forest trees
dimethoate	Barclay Dimethosect	Cereals, peas, ware potatoes, sugarbeet
	BASF Dimethoate 40	Cereals, peas, ware potatoes, sugar beet
	Danadim Dimethoate 40	Cereals, peas, potatoes, beet crops
	Rogor L40	Cereals, peas, potatoes, beet

CHEMICAL	PRODUCT	CROPS/USES
disulfoton	Disyston P 10	Brassicas, beans, sugar beet, carrots
fenitrothion	Dicofen	Cereals, peas
	Unicrop Fenitrothion 50	Cereals, peas
fentin hydroxide	Ashlade Flotin	Potatoes
iprodione	Rovral Flo	Oilseed rape
mancozeb	Agrichem Mancozeb 80	Potatoes
	Ashlade Mancozeb FL	Potatoes, wheat, barley, oats, rye, triticale
	Dithane 945	Potatoes
	Dithane Dry Flowable	Potatoes
	Headland Zebra Flo	Potatoes
	Headland Zebra WP	Potatoes
	Luxan Mancozeb Flowable	Potatoes
	Manzate 200	Potatoes
	Manzate 200 PI	Potatoes
	Opie WP	Potatoes
	Penncozeb WDG	Potatoes
	Stefes Mancozeb DF	Potatoes
	Unicrop Mancozeb 80	Potatoes
mancozeb + metalaxyl	Fubol 75WP	Potatoes
	Osprey 58WP	Potatoes
mancozeb + oxadixyl	Recoil	Potatoes
maneb	Agrichem Maneb 80	Potatoes
	Ashlade Maneb Flowable	Potatoes

CHEMICAL	PRODUCT	CROPS/USES
	Maneb 80	Potatoes
	Unicrop Maneb 80	Barley, potatoes
maneb + zinc oxide	Mazin	Potatoes
metaldehyde	Doff Agricultural Slug Killer with Animal Repellent	All crops
	Doff Horticultural Slug Killer Blue Mini Pellets	All crops
	Doff Metaldehyde Slug Killer Mini Pellets	All crops
	Metarex RG	All crops
	Mifaslug	All crops
	Optimol	All crops
	PBI Slug Pellets	All crops
	Quad Mini Slug Pellets	All crops
	Superflor 6% Metaldehyde Slug Killer	All crops
	Tripart Mini Slug Pellets	All crops
	Unicrop 6% Mini Slug Pellets	All crops
methabenzthiazuron	Tribunil	Barley, winter wheat, autumn sown spring wheat, winter oats, winter rye, triticale, perennial grass, ryegrass
methiocarb	Decoy	All crops
	Draza	All crops
	Draza ST	All crops
	Exit	All crops

CHEMICAL	PRODUCT	CROPS/USES
monolinuron	Arresin	Potatoes, dwarf beans, leeks
phosalone	Zolone Liquid	Brassica seed crops
pirimicarb	Aphox	Cereals, peas, potatoes, sugar beet, beans, leaf brassicas, maize, swedes, oilseed rape, turnips, carrots
	Barclay Pirimisect	Wheat, barley, triticale, oats, rye
	Phantom	Cereals, peas, potatoes, sugar beet, beans, leaf brassicas, maize, sweetcorn, swedes, oilseed rape, turnips, carrots
	Portman Pirimicarb	Wheat, barley, oats
propiconazole	Mantis 250EC	Wheat, barley, rye, oats
	Radar	Wheat, barley, rye, oats, triticale
	Standon Propiconazole	Wheat, barley, rye, oats
	Tilt	Wheat, barley, rye, oats, triticale
sulfur	Headland Sulphur	Wheat, barley, oilseed rape
	Kumulus DF	Wheat, barley, oilseed rape
terbutryn	Prebane SC	Cereals, rye, triticale
triadimefon	Bayleton	Sugar beet, cereals, brassicas (including turnips and swedes)
	Standon Triadimefon	Wheat, barley, rye, oats
tri-allate	Avadex BW Granular	Wheat, barley, winter beans, peas
	Avadex Excel 15G	Winter wheat, barley, winter beans, peas
tridemorph	Calixin	Barley, oats, winter wheat, swedes, turnips

CHEMICAL	PRODUCT	CROPS/USES
	Standon Tridemorph 750	Barley, oats, winter wheat, swedes, turnips
zineb	Unicrop Zineb	Potatoes
zineb-ethylene thiuram disulphide adduct	Polyram DF	Potatoes

SECTION 2
ADJUVANTS

ADJUVANTS

Adjuvants are not themselves classed as pesticides and there is considerable misunderstanding over the extent to which they are legally controlled under the Food and Environment Protection Act. An adjuvant is a substance other than water which enhances the effectiveness of a pesticide with which it is mixed. Consent C(i)5 under the Control of Pesticides Regulations allows that an adjuvant can be used with a pesticide only if that adjuvant is authorised and on a list published from time to time in *The Pesticides Register.* An authorised adjuvant has an *adjuvant number* and may have specific requirements about the circumstances in which it may be used.

Adjuvant product labels must be consulted for full details of authorised use, but the table below provides a summary of the label information to indicate the area of use of the adjuvant. Label precautions refer to the keys given in Appendix 4, A and B, and may include warnings about products harmful or dangerous to fish. The table includes all adjuvants notified by suppliers as available in 1999.

PRODUCT	SUPPLIER	ADJ. NO.	TYPE
Actipron	Bayer	A0013	adjuvant
Contains:	97% highly refined mineral oil		
Use with:	Asulox, Basagran, Benlate, Betanal E, Betanal E + Goltix WG, Checkmate, Checkmate + Goltix WG, Dow Shield + Goltix WG, Goltix WG, Laser, Pilot, Roundup		
Precautions:	U08, U19, U20a, E15, E30a, E31a		
Actirob B	Stefes	A0278	vegetable oil
Contains:	842 g/l esterified rapeseed oil		
Use with:	Cheetah Super and Wildcat on wheat; Sceptre on sugar beet. For details of use with all approved pesticides on specified crops consult label or supplier		
Activator 90	Newman	A0062	non-ionic surfactant/spreader/wetter
Contains:	750 g/l alkylphenyl hydroxypolyoxyethylene		
Use with:	Any spray for which additional wetter is approved and recommended		
Protective clothing:	A, C		
Precautions:	M03, R04a, R04b, U02, U04a, U05a, U10, U11, U19, U20b, C03, E01, E13c, E30a, E31a, E34		

PRODUCT	SUPPLIER	ADJ. NO.	TYPE

Admix-P | Newman | A0301 | wetter

Contains: 80% w/w polyalkylene oxide modified heptamethyltrisiloxane and a maximum of 20 % w/w allyloxypolyethylene glycol methyl ether

Use with: A wide range of pesticides applied as corm, tuber, onion and other bulb treatments in seed production and in seed potato treatment

Precautions: M03, R03a, R03c, R04a, R04b, R04e, U02, U05a, U08, U19, U20b, C03, E01, E13c, E30a, E31a, E34

Agral | Zeneca | A0154 | non-ionic surfactant/spreader/wetter

Contains: 948 g/l alky phenol ethylene oxide

Use with: Any spray for which additional wetter is approved and recommended

Protective clothing: A, C

Precautions: R04a, R04b, R07d, U05a, U08, U20c, C03, E01, E15, E26, E30a, E31a, E34

Agropen | Intracrop | A0143 | vegetable oil

Contains: 95% refined rapeseed oil

Use with: Pesticides which have a recommendation for the addition of a wetter/spreader

Protective clothing: A, C

Precautions: U02, U05a, U08, U19, U20b, C03, E01, E13c, E30a, E31a, E34

Anoint | Intracrop | A0236 | spreader/vegetable oil/wetter

Contains: 95% rapeseed oil

Use with: Pesticides that have a recommendation for the addition of a wetter/spreader

Protective clothing: A, C

Precautions: C03, E01, E13c, E30a, E31a, E34, R04a, R04b, U05a, U08, U19, U20b

---

Aquator | Zeneca | A0265 | surfactant

Contains: A blend of surfactant and ammonium sulphate

FOR FULL CONDITIONS OF USE ALWAYS READ THE PRODUCT LABEL

PRODUCT	SUPPLIER	ADJ. NO.	TYPE
Use with:	Approved formulations of glyphosate-trimesium		
Protective clothing:	C		
Precautions:	C03, E01, E13c, E30a, E31c, U05a, U08, U19, U20b		
Arma	Interagro	A0306	penetrant
Contains:	500 g/l alkoxylated fatty amine + 500 g/l polyoxyethylene monolaurate		
Use with:	Cereal growth regulators, cereal herbicides, cereal fungicides, oilseed rape fungicides and a wide range of other pesticides on specified crops		
Precautions:	U05a, C03, E01, E13c, E30a, E34		
AS Elite	Monsanto	A0230	extender
Contains:	40% w/w ammonium sulphate		
Use with:	Roundup GT, Roundup Elite		
Precautions:	U20a, E30a, E31b		
Ashlade Adjuvant Oil	Ashlade	A0135	mineral oil
Contains:	99% highly refined mineral oil		
Use with:	Goltiw WG, phenmedipham		
Precautions:	R04a, R04b, R04c, U05a, U08, U19, U20a, C03, E01, E13c, E30a, E31b		
Atlas Adherbe	Atlas	A0023	mineral oil
Contains:	83% highly refined mineral oil		
Use with:	Atlas Atrazinc, Atlas Brown, Checkmate, Goltix WG, Goltix WG + Protrum K, Grasp, Laser, Pilot, Protrum K		
Precautions:	R04a, R04b, R04c, U05a, U08, U19, U20a, C03, E01, E13c, E26, E30a, E31b		
Atlas Adjuvant Oil	Atlas	A0021	mineral oil
Contains:	95% highly refined mineral oil		
Use with:	Goltix WG, Goltix WG + Protrum K, phenmedipham, Protrum K		
Precautions:	R04a, R04b, R04c, U05a, U08, U19, U20a, C03, E01, E13c, E26, E30a, E31b		

PRODUCT	SUPPLIER	ADJ. NO.	TYPE
Axiom	Batsons	A0136	mineral oil
Contains:	96.8% w/w paraffinic base oil		
Use with:	Asulox, Basagran, Benlate, Betanal E, Checkmate, Gesaprim 500 SC, Goltix WG, Goltix WG/Betanal E, Goltix WG/Checkmate, Goltix WG/Dow Shield, Goltix WG/Fusilade 5, Laser, Pilot and Roundup		
Precautions:	U09a, U19, U20a, E15, E30a, E31a		
Bandrift Plus	Allied Colloids		drift retardant
Contains:	A non-ionic polyamide dispersed in oil		
Use with:	A wide range of pesticides. See label for restrictions on use with wettable powders, suspension concentrates and water dispersible granules, and other usage limitations		
Protective clothing:	A, C, M		
Precautions:	U05a, U08, U20a, C03, E01, E13c, E30a, E34		
Banka	Interagro	A0245	spreader/sticker/wetter
Contains:	29.2% w/w alkyl pyrrolidones		
Use with:	Potato fungicides and a wide range of other pesticides on specified crops		
Protective clothing:	A, C		
Precautions:	R04b, R04d, U02, U05a, U11, U19, U20a, C03, E01, E13c, E30a, E34		
Barclay Actol	Barclay	A0126	mineral oil
Contains:	99% highly refined paraffinic oil		
Use with:	Any approved pesticide for which the addition of a spraying oil is recommended		
Precautions:	U19, U20b, E13c, E30a, E31b		
Barclay Clinger	Barclay	A0198	sticker/wetter
Contains:	96% w/v poly-1-p-menthene		

FOR FULL CONDITIONS OF USE ALWAYS READ THE PRODUCT LABEL

PRODUCT	SUPPLIER	ADJ. NO.	TYPE

Use with: Barclay Gallup, Barclay Gallup Amenity and a wide range of approved fungicides and insecticides. See label for details

Precautions: U19, U20b, E13c, E30a, E31a

Bazuka Euroagkem A0241 spreader/wetter

Contains: 800 g/l polyalkyleneoxide modified heptamethyltrisiloxane

Use with: All fungicides applied to spring and winter sown wheat, spring and winter sown barley, spring and winter sown oats, triticale, rye and durum wheat where the pesticide label recommends the addition of wetters and spreaders

Protective clothing: A, C

Precautions: C03, E01, E13c, E30a, E31a, E34, M03, R03a, R03c, R04e, U02, U05a, U08, U19, U20b

---

Biodew Intracrop A0253 non-ionic surfactant/spreader/wetter

Contains: 85% w/w alkyl polyglycol ether and fatty acids

Use with: Pesticides where a wetter/spreader is recommended

Protective clothing: A, C

Precautions: C03, E01, E13c, E30a, E31a, E34, M03, R03c, R04a, R07d, U05a, U08, U19, U20c

---

Bond Newman A0184 extender/sticker

Contains: 450 g/l synthetic latex and 100 g/l alkylphenylhydroxy-polyoxyethylene

Use with: Blight sprays on potatoes and other pesticides on a wide range of crops. See label or contact supplier for details

Precautions: M03, R04a, R04b, U05a, U08, U19, U20a, C03, E01, E13c, E30a, E31a

---

Codacide Oil Microcide A0011 vegetable oil

Contains: 95% emulsifiable vegetable oil

Use with: All approved pesticides and tank mixes. See label for details

Protective clothing: [0011] A, C

Precautions: U19, U20b, E13c, E30a, E31b

PRODUCT	SUPPLIER	ADJ. NO.	TYPE
Companion	Allied Colloids	A0239	spreader/sticker/wetter

Contains: 25% w/w polyacrylamide

Use with: Approved herbicides on cereals and any pesticide on non-food crops. See label for restrictions on use with wettable powders suspension concentrates and water dispersible granules, and other usage limitations

Precautions: U05a, U08, C03, E01, E13c, E30a

Comulin	Batsons	A0164	mineral oil

Contains: 96.8% w/w refined paraffinic oil

Use with: Asulox, Basagran, Benlate, Checkmate, Gesaprim 500 SC, Goltix WG, Goltix WG/Checkmate, Goltix WG/Dow Shield, Goltix WG/Fusilade 5, Goltix WG/Laser/Checkmate, Laser, Pilot and Roundup

Precautions: U08, U19, U20a, E15, E30a, E31a

Contact	Interagro	A0191	mineral oil

Contains: 92% w/w paraffinic oil

Use with: Atrazine, Beetomax + metamitron, fenoxaprop-p-ethyl, metamitron, phenmedipham and other approved pesticides which have a label recommendation for use with authorised adjuvant mineral oils

Protective clothing: A, C

Precautions: R04a, R04b, U02, U05a, U08, U19, U20a, C03, E01, E13c, E30a, E31b

---

Cropoil	Chiltern	A0137	mineral oil

Contains: 99% highly refined mineral oil

Use with: Approved pesticides for use on certain specified crops (cereals, combinable break crops, field and dwarf beans, peas, oilseed rape, brassicas, potatoes, carrots, parsnips, sugar beet, fodder beet, mangels, red beet, maize, sweetcorn, onions leek s, horticultural crops, forestry, amenity, grassland)

Protective clothing: A, C

Precautions: E13c, E26, E30a, E31b, R04, R04e, U19, U20b

FOR FULL CONDITIONS OF USE ALWAYS READ THE PRODUCT LABEL

PRODUCT	SUPPLIER	ADJ. NO.	TYPE

Cutback | Tech Nova | A0183 | spreader/wetter

Contains: 800 g/l polyoxyethylene tallow amine

Use with: Glyphosate and other herbicides in agriculture, horticulture, amenity and forestry

Protective clothing: A, C

Precautions: C03, E01, E13c, E30a, E31a, E34, M03, R03c, R04a, R04b, R07d, U02, U05a, U08, U13, U19, U20b

---

Cutinol | Techsol | A0151 | vegetable oil

Contains: 95% emulsifiable rapeseed oil

Use with: All approved pesticides for which addition of a wetter is recommended

Precautions: R04a, R04b, U05a, U08, U20a, C03, E01, E13c, E30a, E31a

---

Cutonic Foliar Booster | Lambson | A0228 | non-ionic surfactant/wetter

Contains: 100% w/w polyether-polymethylsiloxane-copolymer

Use with: All pesticides authorised for use on all cereals and grassland crops where the use of a wetting/spreading/penetrating surfactant is recommended. Can be used with micronutrients and fertilisers when applied as a foliar treatment

Protective clothing: A, C

Precautions: C03, E01, E13c, E26, E30a, E31b, E34, M03, R03a, R03c, R04d, R04e, U02, U05a, U08, U19, U20b

---

Du Pont Adjuvant | DuPont | A0119 | non-ionic surfactant/wetter

Contains: 900 g/l ethylene oxide condensate

Use with: Any spray for which a non-ionic wetter is recommended

Protective clothing: A, C

Precautions: C03, E01, E13c, E30a, E31a, R04a, R04b, R07d, U05a, U08, U20c

---

Emerald | Intracrop | A0031 | anti-transpirant/extender

Contains: 96% di-1-p-menthene

Use with: Alone or with most approved insecticides, fungicides or growth regulators See label for details. Do not use in mixture with adjuvant oils or surfactants

PRODUCT	SUPPLIER	ADJ. NO.	TYPE
Precautions:	U19, U20b, E13c, E30a, E31a		
Enhance	Techsol	A0147	non-ionic surfactant/spreader/wetter
Contains:	900 g/l phenol ethylene oxide condensate		
Use with:	Any spray for which additional wetter is approved and recommended		
Precautions:	R03c, R04a, R04b, R07d, U05a, U08, U13, U19, U20a, C03, E01, E13c, E30a, E31a		
Enhance Low Foam	Techsol	A0148	non-ionic surfactant/spreader/wetter
Contains:	900 g/l alkyl phenol ethylene oxide condensate with silicone anti-foaming agent		
Use with:	Any spray for which wetter is approved and recommended through low volume spraying systems or where low foam is required		
Protective clothing:	A, C		
Precautions:	R04a, R04b, U05a, U08, U20a, C03, E01, E13c, E30a, E31a		
Ethokem	Techsol	A0146	cationic surfactant
Contains:	870 g/l polyoxyethylene tallow amine		
Use with:	A wide range of fungicides, herbicides, insecticides, growth regulators and micronutrients. See leaflet for details		
Precautions:	R04a, R04b, U05a, U08, U20a, C03, E01, E13c, E30a, E31a		
Ethokem C/12	Techsol	A0149	cationic surfactant
Contains:	bis-2 hydroxyethyl coco-amine		
Use with:	Glyphosate, for difficult to kill weeds including rhododendron, bracken and foxglove, especially in forestry		
Protective clothing:	A, C		
Precautions:	R03c, R04a, R04b, R07d, U05a, U08, U13, U19, U20a, C03, E01, E13c, E30a, E31a		
Euroagkem Pen-e-trate	Euroagkem	A0216	acidifier/non-ionic surfactant

FOR FULL CONDITIONS OF USE ALWAYS READ THE PRODUCT LABEL

PRODUCT	SUPPLIER	ADJ. NO.	TYPE

Contains: 350 g/l propionic acid

Use with: Pesticides that recomend the use of a wetter/spreader

Protective clothing: A, C

Precautions: R04a, R04b, R05, R05b, U02, U05a, U08, U19, U20a, C03, E01, E30a, E31a

---

Euroagkem Taktic | Euroagkem | A0210 | extender/sticker

Contains: 450 g/l synthetic latex

Use with: Pesticides that recommend the addition of a sticker/extender

Protective clothing: A, C

Precautions: M03, R04a, R04b, R07d, U02, U05a, U08, U19, U20b, C03, E01, E13c, E30a, E31a

---

Exell | Truchem | A0017 | spreader/wetter

Contains: 64% polyethoxylated tallow amine

Use with: Diquat, glyphosate

Protective clothing: A, C

Precautions: M03, R04a, R04b, U05a, U08, U19, U20a, C03, E01, E13c, E30a, E31a, E34

Felix | Intracrop | A0178 | spreader/wetter

Contains: 60% ethylene oxide condensate

Use with: Mecoprop, 2,4-D in cereals and amenity turf, and a range of grass weedkillers in agriculture. See label for details

Protective clothing: A, C

Precautions: R04a, R04b, R07d, U05a, U08, U20a, C03, E01, E15, E30a, E31a

---

Frigate | Unicrop | A0128 | sticker/wetter

Contains: 800 g/l tallow amine ethoxylate

Use with: Roundup formulations

Protective clothing: A, C

PRODUCT	SUPPLIER	ADJ. NO.	TYPE
Precautions:	M03, R03c, R04a, R04b, R04c, R07d, U05a, U09a, U19, U20a, C03, E01, E13b, E30a, E31a		
Fyzol 11E	AgrEvo	A0172	mineral oil
Contains:	99% highly refined paraffinic oil		
Use with:	Betanal E, Betanal E + Goltix WG, Pilot		
Precautions:	U19, U20b, E13c, E26, E30a, E31b, E34		
Galion	Intracrop	A0162	spreader/wetter
Contains:	60% polyoxyalkylene glycol		
Use with:	Herbicides in grassland and amenity turf and with grass weedkillers in cereals, oilseed rape, sugar beet and other crops		
Protective clothing:	A, C		
Precautions:	R04a, R04b, R07d, U05a, U08, U20b, C03, E01, E15, E30a, E31a		
Grip	Newman	A0211	extender/sticker/wetter
Contains:	450 g/l synthetic latex and 100 g/l alkylphenylhydroxy-polyoxyethylene		
Use with:	Blight sprays on potatoes and other pesticides on a wide range of crops. See label or contact supplier for details		
Protective clothing:	A, C		
Precautions:	C03, E01, E13c, E30a, E31a, E34, M03, R04a, R04b, U02, U05a, U08, U19, U20b		
GS 800	Techsol	A0150	cationic surfactant
Contains:	800 g/l polyoxyethylene tallow amine		
Use with:	A wide range of pesticides. See label for details		
Protective clothing:	A, C		
Precautions:	R03c, R04a, R04b, R07d, U05a, U08, U13, U19, U20a, C03, E01, E13c, E30a, E31a		

PRODUCT	SUPPLIER	ADJ. NO.	TYPE

Guide Newman A0208 acidifier/penetrant

Contains: 350 g/l modified soya lecithin, 100 g/l alkylphenylhydroxypolyoxyethylene and 350 g/l propionic acid

Use with: All plant growth regulators for cereals and oilseed rape, dimethoate and chlorpyrifos for wheat bulb fly control. For other uses see label or contact supplier for details

Protective clothing: A, C

Precautions: M03, R04b, R04d, U02, U04a, U05a, U10, U11, U19, U20a, C03, E01, E13c, E30a, E31a

---

Headland Fortune Headland A0277 penetrant/spreader/vegetable oil/wetter

Contains: 75% w/w mixed methylated fatty acid esters of seed oil and N-butanol

Use with: Herbicides and fungicides in a wide range of crops. See label for details

Protective clothing: A, C

Precautions: C03, E01, E13c, E26, E30a, E31b, E34, R04, R04e, U02, U05a, U14, U20a

---

Headland Guard Headland A0073 spreader/sticker

Contains: 550 g/l organic co-polymer and surfactants

Use with: A wide range of agricultural chemicals and micronutrients

Protective clothing: A, C

Precautions: R04a, R04e, U02, U05a, U08, U20a, C03, E01, E30a, E31a

---

Headland Intake Headland A0074 penetrant

Contains: 450 g/l organic acids and surfactants

Use with: A wide range of fungicides, herbicides, desiccants and growth regulators

Protective clothing: A, C

Precautions: R05, R05b, U02, U05a, U08, U19, U20a, C03, E01, E30a, E31a

---

Headland Quilt Headland A0188 spreader/wetter

Contains: 870 g/l (85% w/w) bis-2 hydroxyethyl coco-amine

Use with: Glyphosate

PRODUCT	SUPPLIER	ADJ. NO.	TYPE

Protective clothing: A, C

Precautions: M03, R03c, R04a, R04b, U02, U05a, U13, U19, U20a, C03, E01, E13c, E30a, E34

Holdtite	Techsol	A0180	extender/sticker

Contains: 52% w/w synthetic latex and 20% alkyl phenol ethylene oxide condensate

Use with: A wide range of agricultural chemicals/micronutrients

Protective clothing: A, C

Precautions: C03, E01, E13b, E26, E30a, E31a, R04a, R04b, R04e, U02, U05a, U08, U19, U20b

---

Hyspray	Fine	A0020	cationic surfactant

Contains: 800 g/l polyethoxylated tallow amine

Use with: All approved formulations of glyphosate

Protective clothing: A, C

Precautions: R03c, R04a, R04b, R07d, U02, U05a, U08, U13, U19, U20a, C03, E01, E13b, E30a, E31b, E34

Intracrop Archer	Intracrop	A0242	spreader/wetter

Contains: 800 g/l polyalkyleneoxide modified heptamethyltrisiloxane

Use with: All fungicides used in cereals where the addition of wetters/spreaders is recommended on the pesticide label

Protective clothing: A, C

Precautions: C03, E01, E13c, E30a, E31a, E34, M03, R03a, R03c, R04e, U02, U05a, U08, U19, U20b

---

Intracrop BLA	Intracrop	A0125	sticker

Contains: 52% synthetic latex and 20% alkyl phenol ethylene oxide condensate

Use with: Any fungicides, herbicides, insecticides or trace elements where addition of an authorised extender/sticker is recommended

FOR FULL CONDITIONS OF USE ALWAYS READ THE PRODUCT LABEL

PRODUCT	SUPPLIER	ADJ. NO.	TYPE
Protective clothing:	A, C		
Precautions:	U02, U05a, U08, U19, U21, C03, E01, E13b, E30a, E31a, E34		
Intracrop Bla-Tex	Intracrop	A0173	extender/sticker
Contains:	450 g/l synthetic latex		
Use with:	Any approved pesticide where the addition of a sticker/extender is required and recommended		
Protective clothing:	A, C		
Precautions:	M03, R04a, R04b, U02, U05a, U08, U19, U20a, C03, E01, E13c, E30a, E31a		
Intracrop Non-Ionic Wetter	Intracrop	A0009	spreader/wetter
Contains:	900 g/l alkyl phenol ethoxylate		
Use with:	Pesticides which have a recommendation for use with a wetter or emulsified oil		
Protective clothing:	A, C		
Precautions:	R04a, R04b, R07d, U02, U05a, U08, U19, U20a, C03, E01, E13c, E30a, E31a		
Intracrop Quiver	Intracrop	A0237	spreader/wetter
Contains:	800 g/l polyalkyleneoxide modified heptamethyltrisiloxane		
Use with:	All approved fungicides applied to wheat, barley, oats, triticale, rye and durum wheat		
Protective clothing:	A, C		
Precautions:	M03, R03a, R03c, R04e, U02, U05a, U08, U19, U20a, C03, E01, E13c, E30a, E31a, E34		
Intracrop Rapide Beta	Intracrop	A0175	acidifier/non-ionic surfactant
Contains:	350 g/l propionic acid and 100 g/l alkylphenyl-hydroxypolyoxyethylene		
Use with:	Any spray where the addition of a wetter/spreader or self-emulsifying oil is required and recommended on the pesticide label		
Protective clothing:	A, C		

PRODUCT	SUPPLIER	ADJ. NO.	TYPE
Precautions:	M03, R05, R05b, U02, U04a, U05a, U08, U10, U11, U19, U20a, C03, E01, E13c, E30a, E31a, E34		
Iona	Intracrop	A0232	spreader/wetter
Contains:	921 g/l alkyl phenol ethoxylate		
Use with:	Any approved pesticide for which the addition of a wetter/spreader is recommended		
Protective clothing:	A, C		
Precautions:	R04a, R04b, U05a, U08, U19, U20a, C03, E01, E13c, E30a, E32a, E34		
Iona Low Foam	Intracrop	A0255	spreader/wetter
Contains:	921 g/l alkyl phenol ethoxylate and silicone antifoam agent		
Use with:	Pesticides that have a recommendation for the use of a wetter/spreader		
Protective clothing:	A, C		
Precautions:	C03, E01, E13c, E30a, E31a, E34, R04a, R04b, U05a, U08, U19, U20b		
Jogral	Intracrop	A0226	cationic surfactant
Contains:	800 g/l tallow amine ethoxylate		
Use with:	Glyphosate		
Protective clothing:	A, C		
Precautions:	M03, R03c, R07d, U08, U19, U20a, C03, E01, E13b, E28, E31b, E32a, E34		
Kandu	Zeneca	A0264	surfactant
Contains:	A blend of surfactant and ammonium sulphate		
Use with:	Approved formulations of glyphosate-trimesium		
Protective clothing:	C		
Precautions:	C03, E01, E13c, E30a, E31c, U05a, U08, U19, U20b		

FOR FULL CONDITIONS OF USE ALWAYS READ THE PRODUCT LABEL

PRODUCT	SUPPLIER	ADJ. NO.	TYPE

Klipper — Service Chemicals — A0260 — spreader/wetter

Contains: 600 g/l ethylene oxide condensate

Use with: Approved salt formulations of mecoprop alone or in mixtures with 2,4-D on managed amenity turf

Protective clothing: A, C

Precautions: C03, E01, E15, E30a, E31a, R04a, R04b, R07d, U05a, U08, U20c

---

LI-700 — Newman — A0176 — acidifier/penetrant

Contains: 350 g/l modified soya lecithin, 100 g/l alkylphenylhydroxypolyoxyethylene and 350 g/l propionic acid

Use with: All plant growth regulators for cereals and oilseed rape, dimethoate and chlorpyrifos for wheat bulb fly control. For other uses see label or contact supplier for details

Protective clothing: A, C

Precautions: M03, R04b, R04d, U02, U04a, U05a, U10, U11, U19, U20a, C03, E01, E13c, E30a, E31a

---

Libsorb — Allied Colloids — A0251 — spreader/wetter

Contains: alkyl alcohol ethoxylate

Use with: Any spray for which additional wetter is approved and recommended

Protective clothing: A, C

Precautions: R04a, R04b, U05a, U08, U20a, C03, E01, E13c, E30a, E31a

---

Lightning — Rigby Taylor — A0280 — mineral oil

Contains: 4.84% w/w paraffinic oil

Use with: Roundup Pro Biactive

Precautions: U08, U19, U20a, E15, E30a, E31a

Lo-Dose — ISK Biosciences — A0129 — cationic surfactant

Contains: 800 g/l tallow amine ethoxylate

Use with: Roundup, Roundup Four 80

PRODUCT	SUPPLIER	ADJ. NO.	TYPE
Protective clothing:	A, C		
Precautions:	M03, R03c, R04a, R04b, R07d, U05a, U08, U19, U20a, C03, E01, E13c, E30a, E31a, E34		
Luxan Non-Ionic Wetter	Luxan	A0139	non-ionic surfactant/spreader/wetter
Contains:	900 g/l alkyl phenol ethylene oxide condensate		
Use with:	Any spray for which additional wetter/spreader is approved and recommended		
Protective clothing:	A, C		
Precautions:	R04a, R04b, R07d, U05a, U08, U20b, C03, E01, E13c, E30a, E31b, E34		
Lyrex	Batsons	A0177	mineral oil
Contains:	96.8% w/w paraffinic base oil		
Use with:	Asulox, Basagran, Benlate, Checkmate, Gesaprim 500 SC, Goltix WG, Goltix WG/Checkmate, Goltix WG/Dow Shield, Goltix WG/Fusilade 5, Goltix WG/Checkmate/Laser, Laser, Pilot, Roundup		
Lyrol	Batsons	A0165	mineral oil
Contains:	96.8% w/w refined paraffinic oil		
Use with:	Asulox, Basagran, Benlate, Checkmate, Gesaprim 500 SC, Goltix WG, Goltix WG/Checkmate, Goltix WG/Dow Shield, Goltix WG/Fusilade 5, Goltix WG/Laser/Checkmate, Laser, Pilot and Roundup		
Precautions:	U08, U19, U20a, E13c, E30a, E31a		
Mangard	Mandops	A0132	adjuvant/anti-transpirant
Contains:	96% w/w di-1-p-menthene		
Use with:	Glyphosate and a wide range of other herbicides, fungicides and insecticides - see label for details		
Precautions:	E15, E30a, E31a, U20c		
Minder	Stoller	A0087	vegetable oil

FOR FULL CONDITIONS OF USE ALWAYS READ THE PRODUCT LABEL

PRODUCT	SUPPLIER	ADJ. NO.	TYPE

Contains: 94% vegetable oil

Use with: Glyphosate

Precautions: E01, E15, E30a, E32a, U20c

Mixture B	Service Chemicals	A0161	non-ionic surfactant/spreader/wetter

Contains: 500 g/l nonyl phenol ethylene oxide condensate and 500 g/l primary alcohol ethylene oxide condensate

Use with: Glyphosate in forestry, amenity and non-crop situations

Protective clothing: A, C

Precautions: M03, R03a, R03c, R04a, R04b, U05a, U09a, C03, E01, E13c, E30a, E31a, E34

MSS Limpet	Mirfield	A0283	spreader/wetter

Contains: 95% rapeseed oil

Use with: All pesticides for which the uses of a wetter or spreader is approved and recommended. See label for latest growth stages recommended for treatment of listed crops

Precautions: U05a, U08, U19, U20b, C03, E01, E13c, E30a, E31a, E34

Newman Cropspray 11E	Newman	A0195	mineral oil

Contains: 99% highly refined mineral oil

Use with: Approved herbicides for which the addition of an adjuvant oil is recommended

Precautions: U19, U20a, E13c, E26, E30a, E31b

Newman's T-80	Newman	A0192	spreader/wetter

Contains: 78% w/w polyoxyethylene tallow amine

Use with: Glyphosate

Protective clothing: A, C

Precautions: M04, R03c, R04a, R04b, U05a, U08, U13, U19, U20a, C03, E01, E13c, E30a, E31a, E34

PRODUCT	SUPPLIER	ADJ. NO.	TYPE
Nion	Service Chemicals	A0258	spreader/wetter
Contains:	90% (900 g/l) ethylene oxide condensate		
Use with:	Pesticides which have a recommendation for use with a wetter/spreader		
Protective clothing:	A, C		
Precautions:	C03, E01, E13c, E30a, E31a, R04a, R04b, U05a, U08, U19, U20b		
Nu Film 17	Intracrop	A0055	sticker/wetter
Contains:	96% di-1-p-menthene		
Use with:	A range of fungicides, insecticides and foliar nutrients. See label for details		
Precautions:	C02 (30 d), E13c, E30a, E32a, U19, U20c		
Nu Film P	Intracrop	A0039	sticker/wetter
Contains:	96% poly-1-p-menthene		
Use with:	Glyphosate and many other pesticides and growth regulators for which a protectant is recommended. Do not use in mixture with adjuvant oils or surfactants		
Precautions:	U19, U20b, E13c, E30a, E31a		
Output	Zeneca	A0163	mineral oil/surfactant
Contains:	60% mineral oil and 40% surfactants		
Use with:	Grasp		
Protective clothing:	A, C		
Precautions:	R04b, U05a, U08, U19, U20a, C03, E01, E13c, E30a, E31b		
Partna	Zeneca	A0246	mineral oil/surfactant
Contains:	A blend of surfactants and mineral oils		
Use with:	Formulations of fluazifop-P-butyl		

FOR FULL CONDITIONS OF USE ALWAYS READ THE PRODUCT LABEL

PRODUCT	SUPPLIER	ADJ. NO.	TYPE

Protective clothing: A

Precautions: C03, E01, E13c, E30a, E31c, U05a, U08, U19, U20b

---

Phase	Newman	A0279	vegetable oil

Contains: 75% w/w methylated mixed fatty acid esters of rapeseed oil and 25% w/w proprietary surfactant bland

Use with: Herbicides and fungicides in sugar beet up to 6 leaves; triazole and morpholine fungicides in winter and spring sown cereals. For other uses see label or contact supplier for details

Protective clothing: A, H

Precautions: U02, U05a, U14, U20b, C03, E01, E13c, E30a, E31a, E34

Pin-o-Film	Intracrop	A0209	sticker/wetter

Contains: 96% di-1-p-menthene

Use with: Approved pesticides on agricultural horticultural and forestry crops. Do not use in tank mixture with adjuvant oils or surfactants

Precautions: U19, U20b, C02 (30 d), E13c, E30a, E32a

Planet	Intracrop	A0224	non-ionic surfactant/spreader/wetter

Contains: 85% alkyl polyglycol ether and fatty acid

Use with: Any spray for which additional wetter is recommended

Precautions: R03c, R04a, R07d, U05a, U08, U19, U20b, E01, E13c, E30a, E31a, E32a

---

Prima	Newman	A0310	mineral oil

Contains: 99% highly refined mineral oil

Use with: All pesticides on all crops when used up to 50% of maximum approved dose for that use (mixtures with Roundup must only be used for treatment of stubbles); all approved pesticides at full dose on listed crops (see label). For other uses see label or contact supplier for details

Precautions: U19, U20b, E13c, E26, E30a, E31b

Q-900 Non-ionic Wetter	Techsol	A0261	spreader/wetter

Contains: 90% w/w (921 g/l) alkyl phenol ethoxylate

PRODUCT	SUPPLIER	ADJ. NO.	TYPE
Use with:	Pesticides which have a recommendation for the use of a wetter/spreader		
Protective clothing:	A, C		
Precautions:	C03, E01, E13c, E30a, E31a, E34, R04a, R04b, U05a, U08, U19, U20b		
Quadrangle Quad-Fast	Quadrangle	A0054	coating agent/surfactant
Contains:	di-1-p-menthene		
Use with:	Glyphosate and other pesticides. Growth regulators and nutrients for which a coating agent is approved and recommended		
Precautions:	U20b, E15, E30a, E31a		
Ranger	Newman	A0296	acidifier/penetrant
Contains:	350 g/l modified soya lecithin, 100 g/l alkylphenylhydroxypolyoxyethylene and 350 g/l propionic acid		
Use with:	All plant growth regulators for cereals and oilseed rape, dimethoate and chlorpyrifos for wheat bulb fly control. For other uses see label or contact supplier for details		
Protective clothing:	A, C		
Precautions:	M03, R04b, R04d, U02, U04a, U05a, U10, U11, U19, U20a, C03, E01, E13c, E30a, E31a		
Rapide	Intracrop	A0116	penetrant/surfactant
Contains:	40% propionic acid		
Use with:	A wide range of pesticides and growth regulators, especially chlormequat		
Protective clothing:	A, C		
Precautions:	R04a, R04b, R05, R05b, U02, U05a, U08, U19, U20a, C03, E01, E30a, E31a		
Retain	Euroagkem	A0263	mineral oil/non-ionic surfactant
Contains:	60% w/w mineral oil and 40% w/w non-ionic surfactants		
Use with:	Tralkoxydim		

FOR FULL CONDITIONS OF USE ALWAYS READ THE PRODUCT LABEL

PRODUCT	SUPPLIER	ADJ. NO.	TYPE
Protective clothing:	A, C		
Precautions:	C03, E01, E13c, E30a, E31b, R04b, U05a, U08, U19		
Ryda	Interagro	A0168	cationic surfactant/wetter
Contains:	800 g/l polyethoxylated tallow amine		
Use with:	Glyphosate		
Protective clothing:	A, C		
Precautions:	M03, R03c, R04a, R04b, R07d, U02, U05a, U08, U13, U19, U20a, C03, E01, E13c, E30a, E31b, E34		
Scout	Newman	A0207	acidifier/non-ionic surfactant/penetrant
Contains:	350 g/l modified soya lecithin, 100 g/l alkylphenylhydroxypolyoxyethylene and 350 g/l propionic acid		
Use with:	All plant growth regulators for cereals and oilseed rape, dimethoate and chlorpyrifos for wheat bulb fly control. For other uses see label or contact supplier for details		
Protective clothing:	A, C		
Precautions:	C03, E01, E13c, E30a, E31a, M03, R04b, R04d, U02, U04a, U05a, U10, U11, U19, U20b		
Silwet L-77	Newman	A0193	wetter
Contains:	80% w/w polyalkylene oxide modified heptamethyltrisiloxane		
Use with:	All approved fungicides on winter and spring sown cereals; all approved pesticides applied at 50% or less of their full approved dose. A wide range of other uses - see label or contact supplier for details		
Protective clothing:	A, C		
Precautions:	M03, U02, U05a, U08, U19, U20a, C03, E01, E13c, E30a, E31a, E34		
Slippa	Interagro	A0206	spreader
Contains:	655 g/l polyalkyleneoxide modified heptamethyltrisiloxane + non-ionic wetters		
Use with:	Cereal fungicides and a wide range of other pesticides and trace elements on specified crops		
Protective clothing:	A, C		

PRODUCT	SUPPLIER	ADJ. NO.	TYPE

Precautions: M03, R03a, R03c, R04b, R04d, R04e, U02, U05a, U08, U11, U19, U20a, C03, E01, E13c, E30a, E31a, E34

Slyx	Euroagkem	A0250	spreader/vegetable oil/wetter

Contains: 95% refined rapeseed oil

Use with: Pesticides that have a recommendation for the addition of a wetter or spreader

Protective clothing: A, C

Precautions: C03, E01, E13c, E30a, E31b, E34, M03, R04a, U04a, U05a, U08, U20b

---

SM99	Newman	A0134	mineral oil

Contains: 99% w/w highly refined paraffinic oil

Use with: Approved herbicides for which the addition of an adjuvant oil is recommended

Precautions: U19, U20a, E15, E30a, E31b

Solar	Intracrop	A0225	activator/non-ionic surfactant/spreader

Contains: 75% polypropoxypropanol and 15% alkyl polyglycol ether

Use with: Foliar applied plant growth regulators

Protective clothing: A, C

Precautions: M03, R04a, R04b, U04a, U05a, U08, U20a, C03, E01, E13c, E30a, E31b, E32a, E34

Sprayfast	Mandops	A0131	extender/sticker/wetter

Contains: di-1-p-menthene and nonyl phenol ethylene oxide condensate

Use with: Glyphosate and other pesticides, growth regulators or nutrients for which a coating agent is approved and recommended

Precautions: U20b, E13c, E30a, E31a

Sprayfix	Newman	A0145	extender/sticker

Contains: 450 g/l synthetic latex and 100 g/l alkylphenylhydroxy-polyoxyethylene

PRODUCT	SUPPLIER	ADJ. NO.	TYPE

Use with: Blight sprays on potatoes and other pesticides on a wide range of crops. See label or contact supplier for details

Protective clothing: A, C

Precautions: M03, R04a, R04b, U02, U05a, U08, U19, U20a, C03, E01, E13c, E30a, E31a

---

Spraymac Newman A0144 acidifier/non-ionic surfactant

Contains: 350 g/l propionic acid and 100 g/l alkylphenyl-hydroxypolyoxyethylene

Use with: All plant growth regulators for cereals and oilseed rape and any spray for which the addition of a wetter/spreader is recommended. See label or contact supplier for details

Protective clothing: A, C

Precautions: M03, R04a, R04b, U02, U04a, U05a, U10, U11, U19, U20a, C03, E01, E13c, E30a, E32a

---

Sprayprover Fine A0238 mineral oil

Contains: 95% highly refined mineral oil

Use with: Any spray for which the addition of a mineral oil is approved and recommended

Protective clothing: A, C

Precautions: R04a, R04b, U05a, U08, U19, U20a, C03, E01, E13c, E30a, E31b

---

Spreader PBI A0189 non-ionic surfactant/spreader/wetter

Contains: nonylphenol ethylene oxide condensate

Use with: Any spray for which additional wetter is approved and recommended

Precautions: U20b, E15, E30a, E31a, E34

Squat Techsol A0179 penetrant

Contains: 40% w/w propionic acid

Use with: Agricultural herbicides, systemic fungicides and desiccants where rapid penetration is required

Protective clothing: A, C

PRODUCT	SUPPLIER	ADJ. NO.	TYPE
Precautions:	C03, E01, E13c, E26, E30a, E31a, R04a, R04b, R05b, U02, U05a, U08, U19, U20b		
Stamina	Interagro	A0202	penetrant
Contains:	100% w/w alkoxylated fatty amine		
Use with:	Glyphosate and a wide range of other pesticides on specified crops		
Protective clothing:	A, C		
Precautions:	M03, R03c, R04b, U02, U05a, U20a, C03, E01, E13b, E30a, E34		
Stefes Abet	Stefes	A0259	spreader/wetter
Contains:	90% w/w ethylene oxide condensate		
Use with:	All approved pesticides which have a recommendation for use with a wetting or spreading agent		
Stefes CAT 80	Stefes	A0182	cationic surfactant
Contains:	800 g/l (78% w/w) polyoxyethylene tallow amine		
Use with:	Glyphosate		
Protective clothing:	A, C		
Precautions:	M03, R03, U05a, U08, U13, U19, U20a, C03, E01, E13c, E30a, E31a, E34		
Stefes Cover	Stefes	A0187	spreader/wetter
Contains:	921 g/l (90% w/w) alkyl phenol ethoxylate		
Use with:	Any spray for which additional wetter or spreader is approved and recommended		
Protective clothing:	A, C		
Precautions:	R04a, R04b, U05a, U08, U19, U20a, C03, E01, E13c, E30a, E31a, E34		
Stefes Eros	Stefes	A0257	vegetable oil
Contains:	95% w/w methylated vegetable oil		

FOR FULL CONDITIONS OF USE ALWAYS READ THE PRODUCT LABEL

PRODUCT	SUPPLIER	ADJ. NO.	TYPE
Use with:	All approved herbicides for cereals, all approved pesticides for forestry and managed amenity turf; all approved formulations of glyphosate; all approved pesticides on all edible and non-edible crops when used at less than 50% of approved dose		
Stefes Oberon	Stefes	A0292	wetter
Contains:	500 g/l alkoxylated fatty amine and 500 g/l polyoxyethylene monolaurate		
Use with:	All approved herbicides on cereals; all approved fungicides and insecticides on cereals up to GS 52 or at any time when used at less than 50% approved dose; all approved pesticides for forestry and managed amenity turf. Consult label or supplier for details of other uses		
Stefes Spread and Seal	Stefes	A0122	anti-transpirant/extender/sticker/wetter
Contains:	di-1-p-menthene and nonyl phenol ethylene oxide condensate		
Use with:	Glyphosate and other pesticides, growth regulators and nutrients for which a coating agent is approved and recommended		
Precautions:	U20b, C02 (4 wk), E13c, E30a, E31a		
Stik-It	Quadrangle	A0071	non-ionic surfactant/spreader/wetter
Use with:	A wide range of fungicides and insecticides for which additional wetter is approved and recommended		
Precautions:	U20b, E15, E30a, E31a, E34		
Sward	Service Chemicals	A0215	spreader/wetter
Contains:	15.2% w/w polyalkylene oxide modified heptomethyltrisiloxane		
Use with:	Specified pesticides in amenity situations		
Protective clothing:	A, C		
Precautions:	C03, E01, E13c, E30a, E31a, R04, R04d, R04e, U05a, U08, U20c		
Swirl	Cyanamid	A0167	mineral oil
Contains:	590 g/l highly refined mineral oil		
Use with:	Aztec, Commando		

PRODUCT	SUPPLIER	ADJ. NO.	TYPE
Precautions:	R07d, U08, U20a, E13b, E30a, E31b, E34		
Talzene	Intracrop	A0233	wetter
Contains:	800 g/l polyoxyethylene tallow amine		
Use with:	Glyphosate		
Protective clothing:	A, C		
Precautions:	C03, E01, E13c, E30a, E31a, E34, M03, R03c, R04a, R04b, U05a, U08, U13, U19, U20b		
Team 2000	Monsanto	A0229	extender
Contains:	40% w/w ammonium sulphate		
Use with:	Roundup 2000 herbicide		
Precautions:	U20a, E30a, E31b		
Team Rodeo	Monsanto	A0231	extender
Contains:	40% w/w ammonium sulphate		
Use with:	Rodeo herbicide		
Precautions:	U20a, E30a, E31b		
Techsol Spreader	Techsol	A0196	spreader/wetter
Contains:	23% w/w (235 gm/litre) alkyl phenol ethoxylate		
Use with:	Pesticides which have a recommendation for the use of a wetter/spreader		
Protective clothing:	A, C		
Precautions:	C03, E01, E13c, E30a, E31a, E34, R04a, R04b, U02, U05a, U08, U19, U20b		
Toil	Interagro	A0248	vegetable oil
Contains:	95% w/w methylated vegetable oil		

PRODUCT	SUPPLIER	ADJ. NO.	TYPE
Use with:	Sugar beet herbicides, oilseed rape herbicides, cereal graminicides and a wide range of other pesticides on specified crops		
Protective clothing:	A		
Precautions:	R04b, U02, U05a, U08, U20a, C03, E01, E15, E30a, E31a		
TopUp Surfactant	FCC	A0080	cationic surfactant
Contains:	800 g/l ethoxylated tallow amine		
Use with:	A wide range of crop protection products. See label for details		
Protective clothing:	A, C		
Precautions:	M03, R03c, R04a, R04b, R07d, U02, U05a, U08, U19, U20a, C03, E01, E13c, E30a, E31a, E34		
Tripart Acer	Tripart	A0097	acidifier/penetrant
Contains:	750 g/l soyal phospholipids		
Use with:	Chlormequat and a wide range of systemic pesticides and trace elements. Contact supplier for details		
Protective clothing:	A, C		
Precautions:	M03, R04a, R04b, U02, U05a, U08, U19, U20b, C03, E01, E13c, E30a, E31a		
Tripart Lentus	Tripart	A0117	extender/sticker
Contains:	450 g/l synthetic latex		
Use with:	A wide range of contact herbicides, fungicides and insecticides. Contact supplier for further details		
Protective clothing:	A, C		
Precautions:	R04a, R04b, U05a, U08, U19, U20a, C03, E01, E13c, E30a, E31a		
Tripart Minax	Tripart	A0108	non-ionic surfactant/spreader/wetter
Contains:	900 g/l alkyl alcohol ethoxylate		
Use with:	Any spray for which additional wetter or spreader is approved and recommended		
Precautions:	E30a, E31a		

PRODUCT	SUPPLIER	ADJ. NO.	TYPE

Tripart Orbis | Tripart | A0204 | mineral oil

Contains: 99% w/w highly refined paraffinic oil

Use with: A wide range of approved pesticides on sugar beet, cereals, oilseed rape, maize, sweetcorn, fodder beet, mangels and red beet (including aerial use where recommended). See label for details

Precautions: U19, U20a, E07, E15, E26, E30a, E31b

Vassgro Non Ionic | Vass | A0190 | spreader/wetter

Contains: 921 g/l alkyl phenol ethoxylate

Use with: Pesticides which have a recommendation for the use of a wetter/spreader

Protective clothing: A, C

Precautions: C03, E01, E13c, E30a, E31a, E34, R04a, R04b, U02, U05a, U08, U19, U20b

---

Vassgro Spreader | Vass | A0035 | non-ionic surfactant/spreader/wetter

Contains: nonyl phenol-ethylene oxide condensates

Use with: Any spray for which additional wetter or spreader is approved and recommended

Precautions: U20a, E15, E30a, E31a, E34

Wayfarer | Hortichem | A0045 | cationic surfactant/spreader/wetter

Contains: 80% tallow amine ethoxylate

Use with: Glyphosate in agriculture

Protective clothing: A, C

Precautions: R04a, R04b, R07d, U05a, U08, U20b, C03, E01, E15, E30a, E31a

---

FOR FULL CONDITIONS OF USE ALWAYS READ THE PRODUCT LABEL

SECTION 3
CROP/PEST GUIDE

CROP INDEX

CROP/PEST GUIDE

Important note: For convenience, some crops and pests have been grouped into generic units in this guide, eg "cereals", "annual grasses". It is essential to check the profile entry in Section 4 *and* the label to ensure that a product is approved for a specific crop/pest combination, eg winter wheat/blackgrass.

Arable crops

Beans, field/broad

Diseases	Ascochyta	benomyl *(seed treatment)*, carbendazim + vinclozolin *(reduction)*, metalaxyl + thiabendazole + thiram *(seed treatment)*, thiabendazole + thiram *(seed treatment)*
	Chocolate spot	benomyl, carbendazim, carbendazim + chlorothalonil, carbendazim + tebuconazole, carbendazim + vinclozolin *(moderate control)*, chlorothalonil, chlorothalonil + cyproconazole, chlorothalonil *(moderate control)*, cyproconazole *(with chlorothalonil)*, iprodione, iprodione + thiophanate-methyl, tebuconazole, thiophanate-methyl, vinclozolin
	Damping off	metalaxyl + thiabendazole + thiram *(seed treatment)*, thiabendazole + thiram *(seed treatment)*, thiram *(seed treatment)*
	Downy mildew	chlorothalonil + metalaxyl, fosetyl-aluminium, metalaxyl + thiabendazole + thiram *(seed treatment)*
	Rust	carbendazim + tebuconazole, chlorothalonil + cyproconazole, cyproconazole, fenpropimorph, tebuconazole
Pests	Aphids	demeton-S-methyl, dimethoate, disulfoton, fatty acids, heptenophos, nicotine, pirimicarb
	Birds	aluminium ammonium sulfate
	Caterpillars	nicotine
	Damaging mammals	aluminium ammonium sulfate
	Leaf miners	oxamyl *(off-label)*
	Mealy bugs	fatty acids
	Pea and bean weevils	cypermethrin, deltamethrin, lambda-cyhalothrin, lambda-cyhalothrin + pirimicarb, zeta-cypermethrin
	Red spider mites	fatty acids
	Scale insects	fatty acids
	Whitefly	fatty acids

Weeds	Annual and perennial weeds	clodinafop-propargyl + diflufenican *(pre-harvest)*, glyphosate (agricultural) *(pre-harvest)*, glyphosate (horticulture, forestry, amenity etc.) *(pre-harvest)*
	Annual dicotyledons	bentazone, carbetamide, chlorpropham + fenuron, cyanazine + pendimethalin, cyanazine *(Scotland only)*, fomesafen + terbutryn, glufosinate-ammonium *(pre-harvest)*, propyzamide, simazine, simazine + trietazine, terbuthylazine + terbutryn, terbutryn + trietazine, trifluralin
	Annual grasses	carbetamide, chlorpropham + fenuron, cyanazine *(Scotland only)*, cycloxydim, diclofop-methyl, fluazifop-P-butyl, fluazifop-P-butyl *(off-label)*, glufosinate-ammonium *(pre-harvest)*, propaquizafop, propyzamide, simazine, terbuthylazine + terbutryn, trifluralin
	Annual meadow grass	cyanazine + pendimethalin, terbutryn + trietazine, tri-allate
	Annual weeds	glyphosate (agricultural) *(post sowing, pre crop emergence)*
	Barren brome	sethoxydim
	Blackgrass	cycloxydim, diclofop-methyl, fluazifop-P-butyl, sethoxydim, tri-allate
	Canary grass	diclofop-methyl
	Chickweed	chlorpropham + fenuron
	Couch	clodinafop-propargyl + diflufenican *(pre-harvest)*, cycloxydim, glyphosate (agricultural) *(pre-harvest)*, glyphosate (horticulture, forestry, amenity etc.) *(pre-harvest)*, sethoxydim
	Creeping bent	cycloxydim, glyphosate (agricultural) *(pre-harvest)*, sethoxydim
	Perennial dicotyledons	glyphosate (agricultural) *(pre-harvest)*
	Perennial grasses	cycloxydim, diclofop-methyl, fluazifop-P-butyl, fluazifop-P-butyl *(off-label)*, glyphosate (agricultural) *(pre-harvest)*, propaquizafop, propyzamide, sethoxydim, tri-allate
	Rough meadow grass	cyanazine + pendimethalin, diclofop-methyl
	Ryegrass	diclofop-methyl
	Volunteer cereals	carbetamide, cycloxydim, fluazifop-P-butyl, glyphosate (agricultural) *(post sowing, pre crop emergence)*, propyzamide, sethoxydim
	Wild oats	cycloxydim, diclofop-methyl, fluazifop-P-butyl, propyzamide, sethoxydim, tri-allate
Crop control	Pre-harvest desiccation	diquat, glufosinate-ammonium
Plant growth regulation	Increasing yield	chlormequat with di-1-p-menthene
	Lodging control	chlormequat with di-1-p-menthene

FOR FULL CONDITIONS OF USE ALWAYS READ THE PRODUCT LABEL

Beet crops

Diseases	Black leg	hymexazol
	Damping off	hymexazol
	Disease control/foliar feed	carbendazim + prochloraz *(off-label)*
	Powdery mildew	copper sulfate + sulfur, cyproconazole, propiconazole, sulfur, triadimefon, triadimenol
	Ramularia leaf spots	cyproconazole, propiconazole
	Rust	cyproconazole, fenpropimorph *(off-label)*, propiconazole
	Seed-borne diseases	thiram *(seed soak)*
Pests	Aphids	aldicarb, carbosulfan, deltamethrin + heptenophos, deltamethrin + pirimicarb, demeton-S-methyl, dimethoate *(excluding Myzus persicae)*, disulfoton, imidacloprid, lambda-cyhalothrin + pirimicarb, oxamyl, pirimicarb, triazamate *(including Myzus persicae)*
	Birds	aluminium ammonium sulfate
	Caterpillars	cypermethrin
	Cutworms	cypermethrin, gamma-HCH, lambda-cyhalothrin, lambda-cyhalothrin + pirimicarb
	Damaging mammals	aluminium ammonium sulfate
	Docking disorder vectors	aldicarb, benfuracarb, carbofuran, oxamyl
	Flea beetles	carbofuran, carbosulfan, deltamethrin, gamma-HCH, imidacloprid, lambda-cyhalothrin, lambda-cyhalothrin + pirimicarb, trichlorfon
	Leaf miners	aldicarb, benfuracarb, dimethoate, lambda-cyhalothrin, lambda-cyhalothrin + pirimicarb, pirimiphos-methyl
	Leatherjackets	chlorpyrifos, gamma-HCH
	Mangold fly	carbofuran, carbosulfan, demeton-S-methyl, dimethoate, disulfoton, imidacloprid, lambda-cyhalothrin, oxamyl, trichlorfon
	Millipedes	aldicarb, benfuracarb, carbofuran, carbosulfan, gamma-HCH, imidacloprid, oxamyl, tefluthrin
	Nematodes	aldicarb, carbofuran, carbosulfan
	Pygmy beetle	aldicarb, benfuracarb, carbofuran, carbosulfan, chlorpyrifos, gamma-HCH, imidacloprid, oxamyl, tefluthrin
	Springtails	benfuracarb, carbofuran, carbosulfan, gamma-HCH, imidacloprid, tefluthrin
	Symphylids	benfuracarb, carbosulfan, gamma-HCH, imidacloprid, tefluthrin
	Tortrix moths	carbofuran
	Virus vectors	deltamethrin + pirimicarb
	Wireworms	carbofuran, carbosulfan, gamma-HCH

Weeds	Annual dicotyledons	carbetamide, chloridazon, chloridazon + ethofumesate, chloridazon + lenacil, clopyralid, desmedipham + ethofumesate + phenmedipham, desmedipham + phenmedipham, ethofumesate, ethofumesate + metamitron + phenmedipham, ethofumesate + phenmedipham, glufosinate-ammonium *(pre-emergence)*, lenacil, lenacil + phenmedipham, metamitron, paraquat, phenmedipham, propyzamide, trifluralin, triflusulfuron-methyl
	Annual grasses	carbetamide, cycloxydim, diclofop-methyl, fluazifop-P-butyl, glufosinate-ammonium *(pre-emergence)*, metamitron, paraquat, propaquizafop, propyzamide, quizalofop-P-ethyl, trifluralin
	Annual meadow grass	chloridazon, chloridazon + ethofumesate, chloridazon + lenacil, desmedipham + ethofumesate + phenmedipham, ethofumesate, ethofumesate + metamitron + phenmedipham, ethofumesate + phenmedipham, lenacil, metamitron, tri-allate
	Annual weeds	glyphosate (agricultural) *(post sowing, pre crop emergence)*, glyphosate (agricultural) *(pre-drilling/pre-emergence)*, glyphosate (horticulture, forestry, amenity etc.) *(pre-drilling/pre-emergence)*
	Barren brome	sethoxydim
	Blackgrass	cycloxydim, diclofop-methyl, ethofumesate, ethofumesate + phenmedipham, sethoxydim, tri-allate
	Canary grass	diclofop-methyl
	Corn marigold	clopyralid
	Couch	cycloxydim, quizalofop-P-ethyl, sethoxydim
	Creeping bent	cycloxydim, sethoxydim
	Creeping thistle	clopyralid
	Fat-hen	metamitron
	Mayweeds	clopyralid
	Perennial dicotyledons	clopyralid
	Perennial grasses	cycloxydim, diclofop-methyl, fluazifop-P-butyl, propaquizafop, propyzamide, quizalofop-P-ethyl, sethoxydim, tri-allate
	Polygonums	sodium chloride (commodity substance)
	Rough meadow grass	diclofop-methyl
	Ryegrass	diclofop-methyl
	Volunteer cereals	carbetamide, cycloxydim, fluazifop-P-butyl, glyphosate (agricultural) *(post sowing, pre crop emergence)*, glyphosate (agricultural) *(pre-drilling/pre-emergence)*, glyphosate (horticulture, forestry, amenity etc.) *(pre-drilling/pre-emergence)*, paraquat, propyzamide, quizalofop-P-ethyl, sethoxydim
	Volunteer potatoes	sodium chloride (commodity substance)

	Volunteer sugar beet	clodinafop-propargyl + diflufenican *(wiper application)*, glyphosate (agricultural) *(wick/wiper)*, glyphosate (agricultural) *(wiper application)*, glyphosate (horticulture, forestry, amenity etc.) *(wick/wiper)*
	Weed beet	glyphosate (agricultural) *(wiper application)*, glyphosate (horticulture, forestry, amenity etc.) *(wiper application)*
	Wild oats	cycloxydim, diclofop-methyl, fluazifop-P-butyl, propyzamide, sethoxydim, tri-allate
Plant growth regulation	Increasing yield	sulfur

Cereals

Diseases	Alternaria	iprodione
	Blue mould	fuberidazole + triadimenol
	Botrytis	carbendazim + chlorothalonil + maneb, fenpropidin + tebuconazole, iprodione, tebuconazole + tridemorph
	Brown foot rot and ear blight	bitertanol + fuberidazole + imidacloprid *(seed treatment - reduction)*, bromuconazole, carbendazim + chlorothalonil + maneb, carboxin + thiram, epoxiconazole + fenpropimorph + kresoxim-methyl *(reduction)*, epoxiconazole + fenpropimorph *(reduction)*, epoxiconazole + kresoxim-methyl *(reduction)*, epoxiconazole *(reduction)*, fenpropidin + tebuconazole, fludioxonil, fuberidazole + triadimenol, guazatine + imazalil *(reduction)*, guazatine *(reduction)*, imazalil *(seed treatment)*, spiroxamine + tebuconazole, tebuconazole, tebuconazole + triadimenol, tebuconazole + tridemorph
	Covered smut	ethirimol + flutriafol + thiabendazole, fludioxonil, fuberidazole + triadimenol
	Crown rust	cyproconazole, fuberidazole + triadimenol, tebuconazole + triadimenol, tebuconazole *(reduction)*, triadimenol + tridemorph
	Eyespot	benomyl, bromuconazole *(reduction)*, carbendazim, carbendazim + chlorothalonil + maneb, carbendazim + cyproconazole, carbendazim + flusilazole, carbendazim + mancozeb, carbendazim + maneb, carbendazim + maneb + tridemorph, carbendazim + prochloraz, carbendazim + propiconazole, chlorothalonil + cyproconazole, cyproconazole, cyproconazole + prochloraz, cyproconazole + quinoxyfen *(reduction)*, cyproconazole + tridemorph, cyprodinil, epoxiconazole + fenpropimorph + kresoxim-methyl *(reduction)*, epoxiconazole + fenpropimorph *(reduction)*, epoxiconazole + kresoxim-methyl *(reduction)*, epoxiconazole *(reduction)*, fenpropidin + prochloraz, fenpropimorph + prochloraz, flusilazole, maneb, prochloraz, prochloraz + tebuconazole, propiconazole *(low levels only)*, propiconazole *(with carbendazim)*
	Foot rot	fludioxonil, guazatine + imazalil

Fusarium root rot	bitertanol + fuberidazole, bitertanol + fuberidazole + imidacloprid *(seed treatment)*
Late ear diseases	azoxystrobin, azoxystrobin + fenpropimorph, chlorothalonil, chlorothalonil + flutriafol, cyproconazole, prochloraz + tebuconazole
Leaf stripe	carboxin + thiram *(partial control)*, ethirimol + flutriafol + thiabendazole, fludioxonil *(partial control)*, fuberidazole + triadimenol, guazatine + imazalil, imazalil *(seed treatment)*, imidacloprid + tebuconazole + triazoxide *(seed treatment)*, tebuconazole + triazoxide
Loose smut	bitertanol + fuberidazole + imidacloprid *(seed treatment - partial control)*, bitertanol + fuberidazole *(partial control)*, carboxin + thiram *(partial control)*, ethirimol + flutriafol + thiabendazole, fuberidazole + triadimenol, imidacloprid + tebuconazole + triazoxide *(seed treatment)*, tebuconazole + triazoxide
Net blotch	azoxystrobin, azoxystrobin + fenpropimorph, bromuconazole, carbendazim + chlorothalonil + maneb, carbendazim + cyproconazole, carbendazim + flusilazole, carbendazim + mancozeb, carbendazim + maneb + sulfur, carbendazim + prochloraz, carbendazim + propiconazole, chlorothalonil + cyproconazole, chlorothalonil + propiconazole, cyproconazole, cyproconazole + prochloraz, cyproconazole + tridemorph, cyprodinil, epoxiconazole, epoxiconazole + fenpropimorph, epoxiconazole + fenpropimorph + kresoxim-methyl, epoxiconazole + kresoxim-methyl, ethirimol + flutriafol + thiabendazole, fenpropidin + prochloraz, fenpropidin + propiconazole, fenpropidin + propiconazole + tebuconazole, fenpropidin + tebuconazole, fenpropimorph + flusilazole, fenpropimorph + flusilazole + tridemorph, fenpropimorph + flusilazole + tridemorph *(reduction)*, fenpropimorph + prochloraz, fenpropimorph + propiconazole, flusilazole, flusilazole + tridemorph, guazatine + imazalil, imazalil *(seed treatment)*, iprodione, mancozeb, maneb, prochloraz, prochloraz + tebuconazole, propiconazole, propiconazole + tebuconazole, propiconazole + tridemorph, spiroxamine + tebuconazole, tebuconazole, tebuconazole + triadimenol, tebuconazole + triazoxide, tebuconazole + tridemorph

Powdery mildew	azoxystrobin, azoxystrobin + fenpropimorph, bromuconazole, carbendazim, carbendazim + chlorothalonil + maneb, carbendazim + cyproconazole, carbendazim + flusilazole, carbendazim + flutriafol, carbendazim + mancozeb, carbendazim + maneb, carbendazim + maneb + sulfur, carbendazim + maneb + tridemorph, carbendazim + prochloraz, carbendazim + propiconazole, carbendazim *(partial control)*, chlorothalonil + cyproconazole, chlorothalonil + flutriafol, chlorothalonil + propiconazole, copper oxychloride + maneb + sulfur, cyproconazole, cyproconazole + prochloraz, cyproconazole + quinoxyfen, cyproconazole + tridemorph, cyprodinil, epoxiconazole, epoxiconazole + fenpropimorph, epoxiconazole + fenpropimorph + kresoxim-methyl, epoxiconazole + kresoxim-methyl, ethirimol + flutriafol + thiabendazole, fenbuconazole + propiconazole *(moderate control)*, fenbuconazole + tridemorph, fenpropidin, fenpropidin + fenpropimorph, fenpropidin + prochloraz, fenpropidin + propiconazole, fenpropidin + propiconazole + tebuconazole, fenpropidin + tebuconazole, fenpropimorph, fenpropimorph + flusilazole, fenpropimorph + flusilazole + tridemorph, fenpropimorph + kresoxim-methyl, fenpropimorph + prochloraz, fenpropimorph + propiconazole, fenpropimorph + quinoxyfen, fenpropimorph + tridemorph, flusilazole, flusilazole + tridemorph, flutriafol, fuberidazole + triadimenol, maneb, prochloraz, prochloraz + tebuconazole, prochloraz *(protection)*, propiconazole, propiconazole + tebuconazole, propiconazole + tridemorph, quinoxyfen, spiroxamine, spiroxamine + tebuconazole, sulfur, tebuconazole, tebuconazole + triadimenol, tebuconazole + tridemorph, triadimefon, triadimefon *(off-label (research/breeding))*, triadimenol, triadimenol + tridemorph, tridemorph
Pyrenophora leaf spot	fuberidazole + triadimenol, guazatine + imazalil

Rhynchosporium	azoxystrobin, azoxystrobin + fenpropimorph, benomyl, bromuconazole, carbendazim, carbendazim + chlorothalonil, carbendazim + chlorothalonil + maneb, carbendazim + cyproconazole, carbendazim + flusilazole, carbendazim + flutriafol, carbendazim + mancozeb, carbendazim + maneb, carbendazim + maneb + sulfur, carbendazim + maneb + tridemorph, carbendazim + prochloraz, carbendazim + propiconazole, chlorothalonil, chlorothalonil + cyproconazole, chlorothalonil + propiconazole, copper oxychloride + maneb + sulfur, cyproconazole, cyproconazole + prochloraz, cyproconazole + tridemorph, cyprodinil, epoxiconazole, epoxiconazole + fenpropimorph, epoxiconazole + fenpropimorph + kresoxim-methyl, epoxiconazole + kresoxim-methyl, ethirimol + flutriafol + thiabendazole, fenbuconazole + tridemorph, fenpropidin, fenpropidin + fenpropimorph *(moderate control)*, fenpropidin + prochloraz, fenpropidin + propiconazole, fenpropidin + propiconazole + tebuconazole, fenpropidin + tebuconazole, fenpropimorph, fenpropimorph + flusilazole, fenpropimorph + flusilazole + tridemorph, fenpropimorph + kresoxim-methyl, fenpropimorph + prochloraz, fenpropimorph + propiconazole, fenpropimorph + tridemorph, flusilazole, flusilazole + tridemorph, flutriafol, mancozeb, maneb, prochloraz, prochloraz + propiconazole, prochloraz + tebuconazole, propiconazole, propiconazole + tebuconazole, propiconazole + tridemorph, spiroxamine + tebuconazole, spiroxamine *(reduction)*, tebuconazole, tebuconazole + triadimenol, tebuconazole + tridemorph, triadimefon, triadimenol, triadimenol + tridemorph

Rust	azoxystrobin, azoxystrobin + fenpropimorph, bromuconazole, carbendazim + chlorothalonil + maneb, carbendazim + cyproconazole, carbendazim + flusilazole, carbendazim + flutriafol, carbendazim + mancozeb, carbendazim + maneb, carbendazim + maneb + sulfur, carbendazim + maneb + tridemorph, carbendazim + propiconazole, chlorothalonil + cyproconazole, chlorothalonil + flutriafol, chlorothalonil + propiconazole, cyproconazole, cyproconazole + prochloraz, cyproconazole + quinoxyfen, cyproconazole + tridemorph, difenoconazole, epoxiconazole, epoxiconazole + fenpropimorph, epoxiconazole + fenpropimorph + kresoxim-methyl, epoxiconazole + kresoxim-methyl, ethirimol + flutriafol + thiabendazole, fenbuconazole + propiconazole, fenbuconazole + tridemorph, fenpropidin, fenpropidin + fenpropimorph, fenpropidin + prochloraz, fenpropidin + propiconazole, fenpropidin + propiconazole + tebuconazole, fenpropidin + tebuconazole, fenpropimorph, fenpropimorph + flusilazole, fenpropimorph + flusilazole + tridemorph, fenpropimorph + prochloraz, fenpropimorph + propiconazole, flusilazole, flusilazole + tridemorph, flutriafol, fuberidazole + triadimenol, mancozeb, maneb, oxycarboxin, prochloraz + tebuconazole, propiconazole, propiconazole + tebuconazole, propiconazole + tridemorph, spiroxamine, spiroxamine + tebuconazole, spiroxamine *(moderate control)*, tebuconazole, tebuconazole + triadimenol, tebuconazole + tridemorph, triadimefon, triadimefon *(off-label (research/breeding))*, triadimenol, triadimenol + tridemorph
Septoria diseases	azoxystrobin, azoxystrobin + fenpropimorph, bitertanol + fuberidazole *(reduction)*, bromuconazole, carbendazim + chlorothalonil, carbendazim + chlorothalonil + maneb, carbendazim + cyproconazole, carbendazim + flusilazole, carbendazim + flutriafol, carbendazim + mancozeb, carbendazim + maneb, carbendazim + maneb + sulfur, carbendazim + prochloraz, carbendazim + propiconazole, carbendazim *(partial control)*, carboxin + thiram, chlorothalonil, chlorothalonil + cyproconazole, chlorothalonil + flutriafol, chlorothalonil + propiconazole, copper oxychloride + maneb + sulfur, cyproconazole, cyproconazole + prochloraz, cyproconazole + quinoxyfen, cyproconazole + tridemorph, difenoconazole, epoxiconazole, epoxiconazole + fenpropimorph, epoxiconazole + fenpropimorph + kresoxim-methyl, epoxiconazole + kresoxim-methyl, fenbuconazole + propiconazole, fenbuconazole + tridemorph, fenpropidin, fenpropidin + prochloraz, fenpropidin + propiconazole, fenpropidin + propiconazole + tebuconazole, fenpropidin + tebuconazole, fenpropimorph + flusilazole, fenpropimorph + flusilazole + tridemorph, fenpropimorph + kresoxim-methyl *(reduction)*, fenpropimorph + prochloraz, fenpropimorph + propiconazole, flusilazole, flusilazole + tridemorph, flutriafol, fuberidazole + triadimenol, guazatine, iprodione, mancozeb, maneb, prochloraz, prochloraz + propiconazole, prochloraz + tebuconazole, propiconazole, propiconazole + tebuconazole, propiconazole + tridemorph, spiroxamine + tebuconazole, tebuconazole, tebuconazole + triadimenol, tebuconazole + tridemorph, triadimenol, triadimenol + tridemorph

	Snow mould	fludioxonil
	Snow rot	propiconazole, triadimefon, triadimenol, triadimenol + tridemorph
	Sooty moulds	bromuconazole, carbendazim, carbendazim + chlorothalonil + maneb, carbendazim + cyproconazole, carbendazim + mancozeb, carbendazim + maneb, carbendazim + maneb + tridemorph, cyproconazole + tridemorph, epoxiconazole + fenpropimorph + kresoxim-methyl *(reduction)*, epoxiconazole + fenpropimorph *(reduction)*, epoxiconazole + kresoxim-methyl *(reduction)*, epoxiconazole *(reduction)*, fenpropidin + tebuconazole, fenpropimorph + propiconazole, mancozeb, maneb, propiconazole, spiroxamine + tebuconazole, tebuconazole, tebuconazole + triadimenol, tebuconazole + tridemorph
	Stinking smut	bitertanol + fuberidazole, bitertanol + fuberidazole + imidacloprid *(seed treatment)*, carboxin + thiram, fludioxonil, fuberidazole + triadimenol, guazatine
Pests	Aphids	alpha-cypermethrin, chlorpyrifos, cypermethrin, deltamethrin, deltamethrin + pirimicarb *(on ears)*, demeton-S-methyl, dimethoate, esfenvalerate, heptenophos, lambda-cyhalothrin, lambda-cyhalothrin + pirimicarb, phosalone, pirimicarb, pirimicarb *(off-label - research/breeding)*, tau-fluvalinate, zeta-cypermethrin
	Barley yellow dwarf virus vectors	cypermethrin, deltamethrin, lambda-cyhalothrin, tau-fluvalinate, zeta-cypermethrin
	Birds	aluminium ammonium sulfate
	Cutworms	gamma-HCH
	Damaging mammals	aluminium ammonium sulfate, sulfonated cod liver oil
	Frit fly	chlorpyrifos, fonofos
	Leatherjackets	chlorpyrifos, fenitrothion, gamma-HCH, methiocarb *(reduction)*
	Saddle gall midge	fenitrothion
	Slugs and snails	metaldehyde, metaldehyde *(seed admixture)*, methiocarb, methiocarb *(seed admixture)*, thiodicarb, thiodicarb *(seed admixture)*
	Thrips	chlorpyrifos, fenitrothion
	Virus vectors	bitertanol + fuberidazole + imidacloprid *(seed treatment)*, imidacloprid + tebuconazole + triazoxide *(seed treatment)*
	Wheat bulb fly	chlorpyrifos, dimethoate, fonofos, pirimiphos-methyl, tefluthrin *(seed treatment)*
	Wheat-blossom midges	chlorpyrifos, fenitrothion
	Wireworms	fonofos, gamma-HCH, gamma-HCH *(seed treatment)*, tefluthrin *(seed treatment)*

	Yellow cereal fly	alpha-cypermethrin, cypermethrin, deltamethrin, lambda-cyhalothrin
Weeds	Annual and perennial weeds	clodinafop-propargyl + diflufenican *(pre-harvest)*, glyphosate (agricultural) *(pre-harvest)*, glyphosate (horticulture, forestry, amenity etc.) *(pre-harvest)*
	Annual dicotyledons	2,4-D, 2,4-DB + MCPA, amidosulfuron, benazolin + bromoxynil + ioxynil, bifenox + dicamba, bromoxynil, bromoxynil + clopyralid + fluroxypyr + ioxynil, bromoxynil + diflufenican + ioxynil, bromoxynil + fluroxypyr, bromoxynil + fluroxypyr + ioxynil, bromoxynil + ioxynil, bromoxynil + ioxynil + mecoprop-P, bromoxynil + ioxynil + triasulfuron, carfentrazone-ethyl + flupyrsulfuron-methyl, carfentrazone-ethyl + isoproturon, carfentrazone-ethyl + mecoprop-P, carfentrazone-ethyl + metsulfuron-methyl, chlorotoluron, clodinafop-propargyl + diflufenican, clodinafop-propargyl + trifluralin, clopyralid, clopyralid + fluroxypyr + ioxynil, cyanazine, cyanazine + pendimethalin, cyanazine + terbuthylazine, dicamba + MCPA + mecoprop, dicamba + MCPA + mecoprop-P, dicamba + mecoprop-P, dichlorprop, dichlorprop + MCPA, diflufenican + flurtamone + isoproturon, diflufenican + isoproturon, diflufenican + terbuthylazine, diflufenican + trifluralin, fluoroglycofen-ethyl + isoproturon, flupyrsulfuron-methyl, flupyrsulfuron-methyl + metsulfuron-methyl, fluroxypyr, fluroxypyr + metosulam, glufosinate-ammonium *(pre-harvest)*, imazamethabenz-methyl, isoproturon, isoproturon + pendimethalin, isoproturon + simazine, isoproturon + trifluralin, linuron, linuron + trifluralin, MCPA, MCPA + MCPB, mecoprop, mecoprop-P, methabenzthiazuron, metoxuron, metsulfuron-methyl, metsulfuron-methyl + thifensulfuron-methyl, pendimethalin, pendimethalin + simazine, pyridate, terbutryn, terbutryn + trifluralin, thifensulfuron-methyl + tribenuron-methyl, triasulfuron, tribenuron-methyl, trifluralin
	Annual grasses	chlorotoluron, cyanazine, diclofop-methyl, diflufenican + isoproturon, glufosinate-ammonium *(pre-harvest)*, isoproturon, isoproturon + pendimethalin, isoproturon + simazine, isoproturon + trifluralin, linuron + trifluralin, metoxuron, pendimethalin, terbutryn, terbutryn + trifluralin, trifluralin
	Annual meadow grass	carfentrazone-ethyl + isoproturon, cyanazine + pendimethalin, cyanazine + terbuthylazine, diflufenican + flurtamone + isoproturon, diflufenican + terbuthylazine, diflufenican + trifluralin, fenoxaprop-P-ethyl + isoproturon, fluoroglycofen-ethyl + isoproturon, isoproturon *(off-label)*, linuron, linuron + trifluralin, methabenzthiazuron, pendimethalin, terbutryn, terbutryn + trifluralin, tri-allate
	Annual weeds	glyphosate (agricultural) *(post sowing, pre crop emergence)*, glyphosate (agricultural) *(pre-drilling/pre-emergence)*, glyphosate (horticulture, forestry, amenity etc.) *(pre-drilling/pre-emergence)*
	Barren brome	metoxuron
	Black bindweed	dichlorprop, dichlorprop + MCPA, fluroxypyr, linuron

Blackgrass	carfentrazone-ethyl + flupyrsulfuron-methyl, chlorotoluron, clodinafop-propargyl, clodinafop-propargyl + diflufenican, clodinafop-propargyl + trifluralin, diclofop-methyl, diclofop-methyl + fenoxaprop-P-ethyl, diflufenican + flurtamone + isoproturon, diflufenican + isoproturon, fenoxaprop-ethyl, fenoxaprop-P-ethyl, fenoxaprop-P-ethyl + isoproturon, flupyrsulfuron-methyl, flupyrsulfuron-methyl + metsulfuron-methyl, imazamethabenz-methyl, isoproturon, isoproturon + pendimethalin, isoproturon + simazine, isoproturon + trifluralin, methabenzthiazuron, metoxuron, pendimethalin, terbutryn, terbutryn + trifluralin, tralkoxydim, tri-allate
Canary grass	diclofop-methyl, fenoxaprop-P-ethyl, fenoxaprop-P-ethyl + isoproturon
Chickweed	bromoxynil + clopyralid + fluroxypyr + ioxynil, bromoxynil + diflufenican + ioxynil, bromoxynil + fluroxypyr + ioxynil, carfentrazone-ethyl + isoproturon, carfentrazone-ethyl + mecoprop-P, clopyralid + fluroxypyr + ioxynil, cyanazine + pendimethalin, dicamba + MCPA + mecoprop, dicamba + MCPA + mecoprop-P, dicamba + mecoprop-P, fenoxaprop-P-ethyl + isoproturon, fluroxypyr, fluroxypyr + metosulam, linuron, mecoprop, mecoprop-P, metsulfuron-methyl, terbutryn + trifluralin, thifensulfuron-methyl + tribenuron-methyl, triasulfuron, tribenuron-methyl
Cleavers	amidosulfuron, bifenox + dicamba, bromoxynil + clopyralid + fluroxypyr + ioxynil, bromoxynil + fluroxypyr + ioxynil, carfentrazone-ethyl + isoproturon, carfentrazone-ethyl + mecoprop-P, clopyralid + fluroxypyr + ioxynil, dicamba + MCPA + mecoprop, dicamba + MCPA + mecoprop-P, dicamba + mecoprop-P, fluoroglycofen-ethyl + isoproturon, fluroxypyr, fluroxypyr + metosulam, mecoprop, mecoprop-P, metsulfuron-methyl + thifensulfuron-methyl, pendimethalin, pyridate
Corn marigold	clopyralid, fenoxaprop-P-ethyl + isoproturon, linuron
Couch	amitrole *(direct-drilled)*, clodinafop-propargyl + diflufenican *(pre-harvest)*, glyphosate (agricultural) *(pre-harvest)*, glyphosate (horticulture, forestry, amenity etc.) *(pre-harvest)*
Creeping bent	glyphosate (agricultural) *(pre-harvest)*
Creeping thistle	clopyralid
Docks	dicamba + MCPA + mecoprop, fluroxypyr
Fat-hen	linuron, MCPA
Field pansy	bifenox + dicamba, cyanazine + pendimethalin, metsulfuron-methyl + thifensulfuron-methyl
Hemp-nettle	bromoxynil + clopyralid + fluroxypyr + ioxynil, bromoxynil + fluroxypyr + ioxynil, clopyralid + fluroxypyr + ioxynil, dichlorprop + MCPA, fenoxaprop-P-ethyl + isoproturon, fluroxypyr, fluroxypyr + metosulam, MCPA

Loose silky bent	carfentrazone-ethyl + isoproturon, diflufenican + flurtamone + isoproturon, fenoxaprop-P-ethyl, imazamethabenz-methyl
Mayweeds	bromoxynil + clopyralid + fluroxypyr + ioxynil, bromoxynil + diflufenican + ioxynil, carfentrazone-ethyl + isoproturon, clopyralid, clopyralid + fluroxypyr + ioxynil, cyanazine + pendimethalin, dicamba + MCPA + mecoprop, dicamba + MCPA + mecoprop-P, dicamba + mecoprop-P, fenoxaprop-P-ethyl + isoproturon, metsulfuron-methyl, terbutryn + trifluralin, thifensulfuron-methyl + tribenuron-methyl, triasulfuron, tribenuron-methyl
Perennial dicotyledons	2,4-D, 2,4-DB + MCPA, clopyralid, dicamba + MCPA + mecoprop, dicamba + MCPA + mecoprop-P, dicamba + mecoprop-P, dichlorprop, dichlorprop + MCPA, glyphosate (agricultural) *(pre-harvest)*, MCPA, MCPA + MCPB, mecoprop, mecoprop-P
Perennial grasses	diclofop-methyl, glyphosate (agricultural) *(pre-harvest)*, imazamethabenz-methyl, pendimethalin + simazine, tri-allate
Perennial ryegrass	linuron + trifluralin, terbutryn
Polygonums	2,4-DB + MCPA, bifenox + dicamba, bromoxynil + diflufenican + ioxynil, dicamba + MCPA + mecoprop, dicamba + MCPA + mecoprop-P, dicamba + mecoprop-P, dichlorprop, dichlorprop + MCPA, linuron, metsulfuron-methyl + thifensulfuron-methyl
Rough meadow grass	carfentrazone-ethyl + isoproturon, chlorotoluron, clodinafop-propargyl, clodinafop-propargyl + diflufenican, clodinafop-propargyl + trifluralin, cyanazine + pendimethalin, diclofop-methyl, fenoxaprop-P-ethyl, isoproturon, linuron + trifluralin, methabenzthiazuron, terbutryn
Ryegrass	diclofop-methyl, diclofop-methyl + fenoxaprop-P-ethyl
Speedwells	bifenox + dicamba, bromoxynil + clopyralid + fluroxypyr + ioxynil, bromoxynil + diflufenican + ioxynil, bromoxynil + fluroxypyr + ioxynil, carfentrazone-ethyl + mecoprop-P, clopyralid + fluroxypyr + ioxynil, metsulfuron-methyl + thifensulfuron-methyl, pendimethalin, pyridate, terbutryn + trifluralin
Volunteer cereals	glyphosate (agricultural) *(post sowing, pre crop emergence)*, glyphosate (agricultural) *(pre-drilling/pre-emergence)*, glyphosate (horticulture, forestry, amenity etc.) *(pre-drilling/pre-emergence)*
Volunteer oilseed rape	fluroxypyr + metosulam, imazamethabenz-methyl
Volunteer potatoes	amitrole *(stubble)*, fluroxypyr

	Wild oats	chlorotoluron, clodinafop-propargyl, clodinafop-propargyl + diflufenican, clodinafop-propargyl + diflufenican *(wiper glove)*, clodinafop-propargyl + trifluralin, diclofop-methyl, diclofop-methyl + fenoxaprop-P-ethyl, difenzoquat, diflufenican + isoproturon, fenoxaprop-ethyl, fenoxaprop-P-ethyl, fenoxaprop-P-ethyl + isoproturon, flamprop-M-isopropyl, glyphosate (agricultural) *(wiper glove)*, glyphosate (horticulture, forestry, amenity etc.) *(wiper glove)*, imazamethabenz-methyl, isoproturon, isoproturon + pendimethalin, pendimethalin, tralkoxydim, tri-allate
Crop control	Pre-harvest desiccation	clodinafop-propargyl + diflufenican, diquat, glufosinate-ammonium, glyphosate (agricultural)
Plant growth regulation	Increasing yield	2-chloroethylphosphonic acid + mepiquat chloride *(low lodging situations)*, chlormequat, chlormequat + choline chloride, chlormequat + choline chloride + imazaquin, sulfur
	Lodging control	2-chloroethylphosphonic acid, 2-chloroethylphosphonic acid + mepiquat chloride, 2-chloroethylphosphonic acid + mepiquat chloride *(some cultivars only)*, chlormequat, chlormequat + 2-chloroethylphosphonic acid, chlormequat + 2-chloroethylphosphonic acid + imazaquin, chlormequat + 2-chloroethylphosphonic acid + mepiquat chloride, chlormequat + choline chloride, chlormequat + choline chloride + imazaquin, chlormequat + mepiquat chloride, trinexapac-ethyl

Cereals undersown

Weeds	Annual dicotyledons	2,4-D, 2,4-DB, 2,4-DB + linuron + MCPA, 2,4-DB + MCPA, benazolin + 2,4-DB + MCPA, bentazone + MCPA + MCPB, bromoxynil + ioxynil, dicamba + MCPA + mecoprop *(grass only)*, dicamba + MCPA + mecoprop-P, dicamba + MCPA + mecoprop-P *(grass only)*, dichlorprop, dichlorprop + MCPA, MCPA, MCPA + MCPB, MCPB
	Black bindweed	dichlorprop, dichlorprop + MCPA
	Chickweed	benazolin + 2,4-DB + MCPA, dicamba + MCPA + mecoprop *(grass only)*, dicamba + MCPA + mecoprop-P, dicamba + MCPA + mecoprop-P *(grass only)*
	Cleavers	benazolin + 2,4-DB + MCPA, dicamba + MCPA + mecoprop-P
	Fat-hen	MCPA
	Hemp-nettle	dichlorprop + MCPA, MCPA
	Mayweeds	dicamba + MCPA + mecoprop *(grass only)*, dicamba + MCPA + mecoprop-P, dicamba + MCPA + mecoprop-P *(grass only)*

	Perennial dicotyledons	2,4-D, 2,4-DB, 2,4-DB + MCPA, benazolin + 2,4-DB + MCPA, dicamba + MCPA + mecoprop *(grass only)*, dicamba + MCPA + mecoprop-P, dicamba + MCPA + mecoprop-P *(grass only)*, dichlorprop, dichlorprop + MCPA, MCPA + MCPB, MCPB
	Polygonums	2,4-DB + MCPA, benazolin + 2,4-DB + MCPA, dicamba + MCPA + mecoprop *(grass only)*, dicamba + MCPA + mecoprop-P, dicamba + MCPA + mecoprop-P *(grass only)*, dichlorprop, dichlorprop + MCPA

Clovers

Weeds	Annual dicotyledons	2,4-DB + MCPA, benazolin + 2,4-DB + MCPA, carbetamide, MCPA + MCPB, MCPB, propyzamide
	Annual grasses	carbetamide, propyzamide
	Chickweed	benazolin + 2,4-DB + MCPA
	Cleavers	benazolin + 2,4-DB + MCPA
	Docks	asulam *(off-label)*
	Perennial dicotyledons	2,4-DB + MCPA, benazolin + 2,4-DB + MCPA, MCPA + MCPB, MCPB
	Perennial grasses	propyzamide
	Polygonums	2,4-DB + MCPA, benazolin + 2,4-DB + MCPA
	Volunteer cereals	carbetamide
Crop control	Pre-harvest desiccation	diquat

Fodder brassica seed crops

Diseases	Alternaria	iprodione
	Botrytis	iprodione
Pests	Aphids	dimethoate *(excluding Myzus persicae)*
	Leaf miners	dimethoate
	Pod midge	phosalone
	Pollen beetles	phosalone
	Seed weevil	phosalone
Weeds	Annual dicotyledons	benazolin + clopyralid *(off-label)*, carbetamide, propyzamide
	Annual grasses	carbetamide, propyzamide
	Chickweed	benazolin + clopyralid *(off-label)*
	Cleavers	benazolin + clopyralid *(off-label)*
	Mayweeds	benazolin + clopyralid *(off-label)*
	Perennial grasses	propyzamide
	Volunteer cereals	carbetamide

Fodder brassicas

Diseases	Alternaria	iprodione, iprodione *(seed treatment)*
	Botrytis	chlorothalonil *(moderate control)*
	Damping off	gamma-HCH + thiram
	Downy mildew	chlorothalonil *(moderate control)*, chlorothalonil *(seedlings only)*
	Ring spot	chlorothalonil, tebuconazole *(off-label)*
Pests	Aphids	alpha-cypermethrin, chlorpyrifos, chlorpyrifos + dimethoate, cypermethrin, deltamethrin + pirimicarb, nicotine, pirimicarb
	Cabbage root fly	carbofuran, chlorpyrifos, chlorpyrifos + dimethoate
	Cabbage stem flea beetle	cypermethrin
	Cabbage stem weevil	carbofuran
	Caterpillars	alpha-cypermethrin, chlorpyrifos, cypermethrin, deltamethrin, deltamethrin + pirimicarb, nicotine
	Cutworms	chlorpyrifos, chlorpyrifos + dimethoate
	Flea beetles	alpha-cypermethrin, carbofuran, cypermethrin, gamma-HCH + thiram
	General insect control	dimethoate *(off-label)*
	Leatherjackets	chlorpyrifos, chlorpyrifos + dimethoate
	Whitefly	chlorpyrifos
	Wireworms	chlorpyrifos + dimethoate
Weeds	Annual dicotyledons	aziprotryne *(off-label)*, chlorthal-dimethyl, clopyralid, cyanazine *(off-label)*, desmetryn, desmetryn *(off-label)*, propachlor, sodium monochloroacetate, tebutam, trifluralin
	Annual grasses	cyanazine *(off-label)*, fluazifop-P-butyl *(off-label)*, fluazifop-P-butyl *(stockfeed only)*, propachlor, tebutam, trifluralin
	Annual meadow grass	aziprotryne *(off-label)*
	Corn marigold	clopyralid
	Creeping thistle	clopyralid
	Fat-hen	desmetryn, desmetryn *(off-label)*
	Mayweeds	clopyralid
	Perennial dicotyledons	clopyralid
	Perennial grasses	fluazifop-P-butyl *(off-label)*, fluazifop-P-butyl *(stockfeed only)*
	Volunteer cereals	fluazifop-P-butyl *(stockfeed only)*, tebutam
	Wild oats	fluazifop-P-butyl *(stockfeed only)*

FOR FULL CONDITIONS OF USE ALWAYS READ THE PRODUCT LABEL

General

Diseases	Soil-borne diseases	dazomet, methyl bromide with amyl acetate, methyl bromide with chloropicrin
Pests	Birds	aluminium ammonium sulfate, quinalbarbitone-sodium, ziram
	Damaging mammals	aluminium ammonium sulfate, ziram
	Nematodes	dazomet, methyl bromide with amyl acetate, methyl bromide with chloropicrin
	Slugs and snails	aluminium sulfate, metaldehyde, methiocarb
	Soil pests	dazomet, methyl bromide with amyl acetate, methyl bromide with chloropicrin
Weeds	Annual and perennial weeds	clodinafop-propargyl + diflufenican, clodinafop-propargyl + diflufenican *(stubble treatment)*, clodinafop-propargyl + diflufenican *(sward destruction/direct drilling)*, glyphosate (agricultural), glyphosate (agricultural) *(stubble treatment)*, glyphosate (agricultural) *(sward destruction/direct drilling)*, glyphosate (agricultural) *(wiper application)*, glyphosate (horticulture, forestry, amenity etc.), glyphosate (horticulture, forestry, amenity etc.) *(stubble treatment)*, glyphosate (horticulture, forestry, amenity etc.) *(sward destruction/direct drilling)*, glyphosate (horticulture, forestry, amenity etc.) *(wiper application)*
	Annual dicotyledons	diquat + paraquat *(autumn stubble, pre-planting/sowing, sward destruction)*, glufosinate-ammonium, glufosinate-ammonium *(pre-cropping situations)*, glufosinate-ammonium *(sward destruction)*, glyphosate (agricultural) *(pre-drilling/pre-crop emergence)*, glyphosate (agricultural) *(stubble treatment)*, paraquat *(conventional cultivation)*, paraquat *(direct drilling)*, paraquat *(minimum cultivation)*, paraquat *(stubble treatment)*
	Annual grasses	diquat + paraquat *(autumn stubble, pre-planting/sowing, sward destruction)*, fluazifop-P-butyl, glufosinate-ammonium, glufosinate-ammonium *(pre-cropping situations)*, glufosinate-ammonium *(sward destruction)*, glyphosate (agricultural) *(pre-drilling/pre-crop emergence)*, glyphosate (agricultural) *(stubble treatment)*, paraquat *(conventional cultivation)*, paraquat *(direct drilling)*, paraquat *(minimum cultivation)*, paraquat *(stubble treatment)*
	Annual weeds	amitrole, amitrole *(pre-sowing, autumn stubble)*, glufosinate-ammonium, glyphosate (agricultural) *(pre-drilling/pre-emergence)*, glyphosate (horticulture, forestry, amenity etc.) *(pre-drilling/pre-emergence)*
	Barren brome	fluazifop-P-butyl, paraquat *(conventional drilling)*, paraquat *(stubble treatment)*

Couch	amitrole, amitrole *(pre-sowing, autumn stubble)*, clodinafop-propargyl + diflufenican *(stubble treatment)*, clodinafop-propargyl + diflufenican *(sward destruction/direct drilling)*, glyphosate (agricultural) *(stubble treatment)*, glyphosate (agricultural) *(sward destruction/direct drilling)*, glyphosate (horticulture, forestry, amenity etc.) *(stubble treatment)*, glyphosate (horticulture, forestry, amenity etc.) *(sward destruction/direct drilling)*
Creeping bent	glyphosate (agricultural) *(stubble treatment)*, paraquat *(conventional drilling)*, paraquat *(direct drilling)*, paraquat *(minimum cultivation)*, paraquat *(stubble treatment)*
Docks	amitrole, amitrole *(pre-sowing, autumn stubble)*
General weed control	dazomet, methyl bromide with amyl acetate, methyl bromide with chloropicrin
Green cover	diquat + paraquat, paraquat
Perennial dicotyledons	glufosinate-ammonium, glufosinate-ammonium *(pre-cropping situations)*, glufosinate-ammonium *(sward destruction)*, glyphosate (agricultural) *(stubble treatment)*
Perennial grasses	diquat + paraquat *(autumn stubble, pre-planting/sowing, sward destruction)*, glufosinate-ammonium, glufosinate-ammonium *(pre-cropping situations)*, glufosinate-ammonium *(sward destruction)*, glyphosate (agricultural) *(pre-harvest)*, glyphosate (agricultural) *(stubble treatment)*
Perennial ryegrass	paraquat *(direct drilling)*
Perennial weeds	amitrole, amitrole *(pre-sowing, autumn stubble)*, glufosinate-ammonium
Rough meadow grass	paraquat *(direct drilling)*
Volunteer cereals	clodinafop-propargyl + diflufenican *(stubble treatment)*, clodinafop-propargyl + diflufenican *(sward destruction/direct drilling)*, diquat + paraquat *(autumn stubble, pre-planting/sowing, sward destruction)*, fluazifop-P-butyl, glyphosate (agricultural) *(pre-drilling/pre-crop emergence)*, glyphosate (agricultural) *(pre-drilling/pre-emergence)*, glyphosate (agricultural) *(stubble treatment)*, glyphosate (agricultural) *(sward destruction/direct drilling)*, glyphosate (horticulture, forestry, amenity etc.) *(pre-drilling/pre-emergence)*, glyphosate (horticulture, forestry, amenity etc.) *(stubble treatment)*, glyphosate (horticulture, forestry, amenity etc.) *(sward destruction/direct drilling)*, paraquat *(conventional drilling)*, paraquat *(minimum cultivation)*, paraquat *(stubble treatment)*
Volunteer potatoes	clodinafop-propargyl + diflufenican *(stubble treatment)*, glyphosate (agricultural) *(stubble treatment)*, glyphosate (horticulture, forestry, amenity etc.) *(stubble treatment)*

	Wild oats	fluazifop-P-butyl, paraquat *(conventional drilling)*, paraquat *(stubble treatment)*

Grass seed crops

Diseases	Crown rust	propiconazole
	Drechslera leaf spot	propiconazole
	Powdery mildew	propiconazole
	Rhynchosporium	propiconazole
Pests	Aphids	deltamethrin *(off-label)*, dimethoate
Weeds	Annual dicotyledons	2,4-D, 2,4-D + mecoprop-P, bromoxynil + ethofumesate + ioxynil, clopyralid *(off-label)*, dicamba + MCPA + mecoprop, dicamba + MCPA + mecoprop-P, MCPA, mecoprop-P
	Annual grasses	ethofumesate
	Annual meadow grass	bromoxynil + ethofumesate + ioxynil
	Blackgrass	ethofumesate
	Chickweed	dicamba + MCPA + mecoprop, dicamba + MCPA + mecoprop-P, ethofumesate, mecoprop-P
	Cleavers	dicamba + MCPA + mecoprop, dicamba + MCPA + mecoprop-P, ethofumesate, mecoprop-P
	Docks	dicamba + MCPA + mecoprop, dicamba + MCPA + mecoprop-P
	Fat-hen	MCPA
	Hemp-nettle	MCPA
	Mayweeds	dicamba + MCPA + mecoprop, dicamba + MCPA + mecoprop-P
	Perennial dicotyledons	2,4-D, 2,4-D + mecoprop-P, clopyralid *(off-label)*, dicamba + MCPA + mecoprop, dicamba + MCPA + mecoprop-P, MCPA, mecoprop-P
	Polygonums	dicamba + MCPA + mecoprop, dicamba + MCPA + mecoprop-P
	Ryegrass	chlorpropham *(off-label)*
	Volunteer cereals	ethofumesate
	Wild oats	difenzoquat

Linseed/flax

Diseases	Alternaria	carboxin + gamma-HCH + thiram, iprodione *(seed treatment)*
	Botrytis	carboxin + gamma-HCH + thiram, tebuconazole *(reduction)*
	Fusarium diseases	carboxin + gamma-HCH + thiram *(partial control)*
	Powdery mildew	tebuconazole
	Seed-borne diseases	prochloraz

Pests	Flea beetles	carboxin + gamma-HCH + thiram
Weeds	Annual and perennial weeds	clodinafop-propargyl + diflufenican *(pre-harvest)*, glyphosate (agricultural) *(pre-harvest)*, glyphosate (horticulture, forestry, amenity etc.) *(pre-harvest)*
	Annual dicotyledons	amidosulfuron, bentazone, bromoxynil, bromoxynil + clopyralid, clopyralid, glufosinate-ammonium *(pre-harvest)*, MCPA, metsulfuron-methyl, trifluralin, trifluralin *(off-label)*
	Annual grasses	cycloxydim, diclofop-methyl, fluazifop-P-butyl, glufosinate-ammonium *(pre-harvest)*, propaquizafop, quizalofop-P-ethyl, sethoxydim, trifluralin, trifluralin *(off-label)*
	Annual weeds	glyphosate (agricultural) *(post sowing, pre crop emergence)*
	Barren brome	sethoxydim
	Blackgrass	cycloxydim, diclofop-methyl, sethoxydim
	Canary grass	diclofop-methyl
	Chickweed	metsulfuron-methyl
	Cleavers	amidosulfuron
	Corn marigold	clopyralid
	Couch	clodinafop-propargyl + diflufenican *(pre-harvest)*, cycloxydim, glyphosate (agricultural) *(pre-harvest)*, quizalofop-P-ethyl
	Creeping bent	cycloxydim
	Creeping thistle	clopyralid
	Mayweeds	clopyralid, metsulfuron-methyl
	Perennial dicotyledons	clopyralid
	Perennial grasses	cycloxydim, diclofop-methyl, fluazifop-P-butyl, propaquizafop, quizalofop-P-ethyl
	Rough meadow grass	diclofop-methyl
	Ryegrass	diclofop-methyl
	Volunteer cereals	cycloxydim, fluazifop-P-butyl, glyphosate (agricultural) *(post sowing, pre crop emergence)*, quizalofop-P-ethyl, sethoxydim
	Wild oats	cycloxydim, diclofop-methyl, fluazifop-P-butyl, sethoxydim
Crop control	Pre-harvest desiccation	diquat, glufosinate-ammonium, glyphosate (agricultural), sulfuric acid (commodity substance)
Plant growth regulation	Lodging control	chlormequat + choline chloride
	Stem shortening	chlormequat + choline chloride

FOR FULL CONDITIONS OF USE ALWAYS READ THE PRODUCT LABEL

Lucerne

Weeds	Annual dicotyledons	2,4-DB, carbetamide, chlorpropham, paraquat *(off-label)*, propyzamide
	Annual grasses	carbetamide, chlorpropham, paraquat *(off-label)*, propyzamide
	Chickweed	chlorpropham
	Perennial dicotyledons	2,4-DB
	Perennial grasses	propyzamide
	Polygonums	chlorpropham
	Volunteer cereals	carbetamide

Lupins

Weeds	Annual dicotyledons	terbuthylazine + terbutryn
	Annual grasses	diclofop-methyl, sethoxydim, terbuthylazine + terbutryn
	Barren brome	sethoxydim
	Blackgrass	diclofop-methyl, sethoxydim
	Canary grass	diclofop-methyl
	Perennial grasses	diclofop-methyl
	Rough meadow grass	diclofop-methyl
	Ryegrass	diclofop-methyl
	Volunteer cereals	sethoxydim
	Wild oats	diclofop-methyl, sethoxydim

Maize/sweetcorn

Diseases	Damping off	thiram *(seed treatment)*
Pests	Aphids	nicotine, pirimicarb, pirimicarb *(off-label)*
	Caterpillars	nicotine
	Cutworms	gamma-HCH
	Frit fly	aldicarb *(off-label)*, bendiocarb, carbofuran, chlorpyrifos, fenitrothion, phorate
	General insect control	demeton-S-methyl *(off-label)*, dimethoate *(off-label)*
	Leatherjackets	fenitrothion, gamma-HCH
	Thrips	fenitrothion *(off-label)*
	Wireworms	bendiocarb, gamma-HCH
Weeds	Annual dicotyledons	atrazine, bromoxynil, bromoxynil + prosulfuron, bromoxynil *(off-label)*, clopyralid, cyanazine, cyanazine + pendimethalin, fluroxypyr, pendimethalin, pendimethalin *(off-label)*, pyridate, rimsulfuron, simazine
	Annual grasses	atrazine, cyanazine, pendimethalin *(off-label)*, simazine

	Annual meadow grass	cyanazine + pendimethalin, pendimethalin
	Black bindweed	bromoxynil + prosulfuron, fluroxypyr
	Black nightshade	pyridate
	Chickweed	bromoxynil + prosulfuron, fluroxypyr
	Cleavers	fluroxypyr, pyridate
	Corn marigold	clopyralid
	Creeping thistle	clopyralid
	Docks	fluroxypyr
	Fat-hen	pyridate
	Hemp-nettle	bromoxynil + prosulfuron, fluroxypyr
	Mayweeds	bromoxynil + prosulfuron, clopyralid
	Perennial dicotyledons	clopyralid
	Polygonums	bromoxynil + prosulfuron
	Rough meadow grass	cyanazine + pendimethalin
	Speedwells	pendimethalin, pyridate
	Volunteer oilseed rape	rimsulfuron
	Volunteer potatoes	fluroxypyr
	Wild oats	difenzoquat

Miscellaneous field crops

Diseases	Fungus diseases	propiconazole *(off-label)*
Pests	Aphids	pirimicarb *(off-label)*
	Flea beetles	deltamethrin *(off-label)*
	Leaf miners	oxamyl *(off-label)*
	Pollen beetles	deltamethrin *(off-label)*
Weeds	Annual dicotyledons	benazolin + clopyralid *(off-label)*, bentazone *(off-label)*, carbetamide, clopyralid + propyzamide *(off-label)*, cyanazine *(off-label)*, MCPA + MCPB, metazachlor *(off-label)*, pendimethalin, pendimethalin *(off-label)*, propyzamide *(off-label)*, trifluralin *(off-label)*
	Annual grasses	carbetamide, clopyralid + propyzamide *(off-label)*, pendimethalin, pendimethalin *(off-label)*, propyzamide *(off-label)*, sethoxydim *(off-label)*, trifluralin *(off-label)*
	Annual meadow grass	pendimethalin, tri-allate
	Blackgrass	pendimethalin, tri-allate
	Chickweed	benazolin + clopyralid *(off-label)*

FOR FULL CONDITIONS OF USE ALWAYS READ THE PRODUCT LABEL

	Cleavers	benazolin + clopyralid *(off-label)*, pendimethalin
	Mayweeds	benazolin + clopyralid *(off-label)*, clopyralid + propyzamide *(off-label)*, clopyralid *(off-label)*
	Perennial dicotyledons	clopyralid *(off-label)*, MCPA + MCPB
	Perennial grasses	propyzamide *(off-label)*, sethoxydim *(off-label)*, tri-allate
	Speedwells	pendimethalin
	Volunteer cereals	carbetamide
	Wild oats	pendimethalin, tri-allate
Crop control	Pre-harvest desiccation	diquat *(off-label)*

Oilseed rape

Diseases	Alternaria	carbendazim + iprodione, carbendazim + prochloraz, carbendazim + tebuconazole, carbendazim + vinclozolin *(reduction)*, difenoconazole, gamma-HCH + fenpropimorph + thiram *(seed treatment)*, iprodione, iprodione + thiophanate-methyl, iprodione *(seed treatment)*, prochloraz, propiconazole, tebuconazole, vinclozolin
	Black scurf and stem canker	difenoconazole, gamma-HCH + fenpropimorph + thiram *(seed treatment)*, iprodione + thiophanate-methyl, tebuconazole
	Botrytis	benomyl, carbendazim + iprodione, carbendazim + prochloraz, carbendazim + vinclozolin *(reduction)*, carbendazim *(reduction)*, chlorothalonil, chlorothalonil *(moderate control)*, iprodione, iprodione + thiophanate-methyl, prochloraz, vinclozolin
	Canker	carbendazim + flusilazole, carbendazim + prochloraz, carbendazim + tebuconazole, carboxin + gamma-HCH + thiram, prochloraz
	Damping off	carboxin + gamma-HCH + thiram, gamma-HCH + thiram, thiram *(seed treatment)*
	Downy mildew	carbendazim + mancozeb, carbendazim + maneb, carbendazim + maneb + sulfur, chlorothalonil, chlorothalonil + metalaxyl, chlorothalonil *(moderate control)*, mancozeb
	Light leaf spot	benomyl, carbendazim, carbendazim + flusilazole, carbendazim + iprodione, carbendazim + maneb, carbendazim + maneb + sulfur, carbendazim + prochloraz, carbendazim + tebuconazole, carbendazim + vinclozolin *(reduction)*, carbendazim *(reduction)*, cyproconazole, difenoconazole, flusilazole, iprodione + thiophanate-methyl, prochloraz, propiconazole, tebuconazole
	Phoma leaf spot	carbendazim + flusilazole, carbendazim + tebuconazole, cyproconazole, tebuconazole
	Ring spot	carbendazim + tebuconazole *(reduction)*, tebuconazole *(reduction)*

	Sclerotinia stem rot	carbendazim + iprodione, carbendazim + prochloraz, carbendazim + tebuconazole, carbendazim + vinclozolin, iprodione, iprodione + thiophanate-methyl, prochloraz, tebuconazole, vinclozolin
	White leaf spot	carbendazim + prochloraz, prochloraz
Pests	Aphids	chlorpyrifos + dimethoate, deltamethrin, deltamethrin + pirimicarb, lambda-cyhalothrin, lambda-cyhalothrin + pirimicarb, pirimicarb, tau-fluvalinate
	Birds	aluminium ammonium sulfate
	Cabbage root fly	carbofuran, chlorpyrifos + dimethoate
	Cabbage seed weevil	alpha-cypermethrin, deltamethrin, lambda-cyhalothrin
	Cabbage stem flea beetle	alpha-cypermethrin, carbofuran, cypermethrin, deltamethrin, deltamethrin *(in store)*, lambda-cyhalothrin, lambda-cyhalothrin + pirimicarb, zeta-cypermethrin
	Cabbage stem weevil	deltamethrin
	Cutworms	chlorpyrifos + dimethoate
	Damaging mammals	aluminium ammonium sulfate, sulfonated cod liver oil
	Flea beetles	carboxin + gamma-HCH + thiram, gamma-HCH + fenpropimorph + thiram *(seed treatment)*, gamma-HCH + thiram, lambda-cyhalothrin, lambda-cyhalothrin + pirimicarb
	Leatherjackets	chlorpyrifos + dimethoate
	Pod midge	alpha-cypermethrin, cypermethrin, deltamethrin, lambda-cyhalothrin, lambda-cyhalothrin + pirimicarb, phosalone, zeta-cypermethrin
	Pollen beetles	alpha-cypermethrin, cypermethrin, deltamethrin, lambda-cyhalothrin, lambda-cyhalothrin + pirimicarb, phosalone, tau-fluvalinate, zeta-cypermethrin
	Rape winter stem weevil	alpha-cypermethrin, carbofuran, cypermethrin, deltamethrin, deltamethrin *(in store)*
	Seed weevil	cypermethrin, lambda-cyhalothrin, lambda-cyhalothrin + pirimicarb, phosalone, zeta-cypermethrin
	Slugs and snails	methiocarb, thiodicarb
	Virus vectors	deltamethrin, lambda-cyhalothrin
	Wireworms	chlorpyrifos + dimethoate
Weeds	Annual and perennial weeds	clodinafop-propargyl + diflufenican *(pre-harvest)*, glyphosate (agricultural) *(pre-harvest)*, glyphosate (horticulture, forestry, amenity etc.) *(pre-harvest)*

Annual dicotyledons	benazolin + clopyralid, benazolin + clopyralid *(off-label)*, carbetamide, chlorthal-dimethyl, clopyralid, clopyralid + propyzamide, cyanazine, glufosinate-ammonium *(pre-harvest)*, metazachlor, metazachlor + quinmerac, napropamide, propachlor, propyzamide, pyridate, pyridate *(off-label)*, tebutam, trifluralin
Annual grasses	carbetamide, clopyralid + propyzamide, cyanazine, cycloxydim, diclofop-methyl, fluazifop-P-butyl, glufosinate-ammonium *(pre-harvest)*, napropamide, propachlor, propaquizafop, propyzamide, quizalofop-P-ethyl, sethoxydim, tebutam, trifluralin
Annual meadow grass	metazachlor, metazachlor + quinmerac
Annual weeds	glyphosate (agricultural) *(post sowing, pre crop emergence)*
Barren brome	clopyralid + propyzamide, sethoxydim
Blackgrass	cycloxydim, diclofop-methyl, metazachlor, metazachlor + quinmerac, sethoxydim
Canary grass	diclofop-methyl, sethoxydim
Chickweed	benazolin, benazolin + clopyralid
Cleavers	benazolin, benazolin + clopyralid, metazachlor + quinmerac, napropamide, pyridate
Corn marigold	clopyralid
Couch	clodinafop-propargyl + diflufenican *(pre-harvest)*, cycloxydim, glyphosate (agricultural) *(pre-harvest)*, glyphosate (horticulture, forestry, amenity etc.) *(pre-harvest)*, quizalofop-P-ethyl
Creeping bent	cycloxydim, glyphosate (agricultural) *(pre-harvest)*
Creeping thistle	clopyralid
Groundsel	napropamide
Groundsel, triazine resistant	metazachlor
Mayweeds	benazolin + clopyralid, clopyralid, clopyralid + propyzamide
Perennial dicotyledons	clopyralid, glyphosate (agricultural) *(pre-harvest)*
Perennial grasses	cycloxydim, diclofop-methyl, fluazifop-P-butyl, glyphosate (agricultural) *(pre-harvest)*, propaquizafop, propyzamide, quizalofop-P-ethyl
Rough meadow grass	diclofop-methyl
Ryegrass	diclofop-methyl
Speedwells	pyridate
Volunteer cereals	carbetamide, cycloxydim, fluazifop-P-butyl, glyphosate (agricultural) *(post sowing, pre crop emergence)*, propyzamide, quizalofop-P-ethyl, sethoxydim, tebutam
Wild oats	cycloxydim, diclofop-methyl, fluazifop-P-butyl, propyzamide, sethoxydim

Crop control	Pre-harvest desiccation	clodinafop-propargyl + diflufenican, diquat, glufosinate-ammonium, glyphosate (agricultural), glyphosate (horticulture, forestry, amenity etc.)
Plant growth regulation	Increasing yield	chlormequat with di-1-p-menthene, sulfur
	Lodging control	chlormequat + choline chloride, chlormequat with di-1-p-menthene
	Stem shortening	chlormequat + choline chloride

Peas

Diseases	Ascochyta	benomyl *(seed treatment)*, carbendazim + chlorothalonil, carbendazim + cymoxanil + oxadixyl + thiram *(seed treatment)*, carbendazim + iprodione, chlorothalonil, chlorothalonil *(moderate control)*, iprodione + thiophanate-methyl, metalaxyl + thiabendazole + thiram *(seed treatment)*, thiabendazole + thiram *(seed treatment)*, vinclozolin
	Botrytis	benomyl, carbendazim + chlorothalonil, carbendazim + iprodione, chlorothalonil, chlorothalonil + cyproconazole, chlorothalonil *(moderate control)*, iprodione + thiophanate-methyl, vinclozolin
	Damping off	carbendazim + cymoxanil + oxadixyl + thiram *(seed treatment)*, metalaxyl + thiabendazole + thiram *(seed treatment)*, thiabendazole + thiram *(seed treatment)*, thiram *(seed treatment)*
	Downy mildew	carbendazim + chlorothalonil, carbendazim + cymoxanil + oxadixyl + thiram *(seed treatment)*, fosetyl-aluminium *(off-label)*, metalaxyl + thiabendazole + thiram *(seed treatment)*
	Mycosphaerella	carbendazim + iprodione, chlorothalonil, chlorothalonil *(moderate control)*, vinclozolin
	Powdery mildew	triadimefon *(off-label (research/breeding))*
	Rust	triadimefon *(off-label (research/breeding))*
	Sclerotinia stem rot	carbendazim + iprodione
	Stem rot	iprodione + thiophanate-methyl
Pests	Aphids	cypermethrin, deltamethrin, deltamethrin + heptenophos, deltamethrin + pirimicarb, demeton-S-methyl, dimethoate, fatty acids, fenitrothion, heptenophos, lambda-cyhalothrin, lambda-cyhalothrin + pirimicarb, nicotine, pirimicarb, pirimicarb *(off-label - research/breeding)*, zeta-cypermethrin
	Birds	aluminium ammonium sulfate
	Caterpillars	nicotine
	Damaging mammals	aluminium ammonium sulfate

	General insect control	chlorpyrifos *(off-label)*
	Leatherjackets	chlorpyrifos
	Mealy bugs	fatty acids
	Pea and bean weevils	cypermethrin, deltamethrin, fenitrothion, lambda-cyhalothrin, lambda-cyhalothrin + pirimicarb, zeta-cypermethrin
	Pea midge	demeton-S-methyl, dimethoate, fenitrothion, lambda-cyhalothrin + pirimicarb
	Pea moth	cypermethrin, deltamethrin, deltamethrin + heptenophos, fenitrothion, lambda-cyhalothrin, lambda-cyhalothrin + pirimicarb, zeta-cypermethrin
	Red spider mites	fatty acids
	Scale insects	fatty acids
	Thrips	dimethoate, fenitrothion
	Whitefly	fatty acids
Weeds	Annual and perennial weeds	clodinafop-propargyl + diflufenican *(pre-harvest)*, glyphosate (agricultural) *(pre-harvest)*, glyphosate (horticulture, forestry, amenity etc.) *(pre-harvest)*
	Annual dicotyledons	atrazine, aziprotryne, bentazone, bentazone + MCPB, chlorpropham + fenuron, cyanazine, cyanazine + pendimethalin, fomesafen + terbutryn *(spring sown)*, glufosinate-ammonium *(pre-harvest)*, isoxaben + terbuthylazine, MCPA + MCPB, MCPB, pendimethalin, prometryn, simazine + trietazine, terbuthylazine + terbutryn, terbutryn + trietazine, trifluralin *(off-label)*
	Annual grasses	atrazine, chlorpropham + fenuron, cyanazine, cycloxydim, diclofop-methyl, fluazifop-P-butyl, glufosinate-ammonium *(pre-harvest)*, pendimethalin, prometryn, propaquizafop, sethoxydim, terbuthylazine + terbutryn, trifluralin *(off-label)*
	Annual meadow grass	aziprotryne, cyanazine + pendimethalin, pendimethalin, terbutryn + trietazine, tri-allate
	Annual weeds	glyphosate (agricultural) *(post sowing, pre crop emergence)*, glyphosate (agricultural) *(pre-drilling/pre-emergence)*, glyphosate (horticulture, forestry, amenity etc.) *(pre-drilling/pre-emergence)*
	Blackgrass	cycloxydim, diclofop-methyl, pendimethalin, sethoxydim, tri-allate
	Canary grass	diclofop-methyl
	Chickweed	chlorpropham + fenuron
	Cleavers	pendimethalin
	Couch	clodinafop-propargyl + diflufenican *(pre-harvest)*, cycloxydim, glyphosate (agricultural) *(pre-harvest)*, glyphosate (horticulture, forestry, amenity etc.) *(pre-harvest)*
	Creeping bent	cycloxydim, glyphosate (agricultural) *(pre-harvest)*
	Perennial dicotyledons	glyphosate (agricultural) *(pre-harvest)*, MCPA + MCPB, MCPB

	Perennial grasses	cycloxydim, diclofop-methyl, fluazifop-P-butyl, glyphosate (agricultural) *(pre-harvest)*, propaquizafop, tri-allate
	Rough meadow grass	cyanazine + pendimethalin, diclofop-methyl
	Ryegrass	diclofop-methyl
	Speedwells	pendimethalin
	Volunteer cereals	cycloxydim, fluazifop-P-butyl, glyphosate (agricultural) *(post sowing, pre crop emergence)*, glyphosate (agricultural) *(pre-drilling/pre-emergence)*, glyphosate (horticulture, forestry, amenity etc.) *(pre-drilling/pre-emergence)*, sethoxydim
	Wild oats	cycloxydim, diclofop-methyl, fluazifop-P-butyl, pendimethalin, sethoxydim, tri-allate
Crop control	Pre-harvest desiccation	diquat, glufosinate-ammonium
Plant growth regulation	Increasing yield	chlormequat with di-1-p-menthene
	Lodging control	chlormequat with di-1-p-menthene

Potatoes

Diseases	Black dot	fenpiclonil
	Black scurf and stem canker	fenpiclonil, imazalil + pencycuron, imazalil + pencycuron *(reduction)*, imazalil + thiabendazole *(reduction)*, iprodione *(seed treatment)*, pencycuron, tolclofos-methyl
	Blight	benalaxyl + mancozeb, Bordeaux mixture, chlorothalonil, chlorothalonil + mancozeb, chlorothalonil + propamocarb hydrochloride, copper oxychloride, copper sulfate + sulfur, cymoxanil + mancozeb, cymoxanil + mancozeb + oxadixyl, dimethomorph + mancozeb, fentin acetate + maneb, fentin hydroxide, ferbam + maneb + zineb, fluazinam, mancozeb, mancozeb + metalaxyl, mancozeb + metalaxyl-M, mancozeb + ofurace, mancozeb + ofurace *(off-label)*, mancozeb + oxadixyl, mancozeb + propamocarb hydrochloride, maneb, zineb, zineb-ethylene thiuram disulphide adduct
	Dry rot	carbendazim + tecnazene, fenpiclonil, imazalil, imazalil + thiabendazole *(reduction)*, tecnazene, tecnazene + thiabendazole, thiabendazole *(post-harvest)*
	Gangrene	2-aminobutane, carbendazim + tecnazene, fenpiclonil, imazalil, imazalil + thiabendazole *(reduction)*, tecnazene + thiabendazole, thiabendazole *(post-harvest)*
	Powdery scab	maneb *(tuber borne)*
	Rhizoctonia	tolclofos-methyl *(off-label)*

	Silver scurf	2-aminobutane, carbendazim + tecnazene, carbendazim + tecnazene *(reduction)*, fenpiclonil, imazalil, imazalil + pencycuron *(reduction)*, imazalil + thiabendazole *(reduction)*, tecnazene + thiabendazole, thiabendazole *(post-harvest)*
	Skin spot	2-aminobutane, carbendazim + tecnazene, carbendazim + tecnazene *(reduction)*, fenpiclonil, imazalil, imazalil + thiabendazole *(reduction)*, tecnazene + thiabendazole, thiabendazole *(post-harvest)*
Pests	Aphids	aldicarb, deltamethrin + heptenophos, deltamethrin + pirimicarb, demeton-S-methyl, dimethoate *(excluding Myzus persicae)*, disulfoton, lambda-cyhalothrin, lambda-cyhalothrin + pirimicarb, malathion, nicotine, oxamyl, phorate, pirimicarb
	Capsids	phorate
	Caterpillars	cypermethrin, nicotine
	Colorado beetle	chlorfenvinphos *(off-label)*, chlorpyrifos *(off-label)*, deltamethrin *(off-label (Statutory Notice))*, lambda-cyhalothrin *(off-label (Statutory Notice))*
	Cutworms	chlorpyrifos, cypermethrin, lambda-cyhalothrin + pirimicarb
	Leaf roll virus vectors	deltamethrin + heptenophos
	Leafhoppers	phorate
	Nematodes	1,3-dichloropropene, aldicarb, oxamyl
	Potato cyst nematode	1,3-dichloropropene, aldicarb, carbofuran, ethoprophos, fosthiazate, oxamyl
	Potato virus vectors	deltamethrin + heptenophos, nicotine
	Slugs and snails	methiocarb, thiodicarb
	Spraing vectors	aldicarb, oxamyl
	Virus vectors	deltamethrin + pirimicarb
	Wireworms	ethoprophos, phorate
Weeds	Annual dicotyledons	atrazine, bentazone, cyanazine + pendimethalin, diquat + paraquat, glufosinate-ammonium *(pre-emergence)*, glufosinate-ammonium *(pre-harvest)*, linuron, metribuzin, monolinuron, paraquat, pendimethalin, prometryn, rimsulfuron, terbuthylazine + terbutryn, terbutryn + trietazine
	Annual grasses	atrazine, cycloxydim, diquat + paraquat, glufosinate-ammonium *(pre-emergence)*, glufosinate-ammonium *(pre-harvest)*, metribuzin, monolinuron, paraquat, pendimethalin, prometryn, propaquizafop, terbuthylazine + terbutryn
	Annual meadow grass	cyanazine + pendimethalin, linuron, monolinuron, pendimethalin, terbutryn + trietazine
	Black bindweed	linuron
	Blackgrass	cycloxydim, pendimethalin, sethoxydim
	Chickweed	linuron
	Cleavers	pendimethalin

	Corn marigold	linuron
	Couch	cycloxydim, sethoxydim
	Creeping bent	cycloxydim, sethoxydim
	Fat-hen	linuron, monolinuron
	Perennial grasses	cycloxydim, diquat + paraquat, propaquizafop, sethoxydim
	Polygonums	linuron, monolinuron
	Rough meadow grass	cyanazine + pendimethalin
	Speedwells	pendimethalin
	Volunteer cereals	cycloxydim, diquat + paraquat, paraquat, sethoxydim
	Volunteer oilseed rape	metribuzin, rimsulfuron
	Wild oats	cycloxydim, pendimethalin, sethoxydim
Crop control	Haulm destruction	sulfuric acid (commodity substance)
	Pre-harvest desiccation	diquat, glufosinate-ammonium
Plant growth regulation	Increasing yield	sulfur
	Sprout suppression	carbendazim + tecnazene, chlorpropham, maleic hydrazide, tecnazene, tecnazene + thiabendazole
	Volunteer suppression	maleic hydrazide

Seed brassicas/mustard

Diseases	Alternaria	iprodione, iprodione *(seed treatment)*
	Botrytis	iprodione
Pests	Cabbage seed weevil	deltamethrin
	Cabbage stem weevil	deltamethrin
	Pod midge	deltamethrin, phosalone
	Pollen beetles	deltamethrin, phosalone
	Seed weevil	phosalone
Weeds	Annual and perennial weeds	clodinafop-propargyl + diflufenican *(pre-harvest)*, glyphosate (agricultural) *(pre-harvest)*, glyphosate (horticulture, forestry, amenity etc.) *(pre-harvest)*
	Annual dicotyledons	benazolin + clopyralid *(off-label)*, chlorthal-dimethyl, propachlor, propyzamide, trifluralin
	Annual grasses	diclofop-methyl, propachlor, propyzamide, quizalofop-P-ethyl, sethoxydim, trifluralin
	Annual weeds	glyphosate (agricultural) *(post sowing, pre crop emergence)*

	Barren brome	sethoxydim
	Blackgrass	diclofop-methyl, sethoxydim
	Canary grass	diclofop-methyl
	Chickweed	benazolin + clopyralid *(off-label)*
	Cleavers	benazolin + clopyralid *(off-label)*
	Couch	clodinafop-propargyl + diflufenican *(pre-harvest)*, glyphosate (agricultural) *(pre-harvest)*, quizalofop-P-ethyl
	Mayweeds	benazolin + clopyralid *(off-label)*
	Perennial grasses	diclofop-methyl, propaquizafop, propyzamide, quizalofop-P-ethyl
	Rough meadow grass	diclofop-methyl
	Ryegrass	diclofop-methyl
	Volunteer cereals	glyphosate (agricultural) *(post sowing, pre crop emergence)*, quizalofop-P-ethyl, sethoxydim
	Wild oats	diclofop-methyl, sethoxydim
Crop control	Pre-harvest desiccation	clodin afop-propargyl + diflufenican, glyphosate (agricultural)

Field vegetables

Asparagus

Diseases	Fungus diseases	thiabendazole *(off-label)*
Pests	Aphids	nicotine
	Caterpillars	nicotine
Weeds	Annual and perennial weeds	clodinafop-propargyl + diflufenican *(off-label)*, glyphosate (agricultural) *(off-label)*
	Annual dicotyledons	aziprotryne *(off-label)*, MCPA, metamitron *(off-label)*, simazine, simazine *(off-label)*, terbacil
	Annual grasses	simazine, simazine *(off-label)*, terbacil
	Annual meadow grass	aziprotryne *(off-label)*
	Cleavers	isoxaben *(off-label)*
	Couch	terbacil
	Fat-hen	isoxaben *(off-label)*
	Perennial dicotyledons	MCPA
	Perennial grasses	terbacil

Beans, french/runner

Diseases	Anthracnose	carbendazim

	Botrytis	carbendazim, iprodione *(off-label)*, thiophanate-methyl, vinclozolin
	Damping off	thiram *(seed treatment)*
	Rust	tebuconazole *(off-label)*
	Sclerotinia	iprodione *(off-label)*
Pests	Aphids	demeton-S-methyl, dimethoate, disulfoton, heptenophos, nicotine, pirimicarb
	Caterpillars	lambda-cyhalothrin *(off-label)*, nicotine
	General insect control	chlorpyrifos *(off-label)*, nicotine *(off-label)*
Weeds	Annual dicotyledons	bentazone, chlorpropham + fenuron *(off-label)*, chlorthal-dimethyl, fomesafen, monolinuron, pendimethalin *(off-label)*, simazine *(off-label)*, trifluralin
	Annual grasses	chlorpropham + fenuron *(off-label)*, cycloxydim, diclofop-methyl, fluazifop-P-butyl *(off-label)*, monolinuron, pendimethalin *(off-label)*, simazine *(off-label)*, trifluralin
	Annual meadow grass	fomesafen, monolinuron
	Blackgrass	cycloxydim, diclofop-methyl
	Canary grass	diclofop-methyl
	Chickweed	chlorpropham + fenuron *(off-label)*
	Couch	cycloxydim
	Creeping bent	cycloxydim
	Fat-hen	monolinuron
	Perennial grasses	cycloxydim, diclofop-methyl, fluazifop-P-butyl *(off-label)*
	Polygonums	monolinuron
	Rough meadow grass	diclofop-methyl
	Ryegrass	diclofop-methyl
	Volunteer cereals	cycloxydim
	Volunteer oilseed rape	fomesafen
	Wild oats	cycloxydim, diclofop-methyl
Crop control	Pre-harvest desiccation	diquat *(off-label)*

Brassica seed crops

Diseases	Alternaria	iprodione
	Botrytis	iprodione
Pests	Pod midge	phosalone
	Pollen beetles	phosalone

FOR FULL CONDITIONS OF USE ALWAYS READ THE PRODUCT LABEL

	Seed weevil	phosalone
Weeds	Annual dicotyledons	carbetamide, propyzamide, trifluralin *(off-label)*
	Annual grasses	carbetamide, propyzamide, trifluralin *(off-label)*
	Volunteer cereals	carbetamide

Brassicas

Diseases	Alternaria	carbendazim *(Must be used in tank mix with Corbel)*, chlorothalonil, chlorothalonil + metalaxyl *(moderate control)*, chlorothalonil *(moderate control)*, difenoconazole, fenpropimorph, iprodione, iprodione *(seed treatment)*, tebuconazole, triadimenol
	Botrytis	chlorothalonil, chlorothalonil *(moderate control)*
	Clubroot	tar acids
	Damping off	etridiazole *(seedlings and transplants)*, fosetyl-aluminium *(off-label)*, gamma-HCH + thiram, propamocarb hydrochloride, thiram *(seed treatment)*
	Damping off and wirestem	chlorothalonil, tolclofos-methyl
	Downy mildew	chlorothalonil, chlorothalonil + metalaxyl, chlorothalonil *(moderate control)*, chlorothalonil *(seedlings only)*, chlorothalonil *(seedlings)*, copper oxychloride + metalaxyl *(off-label)*, dichlofluanid, fosetyl-aluminium *(off-label)*, propamocarb hydrochloride
	Fungus diseases	mancozeb + metalaxyl *(off-label)*
	Light leaf spot	benomyl, carbendazim *(Must be used in tank mix with Corbel)*, fenpropimorph, tebuconazole, triadimenol
	Powdery mildew	copper sulfate + sulfur, tebuconazole, triadimefon, triadimefon *(off-label (research/breeding))*, triadimenol
	Rhizoctonia	chlorothalonil
	Ring spot	benomyl, carbendazim *(Must be used in tank mix with Corbel)*, chlorothalonil, chlorothalonil + metalaxyl *(reduction)*, chlorothalonil *(moderate control)*, difenoconazole, fenpropimorph, tebuconazole, tebuconazole *(off-label)*, triadimenol
	Rust	triadimefon *(off-label (research/breeding))*
	Spear rot	copper oxychloride *(off-label)*
	White blister	chlorothalonil + metalaxyl, mancozeb + metalaxyl *(off-label)*
	Wirestem	quintozene

Pests	Aphids	aldicarb, alpha-cypermethrin, bifenthrin, chlorpyrifos, chlorpyrifos + dimethoate, chlorpyrifos + disulfoton, cypermethrin, deltamethrin + pirimicarb, deltamethrin *(off-label)*, demeton-S-methyl, dimethoate, disulfoton, fatty acids, heptenophos, heptenophos *(off-label)*, lambda-cyhalothrin + pirimicarb, nicotine, phorate, pirimicarb, pirimicarb *(off-label - research/breeding)*
	Birds	aluminium ammonium sulfate
	Cabbage root fly	aldicarb, carbofuran, carbosulfan, chlorfenvinphos, chlorpyrifos, chlorpyrifos + dimethoate, chlorpyrifos + disulfoton, fonofos, fonofos *(off-label)*, phorate, trichlorfon
	Cabbage stem flea beetle	cypermethrin
	Cabbage stem weevil	carbofuran, carbosulfan
	Caterpillars	alpha-cypermethrin, Bacillus thuringiensis, bifenthrin, chlorpyrifos, cypermethrin, deltamethrin, deltamethrin + pirimicarb, deltamethrin *(off-label)*, diflubenzuron, lambda-cyhalothrin, lambda-cyhalothrin + pirimicarb, nicotine, pirimiphos-methyl, trichlorfon
	Cutworms	chlorpyrifos, chlorpyrifos + dimethoate
	Damaging mammals	aluminium ammonium sulfate, sulfonated cod liver oil
	Flea beetles	aldicarb, alpha-cypermethrin, carbofuran, carbosulfan, cypermethrin, deltamethrin, gamma-HCH + thiram
	General insect control	chlorpyrifos *(off-label)*, deltamethrin + heptenophos *(off-label)*, dimethoate *(off-label)*
	Leaf miners	dichlorvos *(off-label)*, nicotine, trichlorfon, trichlorfon *(off-label)*
	Leatherjackets	chlorpyrifos, chlorpyrifos + dimethoate
	Mealy bugs	fatty acids
	Nematodes	aldicarb
	Pollen beetles	alpha-cypermethrin *(off-label)*, lambda-cyhalothrin + pirimicarb
	Red spider mites	fatty acids
	Scale insects	fatty acids
	Slugs and snails	metaldehyde *(seed admixture)*, methiocarb
	Western flower thrips	dichlorvos *(off-label)*
	Whitefly	bifenthrin, chlorpyrifos, cypermethrin, fatty acids, lambda-cyhalothrin + pirimicarb
	Wireworms	chlorpyrifos + dimethoate

Weeds	Annual dicotyledons	aziprotryne, aziprotryne *(off-label)*, carbetamide, chlorthal-dimethyl, clopyralid, cyanazine *(off-label)*, desmetryn, desmetryn *(off-label)*, metazachlor, pendimethalin, propachlor, propachlor *(off-label)*, sodium monochloroacetate, sodium monochloroacetate *(off-label)*, tebutam, trifluralin
	Annual grasses	carbetamide, cyanazine *(off-label)*, cycloxydim, diclofop-methyl, fluazifop-P-butyl *(off-label)*, pendimethalin, propachlor, propachlor *(off-label)*, tebutam, trifluralin
	Annual meadow grass	aziprotryne, aziprotryne *(off-label)*, metazachlor
	Black nightshade	pyridate
	Blackgrass	cycloxydim, diclofop-methyl, metazachlor
	Canary grass	diclofop-methyl
	Cleavers	pyridate
	Corn marigold	clopyralid
	Couch	cycloxydim
	Creeping bent	cycloxydim
	Creeping thistle	clopyralid
	Fat-hen	desmetryn, desmetryn *(off-label)*, pyridate
	Groundsel, triazine resistant	metazachlor
	Mayweeds	clopyralid
	Perennial dicotyledons	clopyralid
	Perennial grasses	cycloxydim, diclofop-methyl, fluazifop-P-butyl *(off-label)*
	Rough meadow grass	diclofop-methyl
	Ryegrass	diclofop-methyl
	Volunteer cereals	carbetamide, cycloxydim, tebutam
	Wild oats	cycloxydim, diclofop-methyl
Plant growth regulation	Increasing yield	sulfur

Carrots/parsnips/parsley

Diseases	Alternaria	iprodione + thiophanate-methyl *(off-label)*, tebuconazole *(off-label)*
	Canker	tebuconazole *(off-label)*
	Cavity spot	mancozeb + metalaxyl, metalaxyl-M
	Crown rot	fenpropimorph *(off-label)*, iprodione + thiophanate-methyl *(off-label)*
	Damping off	thiram *(seed treatment)*
	Fungus diseases	triadimenol *(off-label)*

	Powdery mildew	sulfur *(off-label)*, triadimefon
	Seed-borne diseases	thiram *(seed soak)*
Pests	Aphids	aldicarb, carbofuran, carbosulfan, demeton-S-methyl, dimethoate, disulfoton, lambda-cyhalothrin + pirimicarb, nicotine, pirimicarb, pirimicarb *(off-label)*
	Birds	aluminium ammonium sulfate
	Carrot fly	carbofuran, carbosulfan, chlorfenvinphos, disulfoton, lambda-cyhalothrin *(off-label)*, pirimiphos-methyl, pirimiphos-methyl *(off-label)*, tefluthrin *(off-label)*
	Caterpillars	nicotine
	Cutworms	chlorpyrifos, cypermethrin, lambda-cyhalothrin + pirimicarb
	Damaging mammals	aluminium ammonium sulfate
	General insect control	deltamethrin *(off-label)*, lambda-cyhalothrin *(off-label)*, pyrazophos *(off-label)*
	Nematodes	aldicarb, carbofuran, carbosulfan
Weeds	Annual dicotyledons	atrazine, chlorpropham, chlorpropham + pentanochlor, isoxaben *(off-label)*, linuron, metoxuron, pendimethalin, pendimethalin *(off-label)*, pentanochlor, prometryn, trifluralin
	Annual grasses	atrazine, chlorpropham, cycloxydim, fluazifop-P-butyl, fluazifop-P-butyl *(off-label)*, metoxuron, pendimethalin, pendimethalin *(off-label)*, prometryn, propaquizafop, quizalofop-P-ethyl, trifluralin
	Annual meadow grass	chlorpropham + pentanochlor, linuron, pentanochlor
	Annual weeds	metribuzin *(off-label)*
	Black bindweed	linuron
	Blackgrass	cycloxydim
	Chickweed	chlorpropham, linuron
	Corn marigold	linuron
	Couch	cycloxydim, quizalofop-P-ethyl
	Creeping bent	cycloxydim
	Docks	asulam *(off-label)*
	Fat-hen	linuron
	Mayweeds	metoxuron
	Perennial grasses	cycloxydim, fluazifop-P-butyl, fluazifop-P-butyl *(off-label)*, propaquizafop, quizalofop-P-ethyl
	Polygonums	chlorpropham, linuron
	Volunteer cereals	cycloxydim, fluazifop-P-butyl, quizalofop-P-ethyl

	Wild oats	cycloxydim, fluazifop-P-butyl

Celery

Diseases	Botrytis	carbendazim *(off-label)*
	Celery leaf spot	Bordeaux mixture, carbendazim, chlorothalonil, copper ammonium carbonate, copper oxychloride, zineb
	Damping off	etridiazole *(seedlings and transplants)*
	Seed-borne diseases	thiram *(seed soak)*
	Septoria diseases	chlorothalonil
Pests	Aphids	cypermethrin, heptenophos, nicotine, pirimicarb
	Carrot fly	disulfoton, lambda-cyhalothrin *(off-label)*, phorate
	Caterpillars	cypermethrin, lambda-cyhalothrin *(off-label)*, nicotine
	Cutworms	chlorpyrifos, cypermethrin
	General insect control	deltamethrin *(off-label)*, dimethoate *(off-label)*
	Leaf miners	nicotine
Weeds	Annual dicotyledons	atrazine, chlorpropham, chlorpropham + pentanochlor, linuron, pentanochlor, prometryn
	Annual grasses	atrazine, chlorpropham, prometryn
	Annual meadow grass	chlorpropham + pentanochlor, linuron, pentanochlor
	Black bindweed	linuron
	Chickweed	chlorpropham, linuron
	Fat-hen	linuron
	Polygonums	chlorpropham, linuron
Plant growth regulation	Increasing yield	gibberellins

Cucurbits

Diseases	Black spot	benomyl *(off-label)*
	Botrytis	iprodione *(off-label)*
	Gummosis	benomyl *(off-label)*
	Powdery mildew	bupirimate, bupirimate *(off-label)*, fenarimol *(off-label)*, imazalil *(off-label)*
	Sclerotinia	iprodione *(off-label)*
Pests	Aphids	disulfoton, fatty acids, heptenophos, nicotine, pirimicarb *(off-label)*
	Caterpillars	nicotine
	General insect control	demeton-S-methyl *(off-label)*, dimethoate *(off-label)*

	Leaf miners	deltamethrin *(off-label (fog))*, oxamyl *(off-label)*, trichlorfon *(off-label)*
	Mealy bugs	fatty acids
	Red spider mites	fatty acids
	Scale insects	fatty acids
	Western flower thrips	deltamethrin *(off-label (fog))*
	Whitefly	fatty acids
Weeds	Annual dicotyledons	chlorthal-dimethyl *(off-label)*

General

Diseases	Damping off	propamocarb hydrochloride
	Soil-borne diseases	dazomet, methyl bromide with chloropicrin
Pests	Aphids	nicotine, rotenone
	Capsids	nicotine
	Caterpillars	Bacillus thuringiensis
	Cutworms	Bacillus thuringiensis, trichlorfon
	General insect control	pyrazophos *(off-label)*
	Leafhoppers	nicotine
	Nematodes	dazomet, methyl bromide with chloropicrin
	Sawflies	nicotine
	Slugs and snails	aluminium sulfate, metaldehyde
	Soil pests	dazomet, methyl bromide with chloropicrin
	Thrips	nicotine
Weeds	Annual dicotyledons	diquat, diquat + paraquat, glufosinate-ammonium *(pre-emergence)*, paraquat
	Annual grasses	diquat + paraquat, glufosinate-ammonium *(pre-emergence)*, paraquat
	Annual weeds	ammonium sulfamate *(pre-planting)*
	Barley cover crops	fluazifop-P-butyl
	General weed control	dazomet, methyl bromide with chloropicrin
	Perennial grasses	diquat + paraquat
	Perennial weeds	ammonium sulfamate *(pre-planting)*
	Volunteer cereals	diquat + paraquat, paraquat

Plant growth regulation	Increasing yield	sulfur

Herb crops

Diseases	Alternaria	tebuconazole *(off-label)*
	Crown rot	fenpropimorph *(off-label)*
	Downy mildew	fosetyl-aluminium *(off-label)*
	Fungus diseases	prochloraz *(off-label)*, triadimenol *(off-label)*
	Powdery mildew	sulfur *(off-label)*, tebuconazole *(off-label)*
	Rust	tebuconazole *(off-label)*
	White blister	propamocarb hydrochloride *(off-label)*
Pests	Aphids	pirimicarb *(off-label)*
	Cutworms	trichlorfon
	Leafhoppers	deltamethrin *(off-label)*
Weeds	Annual dicotyledons	chlorpropham + pentanochlor *(off-label)*, chlorthal-dimethyl, chlorthal-dimethyl *(off-label)*, clopyralid *(off-label)*, ethofumesate *(off-label)*, metamitron *(off-label)*, paraquat *(off-label)*, pendimethalin *(off-label)*, pentanochlor *(off-label)*, prometryn *(off-label)*, propachlor, propachlor *(off-label)*, propyzamide *(off-label)*, simazine *(off-label)*, terbacil *(off-label)*, trifluralin *(off-label)*
	Annual grasses	ethofumesate *(off-label)*, paraquat *(off-label)*, pendimethalin *(off-label)*, prometryn *(off-label)*, propachlor, propachlor *(off-label)*, propyzamide *(off-label)*, simazine *(off-label)*, terbacil *(off-label)*, trifluralin *(off-label)*
	Annual meadow grass	chlorpropham + pentanochlor *(off-label)*, pentanochlor *(off-label)*
	Annual weeds	linuron *(off-label)*
	Couch	terbacil *(off-label)*
	Docks	asulam *(off-label)*
	General weed control	monolinuron *(off-label)*
	Perennial dicotyledons	clopyralid *(off-label)*
	Perennial grasses	propyzamide *(off-label)*, terbacil *(off-label)*

Lettuce, outdoor

Diseases	Big vein	carbendazim *(off-label)*
	Botrytis	iprodione, iprodione *(off-label)*, propamocarb hydrochloride *(off-label)*, quintozene, thiram
	Damping off	thiram *(seed treatment)*

	Downy mildew	fosetyl-aluminium *(off-label)*, mancozeb, metalaxyl + thiram, propamocarb hydrochloride *(off-label)*, zineb
	Fungus diseases	prochloraz *(off-label)*
	Rhizoctonia	quintozene, tolclofos-methyl
	Sclerotinia	iprodione *(off-label)*, quintozene
Pests	Aphids	cypermethrin, deltamethrin + heptenophos *(off-label)*, deltamethrin *(off-label)*, demeton-S-methyl, fatty acids, heptenophos, imidacloprid *(off-label)*, lambda-cyhalothrin + pirimicarb, malathion, nicotine, phorate, pirimicarb
	Caterpillars	cypermethrin, deltamethrin, deltamethrin *(off-label)*
	Cutworms	chlorpyrifos, chlorpyrifos *(outdoor only)*, cypermethrin, deltamethrin, lambda-cyhalothrin + pirimicarb
	Damaging mammals	sulfonated cod liver oil
	Leaf miners	trichlorfon *(off-label)*
	Lettuce root aphid	phorate
	Mealy bugs	fatty acids
	Red spider mites	fatty acids
	Scale insects	fatty acids
	Whitefly	fatty acids
Weeds	Annual dicotyledons	chlorpropham, paraquat *(off-label)*, propachlor *(off-label)*, propyzamide, trifluralin
	Annual grasses	chlorpropham, paraquat *(off-label)*, propachlor *(off-label)*, propyzamide, trifluralin
	Chickweed	chlorpropham
	Perennial grasses	propyzamide
	Polygonums	chlorpropham

Miscellaneous field vegetables

Diseases	Botrytis	iprodione *(off-label)*
	Celery leaf spot	chlorothalonil *(off-label)*
	Crown rot	fenpropimorph *(off-label)*
	Damping off	thiram *(off-label seed treatment)*
	Fungus diseases	triadimenol *(off-label)*
	Phytophthora	fosetyl-aluminium *(off-label)*
	Sclerotinia	iprodione *(off-label)*, quintozene *(forcing)*
	Septoria diseases	chlorothalonil *(off-label)*

Pests	Aphids	chlorpyrifos *(off-label)*, deltamethrin *(off-label)*, heptenophos *(off-label)*, nicotine, pirimicarb *(off-label)*
	Carrot fly	chlorfenvinphos *(off-label)*
	Caterpillars	deltamethrin *(off-label)*, nicotine
	Cutworms	chlorpyrifos *(off-label)*
	Flea beetles	deltamethrin *(off-label)*
	General insect control	deltamethrin + heptenophos *(off-label)*, deltamethrin *(off-label)*, dimethoate *(off-label)*, lambda-cyhalothrin *(off-label)*, pyrazophos *(off-label)*
	Leaf miners	trichlorfon *(off-label)*
Weeds	Annual dicotyledons	bentazone *(off-label)*, chlorpropham + pentanochlor, linuron *(off-label)*, pendimethalin *(off-label)*, pentanochlor, prometryn *(off-label)*, propyzamide *(off-label)*, trifluralin *(off-label)*
	Annual grasses	cycloxydim *(off-label)*, linuron *(off-label)*, pendimethalin *(off-label)*, prometryn *(off-label)*, propyzamide *(off-label)*, trifluralin *(off-label)*
	Annual meadow grass	chlorpropham + pentanochlor, pentanochlor
	Annual weeds	linuron *(off-label)*
	Perennial grasses	cycloxydim *(off-label)*, propyzamide *(off-label)*

Onions/leeks

Diseases	Botrytis	carbendazim *(off-label)*, chlorothalonil, chlorothalonil *(moderate control)*, iprodione
	Cladosporium leaf blotch	propiconazole
	Collar rot	iprodione
	Damping off	propamocarb hydrochloride, thiram *(seed treatment)*
	Downy mildew	chlorothalonil + metalaxyl, mancozeb + metalaxyl *(off-label)*
	Fungus diseases	ferbam + maneb + zineb *(off-label)*
	Phytophthora	propamocarb hydrochloride
	Rhynchosporium	chlorothalonil, propiconazole *(off-label)*
	Root rot	propamocarb hydrochloride
	Rust	chlorothalonil, cyproconazole, fenpropimorph, propiconazole, tebuconazole, triadimefon
	White rot	tebuconazole *(off-label)*
	White tip	chlorothalonil + metalaxyl, mancozeb + metalaxyl *(off-label)*
Pests	Aphids	nicotine
	Bean seed flies	tefluthrin *(off-label)*
	Caterpillars	nicotine
	Cutworms	chlorpyrifos

	General insect control	deltamethrin *(off-label)*, dimethoate *(off-label)*
	Leaf miners	deltamethrin *(off-label (fog))*, oxamyl *(off-label)*
	Onion fly	carbofuran, tefluthrin *(off-label)*
	Stem nematodes	aldicarb, aldicarb *(off-label)*, carbofuran, oxamyl *(off-label)*
	Thrips	fenitrothion *(off-label)*, malathion
	Western flower thrips	deltamethrin *(off-label (fog))*
Weeds	Annual dicotyledons	aziprotryne, bentazone *(off-label)*, chloridazon + propachlor, chloridazon + propachlor *(off-label, post-emergence)*, chloridazon *(off-label)*, chlorpropham, chlorpropham + fenuron, chlorthal-dimethyl, clopyralid, cyanazine, ethofumesate *(off-label)*, fluroxypyr *(off-label)*, ioxynil, ioxynil *(off-label)*, linuron *(off-label)*, monolinuron, pendimethalin, pendimethalin *(off-label)*, prometryn, prometryn *(off-label)*, propachlor, sodium monochloroacetate
	Annual grasses	chloridazon + propachlor, chloridazon + propachlor *(off-label, post-emergence)*, chlorpropham, chlorpropham + fenuron, cyanazine, cycloxydim, diclofop-methyl, ethofumesate *(off-label)*, fluazifop-P-butyl, fluazifop-P-butyl *(off-label)*, linuron *(off-label)*, monolinuron, pendimethalin, pendimethalin *(off-label)*, prometryn, prometryn *(off-label)*, propachlor, propaquizafop
	Annual meadow grass	aziprotryne, chloridazon *(off-label)*, monolinuron
	Annual weeds	glyphosate (agricultural) *(post sowing, pre crop emergence)*, glyphosate (agricultural) *(pre-drilling/pre-emergence)*, glyphosate (horticulture, forestry, amenity etc.) *(pre-drilling/pre-emergence)*, linuron *(off-label)*
	Black nightshade	pyridate
	Blackgrass	cycloxydim, diclofop-methyl
	Canary grass	diclofop-methyl
	Chickweed	chlorpropham, chlorpropham + fenuron
	Cleavers	pyridate
	Corn marigold	clopyralid
	Couch	cycloxydim
	Creeping bent	cycloxydim
	Creeping thistle	clopyralid
	Fat-hen	monolinuron, pyridate
	Mayweeds	clopyralid
	Perennial dicotyledons	clopyralid
	Perennial grasses	cycloxydim, diclofop-methyl, fluazifop-P-butyl, fluazifop-P-butyl *(off-label)*, propaquizafop

FOR FULL CONDITIONS OF USE ALWAYS READ THE PRODUCT LABEL

	Polygonums	chlorpropham, monolinuron
	Rough meadow grass	diclofop-methyl
	Ryegrass	diclofop-methyl
	Volunteer cereals	cycloxydim, fluazifop-P-butyl, glyphosate (agricultural) *(post sowing, pre crop emergence)*, glyphosate (agricultural) *(pre-drilling/pre-emergence)*, glyphosate (horticulture, forestry, amenity etc.) *(pre-drilling/pre-emergence)*
	Volunteer potatoes	fluroxypyr *(off-label)*
	Wild oats	cycloxydim, diclofop-methyl, fluazifop-P-butyl
Crop control	Pre-harvest desiccation	sulfuric acid (commodity substance)
Plant growth regulation	Sprout suppression	maleic hydrazide

Peas, mange-tout

Diseases	Botrytis	iprodione *(off-label)*
	Sclerotinia	iprodione *(off-label)*
Pests	Aphids	nicotine
	General insect control	chlorpyrifos *(off-label)*
Weeds	Annual dicotyledons	simazine + trietazine *(off-label)*, simazine *(off-label)*
	Annual grasses	simazine *(off-label)*

Red beet

Diseases	Aphanomyces cochlioides	hymexazol *(off-label)*
	Rust	fenpropimorph *(off-label)*
	Seed-borne diseases	thiram *(seed soak)*
Pests	Aphids	dimethoate *(excluding Myzus persicae)*, nicotine, pirimicarb *(off-label)*
	Beet cyst nematode	aldicarb *(off-label)*
	Caterpillars	cypermethrin, nicotine
	Cutworms	chlorpyrifos, cypermethrin
	Flea beetles	carbofuran, trichlorfon
	General insect control	lambda-cyhalothrin *(off-label)*, pyrazophos *(off-label)*
	Leaf miners	dimethoate
	Mangold fly	trichlorfon
Weeds	Annual dicotyledons	clopyralid, ethofumesate, ethofumesate + phenmedipham, lenacil, metamitron, phenmedipham

	Annual grasses	fluazifop-P-butyl *(off-label)*, metamitron, propaquizafop *(off-label)*, quizalofop-P-ethyl
	Annual meadow grass	ethofumesate, ethofumesate + phenmedipham, lenacil, metamitron
	Blackgrass	ethofumesate, ethofumesate + phenmedipham, tri-allate
	Corn marigold	clopyralid
	Couch	quizalofop-P-ethyl
	Creeping thistle	clopyralid
	Fat-hen	metamitron
	Mayweeds	clopyralid
	Perennial dicotyledons	clopyralid
	Perennial grasses	fluazifop-P-butyl *(off-label)*, propaquizafop *(off-label)*, quizalofop-P-ethyl, tri-allate
	Volunteer cereals	quizalofop-P-ethyl
	Wild oats	tri-allate

Rhubarb

Diseases	Botrytis	benomyl *(off-label)*
	Fungus diseases	benomyl *(off-label)*
Pests	Aphids	heptenophos *(off-label)*
	Rosy rustic moth	chlorpyrifos
Weeds	Annual dicotyledons	propyzamide, propyzamide *(outdoor)*, simazine, simazine *(off-label)*
	Annual grasses	propyzamide, propyzamide *(outdoor)*, simazine, simazine *(off-label)*
	Perennial grasses	propyzamide, propyzamide *(outdoor)*
Plant growth regulation	Increasing yield	gibberellins

Root brassicas

Diseases	Alternaria	iprodione *(seed treatment)*, tebuconazole *(off-label)*
	Damping off	gamma-HCH + thiram, thiram *(seed treatment)*
	Downy mildew	propamocarb hydrochloride *(off-label)*
	Fungus diseases	mancozeb + metalaxyl *(off-label)*
	Light leaf spot	tebuconazole *(off-label)*

FOR FULL CONDITIONS OF USE ALWAYS READ THE PRODUCT LABEL

	Powdery mildew	copper sulfate + sulfur, sulfur, tebuconazole, triadimefon, triadimenol, tridemorph
	Ring spot	tebuconazole *(off-label)*
	White blister	mancozeb + metalaxyl *(off-label)*
Pests	Aphids	aldicarb, carbofuran, deltamethrin + pirimicarb, nicotine, pirimicarb, pirimicarb *(off-label)*
	Cabbage root fly	aldicarb, carbofuran, carbofuran *(off-label)*, carbosulfan, chlorfenvinphos *(off-label)*, chlorpyrifos *(off-label)*, fonofos, trichlorfon *(off-label)*
	Cabbage stem weevil	carbofuran, carbosulfan
	Caterpillars	deltamethrin, deltamethrin + pirimicarb, nicotine
	Cutworms	chlorpyrifos, cypermethrin
	Flea beetles	aldicarb, carbofuran, carbosulfan, deltamethrin, gamma-HCH + thiram
	General insect control	deltamethrin + heptenophos *(off-label)*, dimethoate *(off-label)*
	Nematodes	aldicarb
	Soil pests	carbofuran *(off-label)*
	Turnip root fly	carbofuran
Weeds	Annual dicotyledons	chlorthal-dimethyl, clopyralid, clopyralid *(off-label)*, metazachlor, metazachlor *(off-label)*, prometryn *(off-label)*, propachlor, propachlor *(off-label)*, sodium monochloroacetate *(off-label)*, tebutam, trifluralin, trifluralin *(off-label)*
	Annual grasses	cycloxydim, fluazifop-P-butyl *(off-label)*, fluazifop-P-butyl *(stockfeed only)*, metazachlor *(off-label)*, prometryn *(off-label)*, propachlor, propachlor *(off-label)*, propaquizafop, tebutam, trifluralin, trifluralin *(off-label)*
	Annual meadow grass	metazachlor
	Annual weeds	glyphosate (agricultural) *(post sowing, pre crop emergence)*, glyphosate (agricultural) *(pre-drilling/pre-emergence)*, glyphosate (horticulture, forestry, amenity etc.) *(pre-drilling/pre-emergence)*
	Blackgrass	cycloxydim, metazachlor
	Corn marigold	clopyralid
	Couch	cycloxydim
	Creeping bent	cycloxydim
	Creeping thistle	clopyralid
	Groundsel, triazine resistant	metazachlor
	Mayweeds	clopyralid
	Perennial dicotyledons	clopyralid
	Perennial grasses	cycloxydim, fluazifop-P-butyl *(off-label)*, fluazifop-P-butyl *(stockfeed only)*, propaquizafop

	Volunteer cereals	cycloxydim, fluazifop-P-butyl *(stockfeed only)*, glyphosate (agricultural) *(post sowing, pre crop emergence)*, glyphosate (agricultural) *(pre-drilling/pre-emergence)*, glyphosate (horticulture, forestry, amenity etc.) *(pre-drilling/pre-emergence)*, tebutam
	Wild oats	cycloxydim, fluazifop-P-butyl *(stockfeed only)*
Plant growth regulation	Increasing yield	sulfur

Spinach

Diseases	Downy mildew	copper oxychloride + metalaxyl *(off-label)*, fosetyl-aluminium *(off-label)*
Pests	Aphids	dimethoate, dimethoate *(excluding Myzus persicae)*, nicotine, pirimicarb *(off-label)*
	Caterpillars	nicotine
	Flea beetles	trichlorfon
	General insect control	dimethoate *(off-label)*
	Leaf miners	dimethoate, trichlorfon *(off-label)*
	Mangold fly	trichlorfon
Weeds	Annual dicotyledons	chlorpropham + fenuron, clopyralid *(off-label)*, phenmedipham *(off-label)*
	Annual grasses	chlorpropham + fenuron
	Chickweed	chlorpropham + fenuron

Tomatoes, outdoor

Diseases	Blight	copper ammonium carbonate, copper oxychloride
	Botrytis	iprodione
	Leaf mould	copper ammonium carbonate
	Soil-borne diseases	metam-sodium *(Jersey)*
Pests	Nematodes	metam-sodium *(Jersey)*
	Soil pests	metam-sodium *(Jersey)*
Weeds	General weed control	metam-sodium *(Jersey)*

Watercress

Diseases	Downy mildew	fosetyl-aluminium *(off-label)*

	Phytophthora	etridiazole *(off-label)*, fosetyl-aluminium *(off-label)*, propamocarb hydrochloride *(off-label)*
	Pythium	etridiazole *(off-label)*, fosetyl-aluminium *(off-label)*, propamocarb hydrochloride *(off-label)*
	Rhizoctonia	benomyl *(off-label)*
Pests	Aphids	dimethoate *(off-label)*, fatty acids *(off-label)*
	Flea beetles	malathion *(off-label)*
	Midges	malathion *(off-label)*
	Mustard beetle	malathion *(off-label)*

Flowers and ornamentals

Annuals/biennials

Weeds	Annual dicotyledons	chlorpropham, pentanochlor
	Annual grasses	chlorpropham
	Annual meadow grass	pentanochlor
	Chickweed	chlorpropham
	Polygonums	chlorpropham

Bedding plants

Diseases	Botrytis	carbendazim *(off-label)*, quintozene
	Clubroot	tar acids
	Damping off	furalaxyl, propamocarb hydrochloride
	Phytophthora	propamocarb hydrochloride
	Rhizoctonia	quintozene
	Sclerotinia	quintozene
Pests	Aphids	imidacloprid
	Sciarid flies	imidacloprid
	Vine weevil	imidacloprid
	Whitefly	imidacloprid
Weeds	Annual dicotyledons	pentanochlor
	Annual grasses	sethoxydim *(off-label)*
	Annual meadow grass	pentanochlor
	Perennial grasses	sethoxydim *(off-label)*
Plant growth regulation	Increasing flowering	paclobutrazol

	Stem shortening	chlormequat, chlormequat + choline chloride, daminozide, paclobutrazol

Bulbs/corms

Diseases	Botrytis	carbendazim, carbendazim *(dip - off-label)*, carbendazim *(off-label)*, thiram, zineb
	Fire	carbendazim, dichlofluanid, mancozeb, thiram, zineb
	Fungus diseases	captan *(off-label)*, formaldehyde (commodity substance) *(dip)*
	Fusarium diseases	carbendazim *(dip - off-label)*, carbendazim *(off-label)*, prochloraz *(off-label)*, thiabendazole
	Ink disease	chlorothalonil
	Penicillium rot	carbendazim *(dip - off-label)*, prochloraz *(off-label)*
	Phytophthora	etridiazole, propamocarb hydrochloride
	Pythium	etridiazole, propamocarb hydrochloride
	Rhizoctonia	carbendazim, quintozene
	Sclerotinia	carbendazim, carbendazim *(dip - off-label)*, carbendazim *(off-label)*, quintozene
	Soil-borne diseases	methyl bromide with chloropicrin
	Stagonospora	carbendazim, carbendazim *(off-label)*
Pests	Aphids	nicotine
	Birds	aluminium ammonium sulfate
	Bulb scale mite	endosulfan
	Capsids	nicotine
	Damaging mammals	aluminium ammonium sulfate
	General insect control	chlorpyrifos *(off-label)*
	Large narcissus fly	carbofuran *(off-label)*
	Leaf miners	nicotine
	Leafhoppers	nicotine
	Nematodes	methyl bromide with chloropicrin
	Sawflies	nicotine
	Soil pests	methyl bromide with chloropicrin
	Stem and bulb nematodes	1,3-dichloropropene
	Thrips	nicotine

Weeds	Annual dicotyledons	bentazone, chlorpropham, chlorpropham + linuron, chlorpropham + pentanochlor, cyanazine, diquat + paraquat, paraquat, pentanochlor
	Annual grasses	chlorpropham, chlorpropham + linuron, cyanazine, cycloxydim, diquat + paraquat, paraquat, sethoxydim *(off-label)*
	Annual meadow grass	chlorpropham + pentanochlor, pentanochlor
	Blackgrass	cycloxydim
	Chickweed	chlorpropham
	Couch	cycloxydim
	Creeping bent	cycloxydim, paraquat
	General weed control	methyl bromide with chloropicrin
	Perennial grasses	cycloxydim, diquat + paraquat, sethoxydim *(off-label)*
	Polygonums	chlorpropham
	Volunteer cereals	cycloxydim
	Wild oats	cycloxydim
Crop control	Pre-harvest desiccation	sulfuric acid (commodity substance)
Plant growth regulation	Increasing flowering	paclobutrazol
	Stem shortening	2-chloroethylphosphonic acid, paclobutrazol

Camellias

Plant growth regulation	Stem shortening	chlormequat, chlormequat + choline chloride

Carnations

Diseases	Rust	mancozeb, oxycarboxin, thiram, zineb
Weeds	Annual dicotyledons	pentanochlor
	Annual meadow grass	pentanochlor

Chrysanthemums

Diseases	Botrytis	carbendazim *(off-label)*, quintozene, thiram
	Powdery mildew	bupirimate, copper ammonium carbonate
	Ray blight	mancozeb
	Rhizoctonia	quintozene
	Rust	oxycarboxin, propiconazole *(off-label)*, thiram
	Sclerotinia	quintozene

	Soil-borne diseases	metam-sodium
Pests	Aphids	nicotine
	Capsids	permethrin + thiram
	Caterpillars	permethrin + thiram
	Earwigs	permethrin + thiram
	Leaf miners	nicotine, permethrin + thiram *(partial control)*
	Nematodes	metam-sodium
	Soil pests	metam-sodium
Weeds	Annual dicotyledons	chlorpropham, chlorpropham + pentanochlor, pentanochlor
	Annual grasses	chlorpropham
	Annual meadow grass	chlorpropham + pentanochlor, pentanochlor
	Chickweed	chlorpropham
	General weed control	metam-sodium
	Polygonums	chlorpropham
Plant growth regulation	Stem shortening	daminozide

Container-grown stock

Diseases	Fungus diseases	prochloraz
	Phytophthora	etridiazole, propamocarb hydrochloride
Weeds	Annual dicotyledons	isoxaben, napropamide, oxadiazon
	Annual grasses	napropamide, oxadiazon
	Cleavers	napropamide
	Groundsel	napropamide
Plant growth regulation	Root control	copper hydroxide
	Stem shortening	chlormequat, chlormequat + choline chloride

Dahlias

Diseases	Botrytis	quintozene
	Rhizoctonia	quintozene
	Sclerotinia	quintozene

General

Diseases	Alternaria	iprodione *(seed treatment)*
	Black spot	bupirimate + triforine
	Botrytis	chlorothalonil, dichlofluanid *(off-label)*, prochloraz, thiram *(except Hydrangea)*
	Damping off	copper ammonium carbonate, propamocarb hydrochloride, tolclofos-methyl
	Disease control/foliar feed	difenoconazole *(off-label)*
	Foot rot	tolclofos-methyl
	Phytophthora	propamocarb hydrochloride
	Powdery mildew	bupirimate + triforine, imazalil, pyrifenox
	Pythium	propamocarb hydrochloride
	Root rot	tolclofos-methyl
	Soil-borne diseases	metam-sodium, methyl bromide with chloropicrin
	Verticillium wilt	chloropicrin
Pests	Aphids	cypermethrin, deltamethrin, dimethoate, malathion, nicotine, pirimicarb, rotenone, Verticillium lecanii
	Birds	ziram
	Browntail moth	diflubenzuron, teflubenzuron
	Capsids	cypermethrin, deltamethrin, nicotine
	Caterpillars	Bacillus thuringiensis, cypermethrin, diflubenzuron, nicotine, teflubenzuron
	Cutworms	cypermethrin
	Damaging mammals	bone oil, ziram
	General insect control	aldicarb *(off-label)*, pyrazophos *(off-label)*
	Leaf miners	abamectin, cypermethrin *(off-label)*, dimethoate, heptenophos *(off-label)*, nicotine, oxamyl *(off-label)*, trichlorfon *(off-label)*
	Leafhoppers	malathion, nicotine
	Mealy bugs	malathion
	Nematodes	metam-sodium, methyl bromide with chloropicrin
	Red spider mites	abamectin, bifenthrin, fenazaquin
	Sawflies	nicotine
	Scale insects	deltamethrin
	Sciarid flies	fonofos
	Slugs and snails	aluminium sulfate, metaldehyde, methiocarb
	Soil pests	metam-sodium, methyl bromide with chloropicrin
	Stem nematodes	aldicarb *(off-label)*

	Thrips	cypermethrin, deltamethrin, malathion, nicotine
	Tortrix moths	diflubenzuron
	Vine weevil	chlorpyrifos, fonofos
	Western flower thrips	abamectin
	Whitefly	cypermethrin, malathion, pirimiphos-methyl, teflubenzuron
	Winter moth	diflubenzuron
	Woolly aphid	nicotine
Weeds	Annual and perennial weeds	glyphosate (horticulture, forestry, amenity etc.), glyphosate (horticulture, forestry, amenity etc.) *(pre-planting)*, glyphosate (horticulture, forestry, amenity etc.) *(wiper application)*
	Annual dicotyledons	2,4-D + dicamba + mecoprop, chlorpropham + fenuron, chlorthal-dimethyl, diquat + paraquat, metazachlor, paraquat, propachlor, propyzamide, trifluralin *(off-label)*
	Annual grasses	chlorpropham + fenuron, diquat + paraquat, fluazifop-P-butyl *(off-label)*, paraquat, propachlor, propyzamide, trifluralin *(off-label)*
	Annual meadow grass	metazachlor
	Annual weeds	ammonium sulfamate, ammonium sulfamate *(pre-planting)*
	Blackgrass	metazachlor
	Brambles	2,4-D + dicamba + mecoprop
	Chickweed	chlorpropham + fenuron
	Creeping bent	paraquat
	General weed control	metam-sodium, methyl bromide with chloropicrin
	Groundsel, triazine resistant	metazachlor
	Perennial dicotyledons	2,4-D + dicamba + mecoprop
	Perennial grasses	diquat + paraquat, fluazifop-P-butyl *(off-label)*, propyzamide
	Perennial weeds	ammonium sulfamate, ammonium sulfamate *(pre-planting)*
	Stinging nettle	2,4-D + dicamba + mecoprop
	Woody weeds	2,4-D + dicamba + mecoprop
Plant growth regulation	Rooting of cuttings	2-(1-naphthyl)acetic acid, 4-indol-3-yl-butyric acid, 4-indol-3-yl-butyric acid + 2-(1-naphthyl)acetic acid with dichlorophen, indol-3-ylacetic acid
	Stem shortening	chlormequat, chlormequat + choline chloride

Hardy ornamental nursery stock

Diseases	Fungus diseases	prochloraz
	Phytophthora	etridiazole, fosetyl-aluminium, furalaxyl, propamocarb hydrochloride
	Powdery mildew	pyrifenox
	Pythium	furalaxyl, propamocarb hydrochloride
	Soil-borne diseases	methyl bromide with chloropicrin
Pests	Aphids	imidacloprid, nicotine
	Browntail moth	diflubenzuron
	Capsids	deltamethrin
	Caterpillars	diflubenzuron
	Nematodes	methyl bromide with chloropicrin
	Scale insects	deltamethrin
	Sciarid flies	imidacloprid
	Soil pests	methyl bromide with chloropicrin
	Thrips	deltamethrin
	Tortrix moths	diflubenzuron
	Vine weevil	carbofuran, chlorpyrifos, imidacloprid
	Whitefly	imidacloprid
	Winter moth	diflubenzuron
Weeds	Annual dicotyledons	diuron, diuron + paraquat, glufosinate-ammonium, isoxaben, isoxaben + trifluralin, metazachlor, pentanochlor, simazine
	Annual grasses	diuron, diuron + paraquat, glufosinate-ammonium, simazine
	Annual meadow grass	isoxaben + trifluralin, metazachlor, pentanochlor
	Blackgrass	metazachlor
	Fat-hen	isoxaben + trifluralin
	General weed control	methyl bromide with chloropicrin
	Groundsel, triazine resistant	metazachlor
	Perennial dicotyledons	diuron + paraquat, glufosinate-ammonium
	Perennial grasses	diuron + paraquat, glufosinate-ammonium

Hedges

Pests	Browntail moth	diflubenzuron, trichlorfon
	Caterpillars	diflubenzuron, trichlorfon
	Tortrix moths	diflubenzuron
	Winter moth	diflubenzuron

Weeds	Annual and perennial weeds	glyphosate (agricultural) *(directed spray)*, glyphosate (agricultural) *(pre-planting)*, glyphosate (horticulture, forestry, amenity etc.) *(directed spray)*, glyphosate (horticulture, forestry, amenity etc.) *(pre-planting)*
	Annual weeds	dalapon + dichlobenil
	Bracken	dalapon + dichlobenil
	Perennial grasses	dalapon + dichlobenil
	Rushes	dalapon + dichlobenil
Plant growth regulation	Growth retardation	maleic hydrazide

Perennials

Weeds	Annual dicotyledons	pentanochlor, propachlor
	Annual grasses	propachlor, sethoxydim *(off-label)*
	Annual meadow grass	pentanochlor
	Perennial grasses	sethoxydim *(off-label)*

Roses

Diseases	Black spot	captan, dichlofluanid, mancozeb, myclobutanil
	Powdery mildew	bupirimate, carbendazim, dichlofluanid, dodemorph, fenarimol, imazalil, myclobutanil, pyrazophos
	Rust	bupirimate + triforine, mancozeb, myclobutanil, oxycarboxin, penconazole
	Verticillium wilt	tar acids
Pests	Aphids	dimethoate, malathion
	Caterpillars	trichlorfon
	Leaf miners	dimethoate
	Leafhoppers	malathion
	Mealy bugs	malathion
	Red spider mites	fenpropathrin
	Scale insects	malathion
	Slug sawflies	rotenone
	Thrips	malathion
	Whitefly	malathion
Weeds	Annual dicotyledons	atrazine, pentanochlor, propyzamide, simazine

	Annual grasses	atrazine, propyzamide, simazine
	Annual meadow grass	pentanochlor
	Annual weeds	dalapon + dichlobenil, dichlobenil
	Perennial dicotyledons	dichlobenil
	Perennial grasses	dalapon + dichlobenil, dichlobenil, propyzamide
Plant growth regulation	Increasing flowering	paclobutrazol
	Stem shortening	paclobutrazol

Trees/shrubs

Diseases	Canker	azaconazole + imazalil, octhilinone, tar acids
	Crown gall	tar acids
	Honey fungus	tar acids
	Powdery mildew	penconazole
	Pruning wounds	octhilinone
	Scab	penconazole
	Silver leaf	azaconazole + imazalil, octhilinone
Pests	Aphids	dimethoate, fatty acids
	Browntail moth	diflubenzuron
	Capsids	deltamethrin
	Caterpillars	diflubenzuron
	Leaf miners	dimethoate
	Mealy bugs	fatty acids
	Red spider mites	fatty acids
	Scale insects	deltamethrin, fatty acids
	Thrips	deltamethrin
	Tortrix moths	diflubenzuron
	Whitefly	fatty acids
	Winter moth	diflubenzuron
Weeds	Annual and perennial weeds	glyphosate (agricultural) *(directed spray)*, glyphosate (horticulture, forestry, amenity etc.), glyphosate (horticulture, forestry, amenity etc.) *(directed spray)*, glyphosate (horticulture, forestry, amenity etc.) *(pre-planting/sowing)*, glyphosate (horticulture, forestry, amenity etc.) *(pre-planting)*, glyphosate (horticulture, forestry, amenity etc.) *(wiper application)*
	Annual dicotyledons	clopyralid, diuron, diuron + glyphosate, diuron + paraquat, glufosinate-ammonium, isoxaben, isoxaben + trifluralin, napropamide, oxadiazon, paraquat, propachlor, propyzamide, simazine

	Annual grasses	cycloxydim *(off-label)*, diuron, diuron + glyphosate, diuron + paraquat, glufosinate-ammonium, napropamide, oxadiazon, paraquat, propachlor, propyzamide, sethoxydim *(off-label)*, simazine
	Annual meadow grass	diuron, isoxaben + trifluralin
	Annual weeds	ammonium sulfamate, dalapon + dichlobenil, dichlobenil, dichlobenil *(off-label)*, diuron
	Bindweeds	oxadiazon
	Bracken	glyphosate (horticulture, forestry, amenity etc.)
	Cleavers	napropamide, oxadiazon
	Corn marigold	clopyralid
	Creeping bent	paraquat
	Creeping thistle	clopyralid
	Fat-hen	isoxaben + trifluralin
	General weed control	atrazine *(off-label)*
	Groundsel	isoxaben + trifluralin, napropamide
	Horsetails	propyzamide
	Mayweeds	clopyralid
	Perennial dicotyledons	ammonium sulfamate, clopyralid, dichlobenil, dichlobenil *(off-label)*, diuron + glyphosate, diuron + paraquat, glufosinate-ammonium
	Perennial grasses	cycloxydim *(off-label)*, dalapon + dichlobenil, dichlobenil, dichlobenil *(off-label)*, diuron + glyphosate, diuron + paraquat, glufosinate-ammonium, propyzamide, sethoxydim *(off-label)*
	Polygonums	oxadiazon
	Sedges	propyzamide
	Total vegetation control	glyphosate (horticulture, forestry, amenity etc.)
Plant growth regulation	Sucker inhibition	maleic hydrazide

Water lily

Diseases	Crown rot	carbendazim + metalaxyl *(off-label)*

FOR FULL CONDITIONS OF USE ALWAYS READ THE PRODUCT LABEL

Forestry

Broadleaved trees

Plant growth regulation	Increasing germination	gibberellins

Cut logs/timber

Pests	Ambrosia beetle	chlorpyrifos
	Black pine beetle	permethrin *(off-label)*
	Elm bark beetle	chlorpyrifos
	Larch shoot beetle	chlorpyrifos
	Pine shoot beetle	chlorpyrifos
	Pine weevil	permethrin *(off-label)*

Farm woodland

Diseases	Dutch elm disease	metam-sodium *(off-label)*, thiabendazole *(injection -off-label)*
Weeds	Annual dicotyledons	cyanazine *(off-label)*, lenacil *(off-label)*, metazachlor, napropamide *(off-label)*, pendimethalin *(off-label)*, propyzamide
	Annual grasses	cyanazine *(off-label)*, cycloxydim, fluazifop-P-butyl, metazachlor, napropamide *(off-label)*, pendimethalin *(off-label)*, propaquizafop, propyzamide
	Annual meadow grass	lenacil *(off-label)*
	Perennial grasses	cycloxydim, fluazifop-P-butyl, propaquizafop, propyzamide
	Volunteer cereals	fluazifop-P-butyl
	Wild oats	fluazifop-P-butyl

Forest nursery beds

Pests	Aphids	pirimicarb
	Birds	aluminium ammonium sulfate
	Black pine beetle	permethrin *(off-label)*
	Clay-coloured weevil	chlorpyrifos
	Damaging mammals	aluminium ammonium sulfate
	Pine weevil	permethrin *(off-label)*
Weeds	Annual dicotyledons	paraquat *(stale seedbed)*, pentanochlor, simazine
	Annual grasses	paraquat *(stale seedbed)*, simazine

	Annual meadow grass	pentanochlor

Forestry plantations

Diseases	Canker	azaconazole + imazalil
	Silver leaf	azaconazole + imazalil
Pests	Birds	aluminium ammonium sulfate, ziram
	Black pine beetle	chlorpyrifos, permethrin *(off-label)*
	Caterpillars	diflubenzuron
	Damaging mammals	aluminium ammonium sulfate, aluminium phosphide, sulfonated cod liver oil, ziram
	Grey squirrels	warfarin
	Large pine weevil	carbosulfan, chlorpyrifos
	Pine weevil	permethrin *(off-label)*
	Rodents	aluminium phosphide
	Vine weevil	chlorpyrifos
	Winter moth	diflubenzuron
Weeds	Annual and perennial weeds	clodinafop-propargyl + diflufenican *(directed spray)*, clodinafop-propargyl + diflufenican *(pre-planting)*, glyphosate (agricultural), glyphosate (agricultural) *(directed spray)*, glyphosate (agricultural) *(pre-planting)*, glyphosate (horticulture, forestry, amenity etc.), glyphosate (horticulture, forestry, amenity etc.) *(directed spray)*, glyphosate (horticulture, forestry, amenity etc.) *(post-planting)*, glyphosate (horticulture, forestry, amenity etc.) *(pre-planting)*
	Annual dicotyledons	2,4-D, 2,4-D + dicamba + triclopyr, atrazine, clopyralid *(off-label)*, cyanazine *(off-label)*, diquat + paraquat, isoxaben, metazachlor, paraquat, propyzamide, simazine, simazine *(off-label)*
	Annual grasses	atrazine, cyanazine *(off-label)*, cycloxydim, diquat + paraquat, metazachlor, paraquat, propyzamide, simazine, simazine *(off-label)*
	Annual weeds	ammonium sulfamate, dalapon + dichlobenil, glufosinate-ammonium, glyphosate (agricultural)
	Bracken	asulam, asulam *(off-label)*, clodinafop-propargyl + diflufenican *(directed spray)*, clodinafop-propargyl + diflufenican *(overall dormant spray)*, dalapon + dichlobenil, glyphosate (agricultural), glyphosate (agricultural) *(directed spray)*, glyphosate (agricultural) *(overall dormant spray)*, glyphosate (horticulture, forestry, amenity etc.), glyphosate (horticulture, forestry, amenity etc.) *(directed spray)*, glyphosate (horticulture, forestry, amenity etc.) *(overall dormant spray)*, imazapyr *(site preparation)*

Brambles	2,4-D + dicamba + triclopyr, clodinafop-propargyl + diflufenican *(overall dormant spray)*, glyphosate (agricultural) *(overall dormant spray)*, glyphosate (horticulture, forestry, amenity etc.) *(overall dormant spray)*, triclopyr
Broom	triclopyr
Couch	atrazine, clodinafop-propargyl + diflufenican *(directed spray)*, glyphosate (agricultural) *(directed spray)*, glyphosate (horticulture, forestry, amenity etc.) *(directed spray)*
Creeping bent	paraquat
Docks	2,4-D + dicamba + triclopyr, triclopyr
Firebreak desiccation	paraquat
Gorse	2,4-D + dicamba + triclopyr, triclopyr
Heather	2,4-D, glyphosate (agricultural), glyphosate (horticulture, forestry, amenity etc.), glyphosate (horticulture, forestry, amenity etc.) *(directed spray)*
Horsetails	propyzamide
Japanese knotweed	2,4-D + dicamba + triclopyr
Perennial dicotyledons	2,4-D, 2,4-D + dicamba + triclopyr, ammonium sulfamate, clodinafop-propargyl + diflufenican *(wiper application)*, clopyralid *(off-label)*, glyphosate (agricultural) *(wiper application)*, glyphosate (horticulture, forestry, amenity etc.) *(wiper application)*, triclopyr
Perennial grasses	ammonium sulfamate, atrazine, clodinafop-propargyl + diflufenican *(overall dormant spray)*, cycloxydim, dalapon + dichlobenil, diquat + paraquat, glyphosate (agricultural) *(overall dormant spray)*, glyphosate (horticulture, forestry, amenity etc.) *(overall dormant spray)*, propyzamide
Perennial weeds	glufosinate-ammonium, glyphosate (agricultural)
Rhododendrons	2,4-D + dicamba + triclopyr, ammonium sulfamate, glyphosate (agricultural), glyphosate (horticulture, forestry, amenity etc.), glyphosate (horticulture, forestry, amenity etc.) *(directed spray)*, triclopyr
Rushes	clodinafop-propargyl + diflufenican *(directed spray)*, dalapon + dichlobenil, glyphosate (agricultural) *(directed spray)*, glyphosate (horticulture, forestry, amenity etc.) *(directed spray)*
Sedges	propyzamide
Stinging nettle	2,4-D + dicamba + triclopyr, triclopyr
Total vegetation control	imazapyr *(site preparation)*
Volunteer cereals	diquat + paraquat

	Woody weeds	2,4-D, 2,4-D + dicamba + triclopyr, ammonium sulfamate, clodinafop-propargyl + diflufenican *(directed spray)*, clodinafop-propargyl + diflufenican *(overall dormant spray)*, fosamine-ammonium, fosamine-ammonium *(off-label)*, glyphosate (agricultural), glyphosate (agricultural) *(directed spray)*, glyphosate (agricultural) *(overall dormant spray)*, glyphosate (horticulture, forestry, amenity etc.), glyphosate (horticulture, forestry, amenity etc.) *(directed spray)*, glyphosate (horticulture, forestry, amenity etc.) *(overall dormant spray)*, triclopyr
Crop control	Chemical thinning	clodinafop-propargyl + diflufenican, glyphosate (agricultural), glyphosate (horticulture, forestry, amenity etc.), glyphosate (horticulture, forestry, amenity etc.) *(stem injection)*

Transplant lines

Pests	Black pine beetle	chlorpyrifos
	Pine weevil	chlorpyrifos
	Vine weevil	chlorpyrifos
Weeds	Annual dicotyledons	paraquat, simazine
	Annual grasses	paraquat, simazine
	Creeping bent	paraquat

Fruit and hops

Apples/pears

Diseases	Blossom wilt	vinclozolin
	Botrytis fruit rot	captan, thiram
	Brown rot	captan
	Canker	Bordeaux mixture, carbendazim, carbendazim *(partial control)*, copper oxychloride, octhilinone, thiophanate-methyl
	Collar rot	copper oxychloride + metalaxyl, fosetyl-aluminium
	Crown rot	fosetyl-aluminium
	Gloeosporium	captan, carbendazim, thiram
	Phytophthora fruit rot	captan, mancozeb + metalaxyl *(off-label)*
	Powdery mildew	bupirimate, captan + penconazole, carbendazim, fenarimol, kresoxim-methyl *(reduction)*, myclobutanil, penconazole, pyrazophos, pyrifenox, sulfur, thiophanate-methyl, triadimefon
	Pruning wounds	octhilinone

	Scab	Bordeaux mixture, captan, captan + penconazole, carbendazim, carbendazim *(off-label)*, dithianon, dodine, fenarimol, fenbuconazole, kresoxim-methyl, mancozeb, myclobutanil, pyrifenox, pyrimethanil, sulfur, thiophanate-methyl, thiram
	Silver leaf	octhilinone
	Storage rots	captan, carbendazim, carbendazim + metalaxyl, carbendazim *(off-label)*, thiophanate-methyl
Pests	Aphids	chlorpyrifos, cypermethrin, deltamethrin, dimethoate, fenitrothion, heptenophos, malathion, nicotine, pirimicarb, pirimiphos-methyl, tar oils
	Apple blossom weevil	chlorpyrifos, fenitrothion
	Bryobia mites	dicofol + tetradifon, dimethoate, malathion, phosalone
	Capsids	carbaryl, chlorpyrifos, cypermethrin, deltamethrin, dimethoate, fenitrothion, nicotine
	Caterpillars	chlorpyrifos, cypermethrin, deltamethrin, diflubenzuron, fenitrothion, fenpropathrin, nicotine, pirimiphos-methyl
	Cherry bark tortrix	trichlorfon
	Codling moth	carbaryl, chlorpyrifos, cypermethrin, deltamethrin, diflubenzuron, fenitrothion, malathion, phosalone
	Earwigs	carbaryl, diflubenzuron
	Leafhoppers	malathion
	Red spider mites	amitraz, bifenthrin, chlorpyrifos, clofentezine, dicofol, dicofol + tetradifon, dimethoate, fenazaquin, fenpropathrin, fenpyroximate, malathion, phosalone, tebufenpyrad, tetradifon
	Rust mite	amitraz, diflubenzuron, pirimiphos-methyl
	Sawflies	chlorpyrifos, cypermethrin, deltamethrin, dimethoate, fenitrothion
	Scale insects	tar oils
	Slug sawflies	rotenone
	Suckers	amitraz, chlorpyrifos, cypermethrin, deltamethrin, diflubenzuron, dimethoate, fenitrothion, lambda-cyhalothrin, malathion, tar oils
	Tortrix moths	carbaryl, chlorpyrifos, cypermethrin, deltamethrin, diflubenzuron, fenitrothion, fenoxycarb, phosalone
	Winter moth	carbaryl, chlorpyrifos, cypermethrin, diflubenzuron, fenitrothion, tar oils
	Woolly aphid	chlorpyrifos, heptenophos, malathion
Weeds	Annual and perennial weeds	clodinafop-propargyl + diflufenican, glyphosate (agricultural), glyphosate (horticulture, forestry, amenity etc.)
	Annual dicotyledons	2,4-D, 2,4-D + dichlorprop + MCPA + mecoprop, dicamba + MCPA + mecoprop, dicamba + MCPA + mecoprop-P, diquat + paraquat, diuron, isoxaben, oxadiazon, pendimethalin, pentanochlor, propyzamide, simazine, sodium monochloroacetate, sodium monochloroacetate *(directed spray)*

	Annual grasses	diquat + paraquat, diuron, oxadiazon, pendimethalin, propyzamide, simazine
	Annual meadow grass	pentanochlor
	Annual weeds	amitrole, glufosinate-ammonium
	Bindweeds	oxadiazon
	Chickweed	2,4-D + dichlorprop + MCPA + mecoprop
	Cleavers	2,4-D + dichlorprop + MCPA + mecoprop, oxadiazon
	Couch	amitrole
	Creeping thistle	amitrole
	Docks	amitrole, asulam, dicamba + MCPA + mecoprop, dicamba + MCPA + mecoprop-P
	Perennial dicotyledons	2,4-D, 2,4-D + dichlorprop + MCPA + mecoprop, dicamba + MCPA + mecoprop, dicamba + MCPA + mecoprop-P, glyphosate (agricultural)
	Perennial grasses	diquat + paraquat, glyphosate (agricultural), propyzamide
	Perennial weeds	amitrole, glufosinate-ammonium
	Polygonums	oxadiazon
	Volunteer cereals	diquat + paraquat
Crop control	Control of water shoots	2-(1-naphthyl)acetic acid
	Sucker control	2-(1-naphthyl)acetic acid, clodinafop-propargyl + diflufenican, glyphosate (agricultural)
Plant growth regulation	Controlling vigour	paclobutrazol
	Fruit thinning	carbaryl
	Increasing fruit set	gibberellins, paclobutrazol
	Reducing fruit russeting	gibberellins

Apricots/peaches/nectarines

Diseases	Leaf curl	Bordeaux mixture, copper ammonium carbonate, copper oxychloride
Pests	Aphids	malathion, tar oils
	Red spider mites	tetradifon *(protected crops)*
	Scale insects	tar oils
	Winter moth	tar oils
Weeds	Docks	asulam *(off-label)*

FOR FULL CONDITIONS OF USE ALWAYS READ THE PRODUCT LABEL

	General weed control	amitrole *(off-label)*

Blueberries/cranberries

Weeds	Annual dicotyledons	propachlor *(off-label)*
	Annual grasses	propachlor *(off-label)*
	Docks	asulam *(off-label)*

Cane fruit

Diseases	Botrytis	carbendazim, carbendazim *(off-label)*, chlorothalonil, chlorothalonil *(moderate control)*, dichlofluanid, fenhexamid, iprodione, thiram
	Cane blight	dichlofluanid
	Cane spot	Bordeaux mixture, carbendazim, carbendazim *(off-label)*, chlorothalonil, copper ammonium carbonate, copper oxychloride, dichlofluanid, thiram
	Powdery mildew	bupirimate, carbendazim, carbendazim *(off-label)*, chlorothalonil, dichlofluanid, fenarimol, triadimefon *(off-label)*
	Purple blotch	copper oxychloride
	Root rot	mancozeb + metalaxyl *(off-label)*, mancozeb + oxadixyl *(off-label)*
	Spur blight	Bordeaux mixture, carbendazim, carbendazim *(off-label)*, thiram
Pests	Aphids	chlorpyrifos, dimethoate, pirimicarb, tar oils
	Birds	aluminium ammonium sulfate
	Blackberry mite	endosulfan
	Capsids	dimethoate
	Caterpillars	Bacillus thuringiensis
	Damaging mammals	aluminium ammonium sulfate
	Nematodes	1,3-dichloropropene
	Overwintering pests	tar acids
	Raspberry beetle	chlorpyrifos, deltamethrin, fenitrothion, rotenone
	Raspberry cane midge	chlorpyrifos, fenitrothion
	Red spider mites	chlorpyrifos, clofentezine *(off-label - protected crops)*, clofentezine *(off-label)*, dimethoate, tetradifon
	Scale insects	tar oils
	Virus vectors	1,3-dichloropropene
	Winter moth	tar oils
Weeds	Annual dicotyledons	atrazine, bromacil, chlorthal-dimethyl, diquat + paraquat, glufosinate-ammonium, isoxaben, napropamide, oxadiazon, paraquat, pendimethalin, propyzamide, propyzamide *(England only)*, simazine, trifluralin

	Annual grasses	atrazine, bromacil, diquat + paraquat, fluazifop-P-butyl, glufosinate-ammonium, napropamide, oxadiazon, paraquat, pendimethalin, propyzamide, propyzamide *(England only)*, simazine, trifluralin
	Annual weeds	glufosinate-ammonium
	Barren brome	sethoxydim
	Bindweeds	oxadiazon
	Blackgrass	sethoxydim
	Cleavers	napropamide, oxadiazon
	Couch	bromacil, sethoxydim
	Creeping bent	paraquat, sethoxydim
	Docks	asulam *(off-label)*
	Groundsel	napropamide
	Perennial dicotyledons	bromacil, glufosinate-ammonium
	Perennial grasses	bromacil, diquat + paraquat, fluazifop-P-butyl, glufosinate-ammonium, propyzamide, propyzamide *(England only)*, sethoxydim
	Perennial ryegrass	paraquat
	Perennial weeds	glufosinate-ammonium
	Polygonums	oxadiazon
	Rough meadow grass	paraquat
	Volunteer cereals	diquat + paraquat, fluazifop-P-butyl, sethoxydim
	Wild oats	fluazifop-P-butyl, sethoxydim
Crop control	Sucker control	2-(1-naphthyl)acetic acid, sodium monochloroacetate *(off-label)*

Currants

Diseases	Botrytis	carbendazim, chlorothalonil *(moderate control)*, chlorothalonil *(suppression)*, dichlofluanid, fenhexamid
	Currant leaf spot	Bordeaux mixture, carbendazim, chlorothalonil, copper ammonium carbonate, dodine, mancozeb, zineb
	Leaf spots	pyrifenox
	Powdery mildew	bupirimate, carbendazim, chlorothalonil *(suppression)*, fenarimol, myclobutanil, pyrifenox, triadimefon *(off-label)*
	Rust	copper oxychloride, thiram
Pests	Aphids	chlorpyrifos, dimethoate, malathion, pirimicarb, tar oils
	Big-bud mite	endosulfan
	Capsids	chlorpyrifos, cypermethrin, fenitrothion

	Caterpillars	chlorpyrifos, fenpropathrin
	Earwigs	diflubenzuron
	Gall mite	sulfur
	Overwintering pests	tar acids
	Red spider mites	bifenthrin, chlorpyrifos, dicofol + tetradifon, dimethoate, fenpropathrin, malathion, tetradifon
	Sawflies	cypermethrin, fenitrothion
	Scale insects	tar oils
	Winter moth	diflubenzuron, tar oils
Weeds	Annual and perennial weeds	clodinafop-propargyl + diflufenican *(off-label)*, glyphosate (agricultural) *(off-label)*
	Annual dicotyledons	chlorpropham, chlorthal-dimethyl, diquat + paraquat, diuron *(off-label)*, isoxaben, MCPB, napropamide, oxadiazon, paraquat, pendimethalin, pentanochlor, propachlor *(off-label)*, propyzamide, simazine, sodium monochloroacetate, sodium monochloroacetate *(directed spray)*
	Annual grasses	chlorpropham, diquat + paraquat, diuron *(off-label)*, fluazifop-P-butyl, napropamide, oxadiazon, paraquat, pendimethalin, propachlor *(off-label)*, propyzamide, simazine
	Annual meadow grass	pentanochlor
	Annual weeds	glufosinate-ammonium
	Bindweeds	oxadiazon
	Chickweed	chlorpropham
	Cleavers	napropamide, oxadiazon
	Creeping bent	paraquat
	Docks	asulam, asulam *(off-label)*
	Groundsel	napropamide
	Perennial dicotyledons	MCPB
	Perennial grasses	diquat + paraquat, fluazifop-P-butyl, propyzamide
	Perennial ryegrass	paraquat
	Perennial weeds	glufosinate-ammonium
	Polygonums	chlorpropham, oxadiazon
	Rough meadow grass	paraquat
	Volunteer cereals	diquat + paraquat, fluazifop-P-butyl
	Wild oats	fluazifop-P-butyl

Fruit nursery stock

Weeds	Annual dicotyledons	metazachlor, trifluralin *(off-label)*
	Annual grasses	trifluralin *(off-label)*

	Annual meadow grass	metazachlor
	Blackgrass	metazachlor
	Groundsel, triazine resistant	metazachlor

General

Diseases	Powdery mildew	fenpropimorph *(off-label)*
Pests	Aphids	nicotine, rotenone
	Birds	aluminium ammonium sulfate
	Capsids	nicotine
	Damaging mammals	aluminium ammonium sulfate
	Leafhoppers	nicotine
	Sawflies	nicotine
	Slugs and snails	aluminium sulfate, metaldehyde, methiocarb
	Woolly aphid	nicotine
Weeds	Annual dicotyledons	diquat + paraquat, glufosinate-ammonium, trifluralin *(off-label)*
	Annual grasses	diquat + paraquat, glufosinate-ammonium, trifluralin *(off-label)*
	Annual weeds	dichlobenil
	Perennial dicotyledons	dichlobenil, glufosinate-ammonium
	Perennial grasses	dichlobenil, diquat + paraquat, glufosinate-ammonium
	Volunteer cereals	diquat + paraquat
Plant growth regulation	Increasing yield	sulfur

Gooseberries

Diseases	Botrytis	carbendazim, chlorothalonil *(moderate control)*, dichlofluanid, fenhexamid
	Currant leaf spot	carbendazim, chlorothalonil, dodine, mancozeb
	Leaf spots	pyrifenox
	Powdery mildew	bupirimate, carbendazim, chlorothalonil *(suppression)*, fenarimol, myclobutanil, pyrifenox, sulfur, triadimefon *(off-label)*
	Septoria diseases	chlorothalonil
Pests	Aphids	chlorpyrifos, dimethoate, malathion, pirimicarb, tar oils
	Bryobia mites	lambda-cyhalothrin *(off-label)*

	Capsids	chlorpyrifos, cypermethrin, dimethoate, fenitrothion, lambda-cyhalothrin *(off-label)*
	Caterpillars	chlorpyrifos, lambda-cyhalothrin *(off-label)*, nicotine
	Red spider mites	chlorpyrifos, dimethoate, lambda-cyhalothrin *(off-label)*, malathion, tetradifon
	Sawflies	cypermethrin, fenitrothion, malathion, nicotine, rotenone
	Scale insects	tar oils
	Winter moth	tar oils
Weeds	Annual dicotyledons	chlorpropham, chlorthal-dimethyl, diquat + paraquat, isoxaben, napropamide, oxadiazon, paraquat, pendimethalin, pentanochlor, propachlor *(off-label)*, propyzamide, simazine, sodium monochloroacetate, sodium monochloroacetate *(directed spray)*
	Annual grasses	chlorpropham, diquat + paraquat, fluazifop-P-butyl, napropamide, oxadiazon, paraquat, pendimethalin, propyzamide, simazine
	Annual meadow grass	pentanochlor
	Bindweeds	oxadiazon
	Chickweed	chlorpropham
	Cleavers	napropamide, oxadiazon
	Creeping bent	paraquat
	Docks	asulam *(off-label)*
	Groundsel	napropamide
	Perennial grasses	diquat + paraquat, fluazifop-P-butyl, propyzamide
	Perennial ryegrass	paraquat
	Polygonums	chlorpropham, oxadiazon
	Rough meadow grass	paraquat
	Volunteer cereals	diquat + paraquat, fluazifop-P-butyl
	Wild oats	fluazifop-P-butyl

Grapevines

Diseases	Botrytis	chlorothalonil, dichlofluanid, iprodione *(off-label)*, pyrimethanil *(off-label)*
	Downy mildew	chlorothalonil, copper oxychloride, copper oxychloride + metalaxyl *(off-label)*
	Powdery mildew	copper sulfate + sulfur, sulfur, triadimefon *(off-label)*
Pests	Caterpillars	Bacillus thuringiensis
	Red spider mites	tetradifon
	Scale insects	tar oils
	Tortrix moths	Bacillus thuringiensis

Weeds	Annual and perennial weeds	clodinafop-propargyl + diflufenican *(off-label)*, glyphosate (agricultural) *(off-label)*
	Annual dicotyledons	isoxaben, oxadiazon, paraquat, propyzamide *(off-label)*, simazine *(off-label)*
	Annual grasses	oxadiazon, paraquat, propyzamide *(off-label)*, simazine *(off-label)*
	Annual weeds	glufosinate-ammonium
	Bindweeds	oxadiazon
	Cleavers	oxadiazon
	Perennial grasses	propyzamide *(off-label)*
	Perennial weeds	glufosinate-ammonium
	Polygonums	oxadiazon

Hops

Diseases	Downy mildew	Bordeaux mixture, chlorothalonil, copper oxychloride, copper oxychloride + metalaxyl, copper sulfate + sulfur, fosetyl-aluminium, zineb
	Powdery mildew	bupirimate, copper sulfate + sulfur, fenpropimorph *(off-label)*, myclobutanil *(off-label)*, penconazole, pyrazophos, sulfur, triadimefon
Pests	Aphids	bifenthrin, cypermethrin, deltamethrin, fenpropathrin, imidacloprid, lambda-cyhalothrin, tebufenpyrad
	Caterpillars	fenpropathrin
	Nematodes	1,3-dichloropropene
	Red spider mites	bifenthrin, dicofol, dicofol + tetradifon, fenpropathrin, lambda-cyhalothrin, tebufenpyrad, tetradifon
	Virus vectors	1,3-dichloropropene
Weeds	Annual dicotyledons	diquat + paraquat, isoxaben, oxadiazon, paraquat, pendimethalin, propyzamide *(off-label)*, simazine
	Annual grasses	diquat + paraquat, fluazifop-P-butyl, oxadiazon, paraquat, pendimethalin, propyzamide *(off-label)*, simazine
	Bindweeds	oxadiazon
	Blackgrass	fluazifop-P-butyl
	Cleavers	oxadiazon
	Creeping bent	paraquat
	Docks	asulam
	Perennial grasses	diquat + paraquat, fluazifop-P-butyl, propyzamide *(off-label)*
	Perennial ryegrass	paraquat

	Polygonums	oxadiazon
	Rough meadow grass	paraquat
	Volunteer cereals	diquat + paraquat, fluazifop-P-butyl
	Wild oats	fluazifop-P-butyl
Crop control	Chemical stripping	anthracene oil, diquat, diquat + paraquat, paraquat, sodium monochloroacetate, sodium monochloroacetate *(off-label)*

Plums/cherries/damsons

Diseases	Bacterial canker	Bordeaux mixture, copper oxychloride
	Blossom wilt	myclobutanil *(off-label)*
	Brown rot	carbendazim *(off-label)*
	Pruning wounds	octhilinone
	Silver leaf	octhilinone
Pests	Aphids	chlorpyrifos, cypermethrin, deltamethrin, dimethoate, fenitrothion, malathion, pirimicarb, pirimicarb *(off-label)*, tar oils
	Caterpillars	cypermethrin, deltamethrin, fenitrothion
	Plum fruit moth	deltamethrin, diflubenzuron
	Red spider mites	chlorpyrifos, clofentezine, dimethoate, malathion, phosalone, tetradifon
	Rust mite	diflubenzuron
	Sawflies	deltamethrin, dimethoate
	Scale insects	tar oils
	Tortrix moths	chlorpyrifos, cypermethrin, diflubenzuron
	Winter moth	chlorpyrifos, cypermethrin, diflubenzuron, tar oils
Weeds	Annual and perennial weeds	clodinafop-propargyl + diflufenican, glyphosate (agricultural), glyphosate (horticulture, forestry, amenity etc.)
	Annual dicotyledons	diquat + paraquat, isoxaben, pendimethalin, pentanochlor, propyzamide, sodium monochloroacetate, sodium monochloroacetate *(directed spray)*
	Annual grasses	diquat + paraquat, pendimethalin, propyzamide
	Annual meadow grass	pentanochlor
	Annual weeds	glufosinate-ammonium
	Docks	asulam, asulam *(off-label)*
	General weed control	amitrole *(off-label)*
	Perennial dicotyledons	glyphosate (agricultural)
	Perennial grasses	diquat + paraquat, glyphosate (agricultural), propyzamide
	Perennial weeds	glufosinate-ammonium
	Volunteer cereals	diquat + paraquat

Crop control	Sucker control	2-(1-naphthyl)acetic acid, clodinafop-propargyl + diflufenican, glyphosate (agricultural)
Plant growth regulation	Controlling vigour	paclobutrazol
	Increasing fruit set	paclobutrazol

Quinces

Weeds	Docks	asulam *(off-label)*
	General weed control	amitrole *(off-label)*

Strawberries

Diseases	Botrytis	captan, carbendazim, chlorothalonil, chlorothalonil *(moderate control)*, chlorothalonil *(off-label)*, dichlofluanid, fenhexamid, iprodione, pyrimethanil, thiram
	Crown rot	chloropicrin
	Powdery mildew	bupirimate, carbendazim, fenarimol, myclobutanil, pyrifenox, sulfur, triadimefon *(off-label)*
	Red core	chloropicrin, copper oxychloride + metalaxyl, fosetyl-aluminium, fosetyl-aluminium *(off-label (spring treatment))*
	Soil-borne diseases	methyl bromide with chloropicrin
	Verticillium wilt	chloropicrin
Pests	Aphids	chlorpyrifos, demeton-S-methyl, dimethoate, disulfoton, heptenophos, malathion, nicotine, pirimicarb
	Birds	aluminium ammonium sulfate
	Capsids	cypermethrin
	Caterpillars	Bacillus thuringiensis, cypermethrin, nicotine
	Chafer grubs	gamma-HCH
	Damaging mammals	aluminium ammonium sulfate
	Leatherjackets	gamma-HCH
	Nematodes	1,3-dichloropropene, chloropicrin, methyl bromide with chloropicrin
	Red spider mites	bifenthrin, chlorpyrifos, clofentezine *(off-label - protected crops)*, demeton-S-methyl, dicofol, dicofol + tetradifon, dimethoate, fenbutatin oxide, fenpropathrin, tetradifon
	Sawflies	cypermethrin
	Soil pests	methyl bromide with chloropicrin

FOR FULL CONDITIONS OF USE ALWAYS READ THE PRODUCT LABEL

	Stem nematodes	1,3-dichloropropene
	Strawberry blossom weevil	chlorpyrifos
	Strawberry seed beetle	methiocarb
	Tarsonemid mites	dicofol, dicofol + tetradifon, endosulfan
	Tortrix moths	chlorpyrifos, cypermethrin, fenitrothion, trichlorfon
	Vine weevil	carbofuran, chlorpyrifos
	Virus vectors	1,3-dichloropropene
	Wireworms	gamma-HCH
Weeds	Annual dicotyledons	chlorpropham, chlorpropham + fenuron, chlorthal-dimethyl, clopyralid, diquat + paraquat, isoxaben, napropamide, pendimethalin, phenmedipham, propachlor, propachlor *(off-label)*, propyzamide, simazine, trifluralin
	Annual grasses	chlorpropham, chlorpropham + fenuron, cycloxydim, diquat + paraquat, ethofumesate *(off-label)*, fluazifop-P-butyl, napropamide, pendimethalin, propachlor, propachlor *(off-label)*, propyzamide, simazine, trifluralin
	Annual weeds	glufosinate-ammonium
	Blackgrass	cycloxydim
	Chickweed	chlorpropham, chlorpropham + fenuron
	Cleavers	napropamide
	Clover	ethofumesate *(off-label)*
	Corn marigold	clopyralid
	Couch	cycloxydim
	Creeping bent	cycloxydim
	Creeping thistle	clopyralid
	Docks	asulam *(off-label)*
	General weed control	methyl bromide with chloropicrin
	Groundsel	napropamide
	Mayweeds	clopyralid
	Perennial dicotyledons	clopyralid
	Perennial grasses	cycloxydim, diquat + paraquat, fluazifop-P-butyl, propyzamide
	Perennial weeds	glufosinate-ammonium
	Polygonums	chlorpropham
	Volunteer cereals	cycloxydim, diquat + paraquat, fluazifop-P-butyl
	Wild oats	cycloxydim, fluazifop-P-butyl
Crop control	Runner desiccation	paraquat

Tree fruit

Diseases	Soil-borne diseases	methyl bromide with chloropicrin
	Verticillium wilt	chloropicrin
Pests	Aphids	fatty acids, nicotine, rotenone
	Birds	aluminium ammonium sulfate, ziram
	Browntail moth	Bacillus thuringiensis
	Capsids	nicotine
	Caterpillars	Bacillus thuringiensis
	Damaging mammals	aluminium ammonium sulfate, ziram
	Mealy bugs	fatty acids
	Nematodes	methyl bromide with chloropicrin
	Overwintering pests	tar acids
	Red spider mites	fatty acids
	Sawflies	nicotine
	Scale insects	fatty acids
	Soil pests	methyl bromide with chloropicrin
	Tortrix moths	Bacillus thuringiensis
	Whitefly	fatty acids
	Winter moth	Bacillus thuringiensis
Weeds	Annual and perennial weeds	glyphosate (agricultural), glyphosate (agricultural) *(directed spray)*, glyphosate (horticulture, forestry, amenity etc.) *(directed spray)*
	Annual dicotyledons	glufosinate-ammonium, paraquat
	Annual grasses	glufosinate-ammonium, paraquat
	Annual weeds	dichlobenil, glyphosate (agricultural)
	Creeping bent	paraquat
	General weed control	methyl bromide with chloropicrin
	Perennial dicotyledons	dichlobenil, glufosinate-ammonium, glyphosate (agricultural) *(wiper application)*
	Perennial grasses	dichlobenil, glufosinate-ammonium
	Perennial ryegrass	paraquat
	Perennial weeds	glyphosate (agricultural)
	Rough meadow grass	paraquat

Plant growth regulation	Increasing yield	sulfur

Tree nuts

Weeds	Annual and perennial weeds	clodinafop-propargyl + diflufenican *(off-label)*, glyphosate (agricultural) *(off-label)*
	Annual weeds	glufosinate-ammonium
	Perennial weeds	glufosinate-ammonium

Grain/crop store uses

Food storage

Pests	Food storage pests	methyl bromide with amyl acetate

Stored cabbages

Diseases	Alternaria	iprodione
	Botrytis	iprodione
	Phytophthora	carbendazim + metalaxyl

Stored grain/rapeseed/linseed

Pests	General insect control	magnesium phosphide
	Grain storage pests	chlorpyrifos-methyl, deltamethrin, etrimfos, fenitrothion, fenitrothion + permethrin + resmethrin, gamma-HCH *(empty)*, methyl bromide, methyl bromide with chloropicrin, permethrin, pirimiphos-methyl

Stored products

Pests	Stored product pests	aluminium phosphide

Grassland

Grassland

Diseases	Crown rust	propiconazole, triadimefon
	Damping off	thiram *(seed treatment)*
	Drechslera leaf spot	propiconazole
	Powdery mildew	propiconazole, sulfur, triadimefon

	Rhynchosporium	propiconazole, triadimefon
Pests	Aphids	pirimicarb
	Birds	aluminium ammonium sulfate
	Cutworms	gamma-HCH
	Damaging mammals	aluminium ammonium sulfate, strychnine hydrochloride (commodity substance) *(areas of restricted public access)*
	Frit fly	chlorpyrifos
	Leatherjackets	chlorpyrifos, gamma-HCH
	Slugs and snails	metaldehyde, methiocarb
	Wireworms	gamma-HCH
Weeds	Annual and perennial weeds	clodinafop-propargyl + diflufenican *(pre-cut/graze)*, glyphosate (agricultural) *(pre-cut/graze or sward destruction)*, glyphosate (agricultural) *(pre-cut/graze)*, glyphosate (agricultural) *(sward destruction/direct drilling)*, glyphosate (agricultural) *(sward destruction)*, glyphosate (horticulture, forestry, amenity etc.) *(pre-cut/graze)*
	Annual dicotyledons	2,4-D, 2,4-D + dicamba + triclopyr, 2,4-D + mecoprop-P, 2,4-DB + linuron + MCPA, 2,4-DB + MCPA, benazolin + 2,4-DB + MCPA, benazolin + bromoxynil + ioxynil, bentazone + MCPA + MCPB, bromoxynil + ioxynil + mecoprop-P, clopyralid, dicamba + MCPA + mecoprop, dicamba + MCPA + mecoprop-P, dicamba + mecoprop, dicamba + mecoprop-P, fluroxypyr, glufosinate-ammonium *(sward destruction)*, glyphosate (agricultural) *(sward destruction)*, MCPA, MCPA + MCPB, mecoprop, mecoprop-P, paraquat *(sward desiccation)*, paraquat *(sward destruction/direct drilling)*
	Annual grasses	ethofumesate, glyphosate (agricultural) *(sward destruction)*, paraquat *(sward desiccation)*, paraquat *(sward destruction/direct drilling)*
	Annual weeds	clodinafop-propargyl + diflufenican, glyphosate (agricultural), glyphosate (agricultural) *(sward destruction)*, glyphosate (horticulture, forestry, amenity etc.)
	Black bindweed	fluroxypyr
	Blackgrass	ethofumesate
	Bracken	asulam, glyphosate (agricultural) *(sward destruction)*
	Brambles	2,4-D + dicamba + triclopyr, clopyralid + triclopyr, triclopyr
	Broom	clopyralid + triclopyr, triclopyr
	Buttercups	MCPA
	Chickweed	benazolin + 2,4-DB + MCPA, dicamba, dicamba + MCPA + mecoprop, dicamba + MCPA + mecoprop-P, dicamba + mecoprop-P, ethofumesate, fluroxypyr, mecoprop, mecoprop-P

Cleavers	benazolin + 2,4-DB + MCPA, dicamba + MCPA + mecoprop, dicamba + MCPA + mecoprop-P, dicamba + mecoprop-P, ethofumesate, fluroxypyr, mecoprop-P
Corn marigold	clopyralid
Couch	glyphosate (agricultural) *(sward destruction/direct drilling)*, glyphosate (agricultural) *(sward destruction)*
Creeping bent	paraquat *(sward destruction/direct drilling)*
Creeping thistle	clopyralid, clopyralid + 2,4-D + MCPA, clopyralid *(off-label, weed wiper)*
Destruction of short term leys	clodinafop-propargyl + diflufenican, glyphosate (agricultural), glyphosate (horticulture, forestry, amenity etc.)
Docks	2,4-D + dicamba + triclopyr, asulam, clopyralid + fluroxypyr + triclopyr, clopyralid + triclopyr, dicamba, dicamba + MCPA + mecoprop, dicamba + MCPA + mecoprop-P, dicamba + mecoprop, dicamba + mecoprop-P, fluroxypyr, fluroxypyr + triclopyr, MCPA, thifensulfuron-methyl, triclopyr
Fat-hen	MCPA
Gorse	2,4-D + dicamba + triclopyr, clopyralid + triclopyr, triclopyr
Hemp-nettle	fluroxypyr, MCPA
Japanese knotweed	2,4-D + dicamba + triclopyr
Mayweeds	clopyralid, dicamba + MCPA + mecoprop, dicamba + MCPA + mecoprop-P, dicamba + mecoprop-P
Perennial dicotyledons	2,4-D, 2,4-D + dicamba + triclopyr, 2,4-D + mecoprop-P, 2,4-DB + MCPA, benazolin + 2,4-DB + MCPA, clodinafop-propargyl + diflufenican *(wiper application)*, clopyralid, clopyralid + fluroxypyr + triclopyr, clopyralid + triclopyr, clopyralid + triclopyr *(weed wiper - off-label)*, dicamba + MCPA + mecoprop, dicamba + MCPA + mecoprop-P, dicamba + mecoprop, dicamba + mecoprop-P, glufosinate-ammonium *(sward destruction)*, glyphosate (agricultural) *(sward destruction)*, glyphosate (agricultural) *(wiper application)*, MCPA, MCPA + MCPB, mecoprop, mecoprop-P, triclopyr
Perennial grasses	glufosinate-ammonium *(sward destruction)*, glyphosate (agricultural) *(sward destruction)*
Perennial ryegrass	paraquat *(sward destruction/direct drilling)*
Polygonums	benazolin + 2,4-DB + MCPA, dicamba + MCPA + mecoprop, dicamba + MCPA + mecoprop-P, dicamba + mecoprop-P
Ragwort	MCPA
Rhododendrons	2,4-D + dicamba + triclopyr
Rough meadow grass	paraquat *(sward destruction/direct drilling)*
Rushes	glyphosate (agricultural) *(sward destruction)*, MCPA, triclopyr
Stinging nettle	2,4-D + dicamba + triclopyr, clopyralid + fluroxypyr + triclopyr, clopyralid + triclopyr, triclopyr
Volunteer cereals	ethofumesate

	Volunteer potatoes	fluroxypyr
	Woody weeds	2,4-D + dicamba + triclopyr, triclopyr
Plant growth regulation	Increasing yield	sulfur

Leys

Pests	Frit fly	cypermethrin
Weeds	Annual dicotyledons	2,4-DB + MCPA, benazolin + 2,4-DB + MCPA, bromoxynil + ethofumesate + ioxynil, dicamba + MCPA + mecoprop, dicamba + MCPA + mecoprop-P, dicamba + mecoprop, dicamba + mecoprop-P, fluroxypyr, MCPA, MCPA + MCPB, MCPB, mecoprop, mecoprop-P
	Annual grasses	ethofumesate
	Annual meadow grass	bromoxynil + ethofumesate + ioxynil
	Black bindweed	fluroxypyr
	Blackgrass	ethofumesate
	Chickweed	benazolin + 2,4-DB + MCPA, dicamba, dicamba + MCPA + mecoprop, dicamba + MCPA + mecoprop-P, ethofumesate, fluroxypyr, mecoprop, mecoprop-P
	Cleavers	benazolin + 2,4-DB + MCPA, dicamba + MCPA + mecoprop, dicamba + MCPA + mecoprop-P, ethofumesate, fluroxypyr, mecoprop, mecoprop-P
	Docks	asulam, dicamba, dicamba + MCPA + mecoprop, dicamba + MCPA + mecoprop-P, dicamba + mecoprop, dicamba + mecoprop-P, fluroxypyr
	Fat-hen	MCPA
	Hemp-nettle	fluroxypyr, MCPA
	Mayweeds	dicamba + MCPA + mecoprop, dicamba + MCPA + mecoprop-P
	Perennial dicotyledons	2,4-DB + MCPA, benazolin + 2,4-DB + MCPA, dicamba + MCPA + mecoprop, dicamba + MCPA + mecoprop-P, dicamba + mecoprop, dicamba + mecoprop-P, MCPA, MCPA + MCPB, MCPB, mecoprop, mecoprop-P
	Polygonums	2,4-DB + MCPA, benazolin + 2,4-DB + MCPA, dicamba + MCPA + mecoprop, dicamba + MCPA + mecoprop-P
	Volunteer cereals	ethofumesate
	Volunteer potatoes	fluroxypyr

Non-crop pest control

Farm buildings/yards

Diseases	Fungus diseases	formaldehyde (commodity substance)
Pests	Ants	fenitrothion
	Beetles	dichlorvos, fenitrothion
	Bugs	fenitrothion
	Cockroaches	fenitrothion
	Crickets	fenitrothion
	Damaging mammals	bone oil, sodium cyanide
	Earwigs	fenitrothion
	Fleas	fenitrothion
	Flies	azamethiphos, cyromazine, dichlorvos, fenitrothion, methomyl + (Z)-9-tricosene, phenothrin, phenothrin + tetramethrin, pyrethrins, tetramethrin
	Grey squirrels	warfarin
	Hide beetle	alpha-cypermethrin
	Mealworms	alpha-cypermethrin
	Mites	fenitrothion
	Mosquitoes	dichlorvos, phenothrin + tetramethrin
	Moths	fenitrothion
	Poultry ectoparasites	dichlorvos
	Poultry house pests	alpha-cypermethrin, fenitrothion
	Rodents	alphachloralose, brodifacoum, bromadiolone, calciferol + difenacoum, chlorophacinone, coumatetralyl, difenacoum, diphacinone, sodium cyanide, warfarin, zinc phosphide
	Silverfish	fenitrothion
	Wasps	phenothrin + tetramethrin
Weeds	Annual and perennial weeds	glyphosate (horticulture, forestry, amenity etc.)
	Annual weeds	diuron
	Bracken	imazapyr
	General weed control	glyphosate (agricultural) *(spot treatment)*
	Perennial weeds	diuron
	Total vegetation control	imazapyr

Farmland

Pests	Damaging mammals	aluminium phosphide, bone oil
	Rodents	aluminium phosphide
Weeds	Volunteer potatoes	dichlobenil *(blight prevention)*

Food storage areas

Pests	Ants	fenitrothion
	Beetles	fenitrothion
	Bugs	fenitrothion
	Cockroaches	fenitrothion
	Crickets	fenitrothion
	Earwigs	fenitrothion
	Fleas	fenitrothion
	Flies	fenitrothion
	Food storage pests	methyl bromide
	General insect control	aluminium phosphide
	Mites	fenitrothion
	Moths	fenitrothion
	Silverfish	fenitrothion

Manure heaps

Pests	Flies	cyromazine, permethrin, trichlorfon

Miscellaneous situations

Pests	Birds	carbon dioxide (commodity substance), paraffin oil (commodity substance) *(egg treatment)*
	Damaging mammals	strychnine hydrochloride (commodity substance) *(areas of restricted public access)*
	Rodents	carbon dioxide (commodity substance)
	Wasps	resmethrin + tetramethrin

Refuse tips

Pests	Ants	fenitrothion
	Beetles	fenitrothion

FOR FULL CONDITIONS OF USE ALWAYS READ THE PRODUCT LABEL

	Bugs	fenitrothion
	Cockroaches	fenitrothion
	Crickets	fenitrothion
	Earwigs	fenitrothion
	Fleas	fenitrothion
	Flies	chlorpyrifos-methyl, fenitrothion, pyrethrins, trichlorfon *(off-label)*
	Mites	fenitrothion
	Moths	fenitrothion
	Silverfish	fenitrothion

Stored plant material

Pests	General insect control	magnesium phosphide, methyl bromide

Protected crops

Aubergines

Diseases	Botrytis	dichlofluanid *(off-label)*
	Damping off	propamocarb hydrochloride
	Phytophthora	propamocarb hydrochloride
	Powdery mildew	sulfur *(off-label)*
	Root rot	propamocarb hydrochloride
Pests	Ants	pirimiphos-methyl
	Aphids	nicotine, permethrin, pirimiphos-methyl, Verticillium lecanii
	Capsids	pirimiphos-methyl
	Earwigs	pirimiphos-methyl
	Leaf miners	deltamethrin *(off-label (fog))*, oxamyl *(off-label)*, pirimiphos-methyl, trichlorfon *(off-label)*
	Leafhoppers	nicotine
	Red spider mites	pirimiphos-methyl
	Sawflies	pirimiphos-methyl
	Sciarid flies	permethrin
	Thrips	nicotine, pirimiphos-methyl
	Tomato fruitworm	permethrin
	Western flower thrips	deltamethrin *(off-label (fog))*

	Whitefly	buprofezin, nicotine, permethrin, pirimiphos-methyl, Verticillium lecanii

Beans

Pests	Aphids	Verticillium lecanii
	Red spider mites	tetradifon
	Whitefly	Verticillium lecanii

Chicory

Pests	Leaf miners	cypermethrin *(off-label)*

Cucumbers

Diseases	Botrytis	carbendazim *(off-label)*, chlorothalonil, iprodione
	Damping off	etridiazole *(seedlings and transplants)*, propamocarb hydrochloride
	Damping off and foot rot	propamocarb hydrochloride *(off-label)*
	Downy mildew	copper oxychloride + metalaxyl *(off-label)*
	Phytophthora	propamocarb hydrochloride
	Powdery mildew	bupirimate, carbendazim, carbendazim *(off-label)*, chlorothalonil, copper ammonium carbonate, imazalil, thiophanate-methyl
	Rhizoctonia	quintozene
	Root diseases	carbendazim *(off-label)*
	Root rot	propamocarb hydrochloride
Pests	Ants	pirimiphos-methyl
	Aphids	deltamethrin, fatty acids, nicotine, permethrin, pirimicarb, pirimiphos-methyl, propoxur, pyrethrins + resmethrin, Verticillium lecanii
	Capsids	pirimiphos-methyl
	Caterpillars	Bacillus thuringiensis, deltamethrin
	Earwigs	pirimiphos-methyl
	French fly	pirimiphos-methyl
	Leaf miners	cypermethrin *(off-label)*, deltamethrin *(off-label (fog))*, pirimiphos-methyl, trichlorfon *(off-label)*
	Leafhoppers	heptenophos, nicotine

	Mealy bugs	deltamethrin, fatty acids, petroleum oil
	Mites	dicofol + tetradifon
	Red spider mites	abamectin, dicofol + tetradifon, fatty acids, fenbutatin oxide, petroleum oil, pirimiphos-methyl, tetradifon
	Sawflies	pirimiphos-methyl
	Scale insects	deltamethrin, fatty acids, petroleum oil
	Sciarid flies	permethrin
	Tarsonemid mites	dicofol + tetradifon
	Thrips	deltamethrin, heptenophos, nicotine, pirimiphos-methyl
	Tomato fruitworm	permethrin
	Western flower thrips	abamectin, deltamethrin *(off-label (fog))*, dichlorvos
	Whitefly	buprofezin, cypermethrin, deltamethrin, fatty acids, nicotine, permethrin, pirimiphos-methyl, propoxur, pyrethrins + resmethrin, Verticillium lecanii

General

Diseases	Botrytis	chlorothalonil *(moderate control)*
	Disease control/foliar feed	carbendazim + prochloraz *(off-label)*
	Fungus diseases	formaldehyde (commodity substance)
	Phytophthora	fosetyl-aluminium
	Powdery mildew	imazalil
	Rhizoctonia	tolclofos-methyl *(off-label)*
	Root rot	fosetyl-aluminium
	Soil-borne diseases	dazomet, formaldehyde (commodity substance), metam-sodium, methyl bromide with amyl acetate, methyl bromide with chloropicrin, tar acids
Pests	Ants	gamma-HCH, pirimiphos-methyl, tar acids
	Aphids	deltamethrin, gamma-HCH, heptenophos *(off-label)*, nicotine, permethrin, pirimiphos-methyl, pyrethrins + resmethrin, rotenone
	Capsids	gamma-HCH, nicotine, pirimiphos-methyl
	Caterpillars	Bacillus thuringiensis, deltamethrin, nicotine, permethrin
	Earwigs	gamma-HCH, pirimiphos-methyl
	Leaf miners	cypermethrin *(off-label)*, dichlorvos *(off-label)*, gamma-HCH, heptenophos *(off-label)*, nicotine, pirimiphos-methyl, trichlorfon *(off-label)*
	Leafhoppers	nicotine
	Mealy bugs	deltamethrin
	Mites	dicofol + tetradifon

	Nematodes	dazomet, metam-sodium, methyl bromide with amyl acetate, methyl bromide with chloropicrin
	Red spider mites	dicofol + tetradifon, fenbutatin oxide, pirimiphos-methyl, tetradifon
	Rodents	alphachloralose
	Sawflies	nicotine, pirimiphos-methyl
	Scale insects	deltamethrin
	Sciarid flies	gamma-HCH, permethrin
	Slugs and snails	aluminium sulfate, metaldehyde, methiocarb *(off-label)*, tar acids
	Soil pests	dazomet, metam-sodium, methyl bromide with amyl acetate, methyl bromide with chloropicrin
	Tarsonemid mites	dicofol + tetradifon
	Thrips	gamma-HCH, nicotine, pirimiphos-methyl
	Tortrix moths	Bacillus thuringiensis
	Western flower thrips	dichlorvos *(off-label)*
	Whitefly	buprofezin, cypermethrin, deltamethrin, gamma-HCH, nicotine, permethrin, pirimiphos-methyl, pyrethrins + resmethrin, Verticillium lecanii
	Woodlice	gamma-HCH, tar acids
Weeds	Algae	benzalkonium chloride
	Annual dicotyledons	chlorpropham + fenuron, isoxaben *(off-label)*, paraquat *(off-label)*
	Annual grasses	chlorpropham + fenuron, paraquat *(off-label)*
	Chickweed	chlorpropham + fenuron
	General weed control	dazomet, metam-sodium, methyl bromide with amyl acetate, methyl bromide with chloropicrin
	Lichens	benzalkonium chloride
	Mosses	benzalkonium chloride

Herbs

Pests	Aphids	nicotine
	Leaf miners	dichlorvos *(off-label)*
	Leafhoppers	nicotine
	Thrips	nicotine
	Western flower thrips	dichlorvos *(off-label)*
	Whitefly	nicotine

FOR FULL CONDITIONS OF USE ALWAYS READ THE PRODUCT LABEL

Weeds	Annual dicotyledons	prometryn *(off-label)*, propyzamide *(off-label)*, terbacil *(off-label)*, trifluralin *(off-label)*
	Annual grasses	prometryn *(off-label)*, propyzamide *(off-label)*, terbacil *(off-label)*, trifluralin *(off-label)*
	Couch	terbacil *(off-label)*
	Perennial grasses	propyzamide *(off-label)*, terbacil *(off-label)*

Lettuce

Diseases	Big vein	carbendazim *(off-label)*
	Botrytis	dicloran, iprodione, quintozene, thiram
	Downy mildew	fosetyl-aluminium, mancozeb, metalaxyl + thiram, thiram, zineb
	Rhizoctonia	dicloran, quintozene, tolclofos-methyl
	Sclerotinia	quintozene
Pests	Aphids	deltamethrin + heptenophos *(off-label)*, nicotine, pirimicarb
	Leaf miners	abamectin, cypermethrin *(off-label)*
	Leafhoppers	nicotine
	Sciarid flies	permethrin
	Thrips	nicotine
	Western flower thrips	abamectin
	Whitefly	cypermethrin, nicotine, permethrin, Verticillium lecanii
Weeds	Annual dicotyledons	chlorpropham with cetrimide, pendimethalin *(off-label)*, propyzamide *(off-label)*
	Annual grasses	chlorpropham with cetrimide, pendimethalin *(off-label)*, propyzamide *(off-label)*
	Chickweed	chlorpropham with cetrimide
	Perennial grasses	propyzamide *(off-label)*
	Polygonums	chlorpropham with cetrimide

Mushrooms

Diseases	Bacterial blotch	sodium hypochlorite (commodity substance)
	Bubble	chlorothalonil
	Cobweb	prochloraz
	Dry bubble	chlorothalonil, prochloraz
	Fungus diseases	formaldehyde (commodity substance)
	Mycogone	carbendazim
	Trichoderma	carbendazim *(spawn treatment - off-label)*
	Wet bubble	carbendazim, chlorothalonil, prochloraz

Pests	Mushroom flies	pyrethrins + resmethrin *(Off-label)*
	Sciarid flies	diazinon, dichlorvos, diflubenzuron, methoprene, permethrin, pyrethrins + resmethrin *(Off-label)*

Mustard and cress

Diseases	Damping off	etridiazole *(seedlings and transplants)*

Peppers

Diseases	Botrytis	carbendazim *(off-label)*, chlorothalonil, dichlofluanid *(off-label)*
	Damping off	propamocarb hydrochloride
	Damping off and foot rot	propamocarb hydrochloride *(off-label)*
	Phytophthora	propamocarb hydrochloride
	Powdery mildew	carbendazim *(off-label)*, fenarimol *(off-label)*, sulfur *(off-label)*
	Root rot	propamocarb hydrochloride
Pests	Ants	pirimiphos-methyl
	Aphids	deltamethrin, fatty acids, nicotine, permethrin, pirimicarb, pirimiphos-methyl, Verticillium lecanii
	Capsids	pirimiphos-methyl
	Caterpillars	Bacillus thuringiensis, deltamethrin
	Earwigs	pirimiphos-methyl
	Leaf miners	deltamethrin *(off-label (fog))*, oxamyl *(off-label)*, pirimiphos-methyl, trichlorfon *(off-label)*
	Leafhoppers	nicotine
	Mealy bugs	deltamethrin, fatty acids
	Red spider mites	fatty acids, fenbutatin oxide *(off-label)*, pirimiphos-methyl, tetradifon
	Sawflies	pirimiphos-methyl
	Scale insects	deltamethrin, fatty acids
	Sciarid flies	permethrin
	Thrips	nicotine, pirimiphos-methyl
	Tomato fruitworm	permethrin
	Western flower thrips	deltamethrin *(off-label (fog))*
	Whitefly	buprofezin, deltamethrin, fatty acids, nicotine, permethrin, pirimiphos-methyl, Verticillium lecanii

Pot plants

Diseases	Botrytis	carbendazim, carbendazim *(off-label)*, iprodione, quintozene
	Damping off	propamocarb hydrochloride
	Disease control/foliar feed	difenoconazole *(off-label)*
	Phytophthora	fosetyl-aluminium, furalaxyl, propamocarb hydrochloride
	Powdery mildew	bupirimate, carbendazim, pyrazophos
	Pythium	furalaxyl, propamocarb hydrochloride
	Rhizoctonia	quintozene
	Root rot	fosetyl-aluminium
	Rust	mancozeb, oxycarboxin
	Sclerotinia	quintozene
	Soil-borne diseases	metam-sodium
Pests	Aphids	deltamethrin, heptenophos, imidacloprid, nicotine
	Capsids	permethrin + thiram
	Caterpillars	deltamethrin, permethrin + thiram
	Earwigs	permethrin + thiram
	Leaf miners	abamectin, permethrin + thiram *(partial control)*
	Mealy bugs	deltamethrin, petroleum oil
	Nematodes	metam-sodium
	Red spider mites	abamectin, petroleum oil, tetradifon
	Scale insects	deltamethrin, petroleum oil
	Sciarid flies	imidacloprid
	Soil pests	metam-sodium
	Thrips	heptenophos
	Vine weevil	carbofuran, imidacloprid
	Western flower thrips	abamectin
	Whitefly	deltamethrin, imidacloprid
Weeds	General weed control	metam-sodium
Plant growth regulation	Flower induction	2-chloroethylphosphonic acid
	Flower life prolongation	sodium silver thiosulfate
	Improving colour	paclobutrazol
	Increasing branching	2-chloroethylphosphonic acid
	Increasing flowering	paclobutrazol

	Stem shortening	chlormequat, chlormequat + choline chloride, daminozide, paclobutrazol

Protected asparagus

Pests	Aphids	nicotine

Protected cucurbits

Diseases	Downy mildew	copper oxychloride + metalaxyl *(off-label)*
Pests	Aphids	nicotine, pirimicarb *(off-label)*
	Bean seed flies	bendiocarb *(off-label seed treatment)*
	Leaf miners	cypermethrin *(off-label)*, heptenophos *(off-label)*, trichlorfon *(off-label)*
	Leafhoppers	nicotine
	Red spider mites	tetradifon
	Thrips	nicotine
	Whitefly	nicotine
Weeds	Annual dicotyledons	isoxaben *(off-label)*

Protected cut flowers

Diseases	Botrytis	quintozene
	Powdery mildew	imazalil
	Rhizoctonia	quintozene
	Rust	mancozeb
	Sclerotinia	quintozene
	Soil-borne diseases	metam-sodium
Pests	Aphids	dimethoate, nicotine, pirimicarb, propoxur, Verticillium lecanii
	Leaf miners	dimethoate
	Nematodes	metam-sodium
	Red spider mites	tebufenpyrad
	Soil pests	metam-sodium
	Whitefly	propoxur, Verticillium lecanii
Weeds	General weed control	metam-sodium
Plant growth regulation	Basal bud stimulation	2-chloroethylphosphonic acid

	Flower life prolongation	sodium silver thiosulfate

Protected grapevines

Pests	Mealy bugs	petroleum oil
	Red spider mites	petroleum oil, tetradifon
	Scale insects	petroleum oil

Protected onions/leeks/garlic

Pests	Aphids	nicotine
	General insect control	chlorpyrifos *(off-label)*
	Leafhoppers	nicotine
	Stem nematodes	oxamyl *(off-label)*
	Thrips	nicotine
	Whitefly	nicotine

Protected potatoes

Pests	Aphids	nicotine
	Leafhoppers	nicotine
	Thrips	nicotine
	Whitefly	nicotine

Protected radishes

Diseases	Rhizoctonia	tolclofos-methyl *(off-label)*

Tomatoes

Diseases	Blight	Bordeaux mixture *(outdoor)*, chlorothalonil, copper sulfate + sulfur, zineb
	Botrytis	carbendazim, chlorothalonil, dichlofluanid, dicloran, iprodione, quintozene, thiram
	Damping off	copper oxychloride, etridiazole, etridiazole *(seedlings and transplants)*, propamocarb hydrochloride
	Damping off and foot rot	propamocarb hydrochloride *(off-label)*
	Didymella stem rot	carbendazim, maneb *(off-label)*
	Foot rot	copper oxychloride
	Fusarium wilt	carbendazim

	Leaf mould	carbendazim, chlorothalonil, copper ammonium carbonate, dichlofluanid, zineb
	Phytophthora	copper oxychloride, etridiazole, propamocarb hydrochloride
	Powdery mildew	bupirimate *(off-label)*, fenarimol *(off-label)*, sulfur *(off-label)*
	Rhizoctonia	dicloran, quintozene
	Root diseases	etridiazole *(off-label)*
	Root rot	propamocarb hydrochloride, zineb
	Sclerotinia	quintozene
	Soil-borne diseases	metam-sodium
	Verticillium wilt	carbendazim
Pests	Ants	pirimiphos-methyl
	Aphids	deltamethrin, dimethoate, fatty acids, heptenophos, malathion, nicotine, permethrin, pirimicarb, pirimiphos-methyl, propoxur, pyrethrins + resmethrin, Verticillium lecanii
	Capsids	pirimiphos-methyl
	Caterpillars	Bacillus thuringiensis, deltamethrin, deltamethrin *(off-label)*
	Earwigs	pirimiphos-methyl
	Leaf miners	abamectin, deltamethrin *(off-label (fog))*, deltamethrin *(off-label)*, oxamyl *(off-label)*, pirimiphos-methyl, trichlorfon *(off-label)*
	Leafhoppers	heptenophos, malathion, nicotine
	Mealy bugs	deltamethrin, fatty acids, malathion, petroleum oil
	Mites	dicofol + tetradifon
	Nematodes	metam-sodium
	Red spider mites	abamectin, dicofol + tetradifon, fatty acids, fenbutatin oxide, petroleum oil, pirimiphos-methyl, tetradifon
	Sawflies	pirimiphos-methyl
	Scale insects	deltamethrin, fatty acids, petroleum oil
	Sciarid flies	permethrin
	Soil pests	metam-sodium
	Tarsonemid mites	dicofol + tetradifon
	Thrips	malathion, nicotine, pirimiphos-methyl
	Tomato fruitworm	permethrin
	Tomato moth	Bacillus thuringiensis
	Western flower thrips	abamectin, deltamethrin *(off-label (fog))*

	Whitefly	buprofezin, deltamethrin, fatty acids, malathion, nicotine, permethrin, pirimiphos-methyl, propoxur, pyrethrins + resmethrin, Verticillium lecanii
Weeds	Annual dicotyledons	pentanochlor
	Annual meadow grass	pentanochlor
	General weed control	metam-sodium
Plant growth regulation	Fruit ripening	2-chloroethylphosphonic acid
	Increasing fruit set	(2-naphthyloxy)acetic acid

Total vegetation control

Land temporarily removed from production

Weeds	Annual and perennial weeds	clodinafop-propargyl + diflufenican, glyphosate (agricultural)
	Annual dicotyledons	glufosinate-ammonium
	Annual grasses	glufosinate-ammonium
	Couch	glyphosate (agricultural)
	Docks	thifensulfuron-methyl *(in green cover)*
	Green cover	cycloxydim, diquat + paraquat, fluazifop-P-butyl, glufosinate-ammonium, glyphosate (agricultural), metsulfuron-methyl, paraquat
	Perennial dicotyledons	glufosinate-ammonium
	Perennial grasses	glufosinate-ammonium
	Volunteer oilseed rape	glyphosate (agricultural)
	Volunteer potatoes	glyphosate (agricultural)

Non-crop areas

Diseases	Fungus diseases	dichlorophen
Pests	Grey squirrels	warfarin
Weeds	Algae	benzalkonium chloride, dichlorophen, tar acids
	Annual and perennial weeds	clodinafop-propargyl + diflufenican, glyphosate (agricultural), glyphosate (horticulture, forestry, amenity etc.), glyphosate (horticulture, forestry, amenity etc.) *(pre-planting)*, glyphosate + oxadiazon *(soil surface treatment)*
	Annual dicotyledons	2,4-D + dicamba + mecoprop, 2,4-D + dicamba + triclopyr, diquat + paraquat, diuron, diuron + glyphosate, glufosinate-ammonium, glyphosate (agricultural), MCPA, paraquat, picloram

Annual grasses	diquat + paraquat, diuron, diuron + glyphosate, glyphosate (agricultural), paraquat
Annual meadow grass	diuron
Annual weeds	diuron, glufosinate-ammonium, glyphosate (agricultural)
Bracken	asulam, glyphosate (horticulture, forestry, amenity etc.), glyphosate (horticulture, forestry, amenity etc.) *(pre-planting)*, imazapyr, picloram
Brambles	2,4-D + dicamba + mecoprop, 2,4-D + dicamba + triclopyr, triclopyr
Broom	triclopyr
Couch	amitrole
Creeping bent	amitrole, paraquat
Creeping thistle	amitrole, MCPA
Docks	2,4-D + dicamba + triclopyr, amitrole, asulam, MCPA, triclopyr
Gorse	2,4-D + dicamba + triclopyr, triclopyr
Green cover	glyphosate (agricultural)
Japanese knotweed	2,4-D + dicamba + triclopyr, picloram
Lichens	benzalkonium chloride, tar acids
Liverworts	tar acids
Mosses	benzalkonium chloride, dichlorophen, tar acids
Perennial dicotyledons	2,4-D + dicamba + mecoprop, 2,4-D + dicamba + triclopyr, diuron + glyphosate, MCPA, picloram, triclopyr
Perennial grasses	amitrole, diquat + paraquat, diuron + glyphosate
Perennial ryegrass	paraquat
Perennial weeds	diuron, glufosinate-ammonium, glyphosate (agricultural)
Rhododendrons	2,4-D + dicamba + triclopyr, triclopyr
Rough meadow grass	paraquat
Rushes	triclopyr
Stinging nettle	2,4-D + dicamba + mecoprop, 2,4-D + dicamba + triclopyr, triclopyr
Total vegetation control	amitrole, amitrole + 2,4-D + diuron, amitrole + bromacil + diuron, bromacil, bromacil + diuron, bromacil + picloram, dalapon + dichlobenil, dichlobenil, diuron + paraquat, glyphosate (agricultural), glyphosate (agricultural) *(amenity situations)*, glyphosate (horticulture, forestry, amenity etc.), glyphosate (horticulture, forestry, amenity etc.) *(amenity situations)*, glyphosate + oxadiazon *(soil surface treatment)*, imazapyr, sodium chlorate

	Volunteer cereals	diquat + paraquat
	Woody weeds	2,4-D + dicamba + mecoprop, 2,4-D + dicamba + triclopyr, fosamine-ammonium, picloram, triclopyr
Plant growth regulation	Growth suppression	maleic hydrazide

Stubbles

Weeds	Annual and perennial weeds	glyphosate (agricultural)
	Annual dicotyledons	glyphosate (agricultural) *(pre-drilling/pre-crop emergence)*
	Annual grasses	glyphosate (agricultural) *(pre-drilling/pre-crop emergence)*
	Couch	glyphosate (agricultural)
	Volunteer cereals	glyphosate (agricultural), glyphosate (agricultural) *(pre-drilling/pre-crop emergence)*
	Volunteer potatoes	glyphosate (agricultural)

Weeds in or near water

Weeds	Aquatic weeds	2,4-D, clodinafop-propargyl + diflufenican, dichlobenil, diquat, glyphosate (agricultural), glyphosate (horticulture, forestry, amenity etc.), terbutryn
	Perennial dicotyledons	2,4-D
	Perennial grasses	clodinafop-propargyl + diflufenican, glyphosate (agricultural), glyphosate (horticulture, forestry, amenity etc.)
	Rushes	clodinafop-propargyl + diflufenican, glyphosate (agricultural), glyphosate (horticulture, forestry, amenity etc.)
	Sedges	clodinafop-propargyl + diflufenican, glyphosate (agricultural), glyphosate (horticulture, forestry, amenity etc.)
	Total vegetation control	dalapon + dichlobenil
	Waterlilies	clodinafop-propargyl + diflufenican, glyphosate (agricultural), glyphosate (horticulture, forestry, amenity etc.)
	Woody weeds	fosamine-ammonium
Plant growth regulation	Growth suppression	maleic hydrazide

Turf/amenity grass

Amenity grass

Diseases	Brown patch	iprodione

	Dollar spot	iprodione, thiabendazole
	Fairy rings	oxycarboxin
	Fusarium patch	iprodione, thiabendazole
	Grey snow mould	iprodione
	Melting out	iprodione
	Red thread	iprodione, thiabendazole
Pests	Frit fly	chlorpyrifos
	Leatherjackets	chlorpyrifos
Weeds	Annual and perennial weeds	glyphosate (agricultural) *(destruction/pre-sowing)*, glyphosate (horticulture, forestry, amenity etc.) *(destruction/pre-sowing)*, glyphosate (horticulture, forestry, amenity etc.) *(pre-planting/sowing)*, glyphosate (horticulture, forestry, amenity etc.) *(wiper application)*
	Annual dicotyledons	2,4-D + dicamba + mecoprop, 2,4-D + picloram, dicamba + maleic hydrazide + MCPA, dicamba + MCPA + mecoprop-P, dicamba + paclobutrazol, dichlorprop + MCPA, fluroxypyr + triclopyr, isoxaben, MCPA + mecoprop-P
	Brambles	2,4-D + dicamba + mecoprop, 2,4-D + picloram, clopyralid + triclopyr
	Broom	clopyralid + triclopyr
	Buttercups	dichlorprop + MCPA
	Clovers	dichlorprop + MCPA
	Creeping thistle	2,4-D + picloram
	Daisies	dichlorprop + MCPA
	Docks	2,4-D + picloram, asulam *(not fine turf)*, clopyralid + triclopyr, dicamba + MCPA + mecoprop-P
	Gorse	clopyralid + triclopyr
	Japanese knotweed	2,4-D + picloram
	Perennial dicotyledons	2,4-D + dicamba + mecoprop, 2,4-D + picloram, clopyralid + triclopyr, dicamba + maleic hydrazide + MCPA, dicamba + MCPA + mecoprop-P, dicamba + paclobutrazol, dichlorprop + MCPA, fluroxypyr + triclopyr, MCPA + mecoprop-P
	Ragwort	2,4-D + picloram
	Stinging nettle	2,4-D + dicamba + mecoprop, clopyralid + triclopyr
	Woody weeds	2,4-D + dicamba + mecoprop, 2,4-D + picloram, fluroxypyr + triclopyr
Plant growth regulation	Growth retardation	dicamba + maleic hydrazide + MCPA, dicamba + paclobutrazol, trinexapac-ethyl

	Growth suppression	maleic hydrazide, mefluidide

Turf/amenity grass

Diseases	Anthracnose	carbendazim + chlorothalonil, carbendazim + iprodione, chlorothalonil
	Brown patch	iprodione
	Cladosporium leaf spot	carbendazim + iprodione
	Dollar spot	carbendazim, carbendazim + chlorothalonil, chlorothalonil, fenarimol, iprodione, quintozene, thiabendazole, thiophanate-methyl
	Fairy rings	oxycarboxin, triforine
	Fungus diseases	dichlorophen
	Fusarium patch	carbendazim, carbendazim + chlorothalonil, carbendazim + iprodione, chlorothalonil, fenarimol, iprodione, quintozene, thiabendazole, thiophanate-methyl
	Grey snow mould	chlorothalonil *(reduction)*, iprodione
	Melting out	iprodione
	Red thread	carbendazim + chlorothalonil, carbendazim + iprodione, chlorothalonil, dichlorophen, fenarimol, iprodione, quintozene, thiabendazole, thiophanate-methyl
Pests	Damaging mammals	aluminium ammonium sulfate, aluminium ammonium sulfate *(anti-fouling)*
	Earthworms	carbaryl, carbendazim, carbendazim + chlorothalonil, gamma-HCH + thiophanate-methyl, thiophanate-methyl
	Frit fly	chlorpyrifos
	Leatherjackets	carbaryl, chlorpyrifos, gamma-HCH + thiophanate-methyl
	Nematodes	methyl bromide with chloropicrin
	Soil pests	methyl bromide with chloropicrin
Weeds	Algae	dichlorophen
	Annual and perennial weeds	glyphosate (horticulture, forestry, amenity etc.)
	Annual dicotyledons	2,4-D, 2,4-D + dicamba, 2,4-D + mecoprop, 2,4-D + mecoprop-P, 2,4-D + picloram, clopyralid + diflufenican + MCPA, clopyralid + fluroxypyr + MCPA, dicamba + dichlorprop + ferrous sulfate + MCPA, dicamba + dichlorprop + MCPA, dicamba + maleic hydrazide + MCPA, dicamba + MCPA + mecoprop, dicamba + MCPA + mecoprop-P, fluroxypyr + mecoprop-P, ioxynil, isoxaben, MCPA, mecoprop, mecoprop-P, picloram
	Annual grasses	ethofumesate
	Annual weeds	diuron
	Blackgrass	ethofumesate

	Bracken	picloram
	Brambles	2,4-D + picloram
	Buttercups	clopyralid + diflufenican + MCPA
	Chickweed	dicamba + MCPA + mecoprop-P, ethofumesate, mecoprop, mecoprop-P
	Cleavers	dicamba + MCPA + mecoprop-P, ethofumesate
	Clovers	mecoprop, mecoprop-P
	Corn marigold	clopyralid + diflufenican + MCPA
	Creeping thistle	2,4-D + picloram, clopyralid + diflufenican + MCPA, MCPA
	Daisies	MCPA
	Docks	2,4-D + picloram, asulam, dicamba + MCPA + mecoprop-P, MCPA
	General weed control	methyl bromide with chloropicrin
	Japanese knotweed	2,4-D + picloram, picloram
	Mayweeds	clopyralid + diflufenican + MCPA, dicamba + MCPA + mecoprop-P
	Mosses	dicamba + dichlorprop + ferrous sulfate + MCPA, dichlorophen, dichlorophen + ferrous sulfate, ferrous sulfate, tar acids
	Perennial dicotyledons	2,4-D, 2,4-D + dicamba, 2,4-D + mecoprop, 2,4-D + mecoprop-P, 2,4-D + picloram, clopyralid + diflufenican + MCPA, dicamba + dichlorprop + ferrous sulfate + MCPA, dicamba + dichlorprop + MCPA, dicamba + maleic hydrazide + MCPA, dicamba + MCPA + mecoprop, dicamba + MCPA + mecoprop-P, fluroxypyr + mecoprop-P, MCPA, mecoprop, mecoprop-P, picloram
	Perennial weeds	diuron
	Polygonums	dicamba + MCPA + mecoprop-P
	Ragwort	2,4-D + picloram
	Slender speedwell	chlorthal-dimethyl, fluroxypyr + mecoprop-P
	Total vegetation control	glyphosate (agricultural), glyphosate (horticulture, forestry, amenity etc.)
	Volunteer cereals	ethofumesate
	Weed grasses	amitrole *(off-label)*
	Woody weeds	2,4-D + picloram, picloram
Plant growth regulation	Growth retardation	dicamba + maleic hydrazide + MCPA, trinexapac-ethyl
	Growth suppression	maleic hydrazide, mefluidide

FOR FULL CONDITIONS OF USE ALWAYS READ THE PRODUCT LABEL

Increasing yield	sulfur

SECTION 4
PESTICIDE PROFILES

1 abamectin

A selective acaricide and insecticide for use in ornamentals

Products	Dynamec	Novartis BCM	19 g/l	EC	08701

Uses Leaf miners in FLOWERS, ORNAMENTALS, PROTECTED FLOWERS, PROTECTED LETTUCE, PROTECTED TOMATOES. Two-spotted spider mite in FLOWERS, ORNAMENTALS, PROTECTED CUCUMBERS, PROTECTED FLOWERS, PROTECTED TOMATOES. Western flower thrips in FLOWERS, ORNAMENTALS, PROTECTED CUCUMBERS, PROTECTED FLOWERS, PROTECTED LETTUCE, PROTECTED TOMATOES.

Notes

Efficacy

- Treat at first sign of infestation. Repeat sprays may be required
- For effective control total cover of all plant surfaces is essential, but avoid run-off
- Mites quickly become immobilised but 3-5 d may be required for maximum mortality

Crop Safety/Restrictions

- Number of treatments 6 on protected tomatoes and cucumbers (only 4 of which can be made when flowers or fruit present); 4 on protected lettuce; not restricted on flowers but rotation with other products advised
- Maximum concentration must not exceed 50 ml per 100 l water
- Do not mix with wetters, stickers or other adjuvants
- Treat tomatoes and cucumbers only when in flower or setting fruit and only between 1 Mar and 31 Oct
- Apply to lettuce only between 1 Mar and 31 Oct
- Do not use on ferns (*Adiantum* spp) or Shasta daisies
- Some spotting or staining may occur on carnation, kalanchoe and begonia foliage
- Consult manufacturer for list of plant varieties tested for safety
- There is insufficient evidence to support product compatibility with integrated and biological pest control programmes

Special precautions/Environmental safety

- Harmful if swallowed, in contact with skin and by inhalation
- Irritating to eyes
- Flammable
- Not to be used on food crops
- Extremely dangerous to fish or other aquatic life. Do not contaminate surface waters or ditches with chemical or used container
- High risk to bees. Do not apply to crops in flower or to those in which bees are actively foraging. Do not apply when flowering weeds are present
- Where bumble bees are used in tomatoes as pollinators, keep them out for 24 h after treatment
- Unprotected persons must be kept out of treated areas until the spray has dried

Protective clothing/Label precautions

- A, C, D, H, K, M
- M04, R03a, R03b, R03c, R04a, U02, U05a, U09a, U19, U20a, C03, E01, E12a, E13a, E26, E30b, E31b, E34

2 alachlor

A soil-acting herbicide approvals for which expired on 30 June 1992

3 aldicarb

A soil-applied, systemic carbamate insecticide and nematicide

Products

1 Landgold Aldicarb 10G	Landgold	10% w/w	GR	06036
2 Standon Aldicarb 10G	Standon	10% w/w	GR	05915
3 Temik 10G	RP Agric.	10% w/w	GR	06210

Uses

Aphids in CARROTS, PARSNIPS, POTATOES, SUGAR BEET [1-3]. Aphids in BRASSICAS, ROOT BRASSICAS [3]. Beet cyst nematode in RED BEET *(off-label)* [3]. Cabbage root fly in BRASSICAS, ROOT BRASSICAS [3]. Docking disorder vectors in SUGAR BEET [3]. Flea beetles in BRASSICAS, ROOT BRASSICAS [3]. Free-living nematodes in CARROTS, PARSNIPS, SUGAR BEET [1-3]. Free-living nematodes in BRASSICAS, POTATOES, ROOT BRASSICAS [3]. Frit fly in SWEETCORN *(off-label)* [3]. Insect pests in ORNAMENTALS *(off-label)* [3]. Leaf miners in SUGAR BEET [1-3]. Millipedes in SUGAR BEET [1-3]. Potato cyst nematode in POTATOES [1-3]. Pygmy beetle in SUGAR BEET [1-3]. Spraing vectors in POTATOES [1-3]. Stem nematodes in BULB ONIONS [1-3]. Stem nematodes in LEEKS *(off-label)*, ORNAMENTALS *(off-label)* [3].

Notes

Efficacy

- Must be incorporated into soil by physical means. See label for details of application rates, timing, suitable applicators and techniques of incorporation
- Persistence and activity may be reduced in very wet soils or where pH exceeds 8.0. Do not use within 14 d of liming
- Use in potatoes reduces incidence of spraing disease

Crop Safety/Restrictions

- Maximum number of treatments 1 per crop
- No edible crops other than those listed (see label) should be planted into treated soil for at least 8 wk after application

Special precautions/Environmental safety

- Aldicarb is subject to the Poisons Rules 1982 and the Poisons Act 1972. See notes in Section 1
- This product contains an anticholinesterase carbamate compound. Do not use if under medical advice not to work with such compounds
- Keep in original container, tightly closed, in a safe place, under lock and key
- Toxic in contact with skin, by inhalation and if swallowed
- Dangerous to game, wild birds and animals. Cover granules completely and immediately after application. Bury spillages. Failure to bury granules immediately and completely is hazardous to wildlife
- Dangerous to fish or other aquatic life. Do not contaminate surface waters or ditches with chemical or used container
- Keep unprotected persons out of treated glasshouses for at least 1 d
- Option to return empty container as instructed by supplier [3]

Protective clothing/Label precautions

- A [2, 3]; B [1-3]; C or D+E [2, 3]; C, D [1]; H, K, M
- M02, M04, R02a, R02b, R02c, U02, U04a, U05a, U09a, U13, U19, U20a, C03, E01, E10a, E13b, E30b, E32a, E34 [1-3]; E02 (24 h) [1]; E02, E06a [2]; E06a (13 wk) [1, 3]; E33 [3]

Withholding period

- Keep all livestock out of treated areas for at least 13 wk. Bury or remove spillages

Latest application/Harvest Interval(HI)

- At planting, sowing, drilling or transplanting
- HI potatoes 8 wk; brassicas 10 wk; carrots, parsnips 12 wk; spring sown bulb onions do not harvest until mature bulb stage

Approval

- Off-label approval unlimited for use on outdoor and protected sweetcorn (OLA 2771/96)[3]; to Jan 2001 for use on red beet (OLA 0001/96)[3]; unlimited for use on a range of outdoor and protected ornamentals - see OLA notice for details (OLA 1325/95)[3]; to Sep 1999 for use on outdoor and protected leeks (OLA 1317/95)[3]

Maximum Residue Levels (mg residue/kg food)

- citrus, pecans, cauliflower, Brussels sprouts 0.2; tree nuts (except pecans), pome fruit, stone fruit, berries (except strawberries), cane fruit, miscellaneous fruit (except bananas), root and tuber vegetables (except beetroot, carrots, parsnips), bulb vegetables, fruiting vegetables (except tomatoes), leaf brassicas, leaf vegetables, fresh herbs, legumes, stem vegetables (except leeks), mushrooms, pulses, oilseeds (except linseed, rapeseed, cotton seed), tea, cereals 0.05; products of animal origin 0.01

---

4 aldrin

A persistent organochlorine insecticide, all approvals for which were revoked in 1989. Aldrin is an environmental hazard and is banned under the EC "Prohibition Directive"

---

5 alphachloralose

A narcotic glucofuranose rodenticide used to kill mice

Products	Alphachloralose Pure	Killgerm	100%	CB	00082

Uses Mice in FARM BUILDINGS, GLASSHOUSES.

Notes

Efficacy

- Apply ready-mixed bait or concentrate mixed with suitable bait material in shallow trays where mouse droppings observed
- Lay bait at several points not more than 1.5 m apart
- Leave in position for several days until mouse activity ceases
- After treatment, clear up poison and bury residue

Special precautions/Environmental safety

- A chemical subject to the Poisons Rules 1982 and Poisons Act 1972
- Keep in original container, tighty closed, in a safe place, under lock and key
- Toxic if swallowed

- Do not use outside
- Prevent access by children and animals, particularly cats and dogs
- Should a domestic animal be affected keep the animal warm and quiet

Protective clothing/Label precautions
- R02c, U13, U20a, E01, E30b, V01a, V02, V03a, V04a

6 alpha-cypermethrin

A contact and ingested pyrethroid insecticide for use in arable crops

Products

1 Acquit	DuPont	100 g/l	EC	07000
2 Contest	Cyanamid	15% w/w	WG	09024
3 Fastac	Cyanamid	100 g/l	EC	07008
4 I T Alpha-cypermethrin	I T Agro	100 g/l	EC	08274
5 Littac	Sorex	15 g/l	SC	H5176
6 Standon Alpha-C10	Standon	100 g/l	EC	08823

Uses

Blossom beetles in BROCCOLI *(off-label)*, CALABRESE *(off-label)*, CAULIFLOWERS *(off-label)* [3]. Brassica pod midge in WINTER OILSEED RAPE [1-4, 6]. Brassica pod midge in SPRING OILSEED RAPE [4]. Cabbage aphid in BROCCOLI, BRUSSELS SPROUTS, CABBAGES, CALABRESE, CAULIFLOWERS, KALE [1-3, 6]. Cabbage seed weevil in SPRING OILSEED RAPE, WINTER OILSEED RAPE [1-4, 6]. Cabbage stem flea beetle in WINTER OILSEED RAPE [1-4, 6]. Caterpillars in BROCCOLI, BRUSSELS SPROUTS, CABBAGES, CALABRESE, CAULIFLOWERS, KALE [1-4, 6]. Cereal aphids in BARLEY, WHEAT [1, 6]. Cereal aphids in SPRING BARLEY, SPRING WHEAT, WINTER BARLEY, WINTER WHEAT [2-4]. Flea beetles in BROCCOLI, BRUSSELS SPROUTS, CABBAGES, CALABRESE, CAULIFLOWERS, KALE [1-4, 6]. Hide beetle in POULTRY HOUSES [5]. Lesser mealworm in POULTRY HOUSES [5]. Pollen beetles in SPRING OILSEED RAPE, WINTER OILSEED RAPE [1-4, 6]. Poultry house pests in POULTRY HOUSES [5]. Rape winter stem weevil in WINTER OILSEED RAPE [1-3, 6]. Yellow cereal fly in WINTER BARLEY, WINTER WHEAT [2].

Notes

Efficacy
- For cabbage stem flea beetle control spray oilseed rape when adult or larval damage first seen and about 1 mth later [1-4]
- For flowering pests on oilseed rape apply at any time during flowering, on pollen beetle best results achieved at green to yellow bud stage (GS 3,3-3,7), on seed weevil between 20 pods set stage and 80% petal fall (GS 4,7-5,8)[1-4]
- Spray cereals in autumn for control of cereal aphids, in spring/summer for grain aphids. (See label for details) [1-4]
- For flea beetle, caterpillar and cabbage aphid control on brassicas apply when the pest or damage first seen or as a preventive spray. Repeat if necessary [1-4]
- For lesser mealworm control in poultry houses apply a coarse, low-pressure spray as routine treatment after clean-out and before each new crop. Spray vertical surfaces and ensure an overlap onto ceilings. It is not necessary to treat the floor [5]

Crop Safety/Restrictions

- Maximum number of treatments 4 per crop on edible brassicas, 3 per crop on winter oilseed rape, 2 on spring oilseed rape (only 1 after yellow bud stage - GS 3,7)
- Apply up to 2 sprays on cereals in autumn and spring, 1 in summer between 1 Apr and 31 Aug. See label for details of rates
- Only 1 aphicide treatment may be applied in cereals between 1 Apr and 31 Aug in any one year
- Treatment presents minimal hazard to bees but on flowering crops spray in evening, early morning or in dull weather as a precaution

Special precautions/Environmental safety

- Harmful in contact with skin or if swallowed [1, 3-5]
- Irritating to skin. Risk of serious damage to eyes [1, 3-5]
- Flammable [1, 3-5]
- Irritant. May cause sensitisation by skin contact [2]
- Dangerous to bees. Do not apply at flowering stage except as directed on oilseed rape and cereals. Keep down flowering weeds in all crops [1, 3-5]
- Extremely dangerous to fish or other aquatic life. Do not contaminate surface waters or ditches with chemical or used container
- Do not allow spray from ground sprayers to fall within 6 m or the direct spray from hand-held sprayers to fall within 2 m of surface waters or ditches
- For summer cereal application do not spray within 6 m from edge of crop
- Reduced volume spraying must not be used
- Do not apply directly to poultry; collect eggs before application [5]

Protective clothing/Label precautions

- A [1-6]; C [1, 3-6]; H
- M03, R03a, R03c, R04b, R04d, R07d, U04a, E34 [1, 3, 4, 6]; R04, R04e [2]; U02, U19 [1, 3-6]; U05a, C03, E12c, E16, E26, E31b [1-4, 6]; U09a, C04, C05, C07, C09, E04, E05, E32a [5]; U10 [2-4]; U11 [3, 4]; U20a [1, 6]; U20b [2-5]; E01, E13a, E30a [1-6]

Latest application/Harvest Interval(HI)

- Before the end of flowering for oilseed rape, before 31 Mar in year of harvest for cereals (autumn and spring application), before early dough stage (GS 83) for cereals (summer application).
- HI brassicas 7 d

Approval

- Off-label approval unlimited for use on cauliflowers, calabrese, broccoli (OLA 1750/96)[3]

---

7 aluminium ammonium sulfate

An inorganic bird and animal repellent

Products

1 Guardsman B	Chiltern	83 g/l	SL	05494
2 Guardsman M	Chiltern	83 g/l	SL	05495
3 Guardsman SDP Seed Dressing Powder	Sphere	88% w/w	DS	03606
4 Liquid Curb Crop Spray	Sphere	83 g/l	SC	03164
5 Rezist	Barrettine	88% w/w	WP	08576

Uses Birds in CORMS, FLOWER BULBS, SEEDS [3]. Birds in BEANS, BRASSICAS, BUSH FRUIT, CANE FRUIT, CARROTS, CEREALS, FOREST NURSERY BEDS, FORESTRY

PLANTATIONS, GRASSLAND, OILSEED RAPE, PEAS, STRAWBERRIES, SUGAR BEET, TOP FRUIT [1, 2, 4]. Damaging mammals in CORMS, FLOWER BULBS, SEEDS [3]. Damaging mammals in BEANS, BRASSICAS, BUSH FRUIT, CANE FRUIT, CARROTS, CEREALS, FOREST NURSERY BEDS, FORESTRY PLANTATIONS, GRASSLAND, OILSEED RAPE, PEAS, STRAWBERRIES, SUGAR BEET, TOP FRUIT [1, 2, 4]. Dogs in AMENITY TURF *(anti-fouling)* [5]. Moles in AMENITY TURF [5].

Notes

Efficacy

- Apply as overall spray to growing crops before damage starts or mix powder with seed depending on type of protection required
- Spray deposit protects growth present at spraying but gives little protection to new growth
- Product must be sprayed onto dry foliage to be effective and must dry completely before dew or frost forms. In winter this may require some wind
- Treatments to deter moles should be applied as a planned programme when moles active. Other wild and domestic animals will be repelled during treatment period [5]
- Treatments for dog fouling should be applied before fouling occurs [5]

Protective clothing/Label precautions

- M05, U04b, U10, E31b [1, 2, 4]; U20a [1, 2, 4, 5]; U20b [3]; E15, E30a, E32a [1-5]; S02, S05 [1-4]

Latest application/Harvest Interval(HI)

- HI fruit crops 6 wk

8 aluminium phosphide

A phosphine generating compound used against vertebrates and grain store pests

Products

1 Amos Talunex	Luxan	57% w/w	GE	06567
2 Luxan Talunex	Luxan	57% w/w	GE	06563
3 Phostek	Killgerm	57% w/w	GE	05115
4 Phostek	Killgerm	57% w/w	GE	07921
5 Phostoxin	Rentokil	56% w/w	GE	01775
6 Phostoxin I	Rentokil	56% w/w	GE	05694

Uses

Insect pests in FOOD STORAGE AREAS [6]. Mice in FARMLAND, FORESTRY [4]. Moles in FARMLAND, FORESTRY [1, 2, 4, 5]. Rabbits in FARMLAND, FORESTRY [1, 2, 4, 5]. Rats in FARMLAND, FORESTRY [2, 4]. Stored product pests in STORED PRODUCTS [3].

Notes

Efficacy

- Product releases poisonous hydrogen phosphide gas in contact with moisture
- Place pellets in burrows or runs and seal hole by heeling in or covering with turf. Do not cover pellets with soil. Inspect daily and treat any new or re-opened holes
- Apply pellets by means of Luxan Topex Applicator [1, 2]
- See label for details of fumigation of grain in silos and commodities not stored in bulk [3]

Special precautions/Environmental safety

- Aluminium phosphide is subject to the Poisons Rules 1982 and the Poisons Act 1972. See notes in Section 1
- Only to be used by operators instructed or trained in the use of aluminium phosphide and familiar with the precautionary measures to be taken. See label and HSE Guidance Notes for full precautions
- Very toxic by inhalation, in contact with skin and if swallowed
- Product liberates very toxic, highly flammable gas
- Highly flammable
- Wear suitable protective gloves (synthetic rubber/plastics) when handling product
- Only open container outdoors [1, 2, 5], in well ventilated space [3, 4], and for immediate use. Keep away from liquid or water as this causes immediate release of gas. Do not use in wet weather
- Keep in original container, tightly closed, in a safe place, under lock and key
- Do not use within 3 m of human or animal habitation [1, 2, 4, 5], before application ensure that no humans or domestic animals are in adjacent buildings or structures [3, 4]
- Dangerous to fish or other aquatic life. Do not contaminate surface waters or ditches with chemical or used container
- Pellets must not be placed or allowed to remain on ground surface
- Do not use adjacent to watercourses
- Dust remaining after decomposition is harmless and of no environmental hazard
- Dispose of empty containers as directed on label

Protective clothing/Label precautions

- A
- M04, R01a, R01b, R01c, U07, U13, U19, U20a, C03, E01, E30b, E32b [1-6]; M05, U10 [3, 4]; R07c, U05b [1-5]; U05a, E07 [6]; U14, E27 [1, 2]; U18, E02 [3, 4, 6]; E13b, E34 [1-4, 6]; E15 [5]; E29 [3-6]

Withholding period

- Keep livestock out of treated areas

9 aluminium sulfate

An inorganic salt for slug and snail control

Products	Growing Success Slug Killer	Growing Success	99.5% w/w	SG	04386

Uses Slugs in FIELD CROPS, FRUIT CROPS, ORNAMENTALS, PROTECTED CROPS, VEGETABLES. Snails in FIELD CROPS, FRUIT CROPS, ORNAMENTALS, PROTECTED CROPS, VEGETABLES.

Notes

Efficacy

- Apply granules to soil surface. Product has contact effect
- Best results achieved during mild, damp weather when slugs and snails active

10 amidosulfuron

A post-emergence sulfonylurea herbicide for cleavers and other broad-leaved weed control in cereals

Products

1 Druid	AgrEvo	50% w/w	WG	08714
2 Eagle	AgrEvo	75% w/w	WG	07318
3 Landgold Amidosulfuron	Landgold	75% w/w	WG	09021
4 Pursuit	AgrEvo	75% w/w	WG	07333

Uses

Annual dicotyledons in LINSEED [1, 2, 4]. Annual dicotyledons in DURUM WHEAT, OATS, RYE, SPRING BARLEY, SPRING WHEAT, TRITICALE, WINTER BARLEY, WINTER WHEAT [1-4]. Cleavers in LINSEED [1, 2, 4]. Cleavers in DURUM WHEAT, OATS, RYE, SPRING BARLEY, SPRING WHEAT, TRITICALE, WINTER BARLEY, WINTER WHEAT [1-4].

Notes

Efficacy

- For optimum results apply in spring (from 1 Feb) in warm weather when soil moist and weeds growing actively
- Weed kill is slow, especially under cool, dry conditions. Weeds may sometimes only be stunted but will have little or no competitive effect on crop
- May be used on all soil types
- Spray is rainfast after 1 h
- Cleavers controlled from emergence to flower bud stage. If present at application charlock (up to flower bud), shepherds purse (up to flower bud) and field forget-me-not (up to 6 leaves) will also be controlled

Crop Safety/Restrictions

- Maximum number of treatments 1 per crop
- Apply from 2 leaf stage of crop up to and including first awns visible (GS 12-49)
- Do not apply to crops undersown or due to be undersown with clover or lucerne
- Broadcast crops should be sprayed post-emergence after plants have a well established root system
- Do not spray crops under stress, suffering drought, waterlogged, grazed, lacking nutrients or if soil compacted
- Do not spray if frost expected
- Do not roll or harrow within 1 wk of spraying
- Only cereals, winter oilseed rape, mustard, winter field beans, vetches may be sown in the same year as treatment
- Do not spray in tank mixture, or in sequence, with a product containing any other sulfonylurea except as directed for specific products containing metsulfuron-methyl, metsulfuron-methyl + thifensulfuron-methyl or tribenuron methyl - see label for details. Cereals must be sown as the following crop after use of such mixtures or sequences
- If crop fails cereals may be sown after 15 d
- Avoid drift onto neighbouring broad-leaved plants or onto surface waters or ditches
- Take care to wash out sprayers thoroughly. See label for details

Special precautions/Environmental safety

- Irritating to eyes [1]

- Dangerous to fish or other aquatic life. Do not contaminate surface waters or ditches with chemical or used container

Protective clothing/Label precautions
- C [1]
- R04a, U05a, C03, E01 [1]; U20a [2, 3]; U20b [1, 4]; E13b, E31a [1-4]

Latest application/Harvest Interval(HI)
- Before first spikelets just visible (GS 51)

11 2-aminobutane

A fumigant alkylamine fungicide permitted for use only on stored seed potatoes

Products

Hortichem 2-Aminobutane	Hortichem	720 g/l	VP	06147

Uses

Gangrene in SEED POTATOES. Silver scurf in SEED POTATOES. Skin spot in SEED POTATOES.

Notes

Efficacy
- Treatment must only be carried out by trained operators in suitable fumigation chambers under licence from the British Technology Group
- Fumigate within 21 d of lifting

Crop Safety/Restrictions
- Maximum number of treatments 1 per batch
- Do not treat immature tubers. Allow period of healing before treating damaged tubers

Special precautions/Environmental safety
- Harmful in contact with skin
- Irritating to eyes, skin and respiratory system
- Highly flammable. Keep away from sources of ignition. No smoking
- Keep in original container, tightly closed, in a safe place, under lock and key
- Do not empty into drains
- Do not contaminate surface waters or ditches with chemical or used container
- Do not supply treated potatoes for consumption by humans or lactating dairy cows
- Use must be in accordance with approved Code of Practice for the Control of Substances Hazardous to Health: Fumigation Operations
- The quantity of potatoes to be fumigated in a single stack must not exceed 2000 tonnes

Protective clothing/Label precautions
- A, C
- M03, R03a, R04a, R04b, R04c, R07c, U02, U04a, U05a, U11, U13, U19, U20a, C03, E01, E15, E30b, E31a

Maximum Residue Levels (mg residue/kg food)
- citrus fruits 5; potatoes 1

12 amitraz

An amidine acaricide and insecticide for use in top fruit

Products

Mitac HF	Promark	200 g/l	EC	07358

Uses

Pear sucker in PEARS. Red spider mites in APPLES, PEARS. Rust mite in APPLES.

Notes

Efficacy

- For red spider mites on apples and pears spray at 60-80% egg hatch and repeat 3 wk later
- For pear sucker control spray when significant numbers of nymphs have hatched but before there is significant contamination of the fruit with honeydew, normally Jun/Jul
- Best results achieved in dry conditions, do not spray if rain imminent

Crop Safety/Restrictions

- Maximum total dose 7 l/ha product per yr

Special precautions/Environmental safety

- Harmful in contact with skin and if swallowed
- Harmful to fish. Do not contaminate surface waters or ditches with chemical or used container
- Keep in original container, tightly closed, in a safe place, under lock and key

Protective clothing/Label precautions

- A, C, H
- M03, R03a, R03c, U02, U04a, U05a, U08, U13, U19, U20a, C02 (2 wk), C03, E01, E13c, E26, E30b, E31b, E34

Latest application/Harvest Interval(HI)

- HI apples, pears 2 wk; hops 7 wk

Maximum Residue Levels (mg residue/kg food)

- hops 50; tomatoes 0.5; tea 0.1; all other crops (except tea) 0.02

13 amitrole

A translocated, foliar-acting, non-selective triazole herbicide

Products

1 MSS Aminotriazole Technical	Mirfield	98% w/w	TC	04645
2 Weedazol-TL	Bayer	225 g/l	SL	02979

Uses

Annual weeds in APPLES, FALLOWS, FIELD CROPS *(pre-sowing, autumn stubble)*, HEADLANDS, PEARS [2]. Bent grasses in NON-CROP AREAS [1]. Contaminant grasses in AMENITY TURF - AMITROLE RESISTANT *(off-label)*[2]. Couch in APPLES, FALLOWS, FIELD CROPS *(pre-sowing, autumn stubble)*, HEADLANDS, PEARS, WINTER WHEAT *(direct-drilled)* [2]. Couch in NON-CROP AREAS [1]. Creeping bent in NON-CROP AREAS [1]. Creeping thistle in APPLES, PEARS [2]. Creeping thistle in NON-CROP AREAS [1]. Docks in APPLES, FALLOWS, FIELD CROPS *(pre-sowing, autumn stubble)*, HEADLANDS, PEARS [2]. Docks in NON-CROP AREAS [1]. General weed control in APRICOTS *(off-label)*,

CHERRIES *(off-label)*, PEACHES *(off-label)*, PLUMS *(off-label)*, QUINCES *(off-label)* [2]. Perennial weeds in APPLES, FALLOWS, FIELD CROPS *(pre-sowing, autumn stubble)*, HEADLANDS, PEARS [2]. Total vegetation control in NON-CROP AREAS [1]. Volunteer potatoes in BARLEY *(stubble)* [2].

Notes

Efficacy

- In non-crop land may be applied at any time from Apr to Oct. Best results achieved in spring or early summer when weeds growing actively. For coltsfoot, hogweed and horsetail summer and autumn applications are preferred [1]
- In cropland apply when couch in active growth and foliage at least 7.5 cm high [2]
- In fallows and stubble plough 3-6 wk after application to depth of 20 cm, taking care to seal the furrow [2]

Crop Safety/Restrictions

- Maximum number of treatments 1 per crop for winter wheat, 1 per yr for stone fruit, 2 per yr for amenity turf
- Keep off suckers or foliage of desirable trees or shrubs [1]
- Do not spray areas into which the roots of adjacent trees or shrubs extend [1]
- Do not spray on sloping ground when rain imminent and run-off may occur [1]
- Allow specified interval between treatment and sowing crops (see label) [2]
- Apply in autumn on land to be used for spring barley [2]
- Do not sow direct-drilled winter wheat less than 2 wk after application [2]
- Apply round base of established fruit trees taking care to avoid contaminating trees, especially where bark is damaged. Keep off trees [2]

Special precautions/Environmental safety

- Harmful to fish. Do not contaminate surface waters or ditches with chemical or used container

Protective clothing/Label precautions

- A [1, 2]; C [2]
- U08, U20a, E13c [1]; U09a, C03, E15 [2]; U19, E30a, E31b [1, 2]

Latest application/Harvest Interval(HI)

- At least 2 wk before direct drilling for winter wheat; before drilling for field crops; before end Jun or after harvest for apples, pears; end Oct for fallows [2]

Approval

- Off-label approval unlimited for use on amenity turf (OLA 0693/91) and around stone fruit trees (OLA 0333/92) [2]

Maximum Residue Levels (mg residue/kg food)

- tea, hops 0.1; fruits, vegetables, pulses, oilseeds, potatoes 0.05

---

14 amitrole + bromacil + diuron

A total herbicide mixture of translocated and residual chemicals

Products

BR Destral	RP Amenity	52.8:17.8:17.8% w/w	WP	05184

Uses

Total vegetation control in FENCELINES, INDUSTRIAL SITES, NON-CROP AREAS, RAILWAY TRACKS.

Notes

Efficacy

- Apply as foliage spray at medium to high volume at any time during growing season
- Ensure continuous mechanical or hydraulic agitation during spraying to prevent settling

Crop Safety/Restrictions

- Do not apply or drain or flush equipment on or near young trees, shrubs or other desirable plants or over areas where their roots may extend
- Do not use where chemical may come into contact with roots of desirable plants

Special precautions/Environmental safety

- Irritating to skin, eyes and respiratory system
- Harmful to fish. Do not contaminate surface waters or ditches with chemical or used container

Protective clothing/Label precautions

- A, C
- R04a, R04b, R04c, U04a, U05a, U09a, U20b, C03, E01, E13c, E29, E30a, E32a

Withholding period

- Keep livestock out of treated areas until foliage of poisonous weeds such as ragwort has died and become unpalatable

Maximum Residue Levels (mg residue/kg food)

- see amitrole entry

15 amitrole + 2,4-D + diuron

A total herbicide mixture of translocated and residual chemicals

Products

Trik	Mirfield	26.6:11.0:46.4% w/w	WP	07853

Uses

Total vegetation control in LAND NOT INTENDED TO BEAR VEGETATION.

Notes

Efficacy

- Apply in spring or late summer/early autumn when weeds are growing actively and have sufficient leaf area to absorb chemical
- Apply maintenance treatment if necessary at lower rate when weeds 7-10 cm high
- Increase rate on areas of peat or high carbon content

Crop Safety/Restrictions

- Do not use on ground under which roots of valuable trees or shrubs are growing

Special precautions/Environmental safety

- Harmful to fish. Do not contaminate surface waters or ditches with chemical or used container

Protective clothing/Label precautions

- A, C, D, H, M
- R04a, R04b, R04c, U05a, U08, U19, U20a, C03, E01, E07, E13c, E30a, E32a

Withholding period

- Keep livestock out of treated areas until foliage of poisonous weeds such as ragwort has died and become unpalatable

Maximum Residue Levels (mg residue/kg food)

- see amitrole entry

16 ammonium sulfamate

A non-selective, inorganic, general purpose herbicide and tree-killer

Products					
	1 Amcide	B H & B	99.5% w/w	CR	04246
	2 Root-Out	Dax	98.5% w/w	CR	03510

Uses

Annual weeds in ORNAMENTALS, TREES AND SHRUBS [2]. Annual weeds in FORESTRY, ORNAMENTALS *(pre-planting)*, VEGETABLES *(pre-planting)* [1]. Perennial dicotyledons in TREES AND SHRUBS [2]. Perennial dicotyledons in FORESTRY [1]. Perennial grasses in FORESTRY [1, 2]. Perennial weeds in ORNAMENTALS [2]. Perennial weeds in ORNAMENTALS *(pre-planting)*, VEGETABLES *(pre-planting)* [1]. Rhododendrons in FORESTRY [1, 2]. Woody weeds in FORESTRY [1, 2].

Notes

Efficacy

- Apply as spray to low scrub and herbaceous weeds from Apr to Sep in dry weather when rain unlikely and cultivate after 3-8 wk
- Apply as crystals in frills or notches in trunks of standing trees at any time of year
- Apply as concentrated solution or crystals to stump surfaces within 48 h of cutting. Rhododendrons must be cut level with ground and sprayed to cover cut surface, bark and immediate root area
- Stainless steel or plastic sprayers are recommended. Solutions are corrosive to mild steel, galvanised iron, brass and copper

Crop Safety/Restrictions

- Allow 8-12 wk after treatment before replanting
- Keep spray at least 30 cm from growing plants. Low doses may be used under mature trees with undamaged bark

Special precautions/Environmental safety

- Harmful to fish. Do not contaminate surface waters or ditches with chemical or used container

Protective clothing/Label precautions

- U09b, U20b, E32a [1]; U11, U14, U19, E01, E13c, E30a [1, 2]; U15, U20a [2]

17 anthracene oil

A crop desiccant

Products					
	Sterilite Hop Defoliant	Coventry Chemicals	63.6% w/w	EC	05060

Uses

Chemical stripping in HOPS.

Notes

Efficacy

- Spray when hop bines 1.2 m high and direct spray downward at 45° onto area to be defoliated. Repeat as necessary until cones are formed

Crop Safety/Restrictions

- Do not spray if temperature is above 21°C or after cones have formed
- Do not drench rootstocks
- Do not spray on windy, wet or frosty days

Special precautions/Environmental safety

- Harmful if swallowed. Irritating to eyes, skin and respiratory system
- Dangerous to fish. Do not contaminate surface waters or ditches with chemical or used container

Protective clothing/Label precautions

- A, C, H
- R03c, R04a, R04b, R04c, U05a, U08, U19, U20a, C02, C03, E13a, E25, E26, E29, E30a, E33, S07

18 asulam

A translocated carbamate herbicide for control of docks and bracken

Products	Asulox	RP Agric.	400 g/l	SL	06124

Uses

Bracken in FORESTRY, FORESTRY PLANTATIONS *(off-label)*, NON-CROP AREAS, PERMANENT PASTURE, ROUGH GRAZING. Docks in AMENITY GRASS *(not fine turf)*, APPLES, BLACKCURRANTS, BLUEBERRIES *(off-label)*, CANE FRUIT *(off-label)*, CHERRIES, CRANBERRIES *(off-label)*, CURRANTS *(off-label)*, DAMSONS *(off-label)*, ESTABLISHED LEYS, GOOSEBERRIES *(off-label)*, HOPS, MINT *(off-label)*, NECTARINES *(off-label)*, PARSLEY *(off-label)*, PEARS, PERMANENT PASTURE, PLUMS, QUINCES *(off-label)*, ROAD VERGES, STRAWBERRIES *(off-label)*, TARRAGON *(off-label)*, WASTE GROUND, WHITE CLOVER SEED CROPS *(off-label)*.

Notes

Efficacy

- Spray bracken when fronds fully expanded but not senescent, usually Jul-Aug; docks in full leaf before flower stem emergence
- Bracken fronds must not be damaged by stock, frost or cutting before treatment
- Do not apply in drought or hot, dry conditions
- Uptake and reliability of bracken control may be improved by use of specified additives - see label. Additives not recommended on forestry land
- To allow adequate translocation do not cut or admit stock for 14 d after spraying bracken or 7 d after spraying docks. Preferably leave undisturbed until late autumn
- Complete bracken control rarely achieved by one treatment. Survivors should be sprayed when they recover to full green frond, which may be in the ensuing year but more likely in the second year following initial application

Crop Safety/Restrictions

- Maximum number of treatments 1 per yr

- In forestry areas some young trees may be checked if sprayed directly (see label)
- Allow at least 6 wk between spraying and planting any crop
- Do not use in pasture before mowing for hay
- In fruit crops apply as a directed spray
- Do not treat blackcurrant cuttings, hop sets or weak hills
- Some grasses and herbs will be damaged by full dose. Most sensitive are cocksfoot, Yorkshire fog, timothy, bents, annual meadow-grass, daisies, docks, plaintains, saxifrage
- Apply as spot treatment in parsley, mint and tarragon, not directly to crop

Special precautions/Environmental safety

- Product is approved for use near surface waters. Whilst every care should be taken to avoid contamination of water, any that does occur from normal use should not offer harm to users, consumers of the water, or the environment

Protective clothing/Label precautions

- A, C, D, H, M (for ULV application)
- U19 (ULV), U20b (ULV), U20c, E07 (14 d), E15, E30a, E31a

Withholding period

- Keep livestock out of treated areas for at least 14 d and until foliage of poisonous weeds such as ragwort has died and become unpalatable

Approval

- May be applied through CDA equipment
- Approved for aerial application on bracken. See notes in Section 1
- Approved for use near surface waters. See notes in Section 1
- Off-label Approval unlimited for use on white clover seed crops (OLA 0931/92), strawberries (OLA 0932/92), cane fruit, currants, gooseberries, strawberries, blueberries, cranberries, damsons, nectarines, quinces (OLA 0930/92) [1]; unlimited for use in forestry areas (OLA 1001/92) [1], as spot treatment in parsley, mint, tarragon (OLA 0097/93) [1]

19 atrazine

A triazine herbicide with residual and foliar activity, with restricted permitted uses

Products

1 Alpha Atrazine 50 SC	Makhteshim	500 g/l	SC	04877
2 Atlas Atrazine	Atlas	500 g/l	SC	07702
3 Atrazol	Sipcam	500 g/l	SC	07598
4 Gesaprim	Novartis	500 g/l	SC	08411
5 Gesaprim 500 SC	Ciba Agric.	500 g/l	SC	05845
6 MSS Atrazine 50 FL	Mirfield	500 g/l	SC	01398
7 Unicrop Atrazine FL	Unicrop	500 g/l	SC	08045

Uses

Annual dicotyledons in SWEETCORN [1, 3, 5-7]. Annual dicotyledons in MAIZE [1-3, 5-7]. Annual dicotyledons in CONIFER PLANTATIONS, RASPBERRIES, ROSES [2]. Annual dicotyledons in CARROTS, CELERY, COMBINING PEAS, EARLY POTATOES, PARSLEY, VINING PEAS [4]. Annual grasses in SWEETCORN [1, 3, 5-7]. Annual grasses in MAIZE [1-3, 5-7]. Annual grasses in CONIFER PLANTATIONS, RASPBERRIES, ROSES [2]. Annual grasses in CARROTS, CELERY, COMBINING PEAS, EARLY POTATOES, PARSLEY, VINING PEAS [4]. Couch in CONIFER PLANTATIONS [2]. General weed control in TREES *(off-label)* [7]. Perennial grasses in CONIFER PLANTATIONS [2].

Notes

Efficacy

- May be used pre- or early post-weed emergence in maize and sweetcorn
- Root activity enhanced by rainfall soon after application and reduced on high organic soils. Foliar activity effective on weeds up to 3 cm high
- Not recommended for use on soils with more than 10% organic matter
- In conifers apply as overall spray in Feb-Apr. May be used in first spring after planting
- Apply to raspberries in spring before new cane emergence, not in season of planting
- Application rates vary with crop, soil type and weed problem. See label for details
- Resistant weed strains may develop with repeated use of atrazine or other triazines

Crop Safety/Restrictions

- Maximum number of applications (including other atrazine/simazine products) 1 per crop (or lower doses to 3.0 l/ha total) for maize, sweetcorn; 1 per season for conifer plantations, raspberries, roses
- Do not apply to raspberries in season of planting
- On slopes heavy rainfall soon after application may cause surface run-off
- To reduce soil run-off, especially from forest plantations, users are advised to plant grass strips 6 m wide between treated areas and surface waters
- Do not use on Christmas trees
- After annual weed control only maize or sweetcorn should be sown for at least 7 mth after application. Do not sow oats in autumn following spring treatment

Special precautions/Environmental safety

- Dangerous to fish or other aquatic life and aquatic higher plants. Do not contaminate surface waters or ditches with chemical or used container
- Do not allow direct spray from vehicle-mounted/drawn hydraulic sprayers to fall within 6 m, or from hand-held sprayers to within 2 m, of surface waters or ditches. Direct spray away from water
- Use must be restricted to one product containing atrazine or simazine, and either to a single application at the maximum approved rate or (subject to any existing maximum permitted number of treatments) to several applications at lower doses up to the maximum approved rate for a single application

Protective clothing/Label precautions

- A [1-7]; B [3-5]; C, D, H, M
- U20a [2]; U20b [1, 4, 5, 7]; U20c [3, 6]; E01 [4, 5]; E13b, E16, E30a [1-7]; E26 [1-3, 7]; E31a [2, 3]; E31b [1, 4, 5]; E31c [7]; E32a [6]

Latest application/Harvest Interval(HI)

- 7 mth before a succeeding crop for maize, sweetcorn; Apr for conifer plantations; before cane emergence for raspberries; before weeds at 3 cm for roses

Approval

- Off-label approval unlimited for use in forest situations (OLA 2163/96)[7]

Maximum Residue Levels (mg residue/kg food)

- fruits, vegetables, pulses, oilseeds, potatoes, tea, hops 0.1

FOR FULL CONDITIONS OF USE ALWAYS READ THE PRODUCT LABEL

20 azaconazole

A conazole available only in mixture

21 azaconazole + imazalil

A fungicide mixture for use in horticulture

Products

Nectec Paste	Hortichem	1:2% w/w	PA	08510

Uses Canker in FORESTRY, ORNAMENTAL TREES, SHRUBS. Silver leaf in FORESTRY, ORNAMENTAL TREES, SHRUBS.

Notes

Efficacy
- Paint pruning cuts immediately. If necessary clean wounds and cut back any loose bark
- Ensure whole of cut area is fully covered beyond wound to surrounding healthy bark
- Cut back established cankers to sound healthy wood before treatment. Work paste into all crevices

Crop Safety/Restrictions
- Maximum number of treatments 1 per wound per yr
- Treat only during dry weather and not in frosty conditions
- Use only during dormant periods
- Do not apply to grafting cuts

Special precautions/Environmental safety
- Harmful to fish or other aquatic life. Do not contaminate surface waters or ditches with chemical or used container

Protective clothing/Label precautions
- A
- U05a, U20c, C03, E01, E13c, E26, E30a

Maximum Residue Levels (mg residue/kg food)
- see imazalil entry

22 azamethiphos

A residual organophosphorus insecticide for fly control

Products

1 Alfacron 10 WP	Ciba Agric.	10% w/w	WP	02832
2 Alfacron 10 WP	Novartis A H	10% w/w	WP	08587

Uses Flies in LIVESTOCK HOUSES.

Notes

Efficacy
- Add water as directed to produce paint consistency and paint onto 2.5% of total wall and ceiling surface area at a minimum of 5 points in building or apply as spray to 30% of surface area

Crop Safety/Restrictions

- Only apply to areas out of reach of children and animals

Special precautions/Environmental safety

- This product contains an anticholinesterase organophosphorus compound. Do not use if under medical advice not to work with such compounds
- Irritating to eyes and skin. May cause sensitization by skin contact
- Dangerous to fish. Do not contaminate surface waters or ditches with chemical or used container
- Do not apply directly to livestock and poultry
- Do not apply to surfaces on which food or feed is stored, prepared or eaten. Cover feedstuffs and remove exposed milk and eggs before application

Protective clothing/Label precautions

- A, C
- M01, R04a, R04b, R04e, U05a, U10, U13, U14, U15, U19, U20b, C03, C04, C05, C06, C07, E01, E05, E13b, E30a, E32a

23 aziprotryne

A selective triazine herbicide with foliar and residual activity

Products

Brasoran 50 WP	Novartis	50% w/w	WP	08394

Uses

Annual dicotyledons in ASPARAGUS *(off-label)*, BROCCOLI *(off-label)*, BRUSSELS SPROUTS, CABBAGES, CALABRESE *(off-label)*, CAULIFLOWERS *(off-label)*, CHINESE CABBAGE *(off-label)*, COLLARDS *(off-label)*, COMBINING PEAS, KALE *(off-label)*, LEEKS, ONIONS, VINING PEAS. Annual meadow grass in ASPARAGUS *(off-label)*, BROCCOLI *(off-label)*, BRUSSELS SPROUTS, CABBAGES, CALABRESE *(off-label)*, CAULIFLOWERS *(off-label)*, CHINESE CABBAGE *(off-label)*, COLLARDS *(off-label)*, COMBINING PEAS, KALE *(off-label)*, LEEKS, ONIONS, VINING PEAS.

Notes

Efficacy

- For best results soil should be moist with a fine tilth and the majority of weeds just emerging
- Control lasts 6-8 wk in normal conditions but may be reduced if dry weather follows application
- Emerged weeds not controlled after 2-leaf stage
- Treat peas pre- or post-emergence; brassicas post-emergence only
- Onions and leeks should be treated as part of a weed control programme to ensure weeds are not beyond recommended size when sprayed

Crop Safety/Restrictions

- Maximum number of treatments 1 per crop
- Crops must have reached correct growth stage for treatment. See label for details
- Do not treat crops that are not growing well or are under stress from pest or disease attack
- Do not apply during or before heavy rain, or during frosty weather
- Do not use on brassicas grown under glass

- All varieties of spring sown vining and dried peas may be treated if they are covered by 25 mm of settled soil
- If re-drilling a treated field necessary, any crop may be sown 6 wk after treatment provided correct dose was used and soil has been thoroughly cultivated

Special precautions/Environmental safety

- Dangerous to fish or other aquatic life. Do not contaminate surface waters or ditches with chemical or used container

Protective clothing/Label precautions

- U20b, C02 (brassicas 3 wk; peas 6 wk; onions 6 wk; leeks 6 wk), E13b, E29, E30a, E32a

Latest application/Harvest Interval(HI)

- HI 3 wk Brussels sprouts, cabbages; 6 wk for combining peas, leeks, onions, vining peas

Approval

- Off-label approval unlimited for use on broccoli, calabrese, cauliflowers, Chinese cabbage, collards, kale (OLA 1375/97)[1]; unlimited for use on outdoor asparagus (OLA 1376/97)[1]

24 azoxystrobin

A systemic translaminar and protectant strobilurin fungicide for cereals

Products

Amistar	Zeneca	250 g/l	SC	08517

Uses

Brown rust in SPRING BARLEY, SPRING WHEAT, WINTER BARLEY, WINTER WHEAT. Glume blotch in SPRING WHEAT, WINTER WHEAT. Late ear diseases in SPRING WHEAT, WINTER WHEAT. Leaf spot in SPRING WHEAT, WINTER WHEAT. Net blotch in SPRING BARLEY, WINTER BARLEY. Powdery mildew in SPRING BARLEY, SPRING WHEAT, WINTER BARLEY, WINTER WHEAT. Rhynchosporium in SPRING BARLEY, WINTER BARLEY. Yellow rust in SPRING WHEAT, WINTER WHEAT.

Notes

Efficacy

- Best results obtained from use as a protectant or during early stages of disease establishment
- Control of developed mildew or Rhynchosporium infections can be improved by appropriate tank mixture. See label for details

Crop Safety/Restrictions

- Maximum total dose 3.0 l/ha product
- To discourage development of resistance do not apply more than three foliar treatments of stobilurin products to the same crop
- Avoid sequential use alone against powdery mildew

Special precautions/Environmental safety

- Dangerous to fish or other aquatic life. Do not contaminate surface waters or ditches with chemical or used container

Protective clothing/Label precautions

- U02, U09a, U19, U20a, E13b, E30a, E31c

Latest application/Harvest Interval(HI)

- Grain watery ripe (GS 71)

25 azoxystrobin + fenpropimorph

A protectant and eradicant fungicide mixture

Products	Amistar Pro	Zeneca	100:280 g/l	EC	08871

Uses Brown rust in SPRING BARLEY, SPRING WHEAT, WINTER BARLEY, WINTER WHEAT. Late ear diseases in SPRING WHEAT, WINTER WHEAT. Net blotch in SPRING BARLEY, WINTER BARLEY. Powdery mildew in SPRING BARLEY, SPRING WHEAT, WINTER BARLEY, WINTER WHEAT. Rhynchosporium in SPRING BARLEY, WINTER BARLEY. Septoria diseases in SPRING WHEAT, WINTER WHEAT. Yellow rust in SPRING WHEAT, WINTER WHEAT.

Notes

Efficacy

- Best results obtained from application before infection following a disease risk assessment, or when disease first seen in crop
- Results may be less reliable when used on crops under stress
- Treatments for protection against ear disease should be made at ear emergence

Crop Safety/Restrictions

- Maximum total dose equivalent to two full dose treatments
- Active ingredients have different modes of action and therefore development of resistance less likely. However to discourage development of resistance a maximum of three applications of strobilurin fungicides to the same crop is recommended

Special precautions/Environmental safety

- Irritating to skin
- May cause sensitization by skin contact
- Dangerous to fish or other aquatic life. Do not contaminate surface waters or ditches with chemical or used container
- Do not allow direct spray from vehicle mounted/drawn hydraulic sprayers to fall within 6 m, or from hand-held sprayers to within 2 m, of surface waters or ditches. Direct spray away from water

Protective clothing/Label precautions

- A, H
- M03, R04b, R04e, U02, U05a, U09a, U14, U15, U19, U20b, C03, E01, E13b, E16, E26, E30a, E31c

Latest application/Harvest Interval(HI)

- Before early milk stage (GS 73)
- HI 5 wk

26 Bacillus thuringiensis

A bacterial insecticide for control of caterpillars

Products					
	1 Dipel	English Woodland	16000 IU/mg	WP	08634
	2 Novosol FC	Ashlade	8400 IU/mg	SC	06566

Uses Browntail moth in TOP FRUIT [2]. Cabbage moth in BROCCOLI, BRUSSELS SPROUTS, CABBAGES, CALABRESE, CAULIFLOWERS [2]. Cabbage white butterfly in BROCCOLI, BRUSSELS SPROUTS, CABBAGES, CALABRESE, CAULIFLOWERS [2]. Caterpillars in GRAPEVINES, PROTECTED CROPS, TOP FRUIT, VEGETABLES [2]. Caterpillars in BROCCOLI, BRUSSELS SPROUTS, CABBAGES, CAULIFLOWERS, CUCUMBERS, ORNAMENTALS, PEPPERS, PROTECTED BRASSICA SEEDLINGS, PROTECTED CUCUMBERS, PROTECTED ORNAMENTALS, PROTECTED PEPPERS, PROTECTED TOMATOES, RASPBERRIES, STRAWBERRIES, TOMATOES [1]. Cutworms in VEGETABLES [2]. Diamond-back moth in BROCCOLI, BRUSSELS SPROUTS, CABBAGES, CALABRESE, CAULIFLOWERS [2]. Grape berry moth in GRAPEVINES [2]. Grape leafroller in GRAPEVINES [2]. Omniverous leafroller in GRAPEVINES [2]. Small ermine moth in TOP FRUIT [2]. Tomato moth in TOMATOES [2]. Tortrix moths in GRAPEVINES, PROTECTED CROPS, TOP FRUIT [2]. Winter moth in TOP FRUIT [2].

Notes

Efficacy

- Product affects gut of larvae and must be eaten to be effective. Caterpillars cease feeding and die in 2-3 d (3-5 d for large caterpillars)
- Apply as soon as larvae appear on crop and repeat every 3-14 d for outdoor crops, every 3 wk under glass
- Addition of a wetter recommended for use on brassicas
- Good coverage is essential, especially of undersides of leaves. Use a drop-leg sprayer in field crops

Crop Safety/Restrictions

- No restriction on number of treatments of edible crops
- Do not mix more spray than can be used in a 12 h period

Special precautions/Environmental safety

- Do not contaminate surface waters or ditches with chemical or used container
- Store out of direct sunlight

Protective clothing/Label precautions

- A, C, H [1]
- U20a, E26, E29 [2]; U20c, E01 [1]; E15, E30a, E32a [1, 2]

Latest application/Harvest Interval(HI)

- HI zero for crops

---

27 benalaxyl

A phenylamide (acylalanine) fungicide available only in mixtures

---

28 benalaxyl + mancozeb

A systemic and protectant fungicide mixture

Products Tairel | Sipcam | 8:65% w/w | WB | 07767

Uses Blight in EARLY POTATOES, MAINCROP POTATOES.

Notes

Efficacy

- Apply to potatoes at blight warning prior to crop becoming infected and repeat at 10-21 d intervals depending on risk of infection
- Spray irrigated potatoes after irrigation and at 14 d intervals, crops in polythene tunnels at 10 d intervals
- When active potato growth ceases use a non-systemic fungicide to end of season, starting not more than 10 d after last application
- Do not treat potatoes showing active blight infection
- For reduction of downy mildew in oilseed rape apply at seedling to 7-leaf stage, before end Nov as soon as infection seen

Crop Safety/Restrictions

- Maximum number of treatments 5 per crop (including other phenylamide-based fungicides) for potatoes, 1 per crop for winter oilseed rape

Special precautions/Environmental safety

- Irritating to eyes
- Dangerous to fish or other aquatic life. Do not contaminate surface waters or ditches with chemical or used container

Protective clothing/Label precautions

- A
- R04a, U05a, U08, U20b, C02 (7 d), C03, E01, E13b, E29, E30a

Latest application/Harvest Interval(HI)

- HI 7 d

Maximum Residue Levels (mg residue/kg food)

- (benalaxyl) grapes, onions, tomatoes, peppers 0.2; tea, hops 0.1; fruits (except grapes), vegetables (except onions, tomatoes, peppers), pulses, oilseeds, potatoes, cereals, animal products 0.05. See also mancozeb entry

29 benazolin

A translocated arylacetic acid herbicide for oilseed rape

Products	Galtak 50 SC	AgrEvo	500 g/l	SC	07258

Uses

Chickweed in WINTER OILSEED RAPE. Cleavers in WINTER OILSEED RAPE.

Notes

Efficacy

- For best results spray during mild moist weather when weeds actively growing
- Weeds shaded by crop or grass weeds will not be completely controlled and may grow away later
- Frost after application will not reduce weed control but it should not be on foliage at time of treatment

Crop Safety/Restrictions

- Maximum number of treatments 1 per crop
- Treat from 3 developed crop leaf stage

- Only treat healthy crops and allow time before treatment for recovery from any conditions that reduce wax formation
- Autumn treatment may result in abnormal growth of leaves and stems later in season from which recovery is normally complete
- Do not apply to spring oilseed rape or fodder rape
- Avoid drift outside the target area
- Interval of 14 d must elapse before or after treatment with any other pesticide

Special precautions/Environmental safety

- Irritant, may cause sensitization by skin contact
- Harmful to fish or other aquatic life. Do not contaminate surface waters or ditches with chemical or used container

Protective clothing/Label precautions

- A
- M03, R04, R04e, U04a, U05a, U09a, U10, U14, U20a, C03, E01, E07, E13c, E26, E30a, E31b, E34

Withholding period

- Keep livestock out of treated areas until foliage of any poisonous weeds such as ragwort has died and become unpalatable

Latest application/Harvest Interval(HI)

- Before crop flower buds visible above leaves (GS 3,5)

30 benazolin + bromoxynil + ioxynil

A post-emergence, HBN herbicide mixture for cereal crops and grass

Products	Asset	AgrEvo	50:125:62.5 g/l	EC	07243

Uses

Annual dicotyledons in BARLEY, DURUM WHEAT, NEWLY SOWN GRASS, OATS, WHEAT.

Notes

Efficacy

- Commonly used in mixture with mecoprop and other cereal herbicides to extend the range of weeds controlled. Sequential treatments also recommended
- Best results achieved when weeds small and actively growing and crop competitive
- Weeds should be dry when sprayed

Crop Safety/Restrictions

- Maximum number of treatments 2 per crop or yr
- Use in cereals from 2-leaf stage to first node detectable (GS 12-31), in grass from 2-leaf stage when crop growing vigorously (other times may be advised for mixtures)
- Use on oats only in spring, on other cereals in autumn or spring
- Do not apply to crops undersown with legumes
- Do not use on crops affected by pests, disease, waterlogging or prolonged frost
- Severe frost within 3-4 wk of spraying may scorch crop
- Do not roll or harrow crops for 3 d before or after spraying
- Do not spray grass seed crops later than 5 wk before heading

Special precautions/Environmental safety

- Harmful if swallowed. Irritating to skin and eyes

- Flammable
- Do not apply by hand-held equipment or at concentrations higher than those recommended
- Harmful to bees. Do not apply to crops in flower or to those in which bees are actively foraging. Do not apply when flowering weeds are present
- Dangerous to fish or other aquatic life. Do not contaminate surface waters or ditches with chemical or used container
- Do not allow direct spray from vehicle-mounted/drawn hydraulic sprayers to fall within 6 m of surface waters or ditches. Direct spray away from water

Protective clothing/Label precautions

- A, C
- M03, R03c, R04a, R04b, R07d, U05a, U08, U13, U19, U20a, C03, E01, E07, E12e, E13b, E30a, E31b, E34

Withholding period

- Keep livestock out of treated areas for at least 6 wk and until foliage of poisonous plants such as ragwort has died and become unpalatable

Latest application/Harvest Interval(HI)

- 5 wk before heading (seed crops) for newly sown grass; up to GS 31 for barley, durum wheat, oats, wheat
- HI 6 wk (before grazing)

Maximum Residue Levels (mg residue/kg food)

- see ioxynil entry

31 benazolin + clopyralid

A post-emergence herbicide mixture for use mainly in winter oilseed rape

Products

Benazalox	AgrEvo	30:5% w/w	WP	07246

Uses

Annual dicotyledons in EVENING PRIMROSE *(off-label)*, HONESTY *(off-label)*, SPRING OILSEED RAPE *(off-label)*, SWEDE SEED CROPS *(off-label)*, WHITE MUSTARD *(off-label)*, WINTER OILSEED RAPE. Chickweed in EVENING PRIMROSE *(off-label)*, SWEDE SEED CROPS *(off-label)*, WHITE MUSTARD *(off-label)*, WINTER OILSEED RAPE. Cleavers in EVENING PRIMROSE *(off-label)*, SWEDE SEED CROPS *(off-label)*, WHITE MUSTARD *(off-label)*, WINTER OILSEED RAPE. Mayweeds in EVENING PRIMROSE *(off-label)*, SWEDE SEED CROPS *(off-label)*, WHITE MUSTARD *(off-label)*, WINTER OILSEED RAPE.

Notes

Efficacy

- Weeds are controlled from cotyledon stage to early flowerbud or up to 15 cm high or across in winter oilseed rape
- For best results apply during mild, moist weather when weeds are growing actively and still visible in young crop. Do not spray if rain expected within 4 h
- Do not spray if frost present on foliage; frost after application will not reduce effectiveness
- For grass weed control various tank mixtures are recommended or sequential treatments, allowing specified interval after application of grass-killer. See label for details

Crop Safety/Restrictions

- Maximum number of treatments 1 per crop
- May be used on winter oilseed rape when crop between stages of 3 fully developed leaves to flower buds hidden beneath leaves (GS 1,3-3,1), on spring oilseed rape before green bud
- Oilseed rape may be replanted in the event of crop failure. Wheat, barley or oats may be sown after 1 mth, other crops in the following autumn after ploughing. Winter beans should not be planted in the same year
- Chop and incorporate straw and trash in early autumn to release any clopyralid residues. Ensure remains of plants completely decayed before planting susceptible crops

Special precautions/Environmental safety

- Irritating to eyes and skin
- Harmful to fish or other aquatic life. Do not contaminate surface waters or ditches with chemical or used container

Protective clothing/Label precautions

- A, C
- R04a, R04b, U05a, U09a, U13, U19, U20a, C03, E01, E07, E13c, E30a, E32a

Withholding period

- Keep livestock out of treated areas until foliage of any poisonous weeds such as ragwort has died and become unpalatable

Latest application/Harvest Interval(HI)

- Oilseed rape before green bud stage (GS 3,1) (before end Jan if in tank mix with Butisan S), honesty up to 4-5 pairs true leaves

Approval

- Off-label approval unlimited for use on evening primrose, white mustard and swedes grown for seed (OLA 1596, 1597/96), on honesty (OLA 0610/93), on spring oilseed rape (OLA 1598/97)[1]

32 benazolin + 2,4-DB + MCPA

A post-emergence herbicide for undersown cereals, grass and clover

Products

1 Setter 33	Dow	50:237:43 g/l	SL	05623
2 Stefes Legumex Extra	Stefes	27:237:42.8 g/l	SL	07841

Uses

Annual dicotyledons in CLOVERS, SEEDLING LEYS, UNDERSOWN BARLEY, UNDERSOWN OATS, UNDERSOWN WHEAT [1, 2]. Annual dicotyledons in DIRECT-SOWN SEEDLING CLOVERS [1]. Annual dicotyledons in GRASSLAND [2]. Chickweed in CLOVERS, SEEDLING LEYS, UNDERSOWN BARLEY, UNDERSOWN OATS, UNDERSOWN WHEAT [1, 2]. Chickweed in DIRECT-SOWN SEEDLING CLOVERS [1]. Chickweed in GRASSLAND [2]. Cleavers in CLOVERS, SEEDLING LEYS, UNDERSOWN BARLEY, UNDERSOWN OATS, UNDERSOWN WHEAT [1, 2]. Cleavers in DIRECT-SOWN SEEDLING CLOVERS [1]. Cleavers in GRASSLAND [2]. Knotgrass in CLOVERS, SEEDLING LEYS, UNDERSOWN BARLEY, UNDERSOWN OATS, UNDERSOWN WHEAT [1, 2]. Knotgrass in DIRECT-SOWN SEEDLING CLOVERS [1]. Knotgrass in GRASSLAND [2]. Perennial dicotyledons in CLOVERS, GRASSLAND, SEEDLING LEYS, UNDERSOWN BARLEY, UNDERSOWN OATS, UNDERSOWN WHEAT [2].

Notes

Efficacy

- Best results on annual weeds achieved when treated young and in active growth

- Top growth of established perennials may be checked [1]
- Spray perennial weeds when well developed but before flowering. Higher rates may be used against perennials in established grassland [2]
- Do not spray when rain is imminent or during drought

Crop Safety/Restrictions
- Maximum number of treatments 1 per crop or yr
- Spray undersown cereals before 1st node detectable (GS 31), grass seedlings should have at least 3 leaves
- Spray winter cereals in spring from leaf sheath erect exceeding 5 cm, spring wheat after 5 leaves unfolded (GS 15), spring barley and oats after 2 leaves unfolded (GS 12)
- Spray clovers after the 1-trifoliate leaf stage and before red clover has more than 3 trifoliate leaves. Do not spray clover seed crops in the yr seed is to be taken
- Spray new swards before end Oct when grasses have at least 3 leaves and clovers at stages as above
- Clovers may be damaged if frost occurs soon after spraying
- Do not spray legumes other than clover
- Do not roll, harrow or cut crops for 3 d before or after spraying

Special precautions/Environmental safety
- Harmful to fish or other aquatic life. Do not contaminate surface waters or ditches with chemical or used container

Protective clothing/Label precautions
- U08, U20b, E07 (2 wk), E13c, E26, E30a, E31b [1, 2]; E34 [2]

Withholding period
- Keep livestock out of treated areas for at least 2 wk and until foliage of poisonous weeds such as ragwort has died and become unpalatable

Latest application/Harvest Interval(HI)
- Before first node detectable (GS 31) for undersown cereals

33 bendiocarb

A carbamate insecticide seed dressing

Products

Seedox SC	Uniroyal	500 g/l	FS	07591

Uses

Bean seed flies in PROTECTED COURGETTES *(off-label seed treatment)*, PROTECTED GHERKINS *(off-label seed treatment)*, PROTECTED MARROWS *(off-label seed treatment)*, PROTECTED MELONS *(off-label seed treatment)*, PROTECTED PUMPKINS *(off-label seed treatment)*, PROTECTED SQUASHES *(off-label seed treatment)*. Frit fly in MAIZE, SWEETCORN. Wireworms in MAIZE, SWEETCORN.

Notes

Efficacy
- Apply mixed with Uniroyal Adjuvant Oil using suitable seed treatment machinery

Crop Safety/Restrictions
- Maximum number of treatments 1 per batch of seed

Special precautions/Environmental safety

- This product contains an anticholinesterase carbamate compound. Do not use if under medical advice not to work with such compounds
- Harmful in contact with skin or if swallowed
- Ensure adequate ventilation in confined spaces
- Dangerous to fish or other aquatic life. Do not contaminate surface waters or ditches with chemical or used container
- Dangerous to game, wild birds and animals
- Treated seed harmful to game and wildlife. Bury spillages
- Treated seed not to be used as food or feed

Protective clothing/Label precautions

- A, C
- M02, M04, R03a, R03c, U05a, U09a, U13, U16, U19, U20b, C03, E01, E10a, E13b, E26, E30a, E31a, E34, S01, S02, S04b, S05, S06, S07

Latest application/Harvest Interval(HI)

- Pre-drilling

---

34 benfuracarb

A soil-applied carbamate insecticide and nematicide for beet crops

Products

Oncol 10G	Mirfield	10% w/w	GR	08249

Uses

Docking disorder vectors in FODDER BEET, MANGELS, SUGAR BEET. Leaf miners in FODDER BEET, MANGELS, SUGAR BEET. Millipedes in FODDER BEET, MANGELS, SUGAR BEET. Pygmy beetle in FODDER BEET, MANGELS, SUGAR BEET. Springtails in FODDER BEET, MANGELS, SUGAR BEET. Symphylids in FODDER BEET, MANGELS, SUGAR BEET.

Notes

Efficacy

- Apply at sowing with suitable applicator so that the granules are mixed with the moving soil closing the seed furrow. See label for recommended applicators and calibration

Crop Safety/Restrictions

- Maximum number of treatments 1 per crop

Special precautions/Environmental safety

- Irritating to eyes and skin
- Dangerous to game, wild birds and animals
- Extremely dangerous to fish or other aquatic life. Do not contaminate surface waters or ditches with chemical or used container

Protective clothing/Label precautions

- A, B, C, D, E, H, K, M
- M02, R04a, R04b, U02, U05a, U09a, U13, U19, U20a, C03, E01, E10a, E13a, E30a, E32a

Latest application/Harvest Interval(HI)

- At planting

---

35 benomyl

A systemic, MBC fungicide with protectant and eradicant activity

Products	Benlate Fungicide	DuPont	50% w/w	WP	00229

Uses

Black spot in COURGETTES *(off-label)*. Botrytis in OILSEED RAPE, PEAS, RHUBARB *(off-label)*. Chocolate spot in FIELD BEANS. Eyespot in CEREALS. Fungus diseases in RHUBARB *(off-label)*. Gummosis in COURGETTES *(off-label)*. Leaf and pod spot in FIELD BEANS *(seed treatment)*, PEAS *(seed treatment)*. Light leaf spot in BRUSSELS SPROUTS, OILSEED RAPE. Rhizoctonia in WATERCRESS *(off-label)*. Rhynchosporium in CEREALS. Ring spot in BRUSSELS SPROUTS.

Notes

Efficacy

- Apply as overall spray. One spray effective for some diseases, repeat sprays at 1-4 wk intervals needed for others. Recommended spray programmes vary with disease and crop. See label for details of timing and recommended tank mixes
- Addition of non-ionic wetter recommended for certain uses. See label for details
- To delay appearance of resistant strains in diseases needing more than 2 applications per season, use in programme with fungicide of different mode of action
- Product may lose effectiveness if allowed to become damp during storage

Crop Safety/Restrictions

- Maximum number of treatments (including applications of any product containing benomyl, carbendazim or thiophanate-methyl) 3 per crop for oilseed rape; 2 per crop for barley, Brussels sprouts, field beans, combining and vining peas; 1 per crop for wheat, mange-tout peas

Protective clothing/Label precautions

- A, C, D, H, M
- U20b, E15, E30a, E32a

Latest application/Harvest Interval(HI)

- At full bloom for mange-tout peas; up to and including grain watery ripe stage (GS 71) for barley; before first node detectable (GS 31) for wheat
- HI oilseed rape, Brussels sprouts (with Actipron or non-ionic wetter), combining and vining peas, field and broad beans 3 wk

Approval

- Off-label approval unlimited for use on protected rhubarb (OLA 1806/96), and outdoor courgettes (OLA 1866/96)[1]; unlimited for use on watercress (OLA 0875/97)[1]

36 bentazone

A post-emergence contact diazinone herbicide

Products	1 Basagran SG	BASF	87% w/w	SG	08360
	2 Standon Bentazone	Standon	480 g/l	SL	09204

Uses Annual dicotyledons in BROAD BEANS, LINSEED, NARCISSI, NAVY BEANS, PEAS, POTATOES, RUNNER BEANS, SOYA BEANS *(off-label)*, SPRING FIELD BEANS, WINTER FIELD BEANS [1, 2]. Annual dicotyledons in BULB ONIONS *(off-label)*, EVENING PRIMROSE *(off-label)*, FRENCH BEANS, SALAD ONIONS *(off-label)* [1]. Annual dicotyledons in DWARF BEANS, OUTDOOR DWARF BEANS *(off-label)* [2].

Notes

Efficacy

- Most effective control obtained when weeds are growing actively and less than 5 cm high or across. Good spray cover is essential
- Split dose application may be made in beans, linseed, potatoes and narcissi and generally gives better weed control. See label for details
- Do not apply if rain or frost expected, if unseasonably cold, if foliage wet or in drought
- A minimum of 6 h free from rain is required after application
- Various recommendations are made for use in spray programmes with other herbicides or in tank mixes. See label for details
- Addition of Actipron, Adder or Cropspray 11E adjuvant oils recommended for use in dwarf beans and potatoes to improve fat hen control. Do not use under hot or humid conditions

Crop Safety/Restrictions

- Maximum number of treatments varies with crop and product - see labels
- Crops must be treated at correct stage of growth to avoid danger of scorch. See label for details and for varietal tolerances. Apply to broad beans from 3 to 4 leaf pairs only
- Do not use on crops which have been affected by drought, waterlogging, frost or other stress conditions
- Do not spray at temp above 21°C. Delay spraying until evening if necessary
- Consult processor before using on crops for processing
- A satisfactory wax test must be carried out before use on peas
- Not all varieties are fully tolerant. Do not use on forage peas and other named varieties of peas, beans and potatoes
- May be used on selected varieties of maincrop and second early potatoes (see label for details), not on seed crops or first earlies
- Do not treat narcissi during flower bud initiation

Special precautions/Environmental safety

- Harmful: If swallowed [2]
- Risk of serious damage to eyes [1]
- May cause sensitization by skin contact

Protective clothing/Label precautions

- A [1, 2]; C [1]
- M03, R03c, E31b, E34 [2]; R04, R04d, E32a [1]; R04e, U05a, U08, U19, U20b, C03, E01, E15, E26, E30a [1, 2]

Latest application/Harvest Interval(HI)

- Before shoots exceed 15 cm high for potatoes and spring field beans (or 6-7 leaf pairs), 4 leaf pairs (6 pairs or 15 cm high with split dose) for broad beans, before flower buds visible for French, navy, runner and winter field beans and linseed, before flower buds can be found enclosed in terminal shoot for peas
- HI onions 3 wk

Approval

- Off-label approval unlimited for use on outdoor bulb and salad onions (OLA 1234/97)[1], and evening primrose (OLA 1232/97)[1]; to Apr 2002 for use on soya beans (OLA 0793/97)[1]

37 bentazone + MCPA + MCPB

A post-emergence herbicide for undersown spring cereals and grass

Products	Acumen	BASF	200:80:200 g/l	SL	00028

Uses Annual dicotyledons in NEWLY SOWN GRASS, UNDERSOWN SPRING BARLEY, UNDERSOWN SPRING OATS, UNDERSOWN SPRING WHEAT.

Notes

Efficacy

- Best results when weeds small and actively growing provided crop at correct stage
- A minimum of 6 h free from rain is required after treatment
- Do not apply if frost expected, if crop wet or when temperatures at or above 21°C
- On first year grass leys use alone or in mixture with cyanazine on seedling stage weeds before end of Sep, use mixture with cyanazine on older chickweed in Oct-Nov

Crop Safety/Restrictions

- Maximum number of treatments 1 per crop on undersown cereals; 1 per yr on grassland
- Apply to cereals from 2-fully expanded leaf stage but before first node detectable (GS 12-30) provided clover has reached 1-trifoliate leaf stage
- Do not treat red clover after 3-trifoliate leaf stage
- Apply to newly sown leys after grass has reached 2-leaf stage provided clovers have at least 1 trifoliate leaf and red clover has not passed 3-trifoliate leaf stage. Grasses must have at least 3 leaves before treating with cyanazine mixture
- Do not use on crops suffering from herbicide damage or physical stress
- Do not use on seed crops or on cereals undersown with lucerne
- Do not roll or harrow for 7 d before or after spraying
- Clovers may be scorched and undersown crop checked but effects likely to be outgrown

Special precautions/Environmental safety

- Harmful if swallowed. Irritating to eyes and skin. May cause sensitization by skin contact
- Harmful to fish or other aquatic life. Do not contaminate surface waters or ditches with chemical or used container

Protective clothing/Label precautions

- A, C
- M03, R03c, R04a, R04b, R04e, U05a, U08, U19, U20a, C03, E01, E07, E13c, E30a, E31c, E34

Withholding period

- Keep livestock out of treated areas for at least 2 wk and until foliage of poisonous weeds such as ragwort has died and become unpalatable

Latest application/Harvest Interval(HI)

- Before 1st node detectable for undersown cereals (GS 31), 2 wk before grazing for grassland (4 wk if mixed with cyanazine)

38 bentazone + MCPB

A post-emergence herbicide mixture for tank mixing with cyanazine

Products	Pulsar	BASF	200:200 g/l	SL	04002

Uses Annual dicotyledons in PEAS.

Notes

Efficacy

- To be used in tank mix with cyanazine
- Best results achieved when weeds small and actively growing provided crop at correct stage. Good spray cover is essential
- May be applied as single treatment or as split dose treatment applying first spray when susceptible weeds are not beyond the 2 leaf stage
- Do not apply if rain or frost expected or if foliage wet. A minimum of 24 h free from rain required after treatment
- Do not apply during drought or unseasonably cold weather

Crop Safety/Restrictions

- Maximum number of treatments 1 per crop or 2 per crop with split dose
- Apply only to listed cultivars (see label) from 3 fully expanded leaf (for full dose) or from 2 node stage (for split dose) to before flower buds can be found enclosed in terminal shoot (GS 102 or 103 to before GS 201)
- Do not treat forage pea cultivars or mange-tout peas
- Apply after a satisfactory wax test. Early drilled crops or crops affected by frost or abrasion may not have sufficiently waxy cuticle
- Allow 7 d after spraying before using a grass weed herbicide, or wait 14 d afterwards and test leaf wax before treating
- Do not use as tank mix with any other product than cyanazine, nor after use of TCA
- Do not apply to any crop that may have been subjected to stress conditions, where foliage damaged or under hot, sunny conditions when temperature exceeds 21°C
- Do not apply insecticides within 7 d of treatment

Special precautions/Environmental safety

- Irritating to eyes and skin. May cause sensitization by skin contact
- Harmful to fish or other aquatic life. Do not contaminate surface waters or ditches with chemical or used container

Protective clothing/Label precautions

- A, C
- R04a, R04b, R04e, U05a, U08, U19, U20a, C03, E01, E07, E13c, E30a, E31c, E34

Withholding period

- Keep livestock out of treated areas for 14 d and until foliage of any poisonous weeds such as ragwort has died and become unpalatable

Latest application/Harvest Interval(HI)

- Before enclosed bud stage of crop (GS 10x)

39 benzalkonium chloride

A quaternary ammonium algicide and moss killer for paths, pots etc

Products				
Algaecide	Fargro	12% w/w	SL	H5842

Uses

Algae in GLASSHOUSE STRUCTURES AND SURROUNDS, WALLS. Lichens in GLASSHOUSE STRUCTURES AND SURROUNDS, WALLS. Mosses in GLASSHOUSE STRUCTURES AND SURROUNDS, WALLS.

Notes

Efficacy

- Spray, sprinkle or brush on paths, walls and stonework under dry conditions. Do not use if it has rained within 3-4 d or if rain anticipated within 24 h
- Apply to plant beds or benches before plants are put in place. If applied to pots, plants should be removed during spraying or spray directed at base of pots
- Regrowth of moss normally prevented for whole growing season

Crop Safety/Restrictions

- Avoid contact with leaves of growing plants

Special precautions/Environmental safety

- Irritating to eyes and skin
- Dangerous to fish. Do not contaminate surface waters or ditches with chemical or used container
- Keep unprotected persons away from treated areas for 48 h or until surfaces are dry

Protective clothing/Label precautions

- A, E, H
- R04a, R04b, U02, U05a, U09b, U19, U20b, C03, C04, E01, E02 (48 h), E13b, E20, E30a, E32a

40 bifenox

A diphenyl ether herbicide available only in mixtures

41 bifenox + dicamba

A contact and translocated herbicide mixture for winter cereals

Products				
1 Quickstep	Sandoz	400:40 g/l	SC	07389
2 Quickstep	Novartis	400:40 g/l	SC	08557

Uses

Annual dicotyledons in WINTER BARLEY, WINTER WHEAT. Cleavers in WINTER BARLEY, WINTER WHEAT. Field pansy in WINTER BARLEY, WINTER WHEAT. Polygonums in WINTER BARLEY, WINTER WHEAT. Speedwells in WINTER BARLEY, WINTER WHEAT.

Notes

Efficacy

- Best results achieved by application at early stages of weed growth in warm moist conditions

Crop Safety/Restrictions

- Maximum number of treatments 1 per crop
- Apply from 5 fully expanded leaves (GS 15) until before second node detectable stage (GS 32)
- Do not roll or harrow within 7 d of treatment
- Do not treat crops undersown or to be undersown
- Do not mix with any pesticide formulated as an emulsifiable concentrate nor in a 3-way mixture with mecoprop and isoproturon
- Do not treat crops suffering from any kind of stress or when rain is imminent or falling
- Initial leaf spotting often seen after treatment but is normally outgrown within a few wk without reducing yield

Special precautions/Environmental safety

- Dangerous to fish or other aquatic life. Do not contaminate surface waters or ditches with chemical or used container

Protective clothing/Label precautions

- U08, U20b, E07 (2 wk), E13b, E26, E30a, E31b, E34

Withholding period

- Keep livestock out of treated areas for at least 2 wk and until foliage of poisonous weeds such as ragwort has died and become unpalatable

Latest application/Harvest Interval(HI)

- Before second node detectable stage (GS 32)

---

42 bifenthrin

A contact and residual pyrethroid acaricide/insecticide for use in agricultural and horticultural crops

Products

Talstar	Hortichem	100 g/l	EC	06913

Uses

Aphids in BROCCOLI, BRUSSELS SPROUTS, CABBAGES, CALABRESE, CAULIFLOWERS. Caterpillars in BROCCOLI, BRUSSELS SPROUTS, CABBAGES, CALABRESE, CAULIFLOWERS. Damson-hop aphid in HOPS. Fruit tree red spider mite in APPLES, PEARS. Two-spotted spider mite in BLACKCURRANTS, HOPS, ORNAMENTALS, STRAWBERRIES. Whitefly in BROCCOLI, BRUSSELS SPROUTS, CABBAGES, CALABRESE, CAULIFLOWERS.

Notes

Efficacy

- Timing of application varies with crop and pest. See label for details
- Good spray cover of upper and lower plant surfaces essential to achieve effective pest control

Crop Safety/Restrictions

- Maximum number of treatments 5 per yr for hops; 2 per yr for all other crops

Special precautions/Environmental safety

- Keep in original container, tightly closed, in a safe place under lock and key

- Harmful if swallowed and by inhalation
- Irritating to skin and eyes
- Flammable
- Extremely dangerous to bees. Do not apply to crops in flower or to those in which bees are actively foraging. Do not apply when flowering weeds are present
- Extremely dangerous to fish or other aquatic life. Do not contaminate surface waters or ditches with chemical or used container
- Do not allow direct spray from air-assisted spraying equipment to fall within 18 m, or from ground based vehicle-mounted/drawn sprayers to within 6 m, or from hand-held sprayers to within 2 m, of surface waters or ditches. Direct spray away from water

Protective clothing/Label precautions
- A, C, H
- M03, R03b, R03c, R04a, R04b, R07d, U02, U04a, U05a, U08, U19, U20b, C03, E01, E12b, E13a, E16, E17, E26, E30b, E31b, E34

Latest application/Harvest Interval(HI)
- HI zero for all crops

43 bitertanol

A conazole fungicide available only in mixtures

44 bitertanol + fuberidazole

A broad spectrum fungicide mixture for seed treatment in cereals

Products

Sibutol	Bayer	375:23 g/l	FS	07238

Uses

Bunt in WHEAT. Fusarium root rot in OATS, RYE, TRITICALE, WHEAT. Loose smut in WHEAT *(partial control)*. Septoria seedling blight in WHEAT *(reduction)*.

Notes

Efficacy
- Must be applied simultaneously with water in the ratio 1 part product to 2 parts water in a recommended seed treatment machine
- Treated seed should preferably be drilled in the same season
- Control of loose smut may be inadequate for use on seed for multiplication

Crop Safety/Restrictions
- Maximum number of treatments 1 per batch of seed
- Do not use on seed with more than 16% moisture content, or on sprouted, cracked or skinned seed

Special precautions/Environmental safety
- Dangerous to fish or other aquatic life. Do not contaminate surface waters or ditches with chemical or used container
- Do not use treated seed as food or feed
- Treated seed harmful to game and wildlife

- Product also supplied in returnable containers. See label for guidance on handling, storage, protective clothing and precautions [1]

Protective clothing/Label precautions
- A, H
- U20a, E03, E13b, E30a, E32a, E34, S01, S02, S03, S04b, S05, S06, S07

Latest application/Harvest Interval(HI)
- Before drilling

Maximum Residue Levels (mg residue/kg food)
- (bitertanol) pome fruits, apricots, peaches, nectarines, plums 1; bananas 0.5

45 bitertanol + fuberidazole + imidacloprid

A broad spectrum fungicide and insecticide seed treatment for wheat and oats

Products

Sibutol Secur	Bayer	140:8.6:87.5 g/l	LS	09131

Uses

Bunt in WINTER WHEAT *(seed treatment)*. Fusarium root rot in WINTER OATS *(seed treatment)*, WINTER WHEAT *(seed treatment)*. Loose smut in WINTER WHEAT *(seed treatment - partial control)*. Seedling blight and foot rot in WINTER WHEAT *(seed treatment - reduction)*. Virus vectors in WINTER OATS *(seed treatment)*, WINTER WHEAT *(seed treatment)*.

Notes

Efficacy
- Best applied through recommended seed treatment machines
- Evenness of seed cover improved by simultaneous application of equal volumes of product and water or dilution of product with an equal volume of water
- It is preferable to drill treated seed in the same season although treatment is not known to have any effect on the storage life of treated seed
- In high risk areas where aphid activity is heavy and prolonged a follow-up aphicide treatment may be required
- Protection against foliar air-borne and splash-borne diseases later in the season will require appropriate fungicide follow-up sprays

Crop Safety/Restrictions
- Maximum number of treatments 1 per batch of seed
- Slightly delayed and reduced emergence may occur but this is normally outgrown
- Field emergence which is delayed for any reason may be accentuated by treatment
- Do not use on seed with more than 16% moisture content, or on sprouted, cracked or skinned seed

Special precautions/Environmental safety
- Harmful if swallowed
- Dangerous to fish or other aquatic life. Do not contaminate surface waters or ditches with chemical or used container
- Do not use treated seed as food or feed
- Dangerous to game and wild life. Bury or remove spillages

Protective clothing/Label precautions
- A, H

- M03, R03c, U04a, U05a, U13, U20b, C03, E01, E03, E13b, E26, E30a, E32a, E34, S01, S02, S03, S04b, S05, S06, S07

Latest application/Harvest Interval(HI)
- Before drilling

Maximum Residue Levels (mg residue/kg food)
- see bitertanol + fuberidazole entry

46 bone oil

A ready-to-use animal repellent

Products	Renardine	Roebuck Eyot	33.3% w/w	AL	06769

Uses Badgers in AMENITY AREAS, FARMLAND. Cats in AGRICULTURAL PREMISES, AMENITY AREAS, FARMLAND. Dogs in AGRICULTURAL PREMISES, AMENITY AREAS, FARMLAND. Foxes in AMENITY AREAS, FARMLAND. Moles in AMENITY AREAS, FARMLAND. Rabbits in AMENITY AREAS, FARMLAND.

Notes

Efficacy
- Soak pieces of stick, rags or sand in product and distribute around area to be protected as directed
- Repeat treatment weekly or after heavy rainfall

Crop Safety/Restrictions
- Do not apply directly to animals or crops
- Used as directed product is harmless to animals

47 Bordeaux mixture

A protectant copper sulfate/lime complex fungicide

Products	Wetcol 3	Ford Smith	30 g/l (copper)	SC	02360

Uses Bacterial canker in CHERRIES. Blight in POTATOES, TOMATOES *(outdoor)*. Cane spot in LOGANBERRIES, RASPBERRIES. Canker in APPLES, PEARS. Celery leaf spot in CELERY. Currant leaf spot in BLACKCURRANTS. Downy mildew in HOPS. Leaf curl in APRICOTS, NECTARINES, PEACHES. Scab in APPLES, PEARS. Spur blight in RASPBERRIES.

Notes

Efficacy
- Spray interval normally 7-14 d but varies with disease and crop. See label for details
- Commence spraying potatoes before crop meets in row or immediately first blight period occurs
- For canker control spray monthly from Aug to Oct
- For peach leaf curl control spray at leaf fall in autumn and again in Feb

- Spray when crop foliage dry. Do not spray if rain imminent

Crop Safety/Restrictions
- Do not use on copper sensitive cultivars, including Doyenne du Comice pears

Special precautions/Environmental safety
- Harmful if swallowed. Irritating to eyes, skin and respiratory system
- Harmful to fish or other aquatic life. Do not contaminate surface waters or ditches with chemical or used container
- Harmful to livestock

Protective clothing/Label precautions
- R03c, R04a, R04b, R04c, U13, U15, U20a, E06b, E13c, E30a, E32a

Withholding period
- Keep all livestock out of treated areas for at least 3 wk

Latest application/Harvest Interval(HI)
- HI 7 d

---

48 brodifacoum

An anticoagulant coumarin rodenticide

Products

1 Klerat	Scotts	0.005% w/w	RB	06869
2 Klerat Wax Blocks	Scotts	0.005% w/w	RB	06827

Uses Mice in FARM BUILDINGS, FARMYARDS. Rats in FARM BUILDINGS, FARMYARDS.

Notes

Efficacy
- Effective against rodents resistant to other commonly used anticoagulants
- A single feed representing a fraction of the pest's normal daily food requirement can be lethal
- Use ready-to-use baits in a baiting programme
- Lay small baits about 1 m apart throughout infested areas for mice, larger baits for rats near holes and along runs; place blocks 5-10 m apart depending on severity of infestation
- Cover baits by placing in bait boxes, drain pipes or under boards
- Inspect bait sites frequently and top up as long as there is evidence of feeding

Crop Safety/Restrictions
- Only for use by professional pest contractors
- Cover bait to prevent access by children, animals or birds

Protective clothing/Label precautions
- M05, U13, U20b, C03, E30b, E32a, V01a, V03a, V04a

---

49 bromacil

A soil acting uracil herbicide for non-crop areas and cane fruit

See also amitrole + bromacil + diuron

Products

Hyvar X	DuPont	80% w/w	WP	01105

Uses Annual dicotyledons in BLACKBERRIES, LOGANBERRIES, RASPBERRIES, TAYBERRIES. Annual grasses in BLACKBERRIES, LOGANBERRIES, RASPBERRIES, TAYBERRIES. Couch in BLACKBERRIES, LOGANBERRIES, RASPBERRIES, TAYBERRIES. Perennial dicotyledons in BLACKBERRIES, LOGANBERRIES, RASPBERRIES, TAYBERRIES. Perennial grasses in BLACKBERRIES, LOGANBERRIES, RASPBERRIES, TAYBERRIES. Total vegetation control in NON-CROP AREAS.

Notes

Efficacy

- Apply to established cane fruit as soon as possible in spring after cultivation and before bud break. Avoid further soil disturbance for as long as possible. If inter-row areas are cultivated, only a 30 cm band either side of row need be treated
- Best results achieved when soil is moist at time of application
- For total vegetation control apply to bare ground or standing vegetation. Adequate rainfall is needed to carry chemical into root zone
- Against existing vegetation best results achieved in late winter to early spring but application also satisfactory at other times

Crop Safety/Restrictions

- Maximum number of treatments 1 per yr
- May be used on cane fruit established for at least 2 yr
- In Scotland only may be used on newly planted raspberries at a reduced rate immediately after planting followed by light ridging
- Do not use in last 2 yr before grubbing crop to avoid injury to subsequent crops. Carrots, lettuce, beet, leeks and brassicas are extremely sensitive
- When used non-selectively, take care not to apply where chemical can be washed into root zone of desirable plants
- Treated land should not be cropped within 3 full yr of treatment and then only after obtaining advice from manufacturer

Special precautions/Environmental safety

- Irritating to eyes, skin and respiratory system

Protective clothing/Label precautions

- A, C
- R04a, R04b, R04c, U05a, U09a, U19, U20a, C03, E01, E15, E30a, E32a

Latest application/Harvest Interval(HI)

- Apr for cane fruit

---

50 bromacil + diuron

A root-absorbed residual total herbicide mixture

Products	Borocil K	RP Amenity	0.88:0.88% w/w	GR	05183

Uses Total vegetation control in NON-CROP AREAS.

Notes

Efficacy

- Spray or apply granules in early stage of weed growth at any time of year, provided adequate moisture to activate chemical is supplied by rainfall
- Use higher rates on adsorptive soils or established weed growth
- Do not apply when ground frozen

Crop Safety/Restrictions

- Do not apply on or near trees, shrubs, crops or other desirable plants
- Do not apply where roots of desirable plants may extend or where chemical may be washed into contact with their roots
- Do not use on ground intended for subsequent cultivation

Special precautions/Environmental safety

- Harmful to fish or other aquatic life. Do not contaminate surface waters or ditches with chemical or used container

Protective clothing/Label precautions

- U19, U20b, E13c, E30a, E32a

51 bromacil + picloram

A persistent residual and translocated herbicide mixture

Products

Hydon	Nomix-Chipman	1.08:0.33% w/w	GR	01088

Uses

Total vegetation control in NON-CROP AREAS.

Notes

Efficacy

- Apply at any time from spring to autumn. Best results achieved in Mar-Apr when weeds 50-75 mm high
- Do not apply in very dry weather as moisture needed to carry chemical to roots
- May be used on high fire risk sites

Crop Safety/Restrictions

- Do not apply near crops, cultivated plants and trees
- Do not apply on slopes where run-off to cultivated plants or water courses may occur

Special precautions/Environmental safety

- Harmful to fish or other aquatic life. Do not contaminate surface waters or ditches with chemical or used container

Protective clothing/Label precautions

- U19, U20b, E07, E13c, E30a, E32a

Withholding period

- Keep livestock out of treated area until foliage of any poisonous weeds such as ragwort has died and become unpalatable

52 bromadiolone

An anti-coagulant coumarin-derivative rodenticide

Products

1 Endorats Blue Rat Bait	Irish Drugs	0.005% w/w	GB	08046
2 Slaymor	Ciba Agric.	0.005% w/w	RB	01958
3 Slaymor	Novartis A H	0.005% w/w	RB	08592
4 Slaymor Bait Bags	Ciba Agric.	0.005% w/w	RB	03183
5 Slaymor Bait Bags	Novartis A H	0.005% w/w	RB	08593
6 Tomcat 2	Antec	0.005% w/w	PT	08257
7 Tomcat 2 Blox	Antec	0.005% w/w	BB	08261

Uses

Mice in FARMYARDS [2-5]. Mice in FARM BUILDINGS [1-7]. Rats in FARMYARDS [2-5]. Rats in FARM BUILDINGS [1-7].

Notes

Efficacy
- Ready-to-use baits are formulated on a mould-resistant, whole-wheat base
- Use in baiting programme. Place baits in protected situations, sufficient for continuous feeding between treatments
- Chemical is effective against warfarin- and coumatetralyl-resistant rats and mice and does not induce bait shyness
- Use bait bags where loose baiting inconvenient (eg behind ricks, silage clamps etc)

Special precautions/Environmental safety
- Access to baits by children, birds and animals, particularly cats, dogs, pigs and poultry, must be prevented
- Baits must not be placed where food, feed or water could become contaminated
- Remains of bait and bait containers must be removed after treatment and burned or buried
- Rodent bodies must be searched for and burned or buried. They must not be placed in refuse bins or on rubbish tips

Protective clothing/Label precautions
- M03, U20b, E15, E32a [6, 7]; M04, U20a, E31a [2-5]; U13, E30a, V01a, V02, V04a [1-7]; S06 [1]; V03a [1, 6, 7]

53 bromoxynil

A contact acting HBN herbicide

See also benazolin + bromoxynil + ioxynil

Products

1 Alpha Bromolin 225 EC	Makhteshim	225 g/l	EC	07450
2 Alpha Bromotril P	Makhteshim	250 g/l	LI	07099
3 Barclay Mutiny	Barclay	250 g/l	SC	08933
4 Flagon 400 EC	Makhteshim	400 g/l	LI	08875

Uses

Annual dicotyledons in MAIZE, SWEETCORN *(off-label)* [2]. Annual dicotyledons in LINSEED [1, 3]. Annual dicotyledons in BARLEY, OATS, WHEAT [4]. Annual dicotyledons in FORAGE MAIZE [3].

Notes

Efficacy

- Apply to spring sown maize from 2 fully expanded leaves up to 9 fully expanded leaves [2, 3]
- Apply to spring sown linseed from 1 fully expanded leaf to 20 cm tall [1]
- Spray when main weed flush has germinated and the largest are at the 4 leaf stage
- Weed control can be enhanced by using a split treatment spraying each application when the weeds are seedling to 2 true leaves. Apply the second treatment before the crop canopy covers the ground [2, 3]
- Tank mixture with atrazine recommended for enhanced weed control - see label [2, 3]

Crop Safety/Restrictions

- Maximum number of treatments 1 per yr
- Maximum total dose equivalent to one full dose treatment [2, 3]
- Foliar scorch, which rapidly disappears without affecting growth, will occur if treatment made in hot weather or during rapid growth
- Do not apply with oils or other adjuvants
- Do not apply during frosty weather, drought, when soil is waterlogged, when rain expected within 4 h or to crops under any stress
- Take particular care to avoid drift onto neighbouring susceptible crops or open water surfaces

Special precautions/Environmental safety

- Harmful if swallowed
- Irritating to eyes
- May cause sensitization by skin contact [4]
- Flammable [4]
- Dangerous to fish or other aquatic life. Do not contaminate surface waters or ditches with chemical or used container
- Harmful to bees. Do not apply to crops in flower or to those in which bees are actively foraging. Do not apply when flowering weeds are present
- Do not apply using hand-held equipment or at concentrations higher than those recommended

Protective clothing/Label precautions

- A, C [1-4]; H [1, 4]
- M03, E01 [2-4]; R03c, R04a, U05a, U08, U13, U19, U20a, C03, E12e, E13b, E30a, E31b, E34 [1-4]; R04e, R07d, C02 (6 wk) [4]; U02 [1-3]; U14, U15, E07 (6 wk) [1, 4]; E26 [1, 3, 4]

Withholding period

- Keep livestock out of treated areas for at least 6 wk

Latest application/Harvest Interval(HI)

- Before 10 fully expanded leaf stage of crop for maize [2, 3]; before crop 20 cm tall and before flower buds visible for linseed; before 2nd node detectable (GS 32) for cereals [4]

Approval

- Off-label unlimited for use on outdoor sweetcorn (OLA 2230/97)[2]
- Approval expiry: 31 Dec 99 [1]; replaced by MAFF 08255

54 bromoxynil + clopyralid

A post-emergence contact and translocated herbicide mixture

Products

Vindex	Dow	240:50 g/l	EC	05470

Uses

Annual dicotyledons in LINSEED.

Notes

Efficacy

- Best results achieved when weeds small and growing actively in warm, moist weather
- Vigorous crop competition enhances control of more resistant weeds and prevents those that germinate after treatment becoming a problem
- Do not spray during drought, waterlogging, frost or if rain is imminent
- Weed spectrum can be widened by mixture with bentazone. See label for details

Crop Safety/Restrictions

- Maximum number of treatments 1 per crop
- Avoid spraying during drought, waterlogging, frost, extremes of temperature or if rain is imminent or falling
- Spraying in frosty weather or when hard frost occurs within 3-4 wk may result in leaf scorch
- Do not roll or harrow within 7 d before or after spraying
- Do not apply any other product within 7 d of application (10 d if crop is stressed)
- Wash equipment thoroughly with water and an authorised non-ionic wetting agent according to manufacturer's instructions immediately after use. Traces of product could harm susceptible crops sprayed later

Special precautions/Environmental safety

- Harmful in contact with skin or if swallowed. Irritating to skin and eyes
- Flammable
- Do not apply by hand-held equipment or at concentrations higher than those recommended
- Dangerous to fish or other aquatic life. Do not contaminate surface waters or ditches with chemical or used container
- Harmful to bees. Do not apply to crops in flower or to those in which bees are actively foraging. Do not apply when flowering weeds are present
- Do not harvest crops for animal or human consumption for at least 6 wk after last application

Protective clothing/Label precautions

- A, C
- M03, R03a, R03c, R04a, R04b, R07d, U05a, U08, U13, U19, U20b, C02 (6 wk), C03, E01, E07 (6 wk), E12e, E13b, E26, E30a, E31b, E34

Withholding period

- Keep livestock out of treated areas for at least 6 wk after treatment

Latest application/Harvest Interval(HI)

- Before first flower and before crop exceeds 30 cm tall

55 bromoxynil + clopyralid + fluroxypyr + ioxynil

A contact and translocated herbicide mixture for use in cereals

Products

Crusader S	Dow	75:30:90:87.5 g/l	EC	05174

Uses

Annual dicotyledons in WINTER BARLEY, WINTER WHEAT. Chickweed in WINTER BARLEY, WINTER WHEAT. Cleavers in WINTER BARLEY, WINTER WHEAT. Hemp-nettle in WINTER BARLEY, WINTER WHEAT. Mayweeds in WINTER BARLEY, WINTER WHEAT. Speedwells in WINTER BARLEY, WINTER WHEAT.

Notes

Efficacy

- Best results achieved by application in good growing conditions, when weeds small and growing actively in a strongly competitive crop
- Do not spray if night temperatures are low

Crop Safety/Restrictions

- Maximum number of treatments 1 per crop
- Apply to winter cereals from 3-leaf stage to first node detectable (GS 13-31)
- Do not roll or harrow 10 d before or 7 d after treatment
- Do not spray during prolonged frosty weather or when crop is under stress from drought, waterlogging, nutrient deficiency or pest attack
- Straw from treated crops may contain residues which could damage susceptible crops. See label for detailed guidance on straw disposal or use

Special precautions/Environmental safety

- Harmful if swallowed. Irritating to eyes and skin
- Do not apply by hand-held equipment or at concentrations higher than recommended
- Flammable
- Dangerous to fish or other aquatic life. Do not contaminate surface waters or ditches with chemical or used container
- Harmful to bees. Do not apply to crops in flower or to those in which bees are actively foraging. Do not apply when flowering weeds are present
- Do not harvest crops for animal consumption for at least 6 wk after last application
- Wash spray equipment thoroughly with water and detergent immediately after use. Traces of product can damage susceptible plants sprayed later

Protective clothing/Label precautions

- A, C
- M03, R03c, R04a, R04b, R07d, U05a, U08, U13, U19, U20a, C02 (6 wk), C03, E01, E07 (6 wk), E12e, E13b, E26, E30a, E31b, E34

Withholding period

- Keep livestock out of treated areas for at least 6 wk

Latest application/Harvest Interval(HI)

- Before second node detectable stage (GS 32)

56 bromoxynil + diflufenican + ioxynil

A selective contact and translocated herbicide for cereals

Products Capture RP Agric. 300:50:200 g/l SC 06881

Uses Annual dicotyledons in SPRING BARLEY, SPRING WHEAT, TRITICALE, WINTER BARLEY, WINTER RYE, WINTER WHEAT. Chickweed in SPRING BARLEY, SPRING WHEAT, TRITICALE, WINTER BARLEY, WINTER RYE, WINTER WHEAT. Knotgrass in SPRING BARLEY, SPRING WHEAT, TRITICALE, WINTER BARLEY, WINTER RYE, WINTER WHEAT. Mayweeds in SPRING BARLEY, SPRING WHEAT, TRITICALE, WINTER BARLEY, WINTER RYE, WINTER WHEAT. Speedwells in SPRING BARLEY, SPRING WHEAT, TRITICALE, WINTER BARLEY, WINTER RYE, WINTER WHEAT.

Notes

Efficacy

- Best results obtained on small weeds and when competion removed early
- Good spray coverage essential for good activity

Crop Safety/Restrictions

- Maximum number of treatments 1 per crop
- Do not treat crops undersown or to be undersown
- Do not treat frosted crops or those that are under stress from any cause
- In the event of crop failure winter wheat may be drilled immediately after normal cultivations. Winter barley may be re-drilled after ploughing
- Land must be ploughed and an interval of 12 wk elapse after treatment before planting spring crops of wheat, barley, oilseed rape, peas, field beans, sugar beet, potatoes, carrots, edible brassicas or onions
- Successive treatments of any products containing diflufenican can lead to soil build-up and inversion ploughing must precede sowing any following non-cereal crop. Even where ploughing occurs some crops may be damaged

Special precautions/Environmental safety

- Harmful if swallowed
- Harmful to bees. Do not apply to crops in flower or to those in which bees are actively foraging. Do not apply when flowering weeds are present
- Dangerous to fish or other aquatic life. Do not contaminate surface waters or ditches with chemical or used container
- Do not allow direct spray from vehicle mounted/drawn hydraulic sprayers to fall within 6 m of surface waters or ditches. Direct spray away from water
- Do not apply by hand-held equipment or at concentrations higher than those recommended

Protective clothing/Label precautions

- A, C, H
- M03, R03c, U05a, U08, U13, U19, U20b, C02 (2 wk), C03, E01, E07 (2 wk), E12e, E13b, E16, E26, E30a, E31b, E34

Latest application/Harvest Interval(HI)

- Before 2nd node detectable (GS 32)

Maximum Residue Levels (mg residue/kg food)
- see ioxynil entry

57 bromoxynil + ethofumesate + ioxynil

A post-emergence herbicide for new grass leys

Products

1 Leyclene	AgrEvo	50:200:25 g/l	EC	07263
2 Stefes Leyclene	Stefes	50:200:25 g/l	EC	08173

Uses

Annual dicotyledons in GRASS SEED CROPS, SEEDLING LEYS. Annual meadow grass in GRASS SEED CROPS, SEEDLING LEYS.

Notes

Efficacy
- Best results when weeds small and growing actively in a vigorous crop, soil moist and further rain within 10 d. Mid Oct to end Dec normally suitable
- Annual meadow grass controlled during early crop establishment. Spray weed grasses before fully tillered
- Do not spray in cold conditions or when heavy rain or frost imminent
- Do not use on soils with more than 10% organic matter
- Do not cut for 14 d after spraying or graze in Jan-Feb after spraying in Oct-Dec
- Ash or trash should be burned, buried or removed before spraying

Crop Safety/Restrictions
- Maximum number of treatments 2 per yr
- Apply to healthy ryegrasses or tall fescue after 2-3 leaf stage, to cocksfoot, timothy and meadow fescue at least 60 d after emergence and after 2-3 leaf stage
- Do not use on crops under stress, during periods of very dry weather, prolonged frost or waterlogging
- Do not use where clovers or other legumes are valued components of ley
- Do not roll for 7 d before or after spraying
- Any crop may be sown 5 mth after application following ploughing to at least 15 cm

Special precautions/Environmental safety
- Harmful if swallowed. Irritating to skin and eyes. Flammable
- Do not apply by hand-held equipment or at concentrations higher than those recommended
- Dangerous to fish or other aquatic life. Do not contaminate surface waters or ditches with chemical or used container
- Harmful to bees. Do not apply to crops in flower or to those in which bees are actively foraging. Do not apply when flowering weeds are present

Protective clothing/Label precautions
- A, C
- M03, R03c, R04a, R04b, R07d, U05a, U09a, U13, U19, U20a, C03, E01, E12e, E13b, E30a, E31b, E34

Withholding period
- Keep livestock out of treated areas for at least 6 wk after treatment

Latest application/Harvest Interval(HI)
- 6 wk before cutting or grazing

Maximum Residue Levels (mg residue/kg food)
- see ioxynil entry

58 bromoxynil + fluroxypyr

A post-emergence contact and translocated herbicide for cereals

Products

Sickle	Dow	300:150 g/l	EC	05187

Uses

Annual dicotyledons in SPRING BARLEY, SPRING WHEAT, WINTER BARLEY, WINTER WHEAT.

Notes

Efficacy
- Best results when weeds small and growing actively in a strongly competitive crop
- Spray is rainfast after 2 h

Crop Safety/Restrictions
- Maximum number of treatments 1 per crop
- Apply from 2-fully expanded leaf stage of crop to first node detectable (GS 12-31)
- Crops undersown with grass may be sprayed provided the grasses are tillering. Do not spray cereals undersown with clover or other legume mixtures
- Do not spray when crops are under stress from cold, drought, waterlogging etc, nor when frost imminent
- Do not roll or harrow for 10 d before or 7 d after spraying

Special precautions/Environmental safety
- Harmful if swallowed. Irritating to eyes and skin. Flammable
- Do not apply by hand-held equipment or at concentrations higher than those recommended
- Dangerous to fish or other aquatic life. Do not contaminate surface waters or ditches with chemical or used container
- Harmful to bees. Do not apply to crops in flower or to those in which bees are actively foraging. Do not apply when flowering weeds are present
- Do not allow direct spray from vehicle mounted/drawn hydraulic sprayers to fall within 6 m of surface waters or ditches. Direct spray away from water
- Wash spray equipment thoroughly with water and detergent immediately after use. Traces of product can damage susceptible plants sprayed later

Protective clothing/Label precautions
- A, C
- M03, R03c, R04a, R04b, R07d, U05a, U08, U13, U19, U20b, C02 (6 wk), C03, E01, E07 (6 wk), E12e, E13b, E16, E26, E30a, E31b, E34

Withholding period
- Keep livestock out of treated areas for at least 6 wk

Latest application/Harvest Interval(HI)
- Before second node detectable (GS 32)

FOR FULL CONDITIONS OF USE ALWAYS READ THE PRODUCT LABEL

59 bromoxynil + fluroxypyr + ioxynil

A post-emergence contact and translocated herbicide for cereals

Products

Advance	Dow	100:90:100 g/l	EC	05173

Uses

Annual dicotyledons in SPRING BARLEY, SPRING WHEAT, WINTER BARLEY, WINTER WHEAT. Chickweed in SPRING BARLEY, SPRING WHEAT, WINTER BARLEY, WINTER WHEAT. Cleavers in SPRING BARLEY, SPRING WHEAT, WINTER BARLEY, WINTER WHEAT. Hemp-nettle in SPRING BARLEY, SPRING WHEAT, WINTER BARLEY, WINTER WHEAT. Speedwells in SPRING BARLEY, SPRING WHEAT, WINTER BARLEY, WINTER WHEAT.

Notes

Efficacy

- May be used in autumn or spring on winter or spring sown crops
- Best results from good spray coverage when weeds small and growing actively in warm humid weather
- Do not spray when weed growth is hard from cold or drought

Crop Safety/Restrictions

- Maximum number of treatments 1 per crop
- Apply from 2-leaf stage of crop to first node detectable (GS 12-31)
- Do not spray undersown crops or crops to be undersown with clover or legume mixtures
- Do not spray crops stressed by frost, drought, mineral deficiency pest or disease attack
- Do not roll or harrow for 7 d before or after spraying
- Avoid drift onto non-target crops as damage may occur

Special precautions/Environmental safety

- Harmful if swallowed. Irritating to eyes
- Flammable
- Harmful to bees. Do not apply to crops in flower or to those in which bees are actively foraging. Do not apply when flowering weeds are present
- Dangerous to fish or other aquatic life. Do not contaminate surface waters or ditches with chemical or used container
- Do not apply by hand-held equipment or at concentrations higher than those recommended
- Do not harvest crops for animal or human consumption for at least 6 wk after last application

Protective clothing/Label precautions

- A, C
- M03, R03c, R04a, R07d, U05a, U08, U13, U19, U20b, C02 (6 wk), C03, E01, E07 (6 wk), E12e, E13b, E26, E30a, E31b, E34

Withholding period

- Keep livestock out of treated areas for at least 6 wk following treatment

Latest application/Harvest Interval(HI)

- Before second node detectable (GS 32)

Maximum Residue Levels (mg residue/kg food)

- see ioxynil entry

60 bromoxynil + ioxynil

A contact acting post-emergence HBN herbicide for cereals

Products

1 Alpha Briotril 19/19	Makhteshim	190:190 g/l	EC	04740
2 Alpha Briotril 24/16	Makhteshim	240:160 g/l	EC	04876
3 Deloxil	AgrEvo	190:190 g/l	EC	07313
4 Oxytril CM	RP Agric.	200:200 g/l	EC	06201

Uses

Annual dicotyledons in BARLEY, WHEAT [1, 4]. Annual dicotyledons in OATS [1, 3, 4]. Annual dicotyledons in SPRING BARLEY, SPRING WHEAT, WINTER BARLEY, WINTER WHEAT [2, 3]. Annual dicotyledons in SPRING OATS, WINTER OATS [2]. Annual dicotyledons in UNDERSOWN CEREALS [4]. Annual dicotyledons in RYE, TRITICALE [3, 4]. Annual dicotyledons in DURUM WHEAT [3].

Notes

Efficacy

- Best results achieved on young weeds growing actively in a highly competitive crop
- Do not apply during periods of drought or when rain imminent (some labels say 'if likely within 4 or 6 h')
- Recommended for tank mixture with hormone herbicides to extend weed spectrum. See labels for details

Crop Safety/Restrictions

- Maximum number of treatments 1 per crop
- Apply to winter or spring cereals from 2 fully expanded leaf stage [1, 3], 1-fully expanded leaf [4], but before second node detectable (GS 32)
- Spray oats in spring when danger of frost past. Do not spray winter oats in autumn [1, 3, 4]
- Apply to undersown cereals pre-sowing or pre-emergence of legume provided cover crop is at correct stage. Only spray trefoil pre-sowing [3, 4]
- On crops undersown with grasses alone apply from 2-leaf stage of grass [3, 4]
- Do not spray crops stressed by drought, waterlogging or other factors
- Do not roll or harrow for several days before or after spraying. Number of days specified varies with product. See label for details

Special precautions/Environmental safety

- Harmful if swallowed
- Irritating to skin
- Irritating to eyes [4]
- Do not apply by hand-held equipment or at concentrations higher than those recommended
- Dangerous to fish or other aquatic life. Do not contaminate surface waters or ditches with chemical or used container
- Direct spray from vehicle mounted sprayers must not be allowed to fall within 6 m of surface waters or ditches. Direct spray away from water [1, 3]
- Harmful to bees. Do not apply to crops in flower or to those in which bees are actively foraging. Do not apply when flowering weeds are present

Protective clothing/Label precautions

- A, C

- M03, U20a, E07 (6 wk), E31b [1-3]; R03c, R04a, U05a, U08, U13, U19, C03, E01, E12e, E13b, E30a, E34 [1-4]; R04b, U20b, E32a [4]; R07d, U02, U14, U15 [1, 2]; U04a, E26 [3]; C02 (6 wk), E16 [1, 3]

Withholding period

- Keep livestock out of treated areas for at least 6 wk [2, 3], 14 d [4]

Latest application/Harvest Interval(HI)

- Before 2nd node detectable stage (GS 31).
- HI (animal consumption) 6 wk [2, 3]; 14 d [4]

Maximum Residue Levels (mg residue/kg food)

- see ioxynil entry

61 bromoxynil + ioxynil + mecoprop-P

A post-emergence contact and translocated herbicide

Products

1 Swipe P	Ciba Agric.	56:56:224g/l	EC	08150
2 Swipe P	Novartis	56:56:224 g/l	EC	08452

Uses

Annual dicotyledons in DURUM WHEAT, RYEGRASS, SPRING BARLEY, SPRING OATS, SPRING WHEAT, TRITICALE, WINTER BARLEY, WINTER OATS, WINTER WHEAT.

Notes

Efficacy

- Best results when weeds small and growing actively in a strongly competitive crop
- Do not spray in rain or when rain imminent. Control may be reduced by rain within 6 h
- Application to wet crops or weeds may reduce control
- To achieve optimum control of large over-wintered weeds or in advanced crops increase water volume to aid spray penetration and cover
- Recommended for tank-mixing with approved MCPA-amine for hemp nettle control

Crop Safety/Restrictions

- Maximum number of treatments 1 per crop. The total amount of mecoprop-P applied in a single yr must not exceed the maximum total dose approved for any single product for the crop/situation
- Spray cereals from 3 leaves unfolded (GS 13) to before second node detectable (GS 32) for winter sown cereals, spring wheat and spring barley and before first node detectable (GS 31) for spring oats
- Do not treat durum wheat or winter oats in autumn
- Apply to direct sown ryegrass or cereals undersown with ryegrass from 2-3 leaf stage of grass
- Do not spray crops undersown with legumes or use on winter oats or durum wheat in autumn
- Do not spray crops under stress from frost, waterlogging, drought or other causes
- Do not roll within 5 d after spraying
- Yield of barley may be reduced if frost occurs within 3-4 wk of treatment of low vigour crops on light soils or subject to stress
- Some crop yellowing may follow treatment but yield not normally affected
- Do not mix with manganese sulfate
- Avoid drift onto neighbouring susceptible crops

Special precautions/Environmental safety

- Harmful if swallowed
- Irritating to skin. May cause sensitization by skin contact
- Risk of serious damage to eyes
- Do not apply by hand-held equipment or at concentrations higher than those recommended
- Dangerous to fish or other aquatic life. Do not contaminate surface waters or ditches with chemical or used container
- Do not allow direct spray from ground-based vehicle mounted/drawn sprayers to fall within 6 m of surface waters or ditches. Direct spray away from water
- Harmful to bees. Do not apply to crops in flower or to those in which bees are actively foraging. Do not apply when flowering weeds are present
- Do not harvest crops for animal consumption for at least 6 wk after last application

Protective clothing/Label precautions

- A, C, H, M
- M03, R03c, R04b, R04d, R04e, U05a, U08, U13, U14, U15, U19, U20a, C02 (6 wk), C03, E01, E07 (6 wk), E12e, E13b, E16, E26, E30a, E31b, E34

Withholding period

- Keep livestock out of treated areas for at least 6 wk and until foliage of poisonous weeds such as ragwort has died and become unpalatable

Latest application/Harvest Interval(HI)

- Before first node detectable (GS31) for spring oats; before second node detectable (GS32) for durum wheat, spring barley, spring wheat, triticale, winter barley, winter oats, winter wheat
- HI (animal consumption) 6 wk

Maximum Residue Levels (mg residue/kg food)

- see ioxynil entry

---

62 bromoxynil + ioxynil + triasulfuron

An HBN and sulfonylurea herbicide mixture for cereals

Products

1 Teal	Ciba Agric.	190:190 g/l:20% w/w	KK	06117
2 Teal	Novartis	190:190 g/l:20% w/w	KK	08453

Uses Annual dicotyledons in BARLEY, OATS, RYE, TRITICALE, WHEAT.

Notes

Efficacy

- For best results apply during warm, moist weather when weeds growing actively
- Do not spray when rain imminent. Rainfall within 6 h after spraying may reduce weed control

Crop Safety/Restrictions

- Maximum number of treatments 1 per crop

- Apply in spring, after 1 Feb, from 3 leaves unfolded but before second node detectable (GS 13-32)
- Do not spray undersown crops or those due to be undersown
- Do not spray winter barley of low vigour, on light soil and subject to stress during frosty weather
- Do not use on oats until risk of frost is over
- Do not use during frosty weather, when frost imminent, or on crops under stress from frost, waterlogging or drought
- Do not spray in tank mixture, or in sequence, with a product containing any other sulfonylurea
- See label for details of restrictions on sequential use of other herbicides and sowing of subsequent crops

Special precautions/Environmental safety
- Harmful if swallowed.
- Irritating to eyes
- Do not apply by hand-held equipment or at concentrations higher than those recommended
- Dangerous to fish or other aquatic life and extremely dangerous to aquatic higher plants. Do not contaminate surface waters or ditches with chemical or used container
- Direct spray from vehicle-mounted sprayers must not be allowed to fall within 6 m of surface waters or ditches. Direct spray away from water
- Harmful to bees. Do not apply to crops in flower or to those in which bees are actively foraging. Do not apply when flowering weeds are present

Protective clothing/Label precautions
- A, C, H
- M03, R03c, R04a, U05a, U08, U13, U14, U15, U19, U20a, C03, E01, E12e, E13b, E16, E30a, E32a, E34

Withholding period
- Keep livestock out of treated areas for at least 6 wk

Latest application/Harvest Interval(HI)
- Before second node detectable (GS 32).
- HI (animal consumption) 6 wk

Maximum Residue Levels (mg residue/kg food)
- see ioxynil entry

---

63 bromoxynil + prosulfuron

A contact and residual herbicide mixture for maize and sweetcorn

Products

Jester	Novartis	60:3 % w/w	WG	08681

Uses

Annual dicotyledons in MAIZE, SWEETCORN. Black bindweed in MAIZE, SWEETCORN. Chickweed in MAIZE, SWEETCORN. Hemp-nettle in MAIZE, SWEETCORN. Knotgrass in MAIZE, SWEETCORN. Mayweeds in MAIZE, SWEETCORN.

Notes

Efficacy
- Apply post-emergence up to when the crop has four unfolded leaves
- To minimise the possible development of resistant weeds, use mixtures or sequences with other herbicides with a different mode of action

- Product must be used with Agral

Crop Safety/Restrictions

- Maximum number of treatments 1 per crop
- Apply in cool conditions, or during the evening, to avoid scorch
- Do not treat crops grown for seed production
- Consult before use on crops intended for processing
- Do not use in frosty weather or on crops under stress
- Do not tank mix with organophosphate insecticides or apply in tank mix or sequence with any other sulfonylurea

Special precautions/Environmental safety

- Harmful if swallowed
- Irritating to eyes
- Do not allow direct spray from vehicle mounted/drawn hydraulic sprayers to fall within 6 m of surface waters or ditches. Direct spray away from water
- Do not apply by knapsack sprayer or in volumes less than those recommended
- Take special care to avoid drift outside the target area

Protective clothing/Label precautions

- A, C
- R03c, R04a, U05a, U11, U15, U20c, C03, E01, E16, E29, E30a, E31a

Latest application/Harvest Interval(HI)

- Before 5 crop leaves unfolded for maize, sweetcorn

64 bromuconazole

A systemic conazole fungicide for cereals

Products Granit | RP Agric. | 200 g/l | SC | 08268

Uses Brown rust in SPRING BARLEY, SPRING WHEAT, WINTER BARLEY, WINTER WHEAT. Eyespot in SPRING BARLEY *(reduction)*, SPRING WHEAT *(reduction)*, WINTER BARLEY *(reduction)*, WINTER WHEAT *(reduction)*. Fusarium ear blight in SPRING WHEAT, WINTER WHEAT. Glume blotch in SPRING WHEAT, WINTER WHEAT. Net blotch in SPRING BARLEY, WINTER BARLEY. Powdery mildew in SPRING BARLEY, SPRING WHEAT, WINTER BARLEY, WINTER WHEAT. Rhynchosporium in SPRING BARLEY, WINTER BARLEY. Septoria leaf spot in SPRING WHEAT, WINTER WHEAT. Sooty moulds in SPRING WHEAT, WINTER WHEAT. Yellow rust in SPRING BARLEY, SPRING WHEAT, WINTER BARLEY, WINTER WHEAT.

Notes

Efficacy

- Apply at any time from the tillering stage up to the latest times indicated for each crop
- Best results achieved from applications before disease becomes established
- If disease pressure persists a second treatment may be required to prevent late season attacks
- Best control of ear diseases in wheat obtained from application at or after the ear completely emerged stage

- More consistent control of powdery mildew obtained from tank mixture with fenpropimorph. See label

Crop Safety/Restrictions
- Maximum number of treatments 2 per crop
- Only cereals, oilseed rape, field beans, peas, potatoes, linseed or Italian ryegrass may be sown as following crops

Special precautions/Environmental safety
- Dangerous to fish or other aquatic life. Do not contaminate surface waters or ditches with chemical or used container
- Do not allow direct spray from vehicle mounted/drawn hydraulic sprayers to fall within 6 m, or from hand-held sprayers to within 2 m, of surface waters or ditches. Direct spray away from water

Protective clothing/Label precautions
- A, C, H
- U08, U19, U20b, E13b, E16, E26, E30a, E31b

Latest application/Harvest Interval(HI)
- Varies with dose used and crop - see label

---

65 bupirimate

A systemic pyrimidine fungicide active against powdery mildew

Products

Nimrod	Zeneca	250 g/l	EC	06686

Uses

Powdery mildew in APPLES, BEGONIAS, BLACKCURRANTS, CHRYSANTHEMUMS, COURGETTES, CUCUMBERS, GOOSEBERRIES, HOPS, MARROWS, PEARS, PUMPKINS *(off-label)*, RASPBERRIES, ROSES, SQUASHES *(off-label)*, STRAWBERRIES, TOMATOES *(off-label)*.

Notes

Efficacy
- Apply before or at first signs of disease and repeat at 7-14 d intervals. Timing and maximum dose vary with crop. See label for details
- On apples during periods that favour disease development lower doses applied weekly give better results than higher rates fortnightly
- Product has negligible effect on *Phytoseiulus* and *Encarsia* and may be used in conjunction with biological control of red spider mite

Crop Safety/Restrictions
- Maximum number of treatments or maximum total dose depends on crop and variety (see label for details)
- With apples, hops and ornamentals cultivars may vary in sensitivity to spray. See label for details
- If necessary to spray cucurbits in winter or early spring spray a few plants 10-14 d before spraying whole crop to test for likelihood of leaf spotting problem
- On roses some leaf puckering may occur on young soft growth in early spring or under low light intensity. Avoid use of high rates or wetter on such growth
- Never spray flowering begonias (or buds showing colour) as this can scorch petals
- Do not mix with other chemicals for application to begonias, cucumbers or gerberas

Special precautions/Environmental safety
- Irritating to eyes and skin.
- Flammable
- Harmful to fish or other aquatic life. Do not contaminate surface waters or ditches with chemical or used container

Protective clothing/Label precautions
- A, C
- R04a, R04b, R07d, U05a, U20a, C03, E01, E13c, E30a, E31c, E34

Latest application/Harvest Interval(HI)
- HI depends on crop and variety (see label for details)

Approval
- Off-label approval unlimited for use on protected tomatoes (OLA 0516/96)[1]; to Jan 2001 on outdoor crops of squashes and pumpkins (OLA 0083/96)[1]

66 bupirimate + triforine

A systemic protectant and eradicant fungicide for ornamental crops

Products

1 Nimrod-T	Miracle	62.5:62.5 g/l	EC	07865
2 Nimrod-T	Scotts	62.5:62.5 g/l	EC	09268

Uses Black spot in ORNAMENTALS. Powdery mildew in ORNAMENTALS. Rust in ROSES.

Notes

Efficacy
- To prevent infection on roses spray in early May and repeat every 10-14 d. If infected with blackspot in previous season commence spraying at bud burst

Crop Safety/Restrictions
- Spraying in high glasshouse temperatures may cause temporary leaf damage
- Test varietal susceptibility of roses and other ornamentals by spraying a few plants and allow 14 d for any symptoms to develop

Special precautions/Environmental safety
- Irritating to eyes
- Harmful to fish or other aquatic life. Do not contaminate surface waters or ditches with chemical or used container

Protective clothing/Label precautions
- A, C
- R04a, U05a, U09a, U20b, C03, E01, E13c, E26, E30a, E31b, E34

Maximum Residue Levels (mg residue/kg food)
- see triforine entry

67 buprofezin

A moulting inhibitor, thiadiazine insecticide for whitefly control

Products	Applaud	Zeneca	250 g/l	SC	06900

Uses Glasshouse whitefly in AUBERGINES, CUCUMBERS, PEPPERS, PROTECTED ORNAMENTALS, TOMATOES. Tobacco whitefly in AUBERGINES, CUCUMBERS, PEPPERS, PROTECTED ORNAMENTALS, TOMATOES.

Notes

Efficacy

- Product has contact, residual and some vapour activity
- Whitefly most susceptible at larval stages but residual effect can also kill nymphs emerging from treated eggs and application to pupae reduces emergence
- Adult whitefly not directly affected. Resistant strains of tobacco whitefly are known and where present control likely to be reduced or ineffective
- Product may be used either in IPM programme in association with *Encarsia formosa* or in All Chemical programme
- In IPM programme apply as single application and allow at least 60 d before re-applying
- In All Chemical programme apply twice at 7-14 d interval and allow at least 60 d before re-applying
- Do not leave spray liquid in sprayer for long periods
- Do not apply as fog or mist

Crop Safety/Restrictions

- Maximum number of treatments up to 8 per crop for tomatoes and cucumbers; up to 2 per crop for aubergines and peppers
- Do not apply more than 2 sprays within a 65 d period on tomatoes, or within a 45 d period on cucumbers
- See label for list of ornamentals successfully treated but small scale test advised to check varietal tolerance. This is especially important if spraying flowering ornamentals with buds showing colour
- Do not treat Dieffenbachia or Closmoplictrum
- Do not apply to crops under stress

Protective clothing/Label precautions

- U20c, E15, E30a, E32a, E34

Latest application/Harvest Interval(HI)

- HI edible crops 3 d

68 calciferol

A hypercalcaemic rodenticide available only in mixtures

69 calciferol + difenacoum

A mixture of rodenticides with different modes of action

Products	Sorexa CD	Sorex	0.1:0.0025% w/w	RB	03514

Uses Mice in FARM BUILDINGS.

Notes

Efficacy
- Lay baits in mouse runs in many locations throughout infested area
- It is important to lay many small baits as mice are sporadic feeders
- Cover baits to protect from moisture in outdoor situations
- Inspect baits frequently and replace or top up with fresh material as necessary

Special precautions/Environmental safety
- Protect baits from access by children, domestic or other animals

Protective clothing/Label precautions
- U13, U20a, E30a, E32a, V01a, V02, V03a, V04a

Approval
- Approval expiry: 30 Jan 99 [1]

70 captafol

A protectant dicarboximide fungicide approvals for which expired on 31 December 1990

71 captan

A protectant dicarboximide fungicide with horticultural uses

Products

1 Alpha Captan 80 WDG	Makhteshim	80% w/w	WG	07096
2 Alpha Captan 83 WP	Makhteshim	83% w/w	WP	04806
3 PP Captan 80 WG	Zeneca	80% w/w	SG	06696

Uses Black spot in ROSES [1-3]. Botrytis fruit rot in APPLES, PEARS [3]. Brown rot in APPLES, PEARS [3]. Fungus diseases in FLOWER BULBS *(off-label)* [3]. Gloeosporium rot in PEARS [3]. Gloeosporium rot in APPLES [1-3]. Grey mould in STRAWBERRIES [1, 2]. Phytophthora fruit rot in APPLES, PEARS [3]. Scab in APPLES, PEARS [1-3]. Storage rots in APPLES, PEARS [3].

Notes

Efficacy
- For control of scab apply at bud burst and repeat at 10-14 d intervals until danger of scab infection ceased
- For suppression of fruit storage rots apply from late Jul and repeat at 2-3 wk intervals until 7 d before picking
- To reduce spread of fruit storage rots in store dip or drench fruit immediately after picking
- For black spot control in roses apply after pruning with 3 further applications at 14 d intervals or spray when spots appear and repeat at 7-10 d intervals
- For grey mould in strawberries spray at first open flower and repeat every 7-10 d [1, 2]
- Do not leave diluted material for more than 2 h. Agitate well before and during spraying
- Product is not compatible with adjuvant oils

Crop Safety/Restrictions
- Maximum number of treatments 1 per batch as post-harvest dip on apples and pears [3]
- Do not use on apple cultivars Bramley, Monarch, Winston, King Edward, Spartan, Kidd's Orange or Red Delicious or on pear cultivar D'Anjou
- Do not mix with alkaline materials [1, 2]
- Do not use on strawberries for canning [1, 2]

Special precautions/Environmental safety
- Irritating to eyes, skin and respiratory system [2]; to respiratory system [1]
- Risk of serious damage to eyes [1]
- Irritating to eyes [3]
- May cause sensitization by skin contact [1, 3]
- Harmful to fish or other aquatic life. Do not contaminate surface waters or ditches with chemical or used container

Protective clothing/Label precautions
- A [1-3]; C [3]; D, E [1]; H [1, 3]
- R03, R04d, U09a, U19 [1]; R04a [2, 3]; R04b [2]; R04c, E32a [1, 2]; R04e [1, 3]; U05a, U20b, C03, E01, E13c, E30a [1-3]; E31b [3]

Latest application/Harvest Interval(HI)
- HI apples, pears 7 d

Approval
- Off-label approval unlimited for use on non-edible ornamental bulbs (OLA 1229/95)[3]

Maximum Residue Levels (mg residue/kg food)
- pome fruits, grapes, cane fruits, bilberries, cranberries, currants, gooseberries, tomatoes, peppers, aubergines 3; apricots, peaches, nectarines, plums, lettuce, beans, peas, leeks 2; citrus fruits, bananas, carrots, horseradish, parsnips, parsley root, salsify, swedes, turnips, garlic, onions, shallots, cucumbers, gherkins, courgettes, cauliflowers, Brussels sprouts, head cabbage, celery, rhubarb, mushrooms, potatoes 0.1

72 captan + penconazole

A protectant fungicide for use on apple trees

Products

1 Topas C 50 WP	Ciba Agric.	47.5:2.5% w/w	WB	03232
2 Topas C 50 WP	Novartis	47.5:2.5% w/w	WB	08459

Uses

Powdery mildew in APPLES. Scab in APPLES.

Notes

Efficacy
- Use as a protective spray every 7-14 d from bud burst until extension growth ceases
- High antisporulant activity reduces development of primary mildew and controls spread of secondary mildew
- Does not affect beneficial insects so can be used in integrated control programmes
- Penconazole is recommended alone for mildew control if scab is not a problem

Crop Safety/Restrictions
- Maximum number of treatments 10 per yr

Special precautions/Environmental safety

- Irritating to eyes, skin and respiratory system
- Dangerous to fish or other aquatic life. Do not contaminate surface waters or ditches with chemical or used container

Protective clothing/Label precautions

- A, C
- R04a, R04b, R04c, U02, U05a, U08, U19, U20b, C02 (14 d), C03, E01, E13b, E30a, E32a

Latest application/Harvest Interval(HI)

- HI apples 14 d

Maximum Residue Levels (mg residue/kg food)

- see captan entry

73 carbaryl

A contact carbamate insecticide, with restricted permitted uses

Products

1 Cavalier	RP Amenity	473 g/l	SC	07027
2 Thinsec	Zeneca	450 g/l	SC	06710

Uses

Apple capsid in APPLES [2]. Codling moth in APPLES [2]. Earthworms in AMENITY TURF, FINE TURF [1]. Earwigs in APPLES [2]. Fruit thinning in APPLES [2]. Leatherjackets in AMENITY TURF, FINE TURF [1]. Tortrix moths in APPLES [2]. Winter moth in APPLES [2].

Notes

Efficacy

- Application rate and timing vary with pest. See label for details [2]
- Effective thinning of apples depends on thorough wetting of foliage and fruitlets. Dose and timing vary with cultivar [2]
- Fruit from treated trees may need picking 5-7 d earlier than from untreated trees [2]
- Best results in turf obtained when material placed below thatch and in surface layer of moist soil. Light watering after treatment is beneficial [1]
- On turf commence treatment in autumn or spring when earthworm or leatherjacket activity first seen [1]

Crop Safety/Restrictions

- Maximum number of treatments 2 (amenity turf) or 4 (fine turf) per yr [1]
- Spray concentration must not exceed 3.8 l/500 l water [2]
- Carbaryl is dangerous to pollinating insects and particular care must be taken when spraying close to blossom period or when orchard adjacent to other flowering crops [2]
- Do not spray Laxton's Fortune between pink bud stage and end of first wk in Jun [2]
- Do not use on apples between late pink bud and end of first wk in Jun except for purpose of fruit thinning [2]
- Do not apply to frozen ground [1]

Special precautions/Environmental safety

- This product contains an anticholinesterase carbamate compound. Do not use if under medical advice not to work with such compounds

- Harmful if swallowed and in contact with skin
- Harmful to fish or other aquatic life. Do not contaminate surface waters or ditches with chemical or used container
- Dangerous to bees. Do not apply to crops in flower or to those in which bees are actively foraging. Do not apply when flowering weeds are present [2]
- Spray equipment may only be used where the operator's normal working position is within a closed cab on a tractor or on a self-propelled sprayer. Hand-held or similar application equipment may not be used
- Low level induction bowls, or a closed transfer system must be used when transferring product from container to spray tank [2]

Protective clothing/Label precautions
- A [1, 2]; C [2]; H, K [1, 2]
- M02, M03, R03c, U05a, U08, C03, E01, E13c, E30a, E31a, E34 [1, 2]; R03a, U20b, E12d, E26 [1]; U20a, E12c [2]

Latest application/Harvest Interval(HI)
- HI apples 3 wk

Approval
- All approvals for non-agricultural uses of carbaryl were revoked in November 1995

Maximum Residue Levels (mg residue/kg food)
- apricots, peaches, nectarines, plums, cane fruits, bilberries, cranberries, currants, gooseberries, lettuce 10; citrus fruits, strawberries 7; pome fruits, grapes, bananas, tomatoes, peppers, aubergines, head cabbage, beans, peas 5; cucumbers, gherkins, courgettes, celery, rhubarb 3; carrots, horseradish, parsnips, parsley root, salsify, swedes 2; turnips, garlic, onions, shallots, cauliflowers, Brussels sprouts, leeks, mushrooms 1; potatoes 0.2

74 carbendazim

A systemic benzimidazole fungicide with curative and protectant activity

Products					
	1 Ashlade Carbendazim FL	Ashlade	500 g/l	SC	06213
	2 Bavistin DF	BASF	50% w/w	WG	03848
	3 Carbate Flowable	PBI	500 g/l	SC	08957
	4 Derosal Liquid	AgrEvo	500 g/l	SC	07315
	5 Derosal WDG	AgrEvo	80% w/w	WG	07316
	6 Headland Addstem	Headland	511 g/l	SC	06755
	7 Headland Regain	Headland	50% w/w	WG	08675
	8 Hinge	Quadrangle	500 g/l	SC	04929
	9 Mascot Systemic	Rigby Taylor	500 g/l	SC	08776
	10 MSS Mircarb	Mirfield	500 g/l	SC	08788
	11 Stefes C-Flo 2	Stefes	500 g/l	SC	08059
	12 Stefes Derosal Liquid	Stefes	500 g/l	SC	07649
	13 Stefes Derosal WDG	Stefes	80% w/w	WG	07658
	14 Tripart Defensor FL	Tripart	500 g/l	SC	02752
	15 Turf Systemic Fungicide	PBI	500 g/l	SC	08194
	16 Turfclear	Scotts	500 g/l	SC	07506
	17 Turfclear WDG	Levington	80% w/w	WB	07490
	18 Twincarb	Vitax	500 g/l	SC	08777
	19 UPL Carbendazim 500 FL	United Phosphorus	500 g/l	SC	07472

Uses Alternaria in BRUSSELS SPROUTS *(Must be used in tank mix with Corbel)*[2]. American gooseberry mildew in BLACKCURRANTS, GOOSEBERRIES [1, 2, 4-6, 8, 11-14]. Anthracnose in DWARF BEANS, NAVY BEANS [2]. Big vein in OUTDOOR LETTUCE *(off-label)*, PROTECTED LETTUCE *(off-label)* [2]. Botrytis in BEDDING PLANTS *(off-label)*, BLACKBERRIES *(off-label)*, BULBS/CORMS *(dip - off-label)*, BULBS/CORMS *(off-label)*, CELERY *(off-label)*, CHRYSANTHEMUMS *(off-label)*, CUCUMBERS *(off-label)*, LOGANBERRIES *(off-label)*, NAVY BEANS, ONIONS *(off-label)*, PEPPERS *(off-label)*, POT PLANTS *(off-label)*, RASPBERRIES *(off-label)* [2]. Botrytis in BLACKCURRANTS, GOOSEBERRIES, STRAWBERRIES [1, 2, 4-6, 8, 11-14]. Botrytis in DWARF BEANS [1, 2, 4-6, 10-13]. Botrytis in RASPBERRIES [1, 4, 6, 8, 11-13]. Botrytis in PROTECTED POT PLANTS [4, 6, 11-13]. Botrytis in CANE FRUIT [5, 14]. Botrytis in TOMATOES [5, 8, 13]. Botrytis in POT PLANTS [5]. Botrytis in GLADIOLI, NARCISSI, TULIPS [13]. Botrytis in OILSEED RAPE *(reduction)* [3]. Brown rot in CHERRIES *(off-label)*, PLUMS *(off-label)* [2]. Cane spot in BLACKBERRIES *(off-label)*, LOGANBERRIES *(off-label)*, RASPBERRIES *(off-label)* [2]. Cane spot in RASPBERRIES [1, 4, 6, 8, 11-13]. Cane spot in CANE FRUIT [5, 14]. Canker in APPLES *(partial control)* [8]. Canker in APPLES [1, 14]. Celery leaf spot in CELERY [4-6, 8, 11-13]. Chocolate spot in FIELD BEANS [1-6, 8, 10-14]. Chocolate spot in BROAD BEANS [1, 4-6, 8, 10-14]. Currant leaf spot in GOOSEBERRIES [1, 2, 4-6, 11, 12, 14]. Currant leaf spot in BLACKCURRANTS [1, 2, 4-6, 11-14]. Didymella in TOMATOES [5]. Dollar spot in AMENITY TURF [15]. Dollar spot in TURF [7, 9, 16, 17]. Dollar spot in MANAGED AMENITY TURF [18]. Eye rot in APPLES, PEARS [4, 6, 11-13]. Eyespot in WINTER RYE [2, 3]. Eyespot in WINTER BARLEY, WINTER WHEAT [1-4, 6, 10-12, 14, 19]. Eyespot in BARLEY, WHEAT [5, 13]. Fire in TULIPS [6]. Fusarium patch in AMENITY TURF [15]. Fusarium patch in TURF [7, 9, 16, 17]. Fusarium patch in MANAGED AMENITY TURF [18]. Fusarium wilt in TOMATOES [5]. Fusarium in BULBS/CORMS *(dip - off-label)*, FREESIAS *(off-label)* [2]. Gloeosporium rot in APPLES, PEARS [4, 6, 11, 12]. Gloeosporium in APPLES [2, 8, 13] Gloeosporium in PEARS [13]. Glume blotch in WINTER WHEAT *(partial control)* [8]. Grey bulb rot in TULIPS [6]. Grey mould in TOMATOES [4, 6, 11, 12]. Leaf mould in TOMATOES [4-6, 11-13]. Light leaf spot in BRUSSELS SPROUTS *(Must be used in tank mix with Corbel)*, OILSEED RAPE *(reduction)* [2]. Light leaf spot in OILSEED RAPE [3-6, 8, 10-14]. Mycogone in MUSHROOMS [2]. Penicillium rot in BULBS/CORMS *(dip - off-label)* [2]. Powdery mildew in BARLEY [14]. Powdery mildew in BLACKBERRIES *(off-label)*, CUCUMBERS *(off-label)*, LOGANBERRIES *(off-label)*, PEPPERS *(off-label)*, RASPBERRIES *(off-label)*, ROSES [2]. Powdery mildew in WINTER BARLEY *(partial control)*, WINTER WHEAT *(partial control)* [8]. Powdery mildew in RASPBERRIES [1, 4, 6, 11-13]. Powdery mildew in APPLES [1, 4-6, 8, 11, 12, 14]. Powdery mildew in STRAWBERRIES [1, 4-6, 11-14]. Powdery mildew in PROTECTED POT PLANTS [4, 6, 11-13]. Powdery mildew in PEARS [4, 6, 11, 12]. Powdery mildew in CUCUMBERS [4-6, 11-13]. Powdery mildew in CANE FRUIT [5, 14]. Powdery mildew in POT PLANTS [5]. Rhynchosporium in WINTER BARLEY [2-4, 6, 8, 10-12]. Rhynchosporium in SPRING BARLEY [2-4, 6, 10-12]. Rhynchosporium in BARLEY [1, 5, 13, 14]. Ring spot in BRUSSELS SPROUTS *(Must be used in tank mix with Corbel)* [2]. Root diseases in INERT SUBSTRATE CUCUMBERS *(off-label)* [2]. Scab in PEARS *(off-label)* [2]. Scab in APPLES [1, 2, 4-6, 8, 11-14]. Scab in PEARS [4-6, 11-13]. Sclerotinia in BULBS/CORMS *(dip - off-label)*, BULBS/CORMS *(off-label)* [2]. Sclerotinia in CORMS, FLOWER BULBS, NARCISSI [6]. Sooty moulds in WINTER WHEAT [2, 4, 11, 12]. Sooty moulds in SPRING WHEAT [4]. Sooty moulds in BARLEY, WHEAT [5, 13]. Spur blight in BLACKBERRIES *(off-label)*, LOGANBERRIES *(off-label)*, RASPBERRIES *(off-label)* [2]. Spur blight in RASPBERRIES [1, 8, 13]. Spur blight in CANE FRUIT [5, 14]. Stagonospora in BULBS/CORMS *(off-label)* [2]. Stagonospora in CORMS, FLOWER BULBS, NARCISSI [6]. Storage rots in PEARS *(off-label)* [2]. Storage rots in APPLES [1, 2, 5, 14]. Storage rots in PEARS [5]. Trichoderma in MUSHROOMS *(spawn treatment - off-label)* [2]. Verticillium wilt in

FOR FULL CONDITIONS OF USE ALWAYS READ THE PRODUCT LABEL

TOMATOES [5]. Wet bubble in MUSHROOMS [3]. Wormcast formation in TURF [7, 9, 16, 17]. Wormcast formation in MANAGED AMENITY TURF [18].

Notes

Efficacy

- Products vary in the diseases listed as controlled for several crops. Labels must be consulted for full details and for rates and timings
- Mostly applied as spray or drench. Spray treatments normally applied at first sign of disease and repeated after 10-14 d if required
- On turf apply as preventative treatment in spring or autumn during periods of high disease risk [9, 16-18]
- Allow 4 mth intervals where used for worm cast control. Rain or irrigation after treatment may improve control [9, 16-18]
- Apply as a drench to control soil-borne diseases in cucumbers and tomatoes and as a pre-planting dip treatment for bulbs
- On apples, bush and cane fruit the addition of a non-ionic wetter aids penetration and improves disease control [13]
- Apply by incorporation into casing to control mushroom diseases [2, 3]
- For disease control in Brussels sprouts product must be mixed with fenpropimorph - see label [2]
- To delay appearance of resistant strains alternate treatment with non-MBC fungicide. Eyespot in cereals and *Botrytis cinerea* in many crops is now widely resistant
- Not compatible with alkaline products such as lime sulphur

Crop Safety/Restrictions

- Maximum number of treatments (including applications of any product containing benomyl, carbendazim or thiophanate-methyl) varies with crop treated and product used - see labels for details
- Do not treat crops or turf suffering from drought or other physical or chemical stress
- Do not use on strawberry runner beds
- Apply as drench rather than spray where red spider mite predators are being used
- Consult processors before using on crops for processing
- Use drench for tomatoes, cucumbers and peppers on soil-grown crops only

Special precautions/Environmental safety

- Harmful to fish or other aquatic life. Do not contaminate surface waters or ditches with chemical or used container [2, 4, 6-13, 15, 16, 16-19]
- After use dipping suspension must not be discharged directly into ditches or drains. Preferred disposal method is via a soakaway [13]

Protective clothing/Label precautions

- A [1-19]; B [1-6, 11-14]; C [1-19]; D [2, 5, 7, 13]; H [1-19]; K [1-6, 11-14, 17]; M [1-14, 16-19]
- U20a [8]; U20b [1, 3, 5, 6, 10-13]; U20c [2, 4, 7, 9, 15-19]; U22 [17]; C03 [8, 10]; E01 [1, 8, 10, 14]; E13c [2, 4, 6-13, 15-19]; E15 [1, 3, 5, 17]; E24 [16-18]; E26 [4, 11, 12, 16-19]; E30a [1-4, 6-12, 15-19]; E31a [9, 14, 18]; E31b [1, 3, 4, 6, 11, 12, 15, 19]; E32a [2, 5, 7, 8, 10, 13, 16, 17]; E34 [14, 19]

Latest application/Harvest Interval(HI)

- Varies with crop and product used. See labels for details
- HI 2 d for strawberries, cane fruit, blackcurrants, gooseberries, tomatoes, peppers, cucumbers; 7 d for celery, lettuce; 7-14 d (depends on dose) for apples, pears; 14 d for plums, cherries, onions, mushrooms; 21 d for field beans, broad beans, navy beans, oilseed rape.

Approval

- Approved for aerial application on winter wheat, winter and spring barley [1, 3, 8, 14]; field beans [3]; oilseed rape [1, 3, 14]; See notes in Section 1
- Off-label approval unlimited for use on mushrooms (spawn treatment) (OLA 1144/95)[2]; unlimited for use on protected cucmbers on inert media (OLA1470/94)[2]; unlimited for use on onions, celery, cucumbers, peppers, ornamental bulbs and corms, pears, plums, cherries, raspberries, loganberries, blackberries, freesias, chrysanthemums, pot and bedding plants (OLA 1002/95)[2]; unlimited for use on protected and outdoor crops of lettuce (OLA 1751/96)[2]

Maximum Residue Levels (mg residue/kg food)

- peaches, nectarines, grapes 10; citrus fruits, strawberries, raspberries, currants, tomatoes, lettuce 5; potatoes 3; pome fruits, plums, onions, celery 2; bananas, cultivated mushrooms 1; aubergines, cucumbers, melons, squashes, Brussels sprouts 0.5; soya beans 0.2; tree nuts, bilberries, cranberries, wild berries, miscellaneous fruits (except bananas), beetroot, horseradish, Jerusalem artichokes, parsnips, parsley root, radishes, sweet potatoes, yams, garlic, shallots, spring onions, courgettes, watermelons, sweet corn, Chinese cabbage, kale, kohlrabi, spinach, beet leaves, watercress, chervil, chives, parsley, celery leaves, stem vegetables (except celery), wild mushrooms, lentils, oilseeds (except soya beans), hops, cereals, animal products 0.1

75 carbendazim + chlorothalonil

A systemic and protectant fungicide mixture

Products

1 Bravocarb	Zeneca	100:450 g/l	SC	09105
2 Greenshield	Scotts	100:450 g/l	SC	07988

Uses

Anthracnose in TURF [2]. Ascochyta in PEAS [1]. Botrytis in PEAS [1]. Chocolate spot in FIELD BEANS [1]. Dollar spot in TURF [2]. Downy mildew in PEAS [1]. Fusarium patch in TURF [2]. Red thread in TURF [2]. Rhynchosporium in BARLEY [1]. Septoria in WINTER WHEAT [1]. Wormcast formation in TURF [2].

Notes

Efficacy

- Apply to winter wheat and barley between leaf sheath lengthening and first node detectable stages (GS 30-31) for early Septoria control. Further application up to ear emergence (GS 51) may be needed for Rhynchosporium control [1]
- To protect wheat against Septoria spray immediately infection visible between flag leaf just visible and ear just fully emerged (GS 37-59) [1]
- Treatment gives good control of cereal ear diseases particularly when applied directly to the ear. It also suppresses rusts, mildew and net blotch [1]
- Tank-mix with a specific fungicide if rust/mildew levels become high and with prochloraz if MBC-resistant strains of eyespot are present [1]
- Spray peas and beans at flowering and 2-4 wk later [1]
- Apply at any time of year at the first sign of turf disease. Further treatments may be necessary if conditions remain favourable [2]
- Established or severe infections of Anthracnose will not be controlled [2]

Crop Safety/Restrictions

- Maximum number of treatments (including applications of any product containing benomyl, carbendazim or thiophanate-methyl) 2 per crop for cereals, peas, field beans [1]
- Do not mow or water within 24 h of treatment [2]
- Do not mix with other pesticides, surfactants or fertilisers [2]

Special precautions/Environmental safety

- Irritating to eyes, skin and respiratory system
- Risk of serious damage to eyes [1]
- Dangerous to fish or other aquatic life. Do not contaminate surface waters or ditches with chemical or used container
- Do not allow direct spray from vehicle mounted/drawn hydraulic sprayers to fall within 6 m of surface waters or ditches. Direct spray away from water

Protective clothing/Label precautions

- A, C [1, 2]; H, M [2]
- R04a, R04b, R04c, U05a, U20a, C03, E01, E13b, E16, E26, E30a, E31a [1, 2]; R04d, U08, U14 [1]; U09a, U19 [2]

Latest application/Harvest Interval(HI)

- Up to and including grain watery-ripe stage (GS 71) for cereals [1]
- HI 14 d for peas, 21 d for field beans, oilseed rape [1]

Approval

- Approved for aerial application on winter wheat, barley. See notes in Section 1. Consult firm for details [1]

Maximum Residue Levels (mg residue/kg food)

- see carbendazim and chlorothalonil entries

---

76 carbendazim + chlorothalonil + maneb

A broad-spectrum protectant and eradicant fungicide for cereals

Products

1 Ashlade Mancarb Plus	Ashlade	80:150:200 g/l	SC	08160
2 Tripart Victor	Tripart	80:150:200 g/l	SC	04359

Uses

Botrytis in SPRING BARLEY, SPRING WHEAT, WINTER BARLEY, WINTER WHEAT [2]. Brown foot rot and ear blight in SPRING WHEAT, WINTER WHEAT [2]. Brown rust in SPRING BARLEY, SPRING WHEAT, WINTER BARLEY, WINTER WHEAT [2]. Eyespot in SPRING BARLEY, SPRING WHEAT, WINTER BARLEY, WINTER WHEAT [2]. Net blotch in SPRING BARLEY, WINTER BARLEY [2]. Powdery mildew in SPRING WHEAT, WINTER WHEAT [2]. Rhynchosporium in SPRING BARLEY, WINTER BARLEY [2]. Septoria diseases in SPRING WHEAT, WINTER WHEAT [2]. Septoria in SPRING WHEAT, WINTER WHEAT [1]. Sooty moulds in SPRING WHEAT, WINTER WHEAT [2]. Yellow rust in SPRING BARLEY, SPRING WHEAT, WINTER BARLEY, WINTER WHEAT [2].

Notes

Efficacy

- Correct timing crucial for optimum results. First treatment should be made between when flag leaf just visible and start of ear emergence
- Highly effective against late diseases on flag leaf and ear of winter cereals

Crop Safety/Restrictions

- Maximum number of treatments (including application of any product containing benomyl, carbendazim or thiophanate-methyl) 2 per crop.

Special precautions/Environmental safety

- Irritating to eyes, skin and respiratory system
- Risk of serious damage to eyes [1]
- Dangerous to fish or other aquatic life. Do not contaminate surface waters or ditches with chemical or used container
- Do not allow direct spray from vehicle mounted/drawn hydraulic sprayers to fall within 6 m, or from hand-held sprayers to within 2 m, of surface waters or ditches

Protective clothing/Label precautions

- A, C [1, 2]; H, M [1]
- R04a, R04b, R04c, U05a, U08, U19, U20a, C03, E01, E13b, E30a, E32a [1, 2]; R04d, E16 [1]

Latest application/Harvest Interval(HI)

- Ear emergence complete (GS 59) for wheat [1]; up to and including grain watery-ripe (GS 71) for barley

Maximum Residue Levels (mg residue/kg food)

- see carbendazim, chlorothalonil and maneb entries

77 carbendazim + cymoxanil + oxadixyl + thiram

A broad spectrum fungicide seed treatment for peas

Products Apron Elite | Novartis | 16.7:6.7:16.7:33.4%WG | 08770

Uses Ascochyta in PEAS *(seed treatment)*. Damping off in PEAS *(seed treatment)*. Downy mildew in PEAS *(seed treatment)*.

Notes

Efficacy

- Apply through continuous flow seed treaters which should be calibrated before use

Crop Safety/Restrictions

- Ensure moisture content of treated seed satisfactory and store in a dry place
- Check calibration of seed drill with treated seed before drilling and sow as soon as possible after treatment
- Consult before using on crops for processing

Special precautions/Environmental safety

- May cause sensitization by skin contact
- Dangerous to fish or other aquatic life. Do not contaminate surface waters or ditches with chemical or used container
- Do not use treated seed as food or feed

Protective clothing/Label precautions

- A, C, D, H, M

- M03, R04e, U05a, U08, U20b, C03, E01, E13b, E26, E30a, E31a, E34, S02, S04b, S05, S07

Maximum Residue Levels (mg residue/kg food)

- see carbendazim entry

78 carbendazim + cyproconazole

A systemic protective and curative fungicide for cereals

Products

1 Alto Combi	Sandoz	300:160 g/l	SC	05066
2 Alto Combi	Novartis	300:160 g/l	SC	08465

Uses

Brown rust in SPRING BARLEY, WINTER BARLEY, WINTER WHEAT. Eyespot in WINTER BARLEY, WINTER WHEAT. Net blotch in SPRING BARLEY, WINTER BARLEY. Powdery mildew in SPRING BARLEY, WINTER BARLEY, WINTER WHEAT. Rhynchosporium in SPRING BARLEY, WINTER BARLEY. Septoria diseases in WINTER WHEAT. Sooty moulds in WINTER WHEAT. Yellow rust in SPRING BARLEY, WINTER BARLEY, WINTER WHEAT.

Notes

Efficacy

- Apply at start of disease development or as preventive treatment
- Most effective timing of treatment varies with disease. See label for details
- For eradication of established mildew tank-mix with approved formulation of tridemorph or fenpropimorph
- For high *Septoria tritici* infection tank-mix with approved formulation of chlorothalonil
- When applied early, prior to GS 33, useful reduction of eyespot is obtained

Crop Safety/Restrictions

- Maximum number of treatments (including applications of any product containing benomyl, carbendazim or thiophanate-methyl) 3 per crop
- Application to winter wheat in spring at GS 30-33 may cause straw shortening but does not cause loss of yield

Special precautions/Environmental safety

- Harmful to fish or other aquatic life. Do not contaminate surface waters or ditches with chemical or used container

Protective clothing/Label precautions

- A, C, H, M
- U20c, E13c, E26, E30a, E31b

Latest application/Harvest Interval(HI)

- Up to and including anthesis complete (GS 69) for winter wheat; up to and including emergence of ear complete (GS 59) for barley

Maximum Residue Levels (mg residue/kg food)

- see carbendazim entry

79 carbendazim + flusilazole

A broad-spectrum systemic and protectant fungicide for cereals

Products

1 Contrast	DuPont	125:250 g/l	SC	06150
2 Landgold Flusilazole MBC	Landgold	125:250 g/l	SC	08528
3 Punch C	DuPont	125:250 g/l	SC	06801
4 Standon Flusilazole Plus	Standon	125:250 g/l	SC	07403

Uses

Brown rust in SPRING BARLEY, SPRING WHEAT, WINTER BARLEY, WINTER WHEAT [1-4]. Canker in OILSEED RAPE [1, 3]. Eyespot in SPRING BARLEY [1, 3]. Eyespot in SPRING WHEAT, WINTER BARLEY, WINTER WHEAT [1-4]. Light leaf spot in OILSEED RAPE [1-4]. Mildew in SPRING BARLEY, WINTER BARLEY, WINTER WHEAT [1, 3]. Net blotch in SPRING BARLEY, WINTER BARLEY [1-4]. Phoma leaf spot in OILSEED RAPE [1, 3]. Powdery mildew in SPRING WHEAT [1-4]. Powdery mildew in SPRING BARLEY, WINTER BARLEY, WINTER WHEAT [2, 4]. Rhynchosporium in SPRING BARLEY, WINTER BARLEY [1-4]. Septoria in SPRING BARLEY, WINTER BARLEY [1, 3]. Septoria in SPRING WHEAT, WINTER WHEAT [1-4]. Yellow rust in SPRING BARLEY, SPRING WHEAT, WINTER BARLEY, WINTER WHEAT [1-4].

Notes

Efficacy

- Apply at early stage of disease development or in routine preventive programme
- Most effective timing of treatment varies with disease. See label for details
- Higher rate active against both MBC-sensitive and MBC-resistant eyespot
- Rain occurring within 2-3 h of spraying may reduce effectiveness
- To prevent build-up of resistant strains of mildew tank mix with approved morpholine fungicide [1, 3]
- Treat oilseed rape in autumn when leaf lesions first appear and spring from the start of stem extension when disease appears

Crop Safety/Restrictions

- Maximum number of treatments (including applications of any product containing flusilazole) 3 per winter wheat crop, 2 per crop of winter barley, spring barley, spring wheat or oilseed rape
- Do not apply to crops under stress or during frosty weather

Special precautions/Environmental safety

- Harmful if swallowed
- Irritating to eyes
- Dangerous to fish or other aquatic life. Do not contaminate surface waters or ditches with chemical or used container

Protective clothing/Label precautions

- A, C [1-4]; H, M [1, 2, 4]
- M03, R03c, R04a, U05a, U11, U19, U20b, C03, E01, E13b, E30a, E31b, E34 [1-4]

Latest application/Harvest Interval(HI)

- Normally up to and including grain watery ripe stage (GS 71) for wheat and barley and before first flower opened stage for oilseed rape but timings vary with crop and dose - see labels for details

Maximum Residue Levels (mg residue/kg food)
- see carbendazim entry

80 carbendazim + flutriafol

A systemic fungicide for cereals with protectant and eradicant activity

Products

1 Early Impact	Zeneca	150:94 g/l	SC	06659
2 Pacer	Zeneca	150:94 g/l	SC	06690
3 Palette	Zeneca	150:94 g/l	SC	06691

Uses

Brown rust in SPRING BARLEY, WINTER BARLEY, WINTER WHEAT. Leaf blotch in SPRING BARLEY, WINTER BARLEY. Leaf spot in WINTER WHEAT. Powdery mildew in SPRING BARLEY, WINTER BARLEY, WINTER WHEAT. Yellow rust in SPRING BARLEY, WINTER BARLEY, WINTER WHEAT.

Notes

Efficacy
- Apply at early stage of disease development or as a protectant in high risk situations
- When mildew is established at 5% or more mix with a morpholine fungicide for improved control. See label for details of timing. Seed treatment may be preferable on spring barley
- Control of mildew or leaf blotch is reduced if strains with decreased sensitivity to triazoles or mbc fungicides are present

Crop Safety/Restrictions
- Maximum number of treatments (including applications of any product containing benomyl, carbendazim, thiophanate-methyl or flutriafol) 2 per crop
- Under conditions of stress some wheat varieties can exhibit flag leaf tip scorch, which may be increased by fungicide application

Special precautions/Environmental safety
- Irritating to eyes and skin
- Harmful to fish or other aquatic life. Do not contaminate surface waters or ditches with chemical or used container

Protective clothing/Label precautions
- A, C, H, M
- R04a, R04b, U02, U05a, U08, U19, U20b, C03, E01, E13c, E26, E30a, E31b, E31c, E34

Latest application/Harvest Interval(HI)
- Up to and including grain watery-ripe stage (GS 71)

Maximum Residue Levels (mg residue/kg food)
- see carbendazim entry

81 carbendazim + iprodione

A systemic and contact fungicide mixture

Products

1 Calidan	RP Agric.	87.5:175 g/l	SC	06536
2 Vitesse	RP Amenity	87.5:175 g/l	SC	06537

Uses Alternaria in OILSEED RAPE [1]. Anthracnose in TURF [2]. Ascochyta in COMBINING PEAS, VINING PEAS [1]. Botrytis in COMBINING PEAS, OILSEED RAPE, VINING PEAS [1]. Fusarium patch in TURF [2]. Light leaf spot in OILSEED RAPE [1]. Mycosphaerella in COMBINING PEAS, VINING PEAS [1]. Pink patch in TURF [2]. Red thread in TURF [2]. Sclerotinia stem rot in COMBINING PEAS, OILSEED RAPE, VINING PEAS [1]. Timothy leaf spot in TURF [2].

Notes

Efficacy

- Best results obtained by application at first signs of disease, repeated monthly as necessary [2]
- Maximum efficacy against Anthracnose achieved by treatment at early disease development stage. Curative treatments for well-established Anthracnose are not recommended [2]
- May be used all year round, but is best suited for spring or late summer/early autumn application [2]
- Timing for oilseed rape varies according to disease. See label [1]
- Complete control of Botrytis and Alternaria in rape may require 2 treatments separated by not less than 3 wk [1]
- Treat peas at mid-flowering when first pods 2.5 cm long. On combining peas repeat 2-3 wk later if necessary [1]

Crop Safety/Restrictions

- Maximum number of treatments 2 per crop for oilseed rape and combining peas, 1 per crop for vining peas [1]. (NB Oilseed rape may be treated with 3 applications of products containing benomyl, carbendazim or thiophanate-methyl per season)
- Where grass is being mown, apply after mowing. Delay further mowing for at least 48 h after treatment [2]

Special precautions/Environmental safety

- Harmful to fish or other aquatic life. Do not contaminate surface waters or ditches with chemical or used container

Protective clothing/Label precautions

- A, C, H, M
- U20c, E13c, E26, E30a, E31b

Latest application/Harvest Interval(HI)

- HI 3 wk for oilseed rape, peas [1]

Maximum Residue Levels (mg residue/kg food)

- see carbendazim and iprodione entries

---

82 carbendazim + mancozeb

A broad-spectrum systemic and protectant fungicide for cereals

Products				
Stefes Kombat WDG	Stefes	12.4:63.3% w/w	WG	07985

Uses Downy mildew in WINTER OILSEED RAPE. Eyespot in WINTER WHEAT. Mildew in WINTER BARLEY, WINTER WHEAT. Net blotch in WINTER BARLEY. Rhynchosporium in

WINTER BARLEY. Septoria in WINTER WHEAT. Sooty moulds in WINTER BARLEY, WINTER WHEAT. Yellow rust in WINTER WHEAT.

Notes

Efficacy
- Best results achieved by application at beginning of disease development
- Do not spray when crops wet, if rain imminent or if temperature exceeds 30°C
- Spray wheat between leaf sheath erect and first spikelets visible (GS 30-51) and barley between flag leaf ligule visible and complete ear emergence (GS 39-59)
- Will protect against mildew/rusts but tank-mix with a specific rust/mildew fungicide if these diseases are already present in the crop
- MBC resistant eyespot not controlled. If present use products containing prochloraz
- Spray winter oilseed rape pre-flowering up to yellow bud stage (GS 3,7)

Crop Safety/Restrictions
- Maximum number of treatments (including applications of any product containing benomyl, carbendazim or thiophanate-methyl) 2 per crop for cereals, 3 per crop for oilseed rape
- Do not spray crops suffering stress from drought

Special precautions/Environmental safety
- Irritating to respiratory system
- May cause sensitization by skin contact
- Harmful to fish or other aquatic life. Do not contaminate surface waters or ditches with chemical or used container

Protective clothing/Label precautions
- A, C, D, H, M
- R04c, R04e, U05a, U08, U20a, C03, E01, E13c, E30a, E32a

Latest application/Harvest Interval(HI)
- Up to and including grain watery-ripe stage (GS 71).
- HI 21 d for oilseed rape.

Maximum Residue Levels (mg residue/kg food)
- see carbendazim and mancozeb entries

83 carbendazim + maneb

A broad-spectrum systemic and protectant fungicide

Products

1 Ashlade Mancarb FL	Ashlade	50:320 g/l	SC	07977
2 Headland Dual	Headland	62:400 g/l	SC	03782
3 MC Flowable	United Phosphorus	62:400 g/l	SC	08198
4 Multi-W FL	PBI	50:320 g/l	SC	04131
5 Tripart Legion	Tripart	50:320 g/l	SC	06113

Uses

Brown rust in SPRING BARLEY, SPRING WHEAT, WINTER BARLEY [2]. Brown rust in WINTER WHEAT [1, 2, 4]. Brown rust in BARLEY [3-5]. Brown rust in WHEAT [3, 5]. Downy mildew in OILSEED RAPE [4, 5]. Eyespot in WINTER BARLEY, WINTER WHEAT [2, 3]. Light leaf spot in OILSEED RAPE [4, 5]. Powdery mildew in WINTER WHEAT [2, 4]. Powdery mildew in SPRING BARLEY, SPRING WHEAT, WINTER BARLEY [2]. Powdery mildew in BARLEY [4, 5]. Powdery mildew in WHEAT [5]. Rhynchosporium in SPRING BARLEY, WINTER BARLEY [2]. Rhynchosporium in BARLEY [3-5]. Septoria diseases in WINTER WHEAT [2, 4]. Septoria diseases in SPRING BARLEY, SPRING WHEAT, WINTER BARLEY

[2]. Septoria in WHEAT [5]. Septoria in WINTER WHEAT [1]. Sooty moulds in WINTER WHEAT [2-4]. Sooty moulds in SPRING WHEAT [2]. Sooty moulds in BARLEY [4]. Sooty moulds in WHEAT [5]. Sooty moulds in WINTER BARLEY [3]. Yellow rust in WINTER WHEAT [2, 4]. Yellow rust in BARLEY [2-5]. Yellow rust in WHEAT [2, 3, 5]. Yellow rust in SPRING BARLEY, SPRING WHEAT, WINTER BARLEY [2].

Notes

Efficacy

- Apply in cereals from flag leaf just visible (GS 37) until first ears visible (GS 51), best when flag leaf ligules visible (GS 39) or, if not possible to spray earlier, from ear emergence complete (GS 59) until grain watery ripe (GS 71)
- Treatment gives useful control of late attacks of rust or mildew but for early attacks or established infection tank-mix with specific mildew or rust fungicide
- Apply to oilseed rape as soon as disease appears in autumn or early spring and repeat if necessary between green bud and early flowering (GS 3,3-4,0) [4]
- Do not spray if frost or rain expected

Crop Safety/Restrictions

- Maximum number of treatments (including applications of any product containing benomyl, carbendazim or thiophanate-methyl) 2 per crop for cereals, 3 per crop for oilseed rape

Special precautions/Environmental safety

- Harmful if swallowed [2, 3, 5]
- Irritating to eyes and skin [1, 3-5]
- Irritating to respiratory system [1, 3, 5]
- Harmful to fish or other aquatic life. Do not contaminate surface waters or ditches with chemical or used container

Protective clothing/Label precautions

- A [1-5]; C [1-4]; H, M [1, 2, 4]
- M03, R03c, U04a, E31b, E34 [2, 3, 5]; R04a, R04b, U05a, U08, C03, E01, E13c, E30a [1, 3-5]; R04c, U19 [1, 3, 5]; U02 [3, 5]; U20a [1, 4, 5]; U20b [3]; C02 [2, 5]; E15 [5]; E26 [3, 4]; E32a [1, 4]

Latest application/Harvest Interval(HI)

- Up to and including grain watery-ripe stage (GS 71) for cereals
- HI 21 d for oilseed rape; 7 d for cereals

Approval

- Approved for aerial application on cereals [1]; wheat, barley [2]; winter wheat, winter barley [5]. See notes in Section 1

Maximum Residue Levels (mg residue/kg food)

- see carbendazim and maneb entries

---

84 carbendazim + maneb + sulfur

A broad spectrum protectant fungicide mixture for cereals and oilseed rape

Products	Bolda FL	Atlas	50:320:100 g/l	SC	07653

FOR FULL CONDITIONS OF USE ALWAYS READ THE PRODUCT LABEL

Uses Brown rust in WHEAT. Downy mildew in OILSEED RAPE. Light leaf spot in OILSEED RAPE. Net blotch in BARLEY. Powdery mildew in BARLEY, WHEAT. Rhynchosporium in BARLEY. Septoria leaf spot in WHEAT. Yellow rust in WHEAT.

Notes

Efficacy
- Optimum timing for treatment of wheat and barley between flag leaf just visible stage and ear emergence (GS 39-59)
- Treat oilseed rape in autumn or spring when disease becomes active

Crop Safety/Restrictions
- Maximum number of treatments 2 per crop
- Under severe disease pressure or heavy infection in cereals specific fungicides should be used

Special precautions/Environmental safety
- Irritating to eyes, skin and respiratory system
- Harmful to fish or other aquatic life. Do not contaminate surface waters or ditches with chemical or used container

Protective clothing/Label precautions
- A, C, H
- R04a, R04b, R04c, U05a, U20b, C03, E01, E13c, E30a, E32a, E34

Latest application/Harvest Interval(HI)
- Before grain milky ripe stage (GS 73) for cereals
- HI 21 d for oilseed rape

Maximum Residue Levels (mg residue/kg food)
- see carbendazim and maneb entries

---

85 carbendazim + maneb + tridemorph

A protectant and systemic fungicide for use in cereals

Products				
Cosmic FL	BASF	40:320:90 g/l	SC	03473

Uses Brown rust in BARLEY, WINTER WHEAT. Eyespot in BARLEY, WINTER WHEAT. Powdery mildew in BARLEY, WINTER WHEAT. Rhynchosporium in BARLEY. Sooty moulds in WINTER WHEAT. Yellow rust in BARLEY, WINTER WHEAT.

Notes

Efficacy
- Apply before leaf disease becomes established
- In winter barley autumn application is particularly recommended on disease susceptible cultivars but further treatment may be needed in spring
- Tank mixtures with other fungicides recommended for increased protection against net blotch, rusts, powdery mildew and Septoria. See label for details
- To delay appearance of resistant strains alternate treatment with non-MBC fungicide. Eyespot in cereals is now widely resistant to MBC fungicides
- Do not apply if rain expected or crop wet
- Systemic effect reduced under conditions of severe drought stress

Crop Safety/Restrictions
- Maximum total dose (including applications of any product containing benomyl, carbendazim or thiophanate-methyl) equivalent to three full dose treatments on winter barley and two full dose treatments on winter wheat and spring barley
- Do not apply to wheat during periods of temperature above 21°C or high light intensity. Under such conditions spray in late evening

Special precautions/Environmental safety
- Irritating to eyes, skin and respiratory system
- Harmful to fish or other aquatic life. Do not contaminate surface waters or ditches with chemical or used container

Protective clothing/Label precautions
- A, C, H, M
- R04a, R04b, R04c, U04a, U05a, U08, U20a, C03, E01, E07 (14 d), E13c, E30a, E31c

Withholding period
- Keep all livestock out of treated areas for at least 14 d

Latest application/Harvest Interval(HI)
- Before early milk stage (GS 73)

Approval
- Approved for aerial application on barley and winter wheat. See notes in Section 1

Maximum Residue Levels (mg residue/kg food)
- see carbendazim and maneb entries

86 carbendazim + metalaxyl

A protectant fungicide for use in fruit and cabbage storage

Products

1 Ridomil mbc 60 WP	Ciba Agric.	50:10% w/w	WP	01804
2 Ridomil mbc 60 WP	Novartis	50:10% w/w	WP	08437

Uses

Crown rot in WATER LILY *(off-label)*. Phytophthora in STORED CABBAGES. Storage rots in APPLES, PEARS.

Notes

Efficacy
- Apply as post-harvest drench or dip to prevent spread of storage diseases
- Produce must be treated immediately after harvest and methods must achieve thorough coverage of each fruit/cabbage
- Treated produce should be allowed to drain thoroughly without any further rinsing or cleaning prior to storage
- Little curative activity on crop infected at or before harvest
- Best results achieved when crop stored in a controlled environment
- May be used with calcium chloride for bitter pit control in susceptible apple cultivars
- Other mixtures recommended for broader spectrum control - see label

Crop Safety/Restrictions
- Maximum number of treatments (including applications of any product containing benomyl, carbendazim or thiophanate-methyl) 1 per batch
- No variety restrictions yet found necessary for use as a fruit or cabbage dip or drench

Special precautions/Environmental safety
- Irritating to eyes and skin
- Harmful to fish or other aquatic life. Do not contaminate surface waters or ditches with chemical or used container (remove fish before treating water-lily)

Protective clothing/Label precautions
- A, B, C, D, H, K, M
- R04a, R04b, U05a, U08, U20a, C03, E01, E13c, E29, E30a, E32a

Latest application/Harvest Interval(HI)
- Treated fruit must not be processed or sold for at least 4 wk, cabbage for 7 wk

Approval
- Off-label approval unlimited for use on water-lily (OLA 0912/92)[1]

Maximum Residue Levels (mg residue/kg food)
- see carbendazim and metalaxyl entries

---

87 carbendazim + prochloraz

A broad-spectrum systemic and contact fungicide for cereals and oilseed rape

Products

1 Novak	AgrEvo	100:267 g/l	SE	08020
2 Sportak Alpha HF	AgrEvo	100:267 g/l	SE	07225

Uses

Botrytis in OILSEED RAPE [1, 2]. Canker in OILSEED RAPE [1, 2]. Dark leaf spot in OILSEED RAPE [1, 2]. Disease control/foliar feed in PROTECTED ORNAMENTALS *(off-label)*, SUGAR BEET SEED CROPS *(off-label)*[2]. Eyespot in WINTER BARLEY, WINTER WHEAT [1, 2]. Glume blotch in WINTER WHEAT [1, 2]. Leaf spot in WINTER WHEAT [1, 2]. Light leaf spot in OILSEED RAPE [1, 2]. Net blotch in SPRING BARLEY, WINTER BARLEY [1, 2]. Powdery mildew in SPRING BARLEY, WINTER BARLEY, WINTER WHEAT [1, 2]. Rhynchosporium in SPRING BARLEY, WINTER BARLEY [1, 2]. Sclerotinia stem rot in OILSEED RAPE [1, 2]. White leaf spot in OILSEED RAPE [1, 2].

Notes

Efficacy
- To protect against eyespot in high risk situations apply to winter barley from when leaf sheaths begin to become erect to first node detectable (GS 30-31). May also be used to control eyespot already in crop (up to 10% of tillers affected)
- Applied for eyespot control, product will also control Rhynchosporium and mildew and protect against new infections of net blotch, mildew and Septoria. Where MBC resistance is not a problem product can be used against eyespot and leaf spot but prochloraz alone is preferable where these diseases are resistant
- Tank-mix with tridemorph, fenpropimorph or fenpropidin to control established mildew
- May be applied to winter barley up to full ear emergence (GS 59), but if used earlier in season, prochloraz alone or plus a non-MBC fungicide is preferred treatment
- Apply in oilseed rape at first signs of disease and repeat if necessary. Timing varies with disease - see label for details.

- A period of at least 3 h without rain should follow spraying

Crop Safety/Restrictions

- Maximum number of treatments (including applications of any product containing benomyl, carbendazim or thiophanate-methyl) 2 per crop for cereals; 2 at normal rate or 2 at split plus 1 at normal rate for oilseed rape

Special precautions/Environmental safety

- Harmful in contact with skin [1]
- Irritating to eyes [1, 2]
- Irritating to skin [1]
- May cause sensitization by skin contact [2]
- Flammable [1]
- Dangerous to fish or other aquatic life. Do not contaminate surface waters or ditches with chemical or used container

Protective clothing/Label precautions

- A, C [1, 2]
- M03, R03a, R04b, R07d, U20a, E34 [1]; R04a, U05a, U08, C03, E01, E13b, E30a, E31b [1, 2]; R04e, U02, U20b, E26 [2]

Latest application/Harvest Interval(HI)

- Up to full ear emergence (GS 59) for barley, before flowering (GS 60) for wheat
- HI 6 wk for oilseed rape

Maximum Residue Levels (mg residue/kg food)

- see carbendazim entry

88 carbendazim + propiconazole

A contact and systemic fungicide for winter cereals

Products

1 Hispor 45 WP	Ciba Agric.	20:25% w/w	WP	01050
2 Hispor 45 WP	Novartis	20:25% w/w	WP	08418
3 Sparkle 45 WP	Ciba Agric.	20:25% w/w	WP	04968
4 Sparkle 45 WP	Novartis	20:25% w/w	WP	08450

Uses

Brown rust in WINTER BARLEY, WINTER WHEAT [1-4]. Brown rust in SPRING BARLEY [3, 4]. Eyespot in WINTER BARLEY, WINTER WHEAT [1-4]. Eyespot in SPRING BARLEY [3, 4]. Net blotch in WINTER BARLEY [1-4]. Net blotch in SPRING BARLEY [3, 4]. Powdery mildew in WINTER BARLEY, WINTER WHEAT [1-4]. Powdery mildew in SPRING BARLEY [3, 4]. Rhynchosporium in WINTER BARLEY [1-4]. Rhynchosporium in SPRING BARLEY [3, 4]. Septoria in WINTER WHEAT [1-4]. Yellow rust in WINTER BARLEY, WINTER WHEAT [1-4]. Yellow rust in SPRING BARLEY [3, 4].

Notes

Efficacy

- Major benefit obtained from spring treatment. Control provided for about 30 d
- Apply in spring from stage when leaf sheath begins to lengthen until first node detectable (GS 30-31). Further sprays may be applied up to fully emerged ear (GS 59)

Crop Safety/Restrictions

- Maximum number of treatments (including applications of any product containing benomyl, carbendazim or thiophanate-methyl) 2 per crop

Special precautions/Environmental safety

- Irritating to eyes and skin
- Dangerous to fish or other aquatic life. Do not contaminate surface waters or ditches with chemical or used container
- Harmful to bees. Do not apply to crops in flower or to those in which bees are actively foraging. Do not apply when flowering weeds are present

Protective clothing/Label precautions

- A, C
- R04a, R04b, U05a, U09a, U19, U20a, C02, C03, E01, E12e, E13b, E30a

Latest application/Harvest Interval(HI)

- Up to and including grain watery-ripe stage (GS 71)
- HI 35 d

Approval

- Approved for aerial application in wheat and barley. See notes in Section 1

Maximum Residue Levels (mg residue/kg food)

- see carbendazim and propiconazole entries

---

89 carbendazim + tebuconazole

A systemic fungicide mixture for oilseed rape and field beans

Products Bayer UK 413 | Bayer | 133:167 g/l | SC | 08277

Uses Alternaria in OILSEED RAPE. Canker in OILSEED RAPE. Chocolate spot in FIELD BEANS. Light leaf spot in OILSEED RAPE. Phoma leaf spot in OILSEED RAPE. Ring spot in OILSEED RAPE *(reduction)*. Rust in FIELD BEANS. Sclerotinia stem rot in OILSEED RAPE.

Notes

Efficacy

- Best results against light leaf spot and Phoma achieved from two spray programme starting in early autumn/winter
- Treat Alternaria in oilseed rape at onset of disease
- Treat Sclerotinia in oilseed rape at early flower to first petal fall
- Treat bean rust and chocolate spot at first signs of disease and repeat 3-4 wk later
- Tebuconazole is active against strains of chocolate spot resistant to MBC fungicides

Crop Safety/Restrictions

- Maximum number of treatments (including applications of any product containing benomyl, carbendazim or thiophanate-methyl) 2 per crop for field beans, oilseed rape

Special precautions/Environmental safety

- Harmful to fish or other aquatic life. Do not contaminate surface waters or ditches with chemical or used container

Protective clothing/Label precautions

- A, C, H, M
- U20b, C03, E01, E13c, E15, E30a, E31b, E34

Latest application/Harvest Interval(HI)
- Most seeds green stage (GS 6,3) for oilseed rape
- HI 35 d for field beans

Maximum Residue Levels (mg residue/kg food)
- see carbendazim entry

90 carbendazim + tecnazene

A protectant fungicide and sprout suppressant for stored potatoes

Products

1 Hickstor 6 + MBC	Hickson & Welch	2:6% w/w	DP	04176
2 Hortag Tecnacarb	Hortag	1.8:6% w/w	DS	02929
3 New Arena Plus	Hickson & Welch	2:6% w/w	DS	04598
4 New Hickstor 6 + MBC	Hickson & Welch	2:6% w/w	DP	04599
5 Tripart Arena Plus	Tripart	2:6% w/w	DP	05602

Uses

Dry rot in SEED POTATOES, WARE POTATOES [1-5]. Gangrene in SEED POTATOES, WARE POTATOES [1, 2, 5]. Silver scurf in SEED POTATOES, WARE POTATOES [1, 2, 5]. Silver scurf in SEED POTATOES *(reduction)*, WARE POTATOES *(reduction)* [3, 4]. Skin spot in SEED POTATOES, WARE POTATOES [1, 2, 5]. Skin spot in SEED POTATOES *(reduction)*, WARE POTATOES *(reduction)* [4]. Sprout suppression in WARE POTATOES [1-5].

Notes

Efficacy
- Potatoes should be dormant, have a mature skin and be dry and free from dirt
- Treat tubers as they go into store using a dusting machine
- Ensure even cover of tubers. Cover clamps with straw etc to aid vapour-phase transmission of a.i. Pack boxes as tightly together as possible
- Effectiveness of treatment is reduced if ventilation in store is inadequate or excessive
- Treatment can give protection for 3-4 mth but does not cure blemishes already present on tubers. It will not control sprouting if tubers have already broken dormancy

Crop Safety/Restrictions
- Maximum number of treatments (including applications of any product containing benomyl, carbendazim or thiophanate-methyl) 1 per batch
- Treated tubers must not be removed for sale or processing, including washing, for at least 6 wk after application
- Air seed potatoes for 6 wk and ensure that chitting has commenced before planting out. Treatment may delay emergence and possibly slightly reduce ware yield

Special precautions/Environmental safety
- Dangerous to fish or other aquatic life. Do not contaminate surface waters or ditches with chemical or used container [1, 5]
- Do not contaminate surface waters or ditches with chemical or used container [2-4]

Protective clothing/Label precautions
- A, C, D, H, M [1, 3-5]
- U19 [2-5]; U20a [3]; U20c [1, 2, 4, 5]; E13b [1, 5]; E15 [2-4]; E25 [2]; E30a [1, 3-5]; E32a [1-5]

Latest application/Harvest Interval(HI)
- 6 wk before sale or planting

Maximum Residue Levels (mg residue/kg food)
- see carbendazim and tecnazene entries

---

91 carbendazim + vinclozolin

A protectant and systemic fungicide for oilseed rape and field beans

Products	Konker	BASF	165:250 g/l	SC	03988

Uses

Alternaria in OILSEED RAPE *(reduction)*. Ascochyta in FIELD BEANS *(reduction)*. Botrytis in OILSEED RAPE *(reduction)*. Chocolate spot in FIELD BEANS *(moderate control)*. Light leaf spot in OILSEED RAPE *(reduction)*. Sclerotinia stem rot in OILSEED RAPE.

Notes

Efficacy
- Best results achieved if applied before any disease becomes well established
- In oilseed rape apply from start of rapid spring growth up to and including full flower
- In field beans treat at early to mid-flowering
- Timing depends on diseases present and weather conditions. See label for details

Crop Safety/Restrictions
- Maximum total dose equivalent to two full dose treatments
- Do not apply if rain or frost is expected or when the crop is wet

Special precautions/Environmental safety
- Irritant. May cause sensitisation by skin contact
- Harmful to fish or other aquatic life. Do not contaminate surface waters or ditches with chemical or used container
- Must be applied only by vehicle mounted or trailed hydraulic sprayers
- Vehicles must be fitted with a cab and a forced air filtration unit with a pesticide filter complying with HSE Guidance Note PM74, or equivalent

Protective clothing/Label precautions
- A, C, H, K, M
- R04, R04e, U05a, U08, U19, U20a, C03, E01, E13c, E30a, E31c

Latest application/Harvest Interval(HI)
- HI 7 wk for oilseed rape; 3 wk for field beans

Maximum Residue Levels (mg residue/kg food)
- see carbendazim and vinclozolin entries

---

92 carbetamide

A residual pre- and post-emergence carbamate herbicide

Products	Carbetamex	RP Agric.	70% w/w	WP	06186

Uses Annual grasses in CABBAGE SEED CROPS, COLLARDS, FODDER RAPE SEED CROPS, KALE SEED CROPS, LUCERNE, RED CLOVER, SAINFOIN, SPRING CABBAGE, SUGAR BEET SEED CROPS, SWEDE SEED CROPS, TURNIP SEED CROPS, WHITE CLOVER, WINTER FIELD BEANS, WINTER OILSEED RAPE. Some annual dicotyledons in CABBAGE SEED CROPS, COLLARDS, FODDER RAPE SEED CROPS, KALE SEED CROPS, LUCERNE, RED CLOVER, SAINFOIN, SPRING CABBAGE, SUGAR BEET STECKLINGS, SWEDE SEED CROPS, TURNIP SEED CROPS, WHITE CLOVER, WINTER FIELD BEANS, WINTER OILSEED RAPE. Volunteer cereals in CABBAGE SEED CROPS, COLLARDS, FODDER RAPE SEED CROPS, KALE SEED CROPS, LUCERNE, RED CLOVER, SAINFOIN, SPRING CABBAGE, SUGAR BEET STECKLINGS, SWEDE SEED CROPS, TURNIP SEED CROPS, WHITE CLOVER, WINTER FIELD BEANS, WINTER OILSEED RAPE.

Notes

Efficacy

- Best results achieved pre- or early post-emergence of weeds under cool, moist conditions. Adequate soil moisture is essential
- Dicotyledons controlled include chickweed, cleavers and speedwell
- Weed growth stopped rapidly though full effects may take 6-8 wk to develop
- Do not use on soils with more than 10% organic matter
- Do not apply during prolonged periods of cold weather when weeds fully dormant
- Various tank mixes effective against a wider range of dicotyledons are recommended. See label for details and for mixtures with other pesticides and sequential treatments

Crop Safety/Restrictions

- Maximum number of treatments 1 per crop
- Apply to brassicas from late autumn to late winter provided crop has at least 4 true leaves (spring cabbage, spring greens), 3-4 true leaves (seed crops, oilseed rape)
- Apply to established lucerne and sainfoin from mid-Oct to end Feb, to established red and white clover from Feb to mid-Mar
- After treatment do not sow brassicas or field beans for 2 wk, peas or runner beans for 8 wk, cereals or maize for 16 wk

Protective clothing/Label precautions

- U20c, E15, E30a, E32a

Latest application/Harvest Interval(HI)

- HI 6 wk

93 carbofuran

A systemic carbamate insecticide and nematicide for soil treatment

Products

1 Rampart	Sipcam	5% w/w	GR	05166
2 Tripart Nex	Tripart	5% w/w	GR	05165
3 Yaltox	Bayer	5% w/w	GR	02371

Uses Aphids in CARROTS, PARSNIPS, SWEDES, TURNIPS [1-3]. Beet leaf miner in FODDER BEET, MANGELS, SUGAR BEET [1-3]. Cabbage root fly in KOHLRABI *(off-label)* [3]. Cabbage root fly in BROCCOLI, BRUSSELS SPROUTS, CABBAGES, CALABRESE,

FOR FULL CONDITIONS OF USE ALWAYS READ THE PRODUCT LABEL

CAULIFLOWERS, SWEDES, TURNIPS, WINTER OILSEED RAPE [1-3]. Cabbage root fly in CHINESE CABBAGE, COLLARDS, KALE [1, 2]. Cabbage stem flea beetle in WINTER OILSEED RAPE [1-3]. Cabbage stem weevil in BROCCOLI, BRUSSELS SPROUTS, CABBAGES, CALABRESE, CAULIFLOWERS, SWEDES, TURNIPS [1-3]. Cabbage stem weevil in CHINESE CABBAGE, COLLARDS, KALE [1, 2]. Carrot fly in CARROTS, PARSNIPS [1-3]. Docking disorder vectors in FODDER BEET, MANGELS, SUGAR BEET [1-3]. Flea beetles in BROCCOLI, BRUSSELS SPROUTS, CABBAGES, CALABRESE, CAULIFLOWERS, FODDER BEET, MANGELS, RED BEET, SUGAR BEET, SWEDES, TURNIPS [1-3]. Flea beetles in CHINESE CABBAGE, COLLARDS, KALE [1, 2]. Free-living nematodes in CARROTS, FODDER BEET, MANGELS, PARSNIPS, SUGAR BEET [1-3]. Frit fly in MAIZE, SWEETCORN [1-3]. Large narcissus fly in NARCISSI *(off-label)* [3]. Millipedes in FODDER BEET, MANGELS, SUGAR BEET [1-3]. Onion fly in BULB ONIONS [1-3]. Potato cyst nematode in POTATOES [1-3]. Pygmy beetle in FODDER BEET, MANGELS, SUGAR BEET [1-3]. Rape winter stem weevil in WINTER OILSEED RAPE [1-3]. Soil pests in KOHLRABI *(off-label)* [3]. Springtails in FODDER BEET, MANGELS, SUGAR BEET [1-3]. Stem nematodes in BULB ONIONS [1-3]. Tortrix moths in SUGAR BEET [2]. Turnip root fly in SWEDES, TURNIPS [1-3]. Vine weevil in HARDY ORNAMENTAL NURSERY STOCK, POT PLANTS, STRAWBERRIES [1-3]. Wireworms in FODDER BEET, MANGELS, SUGAR BEET [1-3].

Notes

Efficacy

- Apply through a suitably calibrated granule applicator and incorporate into the soil
- Method of application, rate and timing vary with pest and crop. See label for details
- Performance is reduced in dry soil conditions
- Do not apply to potatoes in very wet or water-logged soils
- Controls only first generation carrot fly, use a follow-up application of a suitable specific insecticide for later attacks
- May be applied in mixture with granular fertilizers

Crop Safety/Restrictions

- Maximum number of treatments 1 per crop or yr
- When applied at drilling do not allow granules to come into contact with crop seed
- Use on carrots is limited to crops growing in mineral soils
- Use only on onions intended for harvest as mature bulbs
- Apply on ornamentals as surface treatment to moist soil after planting or potting up and follow immediately by thorough watering. Do not treat plants grown under cover, see label for details of species which have been treated safely
- On strawberries only apply after last harvest of year

Special precautions/Environmental safety

- This product contains an anticholinesterase carbamate compound. Do not use if under medical advice not to work with such compounds
- Harmful if swallowed
- Dangerous to game, wild birds and animals. Bury or remove spillages
- Keep in original container, tightly closed, in a safe place, under lock and key
- Dangerous to fish or other aquatic life. Do not contaminate surface waters or ditches with chemical or used container
- To minimise risk of enhanced biodegradation do not apply more than once every two yr to any cropping area

Protective clothing/Label precautions

- A, B, C or D+E, H, K, M
- M02, M03, R03c, U02, U04a, U05a, U09a, U13, U19, U20a, C03, E01, E06a, E10a, E13b, E30b, E32a, E34

Latest application/Harvest Interval(HI)

- At drilling when broadcast or used as a band treatment; 6 wk before harvest when applied post-emergence; at least 4 wk before release for sale or supply for ornamentals and pot plants

Approval

- Off-label approval to Jan 2001 for use on outdoor kohlrabi (OLA 1037/96)[3]

Maximum Residue Levels (mg residue/kg food)

- hops 10; radishes 0.5; carrots, parsnips, garlic, onions, shallots 0.3; broccoli, cauliflowers, kohlrabi, tea 0.2; tree nuts (except hazelnuts), grapes, blackberries, dewberries, loganberries, raspberries, bilberries, cranberries, currants, gooseberries, wild berries, miscellaneous fruits, beetroot, horseradish, Jerusalem artichokes, parsley root, salsify, sweet potatoes, yams, spring onions, fruiting vegetables, leafy vegetables and herbs, peas (with and without pods), asparagus, cardoons, fennel, globe artichokes, rhubarb, fungi, lentils, peas, poppy seed, sesame seed, mustard seed, wheat, rye, barley, triticale, maize, animal products 0.1

94 carbon dioxide (commodity substance)

A gas for the control of trapped rodents and other vertebrates

Products carbon dioxide | various | 99.9% | GA

Uses Birds in TRAPS. Mice in TRAPS. Rats in TRAPS.

Notes

Efficacy

- Use to control trapped rodent pests
- Use to control birds covered by general licences issued by the Agriculture and Environment Departments under Section 16(1) of the Wildlife and Countryside Act (1981) for the control of opportunistic bird species, where birds have been trapped or stupefied with alphachloralose/seconal

Special precautions/Environmental safety

- Operators must wear self-contained breathing apparatus when carbon dioxide levels are greater than 0.5% v/v
- Operators must be suitably trained and competent
- Unprotected persons and non-target animals must be excluded from the treatment enclosures and surrounding areas unless the carbon dioxide levels are below 0.5% v/v

Approval

- Only to be used where a licence has been issued in accordance with Section 16(1) of the Wildlife and Countryside Act 1981

95 carbosulfan

A systemic carbamate insecticide for control of soil pests

Products

1 Landgold Carbosulfan 10G	Landgold	10% w/w	GR	09019
2 Marshal 10G	RP Agric.	10% w/w	GR	06165
3 Marshal/suSCon	Fargro	10% w/w	CG	06978
4 Standon Carbosulfan 10G	Standon	10% w/w	GR	05671

Uses

Aphids in FODDER BEET, MANGELS [2]. Aphids in CARROTS, PARSNIPS, SUGAR BEET [1, 2, 4]. Cabbage root fly in COLLARDS [2]. Cabbage root fly in BROCCOLI, BRUSSELS SPROUTS, CABBAGES, CALABRESE, CAULIFLOWERS, SWEDES, TURNIPS [1, 2, 4]. Cabbage stem weevil in COLLARDS [2]. Cabbage stem weevil in BROCCOLI, BRUSSELS SPROUTS, CABBAGES, CALABRESE, CAULIFLOWERS, SWEDES, TURNIPS [1, 2, 4]. Carrot fly in CARROTS, PARSNIPS [1, 4]. Flea beetles in COLLARDS, FODDER BEET, MANGELS [2]. Flea beetles in BROCCOLI, BRUSSELS SPROUTS, CABBAGES, CALABRESE, CAULIFLOWERS, SUGAR BEET, SWEDES, TURNIPS [1, 2, 4]. Free-living nematodes in FODDER BEET, MANGELS [2]. Free-living nematodes in CARROTS, PARSNIPS, SUGAR BEET [1, 2, 4]. Large pine weevil in FORESTRY [3]. Mangold fly in FODDER BEET, MANGELS [2]. Mangold fly in SUGAR BEET [1, 2, 4]. Millipedes in FODDER BEET, MANGELS [2]. Millipedes in SUGAR BEET [1, 2, 4]. Pygmy mangold beetle in FODDER BEET, MANGELS [2]. Pygmy mangold beetle in SUGAR BEET [1, 2, 4]. Springtails in FODDER BEET, MANGELS [2]. Springtails in SUGAR BEET [1, 2, 4]. Symphylids in FODDER BEET, MANGELS [2]. Symphylids in SUGAR BEET [1, 2, 4]. Wireworms in FODDER BEET, MANGELS [2]. Wireworms in SUGAR BEET [1, 2, 4].

Notes

Efficacy

- At recommended rates seed is not damaged by contact with product
- Apply with suitable granule applicator feeding directly into seed furrow or immediately behind drill coulter (behind seed drill boot for brassicas) or use bow-wave technique
- For transplanted brassicas apply sub-surface with 'Leeds' coulter
- See label for details of suitable applicators and settings. Correct calibration is essential
- Where used in forestry, granules should be placed in the planting hole by hand or metered applicator before or after placing the tree [3]
- Where applied annually in intensive brassica growing areas enhanced biodegradation by soil organisms may lead to unsatisfactory control

Crop Safety/Restrictions

- Maximum number of treatments 1 per crop or tree
- Do not mix with compost for blockmaking or use in Hassy trays for brassicas
- Safe to Douglas Fir and Sitka Spruce. Other conifers may vary in their sensitivity [3]
- Forest trees may take 10-15 d to achieve full protection. An additional pre-planting insecticide dip or spray is advised [3]

Special precautions/Environmental safety

- This product contains an anticholinesterase carbamate compound. Do not use if under medical advice not to work with such compounds
- Harmful if swallowed
- Dangerous to game, wild birds and animals. Bury spillages
- Dangerous to fish or other aquatic life. Do not contaminate surface waters or ditches with chemical or used container
- Keep in original container, tightly closed, in a safe place, under lock and key

Protective clothing/Label precautions
- A [3]; B [1-4]; C or D+E [2, 4]; C, D, E [1, 3]; H [1, 2, 4]; K, M
- M02, U02, U04a, U05a, U13, U20a, C03, E01, E10a, E13b [1-4]; M03, R03c, U09a, U19, E30b, E32a, E34 [1, 2, 4]; U10, E30a [3]

Latest application/Harvest Interval(HI)
- HI leaf brassicas 56 d [2, 4]; sugar beet, fodder beet, mangels, swedes, turnips, carrots, parsnips 100 d

Maximum Residue Levels (mg residue/kg food)
- carrots, parsnips, tea 0.1; tree nuts, berries and small fruit, miscellaneous fruits, beetroot, celeriac, horseradish, Jerusalem artichokes, parsley root, radishes, salsify, sweet potatoes, yams, garlic, shallots, spring onions, tomatoes, peppers, aubergines, cucumbers, gherkins, courgettes, sweet corn, leaf vegetables and herbs, legume vegetables, asparagus, cardoons, fennel, globe artichokes, rhubarb, mushrooms, pulses, oilseeds (except sunflower seed, cotton seed), potatoes, cereals, animal products 0.05

96 carboxin

An carboxamide fungicide available only in mixtures

97 carboxin + gamma-HCH + thiram

A fungicide and insecticide dressing for rape seed

Products

Vitavax RS	Uniroyal	45:675:90 g/l	FS	08040

Uses

Alternaria in FLAX, LINSEED. Botrytis in FLAX, LINSEED. Canker in OILSEED RAPE. Damping off in OILSEED RAPE. Flea beetles in FLAX, LINSEED, OILSEED RAPE. Fusarium in FLAX *(partial control)*, LINSEED *(partial control)*.

Notes

Efficacy
- Apply through a conventional seed treater suitable for handling flowable products and equipped with a secondary mixing auger
- Keep at temperatures above 10°C prior to and during application
- Treated seed can affect seed flow in the drill. Always recalibrate before sowing treated seed

Crop Safety/Restrictions
- Maximum number of treatments 1 per batch
- Some varieties of flax and linseed (e.g. Flanders, Laura) may show reduced emergence in certain conditions. Consult manufacturer for information on sensitive varieties
- Treat seed as near to drilling time as possible. Do not treat cracked, split or sprouted seed or with moisture content over 9%
- Do not store treated seed for more than 3 mth
- Do not plant potatoes or carrots within 18 mth of sowing treated flax or linseed

Special precautions/Environmental safety

- Harmful in contact with skin, by inhalation and if swallowed
- Irritating to eyes and skin
- Do not use treated seed as food or feed
- Treated seed harmful to game and wildlife
- Extremely dangerous to fish or other aquatic life. Do not contaminate surface waters or ditches with chemical or used container

Protective clothing/Label precautions

- A, C, D, H
- M03, R03a, R03b, R03c, R04a, R04b, U05a, U08, U20b, C03, E01, E13a, E26, E27, E30a, E31a, E34, S01, S02, S03, S04b, S05, S06, S07

Maximum Residue Levels (mg residue/kg food)

- see gamma-HCH entry

---

98 carboxin + thiram

A fungicide seed dressing for cereals

Products

Anchor	Uniroyal	200:200 g/l	FS	08684

Uses

Bunt in SPRING WHEAT, WINTER WHEAT. Fusarium foot rot and seedling blight in SPRING BARLEY, SPRING OATS, SPRING RYE, SPRING WHEAT, TRITICALE, WINTER BARLEY, WINTER OATS, WINTER RYE, WINTER WHEAT. Leaf stripe in SPRING BARLEY *(partial control)*, WINTER BARLEY *(partial control)*. Loose smut in SPRING BARLEY *(partial control)*, WINTER BARLEY *(partial control)*. Septoria seedling blight in SPRING WHEAT, WINTER WHEAT.

Notes

Efficacy

- Apply through suitable liquid flowable seed treating equipment of the batch treatment or continuous flow type where a secondary mixing auger is fitted
- Do not store treated seed from one season to the next
- Drill flow may be affected by treatment. Always re-calibrate seed drill before use

Crop Safety/Restrictions

- Maximum number of treatments 1 per batch of seed
- Do not treat seed with moisture content above 16%
- Do not apply to cracked, split or sprouted seed

Special precautions/Environmental safety

- Harmful to fish or other aquatic life. Do not contaminate surface waters or ditches with chemical or used container
- Do not use treated seed as food or feed
- Treated seed harmful to game and wildlife

Protective clothing/Label precautions

- A, D, E, H
- R03b, R04a, R04b, R04c, U09a, U20c, E13c, E26, E27, E30a, E31a, E34, S01, S02, S04b, S05, S06, S07

99 carfentrazone-ethyl

A triazolinone herbicide available only in mixture

100 carfentrazone-ethyl + flupyrsulfuron-methyl

A foliar and residual acting herbicide for cereals

Products					
Lexus Class WSB	DuPont	33.3:16.7% w/w	WB	08637

Uses

Annual dicotyledons in TRITICALE, WINTER OATS, WINTER RYE, WINTER WHEAT. Blackgrass in TRITICALE, WINTER OATS, WINTER RYE, WINTER WHEAT.

Notes

Efficacy

- Best results obtained when applied to small actively growing weeds
- Good spray cover of weeds must be obtained
- Increased degradation of active ingredient in high soil temperatures reduces residual activity
- Weed control may be reduced in dry soil conditions but susceptible weeds germinating soon after treatment will be controlled if adequate soil moisture present
- Blackgrass should be treated from 1 leaf but before first node. Strains of blackgrass resistant to other herbicides may not be controlled

Crop Safety/Restrictions

- Maximum number of treatments 1 per crop
- Apply to recommended crops from 2 leaf stage (GS 12) up to 31 Dec on oats, rye, triticale, or before first node detectable (GS 31) on wheat
- Do not use on barley, on crops undersown with grasses or legumes, or any other broad-leaved crop
- Do not apply within 7 d of rolling
- Do not treat any crop suffering from drought, waterlogging, pest or disease attack, nutrient deficiency, or any other stress factors
- Only cereals, oilseed rape, field beans, clover or grass may be sown in the yr of harvest of a treated crop
- In the event of crop failure only winter wheat may be sown 1-3 mth after treatment. Land should be ploughed and cultivated to 15 cm minimum before resowing
- Slight chlorosis and stunting may occur in certain conditions. Recovery is rapid and yield not affected

Special precautions/Environmental safety

- Irritant. May cause sensitization by skin contact
- Dangerous to fish or other aquatic life. Do not contaminate surface waters or ditches with chemical or used container
- Take extreme care to avoid damage by drift outside the target area or onto surface waters or ditches
- Spraying equipment should not be drained or flushed onto land intended for crops other than cereals and should be thoroughly cleansed after use - see label for instructions

Protective clothing/Label precautions

- A

- R04, R04e, U05a, U08, U14, U19, U20b, U22, C03, E01, E13b, E15, E30a

Latest application/Harvest Interval(HI)
- Before 31 Dec in yr of sowing

101 carfentrazone-ethyl + isoproturon

A foliar and residual acting herbicide for cereals

Products

Affinity	DuPont	0.75:50% w/w	WG	08639

Uses

Annual dicotyledons in WINTER BARLEY, WINTER WHEAT. Annual meadow grass in WINTER BARLEY, WINTER WHEAT. Chickweed in WINTER BARLEY, WINTER WHEAT. Cleavers in WINTER BARLEY, WINTER WHEAT. Loose silky bent in WINTER BARLEY, WINTER WHEAT. Mayweeds in WINTER BARLEY, WINTER WHEAT. Rough meadow grass in WINTER BARLEY, WINTER WHEAT.

Notes

Efficacy
- Good weed control depends on burying and dispersing any straw to 15 cm before or during seedbed preparation
- Best results obtained by treatment post weed emergence in crops growing in firm seedbeds with clods no greater than fist size
- Residual activity reduced on soils with more than 10% organic matter. Treat crops on such soils in spring
- Weed control may be reduced if heavy rain falls within a few hours of treatment or if prolonged dry weather follows spring application
- Do not harrow at any time after treatment

Crop Safety/Restrictions
- Maximum number of treatments 1 per crop
- Treat from 2 leaf stage of crop. Do not apply pre-emergence to wheat or barley
- Crops should be drilled to 25 mm and the seeds well covered with soil. Ensure good root system is established on broadcast crops
- Crops drilled in Sep may be damaged if a period of rapid growth follows treatment
- Do not treat undersown crops or those to be undersown
- Transient necrotic spotting may occur after treatment particularly on crops subject to wide diurnal temperature fluctuations. Symptoms disappear within 3-4 wk and yield is not affected
- Do not roll in spring 1 wk before or after application
- In the event of crop failure only wheat, barley, maize, durum wheat, oats or flax may be sown 1 mth after treatment in Dec/Jan. Only maize, wheat or barley may be sown 1 mth after treatment in Feb/Mar. Any crop may be sown 3 mth after treatment

Special precautions/Environmental safety
- Irritant. May cause sensitization by skin contact
- Dangerous to fish or other aquatic life. Do not contaminate surface waters or ditches with chemical or used container
- Do not apply where soils are cracked, to avoid run-off through drains
- Take extreme care to avoid drift onto broad-leaved plants outside the target area or onto ponds waterways or ditches, or onto land intended for cropping

Protective clothing/Label precautions
- A, C, D, H
- R04, R04e, U05a, U08, U13, U14, C03, E01, E13b, E26, E30a, E31b, E34

Latest application/Harvest Interval(HI)
- Before pseudostem erect stage (GS 30)

102 carfentrazone-ethyl + mecoprop-P

A foliar applied herbicide for cereals

Products

Platform S	DuPont	1.5:60% w/w	WG	08638

Uses

Charlock in SPRING BARLEY, SPRING WHEAT, WINTER BARLEY, WINTER WHEAT. Chickweed in SPRING BARLEY, SPRING WHEAT, WINTER BARLEY, WINTER WHEAT. Cleavers in SPRING BARLEY, SPRING WHEAT, WINTER BARLEY, WINTER WHEAT. Red dead-nettle in SPRING BARLEY, SPRING WHEAT, WINTER BARLEY, WINTER WHEAT. Speedwells in SPRING BARLEY, SPRING WHEAT, WINTER BARLEY, WINTER WHEAT.

Notes

Efficacy
- Best results obtained when weeds have germinated and growing vigorously in warm moist conditions
- Treatment of large weeds and poor spray coverage may result in reduced weed control

Crop Safety/Restrictions
- Maximum number of treatments 1 per crop. The total amount of mecoprop-P applied in a single yr must not exceed the maximum total dose approved for any single product for the crop/situation
- Can be used on all varieties of wheat and barley in autumn or spring from the beginning of tillering
- Do not treat crops suffering from stress from any cause
- Early sown crops may be prone to damage if treated after period of rapid growth in autumn
- Do not treat crops undersown or to be undersown
- In the event of crop failure, any cereal, maize, oilseed rape, peas, vetches or sunflowers may be sown 1 mth after a spring treatment. Any crop may be planted 3 mth after treatment

Special precautions/Environmental safety
- Irritant. Risk of serious damage to eyes
- May cause sensitization by skin contact
- Dangerous to fish or other aquatic life. Do not contaminate surface waters or ditches with chemical or used container

Protective clothing/Label precautions
- A, C, H, M
- R04, R04d, R04e, U05a, U08, U11, U13, U14, U20b, C03, E01, E07 (2 wk), E13b, E30a, E32a, E34

Latest application/Harvest Interval(HI)
- Before 3rd node detectable (GS 33)

103 carfentrazone-ethyl + metsulfuron-methyl

A foliar applied herbicide mixture for cereals

Products

1 Ally Express	DuPont	40:10% w/w	WG	08640
2 Standon Carfentrazone MM	Standon	40:10% w/w	WG	09159

Uses

Annual dicotyledons in SPRING BARLEY, SPRING OATS, SPRING WHEAT, WINTER BARLEY, WINTER OATS, WINTER WHEAT.

Notes

Efficacy
- Best results achieved from applications made in good growing conditions
- Good spray cover of weeds must be obtained
- Growth of weeds is inhibited within hours of treatment but the time taken for visible colour changes to appear will vary according to species and weather
- Product has short residual life in soil. Under normal moisture conditions susceptible weeds germinating soon after treatment will be controlled
- Can be used on all soil types
- Apply in the spring from the 3-leaf stage on all crops

Crop Safety/Restrictions
- Maximum number of treatments 1 per crop
- Slight necrotic spotting of crops can occur under certain crop and soil conditions. Recovery is rapid and there is no effect on grain yield or quality
- Do not use on any crop suffering stress from drought, waterlogging, cold, pest or disease attack, nutrient deficiency
- Do not use on crops undersown with grass or legumes or on any broad-leaved crop
- Do not apply within 7 d of rolling
- Do not apply to a crop already treated with any sulfonylurea product except those containing flupyrsulfuron alone or in mixture with carfentrazone-ethyl
- Use recommended procedure for cleaning spray equipment (see label)
- Only cereals, oilseed rape, field beans or grass may be sown as a following crop in the same calendar yr. Any crop may follow in the next spring. In the event of failure of a treated crop, only winter wheat may be sown 1-3 mth later after ploughing and cultivating to at least 15 cm

Special precautions/Environmental safety
- Irritant. May cause sensitization by skin contact
- Extremely dangerous to fish or other aquatic life. Do not contaminate surface waters or ditches with chemical or used container
- Do not allow direct spray from vehicle mounted/drawn hydraulic sprayers to fall within 6 m, or from hand-held sprayers to within 2 m, of surface waters or ditches. Direct spray away from water
- Take extreme care to avoid drift onto broad-leaved plants outside the target area or onto ponds waterways or ditches, or onto land intended for cropping

Protective clothing/Label precautions
- A, H
- R04, R04e, U05a, U08, U14, U19, U20b, C03, E01, E13a, E16, E30a, E32a

Latest application/Harvest Interval(HI)
- Before 3rd node detectable (GS 33)

104 chlordane

A persistent organochlorine earthworm killer, all approvals for which were revoked in 1992. Chlordane is an environmental hazard and is banned under the EC "Prohibition Directive"

105 chlorfenvinphos

A contact and ingested soil-applied organophosphorus insecticide

Products

1 Birlane 24	Cyanamid	240 g/l	EC	07002
2 Sapecron 240 EC	Ciba Agric.	240 g/l	EC	01861
3 Sapecron 240 EC	Novartis	240 g/l	EC	08440

Uses

Cabbage root fly in BROCCOLI, BRUSSELS SPROUTS, CABBAGES, CAULIFLOWERS [1-3]. Cabbage root fly in KOHLRABI *(off-label)* [1]. Carrot fly in CARROTS, PARSNIPS [1-3]. Carrot fly in CELERIAC *(off-label)* [1]. Colorado beetle in POTATOES *(off-label)* [1].

Notes

Efficacy
- Soil incorporation recommended for most treatments. Application method, timing, rate and frequency vary with pest, crop and soil type. See labels for details
- Overall pre-planting sprays are less effective than band treatment and should not be used in areas of heavy cabbage root fly infestation [2, 3]
- Cabbage and cauliflower grown in peat blocks can be protected from cabbage root fly by incorporation into the peat used for blocking [2, 3]
- Efficacy reduced on highly organic or very dry soils unless crop irrigated

Crop Safety/Restrictions
- Maximum number of treatments 1 per crop for most crops, but varies with dose rate and product. See label for details
- On carrots the maximum total dose applied per crop must not exceed the equivalent of 3 (on mineral soils) or 4 (on organic soils) full dose applications
- Danger of scorch if applied to maize during very hot weather or if crop is under stress
- Do not apply in conjunction with seed treatments containing gamma-HCH

Special precautions/Environmental safety
- Chlorfenvinphos is subject to the Poisons Rules 1982 and the Poisons Act 1972. See notes in Section 1
- This product contains an anticholinesterase organophosphorus compound. Do not use if under medical advice not to work with such compounds
- Keep in original container, tightly closed, in a safe place, under lock and key
- Toxic in contact with skin and if swallowed
- Irritating to eyes and skin
- Flammable

- Dangerous to bees. Do not apply to crops in flower or to those in which bees are actively foraging except as directed. Do not apply when flowering weeds are present [1]
- Dangerous (extremely dangerous [1]) to fish or other aquatic life. Do not contaminate surface waters or ditches with chemical or used container [2, 3]
- Do not allow direct spray from ground-based vehicle-mounted/drawn sprayers to fall within 6 m of surface waters or ditches. Direct spray away from water
- Must not be applied by hand-held sprayers

Protective clothing/Label precautions

- A, C, H
- M01, M04, R02a, R02c, R04a, R04b, R07d, U02, U04a, U05a, U13, U19, C03, E01, E16, E30b, E31b, E34 [1-3]; U08, U14, U15, U20a, C02, E13b [2, 3]; U10, U20b, C02 (3 wk), E12d, E13a [1]

Latest application/Harvest Interval(HI)

- HI brassicas, kohlrabi, carrots, celery, celeriac, maize, potatoes 3 wk [1]; parsnips 3 wk [1], 12 wk [2, 3]

Approval

- Off-label approval unlimited for use on kohlrabi, celeriac (OLA 1747/96)[1]

Maximum Residue Levels (mg residue/kg food)

- citrus fruits 1; carrots, horseradish, parsnips, parsley root, salsify, swedes, turnips, garlic, onions, shallots, celery, rhubarb, potatoes 0.5; meat 0.2; tomatoes, peppers, aubergines, cucumbers, gherkins, courgettes, cauliflowers, Brussels sprouts, head cabbage, lettuce, beans, peas, leeks 0.1; pome fruits, apricots, peaches, nectarines, plums, grapes, strawberries, cane fruits, bilberries, cranberries, currants, gooseberries, bananas, mushrooms 0.05; milk 0.008

106 chloridazon

A residual pyridazinone herbicide for beet crops

Products

Product	Company	Content	Formulation	Reg. no.
1 Better DF	Sipcam	65% w/w	SG	06250
2 Better Flowable	Sipcam	430 g/l	SC	04924
3 Gladiator DF	Tripart	65% w/w	SG	06342
4 Luxan Chloridazon	Luxan	430 g/l	SC	06304
5 Portman Weedmaster	Portman	430 g/l	SC	06018
6 Pyramin DF	BASF	65% w/w	SG	03438
7 Questar	BASF	430 g/l	SC	07955
8 Starter Flowable	Truchem	430 g/l	SC	03421
9 Takron	BASF	430 g/l	SC	06237
10 Tripart Gladiator	Tripart	430 g/l	SC	00986

Uses

Annual dicotyledons in BULB ONIONS *(off-label)*, LEEKS *(off-label)*, SALAD ONIONS *(off-label)* [6]. Annual dicotyledons in SUGAR BEET [1-3, 5-10]. Annual dicotyledons in FODDER BEET, MANGELS [1-10]. Annual meadow grass in BULB ONIONS *(off-label)*, LEEKS *(off-label)*, SALAD ONIONS *(off-label)* [6]. Annual meadow grass in SUGAR BEET [1-3, 5-10]. Annual meadow grass in FODDER BEET, MANGELS [1-10].

Notes

Efficacy

- Absorbed by roots of germinating weeds and best results achieved pre-emergence of weeds or crop when soil moist and adequate rain falls after application

- Apply pre-emergence as soon as possible after drilling in mid-Mar to mid-Apr on fine, firm, clod-free seedbed
- Where crop drilled after mid-Apr or soil dry apply pre-drilling and incorporate to 2.5 cm immediately afterwards
- Application rate depends on soil type. See label for details
- Various tank mixes recommended on sugar beet for pre- and (except [7, 9]) post-emergence use and as repeated low dose treatments. See label for details

Crop Safety/Restrictions

- Maximum number of treatments 1 per crop for fodder beet and mangels; 1 per crop [3-5, 7, 9], 1 pre-emergence + 1 post-emergence [1], 1 pre-emergence + 3 post-emergence [1, 6] for sugar beet
- Maximum dose 1.4 kg product/ha for onions and leeks [6]
- Do not use on Coarse Sands, Sands or Fine Sands or where organic matter exceeds 5%
- Crop vigour may be reduced by treatment of crops growing under unfavourable conditions including poor tilth, drilling at incorrect depth, soil capping, physical damage, pest or disease damage, excess seed dressing, trace-element deficiency or a sudden rise in temperature after a cold spell
- In the event of crop failure only sugar or fodder beet, mangels or maize may be re-drilled on treated land after cultivation
- Winter cereals may be sown in autumn after ploughing. Any spring crop may follow treated beet crops harvested normally

Special precautions/Environmental safety

- Irritating to eyes and skin [5, 8, 10]
- Irritating to respiratory system [5, 8, 10]
- May cause sensitization by skin contact
- Harmful to fish or other aquatic life. Do not contaminate surface waters or ditches with chemical or used container

Protective clothing/Label precautions

- A [5, 7-10]; C [5]
- M03, R04, R04e, E31c [7, 9]; R04a, R04b, E31b [5, 8, 10]; R04c [5, 10]; U05a, C03, E01 [5, 7-10]; U08, E13c, E30a [1-10]; U14, U15 [8]; U19 [5, 7-9]; U20a [2-7, 10]; U20b [1, 8, 9]; E26 [5, 8]; E32a [1-4, 6]; E34 [5]

Latest application/Harvest Interval(HI)

- Pre-emergence for fodder beet and mangels; pre-emergence [3, 7, 9], 8-true leaf stage [1-3], before leaves of crop meet in row [4, 6, 8] for sugar beet; up to and including second true leaf stage for onions and leeks [6]

Approval

- Off-label approval unlimited for use on salad and bulb onions and leeks (OLA 0732/97)[6]

107 chloridazon + ethofumesate

A residual pre-emergence herbicide for beet crops

Products	Magnum	BASF	275:170 g/l	SC	08635

FOR FULL CONDITIONS OF USE ALWAYS READ THE PRODUCT LABEL

Uses Annual dicotyledons in FODDER BEET, SUGAR BEET. Annual meadow grass in FODDER BEET, SUGAR BEET.

Notes

Efficacy

- Apply pre- or post-emergence up to cotyledon stage of weeds, on a fine, firm, clod-free seedbed
- Best results achieved when soil moist and adequate rain falls after spraying. Efficacy may be reduced if heavy rain falls just after incorporation
- A reduction in effectiveness may occur under conditions of low pH
- May be used on soil classes Loamy Sand - Silty Clay Loam
- May be applied by conventional or repeat low dose method. See label for details
- Recommended for tank mixing with triallate on sugar beet incorporated pre-drilling and with various other beet herbicides early post-emergence. See labels for details

Crop Safety/Restrictions

- Maximum number of treatments 1 pre-emergence for fodder beet, 1 pre-emergence plus 3 post-emergence for sugar beet
- Crop vigour may be reduced by treatment of crops growing under unfavourable conditions including poor tilth, drilling at incorrect depth, soil capping, physical damage, excess nitrogen, excess seed dressing, trace element deficiency or a sudden rise in temperature after a cold spell. Frost after pre-emergence treatment may check crop growth
- In the event of crop failure only sugar beet, fodder beet or mangels may be re-drilled
- Any crop may be sown 3 mth after spraying following ploughing to 15 cm

Special precautions/Environmental safety

- Harmful to fish or other aquatic life. Do not contaminate surface waters or ditches with chemical or used container

Protective clothing/Label precautions

- A, C
- U08, U19, U20a, E01, E13c, E30a, E31c

Latest application/Harvest Interval(HI)

- Pre-emergence for sugar beet; pre-emergence for fodder beet, mangels; before crop leaves meet between rows for sugar beet

---

108 chloridazon + lenacil

A residual pre-emergence herbicide for beet crops

Products	Advizor	DuPont	200:133 g/l	SC	06571

Uses Annual dicotyledons in FODDER BEET, MANGELS, SUGAR BEET. Annual meadow grass in FODDER BEET, MANGELS, SUGAR BEET.

Notes

Efficacy

- Apply at or immediately after drilling before emergence of crop or weeds
- Best results achieved from application to firm, moist, weed-free seedbed when adequate rain falls afterwards
- Application rate depends on soil type. See label for details
- May be used on Very Light, Light or Medium soils. Not on Sands, stony or gravelly soils, Heavy soils or those with more than 10% organic matter

- Recommended for tank-mixing with Avadex BW incorporated pre-drilling on sugar beet and with various other herbicides pre- or early post-emergence. See label for details

Crop Safety/Restrictions

- Maximum number of treatments 1 per crop pre-drilling (sugar beet) or 1 pre-emergence (sugar beet, fodder beet, mangels) followed in either case by 1 post-emergence (or 3 post-emergence at low dose) per crop
- Crops should be drilled to at least 15 mm and the seed well covered
- Heavy rainfall after spraying may reduce crop stand especially when such rain is followed by very hot weather
- Only use in post-emergence mixtures on crops growing vigorously and not stressed by drought, pest attack, deficiency or other factors
- In the event of crop failure only beet crops may be re-drilled on treated land and no further application should be made for at least 4 mth. Any crop may be sown 4 mth after treatment following ploughing to 15 cm

Special precautions/Environmental safety

- Irritating to eyes, skin and respiratory system
- Harmful to fish or other aquatic life. Do not contaminate surface waters or ditches with chemical or used container

Protective clothing/Label precautions

- R04a, R04b, R04c, U05a, U08, U19, U20a, C03, E01, E13c, E30a, E31a, E34

Latest application/Harvest Interval(HI)

- Before crop plants meet across rows

109 chloridazon + propachlor

A residual pre-emergence herbicide for use in onions and leeks

Products

Ashlade CP	Ashlade	86:400 g/l	SC	06481

Uses

Annual dicotyledons in BULB ONIONS, BULB ONIONS *(off-label, post-emergence)*, LEEKS, LEEKS *(off-label, post-emergence)*, SALAD ONIONS, SALAD ONIONS *(off-label, post-emergence)*. Annual grasses in BULB ONIONS, BULB ONIONS *(off-label, post-emergence)*, LEEKS, LEEKS *(off-label, post-emergence)*, SALAD ONIONS, SALAD ONIONS *(off-label, post-emergence)*.

Notes

Efficacy

- Best results achieved from application to firm, moist, weed-free seedbed when adequate rain falls afterwards
- Apply pre-emergence of sown crops, preferably soon after drilling, before weeds emerge. Loose or fluffy seedbeds must be consolidated before application
- Apply to transplanted crops when soil has settled after planting
- Do not use on soils with more than 10% organic matter

Crop Safety/Restrictions

- Maximum number of treatments 1 per crop

- Ensure crops are drilled to 20 mm depth
- Crops stressed by nutrient deficiency, pests or diseases, poor growing conditions or pesticide damage may be checked by treatment, especially on sandy or gravelly soils
- In the event of crop failure only onions, leeks or maize should be planted
- Any crop can follow a treated onion or leek crop harvested normally as long as the ground is cultivated thoroughly before drilling

Special precautions/Environmental safety

- Irritating to eyes and skin
- Harmful to fish or other aquatic life. Do not contaminate surface waters or ditches with chemical or used container

Protective clothing/Label precautions

- A, C
- M03, R04a, R04b, U02, U04a, U05a, U08, U13, U19, U20a, C03, E01, E13c, E30a, E31b

Latest application/Harvest Interval(HI)

- Pre-emergence of crop; before 2 true leaf stage for onions and leeks (off-label).
- HI 12 wk for chives

Approval

- Off-label Approval unlimited for use post-emergence on bulb onions, leeks (OLA 1362/98)[1]

110 chlormequat

A plant-growth regulator for reducing stem growth and lodging

Products

1 Adjust	Mandops	620 g/l	SL	05589
2 Ashlade 460 CCC	Ashlade	460 g/l	SL	06474
3 Ashlade 700 CCC	Ashlade	700 g/l	SL	06473
4 Atlas 3C:645 Chlormequat	Atlas	645 g/l	SL	07700
5 Atlas Chlormequat 46	Atlas	460 g/l	SL	07704
6 Atlas Chlormequat 700	Atlas	700 g/l	SL	07708
7 Atlas Terbine	Atlas	730 g/l	SL	07709
8 Atlas Tricol	Atlas	670 g/l	SL	07707
9 Barclay Holdup	Barclay	700 g/l	SL	06799
10 Barclay Holdup 600	Barclay	600 g/l	SL	08794
11 Barclay Holdup 640	Barclay	640 g/l	SL	08795
12 Barleyquat B	Mandops	620 g/l	SL	06001
13 Barleyquat B	Mandops	620 g/l	SL	07051
14 BASF 3C Chlormequat 720	BASF	720 g/l	SL	06514
15 Bettaquat B	Mandops	620 g/l	SL	06004
16 Bettaquat B	Mandops	620 g/l	SL	07050
17 Fargro Chlormequat	Fargro	460 g/l	SL	02600
18 Hyquat 70	Agrichem	700 g/l	SL	03364
19 Intracrop Balance	Intracrop	750 g/l	SL	08037
20 Intracrop MCCC	Intracrop	750 g/l	SL	08506
21 Landgold CCC 720	Landgold	720 g/l	SL	08527
22 Mandops Chlormequat 700	Mandops	700 g/l	SL	06002
23 Manipulator	Mandops	620 g/l	SL	05871
24 MSS Chlormequat 40	Mirfield	400 g/l	SL	01401
25 MSS Chlormequat 460	Mirfield	460 g/l	SL	03935
26 MSS Chlormequat 60	Mirfield	600 g/l	SL	03936
27 MSS Chlormequat 70	Mirfield	700 g/l	SL	03937
28 MSS Mirquat	Mirfield	730 g/l	SL	08166
29 Portman Chlormequat 700	Portman	700 g/l	SL	03465
30 Quadrangle Chlormequat 700	Quadrangle	700 g/l	SL	03401
31 Renown	Stefes	640 g/l	SL	09058
32 Sigma PCT	Atlas	460 g/l	SL	08663
33 Stefes CCC 640	Stefes	640 g/l	SL	06993
34 Stefes CCC 700	Stefes	700 g/l	SL	07116
35 Stefes CCC 720	Stefes	720 g/l	SL	05834
36 Tripart Brevis	Tripart	700 g/l	SL	03754
37 Uplift	United Phosphorus	700 g/l	SL	07527

Uses

Increasing yield in WINTER BARLEY [1-3, 9-13, 22, 23, 29, 31, 36]. Increasing yield in SPRING BARLEY [22]. Lodging control in SPRING BARLEY [36]. Lodging control in SPRING WHEAT, WINTER WHEAT [1-11, 14-16, 18-36]. Lodging control in WINTER BARLEY [1-9, 11-14, 19-21, 23-29, 31, 32, 35-37]. Lodging control in WINTER RYE [2, 3, 36]. Lodging control in SPRING OATS [2-11, 14-16, 18, 21, 22, 24-31, 33-36]. Lodging control in WINTER OATS [2-11, 14-16, 19-22, 24-31, 33-36]. Lodging control in RYE, TRITICALE [14]. Lodging control in OATS, WHEAT [37]. Stem shortening in POT PLANTS [17]. Stem shortening in BEDDING PLANTS, CAMELLIAS, HIBISCUS TRIONUM, LILIES [5, 6, 17]. Stem shortening in PELARGONIUMS, POINSETTIAS [5, 6, 14, 17]. Stem shortening in ORNAMENTALS [5, 6]

FOR FULL CONDITIONS OF USE ALWAYS READ THE PRODUCT LABEL

Notes

Efficacy

- Most effective results on cereals normally achieved from application from Apr onwards, on wheat and rye from leaf sheath erect to first node detectable (GS 30-31), on oats at second node detectable (GS 32), on winter barley from mid-tillering to leaf sheath erect (GS 25-30). However, recommendations vary with product. See label for details
- In tank mixes with other pesticides optimum timing for herbicide action may differ from that for growth reduction. See label for details of tank mix recommendations
- Addition of approved non-ionic wetter recommended for products approved on oats
- At least 6 h, preferably 24 h, required before rain for maximum effectiveness. Do not apply to wet crops
- May be used on cereals undersown with grass or clovers

Crop Safety/Restrictions

- Maximum number of treatments 1 per crop for cereals (2 per crop for split application on winter wheat [5-7]); varies with species for ornamentals [5, 6]
- Do not use on very late sown spring wheat or oats or on crops under stress
- Mixtures with liquid nitrogen fertilizers may cause scorch
- Do not use in tank mix with cyanazine [6]
- Do not use on spring barley
- Ornamentals to be treated must be well established and growing vigorously. Do not treat in strong sunlight or when temperatures are likely to fall below 10°C
- Temporary yellow spotting may occur on poinsettias. It can be minimised by use of a non-ionic wetting agent - see label

Special precautions/Environmental safety

- Harmful in contact with skin [3, 19, 20, 24-27, 29, 37]
- Harmful if swallowed
- Harmful to fish or other aquatic life. Do not contaminate surface waters or ditches with chemical or used container [3]
- Wash equipment thoroughly with water and wetting agent immediately after use and spray out. Spray out again before storing or using for another product. Traces can cause harm to susceptible crops sprayed later [5-8]

Protective clothing/Label precautions

- A [1-37]; C [24, 26, 27, 29, 34]; H [27, 34]
- M03, U08, C03, E01, E30a [1-37]; R03a [1, 3, 12, 13, 15, 16, 19, 20, 23-27, 29, 34, 36, 37]; R03c [1-36]; U02, U04a [21, 27, 33-35]; U05a [1-5, 9-27, 29-37]; U05b [6-8, 28]; U13, C01 [2, 9, 17, 18, 22, 29, 30]; U19 [1-4, 6-37]; U20a [2-6, 8, 9, 17, 18, 22, 25, 29, 30]; U20b [1, 7, 10-16, 19-21, 23, 24, 26-28, 31-37]; E08 [2, 9, 17, 18, 22, 29, 30, 36]; E13c [3]; E15 [1, 2, 4-37]; E26 [2, 9-11, 17, 18, 21, 22, 24-27, 29-31, 33-37]; E31a [2, 3, 9, 17, 18, 22, 24, 25, 29, 30, 36]; E31b [1, 4-8, 10-16, 19-21, 23, 26-28, 31-35, 37]; E34 [1-9, 11-22, 26-37]

Latest application/Harvest Interval(HI)

- Varies with product. See label for details

Approval

- Approved for aerial application on wheat, oats [2, 3, 5-8, 14, 18, 22, 30, 32, 36]; wheat [9]; winter barley [2, 3, 8, 14, 17, 18, 22]; rye [2, 3, 14]; triticale [14]. See notes in Section 1

Maximum Residue Levels (mg residue/kg food)

- oats 5; pears 3; wheat, rye, barley, triticale 2; grapes 1; tree nuts, olives, oilseeds, tea, hops 0.1; citrus, quinces, stone fruit, berries and small fruit, miscellaneous fruit (except olives), root and tuber vegetables, bulb vegetables, fruiting vegetables, brassica vegetables, leaf vegetables, fresh herbs, legume vegetables, stem vegetables, fungi, pulses, sorghum, buckwheat, millet, rice 0.05

111 chlormequat + 2-chloroethylphosphonic acid

A plant growth regulator for use in cereals

Products

1 Barclay Banshee XL	Barclay	305:155 g/l	SC	09201
2 Upgrade	RP Agric.	360:180 g/l	SL	06177

Uses

Lodging control in SPRING BARLEY, WINTER BARLEY, WINTER WHEAT.

Notes

Efficacy

- Apply before lodging has started. Best results obtained when crops growing vigorously
- Only crops growing under conditions of high fertility should be treated
- Recommended dose varies with growth stage. See labels for details and recommendations for use of sequential treatments
- Do not spray when crop wet or rain imminent
- Product must always be used with specified wetters - see label [1]

Crop Safety/Restrictions

- Maximum number of treatments 1 per crop; maximum total dose equivalent to one full dose treatment
- Do not spray during cold weather or periods of night frost, when soil is very dry, when crop diseased or suffering pest damage, nutrient deficiency or herbicide stress
- If used on seed crops grown for certification inform seed merchant beforehand
- Do not use on wheat variety Moulin or on any winter varieties sown in spring [1]
- Do not treat barley on soils with more than 10% organic matter [1]
- Do not use straw from treated cereals as a horticultural growth medium or as a mulch [1]
- Do not use in programme with any other product containing 2-chloroethylphosphonic acid [1]

Special precautions/Environmental safety

- Harmful in contact with skin [2]
- Harmful if swallowed
- Irritating to eyes [2]
- Harmful to fish or other aquatic life. Do not contaminate surface waters or ditches with chemical or used container [2]

Protective clothing/Label precautions

- A [1, 2]; C [2]
- M03, R03c, U05a, U08, U19, C03, E01, E26, E30a, E34 [1, 2]; R03a, R04a, U13, U20b, E13c, E31b [2]; U20a, E15, E31c [1]

Latest application/Harvest Interval(HI)

- Before flag leaf ligule/collar just visible (GS 39) or 1st spikelet visible (GS 51) for wheat or barley at top dose; or before flag leaf sheath opening (GS 47) for winter wheat at reduced dose [1]
- Before flag leaf ligule/collar just visible (GS 39) on barley and up to before flag leaf sheath opening (GS 47) on winter wheat [2]

Maximum Residue Levels (mg residue/kg food)

- see chlormequat and 2-chloroethylphosphonic acid entries

112 chlormequat + 2-chloroethylphosphonic acid + imazaquin

A plant growth regulator mixture for cereals

Products

Satellite	Cyanamid	216:480:0.8 g/l	KL	08969

Uses

Lodging control in SPRING BARLEY, WINTER BARLEY, WINTER WHEAT.

Notes

Efficacy

- Best results obtained when crops growing vigorously
- Only crops growing under conditions of high fertility should be treated
- Do not spray when crop wet or rain imminent

Crop Safety/Restrictions

- Maximum total dose equivalent to one full dose treatment
- All varieties of winter wheat, winter barley and spring barley (except Triumph) may be treated. Do not treat spring varieties sown in winter
- Do not spray during cold weather or periods of night frost, when soil is very dry, when crop diseased or suffering pest damage, nutrient deficiency or herbicide stress
- Do not use on crops being grown for seed
- Do not treat undersown crops
- Do not use within 10 d of any herbicide or liquid fertiliser treatment

Special precautions/Environmental safety

- Harmful if swallowed
- Irritating to skin
- Risk of serious damage to eyes
- Harmful to fish or other aquatic life. Do not contaminate surface waters or ditches with chemical or used container

Protective clothing/Label precautions

- A, C, H
- M03, R03c, R04b, R04d, U05a, U08, U13, U19, U20b, C03, E01, E13c, E30a, E31b, E34

Latest application/Harvest Interval(HI)

- Before flag leaf ligule/collar just visible (GS 39)

Maximum Residue Levels (mg residue/kg food)

- see chlormequat and 2-chloroethylphosphonic acid entries

113 chlormequat + 2-chloroethylphosphonic acid + mepiquat chloride

A plant growth regulator for reducing lodging in cereals

Products	Cyclade	BASF	230:155:75 g/l	SL	08958

Uses Lodging control in SPRING BARLEY, WINTER BARLEY, WINTER WHEAT.

Notes

Efficacy
- Best results achieved in a vigorous, actively growing crop with adequate fertility and moisture
- Optimum timing on all crops is from second node detectable stage (GS 32)
- Recommended for use as part of an intensive growing system which includes provision for optimum fertiliser treatment and disease control
- May be applied to crops undersown with grasses or clovers
- Must be used with a non-ionic wetting agent

Crop Safety/Restrictions
- Maximum number of treatments 1 per crop
- Do not apply to stressed crops or those on soils of low fertility unless receiving adequate dressings of fertiliser
- Do not apply in temperatures above 21°C or if crop is wet or if rain expected
- Do not treat variety Moulin nor any winter varieties sown in spring
- Do not use in a programme with any other product containing 2-chloroethylphosphonic acid
- Do not apply to barley on soils with more than 10% organic matter (winter wheat may be treated)
- Treatment may cause some delay in ear emergence

Special precautions/Environmental safety
- Harmful: If swallowed
- Do not contaminate surface waters or ditches with chemical or used container
- Do not use straw from treated crops as a horticultural growth medium
- Notify seed merchant in advance if use on a seed crop is proposed

Protective clothing/Label precautions
- A
- M03, R03c, U05a, U08, U20b, C03, E01, E15, E30a, E31c, E34

Latest application/Harvest Interval(HI)
- Before first spikelet of ear visible (GS 51) using reduced dose on barley; before flag leaf sheath opening (GS 47) using reduced dose on winter wheat

Maximum Residue Levels (mg residue/kg food)
- see chlormequat and 2-chloroethylphosphonic acid entries

FOR FULL CONDITIONS OF USE ALWAYS READ THE PRODUCT LABEL

114 chlormequat + choline chloride

A plant growth regulator for use in cereals and certain ornamentals

Products

1 Ashlade 700 5C	Ashlade	700:32 g/l	SL	07046
2 Atlas 460:46	Atlas	460:46 g/l	SL	07705
3 Atlas 5C Chlormequat	Atlas	460:320 g/l	SL	07701
4 Atlas Quintacel	Atlas	640:64 g/l	SL	07706
5 Barclay Take 5	Barclay	645:- g/l	SL	08524
6 Cropsafe 5C Chlormequat	Hortichem	645 g/l	SL	07897
7 MSS Mircell	Mirfield	640:64 g/l	SL	06939
8 New 5C Cycocel	BASF	645:- g/l	SL	01482
9 Portman Supaquat	Portman	640:- g/l	SL	03466

Uses

Increasing yield in WINTER BARLEY [1, 2, 8, 9]. Lodging control in AUTUMN SOWN SPRING WHEAT, LINSEED, SPRING RYE, WINTER OILSEED RAPE [8]. Lodging control in SPRING OATS, SPRING WHEAT, WINTER OATS, WINTER WHEAT [1-5, 7-9]. Lodging control in WINTER RYE [1, 5, 8]. Lodging control in WINTER BARLEY [3-5, 7]. Lodging control in TRITICALE [5, 8]. Stem shortening in LINSEED, WINTER OILSEED RAPE [8]. Stem shortening in BEDDING PLANTS, CAMELLIAS, HIBISCUS TRIONUM, LILIES [3, 4, 7]. Stem shortening in PELARGONIUMS, POINSETTIAS [3, 4, 6-8]. Stem shortening in ORNAMENTALS [2-4, 7].

Notes

Efficacy

- Influence on growth varies with crop and growth stage. Risk of lodging reduced by application at early stem extension. Root development and yield can be improved by earlier treatment
- Most effective results normally achieved from spring application. On winter barley an autumn treatment may also be useful. Timing of spray is critical and recommendations vary with product. See label for details
- Often used in tank-mixes with pesticides. Recommendations for mixtures and sequential treatments vary with product. See label for details
- Add authorised non-ionic wetter when spraying oats, oilseed rape or linseed
- At least 6 h required before rain for maximum effectiveness. Do not apply to wet crops
- May be used on cereals undersown with grass or clovers

Crop Safety/Restrictions

- Maximum number of treatments 1 per crop for spring wheat, oats, rye, triticale, winter oilseed rape, linseed; 1 or 2 per crop for winter wheat and winter barley (depending on dose); 1-3 per crop for ornamentals. See label for details
- Do not spray very late sown spring crops, crops on soils of low fertility, crops under stress from any cause or if frost expected
- Do not use on spring barley
- Do not use in tank mix with cyanazine [5]
- Mixtures with liquid nitrogen fertilizers may cause scorch

Special precautions/Environmental safety

- Harmful in contact with skin [3, 9]
- Harmful if swallowed [2, 4, 5, 7-9]
- Wash equipment thoroughly with water and wetting agent immediately after use and spray out. Traces can cause harm to susceptible crops sprayed later [3, 4, 7]

Protective clothing/Label precautions

- A [1-5, 7-9]; C [9]

- M03, U05a, U08, U19, C03, E01, E15, E30a [1-5, 7-9]; R03a [1, 3, 9]; R03c [1, 2, 4, 5, 7-9]; U20a [2-4, 7, 9]; U20b [1, 5, 8]; E26 [1, 3]; E31a [1]; E31b [2-5, 7, 9]; E31c [8]; E34 [2-5, 7-9]

Latest application/Harvest Interval(HI)
- Before 3rd node detectable (GS 33) for oats; before 2nd node detectable (GS 32) [2-4, 7, 8] for winter wheat and rye; before 1st node detectable (GS 31) for spring wheat, winter barley, triticale; before first flower bud opens (GS 4,0) for winter oilseed rape; before first flower buds visible for linseed [8]

Approval
- Approved for aerial application on wheat, oats, winter barley [5, 8]; wheat, oats [2-4, 7]; rye, triticale [5, 8]. See notes in Section 1

Maximum Residue Levels (mg residue/kg food)
- see chlormequat entry

115 chlormequat + choline chloride + imazaquin

A plant growth regulator mixture for winter wheat

Products

1 Meteor	Cyanamid	368:28:0.8 g/l	SL	06505
2 Standon Imazaquin 5C	Standon	368:28:0.8 g/l	SL	08813

Uses Increasing yield in WINTER WHEAT. Lodging control in WINTER WHEAT.

Notes

Efficacy
- Apply as single dose from leaf sheath lengthening up to and including 1st node detectable or as split dose, the first from tillers formed to leaf sheath lengthening, the second from leaf sheath erect up to and including 1st node detectable
- Apply to crops during good growing conditions or to those at risk from lodging
- On soils of low fertility, best results obtained where adequate nitrogen fertiliser used
- Do not apply when crop wet or rain imminent

Crop Safety/Restrictions
- Maximum number of applications 1 per crop (2 per crop at split dose)
- Do not treat durum wheat
- Do not apply to undersown crops

Special precautions/Environmental safety
- Harmful if swallowed. Irritating to eyes

Protective clothing/Label precautions
- A, C
- M03, R03c, R04a, U05a, U08, U13, U19, U20a, C03, E01, E15, E30a, E31b, E34

Latest application/Harvest Interval(HI)
- Before second node detectable (GS 31)

Maximum Residue Levels (mg residue/kg food)
- see chlormequat entry

116 chlormequat + mepiquat chloride

A plant growth regulator for reducing lodging in cereals

Products Stronghold | BASF | 345:115 g/l | SL | 09134

Uses Lodging control in WINTER BARLEY, WINTER WHEAT.

Notes

Efficacy
- Apply during good growing conditions at the correct timings - see label
- Optimum timing is when leaf sheaths erect (GS 30)
- On winter barley product should always be used in sequence with Cyclade. This sequence also recommended for winter wheat
- May be applied to crops undersown with grasses or clovers

Crop Safety/Restrictions
- Maximum total dose equivalent to one full dose treatment
- Do not apply to stressed crops or those on soils of low fertility unless receiving adequate dressings of fertiliser
- Do not apply in temperatures above 21°C or if crop is wet or if rain expected
- Do not treat any winter varieties sown in spring
- Do not apply to barley on soils with more than 10% organic matter (winter wheat may be treated)
- Treatment may cause some delay in ear emergence
- Mixtures with liquid fertilisers may cause scorching in some circumstances

Special precautions/Environmental safety
- Harmful: If swallowed
- Do not contaminate surface waters or ditches with chemical or used container
- Do not use straw from treated crops as a horticultural growth medium
- Notify seed merchant in advance if use on a seed crop is proposed

Protective clothing/Label precautions
- A
- M03, R03c, U05a, U08, U19, U20b, C03, E01, E15, E30a, E31c, E34

Latest application/Harvest Interval(HI)
- Before 1st node detectable (GS 31) for winter barley; before 3rd node detectable (GS 33) for winter wheat

Maximum Residue Levels (mg residue/kg food)
- see chlormequat entry

117 chlormequat with di-1-p-menthene

A plant-growth regulator for reducing stem growth and lodging

Products Podquat | Mandops | 470 g/l | SL | 03003

Uses Increasing yield in BROAD BEANS, FIELD BEANS, OILSEED RAPE, PEAS. Lodging control in BROAD BEANS, FIELD BEANS, OILSEED RAPE, PEAS.

Notes

Efficacy

- On winter oilseed rape and beans either apply as soon as possible after 3-leaf stage (GS 1, 3) until growth ceases followed by spring treatment in mid-Mar to early Apr or use a single spring spray. See label for details
- May be used at temperatures down to 1°C provided spray dries on leaves before rain, frost or snow occurs

Crop Safety/Restrictions

- Do not apply to plants covered by frost

Special precautions/Environmental safety

- Harmful if swallowed

Protective clothing/Label precautions

- A
- M03, R03c, U05a, U08, U19, U20a, C03, E01, E15, E30a, E31a, E34

Maximum Residue Levels (mg residue/kg food)

- see chlormequat entry

118 2-chloroethylphosphonic acid

A plant growth regulator for cereals and various horticultural crops

Products

1 Barclay Coolmore	Barclay	480 g/l	SL	07917
2 Cerone	RP Agric.	480 g/l	SL	06185
3 Ethrel C	Hortichem	480 g/l	SL	06995

Uses Basal bud stimulation in PROTECTED ROSES [3]. Fruit ripening in TOMATOES [3]. Increasing branching in PELARGONIUMS [3]. Inducing flowering in BROMELIADS [3]. Lodging control in SPRING BARLEY, TRITICALE, WINTER BARLEY, WINTER RYE, WINTER WHEAT [1, 2]. Stem shortening in NARCISSI [3].

Notes

Efficacy

- Best results achieved on crops growing vigorously under conditions of high fertility
- Optimum timing varies between crops and products. See labels for details
- Do not spray crops when wet or if rain imminent
- Best results on horticultural crops when temperature does not fall below 10°C [3]
- Use on tomatoes 17 d before planned pulling date [3]
- Apply as drench to daffodils when stems average 15 cm [3]
- Apply to glasshouse roses when new growth started after pruning [3]

Crop Safety/Restrictions

- Maximum number of treatments 1 per crop or yr
- Do not spray crops suffering from stress caused by any factor, during cold weather or period of night frost nor when soil very dry

- Do not apply to cereals within 10 d of herbicide or liquid fertilizer application
- Do not spray wheat or triticale where the leaf sheaths have split and the ear is visible

Special precautions/Environmental safety
- Irritating to eyes and skin
- Harmful to fish or other aquatic life. Do not contaminate surface waters or ditches with chemical or used container
- Avoid accidental deposits on painted objects such as cars, trucks, aircraft
- Rinse aircraft windshields after each tank load [2]

Protective clothing/Label precautions
- A, C
- R04a, R04b, U05a, U08, U13, C03, E01, E13c, E30a, E31b [1-3]; U20a, E34 [3]; U20b [2]; S06 [1]

Latest application/Harvest Interval(HI)
- Before 1st spikelet visible (GS 51) for spring barley, winter barley, winter rye; before flag leaf sheath opening (GS 47) for triticale, winter wheat
- HI tomatoes 5 d

Approval
- Approved for aerial application on winter barley [1, 2]. See notes in Section 1

Maximum Residue Levels (mg residue/kg food)
- currants 5; pome fruits, cherries, tomatoes, peppers 3; rye, barley 0.5; wheat, triticale 0.2; tree nuts, tea, hops 0.1; apricots, peaches, plums, strawberries, blackberries, dewberries, loganberries, raspberries, bilberries, cranberries, gooseberries, wild berries, avocados, bananas, dates, kiwi fruit, kumquats, litchis, mangoes, passion fruit, pomegranates, root and tuber vegetables, garlic, shallots, spring onions, aubergines, cucumbers, gherkins, courgettes, melons, squashes, water melons, brassica vegetables, leaf vegetables and herbs, legume vegetables, stem vegetables, fungi, pulses, oilseeds, potatoes, oats, rice, animal products 0.05

119 2-chloroethylphosphonic acid + mepiquat chloride

A plant growth regulator for reducing lodging in cereals

Products

1 Barclay Banshee	Barclay	155:305 g/l	SL	08175
2 Terpal	BASF	155:305 g/l	SL	02103

Uses

Increasing yield in WINTER BARLEY *(low lodging situations)* [1, 2]. Lodging control in SPRING BARLEY, TRITICALE, WINTER BARLEY, WINTER RYE, WINTER WHEAT [1, 2]. Lodging control in AUTUMN SOWN SPRING WHEAT, WINTER OATS *(some cultivars only)* [1].

Notes

Efficacy
- Best results achieved on crops growing vigorously under conditions of high fertility
- Recommended dose and timing vary with crop, cultivar, growing conditions, previous treatment and desired degree of lodging control. See label for details
- Add an authorised non-ionic wetter to spray solution
- May be applied to crops undersown with grass or clovers
- Do not apply to crops if wet or rain expected

Crop Safety/Restrictions
- Maximum number of treatments varies with dose and timing - see labels
- Recommendation for winter oats applies to named varieties only - see label [1]
- Do not spray crops damaged by herbicides or stressed by drought, waterlogging etc
- Do not treat crops on soils of low fertility unless adequately fertilized
- Do not use in a programme with any other product containing 2-chloroethylphosphonic acid
- Late tillering may be increased with crops subject to moisture stress and may reduce quality of malting barley
- Do not apply to winter cultivars sown in spring or treat winter barley, triticale or winter rye on soils with more than 10% organic matter (winter wheat may be treated)
- Do not apply at temperatures above 21°C
- Do not use straw from treated cereals as a mulch or growing medium

Special precautions/Environmental safety
- Do not contaminate surface waters or ditches with chemical or used container

Protective clothing/Label precautions
- U20b, E15, E30a [1, 2]; E31a [1]; E31c [2]

Latest application/Harvest Interval(HI)
- Before ear visible (GS 49) for winter barley, spring barley, winter wheat and triticale; boots swollen (GS 45) for winter oats; flag leaf just visible (GS 37) for winter rye

Maximum Residue Levels (mg residue/kg food)
- see 2-chloroethylphosphonic acid entry

120 chlorophacinone

An anticoagulant rodenticide

Products

1 Drat	RP Amenity	2.5 g/l	CB	05238
2 Endorats	Irish Drugs	0.005% w/w	RB	06503
3 Karate Ready-to-Use Rat & Mouse Bait	Lever Industrial	0.006% w/w	RB	05321
4 Karate Ready-to-Use Rodenticide Sachets	Lever Industrial	0.006% w/w	RB	05890
5 Rat & Mouse Bait	B H & B	0.005% w/w	RB	00764
6 Ruby Rat	Heatherington	0.005% w/w	RB	06059

Uses

Mice in FARM BUILDINGS [1-6]. Rats in FARM BUILDINGS [1-6]. Voles in FARM BUILDINGS [1, 3-6].

Notes

Efficacy
- Chemical formulated with oil, thus improving weather resistance of bait
- Mix concentrate with any convenient bait, such as grain, apple, carrot or potato [1]
- Use in baiting programme. Lay small baits for mice, larger baits for rats
- Replenish baits every few days and remove unused bait when take ceases or after 7-10 d

Special precautions/Environmental safety
- Harmful in contact with skin and if swallowed
- Prevent access to baits by children, domestic animals and birds; see label for other precautions required
- Harmful to game, wild birds and animals

Protective clothing/Label precautions
- A, C [1]
- M03, R03a, R03c, U02, U04a, U05a, U08, U14, U15, C03, E01, E10b, E15, E30b, V05 [1]; U13, V01a, V02, V03a, V04a [1-5]; U20a [1-4]; U20b [5]; E30a [2-5]; E31a [1, 5]; E32a [2-4]

---

121 chloropicrin

A highly toxic horticultural soil fumigant

Products Chloropicrin Fumigant | Dewco-Lloyd | 99.5% | LI | 04216

Uses Crown rot in STRAWBERRIES. Nematodes in STRAWBERRIES. Red core in STRAWBERRIES. Replant disease in HARDY ORNAMENTALS, TOP FRUIT. Verticillium wilt in STRAWBERRIES.

Notes

Efficacy
- Treat pre- or post-planting
- Apply with specialised injection equipment
- For treating small areas or re-planting a single tree, a hand-operated injector may be used. Mark the area to be treated and inject to 22 cm at intervals of 22 cm
- Double roll within 1 h of treatment and leave undisturbed for at least 10 d

Crop Safety/Restrictions
- Polythene sheeting (150 gauge) should be progressively laid over soil as treatment proceeds. The margin of the sheeting around the treated area must be embedded or covered with treated soil. Remove progressively after at least 4 d provided good air movement conditions prevail
- Carry out a cress test before replanting treated soil

Special precautions/Environmental safety
- Chloropicrin is subject to the Poisons Rules 1982 and the Poisons Act 1972. See notes in Section 1
- Very toxic by inhalation, in contact with skin and if swallowed
- Irritating to eyes, respiratory system and skin
- Before use, consult the code of practice for the fumigation of soil with chloropicrin. 2 fumigators must be present
- Avoid treatment or vapour release when persistent still air conditions prevail
- Remove contaminated gloves, boots or other clothing immediately and ventilate them in the open air until all odour is eliminated
- Dangerous to livestock
- Dangerous to game, wild birds and animals
- Dangerous to bees
- Dangerous to fish. Do not contaminate surface waters or ditches with chemical or used container

Protective clothing/Label precautions

- A, G, H, K, M
- M04, R01a, R01b, R01c, R04a, R04b, R04c, U04a, U05a, U10, U13, U14, U15, U19, U20a, C03, E01, E02 (until advised), E06a (until advised), E10a, E12c, E13b, E15, E30b, E32a, E34

Latest application/Harvest Interval(HI)

- Before planting

122 chlorothalonil

A protectant chlorophenyl fungicide for use in many crops

See also carbendazim + chlorothalonil
carbendazim + chlorothalonil + maneb

Products

1 Atlas Cropgard	Atlas	500 g/l	SC	09123
2 Barclay Corrib 500	Barclay	500 g/l	SC	08981
3 Baton SC	Bayer	500 g/l	SC	07945
4 Baton WG	Bayer	75% w/w	WG	07944
5 Bombardier FL	Unicrop	500 g/l	SC	07910
6 Bravo 500	BASF	500 g/l	SC	05637
7 Bravo 500	ISK Biosciences	500 g/l	SC	05638
8 Bravo 500	Zeneca	500 g/l	SC	09059
9 Bravo 720	ISK Biosciences	720 g/l	SC	05544
10 Bravo 720	Zeneca	720 g/l	SC	09104
11 Clortosip	Sipcam	500 g/l	SC	06126
12 Daconil Turf	Miracle	500 g/l	SC	07929
13 Daconil Turf	Scotts	500 g/l	SC	09265
14 ISK 375	ISK Biosciences	75% w/w	WG	07455
15 ISK 375	Zeneca	75% w/w	WG	09103
16 Jupital DG	Zeneca	75% w/w	WG	09181
17 Landgold Chlorothalonil 50	Landgold	500 g/l	SC	06265
18 Mainstay	Quadrangle	500 g/l	SC	05625
19 Miros DF	Sipcam	75% w/w	WG	04966
20 Mycoguard	Chiltern	500 g/l	SC	08115
21 Repulse	Hortichem	500 g/l	SC	07641
22 Sipcam UK Rover 500	Sipcam	500 g/l	SC	04165
23 Standon Chlorothalonil 500	Standon	500 g/l	SC	08597
24 Tripart Faber	Tripart	500 g/l	SC	05505
25 Tripart Faber	Tripart	500 g/l	SC	04549
26 Tripart Ultrafaber	Tripart	720 g/l	SC	05627

Uses

Alternaria in FLOWERHEAD BRASSICAS [22, 25]. Alternaria in LEAF BRASSICAS [22, 24, 25]. Anthracnose in TURF [12, 13]. Ascochyta in PEAS *(moderate control)* [6, 11]. Ascochyta in PEAS [2-5, 7-10, 14-19, 21-24, 26]. Blight in PROTECTED TOMATOES [6-8, 11, 18, 21]. Blight in POTATOES [1-11, 14-26]. Botrytis in FLOWERHEAD BRASSICAS [25]. Botrytis in PEAS [9, 10, 14-18]. Botrytis in PROTECTED TOMATOES [18, 21]. Botrytis in BRUSSELS SPROUTS, GRAPEVINES, ORNAMENTALS, PEPPERS [18]. Botrytis in PROTECTED CUCUMBERS [6-11, 14-16, 18, 21]. Botrytis in CHINESE CABBAGE *(moderate control)*, COLLARDS *(moderate control)*, GOOSEBERRIES *(moderate control)*, KALE *(moderate control)*, PROTECTED ORNAMENTALS *(moderate control)* [6-11, 14-16, 21]. Botrytis in CANE FRUIT *(moderate control)*, PEAS *(moderate control)*, STRAWBERRIES *(moderate control)* [6, 11, 21]. Botrytis in BLACKCURRANTS *(moderate control)*, ONIONS [6, 11]. Botrytis in WINTER OILSEED RAPE *(moderate control)* [6]. Botrytis in CANE FRUIT, STRAWBERRIES, WINTER OILSEED RAPE [7-10, 14-16, 18, 24, 26]. Botrytis in STRAWBERRIES *(off-label)* [7, 8]. Botrytis in OILSEED RAPE *(moderate control)*, REDCURRANTS *(moderate control)*[11]. Botrytis in ONIONS *(moderate control)* [21]. Botrytis in BRUSSELS SPROUTS *(moderate control)*, CABBAGES *(moderate control)*, CAULIFLOWERS *(moderate control)* [5-11, 14-16, 21]. Botrytis in BROCCOLI *(moderate control)* [5, 6, 11]. Botrytis in BLACKBERRIES *(moderate control)*, LOGANBERRIES *(moderate control)*, RASPBERRIES *(moderate control)* [5]. Botrytis in LEAF BRASSICAS [2, 18, 23-26]. Botrytis in BLACKCURRANTS *(suppression)* [2, 7, 8, 23]. Bubble in MUSHROOMS [21]. Cane spot in CANE FRUIT [7-10, 14-16, 18, 26]. Celery leaf spot in CELERIAC *(off-label)* [25]. Celery leaf spot in CELERY [7, 8, 11, 14-16, 18, 24, 25]. Chocolate spot in FIELD BEANS *(moderate control)* [5, 6, 11]. Chocolate spot in FIELD BEANS [1-4, 7-10, 14-20, 22-26]. Currant leaf spot in GOOSEBERRIES [5-8, 11, 14-16, 24, 26]. Currant leaf spot in REDCURRANTS [5, 7-11, 14-16, 18, 21, 24, 26]. Currant leaf spot in BLACKCURRANTS [2, 3, 5-11, 14-16, 18, 21-26]. Damping off and wirestem in LEAF BRASSICAS [24]. Dark leaf spot in FLOWERHEAD BRASSICAS *(moderate control)*, LEAF BRASSICAS *(moderate control)* [3]. Dollar spot in TURF [12, 13]. Downy mildew in FLOWERHEAD BRASSICAS [22, 25]. Downy mildew in BRUSSELS SPROUTS [9, 10, 18]. Downy mildew in GRAPEVINES [18]. Downy mildew in HOPS [6-11, 14-16, 18, 21, 24, 26]. Downy mildew in CHINESE CABBAGE *(moderate control)*, COLLARDS *(moderate control)*, KALE *(moderate control)* [6, 11]. Downy mildew in WINTER OILSEED RAPE *(moderate control)* [6]. Downy mildew in WINTER OILSEED RAPE [7-10, 14-16, 18, 24, 26]. Downy mildew in BRASSICA SEED BEDS [7, 8]. Downy mildew in OILSEED RAPE *(moderate control)* [11]. Downy mildew in BRASSICAS *(seedlings)* [14-16, 18]. Downy mildew in BRUSSELS SPROUTS *(seedlings only)*, CABBAGES *(seedlings only)*, CAULIFLOWERS *(seedlings only)*, CHINESE CABBAGE *(seedlings only)*, COLLARDS *(seedlings only)*, KALE *(seedlings only)* [21]. Downy mildew in BROCCOLI *(moderate control)*, BRUSSELS SPROUTS *(moderate control)*, CABBAGES *(moderate control)*, CAULIFLOWERS *(moderate control)* [5, 6, 11]. Downy mildew in LEAF BRASSICAS [2, 7-10, 14-16, 18, 22-26]. Dry bubble in MUSHROOMS [6-11, 14-16, 18, 21, 24, 26]. Fusarium patch in TURF [12, 13]. Glume blotch in SPRING WHEAT [2-4, 11, 19, 20, 22, 23, 25]. Glume blotch in WINTER WHEAT [1-8, 11, 17, 19, 20, 22-26]. Grey snow mould in TURF *(reduction)* [12, 13]. Ink disease in IRISES [7-11, 14-16, 18, 21]. Late ear diseases in WINTER WHEAT [6]. Leaf blotch in ONIONS [25]. Leaf blotch in SPRING BARLEY, WINTER BARLEY [20]. Leaf mould in PROTECTED TOMATOES [6-8, 11, 18, 21]. Leaf rot in ONIONS [6-11, 14-16, 18, 24, 26]. Leaf rot in ONIONS *(moderate control)* [21]. Leaf spot in GOOSEBERRIES [6-11, 18]. Leaf spot in CELERY [21]. Leaf spot in SPRING WHEAT [2-4, 11, 19, 20, 22, 23, 25]. Leaf spot in WINTER WHEAT [1-4, 6-11, 14-20, 22-26]. Mycosphaerella in PEAS *(moderate control)* [5, 6, 11, 21]. Mycosphaerella in PEAS [1, 3, 4, 7-10, 14-20, 22, 24, 26]. Neck rot in ONIONS [6-11, 14-16, 18, 24, 26]. Neck rot in ONIONS *(moderate control)* [21]. Powdery mildew in GOOSEBERRIES *(suppression)*[6-8, 11]. Powdery mildew in CANE FRUIT [7-10, 26]. Powdery mildew in PROTECTED CUCUMBERS [7-10, 14-16]. Powdery mildew in REDCURRANTS *(suppression)*[11]. Powdery mildew in RASPBERRIES [14-16, 18]. Powdery mildew in BLACKCURRANTS *(suppression)* [2, 6-8, 11, 23]. Red thread in TURF [12, 13]. Rhizoctonia in FLOWERHEAD BRASSICAS, LEAF BRASSICAS [25]. Rhynchosporium in BARLEY [25]. Rhynchosporium in SPRING BARLEY, WINTER BARLEY [1]. Ring spot in FLOWERHEAD BRASSICAS [22, 25]. Ring spot in BRUSSELS SPROUTS [9, 10, 18, 21]. Ring spot in BROCCOLI, CABBAGES, CAULIFLOWERS, CHINESE CABBAGE, COLLARDS, KALE [21]. Ring spot in FLOWERHEAD BRASSICAS *(moderate control)*, LEAF BRASSICAS *(moderate control)* [3]. Ring spot in LEAF BRASSICAS [2, 7-10, 14-16, 18, 22-26]. Rust in LEEKS [18]. Septoria leaf spot in CELERIAC *(off-label)* [7, 8, 26]. Septoria leaf spot in WINTER WHEAT [5]. Wet bubble in MUSHROOMS [6-11, 14-16, 18, 24, 26].

Notes

Efficacy

- For some crops products differ in diseases listed as controlled. See label for details and for application rates, timing and number of sprays
- Apply as protective spray or as soon as disease appears and repeat at 7-21 d intervals
- Activity against Septoria may be reduced where serious mildew or rust present. In such conditions mix with suitable mildew or rust fungicide
- May be used at reduced rate on peas in tank mix with Ronilan FL [6]
- Do not mow or water turf for 24 h after treatment. Do not add surfactant or mix with liquid fertilizer
- For Botrytis control in strawberries important to start spraying early in flowering period and repeat at least 3 times at 10 d intervals

Crop Safety/Restrictions

- Varies with crop and product - see labels for details
- On strawberries some scorching of calyx may occur with protected crops

Special precautions/Environmental safety

- Irritating to eyes [1-3, 7, 9, 11, 12, 17, 18, 20, 22-26]
- Irritating to skin [1-3, 5-7, 9, 11, 12, 17, 18, 20-26]
- Irritating to respiratory system [2-7, 9, 11, 12, 14, 17-19, 21-26]
- Risk of serious damage to eyes [1, 3-7, 9, 11, 12, 14, 17, 19-21, 23, 26]
- Dangerous to fish or other aquatic life. Do not contaminate surface waters or ditches with chemical or used container
- Do not allow direct spray from broadcast air-assisted sprayers to fall within 18 m of surface waters or ditches
- Do not allow direct spray from vehicle-mounted/drawn hydraulic sprayers to fall within 6 m, or from hand-held sprayers to within 2 m, of surface waters or ditches. Direct spray away from water [1-7, 9, 11, 12, 12, 14, 14, 17-20, 22-26]
- Do not spray from the air within 250 m horizontal distance of surface waters or ditches
- Operators must use a vehicle fitted with a cab and a forced air filtration unit with a pesticide filter complying with HSE Guidance Note PM74 or to an equivalent or higher standard when making broadcast or air-assisted applications
- Must only be applied by a pedestrian controlled sprayer or vehicle mounted/drawn equipment [12, 13]

Protective clothing/Label precautions

- A, C [1-26]; D [4, 14-16]; H [1-8, 11-17, 20, 21, 23]; J [2, 3, 5, 6, 11, 23]; M [2-8, 11-17, 21, 23]
- R04a [1, 3, 9-13, 17, 18, 20, 22, 24-26]; R04b [1-3, 5-13, 17, 18, 20-26]; R04c [2-19, 21-26]; R04d [1-17, 19-21, 23]; U02 [1-5, 11, 20, 23, 25]; U05a, U19, U20a, C03, E01, E13b, E30a [1-26]; U09a [1, 2, 4-16, 18-26]; U09b [3, 17]; U13, U14, U15 [1-5, 9-11, 17-20, 22-25]; C02 (blackcurrants 3d; brassicas and field beans 7 d; peas 14 d) [22]; C02 (field beans 7d; celery 7d; onions 7d; blackcurrants 3d) [25]; C02 (field beans and peas 14 d; potatoes 7 d) [1]; C02 (peas 14 d; potatoes 7 d) [20]; E16 [1-20, 23-25]; E17 [2, 3, 5-8, 23]; E18 [2, 3, 5-8]; E26 [1-3, 5, 11, 17, 20, 22, 23, 25]; E31a [7, 8, 12, 13, 21, 26]; E31b [1, 2, 6, 9-11, 17-20, 22-25]; E31c [5]; E32a [3, 4, 14-16]; E34 [1, 6, 9, 10, 12, 13, 18-20, 24]

Latest application/Harvest Interval(HI)

- Before anthesis for spring barley, spring wheat, winter barley, winter wheat [1, 20]; before grain watery ripe (GS 71) for spring wheat [3, 4], winter wheat [3, 4, 6, 7]; ear emergence just complete (GS 59) for spring barley, spring wheat, winter barley, winter wheat [2]; before flowering for winter oilseed rape; end Aug in yr of harvest for blackcurrants, gooseberries, redcurrants, blueberries, loganberries, raspberries
- HI 8 wk for field beans; 6 wk for combining peas; 28 d or before 31 Aug for post-harvest treatment for blackcurrants, gooseberries, redcurrants; 14 d for field beans, onions, strawberries, vining peas; 10 d for hops; 7 d for broccoli, Brussels sprouts, cabbages, cauliflowers, celery, chinese cabbage, collards, field beans, kale, onions, potatoes; 3 d for hops, cane fruit; 48 h for protected cucumbers, protected tomatoes; 24 h for mushrooms

Approval

- Approval for aerial spraying on wheat, barley, field beans, potatoes [2]; wheat, potatoes [22, 25]; potatoes [3, 5]; winter wheat, potatoes [7, 18, 24]; peas, field beans [7, 18]. See notes in Section 1
- Off-label approval to May 2000 for use on celeriac (OLA 0796/95)[7], (OLA 0797/95)[26]; unlimited for use on outdoor strawberries (OLA 2033/97)[7]

Maximum Residue Levels (mg residue/kg food)

- cranberries, tomatoes, peppers, aubergines, peas (with pods) 2; grapes (table), cucumbers 1; garlic, onions, shallots, Brussels sprouts 0.5; tea, wheat, rye, barley, oats, triticale 0.1; lettuce, maize, rice, animal products 0.01

123 chlorothalonil + cyproconazole

A systemic protective and curative fungicide for cereals

Products

1 Alto Elite	Sandoz	375:40 g/l	SC	05069
2 Alto Elite	Novartis	375:40 g/l	SC	08467
3 Octolan	Sandoz	375:40 g/l	SC	06256
4 Octolan	Novartis	375:40 g/l	SC	08480

Uses

Brown rust in SPRING BARLEY, WINTER BARLEY, WINTER WHEAT [1-4]. Chocolate spot in FIELD BEANS [1-4]. Eyespot in WINTER BARLEY, WINTER WHEAT [1-4]. Grey mould in COMBINING PEAS [1]. Grey mould in PEAS [2-4]. Net blotch in SPRING BARLEY, WINTER BARLEY [1-4]. Powdery mildew in SPRING BARLEY, WINTER BARLEY, WINTER WHEAT [1-4]. Rhynchosporium in SPRING BARLEY, WINTER BARLEY [1-4]. Rust in FIELD BEANS [1-4]. Septoria diseases in WINTER WHEAT [1-4]. Yellow rust in SPRING BARLEY, WINTER BARLEY, WINTER WHEAT [1-4].

Notes

Efficacy

- Apply at first signs of infection or as soon as disease becomes active
- A repeat application may be made if re-infection occurs
- For established mildew tank-mix with approved formulation of tridemorph
- When applied prior to third node detectable (GS 33) a useful reduction of eyespot will be obtained
- If infection of eyespot anticipated tank-mix with Sportak

Crop Safety/Restrictions

- Maximum number of treatments 2 per crop
- If applied to winter wheat in spring at GS 30-33 straw shortening may occur but yield is not reduced

Special precautions/Environmental safety
- Irritant. Risk of serious damage to eyes
- Do not apply at concentrations higher than recommended
- Dangerous to fish or other aquatic life. Do not contaminate surface waters or ditches with chemical or used container
- Do not allow direct spray from vehicle mounted/drawn hydraulic sprayers to fall within 6 m, or from hand-held sprayers to within 2 m, of surface waters or ditches

Protective clothing/Label precautions
- A, C, H, M
- R04, R04d, U05a, U11, U19, U20a, C03, E01, E13b, E16, E26, E30a, E31b, E34

Latest application/Harvest Interval(HI)
- Up to and including anthesis complete (GS 69) for winter wheat; up to and including emergence of ear complete (GS 59) for barley

Maximum Residue Levels (mg residue/kg food)
- see chlorothalonil entry

124 chlorothalonil + flutriafol

A systemic eradicant and protectant fungicide for winter wheat

Products

1 Halo	Zeneca	375:47 g/l	SC	06520
2 Impact Excel	Zeneca	300:47 g/l	SC	06680
3 PP 375	Zeneca	375:47 g/l	SC	07898
4 Prospa	Zeneca	375:47 g/l	SC	07899

Uses

Brown rust in WINTER WHEAT. Glume blotch in WINTER WHEAT. Late ear diseases in WINTER WHEAT. Leaf spot in WINTER WHEAT. Powdery mildew in WINTER WHEAT. Yellow rust in WINTER WHEAT.

Notes

Efficacy
- Best results achieved by applying protective spray at flag leaf emergence (GS 37) and full ear emergence (GS 59). If disease already present spray before it reaches the top 2 leaves and ears
- Where mildew serious may be tank-mixed with fenpropidin

Crop Safety/Restrictions
- Maximum number of treatments 2 per crop (including other products containing flutriafol)
- On certain cultivars with erect leaves high transpiration can result in flag leaf tip scorch. This may be increased by treatment but does not affect yield

Special precautions/Environmental safety
- Irritating to eyes and skin. May cause sensitization by skin contact
- Risk of serious damage to eyes [4]
- Dangerous to fish or other aquatic life. Do not contaminate surface waters or ditches with chemical or used container

Protective clothing/Label precautions
- A, C, H
- R04a, R04b, R04d, R04e, U02, U05a, U09a, U19, C03, E01, E13b, E16, E30a [1-4]; U20a, E26, E31a [1, 3, 4]; U20b, E31b, E34 [2]

Latest application/Harvest Interval(HI)
- Before early milk stage (GS 73)

Maximum Residue Levels (mg residue/kg food)
- see chlorothalonil entry

125 chlorothalonil + mancozeb

A protectant fungicide mixture for potatoes

Products

Adagio	PBI	201:274 g/l	SC	07832

Uses

Blight in POTATOES.

Notes

Efficacy
- Start spray treatments immediately after a blight warning or just before the haulm meets in the row
- It is essential to start the programme before blight appears in the crop
- Repeat treatments at 7, 10 or 14 d intervals depending on blight risk and continue until haulm is to be destroyed

Crop Safety/Restrictions
- Maximum number of treatments 5 per crop
- Irrigated crops should be sprayed immediately after irrigation

Special precautions/Environmental safety
- Irritating to eyes, skin and respiratory system
- May cause sensitization by skin contact
- Dangerous to fish or other aquatic life. Do not contaminate surface waters or ditches with chemical or used container
- Do not allow direct spray from broadcast air-assisted sprayers to fall within 18 m, or from vehicle mounted/drawn hydraulic sprayers to within 6 m, or from hand-held sprayers to within 2 m, of surface waters or ditches. Direct spray away from water
- Broadcast air assisted applications must only be made by equipment fitted with a cab with a forced air filtration unit plus a pesticide filter complying with HSE Guidance Note PM 74 or an equivalent or higher standard

Protective clothing/Label precautions
- A, C, H, M
- R04a, R04b, R04c, R04e, U02, U05a, U08, U13, U14, U15, U19, U20a, C03, E01, E13b, E16, E17, E26, E30a, E31b

Latest application/Harvest Interval(HI)
- HI 7 d

Maximum Residue Levels (mg residue/kg food)
- see chlorothalonil and mancozeb entries

126 chlorothalonil + metalaxyl

A systemic and protectant fungicide for various field crops

Products

1 Folio 575 SC	Ciba Agric.	500:75 g/l	SC	05843
2 Folio 575 SC	Novartis	500:75 g/l	SC	08406

Uses

Alternaria in BRUSSELS SPROUTS *(moderate control)*, CALABRESE *(moderate control)*, CAULIFLOWERS *(moderate control)*. Downy mildew in BROAD BEANS, BRUSSELS SPROUTS, BULB ONIONS, CALABRESE, CAULIFLOWERS, FIELD BEANS, SALAD ONIONS, WINTER OILSEED RAPE. Ring spot in BRUSSELS SPROUTS *(reduction)*, CALABRESE *(reduction)*, CAULIFLOWERS *(reduction)*. White blister in BRUSSELS SPROUTS, CALABRESE, CAULIFLOWERS. White tip in LEEKS.

Notes

Efficacy

- Apply at first signs of disease (oilseed rape, beans), at first signs of disease or when weather conditions favourable to disease (Brussels sprouts, onions, leeks)
- Repeat treatment at 14-21 d (14 d for oilseed rape and beans) intervals if necessary
- Oilseed rape crops most likely to benefit from treatment are infected crops between cotyledon and 3-leaf stage (GS 1,0-1,3)

Crop Safety/Restrictions

- Maximum number of treatments 2 per crop for oilseed rape and beans, 3 per crop for Brussels sprouts, onions and leeks

Special precautions/Environmental safety

- Irritating to skin and eyes
- Risk of serious damage to eyes
- Dangerous to fish or other aquatic life. Do not contaminate surface waters or ditches with chemical or used container
- Do not allow direct spray from vehicle mounted/drawn hydraulic sprayers to fall within 6 m, or from hand-held sprayers to within 2 m, of surface waters or ditches. Direct spray away from water

Protective clothing/Label precautions

- A, C
- R04a, R04b, R04d, U05a, U09a, U20a, C02, C03, E01, E13b, E16, E30a, E31a

Latest application/Harvest Interval(HI)

- HI field beans, broad beans, Brussels sprouts, cauliflowers, calabrese, onions, leeks 14 d

Maximum Residue Levels (mg residue/kg food)

- see chlorothalonil and metalaxyl entries

127 chlorothalonil + propamocarb hydrochloride

A contact and systemic fungicide mixture for blight control in potatoes

Products

Merlin	AgrEvo	375:375 g/l	SC	07943

FOR FULL CONDITIONS OF USE ALWAYS READ THE PRODUCT LABEL

Uses Late blight in POTATOES.

Notes

Efficacy
- Commence treatment early in the season as soon as there is risk of infection
- In the absence of a blight warning treatment should start just before potatoes meet along the row
- Use only as a protectant. Stop use when blight readily visible (1% leaf area destroyed)
- Repeat sprays at 10-14 d intervals depending on blight infection risk. See label for details
- Complete blight spray programme after end Aug up to haulm destruction with protectant fungicides preferably fentin based

Crop Safety/Restrictions
- Maximum number of treatments 5 per crop
- Apply to dry foliage. Do not apply if rainfall or irrigation imminent

Special precautions/Environmental safety
- Irritating to eyes
- Risk of serious damage to eyes
- May cause sensitisation by skin contact
- Dangerous to fish or other aquatic life. Do not contaminate surface waters or ditches with chemical or used container
- Do not allow direct spray from ground-based vehicle-mounted/drawn sprayers to fall within 6 m, or from hand-held sprayers to within 2 m, of surface waters or ditches. Direct spray away from water

Protective clothing/Label precautions
- A, C, H
- R04a, R04d, R04e, U05a, U08, U11, U19, U20a, C03, E01, E13b, E16, E30a, E31b, E34

Latest application/Harvest Interval(HI)
- HI 7 d

Maximum Residue Levels (mg residue/kg food)
- see chlorothalonil entry

---

128 chlorothalonil + propiconazole

A systemic and protectant fungicide for winter wheat and barley

Products

1 Sambarin 312.5 SC	Ciba Agric.	250:62.5 g/l	SC	05809
2 Sambarin 312.5 SC	Novartis	250:62.5 g/l	SC	08439

Uses Brown rust in WINTER WHEAT. Mildew in WINTER BARLEY. Net blotch in WINTER BARLEY. Powdery mildew in WINTER WHEAT. Rhynchosporium in WINTER BARLEY. Septoria diseases in WINTER WHEAT. Yellow rust in WINTER WHEAT.

Notes

Efficacy
- On wheat apply from start of flag leaf emergence up and including when ears just fully emerged (GS 37-59), on barley at any time to ears fully emerged (GS 59)
- Best results achieved from early treatment, especially if weather wet, or as soon as disease develops

Crop Safety/Restrictions

- Maximum number of treatments 2 per crop or 1 per crop if other propiconazole based fungicide used in programme

Special precautions/Environmental safety

- Irritating to eyes, skin and respiratory system
- Risk of serious damage to eyes
- Dangerous to fish or other aquatic life. Do not contaminate surface waters or ditches with chemical or used container
- Do not allow direct spray from ground-based vehicle-mounted/drawn sprayers to fall within 6 m, or from hand-held sprayers to within 2 m, of surface waters or ditches. Direct spray away from water

Protective clothing/Label precautions

- A, C
- R04a, R04b, R04c, R04d, U02, U05a, U09a, U11, U19, U20a, C02, C03, E01, E13b, E16, E30a, E31b, E34

Latest application/Harvest Interval(HI)

- Up to and including emergence of ear just complete (GS 59).
- HI 42 d

Maximum Residue Levels (mg residue/kg food)

- see chlorothalonil and propiconazole entries

129 chlorotoluron

A contact and residual urea herbicide for cereals

Products

1 Alpha Chlortoluron 500	Makhteshim	500 g/l	SC	04848
2 Atol	Ashlade	700 g/l	SC	07347
3 Lentipur CL 500	Nufarm	500 g/l	SC	05925
4 Luxan Chlorotoluron 500 Flowable	Luxan	500 g/l	SC	09165
5 MSS Chlorotoluron 500	Mirfield	500 g/l	SC	07871
6 Portman Chlortoluron	Portman	500 g/l	SC	03068
7 Stefes Toluron	Stefes	500 g/l	SC	05779
8 Tolugan 700	Makhteshim	700 g/l	SC	07874
9 Tripart Ludorum	Tripart	500 g/l	SC	03059

Uses

Annual dicotyledons in WINTER BARLEY, WINTER WHEAT [1-9]. Annual dicotyledons in TRITICALE [1-5, 8, 9]. Annual dicotyledons in DURUM WHEAT [3, 4, 9]. Annual grasses in WINTER BARLEY, WINTER WHEAT [1-9]. Annual grasses in TRITICALE [1-5, 8, 9]. Annual grasses in DURUM WHEAT [3, 4, 9]. Blackgrass in WINTER BARLEY, WINTER WHEAT [1-9]. Blackgrass in TRITICALE [1-5, 8, 9]. Blackgrass in DURUM WHEAT [3, 4, 9]. Rough meadow grass in WINTER BARLEY, WINTER WHEAT [1-9]. Rough meadow grass in TRITICALE [1-5, 8, 9]. Rough meadow grass in DURUM WHEAT [3, 4, 9]. Wild oats in WINTER BARLEY, WINTER WHEAT [1-9]. Wild oats in TRITICALE [1-5, 8, 9]. Wild oats in DURUM WHEAT [3, 4, 9].

Notes

Efficacy

- Best results achieved by application soon after drilling. Application in autumn controls most weeds germinating in early spring
- For wild oat control apply within 1 wk of drilling, not after 2-leaf stage. Blackgrass and meadow grasses controlled to 5 leaf, ryegrasses to 3 leaf stage
- Any trash or burnt straw should be buried and dispersed during seedbed preparation
- Do not use on soils with more than 10% organic matter
- Control may be reduced if prolonged dry conditions follow application
- Harrowing after treatment may reduce weed control

Crop Safety/Restrictions

- Maximum number of treatments 1 per crop
- Use only on listed crop varieties. See label. Ensure seed well covered at drilling
- Apply only as pre-emergence spray in durum wheat, pre- or post-emergence in wheat, barley or triticale
- Do not apply pre-emergence to crops sown after 30 Nov
- Do not apply to crops severely checked by waterlogging, pests, frost or other factors
- Do not use on undersown crops or those due to be undersown
- Do not apply post-emergence in mixture with liquid fertilizers
- Do not roll for 7 d before or after application to an emerged crop
- Crops on stony or gravelly soils may be damaged, especially after heavy rain

Special precautions/Environmental safety

- Harmful if swallowed [1, 5, 6, 8]
- Irritating to eyes and skin [1, 2, 5-9]
- Harmful to fish or other aquatic life. Do not contaminate surface waters or ditches with chemical or used container

Protective clothing/Label precautions

- A, C [1, 2, 5, 6, 8]
- M03 [1, 2, 5, 8]; R03c [1, 5, 6, 8]; R04a, R04b, U05a, C03, E01 [1, 2, 5-9]; U08, U19, E30a, E31b, E34 [1-9]; U20a [2, 5-9]; U20b [1, 3, 4]; E13c [1-8]; E15 [9]; E26 [3, 4, 6, 7, 9]; E31a [6, 7, 9]

Latest application/Harvest Interval(HI)

- Pre-emergence for durum wheat. Post emergence timings on other crops vary - see labels

Approval

- May be applied through CDA equipment. See labels for details [2, 7, 9]
- Approved for aerial application on winter wheat, winter barley, durum wheat, triticale [9]. See notes in Section 1

130 chlorpropham

A residual carbamate herbicide and potato sprout suppressant

See also cetrimide + chlorpropham

Products

1 Atlas CIPC 40	Atlas	400 g/l	EC	07710
2 BL 500	Wheatley	500 g/l	HN	00279
3 Luxan Gro-Stop 300 EC	Luxan	300 g/l	EC	08602
4 Luxan Gro-Stop Basis	Luxan	300 g/l	EC	08601
5 Luxan Gro-Stop HN	Luxan	300 g/l	HN	07689
6 MSS CIPC 40 EC	Mirfield	400 g/l	EC	01403
7 MSS CIPC 5 G	Mirfield	5% w/w	GR	01402
8 MSS CIPC 50 LF	Mirfield	500 g/l	EC	03285
9 MSS CIPC 50 M	Mirfield	500 g/l	EC	01404
10 MTM CIPC 40	MTM Agrochem.	400 g/l	EC	05895
11 Standon CIPC 300 HN	Standon	300 g/l	HN	09187
12 Warefog 25	Mirfield	600 g/l	HN	06776

Uses

Annual dicotyledons in LETTUCE [6, 10]. Annual dicotyledons in ANNUAL FLOWERS, BLACKCURRANTS, CELERY, CHRYSANTHEMUMS, GOOSEBERRIES, LUCERNE, PARSLEY, STRAWBERRIES [10]. Annual dicotyledons in FLOWER BULBS, ONIONS [1, 6, 10]. Annual dicotyledons in LEEKS [1, 6]. Annual dicotyledons in CARROTS [1, 10]. Annual grasses in LETTUCE [6, 10]. Annual grasses in ANNUAL FLOWERS, BLACKCURRANTS, CELERY, CHRYSANTHEMUMS, GOOSEBERRIES, LUCERNE, PARSLEY, STRAWBERRIES [10]. Annual grasses in FLOWER BULBS, ONIONS [1, 6, 10]. Annual grasses in LEEKS [1, 6]. Annual grasses in CARROTS [1, 10]. Chickweed in LETTUCE [6, 10]. Chickweed in ANNUAL FLOWERS, BLACKCURRANTS, CELERY, CHRYSANTHEMUMS, GOOSEBERRIES, LUCERNE, PARSLEY, STRAWBERRIES [10]. Chickweed in FLOWER BULBS, LEEKS, ONIONS [1, 6]. Chickweed in CARROTS [1, 10]. Polygonums in LETTUCE [6, 10]. Polygonums in ANNUAL FLOWERS, BLACKCURRANTS, CELERY, CHRYSANTHEMUMS, GOOSEBERRIES, LUCERNE, PARSLEY, STRAWBERRIES [10]. Polygonums in FLOWER BULBS, LEEKS, ONIONS [1, 6]. Polygonums in CARROTS [1, 10]. Sprout suppression in POTATOES [2, 7-9, 12]. Sprout suppression in WARE POTATOES [3-5, 11]. Volunteer ryegrass in GRASS SEED CROPS *(off-label)* [1, 6, 10].

Notes

Efficacy

- Apply weed control sprays to freshly cultivated soil. Adequate rainfall must occur after spraying. Activity is greater in cold, wet than warm, dry conditions
- For sprout suppression apply with suitable fogging or rotary atomiser equipment or sprinkle granules over dry tubers before sprouting commences. Repeat applications may be needed. See labels for details
- Effectiveness of fogging reduced if air spaces between tubers are blocked. best results obtained at 5-10°C and 75-80% humidity
- Cure potatoes according to label instructions before treatment and allow 3 wk between completion of loading into store and first treatment

Crop Safety/Restrictions

- Maximum number of treatments 1 per batch (3 per batch [8]) for fogging potatoes; 1 per yr for grass seed crops [1, 6, 10]; 1 per crop for carrots, onions, leeks [1]; 2 per yr for flower bulbs [1]

- Maximum total dose 2.55 l per 20 tonnes [9], 3 l per 50 tonnes [5], of potatoes; 60 ml per 1000 kg batch of potatoes [11]
- Apply weed control sprays to seeded crops pre-emergence of crop or weeds, to onions as soon as first crop seedlings visible, to planted crops a few days before planting, to bulbs immediately after planting, to fruit crops in late autumn-early winter. See label for further details
- Not to be used on grass seed crops if grass to be grazed or cut for fodder before 31 May following treatment [1, 6, 10]
- Excess rainfall after application may result in crop damage
- Do not use on Sands, Very Light soils or soils low in organic matter
- Poor conditions at drilling or planting, soil compaction, surface capping, waterlogging or attack by pests may result in crop damage
- On crops under glass high temperatures and poor ventilation may cause crop damage
- Only clean, mature, disease-free potatoes should be treated for sprout suppression
- Do not use on potatoes for seed. Do not handle, dry or store seed potatoes or any other seed or bulbs in boxes or buildings in which potatoes are being or have been treated
- Do not fog potatoes with a high level of skin spot [5]
- Do not remove potatoes for sale or processing for at least 21 d (4 wk [5, 11]) after application
- A minimum interval of 45 d (4 wk [5]) must elapse between applications [8, 9]

Special precautions/Environmental safety

- Harmful in contact with skin [1, 6, 10] or by inhalation [2, 5, 9, 11]
- Harmful if swallowed [1, 2, 5, 6, 8-11]
- Irritating to eyes [12] and skin [1, 2, 5, 6, 8, 9, 11]
- Irritating to respiratory system [1, 2, 5, 6, 9, 11, 12]
- May cause sensitization by skin contact and by inhalation [5, 11]
- Highly flammable [2, 9]
- Flammable [8]
- Harmful to fish. Do not contaminate ponds, waterways or ditches with chemical or used container/Harmful to fish or other aquatic life. Do not contaminate surface waters or ditches with chemical or used container [3-5, 11]
- Do not contaminate surface waters or ditches with chemical or used container [1, 2, 6-10, 12]
- Keep unprotected persons out of treated areas for at least 24 h after application [5, 9]

Protective clothing/Label precautions

- A [1-6, 8-12]; C [1-4, 6, 8-10, 12]; D [2, 5, 8, 9, 11, 12]; E [2, 5, 9, 11, 12]; G [8]; H [2-5, 8, 9, 11]; J [2, 8, 9]; M [2, 5, 8, 9, 11]
- M03 [1, 2, 5-11]; R03a [1, 6, 10]; R03b, E02 (24 h) [2, 5, 9, 11]; R03c [1, 2, 5, 6, 8-11]; R04a [1, 2, 5, 6, 8, 9, 11, 12]; R04b [1, 2, 5, 6, 8, 9, 11]; R04c [1, 2, 5, 6, 9, 11, 12]; R04e, R04f, U02, U04a, U13, U14, U17 [5, 11]; R07c [2, 9]; R07d [8]; U05a, C03, E01 [1-12]; U08 [1, 3-5, 8, 9, 11, 12]; U09a [2, 6, 10]; U19 [1, 2, 5, 6, 8-12]; U20a [1, 3-5, 10, 11]; U20b [2, 6-9]; U20c [12]; E13c [3-5, 11]; E15 [1, 2, 6-10, 12]; E26 [7, 10]; E30a [1-4, 6-10, 12]; E31a [1, 2, 8, 10, 12]; E31b [6, 9]; E32a [3-5, 7, 11]; E34 [1, 3-11]; E34, [2]

Latest application/Harvest Interval(HI)

- 14 d after planting for leeks, onions [1]; 21 d before removal for sale or processing for potatoes [2]; 8 wk (4 wk [5]) before marketing or processing for ware potatoes [3, 4]; 48 h before drilling for carrots [1]; before 4 leaf stage (Silt soils only) for onions [1]; before leaves unfurl (tulip); up to 5 cm high (others) for flower bulbs [1]

Approval

- Some products are formulated for application by thermal fogging. See labels for details [2, 5, 11, 12]

- Off-label approval to May 2002 for use on grass seed crops (OLA 0240/98)[6], (OLA 0241/98)[10], OLA 0242/98)[1]

131 chlorpropham with cetrimide

A soil-acting herbicide for lettuce under cold glass

Products

Croptex Pewter	Hortichem	80:80 g/l	SC	02507

Uses

Annual dicotyledons in PROTECTED LETTUCE. Annual grasses in PROTECTED LETTUCE. Chickweed in PROTECTED LETTUCE. Polygonums in PROTECTED LETTUCE.

Notes

Efficacy

- Apply to drilled lettuce under cold glass within 24 h post-drilling, to transplanted crops pre-planting
- Adequate irrigation must be applied before or after treatment
- Best results achieved on firm soil of fine tilth, free from clods and weeds

Crop Safety/Restrictions

- Do not apply to crop foliage or use where seed has chitted
- Excess irrigation may cause temporary check to crop under certain circumstances
- Do not apply where tomatoes or brassicas are growing in the same house
- In the event of crop failure only lettuce should be grown within 2 mth

Special precautions/Environmental safety

- Harmful in contact with skin and if swallowed
- Irritating to eyes, skin and respiratory system
- Highly flammable
- Dangerous to fish or other aquatic life. Do not contaminate surface waters or ditches with chemical or used container

Protective clothing/Label precautions

- A, C
- R03a, R03c, R04a, R04b, R04c, R07c, U05a, U08, U19, U20a, C03, E01, E13b, E26, E30a, E31a

Latest application/Harvest Interval(HI)

- Before crop emergence or pre-planting

132 chlorpropham + fenuron

A residual herbicide for vegetables and ornamentals

Products

Croptex Chrome	Hortichem	80:15 g/l	EC	02415

Uses

Annual dicotyledons in BROAD BEANS, FIELD BEANS, FLOWERS, LEEKS, MAIDEN STRAWBERRIES, ONIONS, PEAS, PROTECTED BULBS, RUNNER BEANS *(off-label)*

SPINACH. Annual grasses in BROAD BEANS, FIELD BEANS, FLOWERS, LEEKS, MAIDEN STRAWBERRIES, ONIONS, PEAS, PROTECTED BULBS, RUNNER BEANS *(off-label)*, SPINACH. Chickweed in BROAD BEANS, FIELD BEANS, FLOWERS, LEEKS, MAIDEN STRAWBERRIES, ONIONS, PEAS, PROTECTED BULBS, RUNNER BEANS *(off-label)*, SPINACH.

Notes

Efficacy

- Apply to soil freshly cultivated and free of established weeds. Adequate rainfall must occur after spraying. Activity is greater in cold, wet than warm, dry conditions

Crop Safety/Restrictions

- Apply to seeded crops pre-emergence of crop or weeds, to transplanted onions and leeks 10-14 d post-planting, to transplanted flowers and strawberries 5 d pre-transplanting
- Do not use on Sands, Very Light soils or soils low in organic matter
- Do not treat protected bulb or flower crops if other crops are being grown in same block of houses
- Poor conditions at drilling or planting, soil compaction, surface capping, waterlogging or attack by pests may result in crop damage
- In the event of crop failure only recommended crops should be replanted in treated soil
- Plough or cultivate to 15 cm after harvest to dissipate any residues

Special precautions/Environmental safety

- Harmful if swallowed and in contact with skin
- Irritating to eyes, skin and respiratory system
- Flammable
- Dangerous to fish or other aquatic life. Do not contaminate surface waters or ditches with chemical or used container

Protective clothing/Label precautions

- A, C
- M03, R03a, R03c, R04a, R04b, R04c, R07d, U05a, U08, U13, U19, U20a, C03, E01, E13b, E26, E30a, E32a, E34

Approval

- Off-label approval unlimited for use on outdoor and covered crops of runner beans (OLA0622/96)[1]

---

133 chlorpropham + linuron

A residual and contact herbicide for use in bulb crops

Products

Profalon	Hortichem	200:100 g/l	EC	01640

Uses

Annual dicotyledons in DAFFODILS, NARCISSI, TULIPS. Annual grasses in DAFFODILS, NARCISSI, TULIPS.

Notes

Efficacy

- Weeds controlled by combined residual action, requiring adequate soil moisture, and contact effect on young seedlings, requiring dry leaf surfaces for good control
- Do not spray during or immediately prior to rainfall
- Do not cultivate after spraying unless necessary

Crop Safety/Restrictions
- Maximum number of treatments 1 per crop
- Spray daffodils and narcissi pre-emergence or post-emergence before flower buds show. Spray tulips pre-emergence only
- Do not apply on Very Light soils or Silts low in humus or clay content

Special precautions/Environmental safety
- Irritating to skin, eyes and respiratory system
- Flammable
- Dangerous to fish or other aquatic life. Do not contaminate surface waters or ditches with chemical or used container
- Do not allow direct spray from vehicle mounted/drawn hydraulic sprayers to fall within 6 m of surface waters or ditches. Direct spray away from water
- Do not apply by hand-held sprayers

Protective clothing/Label precautions
- A, C, H
- R04a, R04b, R04c, R07d, U04a, U05a, U08, U19, U20a, C03, E01, E13b, E16, E26, E30a, E31b, E34

Latest application/Harvest Interval(HI)
- Pre-emergence for tulips

Approval
- Approved for aerial application on bulbs and corms [1]. See notes in Section 1

134 chlorpropham + pentanochlor

A contact and residual herbicide for horticultural crops

Products	Atlas Brown	Atlas	150:300 g/l	EC	07703

Uses Annual dicotyledons in CARROTS, CELERIAC, CELERY, CHRYSANTHEMUMS, FENNEL, NARCISSI, OUTDOOR LEAF HERBS *(off-label)*, PARSLEY, PARSNIPS, TULIPS. Annual meadow grass in CARROTS, CELERIAC, CELERY, CHRYSANTHEMUMS, FENNEL, NARCISSI, OUTDOOR LEAF HERBS *(off-label)*, PARSLEY, PARSNIPS, TULIPS.

Notes

Efficacy
- Apply as pre- or post-weed emergence spray
- Best results by application to weeds up to 2-leaf stage on fine, firm, moist seedbed
- Greatest contact action achieved under warm, moist conditions, the short residual action greatest in earlier part of year

Crop Safety/Restrictions
- Maximum number of treatments 1 per yr for narcissi, tulips; 2 per yr for carrots, celeriac, celery, fennel, parsnips, chrysanthemums, outdoor leaf herbs
- Apply to carrots and related crops pre- or post-emergence after fully expanded cotyledon stage, to narcissi and tulips at any time before emergence

- Apply to chrysanthemums either pre-planting or after planting as carefully directed spray, avoiding foliage
- Any crop may be sown or planted after 4 wk following ploughing and cultivation

Special precautions/Environmental safety

- Irritating to eyes

Protective clothing/Label precautions

- C
- R04a, U05a, U19, U20a, C03, E01, E15, E30a, E31b

Latest application/Harvest Interval(HI)

- Pre-emergence for narcissi, tulips
- HI 28 d for carrots, parsnips

Approval

- Off-label approval unlimited for use on outdoor leaf herbs (OLA 0249/93, 0602/93)[1]

135 chlorpyrifos

A contact and ingested organophosphorus insecticide and acaricide

Products

1 Alpha Chlorpyrifos 48 EC	Makhteshim	480 g/l	EC	04821
2 Barclay Clinch II	Barclay	480 g/l	EC	08596
3 Crossfire 480	RP Amenity	480 g/l	EC	08142
4 Cyren	Cheminova	480 g/l	EC	08358
5 Dursban 4	Dow	480 g/l	EC	07815
6 Lorsban T	Rigby Taylor	480 g/l	EC	07813
7 Maraud	Scotts	480 g/l	EC	09274
8 Pyrinex 48EC	Makhteshim	480 g/l	EC	08644
9 Spannit	PBI	480 g/l	EC	08744
10 Spannit Granules	PBI	6% w/w	GR	08984
11 Standon Chlorpyrifos 48	Standon	480 g/l	EC	08286
12 SuSCon Green Soil Insecticide	Fargro	10.34% w/w	GR	06312
13 Talon	FCC	480 g/l	EC	06017

Uses

Ambrosia beetle in CUT LOGS [1, 4, 6, 8]. Ambrosia beetle in CUT LOGS/TIMBER [2]. Aphids in BARLEY, OATS, WHEAT [1, 2, 4, 5, 8, 9, 11, 13]. Aphids in APPLES, GOOSEBERRIES, PEARS, PLUMS, RASPBERRIES, STRAWBERRIES [1, 2, 4, 5, 8, 9, 11]. Aphids in BRASSICAS [1, 2, 4, 8, 9, 13]. Aphids in BLACKCURRANTS [1, 8, 11]. Aphids in BROCCOLI, BRUSSELS SPROUTS, CABBAGES, CALABRESE, CAULIFLOWERS, CHINESE CABBAGE, COLLARDS [5, 11]. Aphids in FENNEL *(off-label)* [5]. Aphids in REDCURRANTS, WHITECURRANTS [11]. Aphids in KALE [2, 4, 5, 11]. Aphids in CURRANTS [2, 4, 5, 9]. Apple blossom weevil in APPLES [1, 2, 4, 5, 8, 9, 11]. Apple blossom weevil in PEARS [2, 9]. Apple sucker in APPLES [11]. Black pine beetle in FORESTRY TRANSPLANT LINES [1, 2, 5, 6, 8]. Black pine beetle in FORESTRY TRANSPLANTS [4]. Cabbage root fly in BRASSICAS [1, 2, 4, 8, 9, 13]. Cabbage root fly in CALABRESE, CHINESE CABBAGE, COLLARDS [5, 11]. Cabbage root fly in BROCCOLI, BRUSSELS SPROUTS, CABBAGES, CAULIFLOWERS [5, 10, 11]. Cabbage root fly in MOOLI *(off-label)*, RADISHES *(off-label)* [5]. Cabbage root fly in KALE [2, 4, 5, 11]. Capsids in APPLES, GOOSEBERRIES, PEARS [1, 2, 4, 5, 8, 9, 11]. Capsids in CURRANTS [1, 2, 4, 5, 8, 9]. Capsids in BLACKCURRANTS, REDCURRANTS, WHITECURRANTS [11]. Caterpillars in GOOSEBERRIES [1, 2, 4, 5, 8, 9, 11]. Caterpillars in APPLES, CURRANTS, PEARS [1, 2, 4, 5,

8, 9]. Caterpillars in BRASSICAS [1, 2, 4, 8, 9, 13]. Caterpillars in BROCCOLI, BRUSSELS SPROUTS, CABBAGES, CALABRESE, CAULIFLOWERS, CHINESE CABBAGE, COLLARDS [5, 11]. Caterpillars in BLACKCURRANTS, REDCURRANTS, WHITECURRANTS [11]. Caterpillars in KALE [2, 4, 5, 11]. Clay-coloured weevil in FOREST NURSERY BEDS [2]. Codling moth in APPLES, PEARS [1, 2, 4, 5, 8, 9, 11]. Colorado beetle in POTATOES *(off-label)* [5]. Cutworms in CELERY, LEEKS, ONIONS [1, 2, 4, 5, 8, 11]. Cutworms in CARROTS, POTATOES [1, 2, 4, 5, 8, 9, 11, 13]. Cutworms in LEAF BRASSICAS [1, 8]. Cutworms in BROCCOLI, BRUSSELS SPROUTS, CABBAGES, CALABRESE, CAULIFLOWERS, CHINESE CABBAGE, COLLARDS [5, 11]. Cutworms in FENNEL *(off-label)*, LETTUCE *(outdoor only)* [5]. Cutworms in OUTDOOR LETTUCE [4, 5, 11]. Cutworms in PARSNIPS, RED BEET, SWEDES, TURNIPS [4]. Cutworms in KALE [2, 4, 5, 11]. Cutworms in BRASSICAS [2, 4, 9, 13]. Damson-hop aphid in PLUMS [1, 8, 11]. Elm bark beetle in CUT LOGS [4, 6]. Frit fly in AMENITY TURF [1, 3, 7, 8, 11]. Frit fly in BARLEY, GRASSLAND, MAIZE, OATS, WHEAT [1, 2, 4, 5, 8, 9, 11, 13]. Frit fly in SWEETCORN [4, 9]. Frit fly in AMENITY GRASS [2-4, 6, 7, 11]. Frit fly in GOLF COURSES [2, 4, 6, 7, 9]. Insect pests in GREEN BEANS *(off-label)* [5]. Insect pests in EDIBLE PODDED PEAS *(off-label)*, FLOWER BULBS *(off-label)*, PEAS *(off-label)*, PROTECTED BRASSICA SEEDLINGS *(off-label)*, PROTECTED ONIONS/LEEKS/GARLIC *(off-label)* [9]. Larch shoot beetle in CUT LOGS [1, 4-6, 8]. Larch shoot beetle in CUT LOGS/TIMBER [2]. Large pine weevil in FORESTRY TRANSPLANTS [4]. Leatherjackets in AMENITY TURF [1, 3, 7, 8, 11]. Leatherjackets in BARLEY, GRASSLAND, OATS, WHEAT [1, 2, 4, 5, 8, 9, 11, 13]. Leatherjackets in PEAS, SUGAR BEET [1, 2, 4, 5, 8, 9, 11]. Leatherjackets in BROCCOLI, BRUSSELS SPROUTS, CABBAGES, CALABRESE, CAULIFLOWERS, CHINESE CABBAGE, COLLARDS [5, 11]. Leatherjackets in AMENITY GRASS [2-4, 6, 7, 11]. Leatherjackets in GOLF COURSES [2, 4, 6, 7, 9]. Leatherjackets in KALE [2, 4, 5, 11]. Leatherjackets in BRASSICAS [2, 4, 9, 13]. Mealy aphids in PLUMS [1, 8]. Pear sucker in PEARS [11]. Pine shoot beetle in CUT LOGS [1, 4-6, 8]. Pine shoot beetle in CUT LOGS/TIMBER [2]. Pine weevil in FORESTRY TRANSPLANT LINES [1, 2, 5, 6, 8]. Pygmy mangold beetle in SUGAR BEET [2, 4, 5, 9]. Raspberry beetle in RASPBERRIES [1, 2, 4, 5, 8, 9, 11]. Raspberry cane midge in RASPBERRIES [1, 2, 4, 5, 8, 9, 11]. Red spider mites in APPLES, GOOSEBERRIES, PEARS, PLUMS, RASPBERRIES, STRAWBERRIES [1, 2, 4, 5, 8, 9, 11]. Red spider mites in CURRANTS [1, 2, 4, 5, 8, 9]. Red spider mites in BLACKCURRANTS, REDCURRANTS, WHITECURRANTS [11]. Rosy rustic moth in RHUBARB [1, 4, 5, 8, 11]. Sawflies in APPLES [1, 2, 4, 5, 8, 9, 11]. Sawflies in PEARS [2, 5, 9]. Strawberry blossom weevil in STRAWBERRIES [5]. Suckers in APPLES, PEARS [1, 2, 4, 5, 8, 9]. Summer-fruit tortrix moth in APPLES [11]. Thrips in BARLEY, OATS, WHEAT [2, 4, 9, 13]. Tortrix moths in APPLES, PEARS, PLUMS, STRAWBERRIES [1, 2, 4, 5, 8, 9, 11]. Vine weevil in CONIFERS [1, 4-6, 8]. Vine weevil in STRAWBERRIES [1, 2, 4, 5, 8, 9, 11]. Vine weevil in HARDY ORNAMENTAL NURSERY STOCK, ORNAMENTALS [12]. Vine weevil in FORESTRY TRANSPLANT LINES [6]. Wheat bulb fly in BARLEY, OATS, WHEAT [1, 2, 4, 5, 8, 9, 11, 13]. Wheat-blossom midges in BARLEY, OATS, WHEAT [1, 2, 4, 5, 8, 9, 11, 13]. Whitefly in BROCCOLI, BRUSSELS SPROUTS, CABBAGES, CALABRESE, CAULIFLOWERS, CHINESE CABBAGE, COLLARDS, KALE [5]. Whitefly in BRASSICAS [4]. Winter moth in APPLES, PEARS, PLUMS [1, 2, 4, 5, 8, 9, 11]. Woolly aphid in APPLES [1, 2, 4, 5, 8, 9, 11]. Woolly aphid in PEARS [2, 4, 5, 9].

Notes

Efficacy

- Apply as a foliar or soil treatment for most uses, as granules or a drench for soil pests of brassicas, as a drench for vine weevil control, as a dip for forestry transplants
- Brassicas raised in plant-raising beds may require retreatment at transplanting
- Activity may be reduced when soil temperature below 5°C or on organic soils

- In dry conditions the effect of granules applied as a surface band may be reduced [10]
- For vine weevil control in hardy ornamental nursery stock incorporate in growing medium when plants first potted from rooted cutting stage [12]
- For turf pests apply between Nov and Mar where high larval populations detected or damage seen [3, 6]
- Where pear suckers resistant to chlorpyrifos occur control is unlikely to be satisfactory [5, 9]

Crop Safety/Restrictions
- Maximum number of treatments and timing vary with crop, product and pest. See label for details
- On carrots the maximum total dose applied per crop must not exceed the equivalent of 3 (on mineral soils) or 4 (on organic soils) full dose applications
- Do not apply to young lettuce plants [5] or treat potatoes under severe drought stress. The variety Desiree is particularly susceptible
- Do not apply to sugar beet under stress or within 4 d of applying a herbicide
- Do not mix with highly alkaline materials. Not compatible with zineb [5]
- In apples use pre-blossom up to pink/white bud and post-blossom after petal fall
- Cuttings from treated grass should not be used as a mulch for at least 12 mth

Special precautions/Environmental safety
- Products contain an anticholinesterase organophosphorus compound. Do not use if under medical advice not to work with such compounds
- Harmful in contact with skin [2, 3, 5, 6, 9, 11, 13]
- Harmful if swallowed [1-6, 8, 9, 11, 13]
- Irritating to eyes [4, 9, 13]
- Irritating to skin [1-6, 8, 9, 11]
- Risk of serious damage to eyes [1, 8]
- May cause sensitization by skin contact [1, 8]
- Flammable [1-3, 5, 6, 8, 9, 11, 13]
- Dangerous to bees/high risk to bees. Do not apply to crops in flower or to those in which bees are actively foraging. Do not apply when flowering weeds are present [4-6, 9, 13]
- Dangerous to fish or other aquatic life (extremely dangerous [4, 12]). Do not contaminate surface waters or ditches with chemical or used container [5, 6, 9, 10, 12, 13]
- Do not allow direct spray from vehicle mounted sprayers to fall within 6 m of surface waters or ditches, from hand-held sprayers within 2 m. Direct spray away from water
- Do not allow direct spray from orchard air blast sprayers to fall within 18 m of surface waters or ditches [4]

Protective clothing/Label precautions
- A [1-13]; C [1, 4, 8, 9, 13]; H [1, 2, 4-6, 8, 9, 11, 13]
- M01 [1, 3-13]; M03, R03c, U05a, U08, U19, C03, E31b, E34 [1-9, 11, 13]; R03a [2, 3, 5-7, 9, 11, 13]; R04a [4, 9, 13]; R04b [1-9, 11]; R04d, R04e [1, 8]; R05a [13]; R07d [1-3, 5-9, 11, 13]; U02 [1, 8-10, 12, 13]; U20a, C02 [9, 10, 13]; U20b [1-8, 11, 12]; C01 [12]; E01, E30a [1-13]; E07 (14 d) [1, 4, 5, 8, 11]; E07 [2]; E12a, E17 [4]; E12c [1, 5, 6, 8, 9]; E12d [2, 3, 7, 11]; E13a [4, 12]; E13b [1-3, 5-11, 13]; E16 [2-7, 11]; E26 [11, 12]; E32a [9, 10, 12]

Withholding period
- Keep livestock out of treated areas for at least 14 d after treatment [5, 9, 10]

Latest application/Harvest Interval(HI)
- Varies with product. See individual labels

Approval

- Off-label approval unlimited for aerial application to control Colorado beetle in potatoes as required by Statutory Notice (OLA 0796/91) [5]; to Mar 1999 for use on radish and mooli (OLA 0098/98) [5]; unlimited for use on a wide range of seedling brassicas and vegetables (OLA 1199/96)[9]; to Jan 2001 for use on fennel (OLA 0085/96)[5]; unlimited for use on ornamental bulbs (OLA 1498/96)[9]; unlimited for use on phaseolus green beans (kidney beans) harvested dry (OLA 1121/98)[5]
- Approval expiry: 31 Jul 99 [13]

Maximum Residue Levels (mg residue/kg food)

- kiwi fruit 2; pome fruits, grapes, tomatoes, peppers, aubergines 0.5; citrus fruits 0.3; plums, strawberries 0.2; carrots, tea 0.1; tree nuts, apricots, bilberries, cranberries, currants, gooseberries, wild berries, avocados, dates, figs, kumquats, litchis, mangoes, olives, passion fruit, pineapples, pomegranates, beetroot, celeriac, parsnips, horseradish, Jerusalem artichokes, parsley root, radishes, swedes, turnips, salsify, sweet potatoes, yams, garlic, shallots, spring onions, cucurbits, sweetcorn, kohlrabi, cress, spinach, beet leaves, watercress, witloof, mushrooms, pulses, oilseeds, potatoes, cereals, poultry 0.05; hops 0.1; milk, eggs 0.01

136 chlorpyrifos + dimethoate

A contact, systemic and fumigant insecticide for brassica crops

Products

Atlas Sheriff	Atlas	3.6:3.6% w/w	GR	07941

Uses

Aphids in BROCCOLI, BRUSSELS SPROUTS, CABBAGES, CAULIFLOWERS, KALE OILSEED RAPE. Cabbage root fly in BROCCOLI, BRUSSELS SPROUTS, CABBAGES, CAULIFLOWERS, KALE, OILSEED RAPE. Cutworms in BROCCOLI, BRUSSELS SPROUTS, CABBAGES, CAULIFLOWERS, KALE, OILSEED RAPE. Leatherjackets in BROCCOLI, BRUSSELS SPROUTS, CABBAGES, CAULIFLOWERS, KALE, OILSEED RAPE. Wireworms in BROCCOLI, BRUSSELS SPROUTS, CABBAGES, CAULIFLOWERS KALE, OILSEED RAPE.

Notes

Efficacy

- Apply with suitable granule applicator (see label for details) as surface or sub-surface band treatment. Check calibration before applying
- May be used on drilled or transplanted crops and on nursery beds
- On nursery beds apply as overall treatment
- Apply by mid-Apr or at time of drilling or planting and repeat if necessary

Special precautions/Environmental safety

- Product contains organophosphorus compound. Do not use if under medical advice not to work with such compounds
- Dangerous to fish or other aquatic life. Do not contaminate surface waters or ditches with chemical or used container
- Harmful to game, wild birds and animals

Protective clothing/Label precautions

- A, C

- M01, M03, U02, U05a, U20a, C02, C03, E01, E10b, E13b, E30a, E32a

137 chlorpyrifos + disulfoton

A systemic and contact organophosphorus insecticide for brassicas

Products

Twinspan	PBI	4:6% w/w	GR	08983

Uses

Aphids in BROCCOLI, BRUSSELS SPROUTS, CABBAGES, CAULIFLOWERS. Cabbage root fly in BROCCOLI, BRUSSELS SPROUTS, CABBAGES, CAULIFLOWERS.

Notes

Efficacy

- Apply with suitable band applicator as bow-wave treatment at drilling followed if necessary by surface band treatment 2 d after singling, or as sub-surface band treatment at transplanting. See label for details
- Can be used on all mineral and organic soils
- Crops treated in plant-raising beds must be treated again at transplanting
- Effect of surface band application may be reduced in dry weather

Crop Safety/Restrictions

- Do not treat mini-cauliflowers

Special precautions/Environmental safety

- This product contains an anticholinesterase organophosphorus compound. Do not use if under medical advice not to work with such compounds
- Harmful if swallowed and by inhalation
- Keep in original container, tightly closed, in a safe place, under lock and key
- Dangerous to game, wild birds and animals
- Dangerous to fish or other aquatic life. Do not contaminate surface waters or ditches with chemical or used container

Protective clothing/Label precautions

- A, B, C or D+E, H, K, M
- M01, M03, R03b, R03c, U02, U04a, U05a, U09a, U13, U14, U15, U19, U20a, C02, C03, E01, E10a, E13b, E30b, E32a, E34

Latest application/Harvest Interval(HI)

- HI 6 wk

Maximum Residue Levels (mg residue/kg food)

- see chlorpyrifos and disulfoton entries

138 chlorpyrifos-methyl

An organophosphorus insecticide and acaricide for grain store use

Products

1 Reldan 22	Dow	225 g/l	EC	08191
2 Smite	AgrEvo Environ.	300 g/l	EC	H5142

Uses

Flies in REFUSE TIPS [2]. Grain storage pests in STORED GRAIN [1]. Pre-harvest hygiene in GRAIN STORES [1].

Notes

Efficacy
- May be applied pre-harvest to surfaces of empty store and grain handling machinery and as admixture with grain [1]
- Apply to grain after drying to moisture content below 14%, cooling and cleaning [1]
- Insecticide may become depleted at grain surface if grain is being cooled by continuous extraction of air from the base leading to reduced control of grain store pests especially mites [1]
- Resistance to organophosphorus compounds sometimes occurs in insect and mite pests of stored products [1]
- Apply directly onto the surface of exposed refuse. Repeat as necessary [2]

Crop Safety/Restrictions
- Maximum number of treatments 1 per batch [1]

Special precautions/Environmental safety
- This product contains an anticholinesterase organophosphorus compound. Do not use if under medical advice not to work with such compounds
- Irritating to eyes and skin [2]
- Risk of serious damage to eyes [1]
- Flammable [2]
- Extremely dangerous (dangerous [2]) to fish or other aquatic life. Do not contaminate surface waters or ditches with chemical or used container [1]

Protective clothing/Label precautions
- A [1]; B [2]; C, H [1, 2]; M [1]
- M01, R04, R04d, U08, U20b, C03, E01, E13a, E26, E30a, E31b, E34 [1]; M02, R04a, R04b, R07d, U02, C09, E02, E13b, E32a [2]; U05a, U19 [1, 2]

Latest application/Harvest Interval(HI)
- 8 wk before malting barley for stored grain; before use of storage facility for grain stores [1]

Maximum Residue Levels (mg residue/kg food)
- pome fruits, strawberries, tomatoes, peppers, aubergines 0.5; grapes 0.2; tea, hops 0.1; grapefruit, limes, pomelos, tree nuts, apricots, cherries, plums, cane fruits, bilberries, cranberries, currants, gooseberries, wild berries, miscellaneous fruits, root and tuber vegetables, bulb vegetables, cucurbits, brassicas, leaf vegetables, legumes, stem vegetables, mushrooms, pulses, oilseeds, potatoes, rice, meat 0.05; milk, eggs 0.01

139 chlorsulfuron

A sulfonylurea herbicide formulations of which were voluntarily withdrawn in the UK in 1988

140 chlorthal-dimethyl

A residual benzoic acid herbicide for use in horticulture

Products

1 Dacthal W-75	Hortichem	75% w/w	WP	05500
2 Dacthal W-75	ISK Biosciences	75% w/w	WP	05556

FOR FULL CONDITIONS OF USE ALWAYS READ THE PRODUCT LABEL

Uses Annual dicotyledons in BLACKCURRANTS, GOOSEBERRIES, LEEKS, MUSTARD, OILSEED RAPE, ONIONS, RASPBERRIES, STRAWBERRIES, SWEDES, TURNIPS [1, 2]. Annual dicotyledons in BROCCOLI, BRUSSELS SPROUTS, CABBAGES, CALABRESE, CAULIFLOWERS, COURGETTES *(off-label)*, FODDER RAPE, KALE, MARROWS *(off-label)*, ORNAMENTALS, SAGE [1]. Annual dicotyledons in BRASSICAS, OUTDOOR HERBS *(off-label)*, RUNNER BEANS [2]. Slender speedwell in TURF [1, 2].

Notes

Efficacy

- Best results on fine firm weed-free soil when adequate rain or irrigation follows
- Recommended alone on roses, runner beans, various ornamentals, strawberries and turf, in tank-mix with propachlor on brassicas, onions, leeks, sage, ornamentals, established strawberries and newly planted soft fruit
- Apply after drilling or planting prior to weed emergence. Rates and timing vary with crop and soil type. See label for details
- Do not use on organic soils
- For control of slender speedwell in turf apply when weeds growing actively. Do not mow for at least 3 d after treatment

Crop Safety/Restrictions

- Maximum number of treatments 1 per crop for brassicas, oilseed rape, mustard, leeks, onions, sage, newly planted bush and cane fruit; 1 per season for established strawberries
- Apply to brassicas pre-emergence or after 3-4 true leaf stage
- Do not apply mixture with propachlor to newly planted strawberries after rolling or application of other herbicides
- Do not use on strawberries between flowering and harvest
- Many types of ornamental have been treated successfully. See label for details. For species of unknown susceptibility treat a small number of plants first
- Do not use on turf where bent grasses form a major constituent of sward
- Do not plant lettuce within 6 mth of application, seeded turf within 2 mth, other crops within 3 mth. In the event of crop failure deep plough before re-drilling or planting
- Do not use on dwarf French beans

Protective clothing/Label precautions

- R04a, R04b, R04c, U20c, E01, E26 [2]; U20a [1]; E15, E30a, E32a [1, 2]

Latest application/Harvest Interval(HI)

- Before crop emergence for sown crops

Approval

- Off-label approval unlimited for use in outdoor herbs (see OLA notice for list of species) (OLA 1288/93)[2]; to Jun 2000 for use on outdoor courgettes and marrows (OLA 1108/97)[1]

---

141 choline chloride

A plant growth regulator available only in mixtures

See also chlormequat + choline chloride
chlormequat + choline chloride + imazaquin

142 clodinafop-propargyl

A contact acting herbicide for annual grass weed control in cereals

Products

1 Landgold Clodinafop	Landgold	240 g/l	EC	09017
2 Topik	Ciba Agric.	240 g/l	EC	07763
3 Topik	Novartis	240 g/l	EC	08461

Uses

Blackgrass in DURUM WHEAT, RYE, SPRING WHEAT, TRITICALE, WINTER WHEAT. Rough meadow grass in DURUM WHEAT, RYE, SPRING WHEAT, TRITICALE, WINTER WHEAT. Wild oats in DURUM WHEAT, RYE, SPRING WHEAT, TRITICALE, WINTER WHEAT.

Notes

Efficacy

- Spray in autumn, winter or spring from 1 true leaf stage (GS 11) to before second node detectable (GS 32)
- Products contain a herbicide safener (cloquintocet-mexyl) that improves crop tolerance to clodinafop-propargyl
- Spray when majority of weeds have germinated but before competition reduces yield
- Optimum control achieved when all grass weeds emerged. Wait for delayed germination on dry or cloddy seedbed
- A mineral oil additive is recommended to give more consistent control of very high blackgrass populations or for late season treatments. See label for details
- Weed control not affected by soil type, organic matter or straw residues
- Control may be reduced if rain falls within 1 h of treatment

Crop Safety/Restrictions

- Maximum total dose 0.25 l/ha/crop
- Do not treat crops under stress or suffering from waterlogging, pest attack, disease or frost
- Do not treat crops undersown with grass mixtures
- Do not mix with products containing MCPA, mecoprop, 2,4-D or 2,4-DB
- MCPA, mecoprop, 2,4-D or 2,4-DB should not be applied within 21 d before, or 7 d after, treatment
- Only a broad leaved crop may be sown after failure of a treated crop. After normal harvest any broad leaved crop or wheat, durum wheat, rye, triticale or barley should be sown

Special precautions/Environmental safety

- Irritating to skin and eyes
- Flammable
- Harmful to fish or other aquatic life. Do not contaminate surface waters or ditches with chemical or used container

Protective clothing/Label precautions

- A, C, H, K
- R04a, R04b, R07d, U02, U05a, U09a, U20a, C03, E13c, E30a, E32a

Latest application/Harvest Interval(HI)

- Before second node detectable stage (GS 32)

143 clodinafop-propargyl + diflufenican

A contact and residual herbicide mixture for broad spectrum weed control in cereals

Products

1 Amazon	Novartis	30:50 g/l	EC	08128
2 Amazon TP	Ciba Agric.	240:500 g/l	KL	07681
3 Amazon TP	Novartis	240:500 g/l	KL	08384
4 Lucifer	RP Agric.	30:50 g/l	EC	08129

Uses

Annual and perennial weeds in APPLES, ASPARAGUS *(off-label)*, BARLEY *(pre-harvest)*, BLACKCURRANTS *(off-label)*, BROAD-LEAVED TREES *(pre-planting)*, CHERRIES, COMBINING PEAS *(pre-harvest)*, CONIFERS *(pre-planting)*, DAMSONS, FIELD BEANS *(pre-harvest)*, FIELD CROPS *(stubble treatment)*, FIELD CROPS *(sward destruction/direct drilling)*, FORESTRY *(directed spray)*, GRAPEVINES *(off-label)*, GRASSLAND *(pre-cut/graze)*, HEADLANDS, LAND TEMPORARILY REMOVED FROM PRODUCTION, LINSEED *(pre-harvest)*, MUSTARD *(pre-harvest)*, NON-CROP FARM AREAS, OATS *(pre-harvest)*, OILSEED RAPE *(pre-harvest)*, PEARS, PLUMS, TREE NUTS *(off-label)*, WHEAT *(pre-harvest)* [4]. Annual dicotyledons in DURUM WHEAT, TRITICALE, WINTER RYE, WINTER WHEAT [1-3]. Annual weeds in GRASSLAND [4]. Blackgrass in DURUM WHEAT, TRITICALE, WINTER RYE, WINTER WHEAT [1-3]. Bolters in SUGAR BEET *(wiper application)* [4]. Bracken in FORESTRY *(directed spray)*, TOLERANT CONIFERS *(overall dormant spray)* [4]. Brambles in TOLERANT CONIFERS *(overall dormant spray)* [4]. Chemical thinning in FORESTRY [4]. Couch in BARLEY *(pre-harvest)*, COMBINING PEAS *(pre-harvest)*, FIELD BEANS *(pre-harvest)*, FIELD CROPS *(stubble treatment)*, FIELD CROPS *(sward destruction/direct drilling)*, FORESTRY *(directed spray)*, LINSEED *(pre-harvest)*, MUSTARD *(pre-harvest)*, OATS *(pre-harvest)*, OILSEED RAPE *(pre-harvest)*, WHEAT *(pre-harvest)* [4]. Destruction of short term leys in GRASSLAND [4]. Perennial dicotyledons in FORESTRY *(wiper application)*, GRASSLAND *(wiper application)* [4]. Perennial grasses in AQUATIC SITUATIONS, TOLERANT CONIFERS *(overall dormant spray)* [4]. Pre-harvest desiccation in BARLEY, MUSTARD, OATS, OILSEED RAPE, WHEAT [4]. Reeds in AQUATIC SITUATIONS [4]. Rough meadow grass in DURUM WHEAT, TRITICALE, WINTER RYE, WINTER WHEAT [1-3]. Rushes in AQUATIC SITUATIONS, FORESTRY *(directed spray)*[4]. Sedges in AQUATIC SITUATIONS [4]. Sucker control in APPLES, CHERRIES, DAMSONS, PEARS, PLUMS [4]. Volunteer cereals in FIELD CROPS *(stubble treatment)*, FIELD CROPS *(sward destruction/direct drilling)* [4]. Volunteer potatoes in FIELD CROPS *(stubble treatment)* [4]. Waterlilies in AQUATIC SITUATIONS [4]. Wild oats in DURUM WHEAT, TRITICALE, WINTER RYE, WINTER WHEAT [1-3]. Wild oats in CEREALS *(wiper glove)* [4]. Woody weeds in FORESTRY *(directed spray)*, TOLERANT CONIFERS *(overall dormant spray)* [4].

Notes

Efficacy

- Optimum weed control obtained when all grass weeds emerged and broad leaved weeds at susceptible stage before they compete with the crop
- Products contain a herbicide safener (cloquintocet-mexyl) that improves crop tolerance to clodinafop-propargyl
- Spray in autumn, winter or spring from first leaf unfolded stage (GS 11) to end Feb. Soil should be moist at time of treatment
- Delay treatment if dry or cloddy seedbeds favour late weed germination
- Grass weed control unaffected by seedbed condition but control of broad leaved weeds requires a fine, firm seedbed cleared of trash and straw
- Always use a recommended adjuvant. See label

- Speed of action depends on temperature and growing conditions and may appear slow in dry or cold weather

Crop Safety/Restrictions
- Maximum number of treatments 1 per crop
- Do not treat crops under stress or suffering from waterlogging, pest attack, disease or frost
- Do not treat broadcast crops or those undersown with grass mixtures
- Do not mix with products containing MCPA, mecoprop, 2,4-D or 2,4-DB
- MCPA, mecoprop, 2,4-D or 2,4-DB should not be applied within 21 d before, or 7 d after, treatment
- Do not use on Sands, stony or gravelly soils or soil with over 10% organic matter
- Do not roll autumn treated crops until spring. Do not harrow at any time after treatment
- Only listed crops may be sown after failure of a treated crop - see label.
- After normal harvest cereals, field beans, oilseed rape, onions, leaf brassicas or sugar beet seed crops may be sown in the autumn. Land must be ploughed or cultivated to 15 cm except for cereals
- Successive treatments of any products containing diflufenican can lead to soil build up and inversion ploughing must precede sowing any non-cereal crop. Even where ploughing occurs some crops may be damaged - see label

Special precautions/Environmental safety
- Irritating to skin and eyes
- Harmful to fish or other aquatic life. Do not contaminate surface waters or ditches with chemical or used container

Protective clothing/Label precautions
- A, C, H, K
- R04a, R04b, U02, U05a, U09a, C03, E13c, E30a, E32a [1-4]; U13, U20a [2, 3]; U20b, E26 [1, 4]

Latest application/Harvest Interval(HI)
- Before end of Feb in year of harvest

144 clodinafop-propargyl + trifluralin

A contact and residual herbicide for winter wheat

Products					
	1 Hawk	Ciba Agric.	12:383 g/l	EC	08030
	2 Hawk	Novartis	12:383 g/l	EC	08417

Uses

Annual dicotyledons in WINTER WHEAT. Blackgrass in WINTER WHEAT. Rough meadow grass in WINTER WHEAT. Wild oats in WINTER WHEAT.

Notes

Efficacy
- Product must be applied with specified adjuvant - see label
- Product contains a herbicide safener (cloquintocet-mexyl) that improves crop tolerance to clodinafop-propargyl

- Optimum weed control obtained when most grass weeds emerged and broad leaved weeds at susceptible stage before they compete with the crop. Pre-emergence control of annual meadow-grass and broad leaved weeds may not be satisfactory on medium and heavy soils
- Spray in autumn, winter or spring from first to third leaf unfolded stage (GS 11-13) but before ear at 1 cm stage (GS 30)
- Delay treatment if dry or cloddy seedbeds favour late weed germination
- Optimum weed control obtained in crops growing in a fine, firm seedbed cleared of trash and straw
- Wild oats and other weeds germinating from beneath treated zone not controlled
- Always use a recommended adjuvant/wetter. See label
- Speed of action depends on temperature and growing conditions and may appear slow in dry or cold weather
- Rain within 1 h of application may reduce grass weed control

Crop Safety/Restrictions

- Maximum number of treatments 1 per crop
- Do not treat crops under stress or suffering from waterlogging, pest attack, disease or frost
- Do not treat crops undersown with grass mixtures
- Do not mix with products containing MCPA, mecoprop, 2,4-D or 2,4-DB
- MCPA, mecoprop, 2,4-D or 2,4-DB should not be applied within 21 d before, or 7 d after, treatment
- Do not use on Sands, stony or gravelly soils or soils with over 10% organic matter
- After normal harvest any broad leaved crop (except sugar beet), barley, wheat, durum wheat, rye or triticale may be sown. Before drilling or planting subsequent crops soil must be mouldboard ploughed to 15 cm
- See label for details of crops that may be safely drilled or planted within 5 mth of treatment. If a treated crop should fail after 5 mth but before normal harvest, sow only a broad-leaved crop (not sugar beet)
- 12 mth must elapse after treatment before sugar beet is drilled

Special precautions/Environmental safety

- Irritating to eyes and skin
- May cause sensitisation by skin contact
- Harmful to fish or other aquatic life. Do not contaminate surface waters or ditches with chemical or used container
- Crops must not be treated if there is risk of run-off (eg from frozen foliage)

Protective clothing/Label precautions

- A, C, H, K
- R04a, R04b, R04e, U02, U05a, U09a, U20a, C03, E13c, E26, E30a, E32a

Latest application/Harvest Interval(HI)

- Before ear at 1 cm stage (GS 30)

---

145 clofentezine

A selective ovicidal tetrazine acaricide for use in top fruit

Products Apollo 50 SC | Promark | 500 g/l | SC | 07242

Uses Red spider mites in APPLES, BLACKBERRIES *(off-label - protected crops)*, CHERRIES, PEARS, PLUMS, RASPBERRIES *(off-label - protected crops)*, RASPBERRIES *(off-label)*, STRAWBERRIES *(off-label -protected crops)*.

Notes

Efficacy

- Acts on eggs and early motile stages of mites. For effective control total cover of plants is essential, particular care being needed to cover undersides of leaves
- For red spider mite control spray apples and pears between bud burst and pink bud, plums and cherries between white bud and first flower. Rust mite is also suppressed
- On established infestations apply in conjunction with an adult acaricide

Crop Safety/Restrictions

- Maximum number of treatments 1 per yr
- Product safe on predatory mites, bees and other predatory insects

Protective clothing/Label precautions

- U08, U20a, E15, E30a, E32a

Latest application/Harvest Interval(HI)

- 28 d for apple and pear, 8 wk for plum and cherry

Approval

- Off-label approval unlimited for use on blackcurrants, strawberries, raspberries (OLA 1250/95)[1]; unlimited for use on protected raspberries and blackberries (OLA 1645/98)[1]; to Jul 2003 for use on protected strawberries (OLA 1646/98)[1]

146 clopyralid

A foliar translocated picolinic herbicide for a wide range of crops

See also benazolin + clopyralid
bromoxynil + clopyralid
bromoxynil + clopyralid + fluroxypyr
bromoxynil + clopyralid + fluroxypyr + ioxynil

Products

Dow Shield	Dow	200 g/l	SL	05578

Uses

Annual dicotyledons in BROAD-LEAVED TREES *(off-label)*, BROCCOLI, BRUSSELS SPROUTS, CABBAGES, CALABRESE, CAULIFLOWERS, CEREALS, CONIFERS *(off-label)*, ESTABLISHED GRASSLAND, FODDER BEET, FODDER RAPE, GRASS SEED CROPS *(off-label)*, KALE, KOHLRABI *(off-label)*, LINSEED, MAIZE, MANGELS, OILSEED RAPE, ONIONS, OUTDOOR HERBS *(off-label)*, RED BEET, SAGE *(off-label)*, SPINACH BEET *(off-label)*, SPINACH *(off-label)*, STRAWBERRIES, SUGAR BEET, SWEDES, SWEETCORN, TURNIPS, WOODY ORNAMENTALS. Corn marigold in BROCCOLI, BRUSSELS SPROUTS, CABBAGES, CALABRESE, CAULIFLOWERS, CEREALS, ESTABLISHED GRASSLAND, FODDER BEET, FODDER RAPE, KALE, LINSEED, MAIZE, MANGELS, OILSEED RAPE, ONIONS, RED BEET, STRAWBERRIES, SUGAR BEET, SWEDES, SWEETCORN, TURNIPS, WOODY ORNAMENTALS. Creeping thistle in BROCCOLI, BRUSSELS SPROUTS, CABBAGES, CALABRESE, CAULIFLOWERS, CEREALS, ESTABLISHED GRASSLAND, ESTABLISHED GRASSLAND *(off-label, weed wiper)*, FODDER BEET, FODDER RAPE, KALE, LINSEED, MAIZE, MANGELS, OILSEED RAPE, ONIONS, RED BEET, STRAWBERRIES, SUGAR BEET, SWEDES, SWEETCORN, TURNIPS, WOODY ORNAMENTALS. Mayweeds in BROCCOLI, BRUSSELS SPROUTS, CABBAGES, CALABRESE, CAULIFLOWERS, CEREALS, ESTABLISHED GRASSLAND, FODDER BEET, FODDER RAPE, HONESTY *(off-label)*, KALE, LINSEED, MAIZE,

MANGELS, OILSEED RAPE, ONIONS, RED BEET, STRAWBERRIES, SUGAR BEET, SWEDES, SWEETCORN, TURNIPS, WOODY ORNAMENTALS. Perennial dicotyledons in BROAD-LEAVED TREES *(off-label)*, BROCCOLI, BRUSSELS SPROUTS, CABBAGES, CALABRESE, CAULIFLOWERS, CEREALS, CONIFERS *(off-label)*, ESTABLISHED GRASSLAND, FODDER BEET, FODDER RAPE, GRASS SEED CROPS *(off-label)*, KALE, LINSEED, MAIZE, MANGELS, OILSEED RAPE, ONIONS, RED BEET, SAGE *(off-label)*, STRAWBERRIES, SUGAR BEET, SWEDES, SWEETCORN, TURNIPS, WOODY ORNAMENTALS. Thistles in HONESTY *(off-label)*.

Notes

Efficacy

- Best results achieved by application to young actively growing weed seedlings. Treat creeping thistle at rosette stage and repeat 3-4 wk later as directed
- High activity on weeds of Compositae family. For most crops recommended for use in tank mixes. See label for details
- Do not apply when crop damp or when rain expected within 6 h

Crop Safety/Restrictions

- Maximum total dose 2.0 l/ha/crop for winter oilseed rape; 1.5 l/ha/crop for beet crops, vegetable brassicas, fodder rape, swedes, turnips, onions, maize, sweetcorn, spring oilseed rape, strawberries; 1.0 l/ha/yr for established grassland, ornamental plant production; 0.5 l/ha/crop for linseed; 0.35 l/ha/crop for cereals
- Timing of application varies with weed problem, crop and other ingredients of tank mixes. See label for details
- Do not apply to cereals later than the second node detectable stage (GS 32)
- Do not use straw from treated cereals in compost or any other form for glasshouse crops. Straw may be used for strawing down strawberries
- Straw from treated grass seed crops or linseed should be baled and carted away. If incorporated do not plant winter beans in same year
- Do not use on onions at temperatures above 20°C or when under stress
- Do not treat maiden strawberries or runner bed or apply to early leaf growth during blossom period or within 4 wk of picking. Aug or early Sep sprays may reduce yield
- Apply as directed spray in woody ornamentals, avoiding leaves, buds and green stems. Do not apply in root zone of families Compositae or Papilionaceae
- Do not plant susceptible autumn-sown crops in same year as treatment. Do not apply later than Jul where susceptible crops are to be planted in spring. See label for details

Special precautions/Environmental safety

- Wash spray equipment thoroughly with water and detergent immediately after use. Traces of product can damage susceptible plants sprayed later

Protective clothing/Label precautions

- A, C
- U08, U19, U20b, E07 (7 d), E15, E26, E30a, E31b

Withholding period

- Keep livestock out of treated areas for at least 7 d and until foliage of any poisonous weeds such as ragwort has died and become unpalatable

Latest application/Harvest Interval(HI)

- Before 3rd node detectable (GS 33) for cereals; before flower buds visible from above for oilseed rape and linseed.
- HI grassland 7 d; strawberries 4 wk; maize, sweetcorn, onions, Brussels sprouts, broccoli, cabbage, cauliflowers, calabrese, kale, fodder rape, oilseed rape, swedes, turnips, sugar beet, red beet, fodder beet, mangels, sage, honesty 6 wk

Approval

- Off-label approval unlimited for use on established grassland (wiper application), grass seed crops (OLA 0662/92)[1], honesty (OLA 0480/93)[1]; unlimited for use on coniferous and broadleaved trees in forestry (OLA 0757/92)[1]; to Jul 2002 for use on outdoor herbs (see OLA notice for list)(OLA 1359/97)[1]; to Mar 2001 for use on outdoor kohlrabi (OLA 0495/96)[1]; to Oct 2001 for use on spinach and spinach beet (OLA 2360/96)[1]

147 clopyralid + 2,4-D + MCPA

A translocated herbicide mixture for grassland

See also 2,4-D

Products	Lonpar	Dow	35:150:175 g/l	SL	08686

Uses Creeping thistle in ESTABLISHED GRASSLAND.

Notes

Efficacy

- Treatment must be made when weeds and grass are actively growing
- Important to ensure sufficient leaf area for uptake especially on established thistles with extensive root system. Treat at rosette stage
- On large well established thistles and where there is a large soil seed reservoir further treatment in the following year may be needed
- To allow maximum translocation do not cut grass for 28 d after treatment

Crop Safety/Restrictions

- Maximum number of treatments 1 or 2 per yr depending on dose used - see label
- Do not spray in drought, very hot or very cold weather or if rain expected within 6 h
- Do not treat grass less than 1 yr old or sports or amenity turf
- Product kills or severely checks clover and should not be used where clover is an important constituent of the sward
- Very occasionally some yellowing of sward may occur after treatment which is quickly outgrown
- Do not drill grass, grass mixtures, kale, swedes or turnips into the sward within 6 wk of spraying

Special precautions/Environmental safety

- Harmful if swallowed
- Risk of serious damage to eyes
- Harmful to fish or other aquatic life. Do not contaminate surface waters or ditches with chemical or used container
- Wash spray equipment thoroughly with water and detergent immediately after use. Traces of product can damage susceptible plants sprayed later
- Susceptible crops (see label) must not be planted in the same calendar yr as treatment. Ensure that remains of treated crop have completely decayed before planting a subsequent susceptible crop. Where a susceptible crop is to be planted in the spring, product must not be sprayed later than end Jul in previous yr

Protective clothing/Label precautions

- A, C, H, M
- R03c, R04d, U05a, U15, C03, E01, E07 (2 wk), E13c, E26, E34

148 clopyralid + diflufenican + MCPA

A selective herbicide for use in established turf

Products	Spearhead	RP Amenity	20:15:300 g/l	SL	07342

Uses

Annual dicotyledons in TURF. Buttercups in TURF. Corn marigold in TURF. Creeping thistle in TURF. Dandelions in TURF. Mayweeds in TURF. Perennial dicotyledons in TURF.

Notes

Efficacy

- Best results achieved by application when grass and weeds are actively growing
- Treatment during early part of the season is recommended, but not during drought

Crop Safety/Restrictions

- Maximum number of treatments 1 per yr
- Only use when sward is satisfactorily established and regular mowing has begun
- Turf sown in spring or early summer may be ready for treatment after 2 mth. Later sown turf should not be sprayed until growth is resumed in the following spring
- Avoid mowing within 3-4 d before or after treatment
- Do not use cuttings from treated area as a mulch for any crop
- Avoid drift. Small amounts of spray can cause serious injury to herbaceous plants, vegetables, fruit and glasshouse crops

Special precautions/Environmental safety

- Irritating to eyes and skin
- Harmful to fish or other aquatic life. Do not contaminate surface waters or ditches with chemical or used container

Protective clothing/Label precautions

- A, C
- R04a, R04b, U02, U05a, U08, U19, U20b, C03, E01, E07, E13c, E26, E30a, E31b

Withholding period

- Keep livestock out of treated areas

149 clopyralid + fluroxypyr + ioxynil

A contact and translocated herbicide mixture for use in cereals

Products	Hotspur	Dow	45:150:200 g/l	EC	05185

Uses

Annual dicotyledons in BARLEY, DURUM WHEAT, OATS, RYE, TRITICALE, WHEAT. Chickweed in BARLEY, DURUM WHEAT, OATS, RYE, TRITICALE, WHEAT. Cleavers in BARLEY, DURUM WHEAT, OATS, RYE, TRITICALE, WHEAT. Hemp-nettle in BARLEY, DURUM WHEAT, OATS, RYE, TRITICALE, WHEAT. Mayweeds in BARLEY, DURUM WHEAT, OATS, RYE, TRITICALE, WHEAT. Speedwells in BARLEY, DURUM WHEAT, OATS, RYE, TRITICALE, WHEAT.

Notes

Efficacy

- Best results achieved by application in good growing conditions, when weeds small and growing actively in a strongly competitive crop
- Do not spray if rain falling or imminent

Crop Safety/Restrictions

- Maximum number of treatments 1 per crop
- Apply to winter or spring cereals from 2-leaf stage to first node detectable (GS 12-31)
- Crops undersown with grass may be treated from tillering stage of grass. Do not spray crops undersown or about to be undersown with clover or other legumes
- Do not spray during drought, waterlogging, frost or extremes of temperature
- Do not use straw from treated crops in compost or any other form for glasshouse crops

Special precautions/Environmental safety

- Harmful if swallowed. Irritating to eyes and skin
- Do not apply by knapsack sprayer or at concentrations higher than recommended
- Flammable
- Dangerous to fish or other aquatic life. Do not contaminate surface waters or ditches with chemical or used container
- Harmful to bees. Do not apply to crops in flower or to those in which bees are actively foraging. Do not apply when flowering weeds are present
- Wash spray equipment thoroughly with water and detergent immediately after use. Traces of product can damage susceptible plants sprayed later

Protective clothing/Label precautions

- A, C
- M03, R03c, R04a, R04b, R07d, U05a, U08, U13, U19, U20a, C03, E01, E07, E12e, E13b, E26, E30a, E31b, E34

Withholding period

- Keep livestock out of treated areas until foliage of any poisonous weeds such as ragwort has died and become unpalatable

Latest application/Harvest Interval(HI)

- Before second node detectable (GS 32).
- HI 6 wk (animal consumption)

Maximum Residue Levels (mg residue/kg food)

- see ioxynil entry

150 clopyralid + fluroxypyr + MCPA

A translocated herbicide mixture for use in sports and amenity turf

Products	Greenor	Rigby Taylor	20:40:200 g/l	ME	07848

Uses Annual dicotyledons in MANAGED AMENITY TURF.

FOR FULL CONDITIONS OF USE ALWAYS READ THE PRODUCT LABEL

Notes

Efficacy

- Best results achieved when weeds actively growing and turf grass competitive
- Treatment should normally be between Apr-Sep when the soil is moist
- Do not apply during drought unless irrigation is applied
- Allow 3 d before or after mowing established turf to ensure sufficient weed leaf surface present to allow uptake and movement

Crop Safety/Restrictions

- Maximum number of treatments 2 per yr
- Treat young turf only in spring when at least 2 mth have elapsed since sowing
- Allow 5 d after mowing young turf before treatment
- Do not treat grass under stress from frost, drought, waterlogging, trace element deficiency, disease or pest attack
- Do not treat if night temperatures are low, when frost is imminent or during prolonged cold weather
- Product selective on a number of turf grass species (see label) but consultation or testing recommended before treatment of any cultivar

Special precautions/Environmental safety

- Irritating to eyes and skin
- Harmful to fish or other aquatic life. Do not contaminate surface waters or ditches with chemical or used container
- Wash spray equipment thoroughly with water and detergent immediately after use. Traces of product can damage susceptible plants sprayed later

Protective clothing/Label precautions

- A, C
- R04a, R04b, U05a, U08, U19, U20a, C03, E01, E13c, E26, E30a, E31b

151 clopyralid + fluroxypyr + triclopyr

A foliar acting herbicide mixture for grassland

Products

Pastor	Dow	50:75:100 g/l	EC	07440

Uses

Docks in ESTABLISHED GRASSLAND. Stinging nettle in ESTABLISHED GRASSLAND. Thistles in ESTABLISHED GRASSLAND.

Notes

Efficacy

- May be applied in spring or autumn depending on weeds present
- Treatment must be made when weeds and grass are actively growing
- Important to ensure sufficient leaf area for uptake especially on established docks and thistles
- On large well established docks and where there is a large soil seed reservoir further treatment in the following year may be needed
- To allow maximum translocation do not cut grass for 4 wk after treatment

Crop Safety/Restrictions

- Maximum number of treatments 1 per yr at full dose or 2 per yr at half dose
- Application during active growth ensures minimal check to grass
- Product may be used in established grassland which is under non-rotational setaside arrangements

- Do not spray in drought, very hot or very cold weather
- Do not treat grass less than 1 yr old or sports or amenity turf
- Product kills or severely checks clover and should not be used where clover is an important constituent of the sward
- Very occasionally some yellowing of sward may occur after treatment which is quickly outgrown
- Do not roll or harrow 10 d before or 7 d after treatment
- Residues in incompletely decayed plant tissue may affect succeeding susceptible crops such as peas, beans and other legumes, carrots and related crops, potatoes, tomatoes, lettuce
- Do not plant susceptible autumn-sown crops in the same yr as treatment with product. Spring sown crops may follow if treatment was before end Jul in the previous yr
- Do not allow spray or drift to reach other crops, amenity plantings, gardens, ponds, lakes or water courses

Special precautions/Environmental safety
- Irritating to eyes and skin
- May cause sensitisation by skin contact
- Harmful to fish or other aquatic life. Do not contaminate surface waters or ditches with chemical or used container
- Wash spray equipment thoroughly with water and detergent immediately after use. Traces of product can damage susceptible plants sprayed later

Protective clothing/Label precautions
- A, C
- R04a, R04b, R04e, U02, U05a, U08, U19, U20b, C01, C03, E01, E07 (7 d), E13c, E26, E30a, E31b, E34

Withholding period
- Keep livestock out of treated areas for at least 7 d following treatment and until foliage of poisonous weeds such as ragwort has died and become unpalatable

Latest application/Harvest Interval(HI)
- HI 7 d

152 clopyralid + propyzamide

A post-emergence herbicide for winter oilseed rape

Products

1 Matrikerb	PBI	4.3:43% w/w	WP	01308
2 Matrikerb	Rohm & Haas	4.3:43% w/w	WP	02443

Uses

Annual dicotyledons in EVENING PRIMROSE *(off-label)*, WINTER OILSEED RAPE. Annual grasses in EVENING PRIMROSE *(off-label)*, WINTER OILSEED RAPE. Barren brome in WINTER OILSEED RAPE. Mayweeds in EVENING PRIMROSE *(off-label)*, WINTER OILSEED RAPE.

Notes

Efficacy
- Apply from Oct to end Jan. Mayweed and groundsel may be controlled after emergence but before crop large enough to shield seedlings

- Best results achieved on fine, firm, moist soils when weeds germinating or small
- Do not use on soils with more than 10% organic matter
- Effectiveness reduced by surface organic debris, burnt straw or ash
- Do not apply if rainfall imminent or frost present on foliage

Crop Safety/Restrictions
- Maximum number of treatments 1 per crop
- Apply to crop as soon as possible after 3-true leaf stage (GS 1,3)
- Minimum period between spraying and drilling a following crop varies from 10 to 40 wk. See label for details

Protective clothing/Label precautions
- A, C
- U09a, U19, U20a, C02, E01, E15, E30a, E32a, E34

Latest application/Harvest Interval(HI)
- HI oilseed rape 6 wk; evening primrose 14 wk

Approval
- Off-label approval unlimited for use on evening primrose (OLA 0318/92)[1]

Maximum Residue Levels (mg residue/kg food)
- see propyzamide entry

153 clopyralid + triclopyr

A perennial and woody weed herbicide for use in grassland

Products

Grazon 90	Dow	60:240 g/l	EC	05456

Uses

Brambles in AMENITY GRASS, ESTABLISHED GRASSLAND. Broom in AMENITY GRASS, ESTABLISHED GRASSLAND. Docks in AMENITY GRASS, ESTABLISHED GRASSLAND. Gorse in AMENITY GRASS, ESTABLISHED GRASSLAND. Perennial dicotyledons in AMENITY GRASS, ESTABLISHED GRASSLAND, ESTABLISHED GRASSLAND *(weed wiper - off-label)*. Stinging nettle in AMENITY GRASS, ESTABLISHED GRASSLAND. Thistles in AMENITY GRASS, ESTABLISHED GRASSLAND.

Notes

Efficacy
- For good results must be applied to actively growing weeds
- Spray stinging nettle before flowering, docks in rosette stage in spring, creeping thistle before flower stems 15 cm high, brambles, broom and gorse in Jun-Aug
- Allow 2-3 wk regrowth after grazing or mowing before spraying perennial weeds
- Do not cut grass for 21 d before or 28 d after spraying

Crop Safety/Restrictions
- Maximum number of treatments 1 per yr
- Only use on permanent pasture or leys established for at least 1 yr
- Do not apply overall where clover is an important constituent of sward
- Do not roll or harrow within 7 d before or after spraying. Do not direct drill kale, swedes, turnips, grass or grass mixtures within 6 wk of spraying. Do not plant susceptible autumn-sown crops (eg winter beans) in same year as treatment. Do not spray after end Jul where susceptible crops to be planted next spring
- Do not allow drift onto other crops, amenity plantings or gardens

Special precautions/Environmental safety
- Harmful if swallowed
- Irritating to skin. May cause sensitization by skin contact
- Risk of serious damage to eyes
- Not to be used on food crops
- Dangerous to fish or other aquatic life. Do not contaminate surface waters or ditches with chemical or used container
- Do not apply by hand-held rotary atomiser equipment

Protective clothing/Label precautions
- A, C, H, M
- R03c, R04b, R04d, R04e, U02, U05a, U08, U19, U20b, C01, C03, E01, E07 (7 d), E13b, E23, E26, E30a, E31b, E34

Withholding period
- Keep livestock out of treated areas for at least 7 d after spraying and until foliage of any poisonous weeds such as ragwort or buttercup has died down and become unpalatable

Latest application/Harvest Interval(HI)
- 7 d before grazing or harvest

Approval
- Off-label approval unlimited for use on established grassland via a tractor mounted/drawn weed wiper (OLA 0692/95)[1]

154 copper ammonium carbonate

A protectant copper fungicide

Products Croptex Fungex | Hortichem | 8.2% w/w (copper) SL | 02888

Uses Blight in OUTDOOR TOMATOES. Cane spot in LOGANBERRIES, RASPBERRIES. Celery leaf spot in CELERY. Currant leaf spot in BLACKCURRANTS. Damping off in SEEDLINGS OF ORNAMENTALS. Leaf curl in PEACHES. Leaf mould in OUTDOOR TOMATOES, PROTECTED TOMATOES. Powdery mildew in CHRYSANTHEMUMS, CUCUMBERS.

Notes

Efficacy
- Apply spray to both sides of foliage
- With protected crops keep foliage dry before and after spraying

Crop Safety/Restrictions
- Maximum number of treatments 5 per crop for celery; 3 per yr for blackcurrants and loganberries; 2 per yr for peaches and raspberries
- Do not spray plants which are dry at the roots
- Ventilate glasshouse immediately after spraying

Special precautions/Environmental safety
- Harmful if swallowed. Risk of serious damage to eyes

- Harmful to fish or other aquatic life. Do not contaminate surface waters or ditches with chemical or used container
- Harmful to livestock

Protective clothing/Label precautions
- M03, R03c, R04d, U05a, U20a, C03, E01, E06b, E13c, E30a, E31a, E34

Withholding period
- Keep all livestock out of treated areas for at least 3 wk. Bury or remove spillages

155 copper hydroxide

An inorganic root pruning and root development agent

Products

Spin Out	Fargro	71 g/l	SL	07610

Uses Root control in CONTAINER-GROWN STOCK.

Notes

Efficacy
- Product should be painted on inner surfaces of containers prior to planting
- Used containers should be free from any loose soil or dirt
- Apply using conventional painting techniques such as brush, sponge, pad or airless paint sprayer
- Thorough coverage with a single coating to a minimum thickness of 0.075 mm is sufficient for root control

Crop Safety/Restrictions
- Can be used in production of container grown forestry seedlings, hardy ornamental and herbaceous plant species
- Use at any stage of plant development from seedlings to large container grown trees
- Tested on a wide range of ornamental species (see label). Efficacy and tolerance should be tested on a small scale for species not listed
- Performance characteristics may vary under some conditions such as when substrates with a pH of less than 5.0 are used
- Product must not be diluted

Special precautions/Environmental safety
- Harmful to fish or other aquatic life. Do not contaminate surface waters or ditches with chemical or used container
- Knapsack sprayer or similar equipment is not suitable

Protective clothing/Label precautions
- A, D
- U11, U12, U20b, E13c, E26, E30a, E32a

Latest application/Harvest Interval(HI)
- Before planting

156 copper oxychloride

A protectant copper fungicide and bactericide

Products

1 Cuprokylt	Unicrop	50% w/w (copper)	WP	00604
2 Cuprokylt FL	Unicrop	270 g/l (copper)	SC	08299
3 Cuprosana H	Unicrop	6% w/w (copper)	DP	00605
4 Headland Inorganic Liquid Copper	Headland	256 g/l (copper)	SC	07799

Uses

Bacterial canker in CHERRIES [1, 2, 4]. Bacterial canker in PLUMS [1, 2]. Blight in OUTDOOR TOMATOES, POTATOES [1, 2, 4]. Buck-eye rot in TOMATOES [1, 2, 4]. Cane spot in LOGANBERRIES, RASPBERRIES [1, 2]. Canker in APPLES, PEARS [1, 2, 4]. Celery leaf spot in CELERY [1, 2, 4]. Damping off in TOMATOES [1, 2]. Downy mildew in HOPS [1-4]. Downy mildew in GRAPEVINES [1, 2, 4]. Foot rot in TOMATOES [1, 2]. Leaf curl in PEACHES [1, 2]. Purple blotch in BLACKBERRIES [4]. Rust in BLACKCURRANTS [1, 2, 4]. Spear rot in CALABRESE *(off-label)* [1].

Notes

Efficacy

- Spray crops at high volume when foliage dry but avoid run off. Do not spray if rain expected soon
- Spray interval commonly 10-14 d but varies with crop, see label for details
- If buck-eye rot occurs, spray soil surface and lower parts of tomato plants to protect unaffected fruit [1, 2]
- A follow-up spray in the following spring should be made to top fruit severely infected with bacterial canker

Crop Safety/Restrictions

- Maximum number of treatments 3 per crop for apples, blackberries, grapevines, pears [4]. Not specified for other products
- Some peach cultivars are sensitive to copper. Treat non-sensitive varieties only [1, 2]
- Slight damage may occur to leaves of cherries and plums [1, 2]

Special precautions/Environmental safety

- Harmful to livestock.
- Harmful to fish or other aquatic life. Do not contaminate surface waters or ditches with chemical or used container

Protective clothing/Label precautions

- A [4]
- U19, E32a [3]; U20a [1, 4]; U20c [2, 3]; E06b (3 wk), E13c, E30a [1-4]; E26, E31c [2]; E31a [1]

Withholding period

- Keep all livestock out of treated areas for at least 3 wk

Latest application/Harvest Interval(HI)

- Before bud burst for apples and pears.
- HI calabrese 3 d [1]

FOR FULL CONDITIONS OF USE ALWAYS READ THE PRODUCT LABEL

Approval
- Approved for aerial application on potatoes [1]. See notes in Section 1
- Off-label approval unlimited for use on calabrese (OLA 0993/92) [1]

157 copper oxychloride + maneb + sulfur

A protectant fungicide and yield stimulant for wheat and barley

Products

1 Ashlade SMC Flowable	Ashlade	10:160:640 g/l	SC	06494
2 Tripart Senator Flowable	Tripart	10:160:640 g/l	SC	04561

Uses

Glume blotch in WHEAT [2]. Leaf blotch in BARLEY [2]. Leaf spot in WHEAT [2]. Leaf spot in WINTER WHEAT [1]. Mildew in BARLEY, WHEAT [2]. Powdery mildew in WINTER WHEAT [1].

Notes

Efficacy
- Use in a protectant programme covering period from first node detectable to beginning of ear emergence (GS 31-51), see label for details
- Treatment has little effect on established disease
- In addition to fungicidal effects treatment can also give nutritional benefits
- Do not apply when foliage is wet or rain imminent

Crop Safety/Restrictions
- Maximum number of treatments 3 per crop
- Do not apply to crops suffering stress from any cause
- Do not roll or harrow within 7 d of treatment
- Do not apply with any trace elements, liquid fertilizers or wetting agent

Special precautions/Environmental safety
- Harmful by inhalation and if swallowed [2]
- Irritating to eyes and skin [1, 2]
- Harmful to fish or other aquatic life. Do not contaminate surface waters or ditches with chemical or used container

Protective clothing/Label precautions
- A, C [1, 2]
- M03, R03b, R03c, U19, E32a, E34 [2]; R04a, R04b, U05a, U08, U20a, C03, E01, E13c, E30a [1, 2]; E31b [1]

Latest application/Harvest Interval(HI)
- Before ear fully emerged (GS 59)

Maximum Residue Levels (mg residue/kg food)
- see maneb entry

158 copper oxychloride + metalaxyl

A systemic and protectant fungicide mixture

Products

1 Ridomil Plus	Novartis	35:15% w/w	WP	08353
2 Ridomil Plus 50 WP	Ciba Agric.	35:15% w/w	WP	01803

Uses Collar rot in APPLES [1, 2]. Downy mildew in CUCUMBERS *(off-label)*[2]. Downy mildew in BRUSSELS SPROUTS *(off-label)*, CABBAGES *(off-label)*, CALABRESE *(off-label)*, CAULIFLOWERS *(off-label)*, HOPS, PROTECTED COURGETTES *(off-label)*, PROTECTED CUCUMBERS *(off-label)* [1, 2]. Downy mildew in GRAPEVINES *(off-label)*, PROTECTED GHERKINS *(off-label)*, SPINACH BEET *(off-label)*, SPINACH *(off-label)* [1]. Red core in STRAWBERRIES [1, 2].

Notes

Efficacy

- Drench soil at base of apples in Sep to Dec or Mar in first 2 yr after planting only
- Apply drench to maiden strawberries immediately after planting and to established plants immediately new growth begins in autumn
- Spray hops to run-off. Increase dose and volume with hop growth

Crop Safety/Restrictions

- Maximum number of treatments 3 per crop for brassicas; 1 per yr for strawberries; 8 per yr for hops, cucumbers
- Do not use on fruiting apple trees
- Use only on hops when well established

Special precautions/Environmental safety

- Irritating to eyes and skin
- Harmful to fish or other aquatic life. Do not contaminate surface waters or ditches with chemical or used container
- Harmful to livestock

Protective clothing/Label precautions

- A, C
- R04a, R04b, U05a, U09a, U19, U20a, C02, C03, E01, E06b, E13c, E30a, E32a, E34

Withholding period

- Keep all livestock out of treated areas for at least 3 wk. Bury or remove spillages

Latest application/Harvest Interval(HI)

- At planting for maiden strawberries; after harvest but before 30 Nov for established strawberries.
- HI 14 d

Approval

- Off-label Approval unlimited for use in Brussels sprouts, cabbages, calabrese, cauliflowers (OLA 1533/95)[2], (OLA 1383/97)[1]; to Nov 1999 for use on protected cucumbers and courgettes (OLA 1534/95)[2]; unlimited for use on outdoor grapevines (OLA 1362/97)[1]; unlimited for use on outdoor and protected spinach, spinach beet (OLA 1344/98)[1]; to Jun 2003 for use on protect cucumbers, gherkins, courgettes (OLA 1344/98)[1]

Maximum Residue Levels (mg residue/kg food)

- see metalaxyl entry

159 copper sulfate

See Bordeaux Mixture

FOR FULL CONDITIONS OF USE ALWAYS READ THE PRODUCT LABEL

160 copper sulfate + sulfur

A contact fungicide for mildew and blight control

Products					
	Top-Cop	Stoller	6:670 g/l	SC	04553

Uses Blight in POTATOES, TOMATOES. Downy mildew in HOPS. Mildew in GRAPEVINES, LEAF BRASSICAS, SWEDES, TURNIPS. Powdery mildew in HOPS, SUGAR BEET.

Notes

Efficacy
- Apply at first blight warning or when signs of disease appear and repeat every 7-10 d as necessary
- Timing varies with crop and disease, see label for details

Crop Safety/Restrictions
- Do not apply to hops at or after the burr stage

Special precautions/Environmental safety
- Irritating to skin, eyes and respiratory system
- Dangerous to fish or other aquatic life. Do not contaminate surface waters or ditches with chemical or used container
- Harmful to livestock

Protective clothing/Label precautions
- R04a, R04b, R04c, U05a, U08, U13, U14, U15, U20a, C03, E01, E06b, E13b, E30a, E31a

Withholding period
- Keep livestock out of treated areas for at least 3 wk

161 coumatetralyl

An anticoagulant coumarin rodenticide

Products					
	1 Racumin Bait	Bayer	0.0375% w/w	RB	01679
	2 Racumin Tracking Powder	Bayer	0.75% w/w	TP	01681

Uses Rats in FARM BUILDINGS.

Notes

Efficacy
- Chemical is available as a bait and tracking powder and it is recommended that both formulations are used together
- Place bait in runs in sheltered positions near rat holes. Use at least 250 g per baiting point and examine at least every 2 d. Replenish as long as bait being eaten
- When no signs of activity seen for about 10 d remove unused bait
- Apply tracking powder in a 2.5-5 cm wide layer inside and across entrance to holes or blow into holes with dusting machine. Lay a 3 mm thick layer along runs in 30 cm long patches. Only use on exposed runs if well away from buildings

Special precautions/Environmental safety
- Cover bait or powder to prevent access by children, animals or birds

Protective clothing/Label precautions
- U19, U20a, E30a, E31b, E32a, V01a, V02, V03a, V04a

162 cyanazine

A contact and residual triazine herbicide

Products

Fortrol	Cyanamid	500 g/l	SC	07009

Uses

Annual dicotyledons in BROAD BEANS *(Scotland only)*, BROCCOLI *(off-label)*, CABBAGES *(off-label)*, CALABRESE *(off-label)*, CAULIFLOWERS *(off-label)*, COLLARDS *(off-label)*, FARM FORESTRY *(off-label)*, FORESTRY TRANSPLANTS *(off-label)*, KALE *(off-label)*, NARCISSI, ONIONS, PEAS, SWEETCORN, TULIPS, WINTER BARLEY, WINTER OILSEED RAPE, WINTER WHEAT. Annual grasses in BROAD BEANS *(Scotland only)*, BROCCOLI *(off-label)*, CABBAGES *(off-label)*, CALABRESE *(off-label)*, CAULIFLOWERS *(off-label)*, COLLARDS *(off-label)*, FARM FORESTRY *(off-label)*, FORESTRY TRANSPLANTS *(off-label)*, KALE *(off-label)*, NARCISSI, ONIONS, PEAS, SWEETCORN, TULIPS, WINTER BARLEY, WINTER OILSEED RAPE, WINTER WHEAT. Charlock in HONESTY *(off-label)*.

Notes

Efficacy
- Weeds controlled before emergence or at young seedling stage. Chemical must be carried into soil by rainfall
- Best results achieved when applied during mild, bright weather. Avoid applications in dull, cold or wet conditions
- Recommended for use alone, or in range of tank mixtures on cereals, rape, potatoes, linseed and peas. See label for details of tank mix partners and timings
- Residual activity normally persists for about 2 mth but on cereals all previous crop and weed residues must be removed or buried at least 15 cm deep
- Do not use as pre-emergence treatment on soils with more than 10% organic matter
- Some strains of blackgrass have developed resistance to many blackgrass herbicides which may lead to poor control

Crop Safety/Restrictions
- Maximum number of treatments 1 per crop (2 per crop for collards), unspecified for narcissi and tulips
- Apply pre- or post-emergence to winter cereals which must be drilled at least 25 mm deep. Do not spray within 7 d of rolling or harrowing
- Recommended for use alone or as tank mixture on cereals, oilseed rape and peas (some varieties only). See label for details of timings, split applications and rates
- Apply to oilseed rape after 1 Nov from 5-leaf stage when winter hardened. Do not apply after 31 Jan
- Apply pre-emergence to potatoes as soon as possible after planting and at least 7 d before any shoots emerge
- Recommended pre- or post-emergence on some varieties of peas only. See label for details
- On onions only apply post-emergence to crops on fen soils with more than 10% organic matter after 2-true leaf stage

- On maize, sweetcorn, broad beans and field beans use as pre-emergence spray or as a pre-drilling incorporated treatment in maize or sweetcorn
- On flower bulbs treat pre- or early post-emergence
- Do not use on Sands, Very Light or stony soils
- Do not treat crops stressed by frost, waterlogging, drought, chemical damage etc. Heavy rain shortly after treatment may lead to damage, especially on lighter soils

Special precautions/Environmental safety
- Harmful if swallowed and in contact with skin
- Harmful to fish or other aquatic life. Do not contaminate surface waters or ditches with chemical or used container

Protective clothing/Label precautions
- A, C
- M03, R03a, R03c, U05a, U08, U19, U20a, C03, E01, E13c, E30a, E31a, E34

Latest application/Harvest Interval(HI)
- Pre-emergence for broad beans and sweetcorn; before 31 Jan for winter oilseed rape; pre-emergence or before flower buds visible for peas; pre-emergence before end Oct or post-emergence before 2nd node detectable (GS 32) for cereals; before 5 true leaf stage for collards, before flower buds appear above the developing leaves for honesty.
- HI onions, leeks 8 wk, calabrese 11 wk

Approval
- Off-label Approval unlimited for use in farm forestry (OLA 0602/94)[1]; to Feb 2000 for use on broccoli, cabbage, calabrese, cauliflower, collards, kale (OLA 0332/96)[1]; unlimited for use in forest tree establishment (OLA 0317/97)[1]

163 cyanazine + pendimethalin

A contact and residual herbicide mixture

Products

1 Activus	Cyanamid	150:264 g/l	SC	09174
2 Bullet	Cyanamid	150:264 g/l	SC	08049

Uses

Annual dicotyledons in COMBINING PEAS, FODDER MAIZE, SPRING FIELD BEANS, WARE POTATOES [2]. Annual dicotyledons in WINTER BARLEY, WINTER WHEAT [1]. Annual meadow grass in COMBINING PEAS, FODDER MAIZE, SPRING FIELD BEANS, WARE POTATOES [2]. Annual meadow grass in WINTER BARLEY, WINTER WHEAT [1]. Chickweed in WINTER BARLEY, WINTER WHEAT [1]. Field pansy in WINTER BARLEY, WINTER WHEAT [1]. Mayweeds in WINTER BARLEY, WINTER WHEAT [1]. Rough meadow grass in COMBINING PEAS, FODDER MAIZE, SPRING FIELD BEANS, WARE POTATOES [2]. Rough meadow grass in WINTER BARLEY, WINTER WHEAT [1].

Notes

Efficacy
- Spray as soon as possible after planting
- Best results achieved from treatment to moist soil when rain falls afterwards, but before weeds emerge. Do not disturb soil after application
- Weed control may be reduced on soils with a high Kd factor, where OM exceeds 6% or ash content is high. Surface organic crop residues should be dispersed
- Products must be used post emergence of crop. Weeds present at the time of spraying will require treatment with a contact herbicide. See label
- Weeds germinating more than 2 mth after spraying may not be controlled

- Coarse irrigation of potatoes that washes soil particles to the base of the ridges may reduce weed control

Crop Safety/Restrictions
- Maximum number of treatments 1 per crop
- Do not use on Very Light, very stony or gravelly soils or those with more than 10% organic matter
- Do not apply to cereals within 24 h of frost. Frost occurring within 5 d of a post-emergence spray may cause crop damage [1]
- Do not use on crops grown under protection such as polythene sheeting [2]
- Do not treat potato crops grown for certified seed or variety Russet Burbank [2]
- Do not treat pea variety Vedette [2]
- Do not treat after crop emergence or later than 7 d before emergence of most advanced potato shoots or after growing points of peas or beans are within 13 mm of soil surface [2]
- After a dry season land must be ploughed to 150 mm before drilling ryegrass
- Before drilling a winter crop plough or cultivate to 150 mm
- In the event of crop failure land must be ploughed or cultivated to 150 mm and then an interval of 8 wk must elapse before drilling any crop

Special precautions/Environmental safety
- Harmful if swallowed [2]
- Dangerous to fish or other aquatic life. Do not contaminate surface waters or ditches with chemical or used container

Protective clothing/Label precautions
- A [1, 2]
- M03, U20b [1]; R03c, U20a [2]; U05a, U08, U13, U19, C03, E01, E13b, E30a, E31b, E34 [1, 2]

Latest application/Harvest Interval(HI)
- Pre-emergence of broad-leaved crops [2]; before leaf sheath erect (GS 30) for winter barley; before tillering for winter wheat [1]

164 cyanazine + terbuthylazine

A foliar and soil acting herbicide for cereals

Products

1 Angle	Ciba Agric.	306:261 g/l	SC	08066
2 Angle	Novartis	306:261 g/l	SC	08385

Uses Annual dicotyledons in WINTER BARLEY, WINTER WHEAT. Annual meadow grass in WINTER BARLEY, WINTER WHEAT.

Notes

Efficacy
- Best results obtained from application made early post-emergence of weeds
- Weeds germinating after application are controlled by root uptake, those present at application by leaf and root uptake
- Weed control may be reduced if heavy rain falls shortly after application or if there is prolonged dry weather

- Weeds germinating more than 2 mth after application may not be controlled
- Reduced weed control may result from presence of more than 10% organic matter or of surface ash and trash

Crop Safety/Restrictions

- Maximum number of treatments 1 per crop
- May be applied from 1 leaf unfolded on main shoot up to the 2 tiller stage (GS 11-22)
- Plant crop at normal depth of 25 mm; it is important that crop is well covered
- Do not apply to undersown crops or those due to be undersown
- Only spray healthy crops. Do not use on crops under stress or suffering from waterlogging, pest or disease attack, frost or when frost imminent
- Do not use on Sands or Very Light soils. There is a risk of crop damage on stony or gravelly soils, especially if heavy rain follows application
- Early crops (Sep drilled) may be prone to damage if spraying precedes or coincides with period of rapid growth in autumn

Special precautions/Environmental safety

- Harmful in contact with skin and if swallowed
- Irritating to skin and eyes
- May cause sensitization by skin contact
- Harmful to fish or other aquatic life. Do not contaminate surface waters or ditches with chemical or used container

Protective clothing/Label precautions

- A, C
- R03a, R03c, R04a, R04b, R04e, U08, U20a, E13c, E30a, E31b, E34

Latest application/Harvest Interval(HI)

- Before main shoot and 2-tiller stage (GS 22)

165 cycloxydim

A translocated post-emergence oxime herbicide for grass weed control

Products

1 Landgold Cycloxydim	Landgold	200 g/l	EC	06269
2 Laser	BASF	200 g/l	EC	05251
3 Standon Cycloxydim	Standon	200 g/l	EC	08830
4 Stratos	BASF	200 g/l	EC	06891

Uses

Annual grasses in CALABRESE [2, 4]. Annual grasses in BROAD BEANS, BULB ONIONS, CARROTS, DWARF BEANS, FARM FORESTRY, FLOWER BULBS, FORESTRY, LEEKS, LINSEED, PARSNIPS, SALAD ONIONS, SPRING OILSEED RAPE, STRAWBERRIES [2-4]. Annual grasses in AMENITY VEGETATION *(off-label)*, SOYA BEANS *(off-label)*[2]. Annual grasses in BRUSSELS SPROUTS, CABBAGES, CAULIFLOWERS, FIELD BEANS, FODDER BEET, MANGELS, PEAS, POTATOES, SUGAR BEET, SWEDES, WINTER OILSEED RAPE [1-4]. Black bent in CALABRESE [2, 4]. Black bent in BROAD BEANS, BULB ONIONS, CARROTS, DWARF BEANS, FLOWER BULBS, LEEKS, LINSEED, PARSNIPS, SALAD ONIONS, SPRING OILSEED RAPE, STRAWBERRIES [2-4]. Black bent in BRUSSELS SPROUTS, CABBAGES, CAULIFLOWERS, FIELD BEANS, FODDER BEET, MANGELS, PEAS, POTATOES, SUGAR BEET, SWEDES, WINTER OILSEED RAPE [1-4]. Blackgrass in CALABRESE [2, 4]. Blackgrass in BROAD BEANS, BULB ONIONS, CARROTS, DWARF BEANS, FLOWER BULBS, LEEKS, LINSEED, PARSNIPS, SALAD ONIONS, SPRING OILSEED RAPE, STRAWBERRIES [2-4]. Blackgrass in BRUSSELS SPROUTS, CABBAGES, CAULIFLOWERS, FIELD BEANS, FODDER BEET, MANGELS, PEAS, POTATOES,

SUGAR BEET, SWEDES, WINTER OILSEED RAPE [1-4]. Couch in CALABRESE [2, 4]. Couch in BROAD BEANS, BULB ONIONS, CARROTS, DWARF BEANS, FLOWER BULBS, LEEKS, LINSEED, PARSNIPS, SALAD ONIONS, SPRING OILSEED RAPE, STRAWBERRIES [2-4]. Couch in BRUSSELS SPROUTS, CABBAGES, CAULIFLOWERS, FIELD BEANS, FODDER BEET, MANGELS, PEAS, POTATOES, SUGAR BEET, SWEDES, WINTER OILSEED RAPE [1-4]. Creeping bent in CALABRESE [2, 4]. Creeping bent in BROAD BEANS, BULB ONIONS, CARROTS, DWARF BEANS, FLOWER BULBS, LEEKS, LINSEED, PARSNIPS, SALAD ONIONS, SPRING OILSEED RAPE, STRAWBERRIES [2-4]. Creeping bent in BRUSSELS SPROUTS, CABBAGES, CAULIFLOWERS, FIELD BEANS, FODDER BEET, MANGELS, PEAS, POTATOES, SUGAR BEET, SWEDES, WINTER OILSEED RAPE [1-4]. Green cover in SETASIDE [2, 4]. Onion couch in CALABRESE [2, 4]. Onion couch in BROAD BEANS, BRUSSELS SPROUTS, BULB ONIONS, CABBAGES, CARROTS, CAULIFLOWERS, DWARF BEANS, FIELD BEANS, FLOWER BULBS, FODDER BEET, LEEKS, LINSEED, MANGELS, PARSNIPS, PEAS, POTATOES, SALAD ONIONS, SPRING OILSEED RAPE, STRAWBERRIES, SUGAR BEET, SWEDES, WINTER OILSEED RAPE [2-4]. Perennial grasses in FARM FORESTRY, FORESTRY [2-4]. Perennial grasses in AMENITY VEGETATION *(off-label)*, SOYA BEANS *(off-label)* [2]. Volunteer cereals in CALABRESE [2, 4]. Volunteer cereals in BROAD BEANS, BULB ONIONS, CARROTS, DWARF BEANS, FLOWER BULBS, LEEKS, LINSEED, PARSNIPS, SALAD ONIONS, SPRING OILSEED RAPE, STRAWBERRIES [2-4]. Volunteer cereals in BRUSSELS SPROUTS, CABBAGES, CAULIFLOWERS, FIELD BEANS, FODDER BEET, MANGELS, PEAS, POTATOES, SUGAR BEET, SWEDES, WINTER OILSEED RAPE [1-4]. Wild oats in CALABRESE [2, 4]. Wild oats in BROAD BEANS, BULB ONIONS, CARROTS, DWARF BEANS, FLOWER BULBS, LEEKS, LINSEED, PARSNIPS, SALAD ONIONS, SPRING OILSEED RAPE, STRAWBERRIES [2-4]. Wild oats in BRUSSELS SPROUTS, CABBAGES, CAULIFLOWERS, FIELD BEANS, FODDER BEET, MANGELS, PEAS, POTATOES, SUGAR BEET, SWEDES, WINTER OILSEED RAPE [1-4].

Notes

Efficacy

- Best results achieved when weeds small and have not begun to compete with crop. Effectiveness reduced by drought, cool conditions or stress. Weeds emerging after application are not controlled
- Foliage death usually complete after 3-4 wk but longer under cool conditions, especially late treatments to winter oilseed rape
- Perennial grasses should have sufficient foliage to absorb spray and should not be cultivated for at least 14 d after treatment
- On established couch pre-planting cultivation recommended to fragment rhizomes and encourage uniform emergence
- Split applications to volunteer wheat and barley at GS 12-14 will often give adequate control in winter oilseed rape. See label for details
- Apply to dry foliage when rain not expected for at least 2 h
- Must be used with Actipron or alternative adjuvants - see label

Crop Safety/Restrictions

- Maximum number of treatments 1 per crop for spring oilseed rape, early potatoes; 2 per crop (the second at reduced dose) for other crops
- Maximum total dose for winter oilseed rape 2.25 l/ha per crop
- Recommended time of application varies with crop. See label for details
- On peas a crystal violet wax test should be done if leaf wax likely to have been affected by weather conditions or other chemical treatment. The wax test is essential if other products are to be sprayed before or after treatment

- May be used on ornamental bulbs when crop 5-10 cm tall. Product has been used on tulips, narcissi, hyacinths and irises but some subjects may be more sensitive and growers advised to check tolerance on small number of plants before treating the rest of the crop
- May be applied to setaside where the green cover is made up predominantly of tolerant crops listed on label. Use on industrial crops of linseed and oilseed rape on setaside also permitted.
- Do not apply to crops damaged or stressed by adverse weather, pest or disease attack or other pesticide treatment
- Prevent drift onto other crops, especially cereals and grass
- Guideline intervals for sowing succeeding crops after failed treated crop: field beans, peas, sugar beet, rape, kale, swedes, radish, white clover, lucerne 1 wk; onions, dwarf French beans 4 wk; wheat, barley, maize 8 wk
- Oats should not be sown after failure of a treated crop

Special precautions/Environmental safety

- Irritating to skin and eyes
- Harmful to fish or other aquatic life. Do not contaminate surface waters or ditches with chemical or used container [2, 4]

Protective clothing/Label precautions

- A, C
- R04a, R04b, U05a, U08, C03, E01, E30a [1-4]; U20a, E15, E31b [1]; U20b, E13c, E31c [2-4]

Latest application/Harvest Interval(HI)

- Before canopy prevents adequate spray penetration.
- HI cabbage, cauliflower, salad onions 4 wk; peas, dwarf French beans 5 wk; bulb onions, carrots, parsnips, early potatoes, strawberries 6 wk; sugar and fodder beet, broad beans, leeks, mangels, maincrop potatoes, field beans, swedes, Brussels sprouts 8 wk; winter and spring oilseed rape, linseed 12 wk

Approval

- Off-label approval unlimited for use in forestry, amenity vegetation (OLA 2585/96)[2]; to Mar 2002 for use in outdoor soya beans (OLA 0610/97)[2]

---

166 cyhexatin

An organotin acaricide approvals for use of which were revoked in 1988 because of evidence of teratogenicity

---

167 cymoxanil

A urea fungicide available only in mixtures

See also carbendazim + cymoxanil + oxadixyl + thiram

168 cymoxanil + mancozeb

A protectant and systemic fungicide for potato blight control

Products

1 Ashlade Solace	Ashlade	4.5:68% w/w	WP	08087
2 Besiege WSB	DuPont	4.5:68% w/w	WB	08075
3 Curzate M68	DuPont	4.5:68% w/w	WP	08072
4 Curzate M68 WSB	DuPont	4.5:68% w/w	WB	08073
5 Standon Cymoxanil Extra	Standon	4.5:68% w/w	WP	06807
6 Systol M	Quadrangle	4.5:68% w/w	WP	07098
7 Systol M	DuPont	4.5:68% w/w	WP	08085

Uses Blight in POTATOES.

Notes

Efficacy
- Apply immediately after blight warning or as soon as local conditions dictate and repeat at 10-14 d intervals until haulm dies down or is burnt off
- Spray interval should not be more than 10 d in irrigated crops

Crop Safety/Restrictions
- Maximum number of treatments 6 per crop [1, 5, 6]
- At least 10 days must elapse between treatments

Special precautions/Environmental safety
- Irritating to eyes [1-7]
- Irritating to skin and respiratory system [1, 5]
- May cause sensitization by skin contact [2-4, 6, 7]
- Harmful to fish or other aquatic life. Do not contaminate surface waters or ditches with chemical or used container
- Do not allow packs to become wet during storage [2]
- Keep product away from fire or sparks [2]

Protective clothing/Label precautions
- A [1-7]; C [3, 6, 7]
- R04a, U05a, U08, U19, C03, E01, E13c, E30a, E32a [1-7]; R04b, R04c, U20a [1, 5]; R04e, U20b [2-4, 6, 7]; U22 [2, 4]

Latest application/Harvest Interval(HI)
- HI zero

Approval
- Approved for aerial application on potatoes [1, 5, 6]. See notes in Section 1

Maximum Residue Levels (mg residue/kg food)
- see mancozeb entry

169 cymoxanil + mancozeb + oxadixyl

A systemic and contact protective fungicide for potatoes

Products

1 Ripost Pepite	Sandoz	3.2:56:8% w/w	WG	06485
2 Ripost Pepite	Novartis	3.2:56.8% w/w	SG	08485
3 Trustan WDG	DuPont	3.2:56:8% w/w	WG	05050

Uses

Blight in POTATOES.

Notes

Efficacy

- Apply first spray as soon as risk of blight infection or official warning. In absence of warning, spray just before crop meets within row
- Repeat spray every 10-14 d according to blight risk
- Do not treat crops already showing blight infection
- Do not apply within 2-3 h of rainfall or irrigation and only apply to dry foliage
- Use fentin based fungicide for final spray in blight control programme

Crop Safety/Restrictions

- Maximum number of treatments 5 (including other phenylamide-based fungicides) per season

Special precautions/Environmental safety

- Irritating to eyes, skin and respiratory system
- May cause sensitization by skin contact
- Harmful to fish or other aquatic life. Do not contaminate surface waters or ditches with chemical or used container

Protective clothing/Label precautions

- A, C
- R04a, R04b, R04c, R04e, U05a, U10, U11, U19, C03, E01, E13c, E30a, E32a [1-3]; U20a [3]; U20b [1, 2]

Latest application/Harvest Interval(HI)

- Before end of Aug.

Approval

- Approved for aerial application on potatoes. See notes in Section 1

Maximum Residue Levels (mg residue/kg food)

- see mancozeb entry

170 cypermethrin

A contact and stomach acting pyrethroid insecticide

Products

1 Afrisect 10	Stefes	100 g/l	EC	09114
2 Ashlade Cypermethrin 10 EC	Ashlade	100 g/l	EC	06229
3 Cyperkill 10	Chiltern	100 g/l	EC	04119
4 Cyperkill 5	Mitchell Cotts	50 g/l	EC	00625
5 Cypertox	FCC	100 g/l	EC	05122
6 Luxan Cypermethrin 10	Luxan	100 g/l	EC	06283
7 MCC 25 EC	Chiltern	250 g/l	EC	09115
8 Permasect C	Mitchell Cotts	100 g/l	EC	07680
9 Standon Cypermethrin	Standon	100 g/l	EC	06818
10 Stefes Cypermethrin 2	Stefes	100 g/l	EC	05719
11 Toppel 10	United Phosphorus	100 g/l	EC	06516
12 Vassgro Cypermethrin Insecticide	Vass	100 g/l	EC	03240

Uses

Aphids in PEAS [4-7, 9, 11, 12]. Aphids in WINTER CEREALS [5, 6, 9, 11, 12]. Aphids in HOPS [5, 6, 9, 11]. Aphids in BRUSSELS SPROUTS, CABBAGES, CAULIFLOWERS [5, 11]. Aphids in KALE [5]. Aphids in PEARS [2, 5, 6, 9, 11, 12]. Aphids in PLUMS [6, 11, 12]. Aphids in CHERRIES [6, 11]. Aphids in BRASSICAS [6]. Aphids in BROCCOLI, CALABRESE [11]. Aphids in LETTUCE, ORNAMENTALS [9, 11]. Aphids in DURUM WHEAT, RYE, TRITICALE [9]. Aphids in CELERY [1, 3, 4, 9, 11]. Aphids in WINTER BARLEY, WINTER WHEAT [1, 3, 4, 7]. Aphids in OUTDOOR LETTUCE [1, 3, 4]. Aphids in AUTUMN SOWN SPRING BARLEY, AUTUMN SOWN SPRING WHEAT [1, 3, 7]. Aphids in APPLES [1-4, 6, 9, 11, 12]. Apple sucker in APPLES [1, 3, 6]. Barley yellow dwarf virus vectors in WINTER CEREALS [2, 5, 6, 9-12]. Barley yellow dwarf virus vectors in SPRING BARLEY, SPRING WHEAT [8]. Barley yellow dwarf virus vectors in WINTER BARLEY, WINTER WHEAT [1, 3, 4, 7, 8]. Barley yellow dwarf virus vectors in AUTUMN SOWN SPRING BARLEY, AUTUMN SOWN SPRING WHEAT [1, 3, 7]. Cabbage stem flea beetle in BRUSSELS SPROUTS, CABBAGES [5, 11]. Cabbage stem flea beetle in CAULIFLOWERS, KALE [5]. Cabbage stem flea beetle in OILSEED RAPE [2, 5, 9-12]. Cabbage stem flea beetle in BRASSICAS [6]. Cabbage stem flea beetle in WINTER OILSEED RAPE [1, 3, 4, 6-8]. Capsids in APPLES [5, 6, 9-11]. Capsids in STRAWBERRIES [2, 5, 6, 9-11]. Capsids in BLACKCURRANTS, GOOSEBERRIES [2, 6]. Capsids in REDCURRANTS [6]. Capsids in ORNAMENTALS [9, 11]. Capsids in PEARS [1, 3-6, 9-11]. Caterpillars in CALABRESE [4, 7, 11]. Caterpillars in BROCCOLI [4, 7, 9, 11]. Caterpillars in STRAWBERRIES [5, 6, 9, 11]. Caterpillars in PEARS [2, 5, 6, 9, 11, 12]. Caterpillars in BRASSICAS [2, 6]. Caterpillars in PLUMS [6, 10-12]. Caterpillars in CHERRIES [6, 10, 11]. Caterpillars in LETTUCE, ORNAMENTALS [9, 11]. Caterpillars in KALE [1, 3-5, 7, 8, 12]. Caterpillars in BRUSSELS SPROUTS, CABBAGES [1, 3-5, 7-12]. Caterpillars in CAULIFLOWERS [1, 3-5, 7-9, 11]. Caterpillars in CELERY [1, 3, 4, 9, 11]. Caterpillars in FODDER BEET, MANGELS, POTATOES, RED BEET, SUGAR BEET [1, 3, 4, 7]. Caterpillars in OUTDOOR LETTUCE [1, 3, 4]. Caterpillars in APPLES [1-4, 6, 9, 11, 12]. Codling moth in APPLES [1, 3, 4, 6, 9-12]. Cutworms in CARROTS, PARSNIPS, SWEDES [6]. Cutworms in ORNAMENTALS [9, 11]. Cutworms in LETTUCE [9]. Cutworms in POTATOES, SUGAR BEET [1, 3, 4, 6-11]. Cutworms in CELERY [1, 3, 4, 9]. Cutworms in FODDER BEET, MANGELS [1, 3, 4, 7, 8]. Cutworms in RED BEET [1, 3, 4, 7]. Cutworms in OUTDOOR LETTUCE [1, 3, 4]. Damson-hop aphid in HOPS [2, 10, 12]. Flea beetles in BRUSSELS SPROUTS, CABBAGES, KALE [12]. Flea beetles in BRASSICAS [2]. Frit fly in GRASS RE-SEEDS [1, 3, 4, 6, 7, 9-11]. Fruit tree tortrix moth in APPLES [12]. Fruitlet mining tortrix in

APPLES [12]. Leaf miners in ORNAMENTALS *(off-label)*, PROTECTED CELERY *(off-label)*, PROTECTED CHICORY *(off-label)*, PROTECTED COURGETTES *(off-label)*, PROTECTED CUCUMBERS *(off-label)*, PROTECTED ENDIVES *(off-label)*, PROTECTED GHERKINS *(off-label)*, PROTECTED LETTUCE *(off-label)* [4]. Mealy aphids in BRUSSELS SPROUTS, CABBAGES, KALE [12]. Pea and bean weevils in FIELD BEANS [4, 8, 11]. Pea and bean weevils in VINING PEAS [10]. Pea and bean weevils in BEANS [6]. Pea and bean weevils in COMBINING PEAS [8, 10]. Pea and bean weevils in PEAS [1, 3, 4, 6-8, 11]. Pea and bean weevils in SPRING FIELD BEANS, WINTER FIELD BEANS [1, 3, 7, 9, 10]. Pea moth in COMBINING PEAS, VINING PEAS [10]. Pea moth in PEAS [1-4, 6-9, 11, 12]. Pod midge in OILSEED RAPE [11]. Pod midge in SPRING OILSEED RAPE, WINTER OILSEED RAPE [1, 3, 4, 7, 8]. Pollen beetles in OILSEED RAPE [11]. Pollen beetles in SPRING OILSEED RAPE, WINTER OILSEED RAPE [1, 3, 4, 7, 8]. Rape winter stem weevil in OILSEED RAPE [5, 9-12]. Rape winter stem weevil in WINTER OILSEED RAPE [6]. Sawflies in STRAWBERRIES [10]. Sawflies in BLACKCURRANTS, GOOSEBERRIES [2, 6]. Sawflies in APPLES [6, 9-11]. Sawflies in REDCURRANTS [6]. Seed weevil in OILSEED RAPE [11]. Seed weevil in SPRING OILSEED RAPE, WINTER OILSEED RAPE [1, 3, 4, 7, 8]. Strawberry tortrix in STRAWBERRIES [2, 4]. Suckers in APPLES [2, 12]. Suckers in PEARS [2, 5, 6, 9-12]. Summer-fruit tortrix moth in APPLES [12]. Thrips in ORNAMENTALS [9, 11]. Tortrix moths in CHERRIES, PLUMS [9]. Tortrix moths in PEARS [1, 3, 4, 6, 10]. Tortrix moths in STRAWBERRIES [1, 3, 6]. Tortrix moths in APPLES [1-4, 6, 9-11]. Whitefly in BRASSICAS [6]. Whitefly in PROTECTED CELERY, PROTECTED CUCUMBERS, PROTECTED LETTUCE, PROTECTED ORNAMENTALS [11]. Whitefly in BRUSSELS SPROUTS, CABBAGES [9, 10]. Whitefly in ORNAMENTALS [9, 11]. Whitefly in BROCCOLI, CAULIFLOWERS [9]. Winter moth in CHERRIES, PLUMS [9, 10]. Winter moth in APPLES, PEARS [1, 3, 10]. Yellow cereal fly in SPRING BARLEY, SPRING WHEAT [4]. Yellow cereal fly in WINTER CEREALS [2, 5, 6, 9-12]. Yellow cereal fly in WINTER BARLEY, WINTER WHEAT [1, 3, 4, 7]. Yellow cereal fly in AUTUMN SOWN SPRING BARLEY, AUTUMN SOWN SPRING WHEAT [1, 3, 7].

Notes

Efficacy

- Products combine rapid action, good persistence, and high activity on Lepidoptera.
- As effect is mainly via contact good coverage is essential for effective action. Spray volume should be increased on dense crops
- A repeat spray after 10-14 d is needed for some pests of outdoor crops, several sprays at shorter intervals for whitefly and other glasshouse pests
- Rates and timing of sprays vary with crop and pest. See label for details
- Add a non-ionic wetter to improve results on leaf brassicas. In Brussels sprouts use of a drop-leg sprayer may be beneficial
- Where aphids in hops, pear suckers or glasshouse whitefly resistant to cypermethrin occur control is unlikely to be satisfactory

Crop Safety/Restrictions

- Maximum number of treatments varies with crop and product. See label or approval notice for details
- Test spray sample of new or unusual ornamentals before committing whole batches

Special precautions/Environmental safety

- Harmful in contact with skin [2, 5, 9, 11]
- Harmful if swallowed [1, 3-5, 7-9]
- Irritating to eyes and skin [1-6, 8-11]
- May cause sensitization by skin contact [1, 3, 7, 8]
- Flammable
- Dangerous (extremely dangerous [8]) to bees. Do not apply to crops in flower or to those in which bees are actively foraging except as directed on peas. Do not apply when flowering weeds are present [2, 5, 9-11]

- Extremely dangerous to fish or other aquatic life. Do not contaminate surface waters or ditches with chemical or used container
- Do not allow direct spray from ground based sprayer vehicles to fall within 6 m, or from hand-held sprayers within 2 m of surface waters or ditches. Direct spray away from water
- Do not spray cereals after 31 Mar within 6 m of the edge of the growing crop

Protective clothing/Label precautions

- A, C [1-12]; H [12]
- M03, E34 [1-5, 7-10, 12]; R03a [2, 5, 9, 12]; R03c [1, 3-5, 7-9, 12]; R04a, R04b, R07d, U05a, U19, C03, E01, E13a, E30a, E31b [1-12]; R04e, U04a [1, 3, 4, 7, 8]; U02 [1-5, 8-10, 12]; U08, U20a [2, 5, 9, 10]; U09a [12]; U10 [1, 3, 4, 6-8, 11]; U11 [1-5, 7-10]; U20b [1, 3, 4, 6-8, 11, 12]; E12b [8]; E12c [2, 5, 9-11]; E12d [6, 12]; E16 [1, 3, 4, 6-11]; E26 [4, 6, 7, 11]

Latest application/Harvest Interval(HI)

- Varies with product and crop. See labels for details

Approval

- Off-label approval unlimited for use on ornamentals, protected celery, lettuce, endive, chicory, cucumbers, courgettes, gherkins (OLA 1200/96)[4]
- Approval expiry: 30 Nov 99 [9]

Maximum Residue Levels (mg residue/kg food)

- hops 30; citrus fruits, apricots, peaches, nectarines, wild berries, lettuces, herbs 2; Chinese cabbage, kale, wild mushrooms 1; grapes, cane fruits, tomatoes, peppers, aubergines, broccoli, cauliflowers, spinach, beet leaves, beans (with pods), leeks 0.5; cucumbers, gherkins, courgettes, kohlrabi, poppy seed, sesame seed, sunflower seed, rape seed, cotton seed, meat (except poultry) 0.2; garlic, onions, shallots 0.1; tree nuts, cranberries, bilberries, currants, gooseberries, miscellaneous fruits, root and tuber vegetables (except radishes, swedes, turnips), spring onions, sweetcorn, watercress, asparagus, celery, cardoons, rhubarb, cultivated mushrooms, beans, peanuts, soya beans, mustard seed, potatoes, triticale, maize, rice, poultry, eggs 0.05; milk 0.02

171 cyproconazole

A contact and systemic conazole fungicide for cereals and other field crops

See also carbendazim + cyproconazole
chlorothalonil + cyproconazole

Products

1	Alto 100 SL	Sandoz	100 g/l	SL	05065
2	Alto 100 SL	Novartis	100 g/l	SL	08350
3	Alto 240 EC	Novartis	240 g/l	EC	08354
4	Aplan	Sandoz	100 g/l	SL	06121
5	Aplan 240 EC	Novartis	240 g/l	EC	08355
6	Aplan SL	Novartis	100 g/l	SL	08351
7	Barclay Shandon	Barclay	100 g/l	SL	06464
8	Landgold Cyproconazole 100	Landgold	100 g/l	SL	06463
9	Standon Cyproconazole	Standon	100 g/l	SL	07751

Uses Brown rust in SPRING BARLEY, WINTER BARLEY, WINTER WHEAT [1-9]. Brown rust in RYE [3, 5]. Chocolate spot in FIELD BEANS *(with chlorothalonil)* [1-6]. Crown rust in SPRING OATS, WINTER OATS [3, 5]. Eyespot in WINTER BARLEY, WINTER WHEAT [1-9]. Eyespot in SPRING BARLEY [7]. Glume blotch in WINTER WHEAT [8]. Late ear diseases in WINTER WHEAT [3, 5]. Light leaf spot in WINTER OILSEED RAPE [1-6]. Net blotch in SPRING BARLEY, WINTER BARLEY [1-9]. Phoma leaf spot in WINTER OILSEED RAPE [1-6]. Powdery mildew in SPRING BARLEY, WINTER BARLEY, WINTER WHEAT [1-9]. Powdery mildew in SUGAR BEET [2-6]. Powdery mildew in RYE, SPRING OATS, WINTER OATS [3, 5]. Ramularia leaf spots in SUGAR BEET [2-6]. Rhynchosporium in SPRING BARLEY, WINTER BARLEY [1-9]. Rust in FIELD BEANS [1-6]. Rust in SUGAR BEET [2-6]. Rust in LEEKS [3, 5]. Septoria diseases in WINTER WHEAT [1-7, 9]. Septoria leaf spot in WINTER WHEAT [8]. Yellow rust in SPRING BARLEY, WINTER BARLEY, WINTER WHEAT [1-9].

Notes

Efficacy

- Apply at start of disease development or as preventive treatment and repeat as necessary
- Most effective time of treatment varies with disease and use of tank mixes may be desirable. See label for details
- Product alone gives useful reduction of cereal eyespot. Where high infections probable a tank mix with prochloraz recommended
- On oilseed rape a two spray autumn/spring programme recommended for high risk situations and on susceptible varieties

Crop Safety/Restrictions

- Maximum number of treatments 3 per crop
- Application to winter wheat in spring between the start of stem elongation and the third node detectable stage (GS 30-33) may cause straw shortening, but does not cause loss of yield

Special precautions/Environmental safety

- Irritating to eyes and skin [3]
- Harmful (dangerous [3]) to fish or other aquatic life. Do not contaminate surface waters or ditches with chemical or used container [1, 2, 4, 6-9]

Protective clothing/Label precautions

- A [1-9]; C, H [3, 5]
- R04a, R04b, U05a, U08, U20b, C03, E01, E13b [3, 5]; E13c [1, 2, 4, 6-9]; E26 [1-7, 9]; E30a, E31b [1-9]; E34 [1, 2, 4, 6]

Latest application/Harvest Interval(HI)

- Up to and including anthesis complete (GS 69) for wheat [1, 2, 4, 6]; up to and including beginning of anthesis (GS 61) for wheat [7-9]; up to and including emergence of ear complete (GS 59) for barley

172 cyproconazole + prochloraz

A broad spectrum protective and curative fungicide for cereals

Products

1 Profile	AgrEvo	48:320 g/l	EC	08134
2 Sportak Delta 460 HF	AgrEvo	48:320 g/l	EC	07431
3 Tiptor	AgrEvo	80:300 g/l	EC	07295

Uses Brown rust in SPRING BARLEY, WINTER BARLEY, WINTER WHEAT [1-3]. Brown rust in SPRING WHEAT, WINTER RYE [1, 2]. Eyespot in SPRING BARLEY, SPRING WHEAT, WINTER BARLEY, WINTER RYE, WINTER WHEAT [1, 2]. Glume blotch in SPRING WHEAT, WINTER WHEAT [1, 2]. Leaf spot in SPRING WHEAT, WINTER RYE, WINTER WHEAT [1, 2]. Net blotch in SPRING BARLEY, WINTER BARLEY [1-3]. Powdery mildew in SPRING BARLEY, WINTER BARLEY, WINTER WHEAT [1-3]. Powdery mildew in SPRING WHEAT, WINTER RYE [1, 2]. Rhynchosporium in SPRING BARLEY, WINTER BARLEY [1-3]. Rhynchosporium in WINTER RYE [1, 2]. Septoria diseases in WINTER WHEAT [3]. Yellow rust in SPRING BARLEY, WINTER BARLEY, WINTER WHEAT [1-3]. Yellow rust in SPRING WHEAT, WINTER RYE [1, 2].

Notes

Efficacy

- To control foliar diseases apply before infection starts spreading to younger leaves and repeat if necessary
- For established mildew use in tank mixture with an approved morpholine fungicide [1, 2]
- For eyespot control apply in spring from when leaf sheaths start to erect up to and including 3rd node detectable stage (GS 30-33). Effective against eyespot and *Septoria tritici* resistant to MBC fungicides [1, 2]
- Eyespot already in crop on up to 10% tillers also controlled [1, 2]
- When applied to control eyespot may give some control of sharp eyespot and Fusarium if present [1, 2]
- See label for details of timing for foliar disease control. Where infection from Septoria may occur following a rain-splash event treat as soon as possible afterwards to protect second leaf and flag leaf [3]
- Good spray cover of stem bases and leaves is essential [1, 2]
- A period of at least 3 h without rain should follow spraying for full effectiveness [1, 2]

Crop Safety/Restrictions

- Maximum number of treatments 2 per crop
- Application to winter wheat in spring at GS 30-33 may cause straw shortening but does not cause loss of yield [1, 2]

Special precautions/Environmental safety

- Harmful if swallowed [3]
- Irritating to eyes [3]
- Irritating to skin [1, 2]
- Harmful to fish or other aquatic life. Do not contaminate surface waters or ditches with chemical or used container

Protective clothing/Label precautions

- A [1-3]; C [3]
- M03, R03c, R04a, E24, E34 [3]; R04b [1, 2]; U05a, U09a, U20b, C03, E01, E13c, E30a, E31b [1-3]; E26 [2, 3]

Latest application/Harvest Interval(HI)

- Up to and including emergence of ear complete (GS 59) on barley; before beginning of anthesis (GS 60) on wheat, rye

173 cyproconazole + quinoxyfen

A contact and systemic fungicide mixture for cereals

Products

Divora	Novartis	80:75 g/l	SC	08960

Uses

Brown rust in WINTER WHEAT. Eyespot in WINTER WHEAT *(reduction)*. Powdery mildew in WINTER WHEAT. Septoria diseases in WINTER WHEAT. Yellow rust in WINTER WHEAT.

Notes

Efficacy

- Best results are achieved when disease is first detected and seen to be active. A second treatment may be needed under prolonged disease pressure
- Treatment for Septoria control should be made as soon as possible after weather that favours infection to prevent spread to the second and flag leaves
- Useful reduction of eyespot when applied between GS 30-34 to control other diseases

Crop Safety/Restrictions

- Maximum number of treatments equivalent to a total of two full doses
- Application to winter wheat in spring may cause straw shortening but does not cause loss of yield

Special precautions/Environmental safety

- May cause sensitization by skin contact
- Dangerous to fish or other aquatic life. Do not contaminate surface waters or ditches with chemical or used container
- Do not allow direct spray from vehicle mounted/drawn hydraulic sprayers to fall within 6 m of surface waters or ditches. Direct spray away from water

Protective clothing/Label precautions

- A, C, H
- R04e, U05a, U20b, C03, E01, E13b, E16, E26, E30a, E31b

Latest application/Harvest Interval(HI)

- Flag leaf sheath open (GS 49) for winter wheat

174 cyproconazole + tridemorph

A broad spectrum contact and systemic fungicide for cereals

Products

1 Alto Major	Sandoz	80:350 g/l	EC	06979
2 Alto Major	Novartis	80:350 g/l	EC	08468
3 Moot	Sandoz	80:350 g/l	EC	06990
4 Moot	Novartis	80:350 g/l	EC	08479

Uses

Brown rust in BARLEY, WINTER WHEAT. Eyespot in WINTER BARLEY, WINTER WHEAT. Net blotch in BARLEY. Powdery mildew in BARLEY, WINTER WHEAT. Rhynchosporium in BARLEY. Septoria diseases in WINTER WHEAT. Sooty moulds in WINTER WHEAT. Yellow rust in BARLEY, WINTER WHEAT.

Notes

Efficacy

- Apply when disease is first active

- Most effective time of treatment varies with disease. Tank mixtures and/or repeat treatments may be desirable. See label for details
- When applied between GS 30-34 for control of other diseases, useful reduction of eyespot is achieved

Crop Safety/Restrictions

- Maximum number of treatments 2 per crop in the spring/summer (winter wheat, spring barley), 1 in autumn and 2 in spring/summer (winter barley)
- Application to winter wheat in spring at GS 30-33 may cause straw shortening but does not cause loss of yield

Special precautions/Environmental safety

- Harmful if swallowed. Irritating to eyes and skin
- Flammable
- Harmful to fish or other aquatic life. Do not contaminate surface waters or ditches with chemical or used container
- Do not allow direct spray from vehicle mounted/drawn hydraulic sprayers to fall within 6 m, or from hand-held sprayers to within 2 m, of surface waters or ditches. Direct spray away from water

Protective clothing/Label precautions

- A, C
- M03, R03c, R04a, R04b, R07d, U02, U04a, U05a, U09a, U20b, C03, E01, E07 (14 d), E13b, E16, E30a, E31b, E34

Withholding period

- Keep livestock out of treated areas for at least 14 d following treatment

Latest application/Harvest Interval(HI)

- Up to and including emergence of ear complete (GS 59) (winter and spring barley), up to and including anthesis complete (GS 69) (winter wheat)

175 cyprodinil

A systemic broad spectrum fungicide for cereals

Products	Unix	Novartis	75% w/w	WG	08764

Uses Eyespot in WINTER BARLEY, WINTER WHEAT. Net blotch in SPRING BARLEY, WINTER BARLEY. Powdery mildew in SPRING BARLEY, SPRING WHEAT, WINTER BARLEY, WINTER WHEAT. Rhynchosporium in SPRING BARLEY, WINTER BARLEY.

Notes

Efficacy

- Best results obtained from treatment at early stages of disease development
- For best control of eyespot spray before or during the period of stem extension in spring. Control may be reduced if very dry conditions follow treatment

Crop Safety/Restrictions

- Maximum number of treatments equal to one and two thirds full dose on wheat and twice full dose on barley

Special precautions/Environmental safety

- Dangerous to fish or other aquatic life. Do not contaminate surface waters or ditches with chemical or used container
- Do not allow direct spray from vehicle mounted/drawn hydraulic sprayers to fall within 6 m of surface waters or ditches. Direct spray away from water

Protective clothing/Label precautions

- A, H
- U05a, U20a, C03, E01, E13b, E16, E26, E29, E30a

Latest application/Harvest Interval(HI)

- Up to and including first awns visible (GS 49) for spring barley, winter barley; up to and including grain watery ripe (GS 71) for spring wheat, winter wheat

---

176 cyromazine

A triazine insect larvicide

Products

1 Neporex 2SG	Ciba Agric.	2% w/w	SG	06985
2 Neporex 2SG	Novartis A H	2% w/w	SG	08589

Uses

Flies in LIVESTOCK HOUSES, MANURE HEAPS.

Notes

Efficacy

- Apply to surface of manure
- Granules can be scattered dry or applied in water
- Commence treatment when houseflies begin to breed
- Ensure that all breeding sites are treated
- Repeat after 2-3 wk for heavy populations
- Product affects housefly larvae moulting and has no effect on adults. Use an adulticide if large numbers of adult flies present

Crop Safety/Restrictions

- Do not use continuously in intensive or controlled environment animal units. If treatment in subsequent stocking cycles is required use different a.i. and different control method
- Do not feed treated poultry manure to stock
- Spread treated manure only onto grassland or arable land prior to cultivation
- Allow at least 4 wk between spreading and grazing or cropping
- Do not use on manure intended for production of mushroom compost

Special precautions/Environmental safety

- Harmful to fish. Do not contaminate surface waters or ditches with chemical or used container

Protective clothing/Label precautions

- A, B, C, H, M
- U05a, U20a, C03, E01, E13c, E30a, E31a

177 2,4-D

A translocated phenoxy herbicide for cereals, grass and amenity use

See also amitrole + 2,4-D + diuron
clopyralid + 2,4-D + MCPA

Products

1 Agricorn D	FCC	500 g/l	SL	00056
2 Atlas 2,4-D	Atlas	470 g/l	SL	07699
3 Barclay Haybob II	Barclay	490 g/l	SL	08532
4 Depitox	Nufarm	490 g/l	SL	08000
5 Dicotox Extra	RP Amenity	400 g/l	EC	05330
6 Dioweed 50	United Phosphorus	500 g/l	SL	08050
7 Dormone	RP Amenity	465 g/l	SL	05412
8 Forester	Vitax	500 g/l	EC	00914
9 Headland Staff	Headland	470 g/l	SL	07189
10 HY-D	Agrichem	470 g/l	SL	06278
11 MSS 2,4-D Amine	Mirfield	500 g/l	SL	01391
12 MSS 2,4-D Ester	Mirfield	500 g/l	EC	01393
13 Syford	Vitax	500 g/l	SL	02062

Uses

Annual dicotyledons in AMENITY TURF [1, 5-8, 11-13]. Annual dicotyledons in WINTER RYE [1, 2, 6, 9-11]. Annual dicotyledons in ESTABLISHED GRASSLAND [1-4, 6, 8-10, 12, 13] Annual dicotyledons in BARLEY, WHEAT [1-4, 6, 9-12]. Annual dicotyledons in SPRING RYE [1, 6, 11]. Annual dicotyledons in SPORTS TURF [1, 6-8, 11]. Annual dicotyledons in GRASSLAND, UNDERSOWN CEREALS [11]. Annual dicotyledons in GRASS SEED CROPS [13]. Annual dicotyledons in CONIFER PLANTATIONS, FORESTRY [5, 12]. Annual dicotyledons in SPRING OATS [10]. Annual dicotyledons in WINTER OATS [2, 9-12]. Annual dicotyledons in RYE [3, 4, 12]. Annual dicotyledons in APPLES, PEARS [3, 4]. Aquatic weeds in AQUATIC SITUATIONS [11]. Aquatic weeds in WATER OR WATERSIDE AREAS [2, 7]. Heather in CONIFER PLANTATIONS, FORESTRY [5, 12]. Perennial dicotyledons in AMENITY TURF [1, 5-8, 11-13]. Perennial dicotyledons in WINTER RYE [1, 2, 6, 9-11] Perennial dicotyledons in ESTABLISHED GRASSLAND [1-4, 6, 8-10, 12, 13]. Perennial dicotyledons in BARLEY, WHEAT [1-4, 6, 9-12]. Perennial dicotyledons in SPRING RYE [1, 6, 11]. Perennial dicotyledons in SPORTS TURF [1, 6-8, 11]. Perennial dicotyledons in GRASSLAND, UNDERSOWN CEREALS [11]. Perennial dicotyledons in GRASS SEED CROPS [13]. Perennial dicotyledons in CONIFER PLANTATIONS, FORESTRY [5, 12] Perennial dicotyledons in SPRING OATS [10]. Perennial dicotyledons in WATER OR WATERSIDE AREAS [2, 7]. Perennial dicotyledons in WINTER OATS [2, 9-12]. Perennial dicotyledons in RYE [3, 4, 12]. Perennial dicotyledons in APPLES, PEARS [3, 4]. Willows in CONIFER PLANTATIONS, FORESTRY [5, 12]. Woody weeds in CONIFER PLANTATIONS FORESTRY [5, 12].

Notes

Efficacy

- Best results achieved by spraying weeds in seedling to young plant stage when growing actively in a strongly competing crop
- Most effective stage for spraying perennials varies with species. See label for details
- Spray aquatic weeds when in active growth between May and Sep [2, 7]
- Do not spray if rain falling or imminent
- Do not cut grass or graze for at least 7 d after spraying

FOR FULL CONDITIONS OF USE ALWAYS READ THE PRODUCT LABE

Crop Safety/Restrictions
- Maximum number of treatments normally 1 per crop and in forestry or 2 per yr in grassland. Check individual labels
- Spray winter cereals in spring when leaf-sheath erect but before first node detectable (GS 31), spring cereals from 5-leaf stage to before first node detectable (GS 15-31)
- Do not use on newly sown leys containing clover
- Do not spray grass seed crops after ear emergence
- Do not spray within 6 mth of laying turf or sowing fine grass
- Selective treatment of resistant conifers can be made in Aug when growth ceased and plants hardened off, spray must be directed if applied earlier. See label for details
- Do not plant conifers until at least 1 mth after treatment
- Do not spray crops stressed by cold weather or drought or if frost expected
- Do not use shortly before or after sowing any crop
- Do not direct drill brassicas or grass/clover mixtures within 3 wk of application
- Do not roll or harrow within 7 d before or after spraying

Special precautions/Environmental safety
- Harmful in contact with skin and if swallowed
- Irritating to eyes [2, 5-8]
- Irritating to skin [3, 5]
- Risk of serious damage to eyes [3]
- May cause sensitization by skin contact [5]
- Harmful to fish or other aquatic life. Do not contaminate surface waters or ditches with chemical or used container
- May be used to control aquatic weeds in presence of fish if used in strict accordance with directions for waterweed control and precautions needed for aquatic use [2, 7, 11]
- Water containing the herbicide must not be used for irrigation purposes within 3 wk of treatment or until the concentration in water is below 0.05 ppm [2, 7, 11]

Protective clothing/Label precautions
- A, C [1-13]; D [1-3, 6, 10]; H, M [1-3, 5-13]
- M03, R03c, U05a, U08, C03, E01, E13c, E30a, E34 [1-13]; R03a [1, 2, 5-13]; R04a [2, 5, 6]; R04b [3, 5]; R04d, E14b [3]; R04e [5]; U20a [1, 2, 6, 10]; U20b [3-5, 7-9, 11-13]; E07 (2 wk) [3-5, 7-9, 11, 12]; E07 [1, 2, 6, 10, 13]; E19 [2, 7, 10]; E21 (3 wk) [7]; E21 [2, 10]; E26 [3, 4, 9, 11, 12]; E31a [1-3, 6, 8-10]; E31b [1, 2, 4-6, 10-13]

Withholding period
- Keep livestock out of treated areas for at least 2 wk and until foliage of any poisonous weeds such as ragwort has died and become unpalatable

Latest application/Harvest Interval(HI)
- Before first node detectable stage (GS 31] in cereals. Refer to labels for other crops

Approval
- May be applied through CDA equipment (see label for details) [13]. See notes in Section 1 on ULV application
- Approved for aquatic weed control [2, 7, 10]. See notes in Section 1 on use of herbicides in or near water

178 2,4-D + dicamba

A translocated herbicide for use on turf

Products	New Estermone	Vitax	200:35 g/l	EC	06336

Uses Annual dicotyledons in AMENITY TURF, LAWNS. Perennial dicotyledons in AMENITY TURF, LAWNS.

Notes

Efficacy
- Best results achieved by application when weeds growing actively in spring or early summer (later with irrigation and feeding)
- More resistant weeds may need repeat treatment after 3 wk
- Do not use during drought conditions or mow for 3 d before or after treatment

Crop Safety/Restrictions
- Maximum number of treatments 3 per yr
- Do not treat newly sown or turfed areas
- Avoid spray drift onto cultivated crops or ornamentals
- Do not re-seed for 6 wk after application

Special precautions/Environmental safety
- Harmful to fish or other aquatic life. Do not contaminate surface waters or ditches with chemical or used container

Protective clothing/Label precautions
- A, C, H, M
- U08, U20b, E07 (2 wk), E13c, E30a, E31a

Withholding period
- Keep livestock out of treated areas for at least 2 wk and until foliage of any poisonous weeds such as ragwort has died and become unpalatable

179 2,4-D + dicamba + mecoprop

A translocated herbicide for perennial and woody weed control

Products

Weed and Brushkiller (New Formulation)	Vitax	144:32:144 g/l	EC	07072

Uses Annual dicotyledons in AMENITY AREAS, AMENITY GRASS, LAND NOT INTENDED TO BEAR VEGETATION. Brambles in AMENITY AREAS, AMENITY GRASS, LAND NOT INTENDED TO BEAR VEGETATION. Perennial dicotyledons in AMENITY AREAS, AMENITY GRASS, LAND NOT INTENDED TO BEAR VEGETATION. Stinging nettle in AMENITY AREAS, AMENITY GRASS, LAND NOT INTENDED TO BEAR VEGETATION. Woody weeds in AMENITY AREAS, AMENITY GRASS, LAND NOT INTENDED TO BEAR VEGETATION.

Notes

Efficacy
- Mix with water for application as foliar spray. Apply Mar-Jul on perennials when in active growth, Jun-Sep on woody plants, May-Jul on gorse
- Do not spray in drought conditions
- Mix with paraffin or other light oil for application as basal bark treatment on established trees in Jul-Sep or as stump treatment in Jan-Mar

Crop Safety/Restrictions

- Avoid spray drift onto crops or ornamentals. Do not use in hot or windy conditions
- Do not cultivate or replant land for at least 6 wk after treatment
- Do not cut or graze for at least 7 d after treatment

Special precautions/Environmental safety

- Irritating to eyes
- Harmful to fish or other aquatic life. Do not contaminate surface waters or ditches with chemical or used container

Protective clothing/Label precautions

- A, C, H, M
- R04a, U08, U20b, C03, E01, E07, E13c, E30a, E31a

Withholding period

- Keep livestock out of treated areas for at least 2 wk and until foliage of any poisonous weeds such as ragwort has died and become unpalatable

180 2,4-D + dicamba + triclopyr

A translocated herbicide for perennial and woody weed control

Products

1 Broadsword	United Phosphorus	200:85:65 g/l	EC	09140
2 Nufarm Nu-Shot	Nufarm	200:85:65 g/l	EC	09139

Uses

Annual dicotyledons in ESTABLISHED GRASSLAND, FORESTRY, NON-CROP AREAS. Brambles in ESTABLISHED GRASSLAND, FORESTRY, NON-CROP AREAS. Docks in ESTABLISHED GRASSLAND, FORESTRY, NON-CROP AREAS. Gorse in ESTABLISHED GRASSLAND, FORESTRY, NON-CROP AREAS. Japanese knotweed in ESTABLISHED GRASSLAND, FORESTRY, NON-CROP AREAS. Perennial dicotyledons in ESTABLISHED GRASSLAND, FORESTRY, NON-CROP AREAS. Rhododendrons in ESTABLISHED GRASSLAND, FORESTRY, NON-CROP AREAS. Stinging nettle in ESTABLISHED GRASSLAND, FORESTRY, NON-CROP AREAS. Thistles in ESTABLISHED GRASSLAND, FORESTRY, NON-CROP AREAS. Woody weeds in ESTABLISHED GRASSLAND, FORESTRY, NON-CROP AREAS.

Notes

Efficacy

- Apply as foliar spray to herbaceous or woody weeds. Timing and growth stage for best results vary with species. See label for details
- Dilute with water for stump treatment and apply after felling up to the start of regrowth. Treat any regrowth with a spray to the growing foliage
- May be applied at 1/3 dilution in weed wipers or 1/8 dilution with ropewick applicators
- Do not roll or harrow grassland for 7 d before or after spraying

Crop Safety/Restrictions

- Do not use on pasture established less than 1 yr or on grass grown for seed
- Where clover a valued constituent of sward only use as a spot treatment
- Do not graze for 7 d or mow for 14 d after treatment
- Do not direct drill grass, clover or brassicas for at least 6 wk after grassland treatment
- Sprays may be applied in pines, spruce and fir providing drift is avoided. Optimum time is mid-autumn when tree growth ceased but weeds not yet senescent
- Do not plant trees for 1-3 mth after spraying depending on dose applied. See label

- Avoid spray drift into greenhouses or onto crops or ornamentals. Vapour drift may occur in hot conditions

Special precautions/Environmental safety
- Harmful if swallowed
- Irritating to eyes and skin
- Flammable
- Dangerous to fish or other aquatic life. Do not contaminate surface waters or ditches with chemical or used container
- Docks and other weeds may become increasingly palatable after treatment and may be preferentially grazed. Where ragwort or other poisonous weeds are present keep livestock until the weeds have died and become unpalatable

Protective clothing/Label precautions
- A, C, D, H, M
- M03, R03c, R04a, R04b, R07d, U05a, U08, U19, U20b, C03, E01, E07 (7 d), E13b, E26, E30a, E31b, E34

Withholding period
- Keep livestock out of treated area for at least 2 wk or until foliage of any poisonous weeds such as ragwort or buttercup has died and become unpalatable

181 2,4-D + dichlorprop + MCPA + mecoprop

A translocated herbicide for use in apple and pear orchards

Products

Camppex	United Phosphorus	34:133:53:164 g/l	SL	08266

Uses

Annual dicotyledons in APPLES, PEARS. Chickweed in APPLES, PEARS. Cleavers in APPLES, PEARS. Perennial dicotyledons in APPLES, PEARS.

Notes

Efficacy
- Use on emerged weeds in established apple and pear orchards (from 1 yr after planting) as directed application
- Spray when weeds in active growth and at growth stage recommended on label
- Effectiveness may be reduced by rain within 12 h
- Do not roll, harrow or cut grass crops on orchard floor within at least 3 d before or after spraying

Crop Safety/Restrictions
- Applications must be made around, and not directly to, trees. Do not allow drift onto trees
- Do not spray during blossom period. Do not spray to run-off
- Avoid spray drift onto neighbouring crops

Special precautions/Environmental safety
- Harmful in contact with skin and if swallowed
- Irritating to eyes
- Harmful to fish or other aquatic life. Do not contaminate surface waters or ditches with chemical or used container

Protective clothing/Label precautions

- A, C, H, M
- M03, R03a, R03c, R04a, U05a, U08, U20b, C03, E01, E07, E13c, E26, E30a, E31b, E34

Withholding period

- Keep livestock out of treated areas for at least 2 wk and until foliage of any poisonous weeds such as ragwort has died and become unpalatable

Maximum Residue Levels (mg residue/kg food)

- see dichlorprop entry

182 2,4-D + mecoprop

A translocated herbicide for use in turf and grassland

Products

1 Nomix Turf Selective Herbicide	Nomix-Chipman	6.7:13.4% w/w	UL	06777
2 Selective Weedkiller	Yule Catto	12.75:26.95% w/w	SL	06579
3 Supertox 30	RP Amenity	90:190 g/l	SL	05340

Uses

Annual dicotyledons in AMENITY TURF, LAWNS, SPORTS TURF. Perennial dicotyledons in AMENITY TURF, LAWNS, SPORTS TURF.

Notes

Efficacy

- May be applied from Apr to Sep, best results in May-Jun when weeds in active growth
- Some species may need repeat treatment after 4-8 wk
- Do not spray during drought conditions or when rain imminent
- Do not mow for 2-4 d before or 1 d after treatment (3 d after on some labels)
- For best results apply fertiliser 1-2 wk before treatment

Crop Safety/Restrictions

- Do not spray ULV formulation at temperatures above 25-26°C [1]
- Do not use first 4 mowings for mulching unless composted for at least 6 mth (do not use first 2 mowings for composting on some labels)
- Do not resow treated areas for at least 6 wk (8 wk on some labels)
- Avoid spray drift onto nearby trees, shrubs, vegetables or flowers
- Avoid drift onto all broad-leaved plants outside the target area

Special precautions/Environmental safety

- Harmful in contact with skin [2]
- Harmful if swallowed [2]
- Irritating to eyes [2, 3]
- Irritating to eyes, skin and respiratory system [1]
- Harmful to fish or other aquatic life. Do not contaminate surface waters or ditches with chemical or used container [1, 2]

Protective clothing/Label precautions

- A, C [1-3]; D [2]; H, M
- M03, R03a, R03c, U20a, E31a, E34 [2]; R04a, U08, E07 [2, 3]; U02, U09a, U19, E07 (2 wk), E26, E32a [1]; U05a [1, 2]; U20b [1, 3]; C03, E01, E13c, E30a [1-3]; E31b [3]

Withholding period

- Keep livestock out of treated areas for at least 2 wk and until foliage of any poisonous weeds such as ragwort has died and become unpalatable

Approval

- May be applied through CDA equipment (see label for details) [1]. See notes in Section 1 on ULV application

183 2,4-D + mecoprop-P

A translocated herbicide for use in amenity turf

Products

1 Mascot Selective-P	Rigby Taylor	78.5:100 g/l	SL	06105
2 Supertox 30	RP Amenity	93.5:95 g/l	SL	09102
3 Sydex	Vitax	125:125 g/l	SL	06412

Uses

Annual dicotyledons in AMENITY TURF, SPORTS TURF [1, 3]. Annual dicotyledons in GOLF COURSES, LAWNS [1]. Annual dicotyledons in ESTABLISHED GRASSLAND, GRASS SEED CROPS [3]. Annual dicotyledons in MANAGED AMENITY TURF [2]. Perennial dicotyledons in AMENITY TURF, SPORTS TURF [1, 3]. Perennial dicotyledons in GOLF COURSES, LAWNS [1]. Perennial dicotyledons in ESTABLISHED GRASSLAND, GRASS SEED CROPS [3]. Perennial dicotyledons in MANAGED AMENITY TURF [2].

Notes

Efficacy

- May be applied from Apr to Sep, best results in May-Jun when weeds in active growth. Less susceptible weeds may require second treatment not less than 6 wk later
- For best results apply fertiliser 1-2 wk before treatment
- Do not spray during drought conditions or when rain imminent
- Do not close mow infrequently mown turf for 3-4 d before or after treatment. On fine turf do not mow for 24 h before or after treatment

Crop Safety/Restrictions

- Maximum number of treatments 2 per yr. The total amount of mecoprop-P applied in a single yr must not exceed the maximum total dose approved for any single product for the crop/situation
- Do not use first 4 mowings for mulching unless composted for at least 6 mth (do not use first 2 mowings for composting on some labels)
- Newly laid turf or newly sown grass should not be treated for at least 6 mth
- Grass cuttings should not be used for mulching but may be composted and used 6 mth later
- Do not apply during frosty weather
- Do not roll or harrow within 7 d before or after treatment

Special precautions/Environmental safety

- Harmful in contact with skin and if swallowed [1]
- Irritating to eyes [2] and skin [1]
- Harmful to fish or other aquatic life. Do not contaminate surface waters or ditches with chemical or used container
- Avoid drift onto broad-leaved plants outside the target area

Protective clothing/Label precautions
- A, C, H, M
- 6412a) M03, E07, E13b [3]; M03, R04b, U13, E32a [1]; R03a, R03c, E34 [1, 3]; R04a, U05a, U08, U20b, C03, E01, E30a [1-3]; U11, U15, E23 [2]; E07 (2 wk), E13c [1, 2]; E26, E31b [2, 3]

Withholding period
- Keep livestock out of treated areas for at least 2 wk and until foliage of any poisonous weeds such as ragwort has died and become unpalatable

---

184 2,4-D + picloram

A persistent translocated herbicide for non-crop land

Products Atladox HI | Nomix-Chipman | 240:65 g/l | SL | 05559

Uses Annual dicotyledons in AMENITY GRASS, NON-CROP GRASS, ROAD VERGES. Brambles in AMENITY GRASS, NON-CROP GRASS, ROAD VERGES. Creeping thistle in AMENITY GRASS, NON-CROP GRASS, ROAD VERGES. Docks in AMENITY GRASS, NON-CROP GRASS, ROAD VERGES. Japanese knotweed in AMENITY GRASS, NON-CROP GRASS, ROAD VERGES. Perennial dicotyledons in AMENITY GRASS, NON-CROP GRASS, ROAD VERGES. Ragwort in AMENITY GRASS, NON-CROP GRASS, ROAD VERGES. Scrub clearance in AMENITY GRASS, NON-CROP GRASS, ROAD VERGES. Woody weeds in AMENITY GRASS, NON-CROP GRASS, ROAD VERGES.

Notes

Efficacy
- Apply as overall foliar spray during period of active growth when foliage well developed

Crop Safety/Restrictions
- Maximum number of treatments 1 per yr
- Do not apply around desirable trees or shrubs where roots may absorb chemical
- Prevent leaching into areas where desirable plants are present
- Avoid drift of spray onto desirable plants
- Do not use cuttings from treated grass for mulching or composting

Special precautions/Environmental safety
- Irritating to eyes
- Flammable
- Harmful to fish or other aquatic life. Do not contaminate surface waters or ditches with chemical or used container

Protective clothing/Label precautions
- A, C
- R04a, R07d, U05a, U08, U20b, C03, E01, E07, E13c, E26, E30a, E31a

Withholding period
- Keep livestock out of treated areas for at least 2 wk and until foliage of any poisonous weeds such as ragwort has died and become unpalatable

185 dalapon

A translocated chloroalkanoic acid grass herbicide no longer marketed alone but still approved for use

186 dalapon + dichlobenil

A persistent herbicide for use in woody plants and non-crop areas

Products

Fydulan G	Nomix-Chipman	10:6.75% w/w	GR	00958

Uses

Annual weeds in FORESTRY, HEDGES, ROSES, WOODY ORNAMENTALS. Bracken in FORESTRY, HEDGES. Perennial grasses in FORESTRY, HEDGES, ROSES, WOODY ORNAMENTALS. Rushes in FORESTRY, HEDGES. Total vegetation control in NON-CROP AREAS, WATERSIDE AREAS.

Notes

Efficacy

- Apply evenly with suitable granule applicator. Best results achieved by application in late winter/early spring to slightly moist soil, especially if rain falls soon after
- May be used on all soil types. Do not disturb soil after application
- For forestry use apply as a 1 m wide band
- In non-crop situations apply before end of spring

Crop Safety/Restrictions

- Use in ornamentals before onset of spring growth, usually Feb-early Mar. See label for list of tolerant species
- Use in roses established for at least 2 yr before bud growth starts, usually early Feb
- Use in forestry before end of winter. Allow at least 2 mth before planting. Apply after planting as soon as soil settled
- Do not apply to areas underplanted with bulbs or herbaceous stock
- Do not apply near hops or glasshouses, nor if nearby crop foliage wet
- Store well away from bulbs, corms, tubers or seed

Special precautions/Environmental safety

- Irritating to eyes, skin and respiratory system

Protective clothing/Label precautions

- R04a, R04b, R04c, U09a, U20a, C03, E01, E30a, E32a

Approval

- Approved for aquatic weed control [1]. See introductory notes on use of herbicides in or near water

187 daminozide

A hydrazide plant growth regulator for use in certain ornamentals

Products

1 B-Nine	Hortichem	85% w/w	SP	07844
2 Dazide	Fine	85% w/w	SP	02691

Uses

Internode reduction in POT PLANTS [2]. Internode reduction in AZALEAS, BEDDING PLANTS, CHRYSANTHEMUMS, HYDRANGEAS [1, 2]. Internode reduction in POINSETTIAS [1].

Notes

Efficacy

- Best results obtained by application in late afternoon when glasshouse has cooled down
- Spray when foliage dry
- A reduced rate tank mix with chlormequat + choline chloride is recommended for use on poinsettias. See label for details [1]

Crop Safety/Restrictions

- Maximum number of treatments 2 per crop
- Apply only to turgid, well watered plants. Do not water for 24 h after spraying
- Do not use on chrysanthemum Fandango
- Do not mix with other spray chemicals except as recommended above

Special precautions/Environmental safety

- Irritating to eyes [2]
- Harmful to fish or other aquatic life. Do not contaminate surface waters or ditches with chemical or used container [2]

Protective clothing/Label precautions

- A, C [1, 2]; H [2]
- R04a, U05a, U19, C03, E01, E30a [1, 2]; U09a, U20a, E13c, E31a [2]; E15 [1]

Approval

- Sales of daminozide for use on food crops were halted worldwide by the manufacturer in Oct 1989. Sales for use on flower crops were not affected

Maximum Residue Levels (mg residue/kg food)

- tea, hops 0.1; tree nuts, oilseeds, animal products 0.05; citrus fruits, pome fruits, stone fruits, berries and small fruit, miscellaneous fruits, vegetables, pulses, potatoes, cereals 0.02

188 dazomet

A methyl isothiocyanate releasing soil fumigant

Products

Basamid	Hortichem	97% w/w	GR	07204

Uses

Nematodes in FIELD CROPS, PROTECTED CROPS, SOIL LOAM OR TURF FOR COMPOST, VEGETABLES. Soil insects in FIELD CROPS, PROTECTED CROPS, SOIL LOAM OR TURF FOR COMPOST, VEGETABLES. Soil-borne diseases in FIELD CROPS, PROTECTED CROPS, SOIL LOAM OR TURF FOR COMPOST, VEGETABLES. Weed seeds in FIELD CROPS, PROTECTED CROPS, SOIL LOAM OR TURF FOR COMPOST, VEGETABLES.

Notes

Efficacy

- Product acts by releasing methyl isothiocyanate in contact with moist soil
- Soil sterilization is carried out after harvesting one crop and before planting the next
- The soil must be of fine tilth, free of clods and evenly moist to the depth of sterilization
- Soil moisture must not be less than 50% of water-holding capacity or oversaturated. If too dry, water at least 7-14 d before treatment
- Do not treat ground where water table may rise into treated layer
- In order to obtain short treatment times it is recommended to treat soils when soil temperature is above 7°C. Treatment should be used outdoors before winter rains make soil too wet to cultivate - usually early Nov
- For club root control treat only in summer when soil temperature above 10°C
- Where onion white rot a problem unlikely to give effective control where inoculum level high or crop under stress
- Apply granules with suitable applicators, mix into soil immediately to desired depth and seal surface with polythene sheeting, by flooding or heavy rolling. See label for suitable application and incorporation machinery
- With 'planting through' technique polythene seal is left in place to form mulch into which new crop can be planted

Crop Safety/Restrictions

- Maximum number of treatments 1 per crop or batch of soil
- With 'planting through' technique no gas release cultivations are made and safety test with cress is particularly important. Conduct cress test on soil samples from centre as well as edges of bed. Observe minimum of 30 d from application to cress test in soils at 10°C or above, at least 50 d in soils below 10°C. See label for details
- In all other situations 14-28 d after treatment cultivate lightly to allow gas to disperse and conduct cress test after a further 14-28 d (timing depends on soil type and temperature). Do not treat structures containing live plants or any ground within 1 m of live plants

Special precautions/Environmental safety

- Harmful if swallowed
- Do not contaminate surface waters or ditches with chemical or used container

Protective clothing/Label precautions

- A, M
- M03, R03c, U04a, U05a, U09a, U19, U20a, C03, E01, E15, E30a, E32a, E34

Latest application/Harvest Interval(HI)

- Pre-planting

189 2,4-DB

A translocated phenoxy herbicide for use in lucerne

Products					
	DB Straight	United Phosphorus	300 g/l	SL	07523

Uses

Annual dicotyledons in CEREALS UNDERSOWN LUCERNE, LUCERNE. Thistles in CEREALS UNDERSOWN LUCERNE, LUCERNE.

FOR FULL CONDITIONS OF USE ALWAYS READ THE PRODUCT LABEL

Notes **Efficacy**

- Best results achieved on young seedling weeds under good growing conditions. Treatment less effective in cold weather and dry soil conditions
- Rain within 12 h may reduce effectiveness

Crop Safety/Restrictions

- In direct sown lucerne spray when seedlings have reached first trifoliate leaf stage. Optimum time 3-4 trifoliate leaves
- In spring barley and spring oats undersown with lucerne spray from when cereal has 1 leaf unfolded and lucerne has first trifoliate leaf
- In spring wheat undersown with lucerne spray from when cereal has 3 leaves unfolded and lucerne has first trifoliate leaf
- Do not treat any lucerne after fourth trifoliate leaf
- Do not allow spray drift onto neighbouring crops

Special precautions/Environmental safety

- Harmful to fish or other aquatic life. Do not contaminate surface waters or ditches with chemical or used container

Protective clothing/Label precautions

- U08, U20a, E07 (2 wk), E13c, E26, E30a, E31b

Withholding period

- Keep livestock out of treated areas until foliage of any poisonous weeds such as ragwort has died and become unpalatable

190 2,4-DB + linuron + MCPA

A translocated herbicide for undersown cereals and grass

Products

Alistell	Zeneca	220:30:30 g/l	EC	06515

Uses Annual dicotyledons in CEREALS UNDERSOWN CLOVERS, SEEDLING GRASSLAND.

Notes **Efficacy**

- Best results achieved on young seedling weeds growing actively in warm, moist weather
- May be applied at any time of year provided crop at correct stage and weather suitable
- Avoid spraying if rain falling or imminent

Crop Safety/Restrictions

- Maximum number of treatments 1 per crop or yr
- Spray winter cereals when fully tillered but before first node detectable (GS 29-30)
- Spray spring wheat from 5-fully expanded leaf stage (GS 15), barley and oats from 2-fully expanded leaves (GS 12)
- Do not spray cereals undersown with lucerne, peas or beans
- Apply to clovers after 1-trifoliate leaf, to grasses after 2-fully expanded leaf stage
- Do not spray in conditions of drought, waterlogging or extremes of temperature
- In frosty weather clover leaf scorch may occur but damage normally outgrown
- Do not use on sand or soils with more than 10% organic matter
- Do not roll or harrow within 7 d before or after spraying
- Avoid drift of spray or vapour onto susceptible crops

Special precautions/Environmental safety

- Harmful in contact with skin and if swallowed. Irritating to eyes and skin
- Dangerous to fish or other aquatic life. Do not contaminate surface waters or ditches with chemical or used container
- Do not allow direct spray from vehicle mounted/drawn hydraulic sprayers to fall within 6 m of surface waters or ditches. Direct spray away from water
- Do not apply by hand-held sprayers

Protective clothing/Label precautions

- A, C, H
- M03, R03a, R03c, R04a, R04b, U05a, U08, U13, U19, U20b, C03, E01, E07 (2 wk), E13c, E16, E30a, E31b, E34

Withholding period

- Keep livestock out of treated areas for at least 2 wk and until foliage of any poisonous weeds such as ragwort has died and become unpalatable

Latest application/Harvest Interval(HI)

- Before first node detectable (GS 31) for undersown cereals; 2 wk before grazing for grassland

191 2,4-DB + MCPA

A translocated herbicide for cereals, clovers and leys

Products

1 Agrichem DB Plus	Agrichem	243:40 g/l	SL	00044
2 MSS 2,4-DB + MCPA	Mirfield	280 g/l (total)	SL	01392
3 Redlegor	United Phosphorus	244:44 g/l	SL	07519

Uses

Annual dicotyledons in UNDERSOWN CEREALS [1-3]. Annual dicotyledons in CEREALS [1, 2]. Annual dicotyledons in LEYS [1]. Annual dicotyledons in SEEDLING LEYS [2, 3]. Annual dicotyledons in CLOVERS, GRASSLAND [2]. Perennial dicotyledons in UNDERSOWN CEREALS [1-3]. Perennial dicotyledons in CEREALS [1, 2]. Perennial dicotyledons in LEYS [1]. Perennial dicotyledons in SEEDLING LEYS [2, 3]. Perennial dicotyledons in CLOVERS, GRASSLAND [2]. Polygonums in SEEDLING LEYS, UNDERSOWN CEREALS [2, 3]. Polygonums in CEREALS, CLOVERS [2].

Notes

Efficacy

- Best results achieved on young seedling weeds under good growing conditions
- Spray thistles and other perennials when 10-20 cm high provided clover at correct stage
- Effectiveness may be reduced by rain within 12 h, by very cold conditions or drought

Crop Safety/Restrictions

- Apply in spring to winter cereals from leaf sheath erect stage, to spring barley or oats from 2-leaf stage (GS 12), to spring wheat from 5-leaf stage (GS 15)
- Spray clovers as soon as possible after first trifoliate leaf, grasses after 2-3 leaf stage
- Red clover may suffer temporary distortion after treatment
- Do not spray established clover crops or lucerne
- Do not roll or harrow within 7 d before or after spraying

- Do not spray immediately before or after sowing any crop
- Avoid drift onto neighbouring sensitive crops

Special precautions/Environmental safety

- Harmful to fish or other aquatic life. Do not contaminate surface waters or ditches with chemical or used container

Protective clothing/Label precautions

- A, C [2]
- M03, R03a, R03c, R04d, C03 [2]; U05a, E01, E07, E30a, E34 [1, 2]; U08, E13c, E26, E31b [1-3]; U20a [1]; U20b [2, 3]; E07 (2 wk) [3]

Withholding period

- Keep livestock out of treated areas until foliage of any poisonous weeds such as ragwort has died and become unpalatable

Latest application/Harvest Interval(HI)

- Before first node detectable (GS 31) for cereals

192 deltamethrin

A pyrethroid insecticide with contact and residual activity

Products

1 Crackdown	AgrEvo Environ.	10 g/l	SC	H5097
2 Decis	AgrEvo	25 g/l	EC	07172
3 Landgold Deltaland	Landgold	25 g/l	EC	07480
4 Pearl Micro	AgrEvo	6.25% w/w	GR	08620
5 Standon Deltamethrin	Standon	25 g/l	EC	07053
6 Thripstick	Aquaspersions	0.125 g/l	AL	02134

Uses

American serpentine leaf miner in AUBERGINES *(off-label (fog))*, COURGETTES *(off-label (fog))*, CUCUMBERS *(off-label (fog))*, GHERKINS *(off-label (fog))*, ONIONS *(off-label (fog))*, PEPPERS *(off-label (fog))*, TOMATOES *(off-label (fog))* [2]. Aphids in APPLES, CEREALS [2-5]. Aphids in HOPS, ORNAMENTALS, PEAS, PEPPERS, PLUMS, PROTECTED CUCUMBERS, PROTECTED ORNAMENTALS, PROTECTED POT PLANTS, PROTECTED TOMATOES [2, 4]. Aphids in CHINESE CABBAGE *(off-label)*, GRASS SEED CROPS *(off-label)*, LETTUCE *(off-label)*, RADICCHIO *(off-label)* [2]. Aphids in OILSEED RAPE [3, 5]. Barley yellow dwarf virus vectors in CEREALS [2-5]. Beet virus yellows vectors in WINTER OILSEED RAPE [2, 4]. Brassica pod midge in MUSTARD, OILSEED RAPE [2, 4]. Cabbage seed weevil in MUSTARD, OILSEED RAPE [2, 4]. Cabbage stem flea beetle in WINTER OILSEED RAPE [2-5]. Cabbage stem flea beetle in WINTER OILSEED RAPE *(in store)* [1]. Cabbage stem weevil in MUSTARD [2-5]. Cabbage stem weevil in OILSEED RAPE [2, 4]. Capsids in TREES AND SHRUBS [5]. Capsids in APPLES [2-5]. Capsids in NURSERY STOCK, WOODY ORNAMENTALS [2, 4, 5]. Capsids in FLOWERS [2, 4]. Caterpillars in TOMATO HOUSES *(off-label)* [6]. Caterpillars in OUTDOOR LETTUCE, PEPPERS, PROTECTED CUCUMBERS, PROTECTED ORNAMENTALS, PROTECTED POT PLANTS, PROTECTED TOMATOES [2, 4]. Caterpillars in CHINESE CABBAGE *(off-label)*, LETTUCE *(off-label)*, RADICCHIO *(off-label)* [2]. Caterpillars in APPLES, BROCCOLI, BRUSSELS SPROUTS, CABBAGES, CAULIFLOWERS, KALE, PLUMS, SWEDES, TURNIPS [3, 5]. Codling moth in APPLES [2-5]. Colorado beetle in POTATOES *(off-label (Statutory Notice))* [2]. Cutworms in OUTDOOR LETTUCE [3, 5]. Damson-hop aphid in HOPS, PLUMS [3, 5]. Flea beetles in SUGAR BEET [2-5]. Flea beetles in LEAF BRASSICAS, ROOT BRASSICAS [2, 4]. Flea beetles in EVENING PRIMROSE *(off-label)*, SORREL *(off-label)* [2]. Flour beetles in GRAIN STORES [1]. Grain beetles in GRAIN STORES [1]. Grain moths in GRAIN STORES [1]. Grain storage pests in GRAIN STORES [1]. Insect pests in CARROTS *(off-label)*, CELERY *(off-label)*, FENNEL *(off-label)*, GARLIC *(off-label)*, LEEKS *(off-label)*, ONIONS *(off-label)* [2]. Leaf miners in TOMATO HOUSES *(off-label)* [6]. Leafhoppers in MARJORAM *(off-label)* [2]. Mealy bugs in PEPPERS, PROTECTED CUCUMBERS, PROTECTED ORNAMENTALS, PROTECTED POT PLANTS, PROTECTED TOMATOES [2, 4]. Pea and bean weevils in BROAD BEANS, FIELD BEANS, PEAS [2-5]. Pea moth in PEAS [2-5]. Plum fruit moth in PLUMS [2-5]. Pollen beetles in OILSEED RAPE [2-5]. Pollen beetles in MUSTARD [2, 4]. Pollen beetles in EVENING PRIMROSE *(off-label)* [2]. Rape winter stem weevil in WINTER OILSEED RAPE [2-5]. Rape winter stem weevil in WINTER OILSEED RAPE *(in store)*[1]. Raspberry beetle in RASPBERRIES [2-5]. Sawflies in APPLES, PLUMS [2-5]. Scale insects in TREES AND SHRUBS [5]. Scale insects in NURSERY STOCK, WOODY ORNAMENTALS [2, 4, 5]. Scale insects in FLOWERS, PEPPERS, PROTECTED CUCUMBERS, PROTECTED ORNAMENTALS, PROTECTED POT PLANTS, PROTECTED TOMATOES [2, 4]. Suckers in APPLES, PEARS [2-5]. Thrips in PROTECTED CUCUMBERS [6]. Thrips in TREES AND SHRUBS [5]. Thrips in NURSERY STOCK, WOODY ORNAMENTALS [2, 4, 5]. Thrips in FLOWERS [2, 4]. Tortrix moths in APPLES [2-5]. Virus yellows vectors in OILSEED RAPE [3, 5]. Western flower thrips in AUBERGINES *(off-label (fog))*, COURGETTES *(off-label (fog))*, CUCUMBERS *(off-label (fog))*, GHERKINS *(off-label (fog))*, ONIONS *(off-label (fog))*, PEPPERS *(off-label (fog))*, TOMATOES *(off-label (fog))* [2]. Whitefly in PEPPERS, PROTECTED CUCUMBERS, PROTECTED ORNAMENTALS, PROTECTED POT PLANTS, PROTECTED TOMATOES [2, 4]. Yellow cereal fly in CEREALS [2, 4].

Notes

Efficacy

- A contact and stomach poison with 3-4 wk persistence, particularly effective on caterpillars and sucking insects
- Normally applied at first signs of damage with follow-up treatments where necessary at 10-14 d intervals. Rates, timing and recommended combinations with other pesticides vary with crop and pest. See label for details [2, 5]
- Spray is rainfast within 1 h. May be applied in frosty weather provided foliage not covered in ice [2, 5]
- Temperatures above 35°C may reduce effectiveness or persistence
- Spray Thripstick without dilution onto polythene covering on glasshouse floor. Spray before planting cucumbers where thrips are a known problem or when pest problem seen. Repeat every 8 wk to ensure complete control [6]
- Where aphids in hops, pear suckers or glasshouse whitefly resistant to deltamethrin occur control is unlikely to be satisfactory

Crop Safety/Restrictions

- Maximum number of treatments varies with crop and pest, 4 per crop for wheat and barley, only 1 application between 1 Apr and 31 Aug. See label or off-label approval notice for other crops
- Do not apply more than 1 aphicide treatment to cereals in summer
- Do not spray crops suffering from drought or other physical stress
- Consult processer before treating crops for processing

Special precautions/Environmental safety

- Harmful in contact with skin [1-3, 5]
- Harmful if swallowed [1-5]
- Irritating to eyes and skin [1-5]
- Flammable [1-3, 5]

- Dangerous to bees. Do not apply to crops in flower or to those in which bees are actively foraging except as directed (see labels). Do not apply when flowering weeds are present [1-5]
- Extremely dangerous to fish or other aquatic life. Do not contaminate surface waters or ditches with chemical or used container
- Do not apply to a cereal crop if any product containing a pyrethroid insecticide or dimethoate has been applied to that crop after the start of ear emergence (GS 51)
- Do not allow direct spray from ground based spray vehicles to fall within 6 m, or from hand-held sprayers within 2 m, of surface waters or ditches. Direct spray away from water
- Do not spray cereals after 31 Mar in the year of harvest within 6 m of the outside edge of the crop
- Reduced volume spraying must not be used on cereals after 31 Mar in yr of harvest

Protective clothing/Label precautions

- A, C, H
- M03, R03c, R04a, R04b, U04a, U08, E34 [1-5]; R03a, R07d, E31b [1-3, 5]; U02, U05b, U06, U09a, E31a [6]; U05a, U19, C03, E01, E13a, E30a [1-6]; U20a [1, 3, 5]; U20b [2, 4, 6]; E12c [1, 2, 5]; E12d [3, 4]; E16 [2-5]; E26 [1, 2, 4-6]; E27 [1, 5, 6]; E32a [4]

Latest application/Harvest Interval(HI)

- Early dough (GS 83) for barley, oats, wheat [2, 4]; end of flowering for mustard, oilseed rape [2, 4]

Approval

- Off-label approval unlimited for control of Colorado beetle in potatoes as required by Statutory Notice (OLA 1440/97)[2]; unlimited for use in tomato houses (OLA 0180/92)[6]; unlimited for use on carrots (OLA 1265/95)[2]; unlimited for use on onions, garlic (OLA 0506/96)[2]; to Mar 2000 for use on onions, celery, garlic, lettuce, radicchio, sorrel, marjoram (OLA 507/96)[2]; to Mar 2000 for use on fennel (OLA 0508/96)[2]; to Jun 2001 for use on Chinese cabbage (OLA 1396/96)[2]; unlimited for use on lettuce (OLA 1691/96)[2]; unlimited for use on onions, tomatoes, gherkins, aubergines, peppers, cucumbers, courgettes (OLA 1692/96)[2]; unlimited for use on evening primrose, grass seed crops, sorrel, radicchio, marjoram (OLA 1693/96)[2]; to May 1999 for use on leeks (OLA 0931/97)[2]

Maximum Residue Levels (mg residue/kg food)

- tea, hops 5; pulses, cereals 1; Chinese cabbage, kale, lettuces, spinach, beet leaves, herbs 0.5; currants, gooseberries, tomatoes, peppers, aubergines, beans (with pods) 0.2; pome fruits, stone fruits, grapes, garlic, onions, shallots, cucumbers, gherkins, courgettes, broccoli, cauliflowers, Brussels sprouts, head cabbages, peas (with pods), rape seed 0.1; miscellaneous fruit 0.05; citrus fruits, tree nuts, strawberries, bilberries, cranberries, wild berries, root and tuber vegetables, melons, squashes, watermelons, kohlrabi, watercress, peas (without pods), stem vegetables (except globe artichokes), mushrooms, oilseeds (except rape seed), early potatoes, poultry, eggs 0.01

---

193 deltamethrin + heptenophos

A systemic aphicide with knockdown, vapour and anti-feeding activity

Products	Decisquick	AgrEvo	25:400 g/l	EC	07312

Uses Aphids in OUTDOOR LETTUCE *(off-label)*, PEAS, POTATOES, PROTECTED LETTUCE *(off-label)*, SUGAR BEET. Beet mild yellowing virus vectors in SUGAR BEET. Beet virus yellows vectors in SUGAR BEET. Insect pests in BROCCOLI *(off-label)*, BRUSSELS SPROUTS *(off-label)*, CABBAGES *(off-label)*, CALABRESE *(off-label)*, CAULIFLOWERS *(off-label)*, KOHLRABI *(off-label)*, RADICCHIO *(off-label)*, ROSCOFF CAULIFLOWER *(off-label)*. Leaf roll virus vectors in POTATOES. Pea enation virus vectors in PEAS. Pea moth in PEAS. Potato mosaic virus vectors in POTATOES.

Notes

Efficacy

- Works by combination of contact, systemic, vapour and anti-feeding activity and persists for 10-14 d, longer in cool and shorter in hot weather
- Spray peas as soon as aphids seen and repeat once if necessary
- Spray potatoes before population build-up occurs or as soon as aphids are found on varieties where processing quality is important and repeat as necessary. Treat seed crops at 80% emergence whether aphids present or not and repeat up to three times at 7-14 d intervals
- Spray sugar beet when aphids first seen or warning issued and repeat once if necessary
- Product rainfast within 2 h. Do not spray wet crops
- Where aphids resistant to deltamethrin + heptenophos occur repeat treatments are likely to result in lower levels of control

Crop Safety/Restrictions

- Maximum number of treatments 2 per crop for peas and sugar beet; 4 per crop for potatoes
- Do not spray crops suffering from severe moisture stress

Special precautions/Environmental safety

- Toxic if swallowed. Harmful in contact with skin
- Irritating to eyes and skin
- Flammable
- Do not harvest for human or animal consumption for at least 2 d after last application
- Dangerous to bees. Do not apply when flowering weeds are present
- Extremely dangerous to fish or other aquatic life. Do not contaminate surface waters or ditches with chemical or used container
- Do not allow direct spray from vehicle mounted/drawn hydraulic sprayers to fall within 6 m, or from hand-held sprayers to within 2 m, of surface waters or ditches. Direct spray away from water

Protective clothing/Label precautions

- A, C, H
- M01, M04, R02c, R03a, R04a, R04b, R07d, U02, U04a, U05a, U08, U13, U19, U20b, C02 (2 d), C03, E01, E12d, E13a, E16, E30a, E31b, E34

Latest application/Harvest Interval(HI)

- HI 24 h

Approval

- Off-label approval unlimited for use on brassica crops (OLA 1603/96), lettuce (OLA 1604/96), radicchio (OLA 1605/96)[1]

Maximum Residue Levels (mg residue/kg food)

- see deltamethrin entry

194 deltamethrin + pirimicarb

An insecticide mixture combining systemic, contact and stomach activity

Products

1 Evidence	AgrEvo	7.5:100 g/l	EC	06934
2 Patriot EC	AgrEvo	5:100 g/l	EC	08990

Uses

Aphids in BARLEY *(on ears)*, OATS *(on ears)*, WHEAT *(on ears)* [2]. Blackfly in SUGAR BEET [1, 2]. Caterpillars in BROCCOLI, BRUSSELS SPROUTS, CABBAGES, CALABRESE, CAULIFLOWERS, KALE, SWEDES, TURNIPS [1, 2]. Green aphid in BROCCOLI, BRUSSELS SPROUTS, CABBAGES, CALABRESE, CAULIFLOWERS, KALE, POTATOES, SUGAR BEET, SWEDES, TURNIPS [1, 2]. Green aphid in OILSEED RAPE [2]. Mealy aphids in BROCCOLI, BRUSSELS SPROUTS, CABBAGES, CALABRESE, CAULIFLOWERS, KALE, SWEDES, TURNIPS [1, 2]. Mealy aphids in OILSEED RAPE [2]. Pea aphid in PEAS [1, 2]. Virus vectors in POTATOES [1]. Virus yellows vectors in SUGAR BEET [1, 2].

Notes

Efficacy

- Best results obtained from treatment in warm, calm conditions, ideally in morning or evening when uptake by the plant will be at its highest
- Peas should be inspected from beginning of flowering and treated according to current thresholds
- Treat potatoes when first aphids found to control virus spread and preserve processing quality. Other ware crops should be treated from late Jun onwards before population build-up occurs and seed crops must be treated at 80% emergence
- Higher water volume recommended where crop canopy dense
- Treat other crops when pests first seen or when warnings are issued
- Treat cereals when at least two-thirds of heads infested and numbers increasing [2]
- Where aphids resistant to deltamethrin + pirimicarb occur repeat treatments are likely to result in lower levels of control

Crop Safety/Restrictions

- Maximum number of treatments 4 per crop on potatoes, edible brassicas; 2 per crop on peas and sugar beet; 1 per crop on cereals. See label for restrictions on oilseed rape
- Do not spray any crop when wilting or in temperatures above 25°C
- Do not treat wet crops liable to run-off
- Sugar beet may be treated as a band spray provided the pressure and water volume requirements are met

Special precautions/Environmental safety

- Harmful if swallowed
- Risk of serious damage to eyes
- Irritating to respiratory system
- Flammable
- Dangerous to bees. Do not apply to crops in flower or to those in which bees are actively foraging except as directed on peas. Do not apply when flowering weeds are present
- Extremely dangerous to fish or other aquatic life. Do not contaminate surface waters or ditches with chemical or used container.
- Do not allow direct spray from ground-based vehicle-mounted/drawn sprayers to fall within 6 m, or from hand-held sprayers to within 2 m, of surface waters or ditches. Direct spray away from water
- Harmful to livestock
- Must not be applied to cereals if a pyrethroid or dimethoate product has been used at ear emergence (GS 51)

- Product formulation can release traces of other pesticides left in the sprayer. Sprayer should be thoroughly washed out before spraying this product

Protective clothing/Label precautions

- A, D, E, H [1, 2]
- M02, U08, U20a, E06b [1]; M03, R03c, R04c, R04d, R07d, U04a, U05a, U19, C03, E01, E12d, E13a, E16, E30a, E31b, E34 [1, 2]; M05, U11, U20b, E06b (7 d), E26 [2]

Withholding period

- Livestock should be kept out of treated areas for at least 7 d

Maximum Residue Levels (mg residue/kg food)

- see deltamethrin entry

195 demeton-S-methyl

A systemic and contact organophosphorus insecticide and acaricide

Products	Metaphor	United Phosphorus	580 g/l	EC	08017

Uses Aphids in BROAD BEANS, CARROTS, CEREALS, DWARF BEANS, FIELD BEANS, FODDER BEET, FRENCH BEANS, LETTUCE, MANGELS, PARSNIPS, PEAS, POTATOES, RUNNER BEANS, STRAWBERRIES, SUGAR BEET. Cabbage aphid in BROCCOLI, BRUSSELS SPROUTS, CABBAGES, CALABRESE, CAULIFLOWERS. Insect pests in COURGETTES *(off-label)*, MARROWS *(off-label)*, PUMPKINS *(off-label)*, SQUASHES *(off-label)*, SWEETCORN *(off-label)*. Mangold fly in FODDER BEET, MANGELS, SUGAR BEET. Pea midge in PEAS. Red spider mites in STRAWBERRIES.

Notes

Efficacy

- Apply as spray when pest appears and repeat as necessary. Dose, number and timing of sprays vary with pest and crop. See label for details
- May also be applied by soil watering on many ornamentals. Do not apply to dry soil
- Do not spray waxy-leaved plants such as brassicas or sugar beet under hot, dry conditions. Delay until late evening or early morning. Do not spray wilting plants
- For bryobia mite control spray at temperatures above 13°C
- Where strains of red spider mite, peach-potato aphid in sugar beet or other pests resistant to organophosphorus compounds occur, demeton-S-methyl is unlikely to give satisfactory control

Crop Safety/Restrictions

- Maximum number of treatments 1 per crop for mangels, fodder beet, red beet, cereals, cucumbers, lettuce, gooseberries, ornamentals; 2 per crop for peas, apples, pears, plums, cherries, currants, raspberries, tomatoes, chrysanthemums, carnations; 3 per crop for sugar beet; 4 per crop for potatoes; 5 per crop for broccoli, Brussels sprouts, cabbages, calabrese, cauliflowers; 6 per crop for lettuce; 12 per crop for seed potatoes
- On carrots the maximum total dose applied per crop must not exceed the equivalent of 3 (on mineral soils) or 4 (on organic soils) full dose applications
- Do not treat brassicas after the end of Oct. Do not spray cauliflower curds
- Do not spray carnations in bloom as residues may cause unpleasant odours

- Do not mix with fentin compound or alkaline sprays

Special precautions/Environmental safety

- Demeton-S-methyl is subject to the Poisons Rules 1982 and the Poisons Act 1972. See notes in Section 1
- Toxic in contact with skin, by inhalation and if swallowed
- Flammable
- This product contains an anticholinesterase organophosphorus compound. Do not use if under medical advice not to work with such compounds
- Do not apply with hand-held ULV equipment
- Harmful to livestock
- Harmful to game, birds and animals
- Harmful to bees. Do not apply to crops in flower or to those in which bees are actively foraging. Do not apply when flowering weeds are present
- Harmful to fish or other aquatic life. Do not contaminate surface waters or ditches with chemical or used container

Protective clothing/Label precautions

- A, C, H
- M01, M04, R02a, R02b, R02c, R07d, U02, U04a, U05a, U08, U10, U13, U14, U15, U19, U20a, C02 (3 wk), C03, E01, E06b (2 wk), E10b, E12e, E13c, E26, E27, E30b, E31b, E34, S04b

Withholding period

- Keep all livestock out of treated areas for at least 2 wk

Latest application/Harvest Interval(HI)

- Before flowers open for peas.
- HI mangels, fodder beet, red beet 10 d (not to be used as fodder for 21 d); other crops 21 d

Approval

- Off-label approval unlimited for use on courgettes, marrows, squashes, pumpkins (OLA 1826/97)[1]; to Aug 2001 for use on sweetcorn (OLA 1829/97)[1]

196 desmedipham

A contact carbamate herbicide available only in mixtures

197 desmedipham + ethofumesate + phenmedipham

A selective contact and residual herbicide for beet

Products Betanal Progress OF | AgrEvo | 25:151:76 g/l | EC | 07629

Uses Annual dicotyledons in SUGAR BEET. Annual meadow grass in SUGAR BEET.

Notes **Efficacy**

- Best results achieved if weeds are not larger than fully expanded cotyledon stage at spraying
- Where a pre-emergence band spray has been applied treatment must be timed according to size of the untreated weeds between the rows

- Repeat applications as each flush of weeds reaches cotyledon size normally necessary for season long control
- Sequential treatments should be applied when the previous one is still showing an effect on the weeds
- Various mixtures with other beet herbicides are recommended. See label for details

Crop Safety/Restrictions
- Maximum total dose 4.5 l/ha product per crop
- Apply first treatment when majority of crop plants have reached the fully expanded cotyledon stage
- Do not spray crops stressed by nutrient deficiency, wind damage, pest or disease attack, or previous herbicide treatments
- If temperature likely to exceed 21°C spray after 5 pm
- Frost within 7 d of treatment may cause check from which the crop may not recover
- Crystallisation may occur if spray volume exceeds that recommended or spray mixture not used within 2 h, especially if the water temperature is below 5°C
- Before use, wash out sprayer to remove all traces of previous products, especially hormone and sulfonyl urea weedkillers

Special precautions/Environmental safety
- Harmful if swallowed
- Irritating to skin
- Risk of serious damage to eyes
- Harmful to fish or aquatic life. Do not contaminate surface waters or ditches with chemical or used container

Protective clothing/Label precautions
- A, C, H
- M03, R03c, R04b, R04d, U05a, U08, U20a, C03, E01, E13c, E30a, E31b, E34

198 desmedipham + phenmedipham

A mixture of contact herbicides for use in sugar beet

Products

Betanal Compact	AgrEvo	34:129 g/l	EC	07247

Uses

Annual dicotyledons in SUGAR BEET.

Notes

Efficacy
- Best results obtained from treatment when earliest germinating weeds have reached cotyledon stage
- Further treatments must be applied as each flush of weeds reaches cotyledon stage but allowing a minimum of 7 d between each spray
- Where a pre-emergence band spray has been applied, the first treatment should be timed according to the size of the weeds in the untreated area between the rows
- Product should be applied overall as a fine spray to optimise weed cover and spray retention

- Various tank mixtures and sequences recommended to widen weed spectrum and add residual activity - see label for details

Crop Safety/Restrictions
- Maximum number of treatments 3 per crop
- Product safe to use on all soil types
- Spray in evening if daytime temperatures above 21°C expected
- Avoid or delay treatment if frost expected within 7 d
- Avoid or delay treating crops under stress from wind damage, manganese or lime deficiency, pest or disease attack etc
- Check from which recovery may not be complete may occur if treatment made during conditions of sharp diurnal temperature fluctuation

Special precautions/Environmental safety
- Irritant. Risk of serious damage to eyes
- Dangerous to fish or other aquatic life. Do not contaminate surface waters or ditches with chemical or used container
- Keep interval between spray preparation and completion of treatment to 2 h or less to avoid possibility of crystallisation especially if water temperature below 5°C
- Product may cause non-reinforced PVC pipes and hoses to soften and swell. Wherever possible, use reinforced PVC or synthetic rubber hoses

Protective clothing/Label precautions
- A, H
- R04, R04d, U05a, U08, U19, U20a, C03, E01, E13b, E26, E30a, E31b, E34

Latest application/Harvest Interval(HI)
- Before crop leaves meet between rows

---

199 desmetryn

A contact triazine herbicide for use on brassica crops

Products

1 Semeron	Novartis	25% w/w	WP	08441
2 Semeron 25WP	Ciba Agric.	25% w/w	WP	01916

Uses

Annual dicotyledons in BROCCOLI *(off-label)*, BRUSSELS SPROUTS, CABBAGES, CALABRESE *(off-label)*, CAULIFLOWERS *(off-label)*, COLLARDS, FODDER RAPE, KALE *(off-label)*. Fat-hen in BROCCOLI *(off-label)*, BRUSSELS SPROUTS, CABBAGES, CALABRESE *(off-label)*, CAULIFLOWERS *(off-label)*, COLLARDS, FODDER RAPE, KALE *(off-label)*.

Notes

Efficacy
- May be applied from Mar to mid-Nov to weeds up to 5-100 mm high (fat-hen to 350 mm)
- A higher dose required in drier eastern than western areas. See label for details
- Do not apply when rain imminent. Rain within 24 h reduces effectiveness
- May be tank-mixed with aziprotryne or clopyralid. See label for details

Crop Safety/Restrictions
- Maximum number of treatments 1 per crop
- Apply to drilled crops after plants have 3 true leaves and are at least 125 mm high, to transplants at least 2 wk after planting and after reaching above stage

- Do not spray poor, thin crops or seed crops intended for producing Maris Kestrel kale as one of the parent lines is susceptible
- Do not spray in frosty weather or if crops affected by frost or cabbage root fly
- Do not spray for at least 7 d after applying insecticide
- Leaf scorch may occur after spraying but crop recovers quickly without effect on yield

Special precautions/Environmental safety

- Do not contaminate surface waters or ditches with chemical or used container

Protective clothing/Label precautions

- U09a, U20a, C02, E15, E30a, E32a

Latest application/Harvest Interval(HI)

- HI 4 wk (5 wk for calabrese)

Approval

- Off-label approval unlimited for use on broccoli, kale, cauliflowers (OLA 344/93)[2], (OLA 1386/97)[1], calabrese (OLA 0985/93)[2], (OLA 1387/97)[1]

200 diazinon

A contact organophosphorus insecticide for soil and foliar treatment

Products Darlingtons Diazinon Granules | Sylvan | 5% w/w | GR | 05674

Uses Mushroom flies in MUSHROOMS.

Notes

Efficacy

- Incorporate in mushroom compost at spawning for mushroom fly control, it is essential to mix in the granules thoroughly

Crop Safety/Restrictions

- Maximum number of treatments 1 per spawning

Special precautions/Environmental safety

- This product contains an anticholinesterase organophosphorus compound. Do not use if under medical advice not to work with such compounds
- Harmful to game, wild birds and animals
- Harmful to fish or other aquatic life. Do not contaminate surface waters or ditches with chemical or used container

Protective clothing/Label precautions

- A
- M01, U09a, U20a, C02, C03, E01, E10b, E13c, E30a, E32a, E34

Latest application/Harvest Interval(HI)

- HI 2 wk

Maximum Residue Levels (mg residue/kg food)
- meat 0.7; citrus fruit, pome fruits, stone fruits, grapes, cane fruits, bananas, kiwi fruit, olives, beetroot, carrots, celeriac, horseradish, parsnips, radishes, swedes, turnips, bulb vegetables, fruiting vegetables, brassica vegetables, leaf vegetables and fresh herbs, legume vegetables, asparagus, celery, globe artichokes, leeks, cultivated mushrooms, potatoes 0.5; bilberries, currants, gooseberries 0.2; tree nuts, oilseeds, tea, cereals (except buckwheat, millet) 0.05; cranberries, wild berries, avocados, dates, figs, kumquats, litchis, mangoes, passion fruit, pineapples, pomegranates, Jerusalem artichokes, parsley root, salsify, sweet potatoes, yams, cardoons, fennel, rhubarb, buckwheat, millet, milk 0.02

201 dicamba

A translocated benzoic herbicide for control of bracken and perennial weeds

See also 2,4-D + dicamba
2,4-D + dicamba + mecoprop
bifenox + dicamba

Products

Cadence	Novartis	70% w/w	SG	08796

Uses

Chickweed in ESTABLISHED LEYS, PERMANENT PASTURE. Docks in ESTABLISHED LEYS, PERMANENT PASTURE.

Notes

Efficacy
- For weed control in established leys and pasture apply when weeds actively growing before flowering shoots appear. Large well-established docks may require a second treatment in the following yr

Crop Safety/Restrictions
- Maximum total dose equivalent to one full dose per yr
- Avoid drift onto susceptible crops. Tomatoes may be affected by vapour drift at a considerable distance
- Do not roll crops within 1 wk of spraying. Do not treat crops where clover forms an important part of the sward

Special precautions/Environmental safety
- Harmful if swallowed
- Irritating to eyes
- Do not contaminate surface waters or ditches with chemical or used container

Protective clothing/Label precautions
- A, C
- R03c, R04a, U05a, U08, U20b, C02 (21 d), C03, E01, E15, E29, E30a, E31b, E34

Withholding period
- Keep livestock out of treated areas for at least 3 wk and until foliage of any poisonous weeds such as ragwort and buttercups has died and become unpalatable

202 dicamba + dichlorprop + ferrous sulfate + MCPA

A herbicide/fertiliser combination for moss and weed control in turf

Products

1 Longlife Renovator 2	Miracle	0.027:0.446:23.08: 0.223% w/w	GR	08379
2 Renovator 2	Scotts	0.027:0.446:23.08: 0.223% w/w	GR	09272

Uses

Annual dicotyledons in AMENITY TURF, GOLF COURSES. Mosses in AMENITY TURF, GOLF COURSES. Perennial dicotyledons in AMENITY TURF, GOLF COURSES.

Notes

Efficacy
- Apply from mid-Apr to mid-Aug when weeds are growing
- Apply with a suitable calibrated fertilizer distributor
- For best control of moss scarify vigorously after 2 wk to remove dead moss
- Where regrowth of moss or weeds occurs a repeat treatment may be made after 6 wk
- Do not cut grass for at least 3 d before and at least 4 d after treatment
- Avoid treatment of wet grass or during drought. If no rain falls within 48 h water in thoroughly
- Do not apply during freezing conditions or when rain imminent

Crop Safety/Restrictions
- Do not treat new turf until established for 6-9 mth
- The first 4 mowings after treatment should not be used to mulch cultivated plants unless composted at least 6 mth
- Avoid walking on treated areas until it has rained or they have been watered
- Do not re-seed or turf within 8 wk of last treatment

Protective clothing/Label precautions
- U09a, U20a, E15, E30a, E32a

Maximum Residue Levels (mg residue/kg food)
- see dichlorprop entry

203 dicamba + dichlorprop + MCPA

A translocated herbicide mixture for turf

Products

1 Cleanrun 2	Scotts	0.03:0.45:0.22% w/w	MG	09271
2 Intrepid	Miracle	20.8:333:166.5 g/l	SL	07819
3 Intrepid	Scotts	20.8:333:166.5 g/l	SL	09266
4 Longlife Cleanrun 2	Miracle	0.03:0.45:0.22% w/w	MG	07851

Uses

Annual dicotyledons in TURF. Perennial dicotyledons in TURF.

Notes

Efficacy

- Apply between Apr and Sep when weeds growing actively
- If no rain falls within 48 h of treatment irrigate thoroughly [1, 4]
- Do not mow for at least 3 d before and 3-4 d after treatment so that there is sufficient leaf growth for spray uptake and sufficient time for translocation
- If re-growth occurs or new weeds germinate re-treatment recommended

Crop Safety/Restrictions

- Do not use during drought unless irrigation is carried out before and after treatment
- Do not apply during freezing conditions or when heavy rain is imminent
- Do not treat new turf until established for about 6-9 mth after seeding or turfing
- The first 4 mowings after treatment should not be used to mulch cultivated plants unless composted for at least 6 mth
- Avoid walking where possible on treated areas until after rain or irrigation [1, 4]
- Do not re-seed turf within 8 wk of last treatment

Special precautions/Environmental safety

- Harmful if swallowed [2]
- Irritating to skin [2]
- Risk of serious damage to eyes [2]
- May cause sensitisation by skin contact [2]
- Harmful to fish. Do not contaminate surface waters or ditches with chemical or used container [2]

Protective clothing/Label precautions

- A, C [2, 3]
- M03, R03c, R04b, R04d, R04e, U05a, U08, U19, U20a, C03, E01, E13c, E31b, E34 [2, 3]; U09a, E15, E32a, S06 [1, 4]; E30a [1-4]

Maximum Residue Levels (mg residue/kg food)

- see dichlorprop entry

---

204 dicamba + maleic hydrazide + MCPA

A herbicide/plant growth regulator mixture for amenity grass

Products

Mazide Selective	Vitax	6:200:75 g/l	SL	05753

Uses

Annual dicotyledons in AMENITY GRASS, ROAD VERGES. Growth retardation in AMENITY GRASS, ROAD VERGES. Perennial dicotyledons in AMENITY GRASS, ROAD VERGES.

Notes

Efficacy

- Best results achieved by application in Apr-May when grass and weeds growing actively but before weeds have started to flower
- May be used either as one annual spray or as spring spray repeated after 8-10 wk

Crop Safety/Restrictions

- Maximum number of treatments 2 per yr
- Do not use on fine turf

Special precautions/Environmental safety

- Harmful to fish or other aquatic life. Do not contaminate surface waters or ditches with chemical or used container

Protective clothing/Label precautions
- U08, U20b, E07, E13c, E26, E30a, E31a

Withholding period
- Keep livestock out of treated areas until foliage of any poisonous weeds such as ragwort has died and become unpalatable

Maximum Residue Levels (mg residue/kg food)
- see maleic hydrazide entry

205 dicamba + MCPA + mecoprop

A translocated herbicide for cereals, grassland, amenity grass and orchards

Products

1 Barclay Hat-Trick	Barclay	25:200:400 g/l	SL	07579
2 Campbell's Field Marshal	MTM Agrochem.	18:360:160 g/l	SL	08594
3 Campbell's Grassland Herbicide	MTM Agrochem.	25:200:400 g/l	SL	06157
4 Headland Relay	Headland	25:200:400 g/l	SL	07684
5 Quad-Ban	Quadrangle	19.5:245:86.5 g/l	SL	03114
6 Stefes Banlene Plus	Stefes	18:252:84 g/l	SL	07688
7 Stefes Docklene	Stefes	85:85:336 g/l	SL	07816
8 Tribute	Nomix-Chipman	18:252:84 g/l	SL	06921

Uses

Annual dicotyledons in GRASS SEED CROPS, RYE [5, 6]. Annual dicotyledons in AMENITY TURF, SPORTS TURF [8]. Annual dicotyledons in PERMANENT PASTURE [1, 3-7]. Annual dicotyledons in LEYS [1-6]. Annual dicotyledons in APPLES, ESTABLISHED GRASSLAND, PEARS [6]. Annual dicotyledons in APPLE ORCHARDS, ESTABLISHED LEYS [7]. Annual dicotyledons in BARLEY, OATS, WHEAT [2, 5, 6]. Annual dicotyledons in CEREALS UNDERSOWN *(grass only)* [2]. Chickweed in GRASS SEED CROPS, RYE [5, 6]. Chickweed in PERMANENT PASTURE [1, 3-7]. Chickweed in LEYS [1-6]. Chickweed in ESTABLISHED GRASSLAND [6]. Chickweed in ESTABLISHED LEYS [7]. Chickweed in BARLEY, OATS, WHEAT [2, 5, 6]. Chickweed in CEREALS UNDERSOWN *(grass only)* [2]. Cleavers in RYE [5, 6]. Cleavers in GRASS SEED CROPS [5]. Cleavers in PERMANENT PASTURE [3-5, 7]. Cleavers in LEYS [3-5]. Cleavers in ESTABLISHED LEYS [7]. Cleavers in BARLEY, OATS, WHEAT [2, 5, 6]. Docks in GRASS SEED CROPS [5, 6]. Docks in BARLEY, OATS, RYE, WHEAT [5]. Docks in PERMANENT PASTURE [1, 3-7]. Docks in LEYS [1, 3-6]. Docks in APPLES, ESTABLISHED GRASSLAND, PEARS [6]. Docks in ESTABLISHED LEYS [7]. Mayweeds in GRASS SEED CROPS, RYE [5, 6]. Mayweeds in PERMANENT PASTURE [3-7]. Mayweeds in ESTABLISHED GRASSLAND [6]. Mayweeds in ESTABLISHED LEYS [7]. Mayweeds in BARLEY, OATS, WHEAT [2, 5, 6]. Mayweeds in LEYS [2-6]. Mayweeds in CEREALS UNDERSOWN *(grass only)* [2]. Perennial dicotyledons in GRASS SEED CROPS, RYE [5, 6]. Perennial dicotyledons in AMENITY TURF, SPORTS TURF [8]. Perennial dicotyledons in PERMANENT PASTURE [1, 3-7]. Perennial dicotyledons in LEYS [1-6]. Perennial dicotyledons in APPLES, ESTABLISHED GRASSLAND, PEARS [6]. Perennial dicotyledons in APPLE ORCHARDS, ESTABLISHED LEYS [7]. Perennial dicotyledons in BARLEY, OATS, WHEAT [2, 5, 6]. Perennial dicotyledons in CEREALS UNDERSOWN *(grass only)* [2]. Polygonums in RYE [5, 6]. Polygonums in GRASS SEED CROPS [5]. Polygonums in PERMANENT PASTURE [3-5, 7]. Polygonums in ESTABLISHED LEYS [7]. Polygonums in

BARLEY, OATS, WHEAT [2, 5, 6]. Polygonums in LEYS [2-6]. Polygonums in CEREALS UNDERSOWN *(grass only)* [2].

Notes

Efficacy

- Best results achieved on young, actively growing weeds up to 15 cm high in a strongly competitive crop, perennials when well developed but before flowering. See label for details of susceptibility
- Spray grassland and turf from early spring to Oct when grasses growing actively
- Do not spray in rain, when rain imminent or in drought

Crop Safety/Restrictions

- Maximum number of treatments 1 per crop on cereals or 2 per yr on other crops
- Spray winter cereals from main leaf sheath erect and at least 5 cm high but before first node detectable (GS 31), spring cereals from 5-leaf stage to before first node detectable (GS 15-31) (some formulations only recommend treatment in spring)
- Do not spray cereals undersown with clovers or legumes, to be undersown with grass or legumes or grassland where clovers or other legumes are important
- Spray newly sown grass or undersown crops when grass seedlings have 2-3 leaves
- Do not spray leys established less than 18 mth or orchards established less than 3 yr [7]
- Do not spray grass seed crops later than 5 wk before emergence of seed heads
- In orchards avoid drift of spray onto tree foliage. Do not spray during blossom period
- Do not roll, harrow, cut, graze for 3-7 d before or after treatment (products vary)
- The first mowings after treatment should not be used for mulching unless composted for 6 mth. See individual label for details [8]
- Do not use on newly sown turf in the year of establishment. Allow 6-8 wk after treatment before seeding bare patches [8]

Special precautions/Environmental safety

- Harmful in contact with skin and if swallowed [1-3, 5, 8]
- Irritating to eyes [1-3, 5-8]
- Irritating to skin [3, 5, 8]
- Harmful to fish or other aquatic life. Do not contaminate surface waters or ditches with chemical or used container

Protective clothing/Label precautions

- A, C [1-8]; H, M [2, 4]
- M03, R03c [1-3, 5]; R03a [1-3, 5, 8]; R04a [1-3, 5-8]; R04b [3, 5]; U05a, U08, C03, E01, E13c, E30a, E31b [1-8]; U20a [1, 3-6]; U20b [2, 7, 8]; E07 (2 wk) [2, 6, 7]; E07 [1, 3, 5, 8]; E34 [1-5, 8]

Withholding period

- Keep livestock out of treated areas for at least 2 wk and until foliage of any poisonous weeds such as ragwort has died and become unpalatable

Latest application/Harvest Interval(HI)

- Before first node detectable (GS 31) for cereals; 5 wk before head emergence for grass seed crops
- HI normally 7 d before cutting or grazing grass - check labels

206 dicamba + MCPA + mecoprop-P

A translocated herbicide for cereals, grassland, amenity grass and orchards

Products

1 Field Marshal	United Phosphorus	18:360:80	SL	08956
2 Headland Relay P	Headland	25:200:200 g/l	SL	08580
3 Hyprone-P	Agrichem	16:101:92 g/l	SL	09125
4 Hysward-P	Agrichem	16:101:92 g/l	SL	09052
5 Mascot Super Selective-P	Rigby Taylor	15:100:100 g/l	SL	06106
6 MSS Mircam Plus	Mirfield	19.5:245:43.3 g/l	SL	01416
7 Pasturol Plus	FCC	25:200:200 g/l	SL	08581
8 Stefes Banlene Super	Stefes	18:252:42 g/l	SL	07691
9 Stefes Docklene Super	Stefes	85:85:168 g/l	SL	07696
10 Tritox	Scotts	15:178:54 g/l	SL	07764
11 UPL Grassland Herbicide	United Phosphorus	25:200:200	SL	08934

Uses

Annual dicotyledons in GRASSLAND, TURF [6]. Annual dicotyledons in APPLES, PEARS, RYE [8]. Annual dicotyledons in APPLE ORCHARDS, ESTABLISHED LEYS [9]. Annual dicotyledons in SPORTS TURF [2, 5, 7, 10]. Annual dicotyledons in PERMANENT PASTURE [2, 7-9, 11]. Annual dicotyledons in AMENITY GRASS [2, 4, 7]. Annual dicotyledons in LEYS [1, 2, 7, 8, 11]. Annual dicotyledons in BARLEY, OATS, WHEAT [1, 3, 6, 8]. Annual dicotyledons in CEREALS UNDERSOWN *(grass only)* [1]. Annual dicotyledons in AMENITY TURF [4, 5, 10]. Annual dicotyledons in ESTABLISHED GRASSLAND [4, 8]. Annual dicotyledons in GRASS SEED CROPS [3, 6, 8]. Annual dicotyledons in NEWLY SOWN GRASS, UNDERSOWN BARLEY, UNDERSOWN OATS, UNDERSOWN WHEAT [3]. Chickweed in GRASSLAND, TURF [6]. Chickweed in PERMANENT PASTURE [8, 9, 11]. Chickweed in ESTABLISHED GRASSLAND, RYE [8]. Chickweed in ESTABLISHED LEYS [9]. Chickweed in LEYS [1, 8, 11]. Chickweed in BARLEY, OATS, WHEAT [1, 3, 6, 8]. Chickweed in CEREALS UNDERSOWN *(grass only)* [1]. Chickweed in GRASS SEED CROPS [3, 6, 8]. Chickweed in NEWLY SOWN GRASS, UNDERSOWN BARLEY, UNDERSOWN OATS, UNDERSOWN WHEAT [3]. Cleavers in GRASSLAND, TURF [6]. Cleavers in RYE [8]. Cleavers in PERMANENT PASTURE [9, 11]. Cleavers in ESTABLISHED LEYS [9]. Cleavers in LEYS [11]. Cleavers in BARLEY, OATS, WHEAT [1, 3, 6, 8]. Cleavers in GRASS SEED CROPS [3, 6]. Cleavers in NEWLY SOWN GRASS, UNDERSOWN BARLEY, UNDERSOWN OATS, UNDERSOWN WHEAT [3]. Docks in PERMANENT PASTURE [8, 9, 11]. Docks in LEYS [8, 11]. Docks in APPLES, PEARS [8]. Docks in ESTABLISHED LEYS [9]. Docks in ESTABLISHED GRASSLAND [4, 8]. Docks in AMENITY GRASS, AMENITY TURF [4]. Docks in GRASS SEED CROPS [3, 8]. Mayweeds in GRASSLAND, TURF [6]. Mayweeds in PERMANENT PASTURE [8, 9, 11]. Mayweeds in ESTABLISHED GRASSLAND, RYE [8]. Mayweeds in ESTABLISHED LEYS [9]. Mayweeds in LEYS [1, 8, 11]. Mayweeds in BARLEY, OATS, WHEAT [1, 3, 6, 8]. Mayweeds in CEREALS UNDERSOWN *(grass only)* [1]. Mayweeds in GRASS SEED CROPS [3, 6, 8]. Mayweeds in NEWLY SOWN GRASS, UNDERSOWN BARLEY, UNDERSOWN OATS, UNDERSOWN WHEAT [3]. Perennial dicotyledons in GRASSLAND, TURF [6]. Perennial dicotyledons in APPLES, PEARS, RYE [8]. Perennial dicotyledons in APPLE ORCHARDS, ESTABLISHED LEYS [9]. Perennial dicotyledons in SPORTS TURF [2, 5, 7, 10]. Perennial dicotyledons in PERMANENT PASTURE [2, 7-9, 11]. Perennial dicotyledons in AMENITY GRASS [2, 4, 7]. Perennial dicotyledons in LEYS [1, 2, 7, 8, 11]. Perennial dicotyledons in BARLEY, OATS, WHEAT [1, 3, 6, 8]. Perennial dicotyledons in CEREALS UNDERSOWN *(grass only)* [1]. Perennial dicotyledons in AMENITY TURF [4, 5, 10]. Perennial dicotyledons in ESTABLISHED GRASSLAND [4, 8]. Perennial dicotyledons in

GRASS SEED CROPS [3, 6, 8]. Perennial dicotyledons in NEWLY SOWN GRASS, UNDERSOWN BARLEY, UNDERSOWN OATS, UNDERSOWN WHEAT [3]. Polygonums in GRASSLAND, TURF [6]. Polygonums in RYE [8]. Polygonums in PERMANENT PASTURE [9, 11]. Polygonums in ESTABLISHED LEYS [9]. Polygonums in LEYS [1, 8, 11]. Polygonums in BARLEY, OATS, WHEAT [1, 3, 6, 8]. Polygonums in CEREALS UNDERSOWN *(grass only)* [1]. Polygonums in GRASS SEED CROPS [3, 6]. Polygonums in NEWLY SOWN GRASS, UNDERSOWN BARLEY, UNDERSOWN OATS, UNDERSOWN WHEAT [3].

Notes

Efficacy

- Treatment should be made when weeds growing actively. Weeds hardened by winter weather may be less susceptible
- For best results apply in fine warm weather, preferably when soil is moist. Do not spray if rain expected within 6 h or in drought
- Application of fertilizer 1-2 wk before spraying aids weed control in turf

Crop Safety/Restrictions

- Maximum number of treatments (including other mecoprop-P products) 1 per crop on cereals or 2 per yr on other crops and amenity grass. The total amount of mecoprop-P applied in a single yr must not exceed the maximum total dose approved for any single product for the crop/situation
- Apply to winter cereals from the leaf sheath erect stage (GS 30), and to spring cereals from the five expanded leaf stage (GS 15) [3, 6, 8]
- Do not apply to cereals after the first node is detectable (GS 31) [3, 6, 8] or to grass under stress from drought or cold weather [2]
- Do not spray leys established less than 18 mth or orchards established less than 3 yr [9]
- Spray grass seed crops 4-6 wk before flower heads begin to emerge (timothy 6 wk) [3, 6, 8]
- Do not use on turf in year of establishment [2, 4-6, 10]. Allow 6-8 wk after treatment before seeding bare patches
- The first mowings after use should not be used for mulching unless composted for 6 mth [2, 4, 10]
- Do not roll or harrow within 7 d before or after treatment, or graze for at least 7 d afterwards (longer if poisonous weeds present)
- Do not spray cereals undersown with clovers or legumes, to be undersown with grass or legumes or grassland where clovers or other legumes are important
- Fine turf should not be mown for 24 h before or after treatment (3-4 d for closely mown turf) [5, 10]
- Grass cuttings should not be used for mulching but may be composted and used not less than 6 mth later [2, 5]
- Turf containing bulbs may be treated once the foliage has died down completely [10]
- Avoid drift onto all broad-leaved plants outside the target area

Special precautions/Environmental safety

- Harmful in contact with skin and if swallowed [1, 2, 6, 11]
- Irritating to eyes [1-6, 8-11]
- Irritating to skin [3-6]
- Harmful to fish or other aquatic life. Do not contaminate surface waters or ditches with chemical or used container

Protective clothing/Label precautions

- A, C [1-11]; H, M [1-8, 10, 11]

- M03, R03a, R03c [1, 2, 6, 7, 11]; R04a, U08, C03, E01, E13c, E30a [1-11]; R04b [3-6]; U05a [2-11]; U19, E32a [5]; U20b [1, 3-6, 8-11]; E07 (2 wk) [1, 2, 4, 5, 7-11]; E07 [3, 6]; E26 [1-4, 7, 8, 11]; E31a [10]; E31b [1, 2, 6-9, 11]; E31c [3, 4]; E34 [1, 2, 5-7, 10, 11]; E34, [8]; S06 [2, 7]

Withholding period

- Keep livestock out of treated areas for at least 2 wk and until foliage of any poisonous weeds such as ragwort has died and become unpalatable

Latest application/Harvest Interval(HI)

- Before first node detectable (GS 31) for cereals; 4-6 wk before head emergence for grass seed crops
- HI 7 d before cutting or grazing for leys, permanent pasture

207 dicamba + mecoprop

A translocated post-emergence herbicide for cereals and grassland

Products

Condox	Zeneca	112:265 g/l	SL	06519

Uses

Annual dicotyledons in ESTABLISHED LEYS, PERMANENT PASTURE. Docks in ESTABLISHED LEYS, PERMANENT PASTURE. Perennial dicotyledons in ESTABLISHED LEYS, PERMANENT PASTURE. Thistles in ESTABLISHED LEYS, PERMANENT PASTURE.

Notes

Efficacy

- Best results by application to young weed seedlings under good growing conditions
- Do not spray in cold, frosty or windy weather, when rain expected or in drought
- Spray perennial weeds in grassland just before flowering, docks in rosette stage in late spring with repeat spray in Aug-Oct if necessary
- Allow 10-14 d regrowth of docks after grazing

Crop Safety/Restrictions

- Maximum number of treatments 1 per season
- Do not apply to leys established for less than 2 yr
- Do not apply to permanent pastures where clovers are an essential part of the sward
- Do not roll, harrow or cut within 7 d before or after spraying

Special precautions/Environmental safety

- Harmful in contact with skin and if swallowed. Irritating to skin
- Irritating to eyes
- Harmful to fish or other aquatic life. Do not contaminate surface waters or ditches with chemical or used container

Protective clothing/Label precautions

- A, C
- M03, R03a, R04a, R04b, U05a, U08, U20b, C03, E01, E07, E13c, E30a, E31a, E31c

Withholding period

- Keep livestock out of treated areas until foliage of any poisonous weeds such as ragwort has died and become unpalatable

Latest application/Harvest Interval(HI)

- Before first node detectable for cereals
- HI 7 d before cutting or 14 before grazing for permanent pasture

208 dicamba + mecoprop-P

A translocated post-emergence herbicide for cereals and grassland

Products

1 Di-Farmon R	Headland	42:319 g/l	SL	08472
2 Foundation	Sandoz	84:600 g/l	SL	07465
3 Foundation	Novartis	84:600 g/l	SL	08475
4 Hyban-P	Agrichem	18.7:150 g/l	SL	09129
5 Hygrass-P	Agrichem	18.7:150 g/l	SL	09130
6 MSS Mircam	Mirfield	18.7:150 g/l	SL	01415

Uses

Annual dicotyledons in BARLEY, OATS, WHEAT [1-4, 6]. Annual dicotyledons in ESTABLISHED GRASSLAND [1-3]. Annual dicotyledons in RYE, TRITICALE [4]. Annual dicotyledons in LEYS, PERMANENT PASTURE [5]. Chickweed in BARLEY, OATS, WHEAT [1-4, 6]. Chickweed in ESTABLISHED GRASSLAND [1]. Chickweed in RYE, TRITICALE [4]. Cleavers in BARLEY, OATS, WHEAT [1-4, 6]. Cleavers in ESTABLISHED GRASSLAND [1]. Cleavers in RYE, TRITICALE [4]. Docks in LEYS, PERMANENT PASTURE [5]. Mayweeds in BARLEY, OATS, WHEAT [1-4, 6]. Mayweeds in ESTABLISHED GRASSLAND [1]. Mayweeds in RYE, TRITICALE [4]. Perennial dicotyledons in ESTABLISHED GRASSLAND [2, 3]. Perennial dicotyledons in BARLEY, OATS, WHEAT [4, 6]. Perennial dicotyledons in RYE, TRITICALE [4]. Perennial dicotyledons in LEYS, PERMANENT PASTURE [5]. Polygonums in BARLEY, OATS, WHEAT [1-4, 6]. Polygonums in ESTABLISHED GRASSLAND [1]. Polygonums in RYE, TRITICALE [4]. Thistles in LEYS, PERMANENT PASTURE [5].

Notes

Efficacy

- Best results by application in warm, moist weather when weeds are actively growing
- Do not spray in cold or frosty conditions
- Do not spray if rain expected within 6 h
- Ensure any lime is washed off crop and weeds before spraying [6]

Crop Safety/Restrictions

- Maximum number of treatments 1 per crop for cereals or 1 per yr for grass (2 per yr on established grass [5]). The total amount of mecoprop-P applied in a single yr must not exceed the maximum total dose approved for any single product for the crop/situation
- Apply to winter sown crops from 5 expanded leaf stage (GS 15)[1-4], from leaf sheath erect stage [6] but before first node detectable (GS 30-31)
- Apply to spring sown cereals from 5 expanded leaf stage but before first node is detectable (GS 15-31)
- Treat grassland just before perennial weeds flower [1-3]
- Transient crop prostration may occur after spraying but recovery is rapid
- Do not treat undersown grass until tillering begins [6]
- Do not spray cereals undersown with clover or legume mixtures
- Do not roll, harrow within 7 d before or after spraying

- Do not treat crops suffering from stress from any cause
- Avoid treatment when drift may damage neighbouring susceptible crops

Special precautions/Environmental safety
- Harmful if swallowed and in contact with skin [1-3, 6]
- Irritating to eyes [2, 3, 5] and skin [1, 4, 6]
- Harmful to fish or other aquatic life. Do not contaminate surface waters or ditches with chemical or used container

Protective clothing/Label precautions
- A, C [1-6]; H M [1]; H [4-6]; M [5, 6]
- M03 [1-5]; R03a, R03c, U08, E31b [1-3, 6]; R04a, U05a, C03, E01, E07 (2 wk), E13c, E26, E30a [1-6]; R04b [1, 4, 6]; U09a, E31c [4, 5]; U20a [1, 6]; U20b [2-5]; E34 [1-4, 6]

Withholding period
- Keep livestock out of treated areas for at least 2 wk and until foliage of poisonous weeds such as ragwort has died and become unpalatable

Latest application/Harvest Interval(HI)
- Before 1st node detectable for cereals; not specified for established grassland [2, 3, 5]

209 dicamba + paclobutrazol

A plant growth regulator/herbicide mixture for amenity grass

Products				
Holdfast D	Miracle	25:250 g/l	SC	07864

Uses Annual dicotyledons in AMENITY GRASS. Growth retardation in AMENITY GRASS. Perennia dicotyledons in AMENITY GRASS.

Notes

Efficacy
- Uptake mainly via roots and aided by rain after application
- Apply 2 wk before grass growth starts in spring (mid-Mar to mid-May). Can also be applied up to Jul if grass growing rapidly
- Grass growth retarded for about 3 mth
- Do not apply during period of drought or frost
- Where coarse grasses (eg cocksfoot) dominant addition of maleic hydrazide or mefluidide recommended

Crop Safety/Restrictions
- Maximum number of treatments 1 per yr
- Use only on permanent grassland, not on grass within 1 yr after sowing
- Use only on areas of restricted or limited public access
- Do not use where food crops or ornamentals to be grown or grass resown within 5 yr
- Fine-leaved grasses can be retarded more than others leading to an uneven appearance
- Do not use cuttings from treated grass for composting or mulch

Special precautions/Environmental safety
- Harmful if swallowed

- Not to be used on food crops
- Harmful to fish or other aquatic life. Do not contaminate surface waters or ditches with chemical or used container

Protective clothing/Label precautions
- A, H, M
- M03, R03c, U05a, U08, U20a, C01, C03, E01, E07, E13c, E26, E30a, E31b, E34

Withholding period
- Keep livestock out of treated areas for at least 2 yr after treatment

210 dichlobenil

A residual benzonitrile herbicide for woody crops and non-crop areas

See also dalapon + dichlobenil

Products

1 Casoron G	Miracle	6.75% w/w	GR	07926
2 Casoron G	Uniroyal	6.75% w/w	GR	09022
3 Casoron G	Nomix-Chipman	6.75% w/w	GR	09023
4 Casoron G4	Miracle	4% w/w	GR	07927
5 Casoron G4	Uniroyal	4% w/w	GR	09215
6 Casoron G-SR	Miracle	20% w/w	GR	07925
7 Luxan Dichlobenil Granules	Luxan	6.75% w/w	GR	09250
8 Standon Dichlobenil 6G	Standon	6.75% w/w	GR	08874

Uses

Annual weeds in ESTABLISHED WOODY ORNAMENTALS, ROSES [1-5, 7, 8]. Annual weeds in CONIFERS AND BROADLEAVED TREES *(off-label)* [1-3]. Annual weeds in SOFT FRUIT, TOP FRUIT [7, 8]. Aquatic weeds in AQUATIC SITUATIONS [1-3, 6, 7]. Perennial dicotyledons in ESTABLISHED WOODY ORNAMENTALS, ROSES [1-5, 7, 8]. Perennial dicotyledons in CONIFERS AND BROADLEAVED TREES *(off-label)*[1-3]. Perennial dicotyledons in SOFT FRUIT, TOP FRUIT [7, 8]. Perennial grasses in ESTABLISHED WOODY ORNAMENTALS, ROSES [1-5, 7, 8]. Perennial grasses in CONIFERS AND BROADLEAVED TREES *(off-label)* [1-3]. Perennial grasses in SOFT FRUIT, TOP FRUIT [7, 8]. Total vegetation control in NON-CROP AREAS [1-3, 7, 8]. Total vegetation control in LAND NOT INTENDED TO BEAR VEGETATION [1-3]. Volunteer potatoes in POTATO DUMPS *(blight prevention)* [7, 8].

Notes

Efficacy
- Best results achieved by application in winter to moist soil during cool weather, particularly if rain follows soon after. Do not disturb treated soil by hoeing
- Lower rates control annuals, higher rates perennials, see label for details
- Residual activity lasts 3-6 mth with selective, up to 12 mth with non-selective rates
- Apply to potato dump sites before emergence of potatoes for blight prevention [7, 8]
- For control of emergent, floating and submerged aquatics apply to water surface in early spring. Intended for use in still or sluggish flowing water [1-3, 6, 7]
- Do not use on fen peat or moss soils

Crop Safety/Restrictions
- Maximum number of treatments 1 per yr
- Apply to crops in dormant period. See label for details of timing and rates
- Do not treat crops established for less than 2 yr. See label for lists of resistant and sensitive species. Do not treat stone fruit or Norway spruce (Christmas trees)

- Do not apply within 300 m of glasshouses or hops or near areas underplanted with bulbs, annuals or herbaceous stock
- Do not apply to frozen, snow-covered or waterlogged ground or when crop foliage wet
- Do not apply to sites less than 18 mth before replanting or sowing
- Store well away from corms, bulbs, tubers and seed

Special precautions/Environmental safety

- Irritant. May cause sensitization by skin contact [7]
- Harmful to fish or other aquatic life. Do not contaminate surface waters or ditches with chemical or used container
- Do not dump surplus herbicide in water or ditch bottoms [1-3, 6, 7]
- Use not permitted on reservoirs that form part of a water supply system [1-3, 6, 7]
- Do not use treated water for irrigation purposes within 2 wk of treatment or until concentration of dichlobenil falls below 0.3 ppm [1-3, 6, 7]

Protective clothing/Label precautions

- A, H [7]
- M03, R04, R04e, U05a, C03, E01 [7]; U20b, E19 [1-3, 6]; U20c [4, 5, 7, 8]; E13c [1-5, 7, 8]; E21 (2 wk) [1-3]; E21 [6]; E25, E29 [4, 5, 8]; E30a, E32a [1-8]; E34 [1-3, 7]

Latest application/Harvest Interval(HI)

- Early Mar for shrubs and trees; early Apr for apples, pears, currants, gooseberries; before bud movement for cane fruit, roses; before growth starts for potato dumps

Approval

- Approved for aquatic weed control [1-3, 6, 7]. See notes in Section 1 on use of herbicides in or near water
- Off label approval unlimited for use in forestry (OLA 0050/98)[1]

211 dichlofluanid

A protectant trihalomethylthio fungicide for horticultural crops

Products	Elvaron WG	Bayer	50% w/w	WG	04855

Uses Black spot in ROSES. Botrytis in AUBERGINES *(off-label)*, CANE FRUIT, CURRANTS, GOOSEBERRIES, GRAPEVINES, NON-EDIBLE ORNAMENTALS *(off-label)*, PEPPERS *(off-label)*, PROTECTED TOMATOES, STRAWBERRIES. Cane blight in RASPBERRIES. Cane spot in RASPBERRIES. Downy mildew in PROTECTED BRASSICA SEEDLINGS. Fire in TULIPS. Leaf mould in PROTECTED TOMATOES. Mildew in RASPBERRIES, ROSES.

Notes **Efficacy**

- Apply as a protective programme of sprays. See label for recommended timings
- Treatment also gives reduction of mildew and spur blight on raspberries and useful control of mildew on strawberries
- Apply to brassica seedlings under cover as soon as first seedlings appear and repeat after 3, 5, 7 and 10 d using a wetting agent. Thereafter apply 3 sprays at weekly intervals

Crop Safety/Restrictions

- Maximum number of treatments 8 per crop for brassicas under cover, tomatoes, peppers, aubergines, non-edible ornamentals under glass and outdoor grapes for wine-making; 6 per crop for currants; 4 per crop for gooseberries. For other fruit crops number depends on dose applied and restriction on total for season
- Do not use on strawberries under glass or polythene. Plants which have been under cover recently may still be susceptible to leaf scorch
- On raspberries up to 7 applications may be made, up to 6 on currants, up to 8 on grapes for winemaking

Special precautions/Environmental safety

- Harmful to fish or other aquatic life. Do not contaminate surface waters or ditches with chemical or used container

Protective clothing/Label precautions

- U09a, U19, U20a, E13c, E30a, E32a

Latest application/Harvest Interval(HI)

- Before 2 true leaf stage for Brussels sprouts, calabrese; before 8 expanded leaves for planted out cabbage, cauliflower, broccoli.
- HI raspberries 1 wk; strawberries, currants 2 wk; broccoli, cabbage 3 wk; gooseberries, loganberries, blackberries, Rubus hybrids, grapes (fresh consumption) 3 wk; grapes (winemaking) 5 wk; protected tomatoes, peppers, aubergines, non-edible ornamentals 3 d

Approval

- Off-label approval unlimited on protected peppers, aubergines and non-edible ornamentals (OLA 0167/93)[1]

212 dichlorophen

A chlorophenol moss-killer, fungicide, bactericide and algicide

Products

1 50/50 Liquid Mosskiller	Vitax	360 g/l	SL	07191
2 Enforcer	Miracle	360 g/l	SL	07866
3 Enforcer	Scotts	360 g/l	SL	09288
4 Fungo	Dax	340 g/l	SL	H4768
5 Mascot Mosskiller	Rigby Taylor	360 g/l	SL	02439
6 Mossicide	PBI	360 g/l	SL	08183
7 Nomix-Chipman Mosskiller	Nomix-Chipman	340 g/l	SL	06271
8 Panacide M	Coalite	360 g/l	SL	05611
9 Panacide M21	Coalite	165 g/l	SL	H4923
10 Panacide TS	Coalite	480 g/l	SL	05612
11 Panaclean 736	Coalite	45 g/l	SL	H5075
12 Super Mosstox	RP Amenity	340 g/l	SL	05339

Uses

Algae in HARD SURFACES [2-4, 9, 11]. Algae in PATHS AND DRIVES, TURF [4]. Fungus diseases in HARD SURFACES [1, 7]. Fungus diseases in PATHS AND DRIVES, TURF [1]. Mosses in AMENITY TURF, SPORTS TURF [7]. Mosses in TURF [1-5, 8, 10, 12]. Mosses in HARD SURFACES [1-7, 9, 11, 12]. Mosses in PATHS AND DRIVES [1, 4, 5, 12]. Mosses in LAWNS [6, 7]. Red thread in TURF [12].

Notes

Efficacy

- For moss control in turf spray at any time when moss is growing. Supplement spraying with other measures to improve fertility and drainage

- Rake out dead moss 2-3 wk after treatment
- To control mushroom diseases use on trays, floors etc as part of an environmental hygiene programme [4]
- Treat hard surfaces when rain not expected and brush away dead material when treatment fully dried [7]

Crop Safety/Restrictions
- Dose must not exceed 1 litre in 20 litres water
- Must not be applied to growing mushroom crops [4]

Special precautions/Environmental safety
- Harmful if swallowed [4]
- Irritating to eyes and skin
- Keep unprotected persons and animals away while treatment in progress [1, 5, 8] and for 48 h after treatment or until paint is dry [2, 7]
- Prevent surface run-off from entering storm drains [1, 7-11]
- Harmful [5, 12], dangerous [1, 2, 7-11], to fish or other aquatic life. Do not contaminate surface waters or ditches with chemical or used container

Protective clothing/Label precautions
- A [1-5, 7-12]; C [1-5, 7, 8, 10, 12]; E [11]; H [1-5, 7-12]; M [1-5, 7, 8, 10, 12]
- M03 [7, 8, 10]; R03c [4]; R04a, E01, E30a [1-12]; R04b, U05a, C03 [1-3, 5-12]; U02, C09 [1-3, 5-11]; U04a [1-3, 5-8, 10]; U08 [1, 5]; U09a [2-4, 6-12]; U14 [1-3, 5, 7]; U15, U20b [1, 5, 7]; U17 [2, 3, 5, 7]; U19, E32a [9, 11]; U20a [2-4, 6, 8-11]; U20c [12]; E02 (48 h) [1, 6, 9]; E02 (48 hr) [5]; E02 [7, 8, 10, 11]; E13b [1, 5, 7-11]; E13c [4, 12]; E20 [1, 5-9, 11]; E26 [1, 5, 12]; E31a [1, 4, 5, 7, 8, 10, 12]

Withholding period
- Keep unprotected persons out of treated areas for 48 h or until surfaces are dry

213 dichlorophen + ferrous sulfate

A mosskiller/fertilizer mixture for use on turf

Products

1 Aitken's Lawn Sand Plus	Aitken	0.3:10% w/w	GR	04542
2 SHL Lawn Sand Plus	Sinclair	0.3:10% w/w	GR	04439

Uses Mosses in AMENITY TURF.

Notes

Efficacy
- Apply to established turf from late spring to early autumn when soil moist but not when grass wet or damp with dew
- Heavy infestations may need a repeat treatment

Crop Safety/Restrictions
- Maximum number of treatments 2 per yr
- Do not treat newly sown grass or freshly laid turf for the first year
- Do not apply during drought or freezing conditions or when rain imminent
- Avoid walking on treated areas until it has rained or turf has been watered

- Do not mow for 3-4 d before or after application

Special precautions/Environmental safety

- Harmful to fish or other aquatic life. Do not contaminate surface waters or ditches with chemical or used container

Protective clothing/Label precautions

- U09a, U20a, E13c, E30a

214 1,3-dichloropropene

A halogenated hydrocarbon soil nematicide

Products	Telone II	Dow	94% w/w	VP	05749

Uses

Free-living nematodes in POTATOES. Nematodes in HOPS, RASPBERRIES, STRAWBERRIES. Potato cyst nematode in POTATOES. Stem and bulb nematodes in NARCISSI. Stem nematodes in STRAWBERRIES. Virus vectors in HOPS, RASPBERRIES, STRAWBERRIES.

Notes

Efficacy

- Before treatment soil should be in friable seed bed condition above 5°C, with adequate moisture and all crop remains decomposed. Tilling to 30 cm improves results
- Do not use on heavy clays or soils with many large stones
- Apply with sub-surface injector combined with soil-sealing roller or, in hops, with a hollow tined injector. See label for details of suitable machinery
- Leave soil undisturbed for 21 d after treatment. Wet or cold soils need longer exposure
- For control of potato-cyst nematode contact firm's representative, check nematode level and use in integrated control programme
- For use in hops apply in May or Jun following the yr of grubbing and leave as long as possible before replanting
- Do not use water to clean out apparatus nor use aluminium or magnesium alloy containers which may corrode

Crop Safety/Restrictions

- Maximum number of treatments 2 per yr for hops, 1 per yr for other crops
- Do not drill or plant until odour of fumigant is eliminated
- Do not use fertilizer containing ammonium salts after treatment. Use only nitrate nitrogen fertilizers until crop well established and soil temperature above 18°C
- Allow at least 6 wk after treatment before planting raspberries or strawberries

Special precautions/Environmental safety

- Toxic if swallowed. Harmful in contact with skin and by inhalation
- Irritating to eyes, skin and respiratory system
- May cause sensitization by skin contact
- Flammable
- Harmful to fish or other aquatic life. Do not contaminate surface waters or ditches with chemical or used container

Protective clothing/Label precautions

- A, C, H, M
- M04, R02c, R03a, R03b, R04a, R04b, R04c, R04e, R07d, U02, U04a, U05a, U09a, U14, U15, U19, U20a, C03, E01, E13c, E24, E28, E30b, E33, E34

Latest application/Harvest Interval(HI)

- 6 wk before planting for raspberries, strawberries; pre-planting of crop for hops, narcissi, potatoes

215 dichlorprop

A translocated phenoxy herbicide for use in cereals

See also 2,4-D + dichlorprop + MCPA + mecoprop
dicamba + dichlorprop + ferrous sulfate + MCPA
dicamba + dichlorprop + MCPA

Products	MSS 2,4-DP	Mirfield	500 g/l	SL	01394

Uses

Annual dicotyledons in BARLEY, OATS, UNDERSOWN CEREALS, WHEAT. Black bindweed in BARLEY, OATS, UNDERSOWN CEREALS, WHEAT. Perennial dicotyledons in BARLEY, OATS, UNDERSOWN CEREALS, WHEAT. Redshank in BARLEY, OATS, UNDERSOWN CEREALS, WHEAT.

Notes

Efficacy

- Best results achieved by application to young seedling weeds in good growing conditions in a strongly competing crop
- Effectiveness may be reduced by rain within 6 h of spraying, by very cold weather or by drought conditions

Crop Safety/Restrictions

- Spray winter crops in spring from fully tillered stage to before first node detectable (GS 29-30), spring crops from 1-leaf unfolded to before first node detectable (GS 11-30)
- Spray crops undersown with grass as soon as possible after grass begins to tiller
- Do not spray crops undersown with legumes, lucerne, peas or beans
- Do not roll or harrow within 7 d before or after treatment
- Avoid spray drifting onto nearby susceptible crops

Special precautions/Environmental safety

- Harmful in contact with skin and if swallowed

Protective clothing/Label precautions

- A, C, P
- M03, R03a, R03c, U05a, U08, U20b, C03, E01, E07, E13c, E26, E30a, E31b, E34

Withholding period

- Keep livestock out of treated areas until foliage of any poisonous weeds such as ragwort has died and become unpalatable

Maximum Residue Levels (mg residue/kg food)

- tea, hops 0.1; all other products (except cereals and animal products) 0.05

216 dichlorprop + MCPA

A translocated herbicide for use in cereals and turf

Products

1 MSS 2,4-DP+MCPA	Mirfield	500 g/l (total)	SL	01396
2 Redipon Extra	United Phosphorus	350:150 g/l	SL	07518
3 SHL Turf Feed and Weed	Sinclair	0.474:0.366 g/l	DP	04437

Uses

Annual dicotyledons in BARLEY, OATS, WHEAT [1, 2]. Annual dicotyledons in UNDERSOWN CEREALS [1]. Annual dicotyledons in AMENITY GRASS [3]. Black bindweed in BARLEY, OATS, WHEAT [1, 2]. Black bindweed in UNDERSOWN CEREALS [1]. Buttercups in AMENITY GRASS [3]. Clovers in AMENITY GRASS [3]. Daisies in AMENITY GRASS [3]. Dandelions in AMENITY GRASS [3]. Hemp-nettle in BARLEY, OATS, WHEAT [1, 2]. Hemp-nettle in UNDERSOWN CEREALS [1]. Perennial dicotyledons in BARLEY, OATS, WHEAT [1, 2]. Perennial dicotyledons in UNDERSOWN CEREALS [1]. Perennial dicotyledons in AMENITY GRASS [3]. Redshank in BARLEY, OATS, WHEAT [1, 2]. Redshank in UNDERSOWN CEREALS [1].

Notes

Efficacy

- Best results achieved by application to young seedling weeds in active growth
- Do not spray during cold weather, if rain or frost expected, if crop wet or in drought
- Use on turf from Apr-Sep. Avoid close mowing for 3-4 d before or after treatment [3]

Crop Safety/Restrictions

- Apply to winter cereals in spring from fully tillered, leaf-sheath erect stage but before first node detectable (GS 31), to spring barley when crop has 5 leaves unfolded but before first node detectable (GS 15-31)
- Apply to winter and spring cereals from fully tillered stage before 2nd node detectable (GS 32) [1]
- Do not use on undersown crops
- Do not spray crops suffering from herbicide damage or stress
- Do not roll or harrow for 7 d before or after spraying
- Avoid drift of spray onto nearby susceptible crops
- Do not use on turf in year of sowing [3]

Special precautions/Environmental safety

- Harmful if swallowed and in contact with skin
- Harmful to fish or other aquatic life. Do not contaminate surface waters or ditches with chemical or used container [1]

Protective clothing/Label precautions

- A, C [1, 2]; P [1]
- M03, R03a, R03c, U05a, U08, U20b, C03, E01, E30a, E31b, E34 [1-3]; U19 [3]; E07 (2 wk) [2]; E07 [1, 3]; E13c [1]; E15 [2, 3]; E26 [1, 2]

Withholding period

- Keep livestock out of treated areas until foliage of any poisonous weeds such as ragwort has died and become unpalatable

Maximum Residue Levels (mg residue/kg food)

- see dichlorprop entry

217 dichlorvos

A contact and fumigant organophosphorus insecticide

Products

1 Darlingtons Dichlorvos	Sylvan	500 g/l	EC	05699
2 Luxan Dichlorvos 600	Luxan	600 g/l	EC	08297
3 Luxan Dichlorvos Aerosol 15	Luxan	120 g/l	AE	08298
4 Nuvan 500 EC	Ciba Agric.	500 g/l	EC	03861
5 Nuvan 500 EC	Novartis A H	500 g/l	EC	08590

Uses

Beetles in POULTRY HOUSES [4, 5]. Flies in POULTRY HOUSES [4, 5]. Mosquitoes in POULTRY HOUSES [4, 5]. Mushroom flies in MUSHROOM HOUSES [1]. Non-indigenous leaf miners in PROTECTED BRASSICA SEEDLINGS *(off-label)*, PROTECTED HERBS *(off-label)*, PROTECTED ORNAMENTALS *(off-label)*, PROTECTED VEGETABLES *(off-label)* [4, 5]. Poultry ectoparasites in POULTRY HOUSES [4, 5]. Western flower thrips in PROTECTED BRASSICA SEEDLINGS *(off-label)*, PROTECTED HERBS *(off-label)*, PROTECTED ORNAMENTALS *(off-label)*, PROTECTED VEGETABLES *(off-label)* [4, 5]. Western flower thrips in PROTECTED CUCUMBERS [2, 3].

Notes

Efficacy

- Spray walls, roof, floor and ventilators of mushroom houses. Avoid spraying mushroom beds. Use as an aerosol spray or wet spray [1]
- For use in poultry houses apply as surface spray, cold fog or mist using suitable applicator. Direct away from poultry. See label for details [4, 5]
- Product mainly active in vapour phase and is inactive on insect eggs. Best results obtained from programme of treatments on dry crops when temperatures 15-25°C [2, 3]

Crop Safety/Restrictions

- Do not use on chrysanthemums in flower or on roses. Phytotoxicity may occur on cucumbers and other cucurbits
- Do not allow temperature to drop after treatment as condensation may lead to phytotoxicity. Inspect for phytotoxicity before retreatment [2, 3]
- Treatment of cucumbers at flowering may cause flower or fruit abortion [2, 3]
- Do not apply directly to growing cucumbers; apply between rows [2, 3]
- Do not apply other pesticides with product or one day before or after treatment, especially sulfur based formulations [2, 3]

Special precautions/Environmental safety

- Dichlorvos is subject to the Poisons Rules 1982 and the Poisons Act 1972. See notes in Section 1
- This product contains an anticholinesterase organophosphorus compound. Do not use if under medical advice not to work with such compounds
- Toxic in contact with skin or if swallowed
- Very toxic by inhalation
- May cause sensitisation by skin contact
- Flammable, corrosive
- Keep in original container, tightly closed, in a safe place, under lock and key
- Keep unprotected persons out of treated areas for at least 12 h
- Harmful to game, wild birds and animals

- Extremely dangerous to bees. Do not apply to crops in flower or to those in which bees are actively foraging. Do not apply when flowering weeds are present
- Extremely dangerous to fish or other aquatic life. Do not contaminate surface waters or ditches with chemical or used container
- Do not enter treated areas within 12 h and thoroughly ventilate before doing so [2, 3]
- To be used only by operators trained in the use of the chemical and familiar with the precautionary measures to be observed

Protective clothing/Label precautions
- A [1-5]; C [1, 2, 4, 5]; D [2, 3]; H [1-5]; J, K, M [2, 3]
- M01, M04, R02a, R02c, U02, U04a, U05a, U08, U13, U19, U20a, C03, E01, E10b, E12b, E13a, E30b, E34 [1-5]; R01b, R04e, R05, U10, U11, E02 (12 h) [2, 3]; R02b, C02, E02 [1, 4, 5]; R07d, E31b [1, 2, 4, 5]

Latest application/Harvest Interval(HI)
- HI Brussels sprouts, cabbage 14 wk; cauliflower 9 wk; broccoli, calabrese 7 wk; kohlrabi 1 wk; cucurbits 48 h; other edible crops 24 h

Approval
- May be applied as a ULV spray or as a fog (see label for details and for protective clothing requirements for fogging treatment) [1, 4]
- Off-label Approval unlimited for high volume application to protected vegetables for control of western flower thrips (OLA 1590/94)[4], (OLA 0377/95)[1]; unlimited for high volume application to protected non-edible ornamentals and protected vegetables for control of western flower thrips (OLA 0882/92)[4]; unlimited for use as a fog in a range of protected non-edible ornamentals and protected edible crops for control of western flower thrips and non-indigenous leaf miners. See approval notices for details (OLA 1387/93)[4]

Maximum Residue Levels (mg residue/kg food)
- lettuce 1; carrots, horseradish, parsnips, parsley root, salsify, swedes, turnips, garlic, onions, shallots, tomatoes, peppers, aubergines, cucumbers, gherkins, courgettes, cauliflowers, Brussels sprouts, head cabbages, beans (with pods), peas (with pods), celery, leeks, rhubarb, cultivated mushrooms, potatoes 0.5; citrus fruits, pome fruits, apricots, peaches, nectarines, plums, grapes, cane fruits, bilberries, cranberries, currants, gooseberries, bananas 0.1; meat, eggs 0.05; milk 0.02

218 diclofop-methyl

A translocated phenoxypropionic herbicide for grass weed control

Products Hoegrass | AgrEvo | 378 g/l | EC | 07323

Uses Annual grasses in CABBAGES, DURUM WHEAT, DWARF BEANS, LINSEED, LUPINS, MUSTARD, ONIONS, PEAS, SPRING BARLEY, SPRING OILSEED RAPE, SPRING WHEAT, SUGAR BEET, TRITICALE, WINTER BARLEY, WINTER FIELD BEANS, WINTER OILSEED RAPE, WINTER RYE, WINTER WHEAT. Awned canary grass in CABBAGES, DURUM WHEAT, DWARF BEANS, LINSEED, LUPINS, MUSTARD, ONIONS, PEAS, SPRING BARLEY, SPRING OILSEED RAPE, SPRING WHEAT, SUGAR BEET, TRITICALE, WINTER BARLEY, WINTER FIELD BEANS, WINTER OILSEED RAPE, WINTER RYE, WINTER WHEAT. Blackgrass in CABBAGES, DURUM WHEAT, DWARF BEANS, LINSEED, LUPINS, MUSTARD, ONIONS, PEAS, SPRING BARLEY, SPRING OILSEED RAPE, SPRING WHEAT, SUGAR BEET, TRITICALE, WINTER BARLEY, WINTER FIELD BEANS, WINTER OILSEED RAPE, WINTER RYE, WINTER WHEAT. Rough meadow grass in CABBAGES, DURUM WHEAT, DWARF BEANS,

LINSEED, LUPINS, MUSTARD, ONIONS, PEAS, SPRING BARLEY, SPRING OILSEED RAPE, SPRING WHEAT, SUGAR BEET, TRITICALE, WINTER BARLEY, WINTER FIELD BEANS, WINTER OILSEED RAPE, WINTER RYE, WINTER WHEAT. Ryegrass in CABBAGES, DURUM WHEAT, DWARF BEANS, LINSEED, LUPINS, MUSTARD, ONIONS, PEAS, SPRING BARLEY, SPRING OILSEED RAPE, SPRING WHEAT, SUGAR BEET, TRITICALE, WINTER BARLEY, WINTER FIELD BEANS, WINTER OILSEED RAPE, WINTER RYE, WINTER WHEAT. Wild oats in CABBAGES, DURUM WHEAT, DWARF BEANS, LINSEED, LUPINS, MUSTARD, ONIONS, PEAS, SPRING BARLEY, SPRING OILSEED RAPE, SPRING WHEAT, SUGAR BEET, TRITICALE, WINTER BARLEY, WINTER FIELD BEANS, WINTER OILSEED RAPE, WINTER RYE, WINTER WHEAT. Yorkshire fog in CABBAGES, DURUM WHEAT, DWARF BEANS, LINSEED, LUPINS, MUSTARD, ONIONS, PEAS, SPRING BARLEY, SPRING OILSEED RAPE, SPRING WHEAT, SUGAR BEET, TRITICALE, WINTER BARLEY, WINTER FIELD BEANS, WINTER OILSEED RAPE, WINTER RYE, WINTER WHEAT.

Notes

Efficacy

- Best results when applied in autumn, winter or spring to seedling grasses up to 3-4 leaf stage. See label for details of rates and timing. Annual meadow-grass not controlled
- Do not spray if foliage wet or covered in ice. Effectiveness reduced by dry conditions
- Spray is rainfast within 1 h of application
- Do not spray after a sudden drop in temperature or a period of warm days/cold nights
- Recommended for tank mixture with a range of herbicides, growth regulators and other pesticides. See label for details
- Do not mix with hormone herbicides

Crop Safety/Restrictions

- Maximum number of treatments 2 per crop (up to a maximum of 4.5 l/ha) for cereals; 1 per crop for broad-leaved crops
- Apply to wheat, rye and triticale at any time after crop emergence
- Apply to winter barley from emergence to no later than leaf sheath erect (GS 07-30), to spring barley from emergence to 4-fully expanded leaves and 2-tillers (GS 07-22)
- Apply to broad-leaved crops from 100% emergence (cotyledons fully expanded), to onions from 1-true leaf onward (not at crook stage)
- In Scotland and Northern England (north of North Yorkshire) do not treat spring cereals; winter cereals may only be treated until the end of Feb and only at the low rate
- Do not use on cereal crops undersown with grasses
- Do not roll or harrow cereals within 7 d of spraying
- Do not spray crops under stress, suffering from drought or waterlogging or if soil compacted

Special precautions/Environmental safety

- Flammable
- Dangerous to fish or other aquatic life. Do not contaminate surface waters or ditches with chemical or used container

Protective clothing/Label precautions

- A, C, H
- R07d, U08, U20a, C02, E13b, E30a, E31b

Withholding period

- Keep livestock out of treated areas for at least 7 d

Latest application/Harvest Interval(HI)

- Before first node detectable (GS 31) for winter barley; before 5 expanded leaves and 3 tillers (GS 15, 23) for spring barley.
- HI wheat, rye, triticale, broad-leaved crops 6 wk

219 diclofop-methyl + fenoxaprop-P-ethyl

A foliar acting herbicide mixture for grass control in wheat and barley

Products

1 Corniche	AgrEvo	250:20 g/l	EW	08947
2 Tigress Ultra	AgrEvo	250:20 g/l	EW	08946

Uses

Blackgrass in SPRING BARLEY, WINTER BARLEY [1, 2]. Blackgrass in WINTER WHEAT [1]. Ryegrass in SPRING BARLEY, WINTER BARLEY [1, 2]. Ryegrass in WINTER WHEAT [1]. Wild oats in SPRING BARLEY, WINTER BARLEY [1, 2]. Wild oats in WINTER WHEAT [1].

Notes

Efficacy

- Apply from emergence of crop up to before second node detectable (GS 32)
- Wild oats and blackgrass controlled from 2 fully expanded leaves to end of tillering but before 1st node detectable
- Ryegrass from seed controlled from 2 fully expanded leaves up to 3 tillers
- Spray is rainfast from 1 h after spraying. Do not apply to wet or icy foliage
- Can be sprayed in frosty weather provided crop hardened off
- Any conditions resulting in moisture stress may reduce effectiveness especially with later applications to spring barley
- Performance not affected by soils with high OM content or high Kd factor

Crop Safety/Restrictions

- Maximum number of treatments 1 per crop
- Do not apply to undersown crops or those to be undersown
- Broadcast crops should be sprayed post-emergence after root system well established
- Do not roll or harrow within 1 wk
- Do not spray crops under stress from drought, waterlogging etc
- Do not spray immediately before or after a sudden drop in temperature, a period of warm days/cold nights or when extremely low temperatures forecast
- Applications may be followed by transient leaf discoloration

Special precautions/Environmental safety

- Irritating to eyes
- Dangerous to fish or other aquatic life. Do not contaminate surface waters or ditches with chemical or used containers
- Interval of 7 d must elapse before or after application of any other product
- Must not be mixed with weedkillers containing hormones or bifenox

Protective clothing/Label precautions

- A, C, H
- R04a, U05a, U20b, C02 (6 wk), C03, E01, E07 (7 d), E13b, E26, E30a, E31b

Withholding period

- Keep livestock out of treated areas for at least 7 d

Latest application/Harvest Interval(HI)
- Before flag leaf sheath erect (GS 41).
- HI 6 wk

220 dicloran

A protectant nitroaniline fungicide used as a glasshouse fumigant

Products

Fumite Dicloran Smoke	Hortichem	40% w/w	FU	00930

Uses

Botrytis in PROTECTED LETTUCE, PROTECTED TOMATOES. Rhizoctonia in PROTECTED LETTUCE, PROTECTED TOMATOES.

Notes

Efficacy
- Treat at first sign of disease and at 14 d intervals if necessary
- For best results fumigate in late afternoon or evening when temperature is at least 16°C. Keep house closed overnight or for at least 4 h. Do not fumigate in bright sunshine, windy conditions or when temperature too high
- Water plants, paths and straw used as mulches several hours before fumigating. Ensure that foliage is dry before treatment

Crop Safety/Restrictions
- Do not fumigate young seedlings or plants being hardened off
- Treat only crops showing strong growth. Do not treat crops which are dry at the roots

Special precautions/Environmental safety
- Irritating to eyes and respiratory system
- Ventilate glasshouse thoroughly before re-entering
- Harmful to fish or other aquatic life. Do not contaminate surface waters or ditches with chemical or used container

Protective clothing/Label precautions
- R04a, R04c, U05a, U19, U20a, C02, C03, E13c, E30a, E32a

Latest application/Harvest Interval(HI)
- HI tomatoes 2 d; lettuce 14 d

221 dicofol

A non-systemic organochlorine acaricide for horticultural use

Products

Kelthane	Rohm & Haas	18.5% w/w	EC	07449

Uses

Red spider mites in APPLES, HOPS, STRAWBERRIES. Tarsonemid mites in STRAWBERRIES.

Notes

Efficacy

- Spray apples when winter eggs hatched and summer eggs laid and repeat 3 wk later if necessary. Spray other crops at any time before stated harvest interval
- Strawberries may be sprayed after picking where necessary
- Apply fumigant treatment at first sign of infestation and repeat every 14 d
- Do not fumigate in bright sunshine, when foliage wet or roots dry
- Resistance has developed in some areas, in which case sprays are unlikely to give satisfactory control, especially after second or subsequent application

Crop Safety/Restrictions

- Maximum number of treatments 2 per yr for apples and hops, 2 per crop for strawberries, protected cucumbers and tomatoes [1]
- Do not spray apples within 4 wk of petal fall
- Do not spray cucumber or tomato seedlings before mid-May or in bright sunshine

Special precautions/Environmental safety

- Harmful in contact with skin and if swallowed
- Irritating to eyes and skin
- May cause sensitization by skin contact
- Flammable

Protective clothing/Label precautions

- A, C, H
- M03, R03a, R03c, R04a, R04b, R04e, R07d, U02, U05a, U09a, U19, U20b, C03, E01, E15, E30a, E31a, E34

Latest application/Harvest Interval(HI)

- HI tomatoes, cucumbers 2 d [1]; strawberries 7 d [1]; apples 2 wk [1]; hops 4 wk [1]

Maximum Residue Levels (mg residue/kg food)

- hops 50; apricots, peaches, nectarines, plums, currants, garlic, cultivated mushrooms 5; citrus fruit, strawberries, bananas 2; pome fruit, grapes 1; tomatoes, peppers, cucurbits, legume vegetables 0.5; cotton seed 0.1; tree nuts, oilseeds (except cotton seed), eggs 0.05; cane fruit, bilberries, cranberries, gooseberries, wild berries, miscellaneous fruit (except bananas), root and tuber vegetables, bulb vegetables, aubergines, sweet corn, brassica vegetables, leaf vegetables and fresh herbs, peas (without pods), stem vegetables, fungi, pulses, potatoes, cereals 0.02

222 dicofol + tetradifon

A contact acaricide mixture for fruit and glasshouse crops

Products

Childion	Hortichem	170:62.5 g/l	EC	03821

Uses

Broad mite in PROTECTED CUCUMBERS, PROTECTED ORNAMENTALS, PROTECTED TOMATOES. Bryobia mites in APPLES, PEARS. Leaf and bud mite in APPLES, PEARS. Red spider mites in APPLES, BLACKCURRANTS, HOPS, PEARS, PROTECTED CUCUMBERS, PROTECTED ORNAMENTALS, PROTECTED TOMATOES, STRAWBERRIES. Tarsonemid mites in PROTECTED CUCUMBERS, PROTECTED ORNAMENTALS, PROTECTED TOMATOES, STRAWBERRIES.

Notes

Efficacy

- Apply to hops and glasshouse crops as soon as mites appear. Other crops should be sprayed post blossom, strawberries and blackcurrants being treated only if mites appear. Repeat as necessary in all cases

Crop Safety/Restrictions

- Maximum number of treatments 2 per yr for apples, pears; 2 per yr pre-harvest, 1 per yr post-harvest for strawberries, blackcurrants
- A minimum of 3 wk must elapse between treatments for tomatoes, cucumbers, ornamentals
- Do not apply to apples or pears until 3 wk after petal fall, to strawberries during blossoming or to young plants or cucumbers in bright sunshine or before mid-May
- Treat small numbers of new glasshouse subjects before committing whole batches
- Do not apply to cissus, dahlias, ficus, gloxinias, impatiens, kalanchoes, primulas or stephanotis

Special precautions/Environmental safety

- Harmful in contact with skin or if swallowed. Irritating to eyes and skin
- May cause sensitization by skin contact
- Flammable
- Must not be applied by hand-held sprayer

Protective clothing/Label precautions

- A, C
- M03, R03a, R03c, R04a, R04b, R04e, R07d, U05a, U08, U20a, C03, E01, E15, E30a, E31a, E34

Latest application/Harvest Interval(HI)

- HI apples, pears, hops 28 d; blackcurrants 14 d; strawberry 7 d; glasshouse tomatoes and cucumbers 2 d

Maximum Residue Levels (mg residue/kg food)

- see dicofol entry

223 dieldrin

A persistent organochlorine insecticide, all approvals for which were revoked in 1989. Dieldrin is an environmental hazard.

224 difenacoum

An anticoagulant coumarin rodenticide

See also calciferol + difenacoum

Products

1 Deosan Rataway	Deosan	0.005% w/w	RB	05560
2 Deosan Rataway Bait Bags	Deosan	0.005% w/w	RB	05562
3 Killgerm Rat Rods	Killgerm	0.005% w/w	RB	05154
4 Killgerm Wax Bait	Killgerm	0.005% w/w	RB	04096
5 Neokil	Sorex	0.005% w/w	RB	05564
6 Neosorexa	Sorex	0.005% w/w	RB	07756
7 Neosorexa Ratpacks	Sorex	0.005% w/w	RB	04653
8 Ratak	Scotts	0.005% w/w	RB	06832
9 Ratak Wax Blocks	Scotts	0.005% w/w	RB	06829

Uses

Mice in FARM BUILDINGS, FARMYARDS. Rats in FARM BUILDINGS, FARMYARDS.

Notes

Efficacy

- Chemical is a chronic poison and rodents need to feed several times before accumulating a lethal dose. Effective against rodents resistant to other commonly used anticoagulants
- Use ready-to-use baits in baiting programme
- Lay small baits about 1 m apart throughout infested areas for mice, larger baits for rats near holes and along runs; place blocks 5-10 m apart depending on severity of infestation
- Cover baits by placing in bait boxes, drain pipes or under boards
- Inspect bait sites frequently and top up as long as there is evidence of feeding

Special precautions/Environmental safety

- Only for use by farmers, horticulturists and other professional users
- Cover bait to prevent access by children, animals or birds

Protective clothing/Label precautions

- M05 [5, 8, 9]; U13, V01a, V02, V03a, V04a [1-9]; U20a [1, 2, 7]; U20b [3-6, 8, 9]; E30a [1-7]; E30b [8, 9]; E32a [1-5, 7-9]

225 difenoconazole

A diphenyl-ether triazole protectant and curative fungicide

Products

1 Plover	Ciba Agric.	250 g/l	EC	07232
2 Plover	Novartis	250 g/l	EC	08429

Uses

Alternaria in CABBAGES, OILSEED RAPE [1, 2]. Alternaria in BRUSSELS SPROUTS, CALABRESE, CAULIFLOWERS [2]. Brown rust in WINTER WHEAT [1, 2]. Disease control/foliar feed in FLOWERS *(off-label)*, PROTECTED FLOWERS *(off-label)* [1, 2]. Light leaf spot in OILSEED RAPE [1, 2]. Ring spot in CABBAGES [1, 2]. Ring spot in BRUSSELS SPROUTS, CALABRESE, CAULIFLOWERS [2]. Septoria in WINTER WHEAT [1, 2]. Stem canker in OILSEED RAPE [1, 2]. Yellow rust in WINTER WHEAT [1, 2].

Notes

Efficacy

- For most effective control of Septoria, apply as part of a programme of sprays which includes a suitable flag leaf treatment

- Adequate control of yellow rust may require an earlier appropriate treatment
- Treat oilseed rape in autumn from 6 expanded true leaf stage (GS 1,6). A repeat spray may be made in spring at the beginning of stem extension (GS 2,0) if visible symptoms develop
- Improved control of established infections on oilseed rape achieved by mixture with carbendazim. See label
- In cabbages a 3-spray programme should be used starting at the first sign of disease and repeated at 14-21 d intervals
- Product is fully rainfast 2 h after application

Crop Safety/Restrictions

- Maximum number of treatments 3 per crop for cabbages; 2 per crop for oilseed rape; 1 per crop for wheat
- Apply to wheat any time from ear fully emerged stage but before early milk-ripe stage (GS 59-73)
- Recommended for use on all varieties of winter and spring sown oilseed rape and cabbage

Special precautions/Environmental safety

- Irritating to eyes
- May cause sensitization by skin contact
- Harmful to fish or other aquatic life. Do not contaminate surface waters or ditches with chemical or used container
- Do not allow direct spray from ground-based vehicle-mounted/drawn sprayers to fall within 6 m, or from hand-held sprayers to within 2 m, of surface waters or ditches. Direct spray away from water
- May have adverse effect on ladybirds and parasites of insects

Protective clothing/Label precautions

- A, C, H
- R04a, R04e, U05a, U09a, U20b, C03, E01, E13c, E26, E29, E30a, E31a

Latest application/Harvest Interval(HI)

- Before grain early milk-ripe stage (GS 73)

Approval

- Off-label approval unlimited for use on container grown hybrid pinks, sweet williams (OLA 2268/96)[1]

226 difenzoquat

A post-emergence quaternary ammonium herbicide for wild oat control

Products	Avenge 2	Cyanamid	150 g/l	SL	03241

Uses Wild oats in BARLEY, DURUM WHEAT, MAIZE, RYE, RYEGRASS SEED CROPS, TRITICALE, WHEAT.

FOR FULL CONDITIONS OF USE ALWAYS READ THE PRODUCT LABEL

Notes

Efficacy

- Autumn and winter spraying controls wild oats from 2-leaf stage to mid-tillering, spring treatment to end of tillering
- Dose rate varies with season and growth-stage of weed. See label for details
- Spring treatment on barley provides control of powdery mildew as well as wild oats
- Do not apply if rain expected within 6 h
- Recommended for tank mixture with a range of herbicides, growth regulators and other pesticides. See label for details. Other products can be applied after 24 h

Crop Safety/Restrictions

- Maximum number of treatments 2 per crop for winter cereals, durum wheat and ryegrass seed crops; 1 per crop for spring cereals and maize
- Apply to named varieties of winter and spring wheat, durum and triticale (see label) and all varieties of other recommended crops
- Apply to crops from 2-leaf stage to 50% of plants with 3 nodes detectable (GS 12-33)
- Apply to crops undersown with ryegrass and clover either before emergence or after grass has reached 3-leaf stage
- Spray ryegrass seed crops after 3-leaf stage, maize as soon as weeds at susceptible stage
- Do not spray crops suffering stress from waterlogging, drought or other factors
- Temporary yellowing may follow application, especially under extremes of temperature, but is normally outgrown rapidly

Special precautions/Environmental safety

- Harmful if swallowed and in contact with skin
- Irritating to skin
- Risk of serious damage to eyes
- Harmful to fish or other aquatic life. Do not contaminate surface waters or ditches with chemical or used container
- Harmful to livestock

Protective clothing/Label precautions

- A, C
- M03, R03a, R03c, R04b, R04d, U05a, U08, U14, U15, U19, U20a, C03, E01, E06b, E13c, E30a, E31a, E34

Withholding period

- Keep all livestock out of treated areas for at least 6 wk

Latest application/Harvest Interval(HI)

- Before flag leaf just visible (GS 39) for cereals and maize; 6 wk before grazing for ryegrass seed crops

227 diflubenzuron

A selective, persistent, contact and stomach acting insecticide

Products

1 Dimilin Flo	Zeneca	480 g/l	SC	07151
2 Dimilin Flo	Uniroyal	480 g/l	SC	08769

Uses

Browntail moth in AMENITY TREES AND SHRUBS, HEDGES, NURSERY STOCK, ORNAMENTALS [1]. Bud moth in APPLES, PEARS [1]. Carnation tortrix moth in AMENITY TREES AND SHRUBS, HEDGES, NURSERY STOCK, ORNAMENTALS [1]. Caterpillars in BROCCOLI, BRUSSELS SPROUTS, CABBAGES, CALABRESE, CAULIFLOWERS [1].

Clouded drab moth in APPLES, PEARS [1]. Codling moth in APPLES, PEARS [1]. Earwigs in APPLES, BLACKCURRANTS, PEARS [1]. Fruit tree tortrix moth in APPLES, PEARS [1]. Lackey moth in AMENITY TREES AND SHRUBS, HEDGES, NURSERY STOCK, ORNAMENTALS [1]. Oak leaf roller moth in FORESTRY [1]. Pear sucker in PEARS [1]. Pine beauty moth in FORESTRY [1]. Pine looper in FORESTRY [1]. Plum fruit moth in PLUMS [1]. Rust mite in APPLES, PEARS, PLUMS [1]. Sciarid flies in MUSHROOMS [2]. Small ermine moth in AMENITY TREES AND SHRUBS, HEDGES, NURSERY STOCK, ORNAMENTALS [1]. Tortrix moths in PLUMS [1]. Winter moth in AMENITY TREES AND SHRUBS, APPLES, BLACKCURRANTS, FORESTRY, HEDGES, NURSERY STOCK, ORNAMENTALS, PEARS, PLUMS [1].

Notes

Efficacy

- Acts by disrupting chitin synthesis and preventing hatching of eggs. Most active on young caterpillars and most effective control achieved by spraying as eggs start to hatch [1]
- Dose and timing of spray treatments vary with pest and crop. See label for details [1]
- Addition of wetter recommended for use on brassicas and pears [1]
- Where pear suckers resistant to diflubenzuron occur control is unlikely to be satisfactory [1]
- Apply as casing mixing treatment on mushrooms or as post-casing drench [2]

Crop Safety/Restrictions

- Maximum number of treatments 3 per yr for apples, pears; 2 per yr for plums, blackcurrants; 2 per crop for brassicas [1]; 1 per spawning [2]
- Consult supplier regarding range of ornamentals which can be treated and check for varietal differences on a small sample in first instance [1]
- Do not use as a compost drench or incorporated treatment on ornamental crops [1]
- Do not spray protected plants in flower or with flower buds showing colour [1]

Special precautions/Environmental safety

- Do not contaminate surface waters or ditches with chemical or used container
- Do not apply directly to livestock [1]
- Product has negligible effect on many beneficial insects (see label) and may therefore be used with biological control agents and in integrated control programmes [1]

Protective clothing/Label precautions

- U20c, E15, E30a [1, 2]; E05, E32a [1]

Latest application/Harvest Interval(HI)

- HI apples, pears, plums, blackcurrants, brassicas 14 d [1]

Approval

- Approved for aerial application in forestry when average wind velocity does not exceed 18 knots and gusts do not exceed 20 knots [1]. See notes in Section 1

Maximum Residue Levels (mg residue/kg food)

- citrus fruits, pome fruits, plums, tomatoes, peppers, aubergines, Brussels sprouts, head cabbages 1; cultivated mushrooms 0.1; meat, milk, eggs 0.05

228 diflufenican

A shoot absorbed anilide herbicide available only in mixtures

See also bromoxynil + diflufenican + ioxynil
clodinafop-propargyl + diflufenican
clopyralid + diflufenican + MCPA

229 diflufenican + flurtamone + isoproturon

A contact and residual herbicide for winter cereals

Products

Ingot	RP Agric.	27:67:400 g/l	SC	08321

Uses

Annual dicotyledons in WINTER BARLEY, WINTER WHEAT. Annual meadow grass in WINTER BARLEY, WINTER WHEAT. Blackgrass in WINTER BARLEY, WINTER WHEAT. Loose silky bent in WINTER BARLEY, WINTER WHEAT.

Notes

Efficacy

- Apply from when crop has first leaf unfolded before susceptible weeds pass recommended size
- Best results obtained on firm, fine seedbeds with adequate soil moisture
- Speed of control depends on weather conditions and activity will be reduced during prolonged dry periods especially in spring. Weed control, especially of grasses, may also be reduced in wet seasons or where heavy rain falls shortly after treatment
- Do not treat on soils with more than 10% organic matter

Crop Safety/Restrictions

- Maximum number of treatments 1 per crop. Maximum total dose of isoproturon 2.5 kg a.i./ha per crop
- Crops should be drilled to a normal depth of 25 mm and the seed well covered. Do not treat broadcast crops
- Do not apply pre-emergence to wheat or barley
- Do not use on crops being grown for seed or crops undersown or to be undersown
- Do not treat frosted crops or when frost is imminent. Severe frost after application may cause transient discoloration or scorch
- Do not use on Sands or Very Light soils or those that are very stony or gravelly
- Do not harrow at any time after application
- Take particular care to match spray swaths otherwise crop discoloration and biomass reduction may occur which may lead to yield reduction
- Do not broadcast oilseed rape or other brassica crops as a following crop on treated land. See label for details of time and land preparation before sowing other crops in the event of failure of a treated crop
- Successive treatments of any products containing diflufenican can lead to soil build-up and inversion ploughing must precede sowing any following non-cereal crop. Even where ploughing occurs some crops may be damaged

Special precautions/Environmental safety

- Dangerous to fish or other aquatic life. Do not contaminate surface waters or ditches with chemical or used container
- Do not allow direct spray from vehicle mounted/drawn hydraulic sprayers to fall within 6 m, or from hand-held sprayers to within 2 m, of surface waters or ditches. Direct spray away from water

- Do not spray where soils are cracked, to avoid run-off through drains

Protective clothing/Label precautions
- A, C, H
- U20c, E13b, E16, E26, E30a, E31b

Latest application/Harvest Interval(HI)
- Before 2nd node detectable (GS32)

230 diflufenican + isoproturon

A contact and residual herbicide for use in winter cereals

Products

1 Grenadier	RP Agric.	41.7:500 g/l	SC	08136
2 Javelin	RP Agric.	62.5:500 g/l	SC	06192
3 Javelin Gold	RP Agric.	20:500 g/l	SC	06200
4 Landgold DFF 625	Landgold	62.5:500 g/l	SC	06274
5 Panther	RP Agric.	50:500 g/l	SC	06491
6 Standon Diflufenican-IPU	Standon	50:500 g/l	SC	09175
7 Tolkan Turbo	RP Agric.	20:500 g/l	SC	06795

Uses

Annual dicotyledons in WINTER BARLEY, WINTER WHEAT [1-7]. Annual dicotyledons in TRITICALE, WINTER RYE [1-3, 5-7]. Annual grasses in WINTER BARLEY, WINTER WHEAT [1-7]. Annual grasses in TRITICALE, WINTER RYE [1-3, 5-7]. Blackgrass in WINTER BARLEY, WINTER WHEAT [1-7]. Blackgrass in TRITICALE, WINTER RYE [1-3, 5-7]. Wild oats in WINTER BARLEY, WINTER WHEAT [1-7]. Wild oats in TRITICALE, WINTER RYE [1-3, 5-7].

Notes

Efficacy
- May be applied in autumn or spring (but only post-emergence on wheat or barley). Best control normally achieved by early post-emergence treatment
- Best results by application to fine, firm seedbed moist at or after application
- Weeds controlled from before emergence to 6 true leaf stage
- Apply to moist, but not waterlogged, soils
- Any trash or ash should be buried during seedbed preparation

Crop Safety/Restrictions
- Maximum number of treatments 1 per crop. Maximum total dose of isoproturon 2.5 kg a.i./ha per yr
- Spray winter wheat and barley post-emergence before end Feb [2-4], before crop reaches second node detectable stage (GS 32) [1, 5-7]
- On triticale and winter rye only treat named varieties and apply as pre-emergence spray
- Drill crop to normal depth (25 mm) and ensure seed well covered
- Do not use on other cereals, broadcast or undersown crops or crops to be undersown
- Do not use on Sands or Very Light soils, or those that are very stony or gravelly or on soils with more than 10% organic matter
- Do not spray when heavy rain is forecast or on crops suffering from stress, frost, deficiency, pest or disease attack
- Do not harrow after application nor roll autumn-treated crops until spring

- See label for details of time and land preparation needed before sowing other crops (12 wk for most crops). In the event of crop failure winter wheat can be redrilled without ploughing, other crops only after ploughing - see label for details
- Successive treatments of any products containing diflufenican can lead to soil build up and inversion ploughing must precede sowing any non-cereal crop. Even where ploughing occurs some crops may be damaged

Special precautions/Environmental safety
- Irritant. May cause sensitization by skin contact
- Harmful (dangerous [1, 2, 4, 6, 7]) to fish or other aquatic life. Do not contaminate surface waters or ditches with chemical or used container [3, 5]
- Do not spray where soils are cracked, to avoid run-off through drains

Protective clothing/Label precautions
- A, C, H [1, 2, 4, 7]
- R04, R04e [1]; U08, U19 [6]; U20b [3-6]; U20c [1, 2, 7]; E13b [1, 2, 4, 6, 7]; E13c [3, 5]; E26, E30a, E31b [1-7]

Latest application/Harvest Interval(HI)
- Before second node detectable (GS 32) [1, 3, 5-7], before end Feb [2, 4], for wheat and barley; pre-emergence of crop for triticale and rye

231 diflufenican + terbuthylazine

A foliar and residual acting herbicide

Products

1 Bolero	Ciba Agric.	200:400 g/l	SC	07436
2 Bolero	Novartis	200:400	SC	08392

Uses

Annual dicotyledons in WINTER BARLEY, WINTER WHEAT. Annual meadow grass in WINTER BARLEY, WINTER WHEAT.

Notes

Efficacy
- Apply to healthy crops from when crop has 1 leaf unfolded (GS 11) up to the end of tillering. Best results obtained from applications made early post-emergence of weeds
- Product is foliar and soil acting. Weeds germinating after treatment are controlled by root uptake
- Prolonged dry weather or heavy rain after treatment may reduce control
- Weed control may be reduced by presence of high levels of organic matter (over 10%) or surface trash

Crop Safety/Restrictions
- Maximum number of treatments 1 per crop
- Ensure seed covered by 2.5 cm consolidated soil before spraying
- Do not use on crops under stress or suffering from water-logging, pest attack, disease or frost. Do not spray when frost imminent
- Do not apply to undersown crops or those to be undersown
- Sep drilled crops may be damaged if treatment precedes or coincides with a period of rapid growth
- Do not use on Sand or Very Light soils. On stony or gravelly soils crop damage may occur if heavy rain follows treatment

- After harvest treated soils should be inversion ploughed to 15 cm before drilling field beans, leaf brassicas, winter oilseed rape, winter onions or sugar beet seed crops. Autumn sown cereals can be drilled as normal
- Restrictions apply to crops that may be sown after failure of a treated crop. See label for details of crops, cultivations and intervals
- Successive treatments of any products containing diflufenican can lead to soil build up and inversion ploughing must precede sowing any non-cereal crop. Even where ploughing occurs some crops may be damaged - see label

Special precautions/Environmental safety
- Harmful in contact with skin and if swallowed
- Harmful to fish or other aquatic life. Do not contaminate surface waters or ditches with chemical or used container

Protective clothing/Label precautions
- A
- R03a, R03c, U08, U20a, E13c, E26, E30a, E31b, E34

Latest application/Harvest Interval(HI)
- Before ear at 1 cm stage (GS 30)

232 diflufenican + trifluralin

A contact and residual herbicide for use in winter cereals

Products

Ardent	RP Agric.	40:400 g/l	SC	06203

Uses

Annual dicotyledons in TRITICALE, WINTER BARLEY, WINTER RYE, WINTER WHEAT. Annual meadow grass in WINTER BARLEY, WINTER WHEAT.

Notes

Efficacy
- Apply pre- or early post-emergence of crop up to maximum recommended weed size
- Weeds controlled pre-emergence up to 4 leaves (2 leaves for annual meadow grass)
- Best results achieved on firm, fine moist seedbeds
- Rolling after autumn treatment will reduce weed control. Roll in spring if necessary
- Speed of control depends on temperature and growing conditions; activity can be slow under cool conditions
- Do not use on soils with more than 10% organic matter; this may include newly ploughed grassland for a time

Crop Safety/Restrictions
- Maximum number of treatments 1 per crop
- Ensure that crop is evenly drilled and seed well covered. Do not treat broadcast crops or those that are frosted or stressed
- Do not treat undersown crops or those to be undersown
- Do not harrow after application
- Do not use on oats or any spring sown cereals
- Do not use on Sands or Very Light soils, or those that are very stony or gravelly

- Crops occasionally show transient leaf discoloration after treatment. Symptoms are quickly outgrown and yield not affected
- See label for details of time and land preparation needed in the event of crop failure and for normal subsequent cropping
- Successive treatments of any products containing diflufenican can lead to soil build-up and inversion ploughing must precede sowing any following non-cereal crop. Even where ploughing occurs some crops may be damaged

Special precautions/Environmental safety

- Irritating to eyes
- Harmful to fish or other aquatic life. Do not contaminate surface waters or ditches with chemical or used container

Protective clothing/Label precautions

- A, C
- R04a, U05a, U08, U13, U20b, C03, E01, E13c, E26, E30a, E31b

Latest application/Harvest Interval(HI)

- Before 2nd tiller stage (GS 22), or end Nov, whichever is sooner

233 di-1-p-menthene

An antitranspirant and coating agent available only in mixtures

See also chlormequat + di-1-p-menthene

234 dimethoate

A contact and systemic organophosphorus insecticide and acaricide

See also chlorpyrifos + dimethoate

Products

1 Barclay Dimethosect	Barclay	400 g/l	EC	08538
2 BASF Dimethoate 40	BASF	400 g/l	EC	00199
3 Danadim Dimethoate 40	Cheminova	400 g/l	EC	07351
4 P A Dimethoate 40	Portman	400 g/l	EC	01527
5 Portman Sysdim 40	Portman	400 g/l	EC	06902
6 Rogor L40	Isagro	400 g/l	EC	07611

Uses

Aphids in SUGAR BEET SEED CROPS *(excluding Myzus persicae)* [2, 6]. Aphids in GRASS SEED CROPS, SPINACH *(excluding Myzus persicae)*, WATERCRESS *(off-label)* [2]. Aphids in BEET CROPS *(excluding Myzus persicae)* [4-6]. Aphids in APPLES, CANE FRUIT, CARROTS, CHERRIES, CURRANTS, FLOWERS, GOOSEBERRIES, ORNAMENTALS, PEARS, PLUMS, PROTECTED CARNATIONS, ROSES, STRAWBERRIES, TOMATOES, WOODY ORNAMENTALS [4, 5]. Aphids in BROAD BEANS, FRENCH BEANS, RUNNER BEANS [3-5]. Aphids in MANGEL SEED CROPS *(excluding Myzus persicae)*, SPINACH, WHEAT [6]. Aphids in PEAS, POTATOES *(excluding Myzus persicae)* [1-6]. Aphids in CEREALS [1-5]. Aphids in FODDER BEET *(excluding Myzus persicae)*, MANGELS *(excluding Myzus persicae)*, RED BEET *(excluding Myzus persicae)*, SUGAR BEET *(excluding Myzus persicae)* [1-3]. Bryobia mites in APPLES, PEARS [4, 5]. Cabbage aphid in BRASSICAS [1, 3-5]. Capsids in APPLES, CANE FRUIT, GOOSEBERRIES, PEARS [4, 5]. Insect pests in BRASSICA SEED BEDS *(off-label)*, BULB ONIONS *(off-label)*, CELERIAC *(off-label)*, CELERY *(off-label)*, CHINESE CABBAGE *(off-label)*, GARLIC *(off-label)*, KALE *(off-label)*, KOHLRABI *(off-label)*, LEEKS *(off-label)*, MARROWS *(off-label)*, SALAD ONIONS *(off-label)*, SPINACH *(off-label)*, SWEDES *(off-label)*, SWEETCORN *(off-label)*, TURNIPS *(off-label)* [2]. Leaf miners in

SPINACH [2, 6]. Leaf miners in FODDER BEET, MANGELS, RED BEET, SUGAR BEET [2]. Leaf miners in FLOWERS, ORNAMENTALS, PROTECTED CARNATIONS, ROSES, WOODY ORNAMENTALS [4, 5]. Leaf miners in MANGEL SEED CROPS, SUGAR BEET SEED CROPS [6]. Leaf miners in BEET CROPS [1, 3, 6]. Mangold fly in BEET CROPS [4, 5]. Pea midge in PEAS [1-6]. Red spider mites in APPLES, CANE FRUIT, CHERRIES, CURRANTS, GOOSEBERRIES, PEARS, PLUMS, STRAWBERRIES [4, 5]. Sawflies in APPLES, PEARS, PLUMS [4, 5]. Suckers in APPLES, PEARS [4, 5]. Thrips in PEAS [1-6]. Wheat bulb fly in WHEAT [1-6].

Notes

Efficacy

- Chemical has quick knock-down effect and systemic activity lasts for up to 14 d
- With some crops, products differ in range of pests listed as controlled. Uses section above provides summary. See labels for details
- For most pests apply when pest first seen and repeat 2-3 wk later or as necessary. Timing and number of sprays varies with crop and pest. See labels for details
- Best results achieved when crop growing vigorously. Systemic activity reduced when crops suffering from drought or other stress
- In hot weather apply in early morning or late evening
- Do not tank mix with alkaline materials. See label for recommended tank-mixes
- Where aphids or spider mites resistant to organophosphorus compounds occur control is unlikely to be satisfactory and repeat treatments may result in lower levels of control
- Consult processor before spraying crops grown for processing

Crop Safety/Restrictions

- Maximum number of treatments 8 per crop for hops, tomatoes; 7 per crop for seed potatoes; 6 per crop for brassicas, peas, outdoor lettuce, strawberries, swedes, turnips; 4 per crop for cereals, grass seed crops, carrots, apples, pears, plums, cherries, gooseberries, cane fruit; 3 per crop for blackcurrants; 2 per crop for beans, ware potatoes; 1 per crop for beet crops (see also below), celery; 1 per crop for protected lettuce between Oct and Feb
- On beet crops only one treatment per crop may be made for control of leaf miners (max 84 g a.i./ha) and black bean aphid (max 420 g a.i./ha)
- On carrots the maximum total dose applied per crop must not exceed the equivalent of 3 (on mineral soils) or 4 (on organic soils) full dose applications
- In potatoes and beet crops resistant strains of peach-potato aphid (*Myzus persicae*) are common and dimethoate products must not be used to control this pest
- Test for varietal susceptibility on all unusual plants or new cultivars

Special precautions/Environmental safety

- This product contains an anticholinesterase organophosphorus compound. Do not use if under medical advice not to work with such compounds
- Harmful in contact with skin/in contact with skin or if swallowed (products vary - see labels for details)
- Irritating to eyes [1, 4, 5]
- May cause sensitisation by skin contact
- Flammable
- Harmful to game, wild birds and animals. Bury all spillages
- Harmful to livestock
- Likely to cause adverse effects on beneficial arthropods
- Dangerous to bees. Do not apply to crops in flower, except as directed on peas, or to those in which bees are actively foraging. Do not apply when flowering weeds are present

- Dangerous to fish or other aquatic life. Do not contaminate surface waters or ditches with chemical or used container
- Surface residues may also cause bee mortality following spraying
- Do not treat cereals after 1 Apr within 6 m of edge of crop
- Must not be applied to cereals if any product containing a pyrethroid insecticide or dimethoate has been sprayed after the start of ear emergence (GS 51)

Protective clothing/Label precautions
- A, C, H [1-6]; J [1-4, 6]; M
- M01, R03c, R04e, R07d, U02, U19, U20a, E13b, E31b, E34 [1-6]; M03, U08, E30a [1, 3-6]; R03a [2, 6]; R04a [1, 3-5]; R04b [1, 3]; U04a [1-3, 6]; U05a, C03, E06b (7 d), E26 [1, 2, 4-6]; U10, C02, E30b [2]; U13 [2, 4-6]; E01 [1, 4-6]; E06b [3]; E10b, E12d [1-3]; E12c, S04b [4-6]

Withholding period
- Keep all livestock out of treated areas for at least 7 d

Latest application/Harvest Interval(HI)
- Before 30 Jun in yr of harvest for potatoes and beet crops; before 31 Mar in yr of harvest for aerial application to cereals
- HI apples, pears 35 d; protected lettuce, blackcurrants 28 d; field beans, plums, cherries, raspberries, gooseberries, strawberries 21 d; cereals, grass seed crops, carrots, peas, beans, hops 14 d; watercress 10 d; brassicas, outdoor lettuce, tomatoes, celery 7 d

Approval
- Approved for aerial application on cereals, peas, ware potatoes, sugar beet [1-3, 6]. See notes in Section 1
- Off label approval to Apr 2001 for use in celery, leeks, marrows, salad onions, sweetcorn (OLA 0778/96)[2]; unlimited for use on compost for propagation of protected and outdoor watercress (OLA 0394/94)[2]; to Jul 2000 for use on watercress beds (OLA 0993/95)[2]; unlimited for use on outdoor and protected vegetables (see approval notice for details) (OLA 0389/94)[2]; to Apr 2001 for use on protected spring onions, leeks, celery, and on bulb onions, garlic, swedes, turnips (OLA 0778/96)[2]

Maximum Residue Levels (mg residue/kg food)
- citrus fruits, apricots, peaches, nectarines, plums, bilberries, cranberries, currants, gooseberries, cucumbers, gherkins, courgettes, cauliflowers, Brussels sprouts, head cabbages, lettuce, beans (with pods) 2; pome fruits, grapes, cane fruits, bananas, carrots, horseradish, parsnips, parsley root, salsify, swedes, turnips, garlic, onions, shallots, tomatoes, peppers, aubergines, peas (with pods), celery, leeks, rhubarb, cultivated mushrooms 1; potatoes 0.05

235 dimethomorph

A cinnamic acid fungicide with translaminar activity available only in mixtures

236 dimethomorph + mancozeb

A systemic and protectant fungicide for potato blight control

Products	Invader	Cyanamid	7.5:66.7% w/w	WG	06989

Uses Blight in POTATOES.

Notes

Efficacy
- Commence treatment as soon as there is a risk of blight infection
- In the absence of a warning treatment should start before the crop meets along the rows
- Repeat treatments every 10-14 d depending in the degree of infection risk
- Irrigated crops should be regarded as at high risk and treated every 10 d
- For best results good spray coverage of the foliage is essential

Crop Safety/Restrictions
- Maximum number of treatments 8 per crop

Special precautions/Environmental safety
- Irritant. May cause sensitization by skin contact
- Oxidising agent. Contact with combustible material may cause fire
- Dangerous to fish or other aquatic life. Do not contaminate surface waters or ditches with chemical or used container
- Do not allow direct spray from vehicle-mounted/drawn sprayers to fall within 6 m of surface waters or ditches. Direct spray away from water

Protective clothing/Label precautions
- A
- R04e, R08, U02, U04a, U05a, U08, U13, U19, U20a, C03, E01, E13b, E16, E30a

Latest application/Harvest Interval(HI)
- HI 7 d before harvest

Maximum Residue Levels (mg residue/kg food)
- see mancozeb entry

237 dinoseb

A dinitrophenol herbicide and dessicant, all approvals for which were revoked in 1988 because of evidence of teratogenicity and the potential danger to operators

238 diphacinone

An anticoagulant rodenticide

Products

1 Tomcat Rat and Mouse Bait	Antec	0.005%w/w	RB	07171
2 Tomcat Rat and Mouse Blox	Antec	0.005% w/w	BB	07230

Uses Mice in AGRICULTURAL PREMISES. Rats in AGRICULTURAL PREMISES.

Notes

Efficacy
- Products formulated using human food grade ingredients, flavour enhancers and paraffin

- Rodents must consume bait for 3-5 d to produce mortality after 6-15 d
- Maintain uninterrupted supply of fresh bait for 10-15 d or until signs of rodent activity cease
- Remove and replace stale, damp or mouldy bait
- Bait product [1] available loose or in sachets

Special precautions/Environmental safety
- Prevent access to the baits by children, birds and other animals, particularly dogs, cats and pigs
- Do not apply direct to livestock
- Remove exposed milk and collect eggs before application

Protective clothing/Label precautions
- M03, U13, U20a, E15, E30a, E32a, V01a, V03a, V04a

239 diquat

A non-residual bipyridyl contact herbicide and crop desiccant

Products

1 Barclay Desiquat	Barclay	200 g/l	SL	09063
2 Landgold Diquat	Landgold	200 g/l	SL	09020
3 Midstream	Miracle	100 g/l	PC	07739
4 Midstream	Scotts	100 g/l	PC	09267
5 Reglone	Zeneca	200 g/l	SL	06703
6 Standon Diquat	Standon	200 g/l	SL	05587

Uses

Annual dicotyledons in ROW CROPS [5]. Aquatic weeds in AREAS OF WATER [3-5]. Chemical stripping in HOPS [5]. Pre-harvest desiccation in BORAGE *(off-label)*, CLOVER SEED CROPS, HONESTY *(off-label)*, NAVY BEANS *(off-label)*, PEAS [5]. Pre-harvest desiccation in COMBINING PEAS [1, 2, 6]. Pre-harvest desiccation in FIELD BEANS, LAID BARLEY AND OATS, LINSEED, OILSEED RAPE, POTATOES [1, 2, 5, 6].

Notes

Efficacy
- Acts rapidly on green parts of plants and rainfast in 15 min
- Best results achieved by spraying in bright light and low humidity conditions
- Add Agral wetter for improved weed control, not on aquatic weeds [5]
- Apply to potatoes when tubers the desired size, to other crops when mature or approaching maturity. See label for details of timing and of period to be left before harvesting potatoes and for timing of pea and bean desiccation sprays
- Spray linseed when seed matured evenly over whole field; direct combining can normally begin 10-20 d after spraying
- Apply as hop stripping treatment when shoots have reached top wire
- Apply to floating and submerged aquatic weeds in still or slow moving water [5]
- Do not apply in muddy water [3-5]

Crop Safety/Restrictions
- Maximum number of treatments 1 per crop in most situations - see labels for details
- Do not apply haulm destruction treatment when soil dry. Tubers may be damaged if spray applied during or shortly after dry periods. See label for details of maximum allowable soil-moisture deficit and varietal drought resistance scores
- Apply to laid barley and oats to be used only for stock feed, only on peas for harvesting dry, on field beans for pigeon and animal feed only

- Do not add wetters to desiccant sprays for potatoes or processing peas. Agral wetter may be added to spray for use on oilseed rape and peas to be used as animal fodder [5]
- For weed control in crops apply as overall spray before crop emergence or as interrow spray in row crops. Keep off crop foliage
- Do not apply through mist-blower or in windy conditions

Special precautions/Environmental safety

- Harmful if swallowed [1, 2, 5, 6]
- Harmful in contact with skin [3, 4]
- Irritating to skin and eyes
- Harmful to livestock [1, 2, 5, 6]
- Do not use treated straw or haulm as animal feed or bedding within 4 d of spraying [1, 2, 5, 6]
- Do not use treated water for human consumption within 24 h or for overhead irrigation within 10 d of treatment [3-5]
- Do not dump surplus herbicide or containers in water or ditch bottoms
- Do not use on barley, oats and field beans intended for human consumption [1, 2, 5, 6]

Protective clothing/Label precautions

- A, C
- M03, R04a, R04b, U04b, U05a, C03, E01, E19, E30a, E34 [1-6]; R03a, U09a, E21 (10 d), E32a [3, 4]; R03c, U08, U19, E06b, E08, E15, E21, E31a [1, 2, 5, 6]; U02, U20b [3-5]; U20a [1, 2, 6]

Withholding period

- Keep all livestock out of treated areas for at least 24 h [1, 2, 5, 6]

Latest application/Harvest Interval(HI)

- Varies with crop and product - see labels
- HI 14 d before lifting for potatoes; 4 d for laid barley and oats; 7 d for linseed, oilseed rape

Approval

- Approved for aquatic weed control [3-5]. See notes in Section 1 on use of herbicides in or near water
- Off-label Approval unlimited for use on borage (OLA 1216/96)[5], honesty (OLA 0588/93)[5], soya beans (OLA 0614/97)[5], outdoor navy beans (OLA 2604/97)[5]

240 diquat + paraquat

A non-selective non-residual bipyridyl contact herbicide

Products

1 PDQ	Zeneca	80:120 g/l	SL	06518
2 Speedway 2	Scotts	2.5:2.5% w/w	WG	09273

Uses Annual dicotyledons in NON-CROP AREAS [1, 2]. Annual dicotyledons in APPLES, BLACKCURRANTS, BUSH FRUIT, CANE FRUIT, FIELD CROPS *(autumn stubble, pre-planting/sowing, sward destruction)*, FLOWER BULBS, FORESTRY, GOOSEBERRIES, HARDY ORNAMENTALS, HOPS, PEARS, PLUMS/CHERRIES/DAMSONS, POTATOES, RASPBERRIES, ROW CROPS, STRAWBERRIES [1]. Annual grasses in NON-CROP AREAS

FOR FULL CONDITIONS OF USE ALWAYS READ THE PRODUCT LABEL

[1, 2]. Annual grasses in APPLES, BLACKCURRANTS, BUSH FRUIT, CANE FRUIT, FIELD CROPS *(autumn stubble, pre-planting/sowing, sward destruction)*, FLOWER BULBS, FORESTRY, GOOSEBERRIES, HARDY ORNAMENTALS, HOPS, PEARS, PLUMS/CHERRIES/DAMSONS, POTATOES, RASPBERRIES, ROW CROPS, STRAWBERRIES [1]. Chemical stripping in HOPS [1]. Green cover in FIELD MARGINS, LAND TEMPORARILY REMOVED FROM PRODUCTION [1]. Perennial grasses in NON-CROP AREAS [2]. Perennial non-rhizomatous grasses in APPLES, BLACKCURRANTS, BUSH FRUIT, CANE FRUIT, FIELD CROPS *(autumn stubble, pre-planting/sowing, sward destruction)*, FLOWER BULBS, FORESTRY, GOOSEBERRIES, HARDY ORNAMENTALS, HOPS, NON-CROP AREAS, PEARS, PLUMS/CHERRIES/DAMSONS, POTATOES, RASPBERRIES, ROW CROPS, STRAWBERRIES [1]. Volunteer cereals in APPLES, BLACKCURRANTS, BUSH FRUIT, CANE FRUIT, FIELD CROPS *(autumn stubble, pre-planting/sowing, sward destruction)*, FORESTRY, GOOSEBERRIES, HOPS, NON-CROP AREAS, PEARS, PLUMS/CHERRIES/DAMSONS, POTATOES, RASPBERRIES, ROW CROPS, STRAWBERRIES [1].

Notes

Efficacy

- Rapid kill obtained under bright conditions but most effective results on difficult weeds obtained from slower action in winter
- Apply to young emerged weeds less than 15 cm high, annual grasses must have at least 2 leaves when sprayed [1]
- Addition of approved non-ionic wetter recommended for control of certain species and with low dose rates. See label for details
- Interval between spraying and cultivation varies. See labels for details
- For chemical stripping apply in Jul or after hops have reached top wire. Do not use on hops under drought conditions [1]
- Chemical rapidly inactivated in moist soil, activity reduced in dirty or muddy water
- Spray is rainfast in 10 min

Crop Safety/Restrictions

- Maximum number of treatments 2 per yr for grassland destruction; 1 per yr or crop in other crops [1]
- In non-crop areas do not use around green bark [2]
- Apply up to just before sown crops emerge or just before planting [1]
- On sandy or immature peat soils and on forest nursery seedbeds allow 3 d between spraying and planting [1]
- Where trash or dying weeds are left on surface allow at least 3 d before planting
- In potatoes spray earlies up to 10% emergence, maincrop to 40% emergence, provided plants are less than 15 cm high. Do not use post-emergence on potatoes from diseased or small tubers or under very hot, dry conditions [1]
- Use guarded no-drift sprayers to kill inter-row weeds and strawberry runners [1]
- Apply to fruit crops as a directed spray, preferably in dormant season [1]
- If spraying bulbs at end of season ensure all crop foliage is detached from bulbs. Do not use on very sandy soils [1]

Special precautions/Environmental safety

- Paraquat is subject to the Poisons Rules 1982 and the Poisons Act 1972. See notes in Section 1
- Toxic if swallowed. Paraquat can kill if swallowed
- Harmful in contact with skin. Irritating to eyes and skin [1]
- Keep in original container, tightly closed, in a safe place, under lock and key
- Do not put in a food or drinks container
- Products in this guide are for professional use only
- Harmful to animals and livestock

- Paraquat can be harmful to hares, where possible spray stubbles early in the day [1]

Protective clothing/Label precautions
- A, C [1, 2]
- M04, U05a, U19, E01, E15, E31a, E34 [1, 2]; R02c, R03a, R04a, R04b, U04b, U08, U10, U11, U20a, C03, E06b, E11, E30b [1]; U04a, U09a, U20b, E02, E30a [2]

Withholding period
- Keep all livestock out of treated areas for at least 24 h

Maximum Residue Levels (mg residue/kg food)
- see paraquat entry

241 disulfoton

A systemic organophosphorus aphicide and insecticide

See also chlorpyrifos + disulfoton

Products

1 Disulfoton P 10	United Phosphorus	10% w/w	GR	08023
2 Disyston P 10	Bayer	10% w/w	GR	00715

Uses

Aphids in FRENCH BEANS, MARROWS, PARSLEY, RUNNER BEANS, STRAWBERRIES [2]. Aphids in BROAD BEANS, BRUSSELS SPROUTS, CABBAGES, CAULIFLOWERS, FIELD BEANS, POTATOES, SUGAR BEET [1, 2]. Aphids in BROCCOLI, CARROTS, FODDER BEET, MANGELS, PARSNIPS, SUGAR BEET STECKLINGS [1]. Carrot fly in CELERY [2]. Carrot fly in CARROTS, PARSNIPS [1, 2]. Mangold fly in FODDER BEET, MANGELS, SUGAR BEET [1].

Notes

Efficacy
- For effective results granules applied at drilling or planting should be incorporated in soil. See label for details of suitable application machinery
- One application is normally sufficient, a split application may be used on carrots
- Persistence may be reduced in fen peat soils

Crop Safety/Restrictions
- Maximum number of treatments 1 (soil) + 1 (foliage) for Brussels sprouts, cabbages, cauliflowers; 1 (soil) or 2 (foliage) 11 kg/ha (total) for beet crops; 1 per crop for broad beans, celery, field beans, French beans, marrows, parsley, potatoes, runner beans, strawberries; 1 per crop (soil or foliage) for carrots, parsnips
- Consult processor before treating crops grown for processing

Special precautions/Environmental safety
- This product contains an anticholinesterase organophosphorus compound. Do not use if under medical advice not to work with such compounds
- Toxic in contact with skin, by inhalation or if swallowed
- Keep in original container, tightly closed, in a safe place, under lock and key
- Dangerous to livestock
- Dangerous to game, wild birds and animals
- Applied correctly treatment will not harm bees

- Dangerous to fish or other aquatic life. Do not contaminate surface waters or ditches with chemical or used container

Protective clothing/Label precautions
- A, B, C or D + E [1, 2]; C [2]; H, J, K, M [1, 2]
- M01, M04, R02a, R02b, R02c, U04a, U05a, U13, U19, U20a, C03, E01, E10a, E13b, E30b, E32a, E34 [1, 2]; U02, U11, U12, C02 (6 wk), E26 [1]; U09a, U10, C02 [2]

Withholding period
- Keep all livestock out of treated areas for at least 6 wk. Bury or remove spillages

Latest application/Harvest Interval(HI)
- HI 6 wk

Approval
- Approved for aerial application on brassicas, beans, sugar beet, carrots [2]. See notes in Section 1

Maximum Residue Levels (mg residue/kg food)
- barley, sorghum 0.2; wheat 0.1; tea 0.05; all other products 0.02

242 dithianon

A protectant and eradicant dicarbonitrile fungicide for scab control

Products

Dithianon Flowable	Cyanamid	750 g/l	SC	07007

Uses

Scab in APPLES, PEARS.

Notes

Efficacy
- Apply at bud-burst and repeat every 10-14 d until danger of scab infection ceases
- Application at high rate within 48 h of a Mills period prevents new infection
- Spray programme also reduces summer infection with apple canker

Crop Safety/Restrictions
- Maximum number of treatments 8 per crop
- Do not use on Golden Delicious apples after green cluster
- Do not mix with lime sulfur or highly alkaline products

Special precautions/Environmental safety
- Harmful if swallowed.
- Irritating to eyes and skin

Protective clothing/Label precautions
- M03, R03c, R04a, R04b, U05a, U08, U19, U20a, C03, E01, E15, E26, E30a, E31b, E34

Latest application/Harvest Interval(HI)
- HI 4 wk

243 diuron

A residual urea herbicide for non-crop areas and woody crops

See also amitrole + 2,4-D + diuron
amitrole + bromacil + diuron
bromacil + diuron

Products

1 Atlas Diuron	Atlas	500 g/l	SC	08214
2 Chipko Diuron 80	Nomix-Chipman	80% w/w	WP	00497
3 Chipman Diuron 80	Nomix-Chipman	80% w/w	WP	08054
4 Chipman Diuron Flowable	Nomix-Chipman	500 g/l	SC	05701
5 Diuron 80 WP	RP Amenity	80% w/w	WP	05199
6 Freeway	RP Amenity	500 g/l	SC	06047
7 Karmex	DuPont	80% w/w	WP	01128
8 MSS Diuron 50 FL	Mirfield	500 g/l	SC	07160
9 MSS Diuron 500 FL	Mirfield	500 g/l	SC	08171
10 Rescind	RP Amenity	500 g/l	SC	08036
11 Unicrop Flowable Diuron	Unicrop	500 g/l	SC	02270

Uses

Annual dicotyledons in ESTABLISHED WOODY ORNAMENTALS, WOODY NURSERY STOCK [2, 5, 6]. Annual dicotyledons in APPLES, BLACKCURRANTS *(off-label)*, PEARS [11]. Annual dicotyledons in AMENITY TREES AND SHRUBS [3, 6]. Annual dicotyledons in LAND NOT INTENDED TO BEAR VEGETATION [1, 3, 9, 10]. Annual dicotyledons in TREES AND SHRUBS [1, 9, 10]. Annual grasses in ESTABLISHED WOODY ORNAMENTALS WOODY NURSERY STOCK [2, 5, 6]. Annual grasses in APPLES, BLACKCURRANTS *(off-label)*, PEARS [11]. Annual grasses in AMENITY TREES AND SHRUBS [6]. Annual grasses in LAND NOT INTENDED TO BEAR VEGETATION, TREES AND SHRUBS [1, 9, 10]. Annual meadow grass in AMENITY TREES AND SHRUBS, LAND NOT INTENDED TO BEAR VEGETATION [3]. Annual weeds in NON-CROP AREAS [2, 4-8]. Annual weeds in TREES AND SHRUBS [4, 8]. Annual weeds in FARM BUILDINGS, INDUSTRIAL SITES, PATHS RAILWAY TRACKS, ROAD VERGES [4]. Perennial weeds in FARM BUILDINGS INDUSTRIAL SITES, NON-CROP AREAS, PATHS, RAILWAY TRACKS, ROAD VERGES [4].

Notes

Efficacy

- Best results when applied to moist soil and rain falls soon afterwards
- Length of residual activity may be reduced on heavy or highly organic soils or those with ash substrates
- Selective rates must be applied to weed-free soil and activity persists for 2-3 mth
- Around trees and shrubs spray in late winter or early spring [6]. Band spraying is recommended method of application [10]
- Treat hard surfaces between Feb and end May [2-4, 10]
- Application for total vegetation control may be at any time of year, best results obtained in late winter to early spring
- Application to frozen ground not recommended

Crop Safety/Restrictions

- Maximum number of treatments 1 per yr on land not intended for cropping (high rate), 2 per yr around trees and shrubs (low rate) [3, 5, 6]; one per year for apples, pears [11]

- Apply to weed-free soil in apple and pear orchards established for at least 1 yr during Feb-Mar
- Do not treat trees and shrubs less than 5 cm tall or established less than 12 mth. See label for list of sensitive species [1, 3, 6, 10]
- Do not use on Sands or Very Light soils or those that are gravelly or where less than 1% organic matter
- Do not apply to areas intended for replanting during next 12 mth (2 yr for vegetables)
- Do not use on lawns, grass tennis courts or similar areas of turf [3]
- Do not apply non-selective rates on or near desirable plants where chemical may be washed into contact with roots
- Application for amenity use should only take place between the beginning of Feb and end of Apr (May in Scotland)

Special precautions/Environmental safety

- Irritating to eyes and skin [1-8, 10, 11]
- Irritating to respiratory system [2-4, 6, 7, 11]
- Harmful to fish or other aquatic life. Do not contaminate surface waters or ditches with chemical or used container
- Do not apply over drains or in drainage channels or gullies
- Avoid run-off when using on paved and similar surfaces

Protective clothing/Label precautions

- A [1, 3, 4, 6, 7, 10]; C [1, 3, 4, 7]; H, M [6, 10]; N [10]
- R04a, R04b, U05a, E13c, E30a [1-11]; R04c [2, 3, 7, 11]; U08 [1-6, 8-10]; U09a [7, 11]; U19 [1-8, 10, 11]; U20a [7, 9, 10]; U20b [1-6, 8, 11]; C03, E01 [1, 3-11]; E26 [1, 4, 8-11]; E31a [4, 10]; E31b [1, 6, 8, 9]; E31c [11]; E32a [2, 3, 5, 7]; E34 [1, 9, 11]

Approval

- Approved for use through ULV applicators [10]. See notes in Section 1
- Off-label approval unlimited for use on established blackcurrants (OLA 1318/95)[11]

244 diuron + glyphosate

A non-selective residual herbicide mixture for non-crop and amenity use

Products

1 Touché	Nomix-Chipman	217.6:145.3 g/l	RH	06797
2 Xanadu	Monsanto	125:100 g/l	EW	09228

Uses

Annual dicotyledons in AMENITY TREES AND SHRUBS, LAND NOT INTENDED TO BEAR VEGETATION. Annual grasses in AMENITY TREES AND SHRUBS, LAND NOT INTENDED TO BEAR VEGETATION. Perennial dicotyledons in AMENITY TREES AND SHRUBS, LAND NOT INTENDED TO BEAR VEGETATION. Perennial grasses in AMENITY TREES AND SHRUBS, LAND NOT INTENDED TO BEAR VEGETATION.

Notes

Efficacy

- Best results achieved from treatment when weeds are green and actively growing. Symptoms may be slow to appear in poor growing conditions
- Perennial grasses are susceptible when tillering and making new rhizome growth, normally when plants have 4-5 new leaves
- Perennial dicotyledons most susceptible if treated at or near flowering but will be severely checked if treated at other times when growing actively
- Most species of germinating weeds are controlled. See label for more resistant species

- Weed control may be reduced on heavy or highly organic soils and when weeds are suffering stress in any situation
- Treat hard surfaces between Feb and end May
- At least 6 and preferably 24 h rain-free must follow spraying. Rain soon after treatment may reduce initial weed control
- Apply with Nomix System equipment [1]

Crop Safety/Restrictions
- Maximum number of treatments 1 per yr
- Ornamental trees and shrubs should be established for at least 12 mth and be at least 50 mm tall
- Among ornamental trees and shrubs take care to avoid foliage or the stems of young plants
- Do not allow spray to contact desired plants or crops
- Do not use on tree and shrub nurseries or before planting
- Do not treat ornamental trees and shrubs on Sands or Very Light soils or those that are gravelly or with less than 1% organic matter [1]
- Do not decant, connect directly to Nomix applicator [1]

Special precautions/Environmental safety
- Irritating to eyes and skin [1]
- May cause sensitization by skin contact [1]
- Not to be used on food crops [2]
- Harmful to fish or other aquatic life. Do not contaminate surface waters or ditches with chemical or used container
- Do not apply over drains or in drainage channels, gullies or similar structures

Protective clothing/Label precautions
- H, M [1]
- U02, E26 [1]; U09a, U19, U20b, E13c, E30a, E32a [1, 2]; C01 [2]

Approval
- Approved for use through ULV applicators [1]

Maximum Residue Levels (mg residue/kg food)
- see glyphosate entry

245 diuron + paraquat

A total herbicide with contact and residual activity

Products	Dexuron	Nomix-Chipman	300:100 g/l	SC	07169

Uses Annual dicotyledons in ESTABLISHED WOODY ORNAMENTALS, WOODY NURSERY STOCK. Annual grasses in ESTABLISHED WOODY ORNAMENTALS, WOODY NURSERY STOCK. Perennial dicotyledons in ESTABLISHED WOODY ORNAMENTALS, WOODY NURSERY STOCK. Perennial grasses in ESTABLISHED WOODY ORNAMENTALS, WOODY NURSERY STOCK. Total vegetation control in NON-CROP AREAS, PATHS.

Notes

Efficacy

- Apply to emerged weeds at any time of year. Best results achieved by application in spring or early summer
- Effectiveness not reduced by rain soon after treatment

Crop Safety/Restrictions

- Maximum number of treatments 1 per yr
- Avoid contact of spray with green bark, buds or foliage of desirable trees or shrubs
- In nurseries use low dose rate as inter-row spray, not more than once per year

Special precautions/Environmental safety

- Paraquat is subject to the Poisons Rules 1982 and the Poisons Act 1972. See notes in Section 1
- Keep in original container, tightly closed, in a safe place, under lock and key
- Toxic if swallowed
- Harmful in contact with skin. Irritating to eyes and skin
- Harmful to fish or other aquatic life. Do not contaminate surface waters or ditches with chemical or used container

Protective clothing/Label precautions

- A, C
- M04, R02c, R03a, R04a, R04b, U02, U04a, U05a, U09a, U19, U20b, C03, E01, E06b (24 h), E13c, E26, E30b, E31a, E34

Withholding period

- Keep livestock out of treated areas for at least 2 wk and until foliage of any poisonous weeds such as ragwort has died and become unpalatable

Maximum Residue Levels (mg residue/kg food)

- see paraquat entry

---

246 DNOC

A dinitrophenyl insecticide all approvals for which were revoked in December 1989

---

247 dodemorph

A systemic morpholine fungicide for powdery mildew control

Products					
	F238	BASF	385 g/l	EC	00206

Uses

Powdery mildew in ROSES.

Notes

Efficacy

- Spray roses every 10-14 d during mildew period or every 7 d and at increased dose if cleaning up established infection or if disease pressure high
- Add Citowett when treating rose varieties which are difficult to wet
- Product has negligible effect on *Phytoseiulus* spp being used to control red spider mites

Crop Safety/Restrictions

- Do not use on seedling roses

- Do not apply to roses under hot, sunny conditions, particularly under glass, but spray early in the morning or during the evening. Increase the humidity some hours before spraying
- Check tolerance of new varieties before treating rest of crop

Special precautions/Environmental safety
- Irritating to skin. Risk of serious damage to eyes
- Flammable
- Harmful to fish or other aquatic life. Do not contaminate surface waters or ditches with chemical or used container

Protective clothing/Label precautions
- A, C
- R04b, R04d, R07d, U04a, U05a, U09a, U14, U15, U19, U20c, C03, E01, E13c, E26, E30a, E31c

248 dodine

A protectant and eradicant guanidine fungicide

Products Radspor FL | Truchem | 450 g/l | SC | 01685

Uses Currant leaf spot in BLACKCURRANTS, GOOSEBERRIES. Scab in APPLES, PEARS.

Notes

Efficacy
- Apply protective spray on apples and pears at bud-burst and at 10-14 d intervals until late Jun to early Jul
- Apply post-infection spray within 36 h of rain responsible for initiating infection. Where scab already present spray prevents production of spores
- Apply to blackcurrants immediately after flowering, repeat every 10-14 d to within 1 mth of harvest and once or twice post-harvest

Crop Safety/Restrictions
- Do not apply in very cold weather (under 5°C) or under slow drying conditions to pears or dessert apples during bloom or immediately after petal fall
- Do not mix with lime sulfur or tetradifon

Special precautions/Environmental safety
- Harmful: In contact with skin
- Irritating to eyes and skin
- May cause sensitization by skin contact
- Harmful to fish or other aquatic life. Do not contaminate surface waters or ditches with chemical or used container

Protective clothing/Label precautions
- A, C
- M03, R03a, R03c, R04a, R04b, R04e, U05a, U08, U19, U20a, C03, E01, E13c, E30a, E31a, E34

Latest application/Harvest Interval(HI)
- Early Jul for dessert apples; pre-blossom for culinary apples and pears
- HI blackcurrants 1 mth

249 endosulfan

A contact and ingested organochlorine insecticide and acaricide

Products	Thiodan 20	Promark	200 g/l	EC	07335

Uses

Big-bud mite in BLACKCURRANTS. Blackberry mite in BLACKBERRIES. Bulb scale mite in NARCISSI. Tarsonemid mites in STRAWBERRIES.

Notes

Efficacy
- Adjust spray volume to achieve total cover. See label for minimum dilutions
- On blackcurrants apply 3 sprays, at first flower, end of flowering and fruit set
- On blackberries apply 3 sprays at 14 d intervals before flowering
- On strawberries apply immediately after whole crop picked; where a second crop to be picked in autumn within 1 wk of mowing old foliage
- For effective control on strawberries it is essential that the spray penetrates the crowns of the plants
- Apply as drench to boxed narcissi a few days after bringing into greenhouse

Crop Safety/Restrictions
- When timing applications on blackcurrants it is important to consider different cultivars separately. Applications should not be made over a whole plantation of mixed cultivars regardless of variations in maturity

Special precautions/Environmental safety
- Endosulfan is subject to the Poisons Rules 1982 and the Poisons Act 1972. See notes in Section 1
- Toxic if swallowed. Harmful in contact with skin
- Flammable
- Keep in original container, tightly closed, in a safe place, under lock and key
- Keep unprotected persons out of treated areas for at least 1 d
- Dangerous to livestock
- Harmful to bees. Do not apply at flowering stage except as directed on hops, oilseed rape and mustard. Keep down flowering weeds.
- Dangerous to fish or other aquatic life. Do not contaminate surface waters or ditches with chemical or used container

Protective clothing/Label precautions
- A, C, H, J, K, L, M
- M04, R02c, R03a, R07d, U02, U04a, U05a, U09a, U13, U19, U20a, C02 (6 wk), C03, E01, E02 (1 d), E06a (3 wk), E12e, E13b, E30b, E31b, E34

Withholding period
- Keep all livestock out of treated areas for at least 3 wk

Latest application/Harvest Interval(HI)
- HI blackcurrants, blackberries, strawberries 6 wk

Approval

- Off-label approval unlimited for use on outdoor and protected ornamentals (OLA 0656/92)[1]

Maximum Residue Levels (mg residue/kg food)

- strawberries, blackberries, currants, gooseberries 2; citrus fruit, pome fruit, stone fruit, grapes, raspberries, kiwi fruit, olives, onions, fruiting vegetables (except sweet corn), brassica vegetables (except kohlrabi), lettuce, spinach, legume vegetables, cardoons, celery, globe artichokes, leeks, cultivated mushrooms 1; cotton seed 0.3; beetroot, carrots, celeriac, parsley root, swedes, turnips, potatoes, maize 0.2; tree nuts, peanuts, poppy seed, sesame seed, wheat, rye, barley, oats, triticale, meat 0.1; dewberries, loganberries, bilberries, cranberries, wild berries, avocados, dates, figs, kumquats, litchis, mangoes, passion fruit, pineapples, pomegranates, horseradish, Jerusalem artichokes, parsnips, radishes, salsify, sweet potatoes, yams, garlic, shallots, spring onions, sweet corn, kohlrabi, watercress, witloof, herbs, asparagus, fennel, rhubarb, wild mushrooms, pulses, sorghum, buckwheat, millet, rice 0.05; milk, dairy produce, eggs 0.004

250 epoxiconazole

A systemic, protectant and curative triazole fungicide for use in cereals

Products

1 Epic	BASF	125 g/l	SC	08320
2 Landgold Epoxiconazole	Landgold	125 g/l	SC	08631
3 Opus	BASF	125 g/l	SC	08319
4 Standon Epoxiconazole	Standon	125 g/l	SC	08294

Uses

Brown rust in BARLEY, WINTER WHEAT. Eyespot in WINTER BARLEY *(reduction)*, WINTER WHEAT *(reduction)*. Fusarium ear blight in WINTER WHEAT *(reduction)*. Net blotch in BARLEY. Powdery mildew in BARLEY, WINTER WHEAT. Rhynchosporium in BARLEY. Septoria in WINTER WHEAT. Sooty moulds in WINTER WHEAT *(reduction)*. Yellow rust in BARLEY, WINTER WHEAT.

Notes

Efficacy

- Apply at the start of foliar disease attack
- Optimum effect against eyespot achieved by spraying between leaf-sheath erect and second node detectable stages (GS 30-32)
- Best control of ear diseases of wheat obtained by treatment during ear emergence
- For Septoria spray after third node detectable stage (GS 33) when weather favouring disease development has occurred
- Mildew control improved by use of tank mixtures. See label for details

Crop Safety/Restrictions

- Maximum total dose equivalent to two full dose treatments
- Product may cause damage to broad-leaved plant species
- Avoid spray drift onto neighbouring crops

Special precautions/Environmental safety

- Dangerous to fish or other aquatic life. Do not contaminate surface waters or ditches with chemical or used container

- Do not allow direct spray from vehicle mounted/drawn sprayers to fall within 6 m of surface waters or ditches. Direct spray away from water

Protective clothing/Label precautions
- A
- U05a, U20b, C03, E01, E13b, E16, E30a [1-4]; E26, E31b [2]; E31c [1, 3, 4]

Latest application/Harvest Interval(HI)
- Before crop commences flowering (GS 60) in winter wheat; up to and including emergence of ear just complete (GS 59) in barley

Approval
- Approval expiry: 31 Jul 99 [4]

251 epoxiconazole + fenpropimorph

A systemic, protectant and curative fungicide mixture for cereals

Products

1 Barclay Riverdance	Barclay	84:250 g/l	SE	08344
2 Eclipse	BASF	84:250 g/l	SE	07361
3 Opus Team	BASF	84:250 g/l	SE	07362
4 Standon Epoxifen	Standon	84:250 g/l	SE	08972

Uses

Brown rust in BARLEY, WINTER WHEAT [1-4]. Eyespot in WINTER BARLEY *(reduction)*, WINTER WHEAT *(reduction)* [1-4]. Fusarium ear blight in WINTER WHEAT *(reduction)* [1-4]. Net blotch in BARLEY [1-4]. Powdery mildew in BARLEY, WINTER WHEAT [1-4]. Rhynchosporium in BARLEY [1-4]. Septoria in WINTER WHEAT [1-4]. Sooty moulds in WINTER WHEAT *(reduction)* [2, 3]. Yellow rust in BARLEY, WINTER WHEAT [1-4].

Notes

Efficacy
- Apply at the start of foliar disease attack
- Optimum effect against eyespot achieved by spraying between leaf-sheath erect and second node detectable stages (GS 30-32)
- Best control of ear diseases of wheat obtained by treatment during ear emergence
- For Septoria spray after third node detectable stage (GS 33) when weather favouring disease development has occurred

Crop Safety/Restrictions
- Maximum total dose equivalent to two full dose treatments
- Product may cause damage to broad-leaved plant species
- Avoid spray drift onto neighbouring crops

Special precautions/Environmental safety
- Irritating to skin
- Harmful [2, 3] (dangerous [1, 4]) to fish or other aquatic life. Do not contaminate surface waters or ditches with chemical or used container
- Do not allow direct spray from vehicle mounted/drawn sprayers to fall within 6 m of surface waters or ditches. Direct spray away from water

Protective clothing/Label precautions
- A
- R04b, U05a, U20b, C03, E01, E16, E30a [1-4]; E13b, E26, E31b [1, 4]; E13c [2, 3]

Latest application/Harvest Interval(HI)

- Before crop commences flowering (GS 60) in winter wheat; up to and including emergence of ear just complete (GS 59) in barley

Approval

- Approval expiry: 31 Jul 99 [1]

252 epoxiconazole + fenpropimorph + kresoxim-methyl

A protectant, systemic and curative fungicide mixture for cereals

Products					
Products	Mantra	BASF	125:150:125 g/l	SE	08886

Uses Brown rust in SPRING BARLEY, WINTER BARLEY, WINTER WHEAT. Eyespot in WINTER BARLEY *(reduction)*, WINTER WHEAT *(reduction)*. Fusarium ear blight in WINTER WHEAT *(reduction)*. Net blotch in SPRING BARLEY, WINTER BARLEY. Powdery mildew in SPRING BARLEY, WINTER BARLEY, WINTER WHEAT. Rhynchosporium in SPRING BARLEY, WINTER BARLEY. Septoria diseases in WINTER WHEAT. Sooty moulds in WINTER WHEAT *(reduction)*. Yellow rust in SPRING BARLEY, WINTER BARLEY, WINTER WHEAT.

Notes

Efficacy

- For best results spray at the start of foliar disease attack and repeat if infection conditions persist
- Optimum effect against eyespot obtained by treatment between leaf sheaths erect and first node detectable stages (GS 30-32)
- For protection against ear diseases apply during ear emergence

Crop Safety/Restrictions

- Maximum total dose equivalent to two full dose treatments

Special precautions/Environmental safety

- Irritant. May cause sensitization by skin contact
- Dangerous to fish or other aquatic life. Do not contaminate surface waters or ditches with chemical or used container
- Do not allow direct spray from vehicle mounted/drawn hydraulic sprayers to fall within 6 m of surface waters or ditches. Direct spray away from water
- Avoid spray drift onto neighbouring crops. Product may damage broad-leaved species

Protective clothing/Label precautions

- A
- R04, R04e, U05a, U14, U20b, C03, E01, E13b, E16, E30a, E31c

Latest application/Harvest Interval(HI)

- Before start of flowering (GS 60) for winter wheat; completion of ear emergence (GS 59) for barley

253 epoxiconazole + kresoxim-methyl

A protectant, systemic and curative fungicide mixture for cereals

Products Landmark BASF 125:125 g/l SC 08889

Uses Brown rust in SPRING BARLEY, WINTER BARLEY, WINTER WHEAT. Eyespot in WINTER BARLEY *(reduction)*, WINTER WHEAT *(reduction)*. Fusarium ear blight in WINTER WHEAT *(reduction)*. Net blotch in SPRING BARLEY, WINTER BARLEY. Powdery mildew in SPRING BARLEY, WINTER BARLEY, WINTER WHEAT. Rhynchosporium in SPRING BARLEY, WINTER BARLEY. Septoria diseases in WINTER WHEAT. Sooty moulds in WINTER WHEAT *(reduction)*. Yellow rust in SPRING BARLEY, WINTER BARLEY, WINTER WHEAT.

Notes

Efficacy

- For best results spray at the start of foliar disease attack and repeat if infection conditions persist
- Optimum effect against eyespot obtained by treatment between leaf sheaths erect and first node detectable stages (GS 30-32)
- For protection against ear diseases apply during ear emergence

Crop Safety/Restrictions

- Maximum total dose equivalent to two full dose treatments

Special precautions/Environmental safety

- Irritant. May cause sensitization by skin contact
- Dangerous to fish or other aquatic life. Do not contaminate surface waters or ditches with chemical or used container
- Do not allow direct spray from vehicle mounted/drawn hydraulic sprayers to fall within 6 m of surface waters or ditches. Direct spray away from water
- Avoid spray drift onto neighbouring crops. Product may damage broad-leaved species

Protective clothing/Label precautions

- A
- R04, R04e, U05a, U14, U20b, C03, E01, E13b, E16, E30a, E31c

Latest application/Harvest Interval(HI)

- Before start of flowering (GS 60) for winter wheat; completion of ear emergence (GS 59) for barley

254 esfenvalerate

A contact and ingested pyrethroid insecticide

Products Sumi-Alpha Cyanamid 25 g/l EC 07207

Uses Aphids in WINTER BARLEY, WINTER WHEAT.

Notes

Efficacy

- Crops at high risk (e.g. after grass or in areas with history of BYDV) should be treated when aphids first seen or by mid-Oct. Otherwise treat in late Oct-early Nov

- High risk crops will need a second treatment
- Product also recommended between onset of flowering and milky ripe stages (GS 61-73) for control of summer cereal aphids

Crop Safety/Restrictions
- Do not use if another pyrethroid or dimethoate has been applied to crop after start of ear emergence (GS 51)

Special precautions/Environmental safety
- Harmful if swallowed, irritating to skin and eyes
- Extremely dangerous to bees. Do not apply at flowering stage except as directed on oilseed rape and peas. Keep down flowering weeds
- Extremely dangerous to fish or other aquatic life. Do not contaminate surface waters or ditches with chemical or used container
- Do not allow spray from vehicle sprayers to fall within 6 m, or from hand held sprayers to within 2 m of surface waters or ditches. Direct spray away from water
- Store product in dark away from direct sunlight

Protective clothing/Label precautions
- A, C, H
- M03, R03c, R04a, R04b, U04a, U05a, U08, U19, U20b, C03, E01, E12b, E13a, E16, E22a, E26, E30a, E31b, E34

Latest application/Harvest Interval(HI)
- HI 20 d

255 ethirimol

A pyrimidine fungicide available only in mixtures

256 ethirimol + flutriafol + thiabendazole

A systemic liquid fungicide for seed treatment in barley

Products	Ferrax	Bayer	400:30:10 g/l	FS	05284

Uses Brown rust in SPRING BARLEY, WINTER BARLEY. Covered smut in SPRING BARLEY, WINTER BARLEY. Leaf stripe in SPRING BARLEY. Loose smut in SPRING BARLEY, WINTER BARLEY. Net blotch in SPRING BARLEY, WINTER BARLEY. Powdery mildew in SPRING BARLEY, WINTER BARLEY. Rhynchosporium in SPRING BARLEY, WINTER BARLEY.

Notes

Efficacy
- Apply with suitable liquid seed treatment machinery. See label for details
- Controls seed-borne diseases and early attacks of foliar diseases. Additional treatment may be required later
- Recalibrate drill for treated seed
- Do not use treated seed on soils with more than 20% organic matter

- Disease control may be reduced under dry conditions
- May be mixed with Gammasan 30 provided both seed dressings applied together. Do not apply to seed already treated with another seed treatment

Crop Safety/Restrictions
- Maximum number of treatments 1 per batch
- Do not apply to seed with more than 16% moisture
- Emergence of seed may be delayed, especially under poor germination conditions
- Treatment may reduce germination capacity if seed of poor quality
- Treated seed should be stored in cool, well ventilated conditions and drilled as soon as possible. Test germination if stored until next season

Special precautions/Environmental safety
- Irritating to eyes and skin
- Harmful to fish or other aquatic life. Do not contaminate surface waters or ditches with chemical or used container
- Do not use treated seed as food or feed
- Treated seed harmful to game and wildlife
- Seed and stored product must only be treated by means which incorporate engineering controls for workers' protection together with means of accurately dispensing the dose

Protective clothing/Label precautions
- A, C, D, H, M
- M03, R04a, R04b, U02, U05a, U09a, U20a, C03, E01, E13c, E30a, E31a, E34, S01, S02, S03, S04b, S05, S06, S07

Latest application/Harvest Interval(HI)
- Before drilling

---

257 ethofumesate

A benzofuran herbicide for grass weed control in various crops

See also bromoxynil + ethofumesate + ioxynil
chloridazon + ethofumesate
desmedipham + ethofumesate

Products

1 Atlas Thor	Atlas	200 g/l	EC	07732
2 Barclay Keeper 200	Barclay	200 g/l	EC	05266
3 Landgold Ethofumesate 200	Landgold	200 g/l	EC	08980
4 MSS Ethosan	Mirfield	500 g/l	SC	08048
5 MSS Thor	Mirfield	200 g/l	EC	08817
6 Nortron Flo	AgrEvo	500 g/l	SC	08154
7 Standon Ethofumesate 200	Standon	200 g/l	EC	08726
8 Stefes Fumat 2	Stefes	200 g/l	EC	07856

Uses

Annual dicotyledons in SUGAR BEET [1-8]. Annual dicotyledons in RED BEET [1-3, 6-8]. Annual dicotyledons in FODDER BEET, MANGELS [1-3, 5-8]. Annual dicotyledons in GARLIC *(off-label)*, HORSERADISH *(off-label)*, ONIONS *(off-label)* [6]. Annual grasses in GRASS SEED CROPS, LEYS [2, 8]. Annual grasses in AMENITY TURF, ESTABLISHED GRASSLAND [8]. Annual grasses in GARLIC *(off-label)*, HORSERADISH *(off-label)*, ONIONS *(off-label)*, STRAWBERRIES *(off-label)* [6]. Annual meadow grass in SUGAR BEET [1-8]. Annual meadow grass in RED BEET [1-3, 6-8]. Annual meadow grass in FODDER BEET, MANGELS [1-3, 5-8]. Blackgrass in SUGAR BEET [1-8]. Blackgrass in RED BEET [1-3, 6-8].

Blackgrass in FODDER BEET, MANGELS [1-3, 5-8]. Blackgrass in AMENITY TURF, ESTABLISHED GRASSLAND, GRASS SEED CROPS, LEYS [8]. Chickweed in GRASS SEED CROPS, LEYS [2, 8]. Chickweed in AMENITY TURF, ESTABLISHED GRASSLAND [8]. Cleavers in GRASS SEED CROPS, LEYS [2, 8]. Cleavers in AMENITY TURF, ESTABLISHED GRASSLAND [8]. Clover in STRAWBERRIES *(off-label)* [6]. Volunteer cereals in AMENITY TURF, ESTABLISHED GRASSLAND, GRASS SEED CROPS, LEYS [8].

Notes

Efficacy

- May be applied pre- or post-emergence of crop or weeds (pre-emergence only [5])
- Apply in beet crops in tank mixes with other pre- or post-emergence herbicides. Recommendations vary for different mixtures. See label for details
- In grass crops apply to moist soil as soon as possible after sowing or post-emergence when crop in active growth, normally mid-Oct to mid-Dec. See label for details
- Volunteer cereals not well controlled pre-emergence, weed grasses should be sprayed before fully tillered
- Do not use on soils with more than 10% organic matter
- Grass crops may be sprayed during rain or when wet. Not recommended in very dry conditions or prolonged frost
- Do not graze or cut grass for 14 d after, or roll less than 7 d before or after spraying

Crop Safety/Restrictions

- Maximum number of treatments 1 pre-emergence [5]; 1 pre- plus 1 post-emergence (2 at reduced dose) per crop or yr for beet crops and grassland
- Safe timing on beet crops varies with other ingredient of tank mix. See label for details
- May be used in Italian, hybrid and perennial ryegrass, timothy, cocksfoot, meadow fescue and tall fescue. Apply pre-emergence to autumn-sown leys, post-emergence after 2-3 leaf stage. See label for details
- Do not use on swards reseeded without ploughing
- Clovers will be killed or severely checked
- Any crop may be sown 3 mth after application of mixtures in beet crops following ploughing, 5 mth after application in grass crops

Special precautions/Environmental safety

- Flammable
- Harmful to fish or other aquatic life. Do not contaminate surface waters or ditches with chemical or used container

Protective clothing/Label precautions

- M05 [5]; R07d [1-3, 5, 7, 8]; U08 [1-5, 7, 8]; U09a [6]; U19, E13c, E30a [1-8]; U20a, E31b [1-3, 5-8]; U20b [4]; E26 [2, 3, 5, 7]; E34 [3, 4]

Latest application/Harvest Interval(HI)

- Pre-emergence of beet crops [5]; before crops meet across rows for beet crops and mangels; 14 d before cutting or grazing for grass leys; pre-emergence for horseradish

Approval

- Off-label approval unlimited for use on strawberries, onions, garlic, horseradish (OLA 1473/98)[6]

258 ethofumesate + metamitron + phenmedipham

A contact and residual herbicide mixture for sugar beet

Products Betanal Trio WG | AgrEvo | 6.5:28:6.5% w/w | WG | 07537

Uses Annual dicotyledons in SUGAR BEET. Annual meadow grass in SUGAR BEET.

Notes

Efficacy

- Best results obtained from a series of treatments applied as an overall fine spray commencing when earliest germinating weeds are no larger than fully expanded cotyledon
- Apply subsequent sprays as each new flush of weeds reaches early cotyledon and continue until weed emergence ceases (maximum 3 sprays)
- Product must be applied with Actipron
- Product may follow certain pre-emergence treatments and be used in conjunction with other post-emergence sprays - see label for details
- Where a pre-emergence band spray has been applied, the first treatment should be timed according to the size of the weeds in the untreated area between the rows

Crop Safety/Restrictions

- Maximum number of treatments 3 per crop; maximum total dose 6 kg product per ha
- Product may be used on all soil types but residual activity may be reduced on those with more than 5% organic matter
- Crop tolerance may be reduced by stress caused by growing conditions, effects of pests, disease or other pesticides, nutrient deficiency etc
- Only beet crops should be sown within 4 mth of last treatment; winter cereals may be sown after this interval
- Any spring crop may be sown in the year following use
- Mould-board ploughing to 150 mm followed by thorough cultivation recommended before planting any crop

Special precautions/Environmental safety

- Irritating to eyes
- Harmful to fish or other aquatic life. Do not contaminate surface waters or ditches with chemical or used container

Protective clothing/Label precautions

- A, C
- R04a, U05a, U11, U20b, C03, E01, E13c, E30a, E32a

Latest application/Harvest Interval(HI)

- Before crop meets between rows

259 ethofumesate + phenmedipham

A contact and residual herbicide for use in beet crops

Products

1 Barclay Goalpost	Barclay	100:80 g/l	EC	08016
2 Barclay Goldpost	Barclay	100:80 g/l	EC	08964
3 Betosip Combi	Sipcam	100:80 g/l	EC	07235
4 Stefes Medimat 2	Stefes	100:80 g/l	EC	07577
5 Stefes Tandem	Stefes	100:80 g/l	EC	08906
6 Twin	Headland	94:97 g/l	EC	07374

Uses

Annual dicotyledons in RED BEET [1, 4, 5]. Annual dicotyledons in SUGAR BEET [1-6]. Annual dicotyledons in FODDER BEET, MANGELS [1, 2, 4-6]. Annual meadow grass in RED BEET [1, 4]. Annual meadow grass in FODDER BEET, MANGELS, SUGAR BEET [1, 2, 4]. Blackgrass in RED BEET [1, 4]. Blackgrass in FODDER BEET, MANGELS, SUGAR BEET [1, 2, 4].

Notes

Efficacy

- Best results achieved by repeat applications to cotyledon stage weeds. Larger susceptible weeds not killed by first treatment usually checked and controlled by second application
- Apply on all soil types at 7-10 d intervals
- On soils with more than 5% (10% [6]) organic matter residual activity may be reduced
- Do not spray wet foliage or if rain imminent
- Spray must be applied low volume. See label for details

Crop Safety/Restrictions

- Maximum number of treatments normally 3 per crop - see labels for details
- Apply reduced dose from when the crop has fully expanded cotyledons or full dose from 2 fully expanded true leaf stage [1-4]
- Spray in evening if daytime temperatures above 21°C expected
- Avoid or delay treatment if frost expected within 7 d
- Avoid or delay treating crops under stress from wind damage, manganese or lime deficiency, pest or disease attack etc
- Check from which recovery may not be complete may occur if treatment made during conditions of sharp diurnal temperature fluctuation
- Allow 7 d before or after treatment with other herbicides [6]

Special precautions/Environmental safety

- Irritating to eyes [3] and skin [2]. Irritating to skin and respiratory system [1, 4]
- Harmful in contact with skin [1, 4]
- Flammable [1, 4]
- May cause sensitisation by skin contact [3, 6]
- Harmful to fish or aquatic life. Do not contaminate surface waters or ditches with chemical or used container
- Extra care necessary to avoid drift because product is recommended for use as a fine spray
- Spray volumes must not exceed those recommended [1-4]
- Interval between mixing spray and completion of spraying should not exceed 2 h to avoid crystallization [1-4]
- Product may cause non-reinforced PVC pipes and hoses to soften and swell. Wherever possible use reinforced PVC or synthetic rubber [1-4]

Protective clothing/Label precautions
- A [1-6]; C [1-5]
- R03a, R04c [1, 4, 5]; R04 [6]; R04a [2, 3]; R04b, R07d [1, 2, 4, 5]; R04e [3, 6]; U05a, C03, E01, E13c, E30a [1-6]; U08 [1, 3-6]; U10, U11, U20b [2]; U19, E31b [2-6]; U20a, E34 [1, 4-6]; E24 [5]; E26 [2, 3, 5]; E31c [1]

Latest application/Harvest Interval(HI)
- Before crop foliage meets in the rows

260 ethoprophos

An organophosphorus nematicide and insecticide

Products

Mocap 10G	RP Agric.	10% w/w	GR	06773

Uses

Potato cyst nematode in POTATOES. Wireworms in POTATOES.

Notes

Efficacy
- Broadcast shortly before or during final soil preparation with suitable fertilizer spreader and incorporate immediately to 10-15 cm. See label for details
- Treatment can be applied on all soil types. Control of pests reduced on organic soils
- Effectiveness dependent on soil moisture. Drought after application may reduce control

Crop Safety/Restrictions
- Maximum number of treatments 1 per crop

Special precautions/Environmental safety
- This product contains an anticholinesterase organophosphorous compound. Do not use if under medical advice not to work with such compounds
- Harmful by inhalation, in contact with skin and if swallowed
- May cause sensitization by skin contact
- Dangerous to game, wild birds and animals
- Dangerous to fish or other aquatic life. Do not contaminate surface waters or ditches with chemical or used container
- Do not harvest crops for human or animal consumption for at least 8 wk after application

Protective clothing/Label precautions
- A, B, C or D + E, H, K, M
- M01, M03, R03a, R03b, R03c, R04e, U02, U04a, U05a, U08, U13, U19, U20a, C02 (8 wk), C03, E01, E10a, E13b, E30a, E32a, E34

Latest application/Harvest Interval(HI)
- Pre-planting of crop
- HI 8 wk

261 etridiazole

A protective thiadiazole fungicide for soil or compost incorporation

Products

1 Aaterra WP	Zeneca	35% w/w	WP	06625
2 Standon Etridiazole 35	Standon	35% w/w	WP	08778

Uses Damping off in CABBAGES *(seedlings and transplants)*, CAULIFLOWERS *(seedlings and transplants)*, CELERY *(seedlings and transplants)*, CUCUMBERS *(seedlings and transplants)*, MUSTARD AND CRESS *(seedlings and transplants)*, NFT TOMATOES, TOMATOES *(seedlings and transplants)*. Phytophthora in CONTAINER-GROWN STOCK, HARDY ORNAMENTAL NURSERY STOCK, NFT TOMATOES, TULIPS, WATERCRESS *(off-label)*. Pythium in TULIPS, WATERCRESS *(off-label)*. Root diseases in ROCKWOOL TOMATOES *(off-label)*.

Notes

Efficacy

- Best results obtained when incorporated thoroughly into soil or compost. Drench application also recommended
- Do not apply to wet soil
- Treat compost as soon as possible before use

Crop Safety/Restrictions

- Maximum number of treatments depends on application method (see label)
- After drenching wash spray residue from crop foliage
- Do not use on Escallonia, Pyracantha, Gloxinia spp., pansies or lettuces
- Germination of lettuce in previously treated soil may be impaired
- Do not drench seedlings until well established
- When treating compost for blocking reduce dose by 50%
- Test on small numbers of plants in advance when treating subjects of unknown susceptibility or using compost with more than 20% inert material

Special precautions/Environmental safety

- Irritating to eyes and skin

Protective clothing/Label precautions

- A, C [1, 2]
- R04a, R04b, U02, U04a, U05a, U08, U20b, C03, E01, E15, E30a, E32a [1, 2]; U19, E34 [1]

Latest application/Harvest Interval(HI)

- 24 h after seeding watercress.
- HI tomatoes, cucumbers, mustard and cress 3 d

Approval

- Off-label approval unlimited for use on tomatoes grown on rock wool (OLA 0600/94)[1]; unlimited for use on watercress propagation beds (OLA 1213/96)[1]

---

262 etrimfos

A contact organophosphorus insecticide for stored grain crops

Products

1 Satisfar	Nickerson Seeds	525 g/l	EC	04180
2 Satisfar Dust	Nickerson Seeds	2% w/w	DP	04085

Uses Grain beetles in STORED GRAIN, STORED OILSEED RAPE [1, 2]. Grain beetles in GRAIN STORES [1]. Grain storage mites in STORED GRAIN, STORED OILSEED RAPE [1, 2]. Grain storage mites in GRAIN STORES [1]. Grain storage pests in STORED GRAIN, STORED OILSEED RAPE [1, 2]. Grain storage pests in GRAIN STORES [1]. Grain weevils in STORED GRAIN, STORED OILSEED RAPE [1, 2]. Grain weevils in GRAIN STORES [1].

Notes

Efficacy
- Spray internal surfaces of clean store with knapsack or motorized sprayer
- Allow sufficient time for pests to emerge from hiding places and make contact with chemical before grain is stored
- Apply as admixture treatment to grain as it enters store using a suitable stored grain sprayer or automatic seed treater if using dust
- Controls malathion-resistant beetles and gamma-HCH-resistant mites

Crop Safety/Restrictions
- Maximum number of treatments 1 per batch
- Cool grain to below 15°C before treatment and storage. Moisture content of grain should not exceed 16%

Special precautions/Environmental safety
- This product contains an anticholinesterase organophosphorus compound. Do not use if under medical advice not to work with such compounds
- Harmful to game, wild birds and animals
- Dangerous to fish or other aquatic life. Do not contaminate surface waters or ditches with chemical or used container

Protective clothing/Label precautions
- A, C [1, 2]
- M03, R04a, R07d, U05a, U08, U20b, C03, E01 [1]; U09a, U20a [2]; U19, E10b, E13b, E30a, E32a [1, 2]

Latest application/Harvest Interval(HI)
- On entry into store [2]; on removal from store [1]

Maximum Residue Levels (mg residue/kg food)
- cereals (except rice) 5

---

263 fatty acids

A soap concentrate insecticide and acaricide

Products

Savona	Koppert	49% w/w	SL	06057

Uses

Aphids in BEANS, BRUSSELS SPROUTS, CABBAGES, CUCUMBERS, FRUIT TREES, LETTUCE, PEAS, PEPPERS, PUMPKINS, TOMATOES, WATERCRESS *(off-label)*, WOODY ORNAMENTALS. Mealy bugs in BEANS, BRUSSELS SPROUTS, CABBAGES, CUCUMBERS, FRUIT TREES, LETTUCE, PEAS, PEPPERS, PUMPKINS, TOMATOES, WOODY ORNAMENTALS. Red spider mites in BEANS, BRUSSELS SPROUTS, CABBAGES, CUCUMBERS, FRUIT TREES, LETTUCE, PEAS, PEPPERS, PUMPKINS, TOMATOES, WOODY ORNAMENTALS. Scale insects in BEANS, BRUSSELS SPROUTS, CABBAGES, CUCUMBERS, FRUIT TREES, LETTUCE, PEAS, PEPPERS, PUMPKINS, TOMATOES, WOODY ORNAMENTALS. Whitefly in BEANS, BRUSSELS SPROUTS, CABBAGES, CUCUMBERS, FRUIT TREES, LETTUCE, PEAS, PEPPERS, PUMPKINS, TOMATOES, WOODY ORNAMENTALS.

Notes

Efficacy

- Use only soft or rain water for diluting spray
- For glasshouse use apply when insects first seen and repeat as necessary
- To control whitefly spray when required and use biological control after 12 h

Crop Safety/Restrictions

- Do not use on new transplants, newly rooted cuttings or plants under stress
- Do not use on specified susceptible shrubs. See label for details

Special precautions/Environmental safety

- Harmful to fish or other aquatic life. Do not contaminate surface waters or ditches with chemical or used container

Protective clothing/Label precautions

- U20c, E13c, E26, E30a, E31a

Latest application/Harvest Interval(HI)

- HI zero

Approval

- Off-label approval to Apr 1999 for use on watercress grown outdoors (OLA 0735/94)[1]

264 fenarimol

A systemic curative and protective pyrimidine fungicide

Products

1 Rimidin	Rigby Taylor	120 g/l	SC	05907
2 Rubigan	Dow	120 g/l	SC	05489

Uses

Dollar spot in MANAGED AMENITY TURF, SPORTS TURF [1]. Fusarium patch in MANAGED AMENITY TURF, SPORTS TURF [1]. Powdery mildew in APPLES, BLACKCURRANTS, GOOSEBERRIES, PROTECTED PEPPERS *(off-label)*, PROTECTED TOMATOES *(off-label)*, PUMPKINS *(off-label)*, RASPBERRIES, ROSES, SQUASHES *(off-label)*, STRAWBERRIES [2]. Red thread in MANAGED AMENITY TURF, SPORTS TURF [1]. Scab in APPLES [2].

Notes

Efficacy

- Recommended spray interval varies from 7-14 d depending on crop and climatic conditions. See label for timing details [2]
- Efficient coverage and short spray intervals essential, especially for scab control [2]
- Spray strawberry runner beds regularly throughout season [2]
- For turf disease control spray as preventive treatment and repeat as necessary. See label for details [1]
- Do not mow within 24 h after treatment [1]

Crop Safety/Restrictions

- Maximum number of treatments 15 per yr for apples; 3 per yr for raspberries
- Do not use on trees that are under stress from drought, severe pest damage, mineral deficiency or poor soil conditions

Special precautions/Environmental safety
- Dangerous to fish or other aquatic life. Do not contaminate surface waters or ditches with chemical or used container

Protective clothing/Label precautions
- A, C
- U08, U19, C02 (14 d), E13b, E26, E30a, E31b

Latest application/Harvest Interval(HI)
- HI tomatoes, peppers 2 d; other crops 14 d

Approval
- Off-label approval unlimited for use on protected tomatoes, protected peppers (OLA 0645/94) [2]; to Feb 2001 for use on squashes, pumpkins (OLA 0235/96)[2]

Maximum Residue Levels (mg residue/kg food)
- hops 5; currants, goosberries 1; pome fruits, grapes 0.3; tea 0.05; citrus fruits, tree nuts, blackberries, dewberries, loganberries, bilberries, cranberries, wild berries, miscellaneous fruits, root and tuber vegetables, bulb vegetables, sweet corn, brassica vegetables, leaf vegetables and herbs, beans (with and without pods), stem vegetables (except globe artichokes), fungi, pulses, oilseeds, potatoes, rye, oats, triticale, maize, rice, animal products (except liver and kidney) 0.02

265 fenazaquin

A mitochondrial electron transport inhibitor

Products Matador 200 SC | Dow | 200 g/l | SC | 07960

Uses Red spider mites in APPLES. Two-spotted spider mite in ORNAMENTALS.

Notes

Efficacy
- Acts by contact to give rapid knockdown
- Treat apples after petal fall when most overwintered eggs have hatched but before damage is seen
- Product should be used as part of a pest control programme. A further acaricide treatment may be necessary after application
- Control may be reduced where water volumes are reduced or in orchards where water volumes above 750 l/ha are required for good crop cover

Crop Safety/Restrictions
- Maximum number of treatments 1 per yr
- Other mitochondrial electron transport inhibitor (METI) acaricides should not be applied to the same crop in the same calendar yr either separately or in mixture
- Do not apply to roses. Do not treat new ornamental species or varieties without first testing a few plants on a small scale
- Do not use on edible crops other than apples
- Do not treat ornamentals when in blossom or under stress

Special precautions/Environmental safety
- Harmful if swallowed
- Extremely dangerous to fish or other aquatic life. Do not contaminate surface waters or ditches with chemical or used container

- Do not allow direct spray from air-assisted spraying equipment to fall within 15 m, or from hand-held sprayers to within 2 m, of surface waters or ditches. Direct spray away from water

Protective clothing/Label precautions

- A
- M03, R03c, U05a, U19, C02 (30 d), C03, E01, E13a, E16 (hand held sprayers: 2 m), E17 (15 m), E20, E30a, E31a, E34

Latest application/Harvest Interval(HI)

- HI 30 d for apples

266 fenbuconazole

A systemic protectant and curative triazole fungicide for top fruit

Products	Indar 5EW	Headland	50 g/l	EW	07580

Uses Scab in APPLES, PEARS.

Notes

Efficacy

- Most effective when used as part of a routine preventative programme from bud burst to onset of petal fall
- After petal fall, tank mix with other protectant fungicides to enhance scab control
- Safe to use on all main commercial varieties of apples and pears in UK
- See label for recommended spray intervals. In periods of rapid growth or high disease pressure, a 7 d interval should be used

Crop Safety/Restrictions

- Maximum total dose 14 l/ha per yr
- Consult processors before using on pears for processing

Special precautions/Environmental safety

- Irritant. May cause serious damage to eyes
- Harmful to fish or other aquatic life. Do not contaminate surface waters or ditches with chemical or used container
- Do not harvest crops for human or animal consumption for at least 4 wk after last application
- Effect on parasites and predators used in IPM systems not fully established and safety cannot be assumed

Protective clothing/Label precautions

- A, C
- R04, R04d, U05a, U08, U20a, C02 (4 wk), C03, E01, E13c, E32a

Latest application/Harvest Interval(HI)

- 28 d before harvest

267 fenbuconazole + propiconazole

A broad spectrum systemic fungicide mixture with protectant, curative and eradicant properties for use in wheat

Products

1 Graphic	Ciba Agric.	37.5:47 g/l	EC	07585
2 Graphic	Novartis	37.5:47 g/l	EC	08415

Uses

Brown rust in WINTER WHEAT. Powdery mildew in WINTER WHEAT *(moderate control)*. Septoria in WINTER WHEAT. Yellow rust in WINTER WHEAT.

Notes

Efficacy

- Spray at start of disease attack or as part of disease control programme. Apply a second treatment 2-5 wk later if disease pressure remains high or if wet weather continues
- Treat in spring from ear at 1 cm stage (GS 30)

Crop Safety/Restrictions

- Maximum number of treatments 2 per crop
- Use only in spring or summer
- If used in a programme with other products containing propiconazole do not apply more than 500 g a.i./ha to the crop in any one season

Special precautions/Environmental safety

- Irritating to skin
- Risk of serious damage to eyes
- Dangerous to fish or other aquatic life. Do not contaminate surface waters or ditches with chemical or used container
- Do not harvest crops for human or animal consumption for at least 5 wk after application

Protective clothing/Label precautions

- A, C
- R04b, R04d, U02, U05a, U09a, U20a, C02 (5 wk), C03, E01, E13b

Latest application/Harvest Interval(HI)

- Before start of flowering (GS 61)

Maximum Residue Levels (mg residue/kg food)

- see propiconazole entry

268 fenbuconazole + tridemorph

A broad spectrum fungicide mixture for cereals

Products

Unison	PBI	37.5:225 g/l	EC	08318

Uses

Brown rust in SPRING BARLEY, WINTER BARLEY, WINTER WHEAT. Glume blotch in WINTER WHEAT. Leaf spot in WINTER WHEAT. Powdery mildew in SPRING BARLEY, WINTER BARLEY, WINTER WHEAT. Rhynchosporium in SPRING BARLEY, WINTER BARLEY. Yellow rust in WINTER WHEAT.

Notes

Efficacy

- Product has eradicant and protectant activity but best results are obtained from treatments made at the first signs of disease
- First treatment may be applied from the first node detectable stage (GS 31). Repeat applications may be made 3-4 wk later if necessary
- For control of leaf diseases in wheat application soon after flag leaf emergence (GS 37-39) is particularly effective

Crop Safety/Restrictions

- Maximum number of treatments 2 per crop

Special precautions/Environmental safety

- Irritating to skin
- Risk of serious damage to eyes
- Dangerous to fish or other aquatic life. Do not contaminate surface waters or ditches with chemical or used container

Protective clothing/Label precautions

- A, C
- R04b, R04d, U04a, U05a, U08, U19, U20b, C03, E01, E07 (14 d), E13b, E26, E30a, E31b

Latest application/Harvest Interval(HI)

- Before flowering (GS 59)

269 fenbutatin oxide

A selective contact and ingested organotin acaricide

Products	Torq	Fargro	50% w/w	WP	08370

Uses

Two-spotted spider mite in PROTECTED CUCUMBERS, PROTECTED ORNAMENTALS, PROTECTED PEPPERS *(off-label)*, PROTECTED TOMATOES, TUNNEL GROWN STRAWBERRIES.

Notes

Efficacy

- Active on larvae and adult mites. Spray may take 7-10 d to effect complete kill but mites cease feeding and crop damage stops almost immediately
- Apply as soon as mites first appear and repeat as necessary
- On tunnel-grown strawberries apply when mites first appear, usually before flowering starts and repeat 10-14 d later. A post-harvest spray is also recommended

Crop Safety/Restrictions

- Maximum number of treatments 2 per crop pre-harvest + 1 post harvest for tunnel grown strawberries
- Allow at least 10 d between spray applications and do not apply within 10 d of a previous spray
- Do not add wetters or mix with anything other than water

- Do not use white petroleum oil and Torque within 28 d of each other or any other pesticide and Torque within 7 d of each other. Do not apply to crops which are under stress for any reason
- On subjects of unknown susceptibility test treat on a small number of plants in advance

Special precautions/Environmental safety

- Irritating to eyes, skin and respiratory system
- Extremely dangerous to fish or other aquatic life. Do not contaminate surface waters or ditches with chemical or used container
- Not harmful to bees or to *Encarsia* or *Phytoseiulus* being used for biological control

Protective clothing/Label precautions

- A, C, D, H, M
- M03, R04a, R04b, R04c, U05a, U08, U19, U20a, C03, E01, E13a, E30a, E32a, E34

Latest application/Harvest Interval(HI)

- HI glasshouse cucumbers, tomatoes 3 d; tunnel-grown strawberries 7 d

Approval

- Off-label approval unlimited for use on protected peppers (OLA 0857/97)[1]

Maximum Residue Levels (mg residue/kg food)

- pome fruit, grapes 2; cucumbers 0.5; tea 0.1; all other products 0.05

---

270 fenhexamid

A protectant fungicide for soft fruit

Products

Teldor	Bayer	51% w/w	WG	08955

Uses

Botrytis in BLACKBERRIES, BLACKCURRANTS, GOOSEBERRIES, LOGANBERRIES, RASPBERRIES, REDCURRANTS, STRAWBERRIES, WHITECURRANTS.

Notes

Efficacy

- Use as part of a programme of sprays throughout the flowering period to achieve effective control of Botrytis
- To minimise possibility of development of resistance no more than two sprays of the product may be applied consecutively. Other fungicides from a different chemical group should then be used for at least two consecutive sprays
- Complete spray cover of all flowers and fruitlets throughout the blossom period is essential for successful control of Botrytis
- Spray programmes should normally start at the start of flowering

Crop Safety/Restrictions

- Maximum number of treatments 4 per yr but no more than 2 sprays may be applied consecutively

Special precautions/Environmental safety

- Harmful to fish or other aquatic life. Do not contaminate surface waters or ditches with chemical or used container

Protective clothing/Label precautions

- U08, U19, U20b, E13c, E30a, E32a

Latest application/Harvest Interval(HI)

- HI 7 d for raspberries, loganberries, blackberries, blackcurrants, redcurrants, whitecurrants, gooseberries; 1 d for strawberries

271 fenitrothion

A broad spectrum, contact organophosphorus insecticide

Products

1 Dicofen	PBI	500 g/l	EC	00693
2 EC-Kill	Antec	500 g/l	EC	H5468
3 Micromite	Grampian	500 g/l	EC	H4480
4 Unicrop Fenitrothion 50	Unicrop	500 g/l	EC	02267

Uses

Ants in FOOD STORAGE AREAS, LIVESTOCK HOUSES [2, 3]. Ants in REFUSE TIPS [2]. Aphids in APPLES, PEARS, PEAS, PLUMS [1, 4]. Apple blossom weevil in APPLES [1, 4]. Bedbugs in FOOD STORAGE AREAS, LIVESTOCK HOUSES [2, 3]. Bedbugs in REFUSE TIPS [2]. Beetles in FOOD STORAGE AREAS, LIVESTOCK HOUSES [2, 3]. Beetles in REFUSE TIPS [2]. Capsids in APPLES, BLACKCURRANTS, GOOSEBERRIES [1, 4]. Caterpillars in PEARS, PLUMS [1, 4]. Cockroaches in FOOD STORAGE AREAS, LIVESTOCK HOUSES [2, 3]. Cockroaches in REFUSE TIPS [2]. Codling moth in APPLES [1, 4]. Crickets in FOOD STORAGE AREAS, LIVESTOCK HOUSES [2, 3]. Crickets in REFUSE TIPS [2]. Earwigs in FOOD STORAGE AREAS, LIVESTOCK HOUSES [2, 3]. Earwigs in REFUSE TIPS [2]. Fleas in FOOD STORAGE AREAS, LIVESTOCK HOUSES [2, 3]. Fleas in REFUSE TIPS [2]. Flies in FOOD STORAGE AREAS, LIVESTOCK HOUSES [2, 3]. Flies in REFUSE TIPS [2]. Flour beetles in GRAIN STORES [1, 4]. Frit fly in MAIZE, SWEETCORN [1, 4]. Grain beetles in GRAIN STORES [1, 4]. Grain weevils in GRAIN STORES [1, 4]. Leatherjackets in BARLEY, DURUM WHEAT, OATS, RYE, TRITICALE, WHEAT [1, 4]. Leatherjackets in MAIZE, SWEETCORN [4]. Mites in FOOD STORAGE AREAS, LIVESTOCK HOUSES [2, 3]. Mites in REFUSE TIPS [2]. Moths in FOOD STORAGE AREAS, LIVESTOCK HOUSES [2, 3]. Moths in REFUSE TIPS [2]. Pea and bean weevils in PEAS [1, 4]. Pea midge in PEAS [1, 4]. Pea moth in PEAS [1, 4]. Poultry house pests in POULTRY HOUSES [2, 3]. Raspberry beetle in RASPBERRIES [1, 4]. Raspberry cane midge in RASPBERRIES [1, 4]. Saddle gall midge in BARLEY, DURUM WHEAT, OATS, RYE, TRITICALE, WHEAT [1, 4]. Sawflies in APPLES, BLACKCURRANTS, GOOSEBERRIES [1, 4]. Silverfish in FOOD STORAGE AREAS, LIVESTOCK HOUSES [2, 3]. Silverfish in REFUSE TIPS [2]. Suckers in APPLES [1, 4]. Thrips in BARLEY, PEAS, WHEAT [1, 4]. Thrips in LEEKS *(off-label)*, SALAD ONIONS *(off-label)*, SWEETCORN *(off-label)* [1]. Tortrix moths in APPLES, STRAWBERRIES [1, 4]. Wheat-blossom midges in WHEAT [1, 4]. Winter moth in APPLES [1, 4].

Notes

Efficacy

- Number and timing of sprays vary with disease and crop. See label for details
- Apply as bran bait for leatherjacket control in cereals
- For cane midge add suitable spreader and apply to lower 60 cm of young canes
- For control of grain store pests apply to all surfaces of empty stores before filling with grain. Remove dust and debris before applying
- For use in poultry houses apply to breeding sites etc after depopulation [3]
- Where pear suckers resistant to fenitrothion occur control is unlikely to be satisfactory [1, 4]

Crop Safety/Restrictions

- Maximum number of treatments 1 per crop for cereals, maize, sweetcorn; 2 per crop or yr for peas, plums, blackcurrants, gooseberries, strawberries, grain stores; 3 per crop for apples, pears; 4 per crop for raspberries
- Do not mix with magnesium sulfate or highly alkaline materials
- On raspberries may cause slight yellowing of leaves of some varieties which closely resembles that caused by certain viruses

Special precautions/Environmental safety

- This product contains an anticholinesterase organophosphorus compound. Do not use if under medical advice not to work with such compounds
- Harmful in contact with skin or if swallowed [1-4]. Irritating to eyes and skin [1, 4]
- Flammable [1-4]
- Harmful to livestock [1, 4]
- Harmful to game, wild birds and animals [1, 4]
- Harmful to fish or other aquatic life. Do not contaminate surface waters or ditches with chemical or used container
- Dangerous to bees. Do not apply to crops in flower or to those in which bees are actively foraging. Do not apply when flowering weeds are present [1, 4]
- Do not apply directly to livestock and poultry. Remove exposed milk and collect eggs before application. Protect milk machinery and milk containers from contamination [3]

Protective clothing/Label precautions

- A [1-4]; C [1, 3, 4]; D [3]; H [2, 3]
- M01, M03, R03a, R03c, R07d, U05a, U13, U19, C03, E01, E13c, E30a, E31a, E34 [1-4]; R04a, R04b, U02, U08, U14, U15, C02, E10b, E12c [1, 4]; R05a [1]; U09a [2, 3]; U16, U17, U20b, C04, C05, C08, E02, E05 [3]; U20a [1, 2, 4]; E02 (48 h) [2]; E06b (7 d) [4]

Withholding period

- Keep all livestock out of treated areas for at least 7 d [1, 4]
- Keep unprotected persons and animals out of treated areas for 48 h or until surfaces are dry [2]

Latest application/Harvest Interval(HI)

- HI raspberries 7 d; other crops 2 wk

Approval

- Approved for aerial application on cereals, peas [1, 4]. See notes in Section 1
- Off-label approval unlimited for use on leeks, salad onions, sweetcorn (OLA 0352/96)[1]

Maximum Residue Levels (mg residue/kg food)

- cereals 5; citrus fruits 2; pome fruits, apricots, peaches, nectarines, plums, grapes, cane fruits, bilberries, cranberries, currants, gooseberries, bananas, carrots, horseradish, parsnips, parsley root, salsify, swedes, turnips, garlic, onions, shallots, tomatoes, peppers, aubergines, cucumbers, gherkins, courgettes, cauliflowers, Brussels sprouts, head cabbages, lettuce, beans (with pods), peas (with pods), celery, leeks, rhubarb, cultivated mushrooms 0.5; potatoes 0.05

272 fenitrothion + permethrin + resmethrin

An organophosphate/pyrethroid insecticide mixture for grain stores

Products					
	Turbair Grain Store Insecticide	Graincare	10:20:2 g/l	UL	02238

Uses Flour beetles in GRAIN STORES. Grain beetles in GRAIN STORES. Grain moths in GRAIN STORES. Grain weevils in GRAIN STORES.

Notes

Efficacy
- Apply as a surface spray in empty stores using a suitable fan-assisted ULV sprayer
- Clean store thoroughly before applying
- Combines knock-down effect with up to 5 mth residual activity
- Should not be mixed with other sprays

Special precautions/Environmental safety
- This product contains an anticholinesterase organophosphorus compound. Do not use if under medical advice not to work with such compounds
- Irritating to eyes, skin and respiratory system
- Highly flammable
- Extremely dangerous to fish or other aquatic life. Do not contaminate surface waters or ditches with chemical or used container

Protective clothing/Label precautions
- A, C
- M01, R04a, R04b, R04c, R07c, U05a, U08, U19, U20a, C03, E01, E13a, E30a, E31a, E34

Approval
- Product formulated for application as a ULV spray. See label for details [1]

Maximum Residue Levels (mg residue/kg food)
- see fenitrothion and permethrin entries

---

273 fenoxaprop-ethyl

A phenoxypropionic acid herbicide for grass weed control in wheat

Products

Landgold Fenoxaprop	Landgold	60 g/l	EW	06352

Uses Blackgrass in WINTER WHEAT. Wild oats in WINTER WHEAT.

Notes

Efficacy
- Apply post-weed emergence from 2-leaf to flag leaf ligule visible stage of weeds (GS 12-39)
- Treat awned canary grass from 2-leaf stage up to end of tillering but before first node detectable (GS 31)
- Susceptible weeds stop growing within 3 d of spraying and die usually within 2-4 wk
- Spray is rainfast 1 h after application
- Do not spray onto wet foliage or leaves covered with ice
- Dry conditions resulting in moisture stress may reduce effectiveness

Crop Safety/Restrictions
- Maximum number of treatments 2 per crop
- Do not apply to barley, durum wheat, undersown crops or crops to be undersown

- Treat from crop emergence up to and including flag leaf ligule just visible (GS 39)
- Do not mix with products containing bifenox or hormone herbicides
- Do not roll or harrow within 1 wk of spraying
- Do not spray crops under stress, crops suffering from drought, waterlogging or nutrient deficiency or those which have been grazed or if soil compacted

Special precautions/Environmental safety
- Dangerous to fish or other aquatic life. Do not contaminate surface waters or ditches with chemical or used container

Protective clothing/Label precautions
- A, C
- U08, U20a, E13b, E30a, E31b

Latest application/Harvest Interval(HI)
- Before flag leaf sheath extending (GS 41)

274 fenoxaprop-P-ethyl

A phenoxypropionic acid herbicide for use in wheat

See also diclofop-methyl + fenoxaprop-P-ethyl

Products

1 Cheetah Super	AgrEvo	55 g/l	EW	08723
2 Triumph	AgrEvo	120 g/l	EC	08740

Uses

Blackgrass in SPRING WHEAT, WINTER WHEAT. Canary grass in SPRING WHEAT, WINTER WHEAT. Loose silky bent in SPRING WHEAT, WINTER WHEAT. Rough meadow grass in SPRING WHEAT, WINTER WHEAT. Wild oats in SPRING WHEAT, WINTER WHEAT.

Notes

Efficacy
- Treat weeds from 2 fully expanded leaves up to flag leaf ligule just visible; for awned canary-grass from 2 leaves to the end of tillering
- A second application may be made in spring where susceptible weeds emerge after an autumn application
- Spray is rainfast 1 h after application
- Product may be sprayed in frosty weather provided crop hardened off but do not spray wet foliage or leaves covered with ice
- Dry conditions resulting in moisture stress may reduce effectiveness

Crop Safety/Restrictions
- Maximum total dose equivalent to two full dose treatments
- Treat from crop emergence to flag leaf fully emerged (GS 41).
- Do not apply to barley, durum wheat, undersown crops or crops to be undersown
- Do not roll or harrow within 1 wk of spraying
- Do not spray crops under stress, suffering from drought, waterlogging or nutrient deficiency or those grazed or if soil compacted
- Broadcast crops should be sprayed post-emergence after plants have developed well-established root system
- Avoid spraying immediately before or after a sudden drop in temperature or a period of warm days/cold nights
- Do not mix with products containing bifenox or hormones

Special precautions/Environmental safety
- Irritating to eyes and skin
- Dangerous to fish or other aquatic life. Do not contaminate surface waters or ditches with chemical or used container

Protective clothing/Label precautions
- A, C, H [1, 2]
- R04a, U08, E26 [2]; R04b, U05a, U20b, C03, E01, E13c, E30a, E31b [1, 2]; U09a [1]

Latest application/Harvest Interval(HI)
- Before flag leaf sheath extending (GS 41)

275 fenoxaprop-P-ethyl + isoproturon

A foliar and root-acting grass killer for use in wheat

Products

Puma X	AgrEvo	14:300 g/l	SE	08779

Uses

Annual meadow grass in SPRING WHEAT, WINTER WHEAT. Blackgrass in SPRING WHEAT, WINTER WHEAT. Canary grass in SPRING WHEAT, WINTER WHEAT. Chickweed in SPRING WHEAT, WINTER WHEAT. Corn marigold in SPRING WHEAT, WINTER WHEAT. Hemp-nettle in SPRING WHEAT, WINTER WHEAT. Mayweeds in SPRING WHEAT, WINTER WHEAT. Wild oats in SPRING WHEAT, WINTER WHEAT.

Notes

Efficacy
- Annual grass weeds listed are controlled up to and including end of tillering (GS 29) (annual meadow grass up to and including 1 tiller)
- Chickweed and mayweeds controlled up to 6 leaf stage, hemp-nettle and corn marigold up to 2 leaves
- Emerged weeds controlled on all soil types and foliar activity on grass weeds unaffected by high organic matter. Residual activity reduced on soils with more than 10% organic matter
- If prolonged dry weather follows application the speed of action is slower and weed control may be reduced, especially in spring
- In seasons of above average rainfall or where heavy rain falls shortly after application weed control may be reduced
- Foliar activity on grass weeds is unaffected by rainfall from 1 h after application

Crop Safety/Restrictions
- Maximum number of treatments 1 per crop. Maximum total dose of isoproturon 2.5 kg a.i./ha per crop
- Apply from emergence of crop up to and including 1st node stage (GS 31)
- Only treat autumn sown spring wheat after 1 Jan
- Broadcast crops should be sprayed post-emergence after plants have a well-established root system
- Do not apply to barley, durum wheat, undersown crops or those due to be undersown
- Do not spray if frost imminent or after onset of frosty weather
- On free draining, stony or gravelly soils there is a risk of crop damage especially if heavy rain falls soon after application

- Do not spray crops lacking nutrient or under stress from drought, pest or disease attack, waterlogging or soil compaction
- Early sown crops (Sep) may be damaged if spraying precedes or coincides with a period of rapid growth in autumn
- Do not roll within 1 wk before or after or harrow within 1 wk before or at any time after application
- Do not mix with hormone herbicides or bifenox products. See label for permitted mixtures

Special precautions/Environmental safety
- Irritating to skin
- Dangerous to fish or aquatic life. Do not contaminate surface waters or ditches with chemical or used container

Protective clothing/Label precautions
- A, C
- R04b, U04a, U05a, U08, U20a, C03, E01, E13b, E26, E30a, E31b

Latest application/Harvest Interval(HI)
- Before second node detectable (GS 32)

276 fenoxycarb

An insect specific growth regulator for top fruit

Products

Insegar	Novartis	250 g/l	WP	08558

Uses

Summer-fruit tortrix moth in APPLES, PEARS.

Notes

Efficacy
- Best results from application at 5th instar stage before pupation. Product prevents transformation from larva to pupa
- Correct timing best identified from pest warnings
- Because of mode of action rapid knock-down of pest is not achieved and larvae continue to feed for a period after treatment
- Adequate water volume necessary to ensure complete coverage of leaves

Crop Safety/Restrictions
- Maximum number of treatments 2 per crop
- Use on all varieties of apples and pears

Special precautions/Environmental safety
- High risk to bees. Do not apply to crops in flower or to those in which bees are actively foraging. Do not apply when flowering weeds are present
- Dangerous to fish or other aquatic life. Do not contaminate surface waters or ditches with chemical or used container
- Do not allow direct spray from broadcast air-assisted sprayers to fall within 18 m of surface waters or ditches. Direct spray away from water
- Apply to minimise off-target drift to reduce effects on non-target organisms. Some margin of safety to beneficial arthropods is indicated.

Protective clothing/Label precautions
- A, D, H
- U05a, U20a, C03, E01, E12a, E13b, E17, E22b, E29, E30a, E31b, E34

Latest application/Harvest Interval(HI)
- HI 42 d

277 fenpiclonil

A cyanopyrrole fungicide seed treatment for potatoes

Products	Gambit	Novartis	400 g/l	FS	08535

Uses

Black dot in SEED POTATOES. Black scurf and stem canker in SEED POTATOES. Dry rot in SEED POTATOES. Gangrene in SEED POTATOES. Silver scurf in SEED POTATOES. Skin spot in SEED POTATOES.

Notes

Efficacy
- Apply to clean soil-free tubers using conventional hydraulic application equipment, including spinning disc or electrostatic sprayers
- Ensure good even distribution over whole tuber surface
- Product acts primarily against seed-borne diseases. Control of soil-borne diseases may be reduced where soil-borne inoculum present
- Gangrene and dry rot are controlled only during storage

Crop Safety/Restrictions
- Maximum number of treatments 1 per batch of seed potatoes
- Treat seed potatoes at any time from taking out of store to planting. Seed may be treated into store only if there is no possibility that tubers will be used for another purpose
- Treatment recommended pre-chitting as emerged shoots susceptible to mechanical damage resulting in delayed emergence
- Treated tubers may be used only as seed for ware crop or starch production

Special precautions/Environmental safety
- Irritant. May cause sensitization by skin contact
- Dangerous to fish or other aquatic life. Do not contaminate surface waters or ditches with chemical or used container
- Do not use treated seed as food or feed

Protective clothing/Label precautions
- A, D, E, H
- R04, R04e, U20b, E13b, E26, E30a, E31b, S01, S02, S03, S04a, S05, S06

Latest application/Harvest Interval(HI)
- Before drilling

278 fenpropathrin

A contact and ingested pyrethroid acaricide and insecticide

Products	Meothrin	Cyanamid	100 g/l	EC	0720(

Uses Caterpillars in APPLES, BLACKCURRANTS, HOPS. Damson-hop aphid in HOPS. Red spider mites in APPLES, BLACKCURRANTS, HOPS. Two-spotted spider mite in HOPS, ROSES, STRAWBERRIES.

Notes

Efficacy

- Acts on motile stages of mites and gives rapid kill
- Apply on apples as post-blossom spray and repeat 3-4 wk later if necessary
- Apply on hops when hatch of spring eggs complete in May and repeat if necessary
- Apply on roses when pest first seen and repeat as necessary
- Apply on strawberries as a pre-flowering spray and repeat after harvest if necessary
- Where aphids resistant to fenpropathrin occur in hops control is unlikely to be satisfactory and repeat treatments may result in lower levels of control

Crop Safety/Restrictions

- Maximum number of treatments 2 per crop for apples and hops; 2 per yr, 1 pre-flowering and 1 post-harvest for strawberries; 3 per yr for blackcurrants

Special precautions/Environmental safety

- Toxic in contact with skin or if swallowed. Irritating to eyes and skin
- Flammable
- Extremely dangerous to bees. Do not apply to crops in flower or to those in which bees are actively foraging except as directed in blackcurrants. Do not apply when flowering weeds are present
- Extremely dangerous to fish or other aquatic life. Do not contaminate surface waters or ditches with chemical or used container
- Do not operate air-assisted sprayers within 18 m of surface water or ditches, other wheeled sprayers within 6 m, hand-held sprayers within 2 m. Direct spray away from water

Protective clothing/Label precautions

- A, C, H, J
- M04, R02a, R02c, R04a, R04b, R07d, U02, U04a, U05a, U08, U19, U20a, C02, C03, E01, E12b, E13a, E16, E30a, E31a, E34

Latest application/Harvest Interval(HI)

- Pre-flowering for strawberries; 14 d after end of flowering for blackcurrants.
- HI apples and hops 7 d

279 fenpropidin

A systemic, curative and protective piperidine (morpholine) fungicide

Products

Product	Company	Concentration	Formulation	Reg. No.
1 Landgold Fenpropidin 750	Landgold	750 g/l	EC	08973
2 Mallard	Ciba Agric.	750 g/l	EC	07934
3 Mallard	Novartis	750 g/l	EC	08662
4 Patrol	Zeneca	750 g/l	EC	08661
5 Tern	Ciba Agric.	750 g/l	EC	07933
6 Tern	Novartis	750 g/l	EC	08660

Uses Brown rust in BARLEY, WHEAT [1-6]. Glume blotch in WHEAT [1-3, 5, 6]. Leaf blotch in BARLEY [1-6]. Leaf spot in WHEAT [1-3, 5, 6]. Powdery mildew in BARLEY, WHEAT [1-6]. Yellow rust in BARLEY, WHEAT [1-6].

Notes

Efficacy

- Best results obtained when applied at early stage of disease development. See label for details of recommended timing alone and in mixtures
- Disease control enhanced by vapour-phase activity. Control can persist for 4-6 wk
- Alternate with triazole fungicides to discourage build-up of resistance

Crop Safety/Restrictions

- Maximum number of treatments 3 per crop (up to 2 in yr of harvest) for winter crops; 2 per crop for spring crops

Special precautions/Environmental safety

- Harmful in contact with skin and if swallowed
- Irritating to skin
- Risk of serious damage to eyes
- May cause sensitization by skin contact
- Dangerous to fish or other aquatic life. Do not contaminate surface waters or ditches with chemical or used container

Protective clothing/Label precautions

- A, C, H
- M03, R03a, R03c, R04b, R04d, R04e, U02, U04a, U05a, U05b, U10, U20a, C02 (5 wk), C03, E01, E13b, E29, E30a, E31c, E34

Latest application/Harvest Interval(HI)

- Up to and including ear emergence complete (GS 59).
- HI 5 wk

Approval

- Off-label approval unlimited for use on peas, wheat, barley, Brussels sprouts (research/breeding purposes) (OLA 0320/97)[4]

280 fenpropidin + fenpropimorph

A contact and systemic fungicide mixture for use in wheat and barley

Products

1 Agrys	Ciba Agric.	480:270 g/l	EC	08083
2 Agrys	Novartis	480:270 g/l	EC	08382
3 Boscor	Ciba Agric.	188:562 g/l	EC	07416
4 Boscor	Novartis	188:562 g/l	EC	08682

Uses

Brown rust in SPRING BARLEY, SPRING WHEAT, WINTER BARLEY, WINTER WHEAT [1-4]. Powdery mildew in SPRING BARLEY, SPRING WHEAT, WINTER BARLEY, WINTER WHEAT [1-4]. Rhynchosporium in SPRING BARLEY *(moderate control)*, WINTER BARLEY *(moderate control)* [1, 2]. Yellow rust in SPRING BARLEY, SPRING WHEAT, WINTER BARLEY, WINTER WHEAT [1, 2].

Notes

Efficacy

- Spray at start of disease attack or as part of a disease control programme
- Repeat treatment may be needed 3-4 wk later if disease pressure high

- Best results obtained when application made at early stage of disease development. Treatment of established infections will be less effective

Crop Safety/Restrictions
- Maximum total dose 2.0 l/ha

Special precautions/Environmental safety
- Irritating to skin
- Risk of serious damage to eyes
- Flammable [3, 4]
- Harmful to fish or other aquatic life. Do not contaminate surface waters or ditches with chemical or used container
- Do not allow direct spray from ground-based vehicle-mounted/drawn sprayers to fall within 6 m of surface waters or ditches. Direct spray away from water [1, 2, 3, 4]

Protective clothing/Label precautions
- A, C, H
- R04b, R04d, U02, U05a, C03, E01, E13c, E26, E30a, E31b [1-4]; R07d, U04a, U10, E29, E34 [3, 4]; U09a, U19, U20a, E16 [1, 2]

Latest application/Harvest Interval(HI)
- Before start of flowering (GS 60)

281 fenpropidin + prochloraz

A contact and systemic fungicide mixture for cereals

Products Sponsor AgrEvo 250:250 g/l EC 08674

Uses Brown rust in BARLEY, WHEAT. Eyespot in BARLEY, WHEAT. Glume blotch in WHEAT. Leaf blotch in BARLEY. Leaf spot in WHEAT. Net blotch in BARLEY. Powdery mildew in BARLEY, WHEAT. Yellow rust in BARLEY, WHEAT.

Notes

Efficacy
- For eyespot apply in spring from when leaf sheaths erect up to and including third node detectable stage (GS 33)
- Product effective against strains of eyespot resistant to benzimidazole fungicides
- Protection of winter cereals throughout the season usually requires at least 2 treatments. See label for details of timing
- For foliar diseases treat in spring or early summer at first sign of infection on new growth
- Best protection against ear diseases only achieved by spraying at full ear emergence

Crop Safety/Restrictions
- Maximum number of treatments 2 per crop

Special precautions/Environmental safety
- Harmful if swallowed
- Harmful to fish or other aquatic life. Do not contaminate surface waters or ditches with chemical or used container
- Do not allow direct spray from vehicle mounted/drawn hydraulic sprayers to fall within 6 m, or from hand-held sprayers to fall within 2 m, of surface waters or ditches. Direct spray away from water

- Do not harvest crops for human or animal consumption for at least 5 wk after last application

Protective clothing/Label precautions
- A, C, H
- M03, R03c, R04e, U02, U04a, U05a, U09a, U20b, C02 (5 wk), C03, E01, E13c, E26, E30a, E31b, E34

Latest application/Harvest Interval(HI)
- Up to and including ear emergence complete (GS 59)

282 fenpropidin + propiconazole

A systemic curative and protective fungicide for cereals

Products

1 Prophet	Novartis	375:125 g/l	EC	08433
2 Prophet 500 EC	Ciba Agric.	375:125 g/l	EC	07447
3 Sheen	Ciba Agric.	450:100 g/l	EC	07828
4 Sheen	Novartis	450:100 g/l	EC	08442
5 Zulu	Ciba Agric.	450:100 g/l	EC	07829
6 Zulu	Novartis	450:100 g/l	EC	08464

Uses Brown rust in SPRING BARLEY, SPRING WHEAT, WINTER BARLEY, WINTER WHEAT [1-6]. Leaf blotch in SPRING BARLEY, WINTER BARLEY [3-6]. Net blotch in SPRING BARLEY, WINTER BARLEY [1-6]. Powdery mildew in SPRING BARLEY, SPRING WHEAT, WINTER BARLEY, WINTER WHEAT [1-6]. Rhynchosporium in SPRING BARLEY, WINTER BARLEY [1, 2]. Septoria diseases in SPRING WHEAT, WINTER WHEAT [1-6]. Yellow rust in SPRING BARLEY, SPRING WHEAT, WINTER BARLEY, WINTER WHEAT [1-6].

Notes

Efficacy
- Best results obtained when applied at early stage of disease development or as part of a disease control programme. Repeat treatment may be necessary if disease pressure remains high

Crop Safety/Restrictions
- Maximum number of treatments (including any products containing propiconazole) 3 per crop (2 in yr of harvest) for winter crops; 2 per crop for spring crops [3-6]; 2 per crop for all uses [1, 2]
- Transient crop scorch can result on some wheat varieties, particularly those with erect flag leaves. Avoid spraying crops under stress

Special precautions/Environmental safety
- Harmful in contact with skin and if swallowed [3-6]
- Irritating to skin
- Risk of serious damage to eyes
- May cause sensitization by skin contact
- Risk of serious damage to eyes

- Dangerous to fish or other aquatic life. Do not contaminate surface waters or ditches with chemical or used container
- Do not allow direct spray from ground-based vehicle-mounted/drawn sprayers to fall within 6 m, or from hand-held sprayers to within 2 m, of surface waters or ditches. Direct spray away from water

Protective clothing/Label precautions
- A, C [1-6]; H [1, 2]
- M03, R03a, R03c, E30a, E31b, E34 [3-6]; R04b, R04d, R04e, U02, U04a, U05a, U10, U20a, C02 (5 wk), C03, E01, E13b, E16 [1-6]; U11, E26, E29 [1, 2]

Latest application/Harvest Interval(HI)
- Up to and including grain watery ripe (GS 71) for wheat; up to and including emergence of ear just complete (GS 59) for barley.
- HI 5 wk

Maximum Residue Levels (mg residue/kg food)
- see propiconazole entry

283 fenpropidin + propiconazole + tebuconazole

A protectant, curative and eradicant fungicide mixture

Products

Gladio	Novartis	375:125:125 g/l	EC	08413

Uses

Brown rust in SPRING BARLEY, SPRING WHEAT, WINTER BARLEY, WINTER WHEAT. Net blotch in SPRING BARLEY, WINTER BARLEY. Powdery mildew in SPRING BARLEY, SPRING WHEAT, WINTER BARLEY, WINTER WHEAT. Rhynchosporium in SPRING BARLEY, WINTER BARLEY. Septoria diseases in SPRING WHEAT, WINTER WHEAT. Yellow rust in SPRING BARLEY, SPRING WHEAT, WINTER BARLEY, WINTER WHEAT.

Notes

Efficacy
- Spray at first signs of disease and before infection spreads to new growth
- Control of established infections may be less good
- Repeat treatment 3-5 wk later if re-infection occurs or disease pressure remains high

Crop Safety/Restrictions
- Maximum number of treatments equivalent to two full dose treatments

Special precautions/Environmental safety
- Irritating to skin. Risk of serious damage to cyes
- May cause sensitization by skin contact
- Harmful to bees. Do not apply to crops in flower or to those in which bees are actively foraging. Do not apply when flowering weeds are present
- Extremely dangerous to fish or other aquatic life. Do not contaminate surface waters or ditches with chemical or used container

Protective clothing/Label precautions
- A, C, H
- R04b, R04d, R04e, U05a, U09b, U20a, C03, E01, E12e, E13a, E26, E30a, E31b

Latest application/Harvest Interval(HI)
- Before beginning of anthesis (GS 61) for wheat; up to and including ear emergence (GS 59) for barley

Maximum Residue Levels (mg residue/kg food)
- see propiconazole entry

284 fenpropidin + tebuconazole

A broad spectrum fungicide mixture for cereals

Products

Monicle	Bayer	300:200 g/l	EC	07375

Uses

Botrytis in WHEAT. Brown rust in BARLEY, WHEAT. Fusarium ear blight in WHEAT. Glume blotch in WHEAT. Net blotch in BARLEY. Powdery mildew in BARLEY, WHEAT. Rhynchosporium in BARLEY. Septoria leaf spot in WHEAT. Sooty moulds in WHEAT. Yellow rust in BARLEY, WHEAT.

Notes

Efficacy
- Best disease control and yield benefit obtained when applied at early stage of disease development before infection spreads to new growth
- To protect the flag leaf and ear from Septoria diseases apply from flag leaf emergence to ear fully emerged (GS 37-59)
- Applications once foliar symptoms of *Septoria tritici* are already present on upper leaves will be less effective

Crop Safety/Restrictions
- Maximum total dose 2.5 l/ha per crop
- Occasional slight temporary leaf speckling may occur on wheat but this has not been showr to reduce yield response or disease control
- Do not treat durum wheat

Special precautions/Environmental safety
- Harmful if swallowed and in contact with skin
- Irritating to skin
- Risk of serious damage to eyes
- Extremely dangerous to fish or other aquatic life. Do not contaminate surface waters or ditches with chemical or used container

Protective clothing/Label precautions
- A, C, H
- M03, R03a, R03c, R04b, R04d, U02, U04a, U05a, U11, U20a, C02 (5 wk), C03, E01, E13a, E30a, E31b, E34

Latest application/Harvest Interval(HI)
- Before grain milky ripe stage (GS 73) for winter wheat; up to and including ear emergenc just complete (GS 59) for barley and spring wheat
- HI 5 wk for crops for human or animal consumption

285 fenpropimorph

A contact and systemic morpholine fungicide

See also azoxystrobin + fenpropimorph
epoxiconazole + fenpropimorph
epoxiconazole + fenpropimorph + kresoxim-methyl
fenpropidin + fenpropimorph

Products

1 Aura	Novartis	750 g/l	EC	08388
2 Aura 750 EC	Ciba Agric.	750 g/l	EC	05705
3 Corbel	BASF	750 g/l	EC	00578
4 Landgold Fenpropimorph 750	Landgold	750 g/l	EC	06319
5 Mistral	Ciba Agric.	750 g/l	EC	06943
6 Mistral	Novartis	750 g/l	EC	08425
7 Standon Fenpropimorph 750	Standon	750 g/l	EC	08965

Uses

Alternaria in BRUSSELS SPROUTS [3, 7]. Brown rust in FIELD BEANS [5, 6]. Brown rust in BARLEY, WHEAT [1-7]. Brown rust in TRITICALE [1-3, 5-7]. Crown rot in CARROTS *(off-label)*, HORSERADISH *(off-label)*, PARSLEY ROOT *(off-label)*, PARSNIPS *(off-label)*, SALSIFY *(off-label)* [3]. Light leaf spot in BRUSSELS SPROUTS [3, 7]. Powdery mildew in HOPS *(off-label)*, SOFT FRUIT *(off-label)*[3]. Powdery mildew in BARLEY, OATS, WHEAT [1-7]. Powdery mildew in RYE [1-3, 5-7]. Rhynchosporium in BARLEY [1-7]. Ring spot in BRUSSELS SPROUTS [3, 7]. Rust in RED BEET *(off-label)*, SUGAR BEET SEED CROPS *(off-label)*[3]. Rust in FIELD BEANS [1-4, 7]. Rust in LEEKS [1-3, 5-7]. Yellow rust in BARLEY, WHEAT [1-7]. Yellow rust in TRITICALE [1-3, 5-7].

Notes

Efficacy

- On cereals spray at start of disease attack. See labels for recommended tank mixes. Follow-up treatments may be needed if disease pressure remains high
- On field beans a second application may be needed after 2-3 wk
- On Brussels sprouts apply in mixture with carbendazim at start of disease attack - see label
- Product rainfast after 2 h

Crop Safety/Restrictions

- Maximum number of treatments 2 per crop for spring cereals and field beans; 3 per crop at 1.0 l/ha or 4 at 0.75 l/ha (with 1 treatment applied in autumn) for winter cereals; 5 per crop for Brussels sprouts; 6 per crop for leeks
- Scorch may occur if applied during frosty weather or in high temperatures
- Consult processors before using on crops for processing
- An interval of at least 10 d on field beans and 14 d on leeks must elapse between applications

Special precautions/Environmental safety

- Harmful by inhalation [1, 2, 4]
- Irritating to eyes and skin [1-7]
- Irritating to respiratory system [1, 2, 4]
- Dangerous to fish or other aquatic life. Do not contaminate surface waters or ditches with chemical or used container

Protective clothing/Label precautions

- A, C
- M03, E34 [1, 2, 4-6]; R03b, R04c [1, 2, 4]; R04a, R04b, U05a, U08, U14, U15, U19, U20a, C03, E01, E13b, E30a, E31b [1-7]; E26 [3-7]; E29 [5, 6]

Latest application/Harvest Interval(HI)
- HI Brussels sprouts 2 wk; leeks 3 wk; cereals, field beans 5 wk

Approval
- Off-label approval unlimited for use on outdoor carrot, parsnip, parsley root, salsify, horseradish (OLA 2483/96)[3]; unlimited for use on cane and soft fruit (OLA 0787/95)[3]; unlimited for use in red beet (OLA 1246/94)[3], sugar beet seed crops (OLA 1807/96)[3]; to 2001 for use in hops (OLA 2078/96)[3]

---

286 fenpropimorph + flusilazole

A broad-spectrum eradicant and protectant fungicide mixture for cereals

Products	Colstar	DuPont	375:160 g/l	EC	06783

Uses

Brown rust in BARLEY, WINTER WHEAT. Net blotch in BARLEY. Powdery mildew in BARLEY, WINTER WHEAT. Rhynchosporium in BARLEY. Septoria diseases in WINTER WHEAT. Yellow rust in BARLEY, WINTER WHEAT.

Notes

Efficacy
- Disease control is more effective if treatment made at an early stage of disease development
- Treat winter cereals in spring or early summer before diseases spread to new growth
- Spring barley should be treated when diseases are first evident
- Treatment may be repeated after 3-4 wk if necessary

Crop Safety/Restrictions
- Maximum number of treatments (including other products containing flusilazole) 3 per crop (winter wheat), 2 per crop (winter or spring barley)
- Do not apply to crops under stress
- Do not apply during frosty weather

Special precautions/Environmental safety
- Irritating to eyes
- Dangerous to fish or other aquatic life. Do not contaminate surface waters or ditches with chemical or used container

Protective clothing/Label precautions
- A, C
- R04a, U05a, U11, U19, U20b, C03, E01, E13b, E26, E30a, E31b

Latest application/Harvest Interval(HI)
- Before early milk stage (GS 73)

---

287 fenpropimorph + flusilazole + tridemorph

A systemic eradicant and protectant fungicide mixture for cereals

Products

1 DUK 51	DuPont	275:160:100 g/l	EC	06764
2 Justice	DuPont	275:160:100 g/l	EC	07963

Uses

Brown rust in SPRING BARLEY, WINTER BARLEY, WINTER WHEAT. Net blotch in SPRING BARLEY, WINTER BARLEY *(reduction)*. Powdery mildew in SPRING BARLEY, WINTER BARLEY, WINTER WHEAT. Rhynchosporium in SPRING BARLEY, WINTER BARLEY. Septoria diseases in WINTER WHEAT. Yellow rust in SPRING BARLEY, WINTER BARLEY, WINTER WHEAT.

Notes

Efficacy

- Apply at the start of disease attack. Treatment is most effective when disease levels are low at application
- Effectiveness is reduced by rain within 2 h of treatment

Crop Safety/Restrictions

- Maximum number of treatments (including other products containing flusilazole) 3 per crop for winter wheat; 2 per crop for barley
- Do not apply to crops under stress
- Application during frosty weather, high temperatures or drought may cause some crop scorch

Special precautions/Environmental safety

- Harmful if swallowed
- Irritating to skin
- Risk of serious damage to eyes
- Harmful to fish or other aquatic life. Do not contaminate surface waters or ditches with chemical or used container

Protective clothing/Label precautions

- A, C
- M03, R03c, R04b, R04d, U05a, U11, U19, C03, E01, E07 (14 d), E13c, E30a, E31b, E34

Withholding period

- Keep livestock out of treated areas for at least 14 d after treatment

Latest application/Harvest Interval(HI)

- Before early milk stage (GS 73)

288 fenpropimorph + kresoxim-methyl

A protectant and systemic fungicide mixture for cereals

Products

1 Ensign	BASF	300:150 g/l	SE	08362
2 Landgold Strobilurin KF	Landgold	300:150 g/l	SE	09196
3 Standon Kresoxim FM	Standon	300:150 g/l	SE	08922

Uses Powdery mildew in SPRING BARLEY, WINTER BARLEY, WINTER WHEAT. Rhynchosporium in SPRING BARLEY, WINTER BARLEY. Septoria in WINTER WHEAT *(reduction)*.

Notes

Efficacy

- For best results spray at the start of foliar disease attack and repeat if infection conditions persist

Crop Safety/Restrictions

- Maximum total dose for all crops 1.4 litres product/ha

Special precautions/Environmental safety

- Irritant. May cause sensitisation by skin contact
- Dangerous to fish or other aquatic life. Do not contaminate surface waters or ditches with chemical or used container
- Risk to non-target insects or other athropods. Avoid spraying within 6 m of field boundary

Protective clothing/Label precautions

- A
- R04, R04e, U05a, U14, U20b, C03, E01, E13b, E22b, E26, E30a, E31c

Latest application/Harvest Interval(HI)

- Completion of ear emergence (GS 59) for barley; completion of flowering (GS 69) for winter wheat

289 fenpropimorph + prochloraz

A fungicide mixture for late season disease control in cereals

Products

1 Mirage Super 600EC	Makhteshim	300:300 g/l	EC	08531
2 Sprint HF	AgrEvo	375:225 g/l	EC	07292

Uses Brown rust in BARLEY, SPRING WHEAT, WINTER RYE, WINTER WHEAT [2]. Brown rust in WHEAT [1]. Eyespot in BARLEY, SPRING WHEAT, WINTER RYE, WINTER WHEAT [2]. Net blotch in BARLEY [2]. Powdery mildew in SPRING WHEAT, WINTER RYE, WINTER WHEAT [2]. Powdery mildew in BARLEY [1, 2]. Powdery mildew in WHEAT [1]. Rhynchosporium in WINTER RYE [2]. Rhynchosporium in BARLEY [1, 2]. Septoria in SPRING WHEAT, WINTER RYE, WINTER WHEAT [2]. Septoria in WHEAT [1]. Yellow rust in BARLEY, SPRING WHEAT, WINTER RYE, WINTER WHEAT [2].

Notes

Efficacy

- In wheat, if disease present, spray as soon as ligule of flag leaf visible (GS 39). If no disease, delay until first signs appear or use as protectant treatment before flowering
- In barley spray when disease appears on new growth or as protective spray when flag leaf ligule just visible (GS 39)

Crop Safety/Restrictions

- Maximum number of treatments 2 per crop
- Do not apply to crops suffering from stress

Special precautions/Environmental safety

- Harmful if swallowed. Risk of serious damage to eyes [1]
- Irritating to eyes [2]
- May cause sensitization by skin contact [1, 2]
- Dangerous to fish or other aquatic life. Do not contaminate surface waters or ditches with chemical or used container

Protective clothing/Label precautions

- A, C [1, 2]
- M03, U05a, U08, U20b, C03, E01, E13b, E30a, E31b, E34 [1, 2]; R03a, C02 (5 wk), E26 [2]; R03c, R04d, R04e [1]

Latest application/Harvest Interval(HI)

- Up to and including ear emergence just complete (GS 59).
- HI 5 wk

290 fenpropimorph + propiconazole

A fungicide mixture for control of leaf diseases of cereals

Products

1 Belvedere	Makhteshim	375:125 g/l	EC	08084
2 Decade	Novartis	375:125 g/l	EC	08402
3 Decade 500 EC	Ciba Agric.	375:125 g/l	EC	05757
4 Glint	Novartis	375:125 g/l	EC	08414
5 Glint 500 EC	Ciba Agric.	375:125 g/l	EC	04126
6 Mantle	Novartis	250 g/l	EC	08424
7 Mantle 425 EC	Ciba Agric.	300:125 g/l	EC	05715

Uses

Brown rust in WINTER BARLEY, WINTER WHEAT. Mildew in SPRING BARLEY, WINTER BARLEY, WINTER WHEAT. Net blotch in SPRING BARLEY, WINTER BARLEY. Rhynchosporium in SPRING BARLEY, WINTER BARLEY. Septoria in WINTER WHEAT. Sooty moulds in WINTER WHEAT. Yellow rust in WINTER BARLEY, WINTER WHEAT.

Notes

Efficacy

- Apply at start of disease attack
- Spray winter wheat between GS 30-32 against early Septoria, mildew and rusts, at flag leaf against late foliar and ear disease, and repeat 28 d later if necessary [2-5]
- Spray spring barley immediately disease appears in crop
- Spray winter barley at risk from foliar disease at GS 30-31, and repeat at flag leaf (GS 39) if necessary, or spray at awn emergence (GS 49) [2-5]. See label for details

Crop Safety/Restrictions

- Maximum number of treatments 2 per crop [2-5]; 2 per crop for spring wheat and barley, 3 per crop for winter crops (1 in autumn at least 3 mth before 2 in spring/summer) [6, 7]
- If used in programme with other products containing fenpropimorph observe maximum total doses per season permitted for wheat and barley

Special precautions/Environmental safety

- Harmful by inhalation. Irritating to eyes, skin and respiratory system
- Dangerous to fish or other aquatic life. Do not contaminate surface waters or ditches with chemical or used container

Protective clothing/Label precautions
- A, C
- M03, R03b, R04a, R04b, R04c, U05a, U08, U14, U15, U19, U20a, C03, E01, E13b, E30a, E31b, E34

Latest application/Harvest Interval(HI)
- HI 5 wk

Maximum Residue Levels (mg residue/kg food)
- see propiconazole entry

291 fenpropimorph + quinoxyfen

A systemic fungicide mixture for cereals

Products	Orka	Dow	250:66.7 g/l	EW	08879

Uses Powdery mildew in DURUM WHEAT, RYE, SPRING BARLEY, SPRING OATS, SPRING WHEAT, TRITICALE, WINTER BARLEY, WINTER OATS, WINTER WHEAT.

Notes

Efficacy
- For best results treat at early stage of disease development before infection spreads to new crop growth. Further treatment may be necessary if disease pressure remains high
- For control of established infections and broad spectrum disease control use in tank mixtures. See label
- Product rainfast after 1 h
- Systemic activity may be reduced in severe drought

Crop Safety/Restrictions
- Maximum total dose 3.0 l product per ha
- Apply only in the spring from mid-tillering stage (GS 25)
- Crop scorch may occur when treatment made in high temperatures

Special precautions/Environmental safety
- Irritant. May cause sensitization by skin contact
- Dangerous to fish or other aquatic life. Do not contaminate surface waters or ditches with chemical or used container
- Do not allow direct spray from vehicle mounted/drawn hydraulic sprayers to fall within 6 m of surface waters or ditches. Direct spray away from water

Protective clothing/Label precautions
- A, C, H
- R04, R04e, U05a, U14, C03, E01, E13b, E16, E34

Latest application/Harvest Interval(HI)
- First awns visible (GS 49)

292 fenpropimorph + tridemorph

A systemic mixture of morpholine fungicides

Products

1 BAS 46402F	BASF	500:250 g/l	EC	03313
2 Gemini	BASF	500:250 g/l	EC	05684

Uses

Powdery mildew in BARLEY, WINTER WHEAT. Rhynchosporium in BARLEY.

Notes

Efficacy

- Spray when disease first starts to build up
- Control of Rhynchosporium can be improved by mixture with carbendazim

Crop Safety/Restrictions

- Maximum total dose equivalent to two full dose treatments on spring barley, 3 full dose treatments on winter wheat, winter barley
- Crop scorch may follow treatment, especially on winter wheat, particularly if applied in high temperatures or high light intensity or during frosty weather
- Avoid drift on to neighbouring crops

Special precautions/Environmental safety

- Harmful if swallowed. Irritating to skin
- Risk of serious damage to eyes
- Dangerous to fish or other aquatic life. Do not contaminate surface waters or ditches with chemical or used container

Protective clothing/Label precautions

- A, C
- M03, R03c, R04b, R04d, U05a, U08, U14, U15, U19, U20a, C03, E01, E07 (14 d), E13b, E30a, E31c, E34

Withholding period

- Keep livestock out of treated areas for at least 14 d

Latest application/Harvest Interval(HI)

- HI 5 wk

293 fenpyroximate

A mitochondrial electron transport inhibitor acaricide for apples

Products

Sequel	Promark	51.3 g/l	SC	07624

Uses

Fruit tree red spider mite in APPLES.

Notes

Efficacy

- Kills motile stages of fruit tree red spider mite. Best results achieved if applied in warm weather
- Total spray cover of trees essential. Use higher volumes for large trees
- Apply when majority of winter eggs have hatched

Crop Safety/Restrictions
- Maximum number of treatments 1 per yr
- Other mitochondrial electron transport inhibitor (METI) acaricides should not be applied to the same crop in the same calendar yr either separately or in mixture
- Consult processor before use on crops for processing

Special precautions/Environmental safety
- Irritant. Risk of serious damage to eyes
- Dangerous to fish or other aquatic life. Do not contaminate surface waters or ditches with chemical or used container
- Do not allow direct spray from air assisted sprayers to fall within 38 m of surface waters or ditches. Direct spray away from water

Protective clothing/Label precautions
- A, C, H
- R04, R04d, U05a, U08, U14, U20b, C03, E01, E13b, E17 (38 m), E26, E30a, E31b

Latest application/Harvest Interval(HI)
- HI 2 wk

294 fentin acetate

An organotin fungicide available only in mixtures

295 fentin acetate + maneb

A curative and protectant fungicide for use in potatoes

Products	Brestan 60 SP	AgrEvo	54:16% w/w	WB	07305

Uses Blight in POTATOES.

Notes

Efficacy
- Commence spraying before infection occurs, before haulm meets across the rows or at first blight warning, whichever occurs first
- Repeat at 7-14 d intervals until growth ceases or haulm burnt off
- Do not spray if rain imminent

Crop Safety/Restrictions
- Maximum number of treatments 6 per crop

Special precautions/Environmental safety
- Fentin acetate is subject to the Poisons Rules 1982 and the Poisons Act 1972. See notes in Section 1
- Harmful if swallowed. Irritating to eyes, skin and respiratory system
- Keep in original container, tightly closed, in a safe place, under lock and key
- Harmful to livestock

- Harmful to fish or other aquatic life. Do not contaminate surface waters or ditches with chemical or used container

Protective clothing/Label precautions
- A, C, H
- M03, R03c, R04a, R04b, R04c, U02, U04a, U05a, U08, U13, U19, U20b, C03, E01, E06b, E13c, E26, E30a, E32a, E34

Withholding period
- Keep all livestock out of treated areas for at least 1 wk

Latest application/Harvest Interval(HI)
- HI 7 d

Maximum Residue Levels (mg residue/kg food)
- see fentin hydroxide and maneb entries

296 fentin hydroxide

A curative and protectant organotin fungicide

Products

1 Ashlade Flotin	Ashlade	625 g/l	SC	06223
2 Barclay Fentin Flow	Barclay	480 g/l	SC	07914
3 Farmatin 560	AgrEvo	532 g/l	SC	07320
4 Keytin	Chiltern	532 g/l	SC	08894
5 MSS Flotin 480	Mirfield	480 g/l	SC	07616
6 Super-Tin 4L	Chiltern	480 g/l	SC	02995
7 Super-Tin 80 WP	PBI	80% w/w	WB	07605

Uses Blight in POTATOES.

Notes

Efficacy
- Apply to potatoes before haulm meets across rows or on receipt of blight warning and repeat at 7-14 d intervals throughout season. Late sprays protect against tuber blight
- Best results achieved by using for 2 final sprays of blight control programme. Allow at least 14 d after complete death of haulm before lifting unless for immediate sale

Crop Safety/Restrictions
- Maximum number of treatments - see product labels
- Drought stress [1, 3, 4] or mixtures with emulsifiable concentrates [5, 6]may cause localised leaf spotting
- Product may sometimes harden the foliage of young potato plants [2]
- A minimum interval of 7 d (10 d [1, 3]) between applications must be observed [4-6]

Special precautions/Environmental safety
- Fentin hydroxide is subject to the Poisons Rules 1982 and the Poisons Act 1972. See notes in Section 1
- Very toxic by inhalation [7]
- Harmful in contact with skin [1, 3] and if swallowed [2, 4-7]
- Irritating to eyes and skin [1-6]
- Risk of serious damage to eyes [7]
- Irritating to respiratory system [2, 5-7]
- Keep in original container, tightly closed, in a safe place, under lock and key [2, 5-7]

- Harmful (dangerous [1]) to livestock
- Harmful (dangerous [7]) to fish or other aquatic life. Do not contaminate surface waters or ditches with chemical or used container

Protective clothing/Label precautions
- A, C, H [1-7]; K [4]; M [2]
- M03, R04a, E13c, E34 [1-6]; M04, R01b, R04d, U22, E13b [7]; R03a [1, 3, 4, 7]; R03c, U04a, U05a, U08, U19, U20a, C03, E01 [1-7]; R04b [1-5, 7]; R04c, E30b [2, 5-7]; R05a [3]; U02, E06b (7 d) [2-7]; U13, U14 [3-7]; U15 [4]; C02 (7 d) [3, 4]; E06a (7 d) [1]; E26 [2, 5, 6]; E30a [1, 3, 4]; E31b [2, 4-7]; E32a [1, 3]

Withholding period
- Keep all livestock out of treated areas for at least 1 wk

Latest application/Harvest Interval(HI)
- HI potatoes 0 d [6], 1 wk [1, 3, 4]

Approval
- Approved for aerial application on potatoes. See notes in Section 1 [1]

Maximum Residue Levels (mg residue/kg food)
- hops 0.5; potatoes, tea 0.1; all other produce 0.05

297 fenuron

A urea herbicide available only in mixtures

See also chlorpropham + fenuron

298 ferbam + maneb + zineb

A protectant dithiocarbamate complex fungicide

Products	Trimanzone WP	Intracrop	10:65:10% w/w	WP	05860

Uses Blight in POTATOES. Fungus diseases in LEEKS *(off-label).*

Notes

Efficacy
- For blight control apply before haulm meets across the rows or as soon as blight warning received, whichever occurs first. Repeat every 10-14 d

Crop Safety/Restrictions
- Observe minimum period of 10 d between applications

Special precautions/Environmental safety
- Irritating to skin, eyes and respiratory system

Protective clothing/Label precautions
- A, C
- R04a, R04b, R04c, U05a, U08, U20a, C02, C03, E01, E15, E30a, E32a

Latest application/Harvest Interval(HI)
- HI potatoes, leeks 7 d

Approval
- Off-label approval unlimited for use on leeks (OLA 0599/92)[1]

Maximum Residue Levels (mg residue/kg food)
- see maneb entry

299 ferrous sulfate

A herbicide/fertilizer combination for moss control in turf

See also dicamba + dichlorprop + ferrous sulfate + MCPA
dichlorophen + ferrous sulfate

Products

1 Elliott's Lawn Sand	Elliott	17% w/w	SA	04860
2 Elliott's Mosskiller	Elliott	45% w/w	GR	04909
3 Greenmaster Autumn	Scotts	18.1% w/w	GR	07508
4 Greenmaster Mosskiller	Scotts	26.8% w/w	GR	07509
5 Maxicrop Moss Killer & Conditioner	Maxicrop	16.4% w/w	SL	04635
6 SHL Lawn Sand	Sinclair	10% w/w	SA	05254
7 Taylors Lawn Sand	Rigby Taylor	7.35% w/w	SA	04451
8 Vitax Microgran 2	Vitax	20% w/w	MG	04541
9 Vitax Turf Tonic	Vitax	15% w/w	SA	04354
10 Walkover Mosskiller	Allen	16.4% w/w	SL	04662

Uses Mosses in TURF.

Notes

Efficacy
- Apply autumn treatment from Sep onward but not when heavy rain expected or in frosty weather [3]
- Apply from Mar to Sep except during drought or when soil frozen [4]
- For best results apply when light showers expected, mow 3 d before treatment, do not mow or walk on treated area until well watered and water after 2 d if no rain
- Rake out dead moss thoroughly 7-14 d after treatment. See label
- Fertilizer component of autumn treatment encourages root growth, that of mosskiller formulations promotes tillering

Crop Safety/Restrictions
- Maximum number of treatments - see labels
- If spilt on paving, concrete, clothes etc brush off immediately to avoid discolouration
- Observe label restrictions for interval before cutting after treatment

Special precautions/Environmental safety
- Harmful to fish or other aquatic life. Do not contaminate surface waters or ditches with chemical or used container [3, 4]

Protective clothing/Label precautions
- U20a [1, 2, 6, 10]; U20b [3, 4, 8, 9]; U20c [5, 7]; E01 [9]; E13c [3, 4]; E15 [1, 2, 5, 7-10]; E26, E31a [5]; E30a [1-10]; E32a [1-4, 6-10]

300 flamprop-M-isopropyl

A translocated, post-emergence arylalanine wild-oat herbicide

Products

Commando	Cyanamid	200 g/l	EC	07005

Uses

Wild oats in BARLEY, DURUM WHEAT, RYE, TRITICALE, WHEAT.

Notes

Efficacy

- Must be applied when crop and weeds growing actively under conditions of warm, moist days and warm nights
- Timing determined mainly by crop growth stage. Best control at later stages of wild oats but not after weeds visible above crop
- Best results achieved before 3rd node detectable stage (GS 33) in barley, or 4th node detectable (GS 34) in wheat
- Good spray coverage and retention essential. Spray rainfast after 2 h
- Addition of Swirl adjuvant oil recommended for use on winter wheat and winter barley
- Do not treat thin, open crops
- See label for recommended tank mixes and sequential treatments

Crop Safety/Restrictions

- Maximum number of treatments 1 per crop
- May be used on crops undersown with ryegrass and clover
- Do not use on crops under stress or during periods of high temperature
- Do not roll or harrow within 7 d of spraying
- Do not mix with any herbicide other than fluroxypyr

Special precautions/Environmental safety

- Irritating to eyes and skin
- Flammable
- Do not use straw from barley treated after stage GS 33 or wheat treated after stage GS 34 as feed or bedding for animals
- Harmful to fish or other aquatic life. Do not contaminate surface waters or ditches with chemical or used container

Protective clothing/Label precautions

- A, C
- R04a, R04b, R07d, U02, U04a, U05a, U08, U19, U20a, C03, E01, E13c, E26, E30a, E31b, E34

Latest application/Harvest Interval(HI)

- Before flag leaf sheath opening (GS 47) for wheat; before first awns visible (GS 49) for barley; before 3rd node detectable (GS 33) for rye and triticale; before 4th node detectable (GS 34) for durum wheat

FOR FULL CONDITIONS OF USE ALWAYS READ THE PRODUCT LABEI

301 fluazifop-P-butyl

A phenoxypropionic acid grass herbicide for broadleaved crops

Products

1 Citadel	Zeneca	125 g/l	EC	06762
2 Corral	Zeneca	125 g/l	EC	06647
3 Fusilade 250 EW	Zeneca	250 g/l	EW	06531
4 Landgold Fluazifop-P	Landgold	125 g/l	EC	06020
5 Standon Fluazifop-P	Standon	125 g/l	EC	06060

Uses

Annual grasses in BROAD BEANS *(off-label)*, CHINESE CABBAGE *(off-label)*, COLLARDS *(off-label)*, FLOWERS *(off-label)*, GARLIC *(off-label)*, KALE *(off-label)*, NAVY BEANS *(off-label)*, ORNAMENTALS *(off-label)*, PARSNIPS *(off-label)*, RED BEET *(off-label)*, SWEDES *(off-label)*, TURNIPS *(off-label)* [3]. Annual grasses in FODDER BEET, SUGAR BEET, WINTER OILSEED RAPE [1-5]. Annual grasses in BLACKCURRANTS, CARROTS, COMBINING PEAS, FARM FORESTRY, FIELD BEANS, FIELD MARGINS, GOOSEBERRIES, HOPS, KALE *(stockfeed only)*, LINSEED/FLAX, LINSEED/FLAX FOR INDUSTRIAL USE, OILSEED RAPE FOR INDUSTRIAL USE, ONIONS, RASPBERRIES, SPRING OILSEED RAPE, STRAWBERRIES, SWEDES *(stockfeed only)*, TURNIPS *(stockfeed only)*, VINING PEAS [1-3]. Barley cover crops in ROW CROPS [1-3]. Barren brome in FIELD MARGINS [1-3]. Blackgrass in FIELD BEANS, HOPS [1-3]. Green cover in LAND TEMPORARILY REMOVED FROM PRODUCTION, SETASIDE [1-3]. Perennial grasses in BROAD BEANS *(off-label)*, CHINESE CABBAGE *(off-label)*, COLLARDS *(off-label)*, FLOWERS *(off-label)*, GARLIC *(off-label)*, KALE *(off-label)*, NAVY BEANS *(off-label)*, ORNAMENTALS *(off-label)*, PARSNIPS *(off-label)*, RED BEET *(off-label)*, SWEDES *(off-label)*, TURNIPS *(off-label)* [3]. Perennial grasses in FODDER BEET, SUGAR BEET, WINTER OILSEED RAPE [1-5]. Perennial grasses in BLACKCURRANTS, CARROTS, COMBINING PEAS, FARM FORESTRY, FIELD BEANS, GOOSEBERRIES, HOPS, KALE *(stockfeed only)*, LINSEED/FLAX, LINSEED/FLAX FOR INDUSTRIAL USE, OILSEED RAPE FOR INDUSTRIAL USE, ONIONS, RASPBERRIES, SPRING OILSEED RAPE, STRAWBERRIES, SWEDES *(stockfeed only)*, TURNIPS *(stockfeed only)*, VINING PEAS [1-3]. Volunteer cereals in BLACKCURRANTS, CARROTS, COMBINING PEAS, FARM FORESTRY, FIELD BEANS, FIELD MARGINS, FODDER BEET, GOOSEBERRIES, HOPS, KALE *(stockfeed only)*, LINSEED/FLAX, LINSEED/FLAX FOR INDUSTRIAL USE, OILSEED RAPE FOR INDUSTRIAL USE, ONIONS, RASPBERRIES, SPRING OILSEED RAPE, STRAWBERRIES, SUGAR BEET, SWEDES *(stockfeed only)*, TURNIPS *(stockfeed only)*, VINING PEAS, WINTER OILSEED RAPE [1-3]. Wild oats in BLACKCURRANTS, CARROTS, COMBINING PEAS, FARM FORESTRY, FIELD BEANS, FIELD MARGINS, FODDER BEET, GOOSEBERRIES, HOPS, KALE *(stockfeed only)*, LINSEED/FLAX, LINSEED/FLAX FOR INDUSTRIAL USE, OILSEED RAPE FOR INDUSTRIAL USE, ONIONS, RASPBERRIES, SPRING OILSEED RAPE, STRAWBERRIES, SUGAR BEET, SWEDES *(stockfeed only)*, TURNIPS *(stockfeed only)*, VINING PEAS, WINTER OILSEED RAPE [1-3].

Notes

Efficacy

- Best results achieved by application when weed growth active under warm conditions with adequate soil moisture. Agral or other specified adjuvant must always be added to spray. See label
- Spray weeds from 2-expanded leaf stage to fully tillered, couch from 4 leaves when majority of shoots have emerged, with a second application if necessary
- Control may be reduced under dry conditions. Do not cultivate for 2 wk after spraying couch
- Annual meadow grass is not controlled
- May also be used to remove grass cover crops

Crop Safety/Restrictions

- Maximum number of treatments normally 2 per crop for sugar and fodder beet, oilseed rape (including crops for industrial use); 2 per yr for strawberries and farm forestry; 1 per crop or yr for other crops
- Apply to sugar and fodder beet from 1-true leaf to 50% ground cover
- Apply to winter oilseed rape from 1-true leaf to established plant stage
- Apply to spring oilseed rape from 1-true leaf but before 5-true leaves
- Apply in fruit crops after harvest. See label for timing details on other crops
- Before using on onions or peas use crystal violet test to check that leaf wax is sufficient
- Do not sow cereals for at least 8 wk after application of high rate or 2 wk after low rate
- Do not apply through CDA sprayer, with hand-held equipment or from air

Special precautions/Environmental safety

- Irritating to eyes [2, 4, 5] and skin [3]
- Flammable [1, 4, 5]
- Harmful to fish or other aquatic life. Do not contaminate surface waters or ditches with chemical or used container
- Do not apply by hand-held equipment
- Do not treat bush and cane fruit or hops between flowering and harvest
- Oilseed rape, linseed and flax for industrial use must not be harvested for human or animal consumption nor grazed
- Do not use for forestry establishment on land not previously under arable cultivation or improved grassland
- Treated vegetation in field margins, setaside etc, must not be grazed or harvested for human or animal consumption and unprotected persons must be kept out of treated areas for at least 24 h

Protective clothing/Label precautions

- A, C, H
- M03, R04b, U05a, U08, C03, E01, E13c, E30a, E31b [1-5]; R04a [1, 2, 4, 5]; R07d [1, 4, 5]; U20a [1, 3-5]; U20b [2]; E26 [1, 2, 4]; E34 [1-3, 5]

Latest application/Harvest Interval(HI)

- Before 5 leaf stage for spring oilseed rape; after harvest but before flowering for blackcurrants, gooseberries, hops, raspberries, strawberries; before flower buds visible for field beans, peas, linseed, flax, oilseed rape; 2 wk before sowing cereals or grass for field margins, land temporarily removed from production, spring oilseed rape for industrial use, grass crops for farm forestry
- HI beet crops, carrots, kale, 8 wk; onions 4 wk

Approval

- Off-label approval unlimited for use on ornamentals, flowers, garlic, broad beans, dry harvested beans, parsnips, red beet, kale, Chinese cabbage, collards, swedes, turnips (OLA 2768/96)[3]

302 fluazinam

A dinitroaniline fungicide for use in potatoes

Products

1 Barclay Cobbler	Barclay	500 g/l	SC	08349
2 Landgold Fluazinam	Landgold	500 g/l	SC	08060
3 Legacy	ISK Biosciences	500 g/l	SC	07401
4 Salvo	Zeneca	500 g/l	SC	07092
5 Shirlan	Zeneca	500 g/l	SC	07091
6 Shirlan Programme	Zeneca	500 g/l	SC	08761
7 Standon Fluazinam 500	Standon	500 g/l	SC	08670

Uses Blight in POTATOES.

Notes

Efficacy
- Commence treatment at the first blight warning and before blight enters the crop
- In the absence of a warning treatment should start before foliage of adjacent plants meets in the rows
- Spray at 7-14 d intervals depending on severity of risk (see label)
- Ensure complete coverage of the foliage and stems, increasing volume as haulm growth progresses, in dense crops and if blight risk increases

Crop Safety/Restrictions
- Maximum number of treatments 10 per crop

Special precautions/Environmental safety
- Irritant. May cause sensitization by skin contact
- Dangerous to fish or other aquatic life. Do not contaminate surface waters or ditches with chemical or used container
- Do not allow direct spray to fall within 6 m of surface waters or ditches. Direct spray away from water

Protective clothing/Label precautions
- A [1-7]; C [1, 2, 4-7]; H
- M03 [3-6]; R04, U20b [1, 7]; R04e, U05a, C03, E01, E13b, E16, E30a [1-7]; U02, U04a, U08 [1, 3-7]; U20a [2-6]; E26 [1, 2, 4-7]; E31b [1-3, 7]; E31c, E34 [4-6]

Latest application/Harvest Interval(HI)
- HI 7 d before harvest

303 fludioxonil

A cyanopyrrole fungicide seed treatment for wheat and barley

Products

1 Beret Gold	Ciba Agric.	25 g/l	FS	07531
2 Beret Gold	Novartis	25 g/l	FS	08390

Uses Bunt in WINTER WHEAT. Covered smut in SPRING BARLEY, WINTER BARLEY. Foot rot in SPRING BARLEY, WINTER BARLEY. Fusarium foot rot and seedling blight in SPRING BARLEY, WINTER BARLEY, WINTER WHEAT. Leaf stripe in SPRING BARLEY *(partial control)*, WINTER BARLEY *(partial control)*. Snow mould in SPRING BARLEY, WINTER BARLEY, WINTER WHEAT.

Notes

Efficacy

- Apply direct to seed using conventional seed treatment equipment. Continuous flow treaters should be calibrated using product before use
- Effective against benzimidazole-resistant strains of *Fusarium nivale*
- Product may reduce flow rate of seed through drill. Recalibrate with treated seed before drilling

Crop Safety/Restrictions

- Maximum number of treatments 1 per seed batch
- Do not apply to cracked, split or sprouted seed
- Sow treated seed within 6 mth

Special precautions/Environmental safety

- Harmful to fish or other aquatic life. Do not contaminate surface waters or ditches with chemical or used container
- Do not use treated seed as food or feed
- Treated seed harmful to game and wildlife

Protective clothing/Label precautions

- A, H
- U20a, E13c, E26, E32a, S01, S02, S04b, S05, S06, S07

Latest application/Harvest Interval(HI)

- Before drilling

304 fluoroglycofen-ethyl

A diphenyl ether herbicide available only in mixtures

305 fluoroglycofen-ethyl + isoproturon

A contact and residual herbicide for use in winter cereals

Products

1 Competitor	AgrEvo	1.5:60% w/w	WG	07310
2 Stefes Competitor	Stefes	1.5:60% w/w	WG	08753

Uses

Annual dicotyledons in WINTER BARLEY, WINTER WHEAT. Annual meadow grass in WINTER BARLEY, WINTER WHEAT. Cleavers in WINTER BARLEY, WINTER WHEAT.

Notes

Efficacy

- Good weed control depends on burying and dispersing any straw to 15 cm. Seedbed should be firm and not cloddy
- May be applied from emergence of crop up to and including 1st node detectable (GS 31)
- Residual activity reduced on soils with more than 10% organic matter
- Activity may be reduced under dry soil conditions or if heavy rain falls soon after application

Crop Safety/Restrictions
- Maximum number of treatments 1 per crop
- Do not treat durum wheat, undersown crops or those due to be undersown. Do not apply pre-emergence to wheat or barley
- Early (Sep) drilled crops may be prone to damage if spraying should precede or coincide with a period of rapid growth
- Do not spray crops under stress from pests or diseases, waterlogging, drought, frost or other factors. A transient speckled discolouration may occur after treatment
- Do not treat if frost is imminent or after onset of prolonged frosty weather
- On free draining, stony and gravelly soils there is a risk of crop damage, especially if heavy rain falls soon after application
- Do not roll within 1 wk before or after application or harrow within 1 wk before or at any time following

Special precautions/Environmental safety
- Irritant. Risk of serious damage to eyes
- Extremely dangerous to fish or aquatic life. Do not contaminate surface waters or ditches with chemical or used container
- Do not allow direct spray from ground-based sprayer vehicles to fall within 6 m, or from hand-held sprayers within 2 m, of surface waters or ditches. Direct spray away from water
- Do not spray where soils are cracked, to avoid run-off through drains
- Avoid direct or indirect contamination of water by use of buffer strips

Protective clothing/Label precautions
- A, C, H
- R04, R04d, U05a, U08, U20a, C03, E01, E13a, E16, E26, E30a, E31a

Latest application/Harvest Interval(HI)
- Before 2nd node detectable (GS 32)

---

306 flupyrsulfuron-methyl

A sulfonylurea herbicide for winter wheat

See also carfentrazone-ethyl + flupyrsulfuron-methyl

Products Lexus 50 DF | DuPont | 50% w/w | WG | 09026

Uses Annual dicotyledons in WINTER WHEAT. Blackgrass in WINTER WHEAT.

Notes

Efficacy
- Best results achieved from applications made in good growing conditions
- Good spray cover of weeds must be obtained
- Growth of weeds is inhibited within hours of treatment but visible symptoms may not be apparent for up to 4 wk
- Product has moderate residual life in soil. Under normal moisture conditions susceptible weeds germinating soon after treatment will be controlled
- Product may be used on all soil types but residual activity and weed control is reduced on highly alkaline soils
- Apply in the spring from the 3-leaf stage on all crops

Crop Safety/Restrictions
- Maximum number of treatments 1 per crop

- Treat from 2 leaf stage of crop
- Do not use on wheat undersown with grasses or legumes, or any other broad-leaved crop
- Do not apply within 7 d of rolling
- Do not treat any crop suffering from drought, waterlogging, pest or disease attack, nutrient deficiency, or any other stress factors
- Slight chlorosis and stunting may occur in certain conditions. Recovery is rapid and yield not affected
- Use recommended procedure for cleaning spray equipment (see label)
- After treatment other 'ALS-inhibiting' herbicides should not be used on the same crop except metsulfuron-methyl, thifensulfuron-methyl or tribenuron-methyl (alone or in mixtures)
- Only cereals, oilseed rape, field beans, clover or grass may be sown in the yr of harvest of a treated crop
- In the event of crop failure only winter or spring wheat may be sown within 3 mth of treatment. Land should be ploughed and cultivated to 15 cm minimum before resowing

Special precautions/Environmental safety
- Extremely dangerous to fish or other aquatic life. Do not contaminate surface waters or ditches with chemical or used container
- Take extreme care to avoid drift onto broad-leaved plants outside the target area or onto ponds waterways or ditches, or onto land intended for cropping

Protective clothing/Label precautions
- U08, U19, U20b, E13a, E30a, E32a

Latest application/Harvest Interval(HI)
- Before 1st node detectable (GS 31)

307 flupyrsulfuron-methyl + metsulfuron-methyl

A sulfonylurea herbicide mixture for winter wheat

Products

1 Lexus XPE WSB	DuPont	33.3:16.7% w/w	WB	08542
2 Standon Flupyrsulfuron MM	Standon	33.3:16.7% w/w	WG	09098

Uses Annual dicotyledons in WINTER WHEAT. Blackgrass in WINTER WHEAT.

Notes

Efficacy
- Best results obtained when applied to small actively growing weeds
- Good spray cover of weeds must be obtained
- Increased degradation of active ingredient in high soil temperatures reduces residual activity
- Product has moderate residual life in soil. Under normal moisture conditions susceptible weeds germinating soon after treatment will be controlled
- Product may be used on all soil types but residual activity and weed control is reduced on highly alkaline soils
- Blackgrass should be treated from 1 leaf up to mid-tillering. Strains of blackgrass resistant to other herbicides may not be controlled

Crop Safety/Restrictions
- Maximum number of treatments 1 per crop
- Must only be applied in spring after 1 Feb, from 2 leaf stage of crop
- Do not use on wheat undersown with grasses or legumes, or any other broad-leaved crop
- Do not apply within 7 d of rolling
- Do not treat any crop suffering from drought, waterlogging, pest or disease attack, nutrient deficiency, or any other stress factors
- Slight chlorosis and stunting may occur in certain conditions. Recovery is rapid and yield not affected
- Use recommended procedure for cleaning spray equipment (see label)
- After treatment other 'ALS-inhibiting' herbicides should not be used on the same crop except thifensulfuron-methyl or tribenuron-methyl (alone or in mixtures)
- Only cereals, oilseed rape, field beans, clover or grass may be sown in the yr of harvest of a treated crop
- In the event of crop failure only winter wheat may be sown within 3 mth of treatment. Land should be ploughed and cultivated to 15 cm minimum before resowing

Special precautions/Environmental safety
- Extremely dangerous to fish or other aquatic life. Do not contaminate surface waters or ditches with chemical or used container
- Do not allow direct spray from vehicle mounted/drawn hydraulic sprayers to fall within 6 m, or from hand-held sprayers to within 2 m, of surface waters or ditches. Direct spray away from water
- Take extreme care to avoid drift onto broad-leaved plants outside the target area or onto ponds waterways or ditches, or onto land intended for cropping

Protective clothing/Label precautions
- U08, U19, U20b, E13a, E16, E30a, E32a

Latest application/Harvest Interval(HI)
- Beginning of stem extension (GS 30)

---

308 fluroxypyr

A post-emergence aryloxyalkanoic acid herbicide

See also bromoxynil + clopyralid + fluroxypyr
bromoxynil + clopyralid + fluroxypyr + ioxynil
bromoxynil + fluroxypyr
bromoxynil + fluroxypyr + ioxynil
clopyralid + fluroxypyr + ioxynil
clopyralid + fluroxypyr + MCPA
clopyralid + fluroxypyr + triclopyr

Products

1 Barclay Hurler	Barclay	200 g/l	EC	08791
2 Landgold Fluroxypyr	Landgold	200 g/l	EC	06080
3 Standon Fluroxypyr	Standon	200 g/l	EC	08293
4 Starane 2	Dow	200 g/l	EC	05496
5 Tomahawk	Makhteshim	200 g/l	EC	09249

Uses

Annual dicotyledons in BULB ONIONS *(off-label)*, MAIZE [4]. Annual dicotyledons in BARLEY, WHEAT [1-5]. Annual dicotyledons in DURUM WHEAT, ESTABLISHED GRASSLAND, OATS, RYE, SEEDLING LEYS, TRITICALE [1, 3-5]. Annual dicotyledons in FORAGE MAIZE [1, 3, 5]. Black bindweed in MAIZE [4]. Black bindweed in BARLEY,

WHEAT [1-5]. Black bindweed in DURUM WHEAT, ESTABLISHED GRASSLAND, OATS, RYE, SEEDLING LEYS, TRITICALE [1, 3-5]. Black bindweed in FORAGE MAIZE [1, 3, 5]. Chickweed in MAIZE [4]. Chickweed in BARLEY, WHEAT [1-5]. Chickweed in DURUM WHEAT, ESTABLISHED GRASSLAND, OATS, RYE, SEEDLING LEYS, TRITICALE [1, 3-5]. Chickweed in FORAGE MAIZE [1, 3, 5]. Cleavers in MAIZE [4]. Cleavers in BARLEY, WHEAT [1-5]. Cleavers in DURUM WHEAT, ESTABLISHED GRASSLAND, OATS, RYE, SEEDLING LEYS, TRITICALE [1, 3-5]. Cleavers in FORAGE MAIZE [1, 3, 5]. Docks in MAIZE [4]. Docks in BARLEY, WHEAT [1-5]. Docks in DURUM WHEAT, ESTABLISHED GRASSLAND, OATS, RYE, SEEDLING LEYS, TRITICALE [1, 3-5]. Docks in FORAGE MAIZE [1, 3, 5]. Forget-me-not in MAIZE [4]. Forget-me-not in BARLEY, WHEAT [1-5]. Forget-me-not in DURUM WHEAT, ESTABLISHED GRASSLAND, OATS, RYE, SEEDLING LEYS, TRITICALE [1, 3-5]. Forget-me-not in FORAGE MAIZE [1, 3, 5]. Hemp-nettle in MAIZE [4]. Hemp-nettle in BARLEY, WHEAT [1-5]. Hemp-nettle in DURUM WHEAT, ESTABLISHED GRASSLAND, OATS, RYE, SEEDLING LEYS, TRITICALE [1, 3-5]. Hemp-nettle in FORAGE MAIZE [1, 3, 5]. Volunteer potatoes in BULB ONIONS *(off-label)*, MAIZE [4]. Volunteer potatoes in BARLEY, WHEAT [1-5]. Volunteer potatoes in DURUM WHEAT, ESTABLISHED GRASSLAND, OATS, RYE, SEEDLING LEYS, TRITICALE [1, 3-5]. Volunteer potatoes in FORAGE MAIZE [1, 3, 5].

Notes

Efficacy

- Best results achieved under good growing conditions in a strongly competing crop
- A number of tank mixtures with HBN and other herbicides are recommended for use in autumn and spring to extend range of species controlled. See label for details
- Spray is rainfast in 1 h
- Do not spray if frost imminent

Crop Safety/Restrictions

- Maximum number of treatments 1 per crop or yr
- Apply to new leys from 3 expanded leaf stage
- Timing varies in tank mixtures. See label for details
- Do not use on crops undersown with clovers or other legumes
- Crops undersown with grass may be sprayed provided grasses are tillering
- Do not treat crops suffering stress caused by any factor
- Do not roll or harrow for 7 d before or after treatment

Special precautions/Environmental safety

- Flammable
- Harmful to fish or other aquatic life. Do not contaminate surface waters or ditches with chemical or used container
- Wash spray equipment thoroughly with water and detergent immediately after use. Traces of product can damage susceptible plants sprayed later

Protective clothing/Label precautions

- R07d, U08, U19, U20b, E07 (3 d), E13c, E26, E30a, E31b

Withholding period

- Keep livestock out of treated areas for at least 3 d and until foliage of poisonous weeds such as ragwort has died and become unpalatable

FOR FULL CONDITIONS OF USE ALWAYS READ THE PRODUCT LABEL

Latest application/Harvest Interval(HI)

- Before flag leaf sheath opening (GS 47) for winter wheat and barley; before flag leaf sheath extending (GS 41) for spring wheat and barley; before second node detectable (GS 32) for oats, rye, triticale and durum wheat; before 7 leaves unfolded and before buttress roots appear for maize

Approval

- Off-label approval to May 2002 for use in bulb onions (OLA 0943/97)[4]

309 fluroxypyr + mecoprop-P

A post-emergence herbicide for broadleaved weeds in amenity turf

Products

Bastion T	Rigby Taylor	72:300 g/l	EC	06011

Uses

Annual dicotyledons in AMENITY TURF, MANAGED AMENITY TURF. Perennial dicotyledons in AMENITY TURF, MANAGED AMENITY TURF. Slender speedwell in AMENITY TURF, MANAGED AMENITY TURF.

Notes

Efficacy

- Best results achieved when soil moist and weeds in active growth, normally Apr-Sep
- Do not apply in drought period unless irrigation applied
- Avoid mowing for 3 d before or after spraying (5 d before for young turf)
- Do not spray if turf wet

Crop Safety/Restrictions

- Maximum number of treatments 2 per yr. The total amount of mecoprop-P applied in a single yr must not exceed the maximum total dose approved for any single product for the crop/situation
- Young turf must only be treated in spring provided that at least 2 mth have elapsed between sowing and application
- After mowing young turf allow 5 d before treatment
- Do not treat turf under stress from any cause, if night temperatures are low, if ground frost imminent or during prolonged cold weather
- Avoid drift onto all broad-leaved plants outside the target area

Special precautions/Environmental safety

- Harmful in contact with skin or if swallowed, irritating to eyes
- Harmful to fish or other aquatic life. Do not contaminate surface waters or ditches with chemical or used container
- Wash spray equipment thoroughly with water and detergent immediately after use. Traces of product can damage susceptible plants sprayed later

Protective clothing/Label precautions

- A, C
- M03, R03a, R03c, R04a, U05a, U08, U19, U20b, C03, E01, E13c, E26, E30a, E31b, E34

310 fluroxypyr + metosulam

A post-emergence herbicide mixture for cereals

Products

EF1166	Dow	100:10 g/l	SE	07966

Uses

Annual dicotyledons in WINTER BARLEY, WINTER WHEAT. Charlock in WINTER BARLEY, WINTER WHEAT. Chickweed in WINTER BARLEY, WINTER WHEAT. Cleavers in WINTER BARLEY, WINTER WHEAT. Hemp-nettle in WINTER BARLEY, WINTER WHEAT. Volunteer oilseed rape in WINTER BARLEY, WINTER WHEAT.

Notes

Efficacy

- Best results obtained from application in good growing conditions
- Apply after 1 Feb once crop has 3 leaves
- Good spray cover of weeds is necessary.
- Application to wet foliage may result in reduced activity. Do not spray when rain is imminent

Crop Safety/Restrictions

- Maximum number of treatments 1 per crop
- Avoid treatment in frosty or prolonged cold weather
- Product will severely damage clover. Do not apply to crops undersown, or to be undersown, with clover or clover mixtures
- Do not roll or harrow for 7 d before or after application
- Only cereals, grass seed or oilseed rape may be sown as a following crop in the same calendar year. Cereal stubble must be thoroughly cultivated to 15 cm before sowing oilseed rape. Any crop may be sown the following spring

Special precautions/Environmental safety

- Irritating to skin
- Dangerous to fish or other aquatic life. Do not contaminate surface waters or ditches with chemical or used container
- Do not allow direct spray from vehicle mounted/drawn hydraulic sprayers to fall within 6 m of surface waters or ditches. Direct spray away from water
- See label for detailed instructions on tank cleaning

Protective clothing/Label precautions

- A
- R04b, U05a, U08, U20b, C03, E01, E13b, E16, E30a, E31a, E34

Latest application/Harvest Interval(HI)

- Before third node detectable (GS 33)

FOR FULL CONDITIONS OF USE ALWAYS READ THE PRODUCT LABEL

311 fluroxypyr + triclopyr

A foliar acting herbicide for docks in grassland

Products

1 Doxstar	Dow	100:100 g/l	EC	06050
2 Evade	Nomix-Chipman	20:60 g/l	ME	08071

Uses

Annual dicotyledons in AMENITY GRASS [2]. Docks in ESTABLISHED GRASSLAND [1]. Perennial dicotyledons in AMENITY GRASS [2]. Woody weeds in AMENITY GRASS [2].

Notes

Efficacy

- For dock control apply in spring or autumn or, at lower dose, in spring and autumn [1]
- Treat docks in rosette stage, up to 200 mm high or across [1]
- Allow 2-3 wk growth after cutting or grazing to allow sufficient regrowth to occur before spraying [1]
- For woody weed control apply between May and end Aug when weeds are actively growing [2]
- Do not spray in drought, very hot or very cold weather
- Control may be reduced if rain falls within 2 h of application
- Do not cut grass for 28 d after spraying [1]

Crop Safety/Restrictions

- Maximum number of treatments 2 per yr
- Do not spray grass less than 1 yr old
- Clover will be killed or severely checked by treatment
- Do not roll or harrow for 10 d before or 7 d after spraying
- Avoid contact with foliage, stems or branches of desirable vegetation either directly or by drift. Use of a tree guard recommended when applying near trees and shrubs [2]
- Do not sow kale, turnips, swedes or grass mixtures containing clover by direct drilling or minimum cultivation techniques within 6 wk of application [1]
- Do not plant winter beans or legumes in same yr as treatment. Restrictions apply to many other crops planted in spring following treatment. See label for details of cultivations and intervals required [2]

Special precautions/Environmental safety

- Irritating to skin
- Irritating to eyes [2]
- Flammable [1]
- Harmful to fish or other aquatic life. Do not contaminate surface waters or ditches with chemical or used container
- Do not allow direct spray from vehicle mounted/drawn hydraulic sprayers to fall within 6 m, or from hand-held sprayers to within 2 m, of surface waters or ditches. Direct spray away from water [2]
- Do not allow drift to come into contact with crops, amenity plantings, gardens, ponds, lakes or watercourses
- Wash spray equipment thoroughly with water and detergent immediately after use. Traces of product can damage susceptible plants sprayed later

Protective clothing/Label precautions

- A, C [1, 2]; H, M [2]
- R04a, E16 [2]; R04b, U05a, U08, U20b, C03, E01, E13c, E26, E30a, E31b, E34 [1, 2]; R07d, U02, U19, C01, E07 (7 d) [1]

Withholding period
- Keep livestock out of treated areas for at least 7 d and until foliage of any poisonous weeds such as ragwort has died and become unpalatable [1]

Latest application/Harvest Interval(HI)
- 7 d before grazing or harvest of grass

312 flurtamone

A carotenoid synthesis inhibitor available only in mixtures

See also diflufenican + flurtamone + isoproturon

313 flusilazole

A systemic, protective and curative conazole fungicide for cereals, oilseed rape and apples

See also carbendazim + flusilazole
fenpropimorph + flusilazole
fenpropimorph + flusilazole + tridemorph

Products

1 DUK 747	DuPont	400 g/l	EC	08239
2 Genie	DuPont	400 g/l	EC	08238
3 Lyric	DuPont	250 g/l	EC	08252
4 Sanction	DuPont	400 g/l	EC	08237

Uses

Brown rust in SPRING BARLEY, WINTER BARLEY, WINTER WHEAT. Eyespot in SPRING BARLEY, WINTER BARLEY, WINTER WHEAT. Light leaf spot in OILSEED RAPE. Net blotch in SPRING BARLEY, WINTER BARLEY. Powdery mildew in SPRING BARLEY, WINTER BARLEY, WINTER WHEAT. Rhynchosporium in SPRING BARLEY, WINTER BARLEY. Septoria in SPRING BARLEY, WINTER BARLEY, WINTER WHEAT. Yellow rust in SPRING BARLEY, WINTER BARLEY, WINTER WHEAT.

Notes

Efficacy
- Use as a routine preventative spray or when disease first develops
- Best control of eyespot achieved by spraying between leaf-sheath erect and second node detectable stages (GS 30-32)
- Product active against both MBC-sensitive and MBC-resistant strains of eyespot
- Treat oilseed rape in autumn and/or spring at stem extension stage
- Rain occurring within 2 h after application may reduce effectiveness
- See label for recommended tank-mixes

Crop Safety/Restrictions
- Maximum number of treatments (including other products containing flusilazole) on cereals depends on dose and timing - see labels for details. High rate must not be used more than once in any crop
- Do not apply to crops under stress or during frosty weather

Special precautions/Environmental safety
- Harmful if swallowed
- Dangerous to fish or other aquatic life. Do not contaminate surface waters or ditches with chemical or used container

Protective clothing/Label precautions
- A, C
- M03, R03c, U05a, U11, U19, U20a, C03, E01, E13b, E26, E30a, E31b, E34

Latest application/Harvest Interval(HI)
- High dose before 3rd node detectable (GS 33) plus reduced dose before early milk stage (GS 72) on winter wheat and barley; before first flowers open (GS 4,0) on oilseed rape

314 flusilazole + tridemorph

A systemic protective and curative fungicide for cereals

Products

1 Fusion	DuPont	160:350 g/l	EC	04908
2 Option	DuPont	160:350 g/l	EC	07951

Uses

Brown rust in SPRING BARLEY, WINTER BARLEY, WINTER WHEAT. Net blotch in SPRING BARLEY, WINTER BARLEY. Powdery mildew in SPRING BARLEY, WINTER BARLEY, WINTER WHEAT. Rhynchosporium in SPRING BARLEY, WINTER BARLEY. Septoria diseases in WINTER WHEAT. Yellow rust in SPRING BARLEY, WINTER BARLEY, WINTER WHEAT.

Notes

Efficacy
- Apply at start of disease attack. Treatment most effective when disease levels low at application
- Rain within 2 h of application may reduce effectiveness

Crop Safety/Restrictions
- Maximum number of treatments (including other products containing flusilazole) 3 per crop for winter wheat, 2 per crop for winter or spring barley
- Some crop scorch may occur if treatment applied during frosty weather, high temperatures or drought

Special precautions/Environmental safety
- Irritating to eyes
- Dangerous to fish or other aquatic life. Do not contaminate surface waters or ditches with chemical or used container

Protective clothing/Label precautions
- A, C
- R04a, U05a, U11, U19, U20a, C03, E01, E07, E13b, E30a, E31b

Withholding period
- Keep livestock out of treated areas for at least 14 d

Latest application/Harvest Interval(HI)
- Before early milk stage of grain (GS 73)

315 flutriafol

A conazole fungicide for broad spectrum disease control in wheat and barley

See also carbendazim + flutriafol
chlorothalonil + flutriafol
ethirimol + flutriafol + thiabendazole

Products

1 Landgold Flutriafol	Landgold	125 g/l	SC	06244
2 Pointer	Zeneca	125 g/l	SC	06695

Uses

Brown rust in SPRING BARLEY, WINTER BARLEY, WINTER WHEAT. Leaf blotch in SPRING BARLEY, WINTER BARLEY. Powdery mildew in SPRING BARLEY, WINTER BARLEY, WINTER WHEAT. Septoria leaf spot in WINTER WHEAT. Yellow rust in SPRING BARLEY, WINTER BARLEY, WINTER WHEAT.

Notes

Efficacy

- Best results obtained from application at early stage of disease development
- See label for details of recommended timing, need for repeat treatments and tank mixes

Crop Safety/Restrictions

- Maximum number of treatments 2 per crop (including any other products containing flutriafol)
- Under conditions of stress some wheat varieties can exhibit flag leaf tip scorch which may be increased by fungicide treatment

Special precautions/Environmental safety

- Irritating to eyes and skin
- Harmful to fish or other aquatic life. Do not contaminate surface waters or ditches with chemical or used container

Protective clothing/Label precautions

- A, C [1, 2]
- M03, E34 [1]; R04a, R04b, U02, U05a, U09a, U19, U20a, C03, E01, E13c, E30a, E31c [1, 2]

Latest application/Harvest Interval(HI)

- Up to grain watery ripe (GS 71)

316 fomesafen

A contact and residual diphenyl ether herbicide for use in leguminous crops

Products

Flex	Novartis	250 g/l	SL	08885

Uses

Annual dicotyledons in GREEN BEANS. Annual meadow grass in GREEN BEANS. Voluntee oilseed rape in GREEN BEANS.

Notes

Efficacy

- Best results obtained when weeds actively growing in warm moist conditions with adequate soil moisture
- Weeds should be treated at seedling stage. Larger weeds, and those hardened off by adverse conditions, may be less well controlled
- Residual activity gives control of later germinating weed seedlings
- Product may be used on all soil types but residual control of weeds germinating after treatment may be reduced on soils with high organic matter
- Weed spectrum may be widened by use in a programme with other herbicides. See label for details

Crop Safety/Restrictions

- Maximum total dose 0.9 l/ha
- Apply to healthy crops only. Crops growing under stress from drought, waterlogging etc should not be treated
- Some crop damage, such as leaf crinkling, may occur. Effect is transient and should not affect yield

Special precautions/Environmental safety

- Irritating to eyes
- May cause sensitisation by skin contact
- Do not contaminate surface waters or ditches with chemical or used container

Protective clothing/Label precautions

- A, C, H
- M03, R04a, R04e, U04a, U05a, U11, U14, U19, U20b, C03, E01, E15, E26, E30a, E31a

Latest application/Harvest Interval(HI)

- HI 4 wk

Approval

- Off-label approval to Jul 2003 for use on outdoor soya beans (OLA 1365/97)[1]

317 fomesafen + terbutryn

A residual herbicide for use in leguminous crops

Products	Reflex T	Novartis	80:400 g/l	SC	08884

Uses

Annual dicotyledons in PEAS *(spring sown)*, SPRING BROAD BEANS, SPRING FIELD BEANS.

Notes

Efficacy

- Apply after planting, but before the crop emerges
- Best results achieved when applied to moist, fine, firm tilth
- Rain soon after application is essential for optimum weed control
- Results may be unsatisfactory in dry, or excessively wet, soil conditions
- Weed control may be reduced on soils with more than 10% organic matter

Crop Safety/Restrictions

- Maximum number of treatments 1 per crop
- Use only once every 5 yr

- All varieties of spring sown peas may be treated but forage varieties may be damaged from which recovery may not be complete. All varieties of spring sown field and broad beans may be treated
- Emerged crop leaves may be severely scorched or killed
- Do not use on Sands
- Plough or cultivate to at least 150 mm before drilling or planting another crop. Only cereals should be planted in the calendar year of use with a minimum interval of 4 mth after treatment

Special precautions/Environmental safety
- Irritant. May cause sensitisation by skin contact
- Harmful to fish or other aquatic life. Do not contaminate surface waters or ditches with chemical or used container

Protective clothing/Label precautions
- A, C, H
- R04, R04e, U04a, U05a, U08, U19, U20b, C03, E01, E13c, E26, E30a, E31a, E34

Latest application/Harvest Interval(HI)
- Pre-emergence of crop

318 fonofos

An organophosphorus insecticide for soil and seed treatment

Products

1 Cudgel	Zeneca	433 g/l	CS	06648
2 Fonofos Seed Treatment	Zeneca	433 g/l	CS	06664

Uses

Cabbage root fly in CHINESE CABBAGE *(off-label)*, LEAF BRASSICAS, ROOT BRASSICA [1]. Frit fly in SPRING WHEAT, WINTER BARLEY, WINTER WHEAT [2]. Sciarid flies i HARDY ORNAMENTALS [1]. Vine weevil in HARDY ORNAMENTALS [1]. Wheat bulb fl in SPRING WHEAT, WINTER BARLEY, WINTER WHEAT [2]. Wireworms in SPRIN WHEAT, WINTER BARLEY, WINTER WHEAT [2].

Notes

Efficacy
- May be applied either by incorporation into compost, as a compost or soil drench, or as a hygiene spray treatment under glass against adults and larvae of sciarid flies [1]
- Should be incorporated into compost before striking cuttings, pricking out or potting on an followed with a drench treatment 6 wk later [1]
- When drenching use sufficient water to ensure thorough wetting of the growing medium b without excessive run-through [1]
- When used for cabbage root fly control in modular raised brassicas, slowly maturing crops should be re-treated in the field with a suitable granular insecticide to prevent re-infestatio [1]
- Apply seed treatment with suitable liquid seed treatment machinery [2]
- For best results drill into a well-prepared seed-bed no more than 25 mm deep. Do not broadcast treated seed [2]
- Recommended for all crops sown from early Sep until mid-Mar [2]
- Check drill calibration after each seed lot [2]

- Where wheat bulb fly egg counts are high, insecticide sprays at egg-hatch or dead-heart may be necessary [2]

Crop Safety/Restrictions

- Maximum number of treatments 1 per batch of seed [2]; 3 per season for container nursery stock [1]
- Do not drench under hot sunny conditions and rinse off plant foliage immediately with plain water [1]
- Check susceptibility of any one variety by treating a small number of plants first [1]
- Do not treat seed with more than 16% moisture [2]
- Treated seed may be stored for up to 3 mth in cool, dry, well ventilated place [2]
- All varieties of winter wheat, spring wheat and winter barley seed may be treated [2]

Special precautions/Environmental safety

- Fonofos is subject to the Poisons Rules 1982 and the Poisons Act 1972. See notes in Section 1
- Contains an anticholinesterase organophosphorus compound. Do not use if under medical advice not to work with such compounds
- Toxic in contact with skin. Harmful if swallowed [1, 2]
- May cause sensitization by skin contact [2]
- Dangerous to game, wild birds and animals [1]
- Extremely dangerous to fish or other aquatic life. Do not contaminate surface waters or ditches with chemical or used container
- Treated foliage must not be handled until the spray deposit has dried [1]
- Do not harvest crops for human or animal consumption for at least 6 wk after sowing treated seed [2]
- Do not use treated seed as food or feed [2]
- Treated seed harmful to game and wildlife [2]

Protective clothing/Label precautions

- A [2]; B [1]; C [1, 2]; D [2]; H [1, 2]; J [1]; K [1, 2]; L [1]; M/ [2]; M [1]
- M01, M03, U02, U04a, U05a, U09a, U13, U19, C03, E01, E10a, E13a, E30b, E31a, E34 [1, 2]; R02a, R03c, R04e, U20a, C02, E06a, S01, S02, S03, S04b, S05, S06, S07 [2]; R03a, U14, U15, U20b, E26 [1]

Withholding period

- Keep all livestock out of treated areas for at least 6 wk. Bury or remove spillages [2]

Latest application/Harvest Interval(HI)

- HI 6 wk

Approval

- Off-label approval unlimited for use on outdoor Chinese cabbage seedlings (OLA1202/96)[1]

---

319 formaldehyde (commodity substance)

An agricultural/horticultural and animal husbandry fungicide

Products				
	1 formaldehyde	various	38-40%	SL
	2 paraformaldehyde	various		

Uses Fungus diseases in FLOWER BULBS *(dip)*, GLASSHOUSES, MUSHROOM HOUSES [1]. Fungus diseases in LIVESTOCK HOUSES [2]. Soil-borne diseases in SOIL AND COMPOST [1].

Notes

Efficacy
- Use as a dip to sterilize flower bulbs [1]
- Use as drench to sterilize soil and compost, indoors and outdoors [1]
- Use as spray or fumigant in mushroom houses [1]
- Use as spray, dip or fumigant for glasshouse hygiene [1]
- Use as fumigant to sterilize animal houses [2]

Crop Safety/Restrictions
- Observe maximum permitted concentrations. See PR 12, 1990, pp 9-10

Special precautions/Environmental safety
- Formaldehyde is subject to the Poisons Rules 1982 and the Poisons Act 1972. See notes in Section 1
- Operators must observe Occupational Exposure Standard as set out in HSE Guidance Note EH40/90 and ACOP 30. Control of Substances Hazardous to Health in Fumigation
- Operators must be supplied with a Section 6 (HSW) Safety Data Sheet before commencing work

320 fosamine-ammonium

A contact phosphonic acid herbicide for woody weed control

Products

Krenite	DuPont	480 g/l	SL	01165

Uses Woody weeds in CONIFER PLANTATIONS *(off-label)*, FORESTRY, NON-CROP AREAS, WATERSIDE AREAS.

Notes

Efficacy
- Apply as overall spray to foliage during Aug-Oct when growth ceased but before leaves have started to change colour. Effects develop in following spring
- Thorough coverage of leaves and stems needed for effective control
- Addition of non-ionic wetter or Actipron recommended. Rain within 24 h may reduce effectiveness
- Most deciduous species controlled or suppressed but evergreens resistant
- Little effect on underlying herbaceous vegetation

Crop Safety/Restrictions
- Do not use in conifer plantations

Protective clothing/Label precautions
- U08, U20a, E15, E26, E30a, E31a

Approval
- May be used alongside river, canal and reservoir banks but should not be sprayed into water. See notes in Section 1 on use of herbicides in or near water

- Off-label approval unlimited for use in conifer plantations (OLA 1437/96)[1]

321 fosetyl-aluminium

A systemic phosphonic acid fungicide for various horticultural crops

Products

Aliette	Hortichem	80% w/w	WP	05648

Uses

Collar rot in APPLES. Crown rot in APPLES. Damping off in BRASSICAS *(off-label)*. Downy mildew in BRASSICAS *(off-label)*, BROAD BEANS, COMBINING PEAS *(off-label)*, HERBS *(off-label)*, HOPS, LETTUCE *(off-label)*, PROTECTED LETTUCE, SPINACH BEET *(off-label)*, SPINACH *(off-label)*, VINING PEAS *(off-label)*, WATERCRESS *(off-label)*. Phytophthora in CAPILLARY BENCHES, CHICORY *(off-label)*, HARDY ORNAMENTAL NURSERY STOCK, PROTECTED POT PLANTS, WATERCRESS *(off-label)*. Pythium in WATERCRESS *(off-label)*. Red core in STRAWBERRIES, STRAWBERRIES *(off-label (spring treatment))*. Root rot in CAPILLARY BENCHES, PROTECTED POT PLANTS.

Notes

Efficacy

- Spray young orchards for crown rot protection after blossom and repeat after 3-6 wk. Apply as paste to bark of apples to control collar rot
- Apply to broad beans at early flowering and 14 d later
- Spray autumn-planted strawberries 2-3 wk after planting or use dip treatment at planting. Spray established crops in late summer/early autumn after picking
- Apply to hops as early season band spray and as foliar spray every 10-14 d
- May be used incorporated in blocking compost for protected lettuce
- Apply as drench to rooted cuttings of hardy nursery stock after first potting and repeat mthly. Up to 6 applications may be needed
- See label for details of application to capillary benches and list of tolerant plants

Crop Safety/Restrictions

- Maximum number of treatments 1 per batch of compost for lettuce; 1 per yr for strawberries (root dip or foliar spray); 1 per yr for apples (bark paste), 6 per yr for apples (foliar spray); 2 per crop for broad beans, peas; 2 per yr for hops (basal spray); 6 per yr for hops (foliar spray); 1 per crop for brassicas; 1 per crop during propagation, 2 per crop after planting out for lettuce
- When using on strawberries do not mix with any other product
- Use in protected lettuce only from Sep to Apr
- Do not apply to ornamentals in mixture with nutrient solutions
- Apply test treatment on lettuce cultivars or ornamentals of unknown tolerance before committing whole batches

Protective clothing/Label precautions

- A, C, H
- U20b, C02, C03, E01, E15, E30a, E32a

Latest application/Harvest Interval(HI)

- Up to 31 Dec for autumn treatment on strawberries; pre-emergence for brassicas.
- HI hops 4 d; lettuce 14 d; broad beans, peas 17 d; apples 14 d; strawberries 3 mth; applied as bark paste 5 mth

Approval

- Off-label approval unlimited for use on strawberries (OLA 0210/96)[1]; to Feb 2002 for use on leaf and flowerhead brassicas (before transplanting or pre-emergence)(OLA 0253/97, 0254/97)[1]; to Feb 2002 for use on outdoor and protected lettuce (OLA 0255/97)[1]; to Apr 2002 for use on watercress during propagation (OLA 0810/97)[1]; to Jun 2002 for use on outdoor spinach and spinach beet (OLA 1190/97)[1]; unlimited for use on vining and combining peas (OLA 2390/97)[1]; to Jul 2003 for use on chicory in forcing sheds (OLA 1206/98)[1]

322 fosthiazate

A contact nematicide for potatoes

Products

Nemathorin 10G	Zeneca	10% w/w	FG	08915

Uses

Potato cyst nematode in POTATOES.

Notes

Efficacy

- Apply and incorporate granules in one operation to a uniform depth of 10-15 cm. Deeper incorporation will reduce control
- Application best achieved using equipment such as Horstine Farmery Microband Applicator, Matco or Stocks Micrometer applicators together with a rear mounted powered rotary cultivator
- Granules must not become wet or damp before use

Crop Safety/Restrictions

- Maximum number of treatments 1 per crop
- Product must only be applied using tractor-mounted/drawn direct placement machinery. Do not use air assisted broadcast machinery
- Do not apply more than once every four years on the same area of land
- Consult before using on crops intended for processing

Special precautions/Environmental safety

- Harmful if swallowed
- Dangerous to game, wild birds and animals
- Harmful to fish or other aquatic life. Do not contaminate surface waters or ditches with chemical or used container
- Incorporation to 10-15 cm and ridging up of treated soil must be carried out immediately after application
- Failure completely to bury granules immediately after application is hazardous to wildlife

Protective clothing/Label precautions

- A, E, G, H, K, M
- M01, M03, R03c, U02, U04a, U05a, U09a, U13, U19, U20a, C03, E01, E06a, E10a, E13c E26, E29, E32a, E34

Latest application/Harvest Interval(HI)

- Before planting

323 fuberidazole

A benzimidazole (MBC) fungicide available only in mixtures

See also bitertanol + fuberidazole
bitertanol + fuberidazole + imidacloprid

324 fuberidazole + triadimenol

A broad spectrum systemic fungicide seed treatment for cereals

Products

Baytan Flowable	Bayer	22.5:187.5 g/l	FS	02593

Uses

Blue mould in TRITICALE, WHEAT. Brown foot rot in BARLEY, OATS, RYE, TRITICALE, WHEAT. Brown rust in BARLEY, RYE, WHEAT. Bunt in WHEAT. Covered smut in BARLEY. Crown rust in OATS. Leaf stripe in SPRING BARLEY. Loose smut in BARLEY, OATS, WHEAT. Mildew in BARLEY, WHEAT. Pyrenophora leaf spot in OATS. Septoria in WHEAT. Yellow rust in BARLEY, WHEAT.

Notes

Efficacy

- Apply through suitable seed treatment machinery. See label for details
- Calibrate drill for treated seed and drill at 2.5-4 cm into firm, well prepared seedbed
- Ferrax recommended as preferable seed treatment for barley
- In addition to seed-borne diseases early attacks of various foliar, air-borne diseases are controlled or suppressed. See label for details
- Product may be used in conjunction with Gammasan 30

Crop Safety/Restrictions

- Maximum number of treatments 1 per batch of seed
- Do not use on naked oats
- Do not drill treated winter wheat or rye seed after end of Nov. Seed rate should be increased as drilling season progresses
- Treated spring wheat may be drilled in autumn up to end of Nov or from Feb onward
- Germination tests should be done on all batches of seed to be treated to ensure seed viability and suitability for treatment
- Do not use on seed with moisture content above 16%, on sprouted, cracked or skinned seed or on seed already treated with another seed treatment
- Store treated seed in cool, dry, well-ventilated store and drill as soon as possible, preferably in season of purchase
- Treatment may accentuate effects of adverse seedbed conditions on crop emergence

Special precautions/Environmental safety

- Dangerous to fish or other aquatic life. Do not contaminate surface waters or ditches with chemical or used container
- Do not use treated seed as food or feed

Protective clothing/Label precautions

- A
- U20b, E03, E13b, E26, E30a, E32a, E34, S01, S02, S03, S04a, S05, S06, S07

Latest application/Harvest Interval(HI)

- Before drilling

325 furalaxyl

A protective and curative phenylamide (acylalanine) fungicide for ornamentals

Products

1 Fongarid	Novartis	25% w/w	WP	08407
2 Fongarid 25 WP	Ciba Agric.	25% w/w	WP	03595

Uses

Damping off in BEDDING PLANTS. Phytophthora in HARDY ORNAMENTAL NURSERY STOCK, POT PLANTS. Pythium in HARDY ORNAMENTAL NURSERY STOCK, POT PLANTS.

Notes

Efficacy
- Apply by incorporation into compost or as a drench to obtain at least 12 wk protection
- Apply drench within 3 d of seeding or planting out as a protective treatment or as soon as first signs of root disease appear as curative treatment
- Do not apply to field or border soil

Crop Safety/Restrictions
- Manufacturer's literature lists genera which have been treated successfully. With all subjects of unknown susceptibility treat a few plants before committing whole batches

Special precautions/Environmental safety
- Irritating to eyes and skin
- Harmful to fish or other aquatic life. Do not contaminate surface waters or ditches with chemical or used container

Protective clothing/Label precautions
- A, C
- R04a, R04b, U02, U05a, U09a, U19, U20a, C03, E01, E13c, E30a, E31b, E34

326 gamma-HCH

A contact, ingested and fumigant organochlorine insecticide

See also carboxin + gamma-HCH + thiram

Products

1 Atlas Steward	Atlas	560 g/l	SC	07728
2 Fumite Lindane 10	Hortichem	20.8 g a.i.	FU	00933
3 Fumite Lindane 40	Hortichem	84.4 g a.i.	FU	00934
4 Fumite Lindane Pellets	Hortichem	2.7 g or 7.6 g a.i.	FW	00937
5 Gamma-Col	Zeneca	800 g/l	SC	06670
6 Gammasan 30	Zeneca	30% w/w	FS	06671
7 Kotol FS	RP Agric.	125 g/l	FS	06968
8 Lindane Flowable	PBI	800 g/l	SC	02610

Uses

Ants in PROTECTED CROPS [4]. Aphids in PROTECTED CROPS [4]. Capsids in PROTECTED CROPS [4]. Chafer grubs in STRAWBERRIES [1, 5]. Cutworms in CEREALS GRASSLAND, MAIZE, SUGAR BEET [1, 5, 8]. Earwigs in PROTECTED CROPS [4]. Flea beetles in SUGAR BEET [1, 5]. Grain beetles in GRAIN STORES *(empty)* [2, 3]. Grain storage pests in GRAIN STORES *(empty)* [2, 3]. Grain weevils in GRAIN STORES *(empty)* [2, 3]. Leaf miners in PROTECTED CROPS [4]. Leatherjackets in CEREALS, GRASSLAND, MAIZE, SUGAR BEET [1, 5, 8]. Leatherjackets in STRAWBERRIES [1, 5]. Millipedes in SUGAR BEET [1, 5, 8]. Mushroom flies in PROTECTED CROPS [4]. Pygmy mangold beetle in SUGAR BEET [1, 5, 8]. Springtails in SUGAR BEET [1, 5, 8]. Symphylids in SUGAR BEET [1, 5, 8]. Thrips in PROTECTED CROPS [4]. Whitefly in PROTECTED CROPS [4]. Wireworms in BARLEY *(seed treatment)*, OATS *(seed treatment)*, WHEAT *(seed treatment)* [6, 7]. Wireworms in RYE *(seed treatment)* [7]. Wireworms in CEREALS, GRASSLAND, MAIZE, SUGAR BEET [1, 5, 8]. Wireworms in STRAWBERRIES [1, 5]. Woodlice in PROTECTED CROPS [4].

Notes

Efficacy

- Method of application, dose, timing and number of applications vary with formulation, crop and pest. See label for details
- Where glasshouse whitefly resistant to gamma-HCH occur control is unlikely to be satisfactory

Crop Safety/Restrictions

- Maximum number of treatments 1 per crop for cereals, maize, sugar beet, strawberries; 1 per yr for grassland; 1 per batch of seed [7]
- Do not use if potatoes or carrots are to be planted within 18 mth [1, 8]
- Do not fumigate newly pricked-out seedlings until root action has started again. Advisable to cut blooms before fumigating. Do not fumigate rare or unusual plants without first testing on small scale

Special precautions/Environmental safety

- Toxic if swallowed [5]
- Harmful if swallowed [1, 6-8]; irritating to eyes and respiratory system [2, 3]
- Extremely dangerous to fish [1, 5-8] dangerous to fish [2-4] or aquatic life. Do not contaminate surface waters or ditches with chemical or used container
- Do not allow direct spray from vehicle mounted/drawn hydraulic sprayers to fall within 6 m, or from hand-held sprayers to within 2 m, of surface waters or ditches. Direct spray away from water [1, 5, 8]
- Harmful to livestock [1, 5]
- Do not use treated seed as food or feed [6, 7]
- Treated seed harmful to game and wildlife [6, 7]
- Must be incorporated into soil following application [1, 5, 8]

Protective clothing/Label precautions

- A, C [1-8]; D [6]; H [1, 5, 6, 8]; K [1, 8]; M [8]
- M03 [2-4, 6-8]; M04 [1, 5]; R02c, E07, E17 [5]; R03a [1, 6, 7]; R03b, U04a, U20c, E13c, S04b, S07 [7]; R03c [1, 6-8]; R04a, R04c, U09a, U20a, E06b, E13b, E31b [2-4]; U02 [2-4, 8]; U05a, C03, E01, E30a, E34 [1-8]; U08 [1, 5-8]; U19 [2-4, 7, 8]; U20b, E13a, E31a [1, 5, 6, 8]; C02 (14 d) [1]; C02 [2-5]; E12c [2-5, 8]; E16 [1, 5, 8]; E26 [1-5, 8]; E32a [2-4, 7]; S01, S02, S05, S06 [6, 7]

Withholding period

- Keep all livestock out of treated areas for at least 14 d [5, 8]

Latest application/Harvest Interval(HI)

- HI 2 d [4]

Maximum Residue Levels (mg residue/kg food)

- strawberries, blackberries, loganberries, raspberries, bilberries, cranberries, currants, gooseberries 3; tomatoes, peppers, aubergines, cauliflowers, Brussels sprouts, head cabbages, lettuce, meat (sheep meat) 2; citrus fruits, pome fruits, apricots, peaches, nectarines, plums, bananas, swedes, turnips, garlic, onions, shallots, cucumbers, gherkins, courgettes, beans (with pods), celery, leeks, rhubarb, cultivated mushrooms, meat (except sheep meat) 1; grapes 0.5; carrots, horseradish, parsnips, parsley root, salsify, tea 0.2; peas (with pods), cereals, eggs 0.1; potatoes 0.05

327 gamma-HCH + fenpropimorph + thiram

An insecticide and fungicide seed dressing for oilseed rape

See also fenpropimorph

Products

Lindex-Plus FS Seed Treatment	Uniroyal	545:43:73 g/l	FS	08212

Uses

Alternaria in OILSEED RAPE *(seed treatment)*. Flea beetles in OILSEED RAPE *(seed treatment)*. Stem canker in OILSEED RAPE *(seed treatment)*.

Notes

Efficacy

- Use product within 6 mth of manufacture and apply via a conventional seed treatment machine fitted with secondary mixers (augers)
- Flea beetle normally reduced up to first crop true leaf stage. Heavy attacks may require foliar insecticide spray

Crop Safety/Restrictions

- Maximum number of treatments 1 per batch of seed
- Treatment may reduce germination capacity if seed is grown, harvested and stored under adverse conditions
- Do not treat cracked, split or sprouted seed or seed with moisture above 9%

Special precautions/Environmental safety

- Harmful if swallowed. Irritating to eyes
- Extremely dangerous to fish or other aquatic life. Do not contaminate surface waters or ditches with chemical or used container
- Do not store treated seed from one season to the next
- Do not use treated seed as food or feed
- Treated seed harmful to game and wildlife
- Do not mix with other formulations and flush machine pumps and pipelines after every use

Protective clothing/Label precautions

- A, C, D, H
- M04, R03c, R04a, U05a, U08, U20b, C03, E01, E03, E13a, E26, E27, E29, E30a, E31a, E34, S01, S02, S03, S04b, S05, S06

Latest application/Harvest Interval(HI)

- Before drilling

Maximum Residue Levels (mg residue/kg food)
- see gamma-HCH entry

328 gamma-HCH + thiophanate-methyl

An insecticide and lumbricide for use on turf

Products

Castaway Plus	RP Amenity	60:500 g/l	SC	05327

Uses Earthworms in TURF. Leatherjackets in TURF.

Notes

Efficacy
- Apply by spraying in spring or autumn. Drenching is not required
- Do not mow for 2 d after spraying. If mown beforehand leave clippings
- Do not collect first clippings after treatment
- Do not spray during drought or if ground frozen
- Effectiveness is not impaired by rain or irrigation immediately after application
- Do not mix with any other product

Special precautions/Environmental safety
- Irritating to eyes and skin
- Harmful to bees. Do not apply to crops in flower or to those in which bees are actively foraging. Do not apply when flowering weeds are present
- Harmful to fish or other aquatic life. Do not contaminate surface waters or ditches with chemical or used container
- Do not allow direct spray from vehicle mounted/drawn hydraulic sprayers to fall within 6 m, or from hand-held sprayers to within 2 m, of surface waters or ditches. Direct spray away from water

Protective clothing/Label precautions
- A, C, H, M
- M03, R04a, R04b, U02, U05a, U08, U19, U20c, C03, E01, E13c, E16, E26, E30a, E34

Withholding period
- Keep livestock out of treated areas for at least 14 d

Maximum Residue Levels (mg residue/kg food)
- see gamma-HCH entry

329 gamma-HCH + thiram

An insecticide and fungicide seed dressing for brassica crops

Products

Hydraguard	Agrichem	533:200 g/l	LS	08877

Uses Damping off in BRUSSELS SPROUTS, CABBAGES, KALE, OILSEED RAPE, SWEDES, TURNIPS. Flea beetles in BRUSSELS SPROUTS, CABBAGES, KALE, OILSEED RAPE, SWEDES, TURNIPS.

Notes **Efficacy**

- Apply in batch or continuous flow type seed treatment machinery
- Normally used undiluted but may be diluted with up to an equal volume of water in which case seed may require drying before bagging
- Seed should be evenly treated for best pest and disease control
- Control of seed-borne Alternaria can be achieved by co-application with iprodione (see label)

Crop Safety/Restrictions

- Maximum number of treatments 1 per batch

Special precautions/Environmental safety

- Harmful in contact with skin and if swallowed
- Irritating to skin, eyes and respiratory system
- May cause sensitisation by skin contact
- Harmful to fish or other aquatic life. Do not contaminate surface waters or ditches with chemical or used container
- Do not use treated seed as food or feed
- Treated seed harmful to game and wildlife
- For use only by agricultural contractors using batch or continuous flow type seed treatment machinery
- Must be dispensed from 20 l container with a suitable probe and pump mechanism

Protective clothing/Label precautions

- A, C, D, H
- M03, R03a, R03c, R04a, R04b, R04c, R04e, U05a, U08, U19, U20b, C03, E01, E03, E13a, E26, E30a, E31c, E34, S01, S02, S03, S04b, S05, S06, S07

Latest application/Harvest Interval(HI)

- Pre-drilling

Maximum Residue Levels (mg residue/kg food)

- see gamma-HCH entry

330 gibberellins

A plant growth regulator for use in apples, pears etc

Products

1 Berelex	Whyte	1 g/tablet	TB	06637
2 Novagib	Fine	10 g/l	SL	08954
3 Regulex	Zeneca	10 g/l	SL	07997

Uses Increasing fruit set in PEARS [1]. Increasing germination in NOTHOFAGUS [3]. Increasing yield in CELERY, RHUBARB [1]. Reducing fruit russeting in APPLES [2, 3].

Notes **Efficacy**

- Apply to apples at completion of petal fall and repeat 3 or 4 times at 7-10 d intervals. Number of sprays and spray interval depend on dose (see labels) [2, 3]

FOR FULL CONDITIONS OF USE ALWAYS READ THE PRODUCT LABEL

- Good coverage of fruitlets is essential for successful results. Best results achieved by spraying under humid, slow drying conditions [3]
- Useful in pears when blossom sparse, setting conditions poor or where frost has killed many flowers. Apply as single or split application. See label for details. Resulting fruit may be seedless [1]
- May be used on pears if 80% of flowers frosted but not effective on very severely frosted blossom. Conference pear responds well, Beurre Hardy only in some seasons. Young trees generally less responsive [1]
- Apply to Nothofagus seed for 24 h as a soak treatment and sow immediately [3]
- Apply to celery 3 wk before harvest to increase head size, to rhubarb crowns on transfer to forcing shed or as drench at first signs of growth in field [1]

Crop Safety/Restrictions

- Maximum number of treatments 1 per crop or yr on celery and rhubarb; 2 per yr on pears [1]
- Maximum total dose 2.0 l/ha per yr for apples
- Good results achieved on Cox's Orange Pippin, Discovery, Golden Delicious and Karmijn apples. For other cultivars test on a small number of trees [3]
- Return bloom may be reduced in yr following treatment [2]
- Do not apply to pears after petal fall [1]
- Consult before treating crops grown for processing [2]

Special precautions/Environmental safety

- Do not contaminate surface waters or ditches with chemical or used container

Protective clothing/Label precautions

- U20a, E27, E31a, E34 [1, 3]; U20b, E01, E24, E29 [2]; E15, E26, E30a [1-3]; E31b [1, 2]; E31c [3]

331 glufosinate-ammonium

A non-selective, non-residual phosphinic acid contact herbicide

Products

1 Challenge	AgrEvo	150 g/l	SL	07306
2 Challenge 60	Fargro	60 g/l	SL	08236
3 Dash	Nomix-Chipman	120 g/l	SL	05177
4 Harvest	AgrEvo	150 g/l	SL	07321

Uses

Annual dicotyledons in NURSERY STOCK, WOODY ORNAMENTALS [3]. Annual dicotyledons in BARLEY *(pre-harvest)*, COMBINING PEAS *(pre-harvest)*, FIELD BEANS *(pre-harvest)*, FIELD CROPS *(pre-cropping situations)*, GRASSLAND *(sward destruction)*, LINSEED *(pre-harvest)*, NON-CROP AREAS, OILSEED RAPE *(pre-harvest)*, POTATOES *(pre-emergence)*, POTATOES *(pre-harvest)*, SUGAR BEET *(pre-emergence)*, VEGETABLES *(pre-emergence)*, WHEAT *(pre-harvest)* [1, 4]. Annual dicotyledons in FIELD CROPS *(sward destruction)*, LAND TEMPORARILY REMOVED FROM PRODUCTION [4]. Annual dicotyledons in BUSH FRUIT, CANE FRUIT, CULTIVATED LAND/SOIL, FRUIT TREES, ORNAMENTAL TREES, SHRUBS [2]. Annual grasses in NURSERY STOCK, WOODY ORNAMENTALS [3]. Annual grasses in BARLEY *(pre-harvest)*, COMBINING PEAS *(pre-harvest)*, FIELD BEANS *(pre-harvest)*, FIELD CROPS *(pre-cropping situations)*, FIELD CROPS *(sward destruction)*, LAND TEMPORARILY REMOVED FROM PRODUCTION, LINSEED *(pre-harvest)*, OILSEED RAPE *(pre-harvest)*, POTATOES *(pre-emergence)*, POTATOES *(pre-harvest)*, SUGAR BEET *(pre-emergence)*, VEGETABLES *(pre-emergence)*, WHEAT *(pre-harvest)* [1, 4]. Annual grasses in BUSH FRUIT, CANE FRUIT, CULTIVATED LAND/SOIL, FRUIT TREES, ORNAMENTAL TREES, SHRUBS [2]. Annual weeds in

APPLES, CANE FRUIT, CHERRIES, CURRANTS, DAMSONS, FORESTRY, GRAPEVINES, HEADLANDS, NON-CROP AREAS, PEARS, PLUMS, STRAWBERRIES, TREE NUTS [1, 4]. Green cover in LAND TEMPORARILY REMOVED FROM PRODUCTION [1, 4]. Perennial dicotyledons in NURSERY STOCK, WOODY ORNAMENTALS [3]. Perennial dicotyledons in FIELD CROPS *(pre-cropping situations)*, GRASSLAND *(sward destruction)*, LAND TEMPORARILY REMOVED FROM PRODUCTION [1, 4]. Perennial dicotyledons in FIELD CROPS *(sward destruction)* [4]. Perennial dicotyledons in BUSH FRUIT, CANE FRUIT, CULTIVATED LAND/SOIL, FRUIT TREES, ORNAMENTAL TREES, SHRUBS [2]. Perennial grasses in NURSERY STOCK, WOODY ORNAMENTALS [3]. Perennial grasses in FIELD CROPS *(pre-cropping situations)*, GRASSLAND *(sward destruction)*, LAND TEMPORARILY REMOVED FROM PRODUCTION [1, 4]. Perennial grasses in FIELD CROPS *(sward destruction)* [4]. Perennial grasses in BUSH FRUIT, CANE FRUIT, CULTIVATED LAND/SOIL, FRUIT TREES, ORNAMENTAL TREES, SHRUBS [2]. Perennial weeds in APPLES, CANE FRUIT, CHERRIES, CURRANTS, DAMSONS, FORESTRY, GRAPEVINES, HEADLANDS, NON-CROP AREAS, PEARS, PLUMS, STRAWBERRIES, TREE NUTS [1, 4]. Pre-harvest desiccation in BARLEY, COMBINING PEAS, FIELD BEANS, LINSEED, OILSEED RAPE, POTATOES, WHEAT [1, 4].

Notes

Efficacy

- Activity quickest under warm, moist conditions. Light rainfall 3-4 h after application will not affect activity. Do not spray wet foliage or if rain likely within 6 h
- For weed control uses treat when weeds growing actively. Deep rooted weeds may require second treatment [2]
- Ploughing or other cultivations can follow 4 h after spraying
- On uncropped headlands apply in May/Jun to prevent weeds invading field
- Crops can normally be sown/planted immediately after spraying or sprayed post-drilling. On sand, very light or immature peat soils allow at least 3 d before sowing/planting or expected emergence
- For weed control in potatoes apply pre-emergence or up to 10% emergence on earlies and seed crops, up to 40% on maincrop, on plants up to 15 cm high
- In sugar beet and vegetables apply just before crop emergence, using stale seedbed technique
- In top and soft fruit, grapevines, woody ornamentals, nursery stock and forestry apply up to 3 treatments (2 treatments [2]) between 1 Mar and 30 Sep as directed sprays
- For grass destruction apply before winter dormancy occurs. Heavily grazed fields should show active regrowth. Plough from the day after spraying
- Apply pre-harvest desiccation treatments 10-21 d before harvest (14-21 d for oilseed rape). See label for timing details on individual crops
- For potato haulm desiccation apply to listed varieties (not seed crops) at onset of senescence, 14-21 d before harvest

Crop Safety/Restrictions

- Maximum number of treatments 4 per crop for potatoes (including 2 desiccant uses); 2 per crop (including 1 desiccant use) for oilseed rape, dried peas, field beans, linseed, wheat and barley; 3 (2 [2]) per yr for fruit and forestry; 2 per yr for strawberries, ornamental trees and shrubs, non-crop land and setaside; 1 per crop for sugar beet, vegetables and other crops; 1 per yr for grassland destruction, on cultivated land prior to planting edible crops
- Pre-harvest desiccation sprays should not be used on seed crops of wheat, barley, peas or potatoes but may be used on seed crops of oilseed rape, field beans and linseed

- Do not desiccate potatoes in exceptionally wet weather or in saturated soil. See label for details
- Do not spray potatoes after emergence if grown from small or diseased seed or under very dry conditions
- For weed control uses application must be between 1 Mar and 30 Sep
- Do not spray hedge bottoms
- Do not allow spray to contact dormant or green buds, suckers, damaged or green bark and foliage of wanted plants [2]

Special precautions/Environmental safety

- Harmful if swallowed and in contact with skin. Irritating to eyes [1, 3, 4]
- Treated pea haulm may be fed to livestock from 7 d after spraying, treated grain from 14 d
- Do not use straw from treated crops as animal feed or bedding
- Harmful to fish or other aquatic life. Do not contaminate surface waters or ditches with chemical or used container
- Product supplied in small volume returnable container - see label for filling and mixing instructions [4]

Protective clothing/Label precautions

- A, C, H, M
- M03, R03a, R03c, U04a, U13, U14, U15, U19, E09, E34 [1, 4]; R04a, U05a, C03, E01 [1, 3, 4]; U02, E31a [2]; U08, E07, E13c, E30a [1-4]; U20a [1, 2, 4]; U20b, E26 [3]

Withholding period

- Keep livestock out of treated areas until foliage of any poisonous weeds such as ragwort has died and become unpalatable

Latest application/Harvest Interval(HI)

- 30 Sep for use on non-crop land, top fruit, forestry, pre-drilling, pre-planting, pre-emergence in sugar beet, vegetables and other crops; before winter dormancy for grassland destruction.
- HI potatoes, oilseed rape, drying peas, field beans, linseed 7 d; wheat, barley 14 d

332 glyphosate (agricultural)

A translocated non-residual phosphonic acid herbicide

See also diuron + glyphosate

Products

1 Alpha Glyphogan	Makhteshim	360 g/l	SL	05784
2 Apache	Zeneca	220 g/l	SL	06748
3 Barbarian	Barclay	360 g/l	SL	07980
4 Barclay Cleanup	Barclay	180 g/l	SL	09179
5 Barclay Dart	Barclay	180 g/l	SL	05129
6 Barclay Gallup	Barclay	360 g/l	SL	05161
7 Barclay Garryowen	Barclay	360 g/l	SL	08599
8 Clinic	Nufarm	360 g/l	SL	08579
9 Glyfos	Cheminova	360 g/l	SL	07109
10 Glyper	PBI	360 g/l	SL	07968
11 Helosate	Helm	360 g/l	SL	06499
12 Hilite	Nomix-Chipman	144 g/l	RH	06261
13 I T Glyphosate	I T Agro	360 g/l	SL	07212
14 Landgold Glyphosate 360	Landgold	360 g/l	SL	05929
15 MON 240	Monsanto	240 g/l	SL	04538
16 Poise	Unicrop	480 g/l	SL	08276
17 Portman Glider	Portman	480 g/l	SL	04695
18 Portman Glyphosate	Portman	360 g/l	SL	05891
19 Portman Glyphosate 360	Portman	360 g/l	SL	04699
20 Portman Glyphosate 480	Portman	480 g/l	SL	07194
21 Roundup	Monsanto	360 g/l	SL	01828
22 Roundup 2000	Monsanto	400 g/l	SL	08069
23 Roundup Biactive	Monsanto	360 g/l	SL	06941
24 Roundup Biactive Dry	Monsanto	42% w/w	WG	06942
25 Roundup Four 80	Monsanto	480 g/l	SL	03176
26 Roundup GT	Monsanto	400 g/l	SL	08068
27 Roundup Pro	Monsanto	360 g/l	SL	04146
28 Roundup Pro Biactive	Monsanto	360 g/l	SL	06954
29 Roundup Rapide	Monsanto	400 g/l	SL	08067
30 Stacato	Sipcam	360 g/l	SL	05892
31 Standon Glyphosate 360	Standon	360 g/l	SL	05582
32 Stefes Glyphosate	Stefes	360 g/l	SL	05819
33 Stefes Kickdown	Stefes	360 g/l	SL	06329
34 Sting CT	Monsanto	120 g/l	SL	04754
35 Sting ECO	Monsanto	120 g/l	SL	08291
36 Stirrup	Nomix-Chipman	144 g/l	RH	06132
37 Stride	Zeneca	440 g/l	SL	06750
38 Touchdown	Zeneca	330 g/l	SL	06326

Uses Annual and perennial weeds in STUBBLES [15]. Annual and perennial weeds in NON-CROP FARM AREAS [6-8, 15, 19, 21, 23, 24, 30, 32]. Annual and perennial weeds in CULTIVATED LAND/SOIL [6-8, 32, 33]. Annual and perennial weeds in DURUM WHEAT *(pre-harvest)* [6-8, 10, 20, 23, 24]. Annual and perennial weeds in FIELD CROPS *(wiper application)* [6-8]. Annual and perennial weeds in DAMSONS [1, 6-11, 21, 23, 24, 32, 33]. Annual and perennial weeds in CHERRIES, PLUMS [1, 6-11, 19, 21, 23, 24, 32, 33]. Annual and perennial weeds in FIELD BEANS *(pre-harvest)* [1, 6-11, 14, 16, 19-26, 29, 31-33]. Annual and perennial weeds in PEARS

FOR FULL CONDITIONS OF USE ALWAYS READ THE PRODUCT LABEL

[1, 6-8, 10, 11, 15, 19, 21, 23, 24, 32, 33]. Annual and perennial weeds in GRASSLAND *(pre-cut/graze)* [1, 6-8, 10, 11, 15, 19, 21, 23]. Annual and perennial weeds in APPLES [1, 6-8, 10, 11, 19, 21, 23, 24, 32, 33]. Annual and perennial weeds in OATS *(pre-harvest)*[1, 6-8, 10, 11, 13-17, 19-26, 29-31]. Annual and perennial weeds in OILSEED RAPE *(pre-harvest)* [1, 2, 6-11, 14-16, 19-26, 29-33, 37, 38]. Annual and perennial weeds in COMBINING PEAS *(pre-harvest)* [1, 2, 6-8, 10, 11, 14, 16, 19-26, 29, 31-33, 37, 38]. Annual and perennial weeds in BARLEY *(pre-harvest)*, WHEAT *(pre-harvest)* [1, 2, 6-8, 10, 11, 13-17, 19-26, 29-33, 37, 38]. Annual and perennial weeds in LINSEED *(pre-harvest)* [1, 2, 8-11, 20, 21, 23, 24, 37, 38]. Annual and perennial weeds in MUSTARD *(pre-harvest)* [1, 2, 8, 10, 11, 15, 16, 20-26, 29, 37, 38]. Annual and perennial weeds in FIELD CROPS *(stubble treatment)* [1, 2, 4-11, 13, 14, 16, 17, 19-26, 29-33, 37, 38]. Annual and perennial weeds in FIELD CROPS *(sward destruction/direct drilling)* [1, 4-8, 10, 11, 13, 21, 23, 30]. Annual and perennial weeds in GRASSLAND *(pre-cut/graze or sward destruction)* [32]. Annual and perennial weeds in GRASSLAND *(sward destruction/direct drilling)* [14, 15, 24, 31]. Annual and perennial weeds in HEDGES *(directed spray)*, HEDGES *(pre-planting)*, TOP FRUIT *(directed spray)* [12, 36]. Annual and perennial weeds in OILSEED RAPE FOR INDUSTRIAL USE *(pre-harvest)* [2, 37]. Annual and perennial weeds in GRASSLAND *(sward destruction)* [2, 9, 19, 33, 37, 38]. Annual and perennial weeds in LAND TEMPORARILY REMOVED FROM PRODUCTION [2, 9, 23, 24, 37, 38]. Annual and perennial weeds in CEREALS *(pre-harvest)*, ORCHARDS, PEAS *(pre-harvest)* [9]. Annual dicotyledons in CULTIVATED LAND/SOIL *(pre-drilling/pre-crop emergence)*, NON-CROP FARM AREAS, STUBBLES *(pre-drilling/pre-crop emergence)* [34, 35]. Annual dicotyledons in FIELD CROPS *(stubble treatment)* [18]. Annual dicotyledons in GRASSLAND *(sward destruction)* [3, 9, 18]. Annual grasses in CULTIVATED LAND/SOIL *(pre-drilling/pre-crop emergence)*, NON-CROP FARM AREAS, STUBBLES *(pre-drilling/pre-crop emergence)* [34, 35]. Annual grasses in GRASSLAND *(sward destruction)* [3, 18]. Annual grasses in FIELD CROPS *(stubble treatment)* [3, 9, 18]. Annual weeds in GRASSLAND *(sward destruction)* [34, 35]. Annual weeds in GRASSLAND [1, 4-8, 10, 11, 16, 20-23, 25, 26, 29, 30, 32, 33]. Annual weeds in BARLEY *(post sowing, pre crop emergence)*, DURUM WHEAT *(post sowing, pre crop emergence)*, FIELD BEANS *(post sowing, pre crop emergence)*, LINSEED *(post sowing, pre crop emergence)*, MUSTARD *(post sowing, pre crop emergence)*, OATS *(post sowing, pre crop emergence)*, OILSEED RAPE *(post sowing, pre crop emergence)*, PEAS *(post sowing, pre crop emergence)*, SUGAR BEET *(post sowing, pre crop emergence)*, SWEDES *(post sowing, pre crop emergence)*, TURNIPS *(post sowing, pre crop emergence)*, WHEAT *(post sowing, pre crop emergence)* [23, 24]. Annual weeds in NON-CROP FARM AREAS [3, 18]. Annual weeds in ORCHARDS [3]. Annual weeds in FIELD CROPS *(pre-drilling/pre-emergence)*, PEAS *(pre-drilling/pre-emergence)*, SPRING CEREALS *(pre-drilling/pre-emergence)*, SUGAR BEET *(pre-drilling/pre-emergence)*, SWEDES *(pre-drilling/pre-emergence)*, TURNIPS *(pre-drilling/pre-emergence)* [4, 5]. Black bent in OATS *(pre-harvest)* [18]. Black bent in OILSEED RAPE *(pre-harvest)* [2, 3, 18]. Black bent in OILSEED RAPE FOR INDUSTRIAL USE *(pre-harvest)* [2]. Black bent in COMBINING PEAS *(pre-harvest)* [37]. Black bent in FIELD CROPS *(stubble treatment)* [9]. Black bent in BARLEY *(pre-harvest)*, WHEAT *(pre-harvest)* [3, 18]. Black bent in FIELD BEANS *(pre-harvest)*, FIELD CROPS *(pre-harvest)*, PEAS *(pre-harvest)* [3]. Bolters in SUGAR BEET *(wiper application)* [21, 23]. Bolters in SUGAR BEET *(wick/wiper)* [3, 6-8, 24, 32, 33]. Bracken in GRASSLAND *(sward destruction)* [2, 37, 38]. Couch in STUBBLES [15]. Couch in OATS *(pre-harvest)* [1, 6-8, 10, 11, 13-26, 29-31]. Couch in COMBINING PEAS *(pre-harvest)* [1, 2, 6-8, 10, 11, 14-16, 19-26, 29, 31-33, 37]. Couch in LINSEED *(pre-harvest)*[1, 2, 9-11, 20, 21, 23, 24, 37, 38]. Couch in MUSTARD *(pre-harvest)*[1, 2, 10, 11, 15, 21, 23, 24, 37, 38]. Couch in OILSEED RAPE *(pre-harvest)* [1-3, 6-11, 14-16, 18-26, 29-33, 37, 38]. Couch in BARLEY *(pre-harvest)*, WHEAT *(pre-harvest)* [1-3, 6-8, 10, 11, 13-26, 29-33, 37, 38]. Couch in FIELD CROPS *(stubble treatment)* [1-11, 13, 14, 16, 17, 19-26, 29-33, 37, 38]. Couch in FIELD BEANS *(pre-harvest)* [1, 3, 6-11, 14-16, 19-26, 29, 31-33]. Couch in FIELD CROPS *(sward destruction/direct drilling)* [1, 4-8, 10, 11, 13, 21, 23, 30]. Couch in GRASSLAND *(sward destruction/direct drilling)*[14, 15, 24, 31, 32]. Couch in GRASSLAND *(sward destruction)*, LAND TEMPORARILY REMOVED FROM PRODUCTION [38]. Couch in OILSEED RAPE FOR INDUSTRIAL USE *(pre-harvest)* [2]. Couch in CEREALS *(pre-harvest)* [9]. Couch in DURUM WHEAT *(pre-harvest)* [10, 20, 23, 24]. Couch in PEAS *(pre-harvest)* [3, 9]. Creeping bent in OATS *(pre-harvest)* [18]. Creeping bent in OILSEED RAPE *(pre-harvest)* [2, 3, 18]. Creeping bent in OILSEED RAPE FOR INDUSTRIAL USE *(pre-harvest)* [2]. Creeping bent in COMBINING PEAS *(pre-harvest)* [37]. Creeping bent in BARLEY *(pre-harvest)*, WHEAT *(pre-harvest)* [3, 18]. Creeping bent in FIELD CROPS *(stubble treatment)* [3, 9]. Creeping bent in FIELD BEANS *(pre-harvest)*, PEAS *(pre-harvest)* [3]. Destruction of short term leys in GRASSLAND [1, 4-8, 10, 11, 14-16, 20-26, 29-32]. Green cover in LAND TEMPORARILY REMOVED FROM PRODUCTION [2, 9, 37, 38]. Perennial dicotyledons in GRASSLAND *(wiper application)* [21, 23, 24, 32]. Perennial dicotyledons in COMBINING PEAS *(pre-harvest)* [15, 37]. Perennial dicotyledons in APPLES, CHERRIES, DAMSONS, PEARS, PLUMS [15]. Perennial dicotyledons in OATS *(pre-harvest)* [18]. Perennial dicotyledons in OILSEED RAPE *(pre-harvest)* [2, 3, 18]. Perennial dicotyledons in OILSEED RAPE FOR INDUSTRIAL USE *(pre-harvest)* [2]. Perennial dicotyledons in ORCHARDS *(wiper application)* [24]. Perennial dicotyledons in FIELD BEANS *(pre-harvest)* [3, 15]. Perennial dicotyledons in BARLEY *(pre-harvest)*, GRASSLAND *(sward destruction)*, WHEAT *(pre-harvest)* [3, 18]. Perennial dicotyledons in FIELD CROPS *(stubble treatment)* [3, 9]. Perennial dicotyledons in PEAS *(pre-harvest)* [3]. Perennial grasses in APPLES, CHERRIES, COMBINING PEAS *(pre-harvest)*, DAMSONS, FIELD BEANS *(pre-harvest)*, PEARS, PLUMS [15]. Perennial grasses in GRASSLAND *(sward destruction)* [3, 9, 18]. Perennial weeds in NON-CROP FARM AREAS [3, 18]. Perennial weeds in ORCHARDS [3]. Pre-harvest desiccation in OILSEED RAPE [1, 6-8, 10, 11, 13-15, 21, 23, 24, 30-33]. Pre-harvest desiccation in BARLEY, WHEAT [1, 10, 11, 15, 21, 23, 24, 30, 33]. Pre-harvest desiccation in OATS [1, 10, 11, 15, 21, 23, 24, 30]. Pre-harvest desiccation in MUSTARD [1, 10, 11, 13, 15, 21, 23, 24]. Pre-harvest desiccation in DURUM WHEAT, LINSEED [24]. Rushes in GRASSLAND *(sward destruction)* [2, 37, 38]. Sucker control in APPLES, CHERRIES, DAMSONS, PEARS, PLUMS [21, 23, 24]. Volunteer cereals in STUBBLES [15]. Volunteer cereals in CULTIVATED LAND/SOIL *(pre-drilling/pre-crop emergence)*, STUBBLES *(pre-drilling/pre-crop emergence)* [34, 35]. Volunteer cereals in FIELD CROPS *(stubble treatment)* [1-11, 13, 14, 17-19, 21, 23, 24, 30-33, 37, 38]. Volunteer cereals in FIELD CROPS *(sward destruction/direct drilling)* [1, 4-8, 10, 11, 13, 15, 21, 24, 30, 32, 33]. Volunteer cereals in BARLEY *(post sowing, pre crop emergence)*, DURUM WHEAT *(post sowing, pre crop emergence)*, FIELD BEANS *(post sowing, pre crop emergence)*, LINSEED *(post sowing, pre crop emergence)*, MUSTARD *(post sowing, pre crop emergence)*, OATS *(post sowing, pre crop emergence)*, OILSEED RAPE *(post sowing, pre crop emergence)*, PEAS *(post sowing, pre crop emergence)*, SUGAR BEET *(post sowing, pre crop emergence)*, SWEDES *(post sowing, pre crop emergence)*, TURNIPS *(post sowing, pre crop emergence)*, WHEAT *(post sowing, pre crop emergence)* [23, 24]. Volunteer cereals in FIELD CROPS *(pre-drilling/pre-emergence)*, PEAS *(pre-drilling/pre-emergence)*, SPRING CEREALS *(pre-drilling/pre-emergence)*, SUGAR BEET *(pre-drilling/pre-emergence)*, SWEDES *(pre-drilling/pre-emergence)*, TURNIPS *(pre-drilling/pre-emergence)* [4, 5]. Volunteer oilseed rape in LAND TEMPORARILY REMOVED FROM PRODUCTION [38]. Volunteer potatoes in STUBBLES [15]. Volunteer potatoes in FIELD CROPS *(stubble treatment)* [1-11, 13, 14, 16-26, 29-33, 37, 38]. Volunteer potatoes in LAND TEMPORARILY REMOVED FROM PRODUCTION [38]. Weed beet in SUGAR BEET *(wiper application)* [6-8]. Wild oats in CEREALS *(wiper glove)* [6-8, 21, 23].

Notes

Efficacy

- For best results apply to actively growing weeds with enough leaf to absorb chemical
- Products are formulated as either isopropylamine [1, 3-7, 9-36], or trimesium [2, 37, 38] salts of glyphosate and may vary in the details of efficacy claims. See individual product labels
- Annual weed grasses should have at least 5 cm of leaf and annual broad-leaved weeds at least 2 expanded true leaves

- Perennial grass weeds should have 4-5 new leaves and be at least 10 cm long when treated. Perennial broad-leaved weeds should be treated at or near flowering but before onset of senescence
- Volunteer potatoes and polygonums are not controlled by harvest-aid rates
- In order to allow translocation, do not cultivate before treating perennials and do not apply other pesticides, lime, fertiliser or farmyard manure within 5 d of treatment
- Recommended intervals after treatment and before cultivation vary. See labels
- A rainfree period of at least 6 h (preferably 24 h) should follow spraying
- Do not tank-mix with other pesticides or fertilisers as such mixtures may lead to reduced control. Adjuvants are obligatory for some products and recommended for some uses with others. See labels
- Do not spray weeds affected by drought, waterlogging, frost or high temperatures

Crop Safety/Restrictions

- Maximum number of applications normally 1 per crop (pre-harvest), 1 per yr (grassland destruction, non-crop land, green cover treatment) and 1 in other situations but check label for details
- Do not treat cereals grown for seed or undersown crops
- Disperse decaying matter by thorough cultivation before sowing or planting a following crop
- Consult grain merchant before treating crops grown on contract or intended for malting
- With wiper application weeds should be at least 10 cm taller than crop

Special precautions/Environmental safety

- Harmful if swallowed [2, 4, 5, 33, 37, 38]
- Irritating to skin [1, 3, 5-7, 10, 11, 13, 14, 16-21, 25, 27, 30-33] and eyes [2, 8, 37, 38]
- Risk of serious damage to eyes [4, 5, 15, 33]
- Harmful (dangerous [4, 5, 15]) to fish or other aquatic life. Do not contaminate surface waters or ditches with chemical or used container [1-3, 6-14, 16-22, 25-27, 29-35, 37, 38]
- Do not use on oilseed rape to be sold for human or animal consumption [2, 37, 38]
- Do not mix, store or apply in galvanized or unlined mild steel containers or spray tanks
- Do not leave spray in spray tanks for long period and make sure tanks are well vented
- Take extreme care to avoid drift
- Treated poisonous plants must be removed before grazing or conserving [1, 6, 7, 10, 21, 23-25, 32]
- Do not use in covered areas such as greenhouses or under polythene
- For field edge treatment direct spray away from hedge bottoms

Protective clothing/Label precautions

- A, C [1-21, 23-25, 27, 28, 30-33, 36-38]; D [9]; F [1, 8, 10, 11, 21, 27]; H [1, 2, 8-11, 15, 19, 21, 23, 24, 27, 28, 37, 38]; M [1, 2, 8-12, 15, 18, 19, 21, 23, 24, 27, 28, 36-38]; N [1, 8, 10-12, 18, 21, 27, 36]
- M03 [2, 4, 5, 33, 37, 38]; R03c [1, 2, 4, 5, 33, 37, 38]; R04 [15]; R04a [1-3, 5-8, 10, 11, 13, 14, 16-21, 25, 27, 30-33, 36-38]; R04b [1, 3, 5-8, 10, 11, 13, 14, 16-21, 25, 27, 30-33]; R04d [4, 5, 15, 33]; R05a [5]; U02 [1, 4-14, 16-21, 25, 27, 30-33, 36]; U05a [1-8, 10-21, 25, 27, 30-33, 36-38]; U08 [1, 3, 5-21, 25, 27, 30-33, 36]; U09a [4]; U19 [1, 3-21, 25, 27, 30-33, 36]; U20a [2, 3, 5, 6, 9, 10, 12-14, 17-20, 23, 28, 31-33, 37, 38]; U20b [1, 4, 7, 8, 11, 15, 16, 21, 22, 24-27, 29, 30, 36]; U20c [34, 35]; C03 [1-8, 10-21, 23-25, 27, 28, 30-33, 36-38]; E01 [1-21, 23-25, 27, 28, 30-33, 36-38]; E07 [2, 3, 37, 38]; E09 [2, 37, 38]; E13b [4, 15]; E13c [1-3, 6-14, 16-22, 25-27, 29-38]; E15 [23, 24, 28]; E19 [8-11, 17, 19, 21]; E26 [1, 2, 4-7, 10-12, 14, 16, 17, 19, 21, 22, 25, 26, 29, 31-36]; E30a [1-38]; E31a [7]; E31b [1, 3, 4, 8-11, 13, 15, 18, 20-24, 26-30, 34, 35]; E31c [16, 25, 37, 38]; E32a [2, 12, 36-38]; E34 [2, 4-6, 14, 15, 17, 19, 31-35, 37, 38]

Withholding period

- Keep livestock out of treated areas until foliage of any poisonous weeds such as ragwort has died and become unpalatable [2, 3, 37, 38]

Latest application/Harvest Interval(HI)

- 24 h-5 d before cultivating stubbles depending on product and dose (see labels for details); pre-emergence for autumn crops and spring cereals; 72 h post-drilling for sugar beet, peas, swedes, turnips, onions and leeks; post-leaf fall but before white bud or green cluster for top fruit.
- HI grass 5 d; wheat, barley, oats, field beans, combining peas, linseed 7 d; mustard 8 d; oilseed rape 14 d

Approval

- May be applied through CDA equipment. See label for details. When applying through rotary atomisers the spray droplet spectra must have a minimum Volume Median Diameter (VMD) of 200 microns

Maximum Residue Levels (mg residue/kg food)

- wild mushrooms 50; barley, oats 20; linseed, rape seed 10; wheat, rye, triticale 5; meat (cattle, goat, sheep kidney) 2; meat (pig kidney) 0.5; all other products (except beans, peas) 0.1

333 glyphosate (horticulture, forestry, amenity etc.)

A translocated non-residual phosphonic acid herbicide

See also diuron + glyphosate

Products

1 Alpha Glyphogan	Makhteshim	360 g/l	SL	05784
2 Barbarian	Barclay	360 g/l	SL	07980
3 Barclay Cleanup	Barclay	180 g/l	SL	09179
4 Barclay Dart	Barclay	180 g/l	SL	05129
5 Barclay Gallup	Barclay	360 g/l	SL	05161
6 Barclay Gallup Amenity	Barclay	360 g/l	SL	06753
7 Barclay Garryowen	Barclay	360 g/l	SL	08599
8 CDA Vanquish	RP Amenity	120 g/l	RH	08577
9 Clinic	Nufarm	360 g/l	SL	08579
10 Glyfos	Cheminova	360 g/l	SL	07109
11 Glyfos ProActive	Nomix-Chipman	360 g/l	SL	07800
12 Glyper	PBI	360 g/l	SL	07968
13 Helosate	Helm	360 g/l	SL	06499
14 Hilite	Nomix-Chipman	144 g/l	RH	06261
15 MSS Glyfield	Mirfield	360 g/l	SL	08009
16 Portman Glyphosate	Portman	360 g/l	SL	05891
17 Rival	Monsanto	360 g/l	SL	09220
18 Roundup	Monsanto	360 g/l	SL	01828
19 Roundup Amenity	Monsanto	360 g/l	SL	08721
20 Roundup Biactive	Monsanto	360 g/l	SL	06941
21 Roundup Biactive Dry	Monsanto	42% w/w	WG	06942
22 Roundup Pro	Monsanto	360 g/l	SL	04146
23 Roundup Pro Biactive	Monsanto	360 g/l	SL	06954
24 Spasor	RP Amenity	360 g/l	SL	07211
25 Spasor Biactive	RP Amenity	360 g/l	SL	07651
26 Stefes Glyphosate	Stefes	360 g/l	SL	05819
27 Stefes Kickdown	Stefes	360 g/l	SL	06329
28 Sting CT	Monsanto	120 g/l	SL	04754
29 Stirrup	Nomix-Chipman	144 g/l	RH	06132
30 Touchdown LA	Miracle	330 g/l	SL	07747
31 Touchdown LA	Scotts	330 g/l	SL	09270

Uses

Annual and perennial weeds in ASPARAGUS *(off-label)*, BLACKCURRANTS *(off-label)*, GRAPEVINES *(off-label)*, TREE NUTS *(off-label)* [18]. Annual and perennial weeds in FORESTRY *(directed spray)* [1, 5-7, 9, 12-14, 17, 19, 22, 23, 26, 27, 29]. Annual and perennial weeds in AMENITY GRASS *(destruction/pre-sowing)*, FORESTRY *(pre-planting)* [14, 29]. Annual and perennial weeds in AMENITY TREES AND SHRUBS *(directed spray)* [14, 19, 22, 29]. Annual and perennial weeds in WOODY ORNAMENTALS *(directed spray)* [14, 19, 22, 24, 25, 29]. Annual and perennial weeds in PATHS AND DRIVES [6, 17, 25, 30, 31]. Annual and perennial weeds in ROAD VERGES [6, 17, 25]. Annual and perennial weeds in INDUSTRIAL SITES [6, 17]. Annual and perennial weeds in AMENITY AREAS, AMENITY AREAS *(wiper application)*, AMENITY TREES AND SHRUBS *(pre-planting)*, BROAD-LEAVED TREES *(post-planting)* [6]. Annual and perennial weeds in AMENITY GRASS *(pre-planting/sowing)*, AMENITY GRASS *(wiper application)*, AMENITY TREES AND SHRUBS *(pre-planting/sowing)*, AMENITY TREES AND SHRUBS *(wiper application)* [23, 24]. Annual and perennial weeds in FORESTRY [10, 11, 15]. Annual and perennial weeds in AMENITY AREAS *(pre-planting)* [25]. Annual and perennial weeds in AMENITY TREES AND SHRUBS, FARM BUILDINGS/YARDS, LAND CLEARANCE *(pre-planting)* [30, 31]. Annual and perennial weeds in NON-CROP AREAS [12, 13, 17, 25, 30, 31]. Annual and perennial weeds in LAND NOT INTENDED TO BEAR VEGETATION [8, 10, 11, 15, 27]. Annual and perennial weeds in

AMENITY VEGETATION [8]. Annual and perennial weeds in FENCELINES [17, 25, 30, 31]. Annual and perennial weeds in HARD SURFACES, WALLS [17, 25]. Annual and perennial weeds in BROAD-LEAVED TREES *(pre-planting)*, CONIFERS *(pre-planting)* [17, 19, 22, 23]. Annual weeds in LEEKS *(post sowing, pre crop emergence)*, ONIONS *(post sowing, pre crop emergence)* [20, 21]. Annual weeds in FORESTRY [2, 16]. Annual weeds in LEEKS *(pre-drilling/pre-emergence)*, ONIONS *(pre-drilling/pre-emergence)* [3, 4]. Bracken in FORESTRY *(directed spray)* [1, 5-7, 9, 12, 13, 17, 19, 22, 23, 26, 27]. Bracken in FORESTRY [10, 11, 15]. Bracken in AMENITY TREES AND SHRUBS, FENCELINES, LAND CLEARANCE *(pre-planting)*, NON-CROP AREAS, PATHS AND DRIVES [30, 31]. Bracken in TOLERANT CONIFERS *(overall dormant spray)*[17, 19, 22, 23]. Brambles in TOLERANT CONIFERS *(overall dormant spray)* [17, 19, 22, 23]. Chemical thinning in FORESTRY [1, 5, 7, 9, 12, 13, 17, 19, 22, 23]. Chemical thinning in FORESTRY *(stem injection)*[6]. Couch in FORESTRY *(directed spray)* [1, 5-7, 9, 12, 13, 17, 19, 22, 23, 26, 27]. General weed control in FARMYARDS *(spot treatment)*[21]. Heather in FORESTRY *(directed spray)* [6]. Heather in FORESTRY [10, 11, 15]. Perennial dicotyledons in FORESTRY *(wiper application)*[23, 26]. Perennial grasses in AQUATIC SITUATIONS [1, 6, 9-13, 15, 17, 19, 21-25]. Perennial grasses in TOLERANT CONIFERS *(overall dormant spray)*[10, 11, 15, 17, 19, 22, 23]. Perennial weeds in FORESTRY [2, 16]. Reeds in AQUATIC SITUATIONS [1, 6, 9-13, 15, 17, 19, 21-25]. Rhododendrons in FORESTRY *(directed spray)* [6]. Rhododendrons in FORESTRY [10, 11, 15]. Rushes in AQUATIC SITUATIONS [1, 6, 9-13, 15, 17, 19, 21-25]. Rushes in FORESTRY *(directed spray)* [1, 12, 13, 23, 26, 27]. Sedges in AQUATIC SITUATIONS [1, 6, 9-13, 15, 17, 19, 21-25]. Total vegetation control in NON-CROP AREAS *(amenity situations)* [14, 23, 24, 26, 29]. Total vegetation control in HARD SURFACES [14, 19, 22, 29]. Total vegetation control in INDUSTRIAL SITES [14, 19, 22, 24, 25, 29]. Total vegetation control in AMENITY VEGETATION, LAND NOT INTENDED TO BEAR VEGETATION [8]. Total vegetation control in FENCELINES, ROAD VERGES [19, 22, 24]. Total vegetation control in NON-CROP AREAS, PATHS AND DRIVES, WALLS [19, 22]. Volunteer cereals in LEEKS *(post sowing, pre crop emergence)*, ONIONS *(post sowing, pre crop emergence)* [20, 21]. Volunteer cereals in LEEKS *(pre-drilling/pre-emergence)*, ONIONS *(pre-drilling/pre-emergence)* [3, 4]. Waterlilies in AQUATIC SITUATIONS [1, 6, 9-13, 15, 17, 19, 21-25]. Woody weeds in FORESTRY *(directed spray)* [1, 5-7, 9, 12-14, 17, 19, 22, 23, 26, 27, 29]. Woody weeds in FORESTRY [2, 10, 11, 15, 16]. Woody weeds in TOLERANT CONIFERS *(overall dormant spray)* [17, 19, 22, 23].

Notes

Efficacy

- For best results apply to actively growing weeds with enough leaf to absorb chemical
- Products are formulated as either isopropylamine [1, 2, 4-8, 10-16, 18-29]or trimesium [30, 31] salts of glyphosate and may vary in the details of efficacy claims. See individual product labels
- Annual weed grasses should have at least 5 cm of leaf and annual broad-leaved weeds at least 2 expanded true leaves
- Perennial grass weeds should have 4-5 new leaves and be at least 10 cm long when treated. Perennial broad-leaved weeds should be treated at or near flowering but before onset of senescence
- Bracken must be treated at full frond expansion
- In order to allow translocation, do not cultivate before treating perennials and do not apply other pesticides, lime, fertiliser or farmyard manure within 5 d of treatment
- Recommended intervals after treatment and before cultivation vary. See labels
- A rainfree period of at least 6 h (preferably 24 h) should follow spraying
- Most products should not be tank-mixed with other pesticides or fertilisers as such mixtures may lead to reduced control, but some [30, 31] are compatible with listed diuron-containing products. See labels for details

FOR FULL CONDITIONS OF USE ALWAYS READ THE PRODUCT LABEL

- Adjuvants are obligatory for some products and recommended for some uses with others. See labels
- Do not spray weeds affected by drought, waterlogging, frost or high temperatures
- Fruit tree suckers best treated in late spring
- Chemical thinning treatment can be applied as stump spray or stem injection
- For rhododendron control apply to stumps or regrowth. Addition of adjuvant Mixture B recommended for application to foliage by knapsack sprayer [18, 22]
- Product formulated ready for use without dilution. Apply only through specified applicator (see label for details) [8, 14, 29]

Crop Safety/Restrictions

- Maximum number of treatments normally 1 per crop or season on field and edible crops and unrestricted for non-crop uses. Check labels for details
- Decaying remains of plants killed by spraying must be dispersed before direct drilling
- Do not use treated straw as a mulch or growing medium for horticultural crops
- For use in orchards, grapevines and tree nuts care must be taken to avoid contact with the trees. Do not use in orchards established less than 2 yr and keep off low-lying branches
- Do not spray root suckers in orchards in late summer or autumn
- Do not use under glass or polythene as damage to crops may result
- With wiper application weeds should be at least 10 cm taller than crop
- Do not use wiper techniques in soft fruit crops
- Certain conifers may be sprayed overall in dormant season. See label for details [10, 11, 15, 18, 20-23]
- Use a tree guard when spraying in established forestry plantations

Special precautions/Environmental safety

- Harmful if swallowed [4, 27, 30]
- Irritating to eyes and skin [1, 2, 4-6, 12, 13, 16-19, 22, 24, 27]
- Risk of serious damage to eyes [4]
- Harmful (dangerous [4]) to fish or other aquatic life. Do not contaminate surface waters or ditches with chemical or used container [1, 2, 5, 6, 10-19, 22, 24, 26-28, 30]
- Maximum permitted concentration in treated water 0.2 ppm [1, 6, 10, 11, 13, 15, 17, 18, 22]
- The Environment Agency or Local River Purification Authority must be consulted before use in or near water
- Do not mix, store or apply in galvanised or unlined mild steel containers or spray tanks
- Do not leave spray in spray tanks for long period and make sure that tanks are well vented
- Take extreme care to avoid drift onto other crops
- For field edge treatment direct spray away from hedge bottoms
- Do not dump surplus herbicide in water or ditch bottoms [10-13, 15]

Protective clothing/Label precautions

- A [1-27, 29-31]; C [1-7, 9-24, 26, 27, 29-31]; D [10, 11, 15]; F [9, 12, 13, 17-19, 22]; H [8-13, 15, 17-23, 25, 30, 31]; M [8-25, 29-31]; N [9, 12-14, 16-19, 22, 24, 29]
- M03 [3, 4, 27, 28, 30, 31]; R03c [3, 4, 27, 30, 31]; R04a [1, 2, 4-7, 9, 12, 13, 16-19, 22, 24, 27, 29]; R04b [1, 2, 4-7, 9, 12, 13, 16-19, 22, 24]; R04d [3, 4]; R05a [4]; U02 [1, 3-7, 9-19, 22, 24, 26, 27, 29]; U05a [1-7, 9, 12-14, 16-19, 22, 24, 26, 27, 29-31]; U08 [1, 2, 4-7, 9, 10, 12-14, 16-19, 22, 24, 26, 27, 29]; U09a, E13b [3]; U19 [1-19, 22, 24, 26, 27, 29]; U20a [1, 2, 4-6, 10-13, 15, 16, 18, 20-22, 25-27, 30, 31]; U20b [3, 7-9, 14, 17, 19, 23, 24, 28, 29]; C03 [1-9, 12-14, 16-27, 29-31]; E01 [1-27, 29-31]; E07 [2]; E13c [1, 2, 5-7, 9-19, 22, 24, 26-31]; E15 [8, 20, 21, 23, 25]; E19 [9-13, 15, 24]; E26 [1, 3-8, 11-15, 26-29]; E30a [1-31]; E31a [7]; E31b [2, 3, 8-13, 15-23, 25, 28]; E32a [14, 29-31]; E34 [1, 3-6, 13, 26-28, 30, 31]

Latest application/Harvest Interval(HI)

- 5 d before drilling for grassland; stubbles 24 h-5 d depending on product and dose (see labels for details); post-leaf fall but before white bud or green cluster for top fruit

Approval

- Approved for aquatic weed control. See notes in Section 1 on use of herbicides in or near water [1, 6, 10, 11, 13, 15, 18-25]
- May be applied through CDA equipment. See label for details. When applying through rotary atomisers the spray droplet spectra must have a minimum Volume Median Diameter (VMD) of 200 microns
- Off-label approval unlimited for use on grapevines and tree nuts (OLA 0337/92)[18]; unlimited for use on outdoor asparagus (OLA 0584/94)[18]; unlimited for use on blackcurrants (OLA 1407/96)[18]

Maximum Residue Levels (mg residue/kg food)

- see glyphosate (agriculture) entry

334 glyphosate + oxadiazon

A non-selective herbicide for total vegetation control

Products	Zapper	RP Amenity	10.8:30% w/w	WB	08605

Uses Annual and perennial weeds in LAND NOT INTENDED TO BEAR VEGETATION *(soil surface treatment)*. Total vegetation control in LAND NOT INTENDED TO BEAR VEGETATION *(soil surface treatment)*.

Notes

Efficacy

- Apply post-emergence of weeds at any time when they are actively growing from Mar to end Sep. Control may be reduced if weeds are under stress when treated
- Best results on perennials including docks, perennial sowthistle and willowherb obtained from treatment just before flowering or seed set
- A rainfree period of at least 6 h (preferably 24 h) should follow spraying
- Perennials such as dandelions or docks emerging after application from established rootstocks will not be controlled
- Pre-emergence activity is reduced on soils with more than 10% organic matter or where leaves have collected or a mat of organic matter has built up
- Use of wetting agents or adjuvants may reduce activity

Crop Safety/Restrictions

- When used on sites that are to be cleared or grubbed 2 yr should elapse before sowing or planting anything. Soil should be ploughed or dug after clearance
- Take care to prevent heavy rain after application washing product onto sensitive areas such as newly sown grass or areas about to be planted

Special precautions/Environmental safety

- Dangerous to fish or other aquatic life. Do not contaminate surface waters or ditches with chemical or used container

- Do not allow direct spray from vehicle mounted/drawn hydraulic sprayers to fall within 6 m, or from hand-held sprayers to within 2 m, of surface waters or ditches. Direct spray away from water

Protective clothing/Label precautions
- A, C
- U20c, U22, E13b, E16, E30a, E32a

Maximum Residue Levels (mg residue/kg food)
- see glyphosate (agricultural) entry

---

335 guazatine

A guanidine fungicide seed dressing for cereals

Products

Panoctine	RP Agric.	300 g/l	LS	06207

Uses

Bunt in WHEAT. Fusarium foot rot and seedling blight in BARLEY *(reduction)*, OATS *(reduction)*, WHEAT *(reduction)*. Septoria seedling blight in WHEAT.

Notes

Efficacy
- Apply with conventional seed treatment machinery
- After treating, bag seed immediately and keep in dry, draught-free store

Crop Safety/Restrictions
- Maximum number of treatments 1 per batch
- Do not treat grain with moisture content above 16% and do not allow moisture content of treated seed to exceed 16%
- Do not apply to cracked, split or sprouted seed
- Treatment may lower germination capacity, particularly if seed grown, harvested or stored under adverse conditions

Special precautions/Environmental safety
- Harmful if swallowed and in contact with skin. Risk of serious damage to eyes
- Dangerous to fish or other aquatic life. Do not contaminate surface waters or ditches with chemical or used container
- Do not use treated seed as food or feed

Protective clothing/Label precautions
- A, C, D, H
- M03, R03a, R03c, R04d, U05a, U20b, C03, E01, E13b, E26, E30a, E31a, E34, S01, S02, S04b, S05, S06, S07

Latest application/Harvest Interval(HI)
- Pre-drilling

---

336 guazatine + imazalil

A fungicide seed treatment for barley and oats

Products

Panoctine Plus	RP Agric.	300:25 g/l	LS	06208

Uses Brown foot rot in BARLEY *(reduction)*, OATS *(reduction)*. Foot rot in BARLEY. Fusarium foot rot and seedling blight in BARLEY *(reduction)*, OATS *(reduction)*. Leaf stripe in BARLEY. Net blotch in BARLEY. Pyrenophora leaf spot in OATS.

Notes

Efficacy

- Apply with conventional seed treatment machinery
- After treating, bag seed immediately and keep in dry, draught-free store

Crop Safety/Restrictions

- Maximum number of treatments 1 per batch
- Do not treat grain with moisture content above 16% and do not allow moisture content of treated seed to exceed 16%
- Do not apply to cracked, split or sprouted seed
- Do not store treated seed for more than 6 mth
- Treatment may lower germination capacity, particularly if seed grown, harvested or stored under adverse conditions

Special precautions/Environmental safety

- Harmful if swallowed and in contact with skin. Risk of serious damage to eyes
- Dangerous to fish or other aquatic life. Do not contaminate surface waters or ditches with chemical or used container
- Do not use treated seed as food or feed

Protective clothing/Label precautions

- A, C, D, H
- M03, R03a, R03c, R04d, U05a, U09a, U20b, C03, E01, E13b, E26, E30a, E31a, E34, S01, S02, S04b, S05, S06, S07

Latest application/Harvest Interval(HI)

- Pre-drilling

Maximum Residue Levels (mg residue/kg food)

- see imazalil entry

337 heptenophos

A contact, systemic and fumigant organophosphorus insecticide

See also deltamethrin + heptenophos

Products					
	Hostaquick	AgrEvo	550 g/l	EC	07326

Uses American serpentine leaf miner in NON-EDIBLE ORNAMENTALS *(off-label)*, PROTECTED CUCURBITS *(off-label)*, PROTECTED VEGETABLES *(off-label)*. Aphids in APPLES, BRASSICAS, BROAD BEANS, CELERY, CEREALS, CHICORY *(off-label)*, COURGETTES, ENDIVES *(off-label)*, FIELD BEANS, FLOWERHEAD BRASSICAS *(off-label)*, FRENCH BEANS, LEAF BRASSICAS *(off-label)*, LETTUCE, MARROWS, PEARS, PEAS, PROTECTED POT PLANTS, PROTECTED VEGETABLE SEEDLINGS *(off-label)*, RHUBARB *(off-label)*, RUNNER BEANS, STRAWBERRIES, TOMATOES. Leafhoppers in

CUCUMBERS, TOMATOES. Thrips in CUCUMBERS, PROTECTED POT PLANTS. Woolly aphid in APPLES.

Notes

Efficacy

- Spray when pests first seen and repeat as necessary. Knock-down is rapid
- Ensure good spray cover and use high volume on glasshouse crops
- May be used in integrated control programmes with *Encarsia* or *Phytoseiulus* in glasshouses. Allow at least 4 d after treatment before introducing any new beneficial insects
- Where aphids resistant to heptenophos occur repeat treatments are likely to result in lower levels of control

Crop Safety/Restrictions

- In glasshouses do not apply early in morning after a cold night nor in daytime temperatures above 30°C

Special precautions/Environmental safety

- This product contains an anticholinesterase organophosphorus compound. Do not use if under medical advice not to work with such compounds
- Toxic if swallowed. Harmful in contact with skin
- Flammable
- Do not apply to crops in flower or to those in which bees are actively foraging. Do not apply when flowering weeds are present
- Harmful to fish or other aquatic life. Do not contaminate surface waters or ditches with chemical or used container

Protective clothing/Label precautions

- A, C
- M01, M04, R02c, R03a, R07d, U02, U04a, U05a, U08, U13, U19, U20a, C02, C03, E01, E12c, E13c, E26, E30a, E31b, E34

Latest application/Harvest Interval(HI)

- HI 24 h

Approval

- Off-label approval unlimited for use on rhubarb, endives, chicory, leaf brassicas, flowerhead brassicas (OLA 1304/96)[1]; unlimited for use on ornamentals, aubergines, peppers, cucurbits (OLA 1606/96)[1]

---

338 hymexazol

A systemic isoxazole fungicide for pelleting sugar beet seed

Products

Tachigaren 70 WP	Sumitomo	70% w/w	WP	02649

Uses

Aphanomyces cochlioides in BEETROOT *(off-label)*. Black leg in SUGAR BEET. Damping off in SUGAR BEET.

Notes

Efficacy

- Incorporate into pelleted seed using suitable seed pelleting machinery

Crop Safety/Restrictions

- Maximum number of treatments 1 per batch of seed

Special precautions/Environmental safety
- Irritating to eyes, skin and respiratory system
- Harmful to fish of aquatic life. Do not contaminate surface waters or ditches with chemical or used container
- Do not use treated seed as food or feed
- Treated seed harmful to game and wildlife

Protective clothing/Label precautions
- A, C, F
- R04a, R04b, R04c, U05a, U20a, C03, E01, E13c, E30a, E32a, S01, S02, S05, S06

Approval
- Off-label Approval unlimited for use on pelleted beetroot seed to be sown outdoors (OLA 0565/98)[1]

339 imazalil

A systemic and protectant conazole fungicide

See also azaconazole + imazalil
guazatine + imazalil

Products

1 Fungaflor	Hortichem	200 g/l	EC	05967
2 Fungaflor Smoke	Hortichem	15% w/w	FU	05968
3 Fungazil 100 SL	RP Agric.	100 g/l	LS	06202
4 Stryper	Uniroyal	50 g/l	LS	08310

Uses

Brown foot rot in SPRING BARLEY *(seed treatment)*, WINTER BARLEY *(seed treatment)* [4]. Dry rot in SEED POTATOES, WARE POTATOES [3]. Gangrene in SEED POTATOES, WARE POTATOES [3]. Leaf stripe in SPRING BARLEY *(seed treatment)*, WINTER BARLEY *(seed treatment)* [4]. Net blotch in SPRING BARLEY *(seed treatment)*, WINTER BARLEY *(seed treatment)* [4]. Powdery mildew in PROTECTED ORNAMENTALS, PROTECTED ROSES [1, 2]. Powdery mildew in COURGETTES *(off-label)*, CUCUMBERS, GHERKINS *(off-label)*, ORNAMENTALS, ROSES [1]. Powdery mildew in PROTECTED CUCUMBERS [2]. Silver scurf in SEED POTATOES, WARE POTATOES [3]. Skin spot in SEED POTATOES, WARE POTATOES [3].

Notes

Efficacy
- Treat cucurbits before or as soon as disease appears and repeat every 10-14 d or every 7 d if infection pressure great or with susceptible cultivars [1, 2]
- Apply to clean soil-free potatoes post-harvest before putting into store or at first grading. A further treatment may be applied in early spring before planting [3]
- For best control of skin and wound diseases of ware potatoes treat as soon as possible after harvest, preferably within 7-10 d, before any wounds have healed [3]
- Apply through canopied hydraulic or spinning disc equipment preferably diluted with up to 2 l water per tonne of potatoes to obtain maximum skin cover and penetration [3]
- Apply seed treatment via conventional machine and sow treated seed as soon as possible [4]

Crop Safety/Restrictions

- Maximum number of treatments 1 for ware potatoes and 2 per batch of seed potatoes [3]; 1 per batch of barley seed [4]
- Where possible apply to ware potatoes at least 6 wk before washing for sale or use. Consult processor before treating potatoes for processing [3]
- Do not spray cucurbits or ornamentals in full, bright sunshine. When spraying in the evening the spray should dry before nightfall. May cause damage if open flowers are sprayed. Do not use on rose cultivar Dr A.J. Verhage [1]
- With ornamentals of unknown tolerance test on a few plants in first instance [1]
- Do not treat grain over 16% moisture content and keep seed dry and draught free after treatment [4]
- Do not apply to cracked, split or sprouted seed [4]

Special precautions/Environmental safety

- Harmful if swallowed [1, 3]
- Irritating to eyes and skin [1, 3]; irritating to eyes [2, 4]
- Causes burns [1]
- Flammable [1]; highly flammable [3]
- Keep unprotected persons out of treated areas for at least 2 h [2]
- Harmful to bees. Do not apply to crops in flower or to those in which bees are actively foraging. Do not apply when flowering weeds are present [1]
- Harmful to fish or other aquatic life. Do not contaminate surface waters or ditches with chemical or used container [1, 3, 4]
- Do not use treated seed as food or feed [3, 4]

Protective clothing/Label precautions

- A, C [1, 3, 4]; D [3]; H [1, 4]
- M03, E26, E31a, S01, S02, S05, S06 [3, 4]; R03c, R04b, U19, U20a, E34 [1, 3]; R04a, E30a [1-4]; R05b, R07d, U08, C02 (1 d), E12e [1]; R07c, C02, E02 [2]; U04a [3]; U05a, C03, E01 [2-4]; U09a, U20b, E03, S03, S04a, S07 [4]; E13c [1, 3, 4]; E32a [1, 2]

Latest application/Harvest Interval(HI)

- During storage and/or prior to chitting for potatoes [3].
- HI cucumbers 1 d [1, 2]

Approval

- Off-label approval unlimited for use on courgettes and gherkins (OLA 0939/95, OLA 0841/96)[1]

Maximum Residue Levels (mg residue/kg food)

- citrus fruits; pome fruits, ware potatoes 5; bananas 2; cucumbers, gherkins, courgettes 0.2; tea, hops 0.1; all other products 0.02

---

340 imazalil + pencycuron

A fungicide mixture for treatment of seed potatoes

Products

1 Monceren IM	Bayer	0.6:12.5% w/w	DS	06259
2 Monceren IM Flowable	Bayer	20:250 g/l	FS	06731

Uses

Black scurf in POTATOES [1]. Black scurf in SEED POTATOES [2]. Silver scurf in POTATOES *(reduction)* [1]. Silver scurf in SEED POTATOES *(reduction)* [2]. Stem canker in POTATOES *(reduction)* [1]. Stem canker in SEED POTATOES *(reduction)* [2].

Notes **Efficacy**

- Apply to seed tubers during planting (see label for suitable method) or sprinkle over tubers in chitting trays before loading into planter. It is essential to obtain an even distribution over tubers [1]
- Apply to clean seed tubers at any time, into or out of store or in hopper during planting [2]
- If seed tubers become damp from light rain distribution of product should not be affected. Tubers in the hopper should be covered if a shower interrupts planting
- Use suitable misting equipment. See label for details [2]
- Treatment usually most conveniently carried out over roller table at the end of grading out [2]

Crop Safety/Restrictions

- Maximum number of treatments 1 per batch
- May be used on seed tubers previously treated with a liquid fungicide but not if this contained imazalil
- Do not use on tubers which have previously been treated with a dry powder seed treatment or hot water
- Do not tank mix with other potato fungicides or storage products [2]
- Treated tubers may only be used as seed. They must not be used for human or animal consumption

Special precautions/Environmental safety

- Operators must wear suitable respiratory equipment and gloves when handling product and when riding on planter. Wear gloves when handling treated tubers
- Irritating to eyes [2]
- Harmful to fish or other aquatic life. Do not contaminate surface waters or ditches with chemical or used container
- Do not use treated seed as food or feed
- Treated seed harmful to game and wildlife

Protective clothing/Label precautions

- A [1, 2]; C [2]; D [1]
- R04a [2]; U20a, E03, E13c, E30a, E32a, S01, S02, S03, S04b, S05, S06 [1, 2]

Latest application/Harvest Interval(HI)

- Immediately before planting

Maximum Residue Levels (mg residue/kg food)

- see imazalil entry

341 imazalil + thiabendazole

A fungicide mixture for treatment of seed potatoes

Products	Extratect Flowable	Seedcote	100:300 g/l	FS	08704

Uses Dry rot in SEED POTATOES *(reduction)*. Gangrene in SEED POTATOES *(reduction)*. Silve scurf in POTATOES *(reduction)*, SEED POTATOES *(reduction)*. Skin spot in POTATOE *(reduction)*, SEED POTATOES *(reduction)*. Stem canker in POTATOES *(reduction)*.

FOR FULL CONDITIONS OF USE ALWAYS READ THE PRODUCT LABEI

Notes

Efficacy

- Apply immediately after lifting with suitable spinning disc or low volume hydraulic applicator for disease reduction during storage
- Apply within 2 mth of planting for maximum disease reduction in the progeny crop

Crop Safety/Restrictions

- Maximum number of treatments 1 per batch
- Do not mix with any other product
- May not be used on seed tubers previously treated with a product containing imazalil
- Delayed emergence noted under certain conditions but with no final effect on yield

Special precautions/Environmental safety

- Irritating to respiratory system. Risk of serious damage to eyes
- Harmful to fish or other aquatic life. Do not contaminate surface waters or ditches with chemical or used container
- Do not use treated seed as food or feed

Protective clothing/Label precautions

- A, C, D, H
- R04c, R04d, U05a, U09a, U11, U19, U20c, C03, E01, E13c, E30a, E31a, S01, S02, S05, S06

Latest application/Harvest Interval(HI)

- Pre-planting

Approval

- Approved for use throught ULV equipment [1]

Maximum Residue Levels (mg residue/kg food)

- see imazalil and thiabendazole entries

---

342 imazamethabenz-methyl

A post-emergence imidazolinone grass weed herbicide for use in winter cereals

Products

Dagger	Cyanamid	300 g/l	SC	03737

Uses

Blackgrass in WINTER BARLEY, WINTER WHEAT. Charlock in WINTER BARLEY, WINTER WHEAT. Loose silky bent in WINTER BARLEY, WINTER WHEAT. Onion couch in WINTER BARLEY, WINTER WHEAT. Volunteer oilseed rape in WINTER BARLEY, WINTER WHEAT. Wild oats in WINTER BARLEY, WINTER WHEAT.

Notes

Efficacy

- Has contact and residual activity. Wild oats controlled from pre-emergence to 4 leaves + 3 tillers stage, charlock and volunteer oilseed rape to 10 cm. Good activity on onion couch up to 7.5 cm and useful suppression of cleavers
- When used alone add Agral for improved contact activity
- Best results achieved when applied to fine, firm, clod-free seedbed when soil moist
- Do not use on soils with more than 10% organic matter
- Effects of autumn/winter treatment normally persist to control spring flushes of weeds
- Tank mixture with isoproturon recommended for control of tillered blackgrass, with pendimethalin for general grass and dicotyledon control

Crop Safety/Restrictions

- Maximum number of treatments 2 per crop (as split dose treatment)
- Apply from 2-fully expanded leaf stage of crop to before second node detectable (GS 12-32)
- Do not use on durum wheat
- Do not use on soils where surface water is likely to accumulate
- See label for guidance on following crops after normal harvesting
- In case of crop failure land must be ploughed to 15 cm and re-drilled in spring but only with specified crops - see label for details

Special precautions/Environmental safety

- Irritating to eyes

Protective clothing/Label precautions

- A, C
- M03, R04a, U05a, U08, U20a, C02, E01, E15, E26, E30a, E31b, E34

Latest application/Harvest Interval(HI)

- Before 2nd node detectable (GS 32)

343 imazapyr

A non-selective translocated and residual imidazolinone herbicide for use on non-crop land

Products

1 Arsenal	Nomix-Chipman	250 g/l	SL	05537
2 Arsenal	Cyanamid	250 g/l	SL	04064
3 Arsenal 50	Nomix-Chipman	50 g/l	SL	05567
4 Arsenal 50	Cyanamid	50 g/l	SL	04070

Uses

Bracken in NON-CROP AREAS [1-4]. Bracken in FARM BUILDINGS/YARDS, FENCELINES, FORESTRY *(site preparation)*, INDUSTRIAL SITES, RAILWAY TRACKS [3, 4]. Total vegetation control in NON-CROP AREAS [1-4]. Total vegetation control in FARM BUILDINGS/YARDS, FENCELINES, FORESTRY *(site preparation)*, INDUSTRIAL SITES, RAILWAY TRACKS [3, 4].

Notes

Efficacy

- Chemical is absorbed through roots and foliage, kills underground storage organs and gives long term residual control. Complete kill may take several wk
- May be applied before weed emergence but gives best results from application at any time of year when weeds are growing actively
- Apply from beginning Jul to end Oct as a conifer site preparation treatment [3, 4]

Crop Safety/Restrictions

- Maximum number of treatments 1 per yr
- Avoid drift onto desirable plants
- Do not apply to soil which may later be used to grow desirable plants
- Do not apply on or near desirable trees or plants or on areas where their roots may extend or in locations where the chemical may be washed or move into contact with their roots

- At least 5 mth must elapse between application and planting of named conifers only (Sitka spruce, Lodgepole pine, Corsican pine)
- Must not be used as site preparation treatment for broad-leaved tree species

Special precautions/Environmental safety
- Irritating to eyes
- Not to be used on food crops
- Dangerous to fish or other aquatic life. Do not contaminate surface waters or ditches with chemical or used container [3, 4]
- Do not contaminate surface waters or ditches with chemical or used container [1, 2]
- Do not allow direct spray from vehicle mounted/drawn hydraulic sprayers to fall within 6 m, or from hand-held sprayers to within 2 m, of surface waters or ditches. Direct spray away from water

Protective clothing/Label precautions
- A, C [1-4]; M [1, 2]
- R04a, U02, U05a, U08, U09a, U20b, C03, E01, E16, E26, E30a, E31b [1-4]; E13b [3, 4]; E15 [1, 2]

Approval
- May be applied through CDA equipment. See label for details

344 imazaquin

An imidazolinone herbicide and plant growth regulator available only in mixtures

See also chlormequat + 2-chloroethylphosphonic acid + imazaquin
chlormequat + choline chloride + imazaquin

345 imidacloprid

A nitroimidazolidinimine insecticide for seed, soil or foliar treatment

See also bitertanol + fuberidazole + imidacloprid

Products

1 Admire	Bayer	70% w/w	WG	07481
2 Gaucho	Bayer	70% w/w	WS	06590
3 Intercept 5GR	Scotts	5% w/w	GR	08126
4 Intercept 70WG	Scotts	70% w/w	WG	08585

Uses

Aphids in LETTUCE *(off-label)* [2]. Aphids in BEDDING PLANTS, HARDY ORNAMENTAL NURSERY STOCK, POT PLANTS [3, 4]. Beet virus yellows vectors in SUGAR BEET [2]. Damson-hop aphid in HOPS [1]. Flea beetles in SUGAR BEET [2]. Mangold fly in SUGAR BEET [2]. Millipedes in SUGAR BEET [2]. Pygmy beetle in SUGAR BEET [2]. Sciarid flies in BEDDING PLANTS, HARDY ORNAMENTAL NURSERY STOCK, POT PLANTS [3, 4]. Springtails in SUGAR BEET [2]. Symphylids in SUGAR BEET [2]. Vine weevil in BEDDING PLANTS, HARDY ORNAMENTAL NURSERY STOCK, POT PLANTS [3, 4]. Whitefly in BEDDING PLANTS, HARDY ORNAMENTAL NURSERY STOCK, POT PLANTS [3, 4].

Notes

Efficacy
- Apply to sugar beet seed as part of the normal commercial pelleting process using special treatment machinery [2]

- Treated seed should be drilled within the season of purchase [2]
- Apply as a directed stem base spray before most bines reach a height of 2 m. If necessary treat both sides of the crown at half the normal concentration [1]
- Base of hop plants should be free of weeds and debris at application [1]
- Bines emerging away from the main stock or adjacent to poles may require a separate application [1]
- Uptake and movement within hops requires soil moisture and good growing conditions [1]
- Control may be impaired in plantations greater than 3640 plants/ha [1]
- When applied as drench or incorporated as granules in moist compost compound is readily absorbed and translocated to aerial parts of plant [3, 4]

Crop Safety/Restrictions

- Maximum number of treatments 1 per batch of seed [2] or growing medium [3], 1 per yr [1], 1 per plant per yr [4]
- To minimise likelihood of resistance do not treat all the hop crop in any one yr [1] and adopt a planned programme to alternate with pesticides of different types or use other measures when using in compost [3, 4]
- Product must not be used in compost that has already been treated with an imidacloprid-containing product [4]
- Product formulated for use only as a compost incorporation treatment into peat-based growing media [3]
- For use only on container grown ornamentals [3, 4]
- The safety of seeds sown into treated compost should not be assumed. Test treat before full-scale use [3]
- Product must not be used on crops for human or animal consumption and treated compost must not be re-used for this purpose [3, 4]

Special precautions/Environmental safety

- Irritant. May cause sensitization by skin contact [1, 2]
- Extremely dangerous to bees. Do not apply to crops in flower or those in which bees are actively foraging. Do not apply when flowering weeds are present [1]
- High risk to bees. Do not apply to crops in flower or to those in which bees are actively foraging. Do not apply when flowering weeds are present [4]
- Do not contaminate surface waters or ditches with chemical or used container
- Avoid spillage or other environmental contamination when incorporating into compost [3]
- Do not use treated seed as food or feed [2]
- Treated seed harmful to game and wildlife [2]

Protective clothing/Label precautions

- A [1-3]; C, D, H [2]
- R04, E12b [1]; R04e, U20a [1, 2]; U05a, C03, E01, E34 [1, 4]; U20b [3, 4]; E12a [4]; E15, E30a, E32a [1-4]; S02, S04b, S05, S06, S07 [2]

Latest application/Harvest Interval(HI)

- Before bines reach 2 m or before end 1st wk Jun for hops [1]; Before drilling for sugar beet [2]; before sowing or planting for bedding plants, hardy ornamental nursery stock, pot plants [3]

Approval

- Off-label approval to May 2001 for use on lettuce seed (OLA 1041/96)[2]

346 imidacloprid + tebuconazole + triazoxide

A broad spectrum fungicide and insecticide seed treatment for barley

Products

Raxil Secur	Bayer	233:20:20 g/l	LS	08966

Uses

Leaf stripe in WINTER BARLEY *(seed treatment)*. Loose smut in WINTER BARLEY *(seed treatment)*. Virus vectors in WINTER BARLEY *(seed treatment)*.

Notes

Efficacy

- Best applied through recommended seed treatment machines
- Evenness of seed cover improved by simultaneous application of equal volumes of product and water or dilution of product with an equal volume of water
- Drill treated seed in the same season
- In high risk areas where aphid activity is heavy and prolonged a follow-up aphicide treatment may be required
- Protection against foliar air-borne and splash-borne diseases later in the season will require appropriate fungicide follow-up sprays

Crop Safety/Restrictions

- Maximum number of treatments 1 per batch of seed
- Slightly delayed and reduced emergence may occur but this is normally outgrown
- Any delay in field emergence, for whatever reason, may be accentuated by treatment
- Do not use on seed with more than 16% moisture content, or on sprouted, cracked or skinned seed

Special precautions/Environmental safety

- Harmful if swallowed
- May cause sensitization by skin contact
- Harmful to fish or other aquatic life. Do not contaminate surface waters or ditches with chemical or used container
- Do not use treated seed as food or feed
- Dangerous to game and wild life. Bury spillages/Bury or remove spillages

Protective clothing/Label precautions

- A, H
- M03, R03c, R04e, U04a, U05a, U13, U20b, C03, E01, E03, E13c, E26, E30a, E32a, E34, S01, S02, S03, S04b, S05, S06, S07

Latest application/Harvest Interval(HI)

- Before drilling

347 indol-3-ylacetic acid

A plant growth regulator for promoting rooting of cuttings

Products

1 Rhizopon A Powder	Fargro	1% w/w	DP	07131
2 Rhizopon A Tablets	Fargro	50 mg a.i.	TB	07132

Uses

Rooting of cuttings in ORNAMENTALS.

Notes

Efficacy
- Apply by dipping end of prepared cuttings into powder or dissolved tablet solution
- Shake off excess powder and make planting holes to prevent powder stripping off [1]
- Consult manufacturer for details of application by spray or total immersion

Crop Safety/Restrictions
- Maximum number of treatments 1 per cutting
- Store product in a cool, dark and dry place
- Use solutions once only. Discard after use [2]
- Use plastic, not metallic, container for solutions [2]

Protective clothing/Label precautions
- U19 [1]; U20a, E15, E30a, E32a [1, 2]

Latest application/Harvest Interval(HI)
- Before cutting insertion

348 4-indol-3-yl-butyric acid

A plant growth regulator promoting the rooting of cuttings

See also 4-indol-3-yl-butyric acid + 2-(1-naphthyl)acetic acid with dichlorophen

Products

1 Chryzoplus Grey	Fargro	0.8% w/w	DP	07984
2 Chryzopon Rose	Fargro	0.1% w/w	DP	07982
3 Chryzosan White	Fargro	0.6% w/w	DP	07983
4 Chryzotek Beige	Fargro	0.4% w/w	DP	07125
5 Chryzotop Green	Fargro	0.25% w/w	DP	07129
6 Rhizopon AA Powder (0.5%)	Fargro	0.5% w/w	DP	07126
7 Rhizopon AA Powder (1%)	Fargro	1% w/w	DP	07127
8 Rhizopon AA Powder (2%)	Fargro	2% w/w	DP	07128
9 Rhizopon AA Tablets	Fargro	50 mg a.i.	TB	07130
10 Seradix 1	Hortichem	0.1% w/w	DP	06191
11 Seradix 2	Hortichem	0.3% w/w	DP	06193
12 Seradix 3	Hortichem	0.8% w/w	DP	06194

Uses

Rooting of cuttings in ORNAMENTALS.

Notes

Efficacy
- Dip base of cuttings into powder immediately before planting
- Powders or solutions of different concentration are required for different types of cutting. Lowest concentration for softwood, intermediate for semi-ripe, highest for hardwood
- See label for details of concentration and timing recommended for different species
- Use of planting holes recommended for powder formulations to ensure product is not removed on insertion of cutting. Cuttings should be watered in if necessary

Crop Safety/Restrictions
- Use of too strong a powder or solution may cause injury to cuttings
- No unused moistened powder should be returned to container

Special precautions/Environmental safety
- Do not contaminate surface waters or ditches with chemical or used container

Protective clothing/Label precautions
- U19, U20a, E15, E30a, E32a

349 4-indol-3-yl-butyric acid + 2-(1-naphthyl)acetic acid with dichlorophen

A plant growth regulator for promoting rooting of cuttings

Products	Synergol	Hortichem	5.0:5.0 g/l	SL	07386

Uses Rooting of cuttings in ORNAMENTALS.

Notes

Efficacy
- Dip base of cuttings into diluted concentrate immediately before planting
- Suitable for hardwood and softwood cuttings
- See label for details of concentration and timing for different species

Special precautions/Environmental safety
- Harmful if swallowed

Protective clothing/Label precautions
- A, C
- M03, R03c, U05a, U13, U20b, C03, E01, E15, E30a, E32a, E34

350 ioxynil

A contact acting HBN herbicide for use in turf and onions

See also benazolin + bromoxynil + ioxynil
bromoxynil + clopyralid + fluroxypyr + ioxynil
bromoxynil + diflufenican + ioxynil
bromoxynil + ethofumesate + ioxynil
bromoxynil + fluroxypyr + ioxynil
bromoxynil + ioxynil
bromoxynil + ioxynil + mecoprop-P
bromoxynil + ioxynil + triasulfuron
clopyralid + fluroxypyr + ioxynil

Products	1 Actrilawn 10	RP Amenity	100 g/l	SL	05247
	2 Totril	RP Agric.	225 g/l	EC	06116

Uses Annual dicotyledons in NEWLY SOWN TURF [1]. Annual dicotyledons in CHIVES *(off-label)*, GARLIC, LEEKS, ONIONS, SHALLOTS [2].

Notes

Efficacy
- Best results on seedling to 4-leaf stage weeds in active growth during mild weather
- In newly sown turf apply after first flush of weed seedlings, in spring normally 4 wk after sowing

- May be used on established turf from May to Sep under suitable conditions. Do not mow within 7 d of treatment

Crop Safety/Restrictions
- Maximum number of treatments 1 per yr for turf; 1 per crop (more at split doses) for spring or autumn sown bulb onions, spring sown leeks and autumn sown salad onions; 1 per crop (more at split doses) for spring sown pickling and salad onions, transplanted onions, leeks, garlic and shallots. See label for details of split applications
- Apply to sown onion crops as soon as possible after plants have 3 true leaves or to transplanted crops when established
- Apply to newly sown turf after the 2-leaf stage of grasses
- Do not use on crested dogstail

Special precautions/Environmental safety
- Harmful if swallowed [1], if swallowed or in contact with skin [2]
- Irritating to eyes and skin [2]
- Do not apply by hand-held equipment or at concentrations higher than those recommended
- Dangerous to fish or other aquatic life. Do not contaminate surface waters or ditches with chemical or used container
- Harmful to bees. Do not apply to crops in flower or to those in which bees are actively foraging. Do not apply when flowering weeds are present

Protective clothing/Label precautions
- A, C [1, 2]
- M03, R03c, U05a, U08, U13, U19, U20b, C03, E01, E07 (6 wk), E26, E30a, E31b, E34 [1, 2]; R03a, R04a, E12e, E13b [2]; E13c [1]

Withholding period
- Keep livestock out of treated areas for at least 6 wk after treatment and until foliage of any poisonous weeds such as ragwort has died and become unpalatable

Latest application/Harvest Interval(HI)
- HI onions, shallots, garlic, leeks 14 d [2]

Maximum Residue Levels (mg residue/kg food)
- garlic, onions, shallots 0.1

351 iprodione

A protectant dicarboximide fungicide with some eradicant activity

See also carbendazim + iprodione

Products

1 I T Iprodione	I T Agro	50% w/w	WP	08267
2 Landgold Iprodione 250	Landgold	255 g/l	SC	06465
3 Rovral Flo	RP Agric.	255 g/l	SC	06328
4 Rovral Green	RP Amenity	250 g/l	SC	05702
5 Rovral Liquid FS	RP Agric.	500 g/l	FS	06366
6 Rovral WP	RP Agric.	50% w/w	WP	06091

Uses Alternaria in BRASSICAS *(seed treatment)*, FLOWER SEEDS *(seed treatment)* [6]. Alternaria in KALE SEED CROPS, LEAF BRASSICAS, MUSTARD, OILSEED RAPE, STUBBLE TURNIPS, WINTER WHEAT [2, 3]. Alternaria in STORED CABBAGES [1, 6]. Alternaria in LINSEED *(seed treatment)*, OILSEED RAPE *(seed treatment)* [1, 5, 6]. Alternaria in BRASSICA SEED CROPS [1-3, 6]. Alternaria in FODDER RAPE *(seed treatment)*, LEAF BRASSICAS *(seed treatment)*, MUSTARD *(seed treatment)*, ORNAMENTALS *(seed treatment)*, STUBBLE TURNIPS *(seed treatment)*, SWEDES *(seed treatment)*, TURNIPS *(seed treatment)* [1]. Black scurf and stem canker in POTATOES *(seed treatment)* [6]. Black scurf in SEED POTATOES *(seed treatment)*[1, 5]. Botrytis in COURGETTES *(off-label)*, DWARF BEANS *(off-label)*, ENDIVES *(off-label)*, FENNEL *(off-label)*, GREEN BEANS *(off-label)*, MANGE-TOUT PEAS *(off-label)*, RUNNER BEANS *(off-label)* [6]. Botrytis in RADICCHIO *(off-label)* [3, 6]. Botrytis in GRAPEVINES *(off-label)*, OUTDOOR LETTUCE *(off-label)* [3]. Botrytis in BRASSICA SEED CROPS, KALE SEED CROPS, MUSTARD, OILSEED RAPE, WINTER WHEAT [2, 3]. Botrytis in CUCUMBERS, OUTDOOR LETTUCE, OUTDOOR TOMATOES, POT PLANTS, PROTECTED LETTUCE, PROTECTED TOMATOES, RASPBERRIES, STORED CABBAGES [1, 6]. Botrytis in STRAWBERRIES [1-3, 6]. Brown patch in AMENITY GRASS, TURF [4]. Chocolate spot in FIELD BEANS [2, 3]. Collar rot in ONIONS, SALAD ONIONS [2, 3]. Dollar spot in AMENITY GRASS, TURF [4]. Fusarium patch in AMENITY GRASS, TURF [4]. Glume blotch in WINTER WHEAT [2, 3]. Grey snow mould in AMENITY GRASS, TURF [4]. Leaf rot in ONIONS, SALAD ONIONS [2, 3]. Melting out in AMENITY GRASS, TURF [4]. Net blotch in BARLEY [2, 3]. Red thread in AMENITY GRASS, TURF [4]. Sclerotinia stem rot in OILSEED RAPE [2, 3]. Sclerotinia in COURGETTES *(off-label)*, DWARF BEANS *(off-label)*, ENDIVES *(off-label)*, FENNEL *(off-label)*, GREEN BEANS *(off-label)*, MANGE-TOUT PEAS *(off-label)*, RUNNER BEANS *(off-label)* [6]. Sclerotinia in RADICCHIO *(off-label)*[3, 6]. Sclerotinia in OUTDOOR LETTUCE *(off-label)* [3].

Notes

Efficacy

- Many diseases require a programme of 2 or more sprays at intervals of 2-4 wk. Recommendations vary with disease and crop - see label for details
- Use as a drench to control cabbage storage diseases. Spray ornamental pot plants and cucumbers to run-off [1, 6]
- Apply turf treatments after mowing and do not mow again for 24 h [3]
- Treatment harmless to *Encarsia* or *Phytoseiulus* being used for integrated pest control
- Apply to dry rapeseed and linseed prior to sowing [5]
- Best results on seed potatoes achieved using hydraulic sprayer with solid or hollow cone jets with air or liquid pressure atomisation [5]
- Apply to seed potatoes after harvest or after grading out and before traying out for chitting [5]

Crop Safety/Restrictions

- Maximum number of treatments 1 per batch for seed treatments [5]; 1 per crop on cereals, vining peas, cabbage (as drench), chicory, borage; 2 per crop on combining peas, field beans and stubble turnips; 3 per crop on brassicas (including seed crops), oilseed rape, protected winter lettuce (Oct-Feb); 4 per crop on strawberries, grapevines, salad onions, cucumbers; 5 per crop on raspberries; 6 per crop on bulb onions, tomatoes, curcurbits, peppers, aubergines and turf; 7 per crop on lettuce (Mar-Sep)
- Not to be used on protected lettuce or radicchio [3]
- A minimum of 3 wk must elapse between treatments on leaf brassicas [3]
- See label for pot plants showing good tolerance. Check other species before applying on a large scale [1, 6]
- Do not treat oilseed rape seed that is cracked or broken or of low viability [5]
- Do not excessively wet seed potato skins [5]
- Do not treat oats [3]

Special precautions/Environmental safety

- Irritating to eyes and skin [4]
- Harmful to fish or other aquatic life. Do not contaminate surface waters or ditches with chemical or used container
- Treated brassica seed crops not to be used for human or animal consumption [1, 6]
- See label for guidance on disposal of spent drench liquor [1, 6]
- Do not use treated seed as food or feed [1, 5]

Protective clothing/Label precautions

- A [2-4]; C [4]
- R04a, R04b, U04a [4]; U05a, C03, E01 [3, 4]; U08, E31b [2-4]; U19 [1, 6]; U20a [2]; U20b [3]; U20c, E34 [1, 4-6]; E13c, E30a [1-6]; E26 [2, 4, 5]; E32a [1, 5, 6]; S01, S02, S05, S06 [5, 6]; S07 [5]

Latest application/Harvest Interval(HI)

- Before grain watery ripe (GS 69) for wheat and barley; pre-planting for seed treatments.
- HI strawberries, protected tomatoes 1 d; outdoor tomatoes, cucurbits, peppers, aubergines 2 d; lettuce (Mar-Sep), onions, raspberries 7 d; grapevines 2 wk; brassicas, brassica seed crops, oilseed rape, beans, stubble turnips, borage 21 d; protected winter lettuce (Oct-Feb) 4 wk; cabbage (drench treatment) 2 mth

Approval

- Approved for aerial application on oilseed rape [3]. See notes in Section 1
- Off-label Approval unlimited for use on outdoor crops of lettuce, cress, lamb's lettuce, scarole, radicchio (OLA 0715/95)[3]; unlimited for use on grapevines (OLA 0478/93)[3]; to Mar 2003 for use on pears as a post-harvest dip (OLA 0693/98)[6]; unlimited for use on French, dwarf and runner beans, mange-tout peas (OLA 1565/98)[6], and to Jul 2003 on the same crops with a shorter harvest interval (OLA 1501/95)[6]

Maximum Residue Levels (mg residue/kg food)

- pome fruits, bilberries, currants, gooseberries, lettuces, herbs 10; stone fruits, cane fruits, kiwi fruit, garlic, onions, shallots, tomatoes, peppers, aubergines, head cabbages 5; cucumbers, gherkins, courgettes 2; witloof 1; rape seed, wheat 0.5; pulses 0.2; kohlrabi, tea, hops 0.1; animal products 0.05; citrus fruits, tree nuts, cranberries, wild berries, miscellaneous fruit (except bananas and kiwi fruit), root and tuber vegetables (except carrots, radishes), squashes, water melons, sweetcorn, spinach, beet leaves, water cress, stem vegetables (except fennel), mushrooms, oilseeds (except linseed, rape seed, mustard seed), rye, oats, triticale, maize 0.02

352 iprodione + thiophanate-methyl

A protectant and systemic fungicide for oilseed rape, field beans and peas

Products

1 Compass	RP Agric.	167:167 g/l	SC	06190
2 Snooker	RP Agric.	150:200 g/l	SC	07940

Uses Alternaria in CARROTS *(off-label)*, OILSEED RAPE [1]. Alternaria in WINTER OILSEED RAPE [2]. Chocolate spot in FIELD BEANS [1, 2]. Crown rot in CARROTS *(off-label)* [1]. Grey mould in COMBINING PEAS, OILSEED RAPE, VINING PEAS [1]. Grey mould in WINTER OILSEED RAPE [2]. Leaf and pod spot in COMBINING PEAS, VINING PEAS [1]. Light lea spot in OILSEED RAPE [1]. Light leaf spot in WINTER OILSEED RAPE [2]. Sclerotinia stem rot in OILSEED RAPE [1]. Sclerotinia stem rot in WINTER OILSEED RAPE [2]. Stem canker in OILSEED RAPE [1]. Stem rot in COMBINING PEAS, VINING PEAS [1].

Notes

Efficacy

- Timing of sprays on oilseed rape varies with disease, see label for details
- Apply to field beans at early flowering stage before disease enters aggressive stage and repeat after 3 wk if necessary
- Apply to peas at mid flowering stage. For combining peas repeat treatment in 2-4 wk [1]

Crop Safety/Restrictions

- Maximum number of treatments (refers to total sprays containing benomyl, carbendazim and thiophanate methyl) 2 per crop for winter oilseed rape, field beans and combining peas; 1 per crop for vining peas
- Under conditions of crop stress, e.g. after prolonged dry conditions on light soils do not spray winter oilseed rape from aircraft in less than 100 l/ha in strong direct sunlight. Under these conditions aerial application should be made in early morning or evening [1]
- Treatment may extend duration of green leaf in winter oilseed rape [2]

Special precautions/Environmental safety

- Irritating to eyes and skin. May cause sensitization by skin contact [2]
- Treated haulm must not be fed to livestock [1]
- Harmful to fish or other aquatic life. Do not contaminate surface waters or ditches with chemical or used container

Protective clothing/Label precautions

- A, C, H, M [1, 2]
- R04a, R04b, R04e, U05a, U09a, U20b, C03, E01, E26 [2]; U20c [1]; E13c, E30a, E31b [1, 2]

Latest application/Harvest Interval(HI)

- Before end of flowering for winter oilseed rape and field beans.
- HI 3 wk

Approval

- Off-label Approval to Aug 1999 for use on carrots (OLA 1264/94)[1]

Maximum Residue Levels (mg residue/kg food)

- see iprodione entry

353 isoproturon

A residual urea herbicide for use in cereals

See also carfentrazone-ethyl + isoproturon
diflufenican + flurtamone + isoproturon
diflufenican + isoproturon
fenoxaprop-P-ethyl + isoproturon

Products

1 Alpha Isoproturon 500	Makhteshim	500 g/l	SC	05882
2 Alpha Isoproturon 650	Makhteshim	650 g/l	SC	07034
3 Arelon 500	AgrEvo	500 g/l	SC	08100
4 Atlas Fieldgard	Atlas	500 g/l	SC	08582
5 Atum WDG	RP Agric.	83% w/w	WG	07778
6 Auger	RP Agric.	500 g/l	SC	06581
7 Barclay Guideline	Barclay	500 g/l	SC	06743
8 Isoguard	Chiltern	500 g/l	SC	08497
9 Isoproturon 500	AgrEvo	500 g/l	SC	06718
10 Landgold Isoproturon	Landgold	500 g/l	SC	06012
11 Luxan Isoproturon 500 Flowable	Luxan	500 g/l	SC	07437
12 MSS Iprofile	Mirfield	500 g/l	SC	06341
13 Mysen	Portman	500 g/l	SC	07695
14 Portman Isotop SC	Portman	500 g/l	SC	07663
15 Sabre 500	AgrEvo	500 g/l	SC	08101
16 Standon IPU	Standon	500 g/l	SC	08671
17 Stefes IPU 500	Stefes	500 g/l	SC	08102
18 Tolkan Liquid	RP Agric.	500 g/l	SC	06172
19 Tripart Pugil	Tripart	500 g/l	SC	06153

Uses

Annual dicotyledons in WINTER BARLEY, WINTER WHEAT [1-19]. Annual dicotyledons in AUTUMN SOWN SPRING WHEAT [5, 6, 18]. Annual dicotyledons in SPRING WHEAT [3, 7, 9, 13, 15-17, 19]. Annual dicotyledons in WINTER RYE [3, 5-7, 9, 13, 15, 17-19]. Annual dicotyledons in TRITICALE [3, 5-7, 9, 13, 15-19]. Annual grasses in WINTER BARLEY, WINTER WHEAT [1-19]. Annual grasses in AUTUMN SOWN SPRING WHEAT [5, 6, 18]. Annual grasses in SPRING WHEAT [3, 7, 9, 13, 15-17, 19]. Annual grasses in WINTER RYE [3, 5-7, 9, 13, 15, 17-19]. Annual grasses in TRITICALE [3, 5-7, 9, 13, 15-19]. Annual meadow grass in SPRING BARLEY *(off-label)* [18]. Blackgrass in WINTER BARLEY, WINTER WHEAT [1-19]. Blackgrass in AUTUMN SOWN SPRING WHEAT [5, 6, 18]. Blackgrass in SPRING WHEAT [3, 7, 9, 13, 15-17, 19]. Blackgrass in WINTER RYE [3, 5-7, 9, 13, 15, 17-19]. Blackgrass in TRITICALE [3, 5-7, 9, 13, 15-19]. Rough meadow grass in WINTER BARLEY, WINTER WHEAT [1-19]. Rough meadow grass in AUTUMN SOWN SPRING WHEAT [5, 6, 18]. Rough meadow grass in SPRING WHEAT [3, 7, 9, 13, 15-17, 19]. Rough meadow grass in WINTER RYE [3, 5-7, 9, 13, 15, 17-19]. Rough meadow grass in TRITICALE [3, 5-7, 9, 13, 15-19]. Wild oats in WINTER BARLEY, WINTER WHEAT [1-19]. Wild oats in AUTUMN SOWN SPRING WHEAT [5, 6, 18]. Wild oats in SPRING WHEAT [3, 7, 9, 13, 15-17, 19]. Wild oats in WINTER RYE [3, 5-7, 9, 13, 15, 17-19]. Wild oats in TRITICALE [3, 5-7, 9, 13, 15-19].

Notes

Efficacy

- May be applied in autumn or spring (but only post-emergence on wheat or barley). Best control normally achieved by early post-emergence treatment

- See label for details of rates, timings and tank mixes for different weed problems
- Apply to moist soils. Effectiveness may be reduced in seasons of above average rainfall or by prolonged dry weather
- Residual activity reduced on soils with more than 10% organic matter. Only use on such soils in spring

Crop Safety/Restrictions

- Maximum total dose 2.5 kg a.i./ha for any crop. The IPU stewardship programme guidelines recommend that where possible this should be reduced to 1.5 kg a.i./ha using mixtures or sequences with other herbicides
- Recommended timing of treatment varies depending on crop to be treated, method of sowing, season of application, weeds to be controlled and product being used. See label for details
- Do not apply pre-emergence to wheat or barley. On triticale and winter rye use pre-emergence only and on named varieties
- Do not use on durum wheats, undersown cereals or crops to be undersown
- Do not apply to very wet or waterlogged soils, or when heavy rainfall is forecast
- Do not use on very cloddy soils or if frost is imminent or after onset of frosty weather
- Crop damage may occur on free draining, stony or gravelly soils if heavy rain falls soon after spraying. Early sown crops may be damaged if spraying precedes or coincides with a period of rapid growth
- Do not roll for 1 wk before or after treatment or harrow for 1 wk before or any time after treatment

Special precautions/Environmental safety

- Irritant. May cause sensitisation by skin contact [5]
- Irritating to eyes and skin [4, 8, 14]
- Dangerous (harmful [1, 2, 7, 11, 12, 19]) to fish or other aquatic life. Do not contaminate surface waters or ditches with chemical or used container [3-5, 8-10, 14-17]
- Do not spray where soils are cracked, to avoid run-off through drains
- Product available in refillable container - see label for instructions [3, 9, 15]

Protective clothing/Label precautions

- A [1-5, 7-10, 12, 14-17]; C [1-5, 8-10, 12, 14-17]; D [5]; H [1-5, 8-10, 12, 15-17]
- R04, R04e, E32a [5]; R04a, R04b [4, 8, 14]; U04a [16, 19]; U05a [1, 2, 4, 8, 12-14, 16]; U08 [1-6, 8, 9, 12-18]; U10 [3]; U19 [1, 2, 4-6, 8, 12, 14, 16, 18]; U20a [1, 7, 10, 14]; U20b [2-6, 8, 9, 11-13, 15-19]; C03, E01 [1, 2, 4, 8, 12]; E13b [3-5, 8-10, 12, 14-17]; E13c [1, 2, 7, 11, 19]; E15 [6, 13, 18]; E26 [1, 2, 4-6, 8, 10-14, 16-19]; E30a [1-19]; E31b (C), E33 (R), E34 (R), E36 (R) [3, 15]; E31b [1, 2, 4, 6-14, 16-19]; E34 [19]; E36 [17]

Latest application/Harvest Interval(HI)

- Not later than second node detectable stage (GS 32)

Approval

- May be applied through CDA equipment (see labels for details) [6, 18]
- All approvals for aerial use of isoproturon were revoked in 1995
- Off-label approval unlimited for use on spring barley (OLA 0846/96)[18]

354 isoproturon + pendimethalin

A contact and residual herbicide for use in winter cereals

Products

1 Encore	Cyanamid	125:250 g/l	SC	04737
2 Jolt	Cyanamid	125:250 g/l	SC	05488
3 Trump	Cyanamid	236:236 g/l	SC	03687

Uses

Annual dicotyledons in AUTUMN SOWN SPRING WHEAT [3]. Annual dicotyledons in TRITICALE, WINTER BARLEY, WINTER RYE, WINTER WHEAT [1-3]. Annual grasses in AUTUMN SOWN SPRING WHEAT [3]. Annual grasses in TRITICALE, WINTER BARLEY, WINTER RYE, WINTER WHEAT [1-3]. Blackgrass in AUTUMN SOWN SPRING WHEAT [3]. Blackgrass in TRITICALE, WINTER BARLEY, WINTER RYE, WINTER WHEAT [1-3]. Wild oats in AUTUMN SOWN SPRING WHEAT [3]. Wild oats in TRITICALE, WINTER BARLEY, WINTER RYE, WINTER WHEAT [1-3].

Notes

Efficacy

- May be applied in autumn or spring (but only post-emergence on wheat or barley)
- Annual grasses controlled from pre-emergence to 3-leaf stage, extended to 3-tiller stage with blackgrass by tank mixing with additional isoproturon. Best results on wild oats post-emergence in autumn
- Annual dicotyledons controlled pre-emergence and up to 4-leaf stage [1, 2], up to 8-leaf stage [3]. See label for details
- Apply to moist soils. Best results achieved by application to fine, firm seedbed. Trash, ash or straw should have been incorporated evenly
- Contact activity is reduced by rain within 6 h of application
- Activity may be reduced on soil with more than 6% organic matter or ash. Do not use on soils with more than 10% organic matter

Crop Safety/Restrictions

- Maximum number of treatments 1 per crop. Maximum total dose of isoproturon 2.5 kg a.i./ha per crop
- Apply before first node detectable (GS 31) [1, 2], to before leaf sheath erect (GS 30) [3]
- Do not apply pre-emergence to wheat or barley. On winter rye and triticale use pre-emergence only on named cultivars
- Do not use on durum wheat or spring cereals
- May be used on autumn sown spring wheat [3]
- Do not use pre-emergence on winter rye or triticale drilled after 30 Nov [3]
- Do not use on crops suffering stress from disease, drought, waterlogging, poor seedbed conditions or other causes or apply post-emergence when frost imminent
- Do not apply to very wet or waterlogged soils, or when heavy rain is forecast
- Do not undersow treated crops
- Do not roll or harrow after application
- In the event of autumn crop failure spring wheat, spring barley, maize, potatoes, beans or peas may be grown following ploughing to at least 150 mm

Special precautions/Environmental safety

- Dangerous to fish or other aquatic life. Do not contaminate surface waters or ditches with chemical or used container

- Do not spray where soils are cracked, to avoid run-off through drains

Protective clothing/Label precautions

- A [1-3]; C [1, 2]
- U08, U13, U19, U20a, E13b, E30a, E31b, E34 [1-3]

Latest application/Harvest Interval(HI)

- Pre-emergence for winter rye and triticale; before leaf sheath erect (GS 30)[3] or before 1st node detectable (GS 31)[1, 2] for winter wheat and barley

355 isoproturon + simazine

A contact and residual herbicide for winter wheat and barley

Products

1 Harlequin 500 SC	Ciba Agric.	450:50 g/l	SC	05847
2 Harlequin 500 SC	Novartis	450:50 g/l	SC	08416

Uses

Annual dicotyledons in WINTER BARLEY, WINTER WHEAT. Annual grasses in WINTER BARLEY, WINTER WHEAT. Blackgrass in WINTER BARLEY, WINTER WHEAT.

Notes

Efficacy

- May be applied in autumn or spring (but only post-emergence of the crop)
- Annual grasses controlled up to early tillering, annual dicotyledons to 50 mm
- Apply to fine, firm, even seedbed. Any trash or burnt straw should be dispersed in preparing seedbed
- Apply to moist soils. Weed control may be reduced in excessively wet autumns or if prolonged dry weather follows application to dry soil

Crop Safety/Restrictions

- Maximum number of treatments 1 per crop. Maximum total dose of isoproturon 2.5 kg a.i./ha per crop
- Apply from 2-leaf stage of crop to before leaf sheath erect (GS 12-30)
- Do not use on durum wheat, undersown crops or those due to be undersown
- With direct drilled crops soil surface should be broken by surface cultivation and seed covered by 12-25 mm soil
- Do not use on sand or on soils with more than 10% organic matter
- On stony or gravelly soils there is risk of crop damage, especially with heavy rain soon after application
- Do not apply when frost or heavy rain is forecast or to crops severely checked by frost, waterlogging, pest or disease attack.
- Early sown crops may be prone to damage if spraying precedes or coincides with period of rapid growth in autumn
- Do not harrow for 7 d before treatment or roll for 7 d before or after treatment in spring
- In the event of crop failure land should be inverted by mouldboard ploughing to at least 150 mm and harrowed before drilling/planting another crop
- Do not harrow after application

Special precautions/Environmental safety

- Irritating to skin
- Do not spray where soils are cracked, to avoid run-off through drains
- Do not contaminate ponds, waterways or ditches with chemical or used container/Do not contaminate surface waters or ditches with chemical or used container

- Do not allow direct spray from vehicle mounted/drawn hydraulic sprayers to fall within 6 m, or from hand-held sprayers to within 2 m, of surface waters or ditches. Direct spray away from water [1]

Protective clothing/Label precautions
- R04b, U05a, U08, U19, U20a, C03, E01, E15, E16, E30a, E31b, E34

Latest application/Harvest Interval(HI)
- Leaf sheath erect (GS 30)

356 isoproturon + trifluralin

A residual early post-emergence herbicide for cereals

Products

Autumn Kite	AgrEvo	300:200 g/l	EC	07119

Uses

Annual dicotyledons in WINTER BARLEY, WINTER WHEAT. Annual grasses in WINTER BARLEY, WINTER WHEAT. Blackgrass in WINTER BARLEY, WINTER WHEAT.

Notes

Efficacy
- Provides contact and residual control. Best results achieved by application to dicotyledons pre-emergence or up to 2-leaf stage, to blackgrass up to 2-3 tillers
- Apply post-emergence when leaves dry, weeds are actively growing and rain not expected for 2 h
- Effectiveness may be reduced by prolonged dry or sunny weather after application
- Do not use on soils with more than 10% organic matter
- If used in conjunction with minimum cultivation ensure that all trash and burnt straw is removed, buried or dispersed before spraying

Crop Safety/Restrictions
- Maximum number of treatments 1 per crop. Maximum total dose of isoproturon 2.5 kg a.i./ha per crop
- Apply post-emergence to 4-leaf and 2 tillers stage and to broadcast crops after 3-leaf stage (GS 13)
- Do not use on durum wheat or crops to be undersown
- Do not use on Sands and do not incorporate in soil. On very stony, gravelly or other free draining soils crops may be damaged if heavy rain falls soon after treatment
- Do not spray when heavy rain is forecast or to crops stressed by frost, waterlogging, deficiency or pest attack
- Do not roll after treatment until following spring
- In case of crop failure only sow carrots, peas or sunflowers within 5 mth and plough to at least 15 cm. Do not sow sugar beet in spring following treatment
- Do not harrow treated crops

Special precautions/Environmental safety
- Irritating to skin and eyes. Flammable
- Keep in original container, tightly closed, in a safe place, under lock and key
- Harmful to fish or other aquatic life. Do not contaminate surface waters or ditches with chemical or used container

- Do not spray where soils are cracked, to avoid run-off through drains

Protective clothing/Label precautions
- A, C
- R04a, R04b, U05a, U08, U13, U19, U20a, C03, E01, E13b, E26, E30a, E31b, E34

Latest application/Harvest Interval(HI)
- Before 5 leaf stage (GS 15)

357 isoxaben

A soil-acting amide herbicide for use in grass and fruit

Products

1 Flexidor 125	Dow	125 g/l	SC	05104
2 Gallery 125	Rigby Taylor	125 g/l	SC	06889
3 Knot Out	Vitax	125 g/l	SC	05163

Uses

Annual dicotyledons in FORESTRY TRANSPLANTS [1, 2]. Annual dicotyledons in APPLES, BLACKBERRIES, BLACKCURRANTS, CHERRIES, CONTAINER-GROWN WOODY ORNAMENTALS, GOOSEBERRIES, GRAPEVINES, HARDY ORNAMENTAL NURSERY STOCK, HOPS, OUTDOOR CARROTS *(off-label)*, PEARS, PLUMS, PROTECTED COURGETTES *(off-label)*, PROTECTED MARROWS *(off-label)*, PROTECTED ORNAMENTALS *(off-label)*, PROTECTED PUMPKINS *(off-label)*, PROTECTED SQUASHES *(off-label)*, RASPBERRIES, STRAWBERRIES [1]. Annual dicotyledons in AMENITY GRASS [2, 3]. Annual dicotyledons in AMENITY TREES AND SHRUBS, MANAGED AMENITY TURF [2]. Cleavers in ASPARAGUS *(off-label)* [1]. Fat-hen in ASPARAGUS *(off-label)* [1].

Notes

Efficacy
- When used alone apply pre-weed emergence
- Effectiveness is reduced in dry conditions. Weed seeds germinating at depth are not controlled
- Activity reduced on soils with more than 10% organic matter. Do not use on peaty soils
- Various tank mixtures are recommended for early post-weed emergence treatment (especially for grass weeds). See label for details
- Best results on turf achieved by applying to firm, moist seedbed within 2 d of sowing. Avoid disturbing soil surface after application [3]

Crop Safety/Restrictions
- Maximum number of treatments 1 per crop, sowing or yr for amenity grass, ornamentals, strawberries, 2 per yr for hardy ornamental nursery stock and forestry transplants
- See label for details of crops which may be sown in the event of failure of a treated crop

Special precautions/Environmental safety
- Do not contaminate surface waters or ditches with chemical or used container
- Dangerous to livestock

Protective clothing/Label precautions
- A, C [1-3]; H [2]
- U20b, E07 (50 d), E15, E30a, E32a [1-3]; E26, E31a, E34 [3]

Withholding period
- Keep all livestock out of treated areas for at least 50 d

Latest application/Harvest Interval(HI)

- Before 1 Apr in yr of harvest for edible crops

Approval

- Off-label approval unlimited for use in protected ornamentals (OLA 0605/94)[1]; unlimited for use on outdoor carrots (OLA 2209/96)[1]; unlimited for use on outdoor asparagus (OLA 2318/97)[1]; unlimited for use on courgettes, marrows, squashes, pumpkins under crop covers (OLA 1595/98)[1]

358 isoxaben + terbuthylazine

A contact and residual herbicide for use in peas

Products

1 Skirmish	Ciba Agric.	75:420 g/l	SC	08079
2 Skirmish	Novartis	75:420 g/l	SC	08444

Uses Annual dicotyledons in COMBINING PEAS, VINING PEAS.

Notes

Efficacy

- May be applied pre- or post-emergence of crop but before second node stage (GS 102)
- Product will give residual control of germinating weeds on mineral soils for up to 8 wk
- Product is slow acting and control may not be evident for 7-10 d or more after spraying
- Best results achieved when soil surface damp and with a fine, firm tilth. Do not use on very cloddy or stony soil
- Rain after spraying will normally improve weed control but excessive rainfall, very dry conditions or unusually low soil temperatures may lead to unsatisfactory control

Crop Safety/Restrictions

- Maximum number of treatments 1 per crop
- Product may be used on all varieties of spring sown vining and combining peas but Vedette and Printana may be damaged
- Do not use on forage peas
- Do not use on soils lighter than coarse sandy loam, on very stony soils or soils with more than 10% organic matter
- Pea seed should be covered by at least 25 mm of soil
- Heavy rain after application may cause some crop damage, especially on light soils. Do not use on soils where surface water is likely to accumulate
- For post-emergence treatment a crystal violet test for cuticle wax is advised

Special precautions/Environmental safety

- Irritant. May cause sensitization by skin contact
- Harmful to fish or other aquatic life. Do not contaminate surface waters or ditches with chemical or used container

Protective clothing/Label precautions

- A
- R04e, U05a, U20a, C03, E01, E15, E30a, E31b

Latest application/Harvest Interval(HI)
- Before second node stage (GS 102)

359 isoxaben + trifluralin

A residual herbicide mixture for amenity use

Products

1 Axit GR	Fargro	0.55:2.06% w/w	GR	08892
2 Premiere Granules	Dow	0.55:2.06% w/w	GR	07987

Uses

Annual dicotyledons in AMENITY TREES AND SHRUBS, AMENITY VEGETATION [2]. Annual dicotyledons in HARDY ORNAMENTAL NURSERY STOCK [1]. Annual meadow grass in AMENITY TREES AND SHRUBS, AMENITY VEGETATION [2]. Annual meadow grass in HARDY ORNAMENTAL NURSERY STOCK [1]. Fat-hen in AMENITY TREES AND SHRUBS, AMENITY VEGETATION [2]. Fat-hen in HARDY ORNAMENTAL NURSERY STOCK [1]. Groundsel in AMENITY TREES AND SHRUBS, AMENITY VEGETATION [2]. Sowthistle in AMENITY TREES AND SHRUBS, AMENITY VEGETATION [2]. Sowthistle in HARDY ORNAMENTAL NURSERY STOCK [1].

Notes

Efficacy
- Best results obtained when granules applied to firm, moist soil, free from clods. 20-30 mm irrigation or rainfall required within 3 d after treatment
- Weed control reduced under dry soil conditions
- Treatment should be made before mulching and before weed emergence
- Uniform cover with granules optimises results. Improved uniformity of cover may be achieved by spreading the granules twice at half dose at right angles
- Control may be reduced on soils with high organic matter or where organic manure has been applied

Crop Safety/Restrictions
- Maximum number of treatments 1 per yr
- Ensure soil has settled and no cracks are present before application to transplanted trees and shrubs [2]
- May be used on Light, Medium and Heavy soils, but on Light soils early growth may be reduced under adverse conditions
- Do not use on soils with more than 10% organic matter
- For newly planted containerised stock allow compost to settle 7-10 d before treatment [1]
- Do not use in the 12 mth prior to lifting field grown nursery stock if edible crops are to be grown as a following crop
- Land must be mouldboard ploughed to at least 20 cm before drilling or planting succeeding plants except well rooted forestry trees and ornamentals

Special precautions/Environmental safety
- Harmful to fish or other aquatic life. Do not contaminate surface waters or ditches with chemical or used container
- Do not allow direct applications of granules from ground based/vehicle mounted applicators to fall within 6 m, or from hand-held applicators to within 2 m, of surface waters or ditches. Direct applications away from water
- Product contains up to 1% crystalline silica

Protective clothing/Label precautions
- U20b, E01, E13c, E16, E30a, E32a, E34

360 kresoxim-methyl

A systemic strobilurin fungicide for apples

See also epoxiconazole + fenpropimorph + kresoxim-methyl
fenpropimorph + kresoxim-methyl

Products

Stroby WG	BASF	50% w/w	WG	08653

Uses

Powdery mildew in APPLES *(reduction)*. Scab in APPLES.

Notes

Efficacy
- Best results achieved from treatments starting at bud burst and repeated at 10-14 d intervals depending on severity of disease pressure
- Do not spray product more than three times consecutively. To minimise risk of development of resistance apply in a programme with products from different chemical groups.
- Product should not be used as final spray of the season
- Product may be applied in ultra low volumes (ULV) but disease control may be reduced

Crop Safety/Restrictions
- Maximum number of treatments 5 per yr, but not more than three consecutively
- Consult before using on crops intended for processing

Special precautions/Environmental safety
- Dangerous to fish or other aquatic life. Do not contaminate surface waters or ditches with chemical or used container
- Do not allow direct spray from broadcast air-assisted sprayers to fall within 18 m of surface waters or ditches. Direct spray away from water
- Harmless to ladybirds and predatory mites
- Harmless to honey bees and may be applied during flowering. Nevertheless local beekeepers should be notified when treatment of orchards in flower is to occur

Protective clothing/Label precautions
- U20b, E13b, E17, E30a, E31c

Latest application/Harvest Interval(HI)
- HI 35 d

Approval
- Product approved for use in ULV systems

361 lambda-cyhalothrin

A quick-acting contact and ingested pyrethroid insecticide

Products

1 Hallmark	Zeneca	50 g/l	EC	06434
2 Landgold Lambda-C	Landgold	50 g/l	EC	09205
3 Standon Lambda-C	Standon	50 g/l	EC	08831

Uses Aphids in DURUM WHEAT, POTATOES, WINTER BARLEY, WINTER WHEAT [1-3]. Aphids in WINTER OILSEED RAPE [2, 3]. Barley yellow dwarf virus vectors in DURUM WHEAT, WINTER BARLEY, WINTER WHEAT [1-3]. Beet leaf miner in SUGAR BEET [1]. Beet virus yellows vectors in SPRING OILSEED RAPE, WINTER OILSEED RAPE [1]. Bryobia mites in GOOSEBERRIES *(off-label)* [1]. Cabbage seed weevil in SPRING OILSEED RAPE, WINTER OILSEED RAPE [3]. Cabbage stem flea beetle in WINTER OILSEED RAPE [1-3]. Cabbage stem flea beetle in SPRING OILSEED RAPE [1]. Capsids in GOOSEBERRIES *(off-label)* [1]. Carrot fly in CELERY *(off-label)*, PARSNIPS *(off-label)* [1]. Caterpillars in BROCCOLI, BRUSSELS SPROUTS, CABBAGES, CALABRESE, CAULIFLOWERS [1-3]. Caterpillars in GOOSEBERRIES *(off-label)* [1]. Colorado beetle in POTATOES *(off-label (Statutory Notice))* [1]. Cutworms in SUGAR BEET [1-3]. Damson-hop aphid in HOPS [1-3]. Flea beetles in SUGAR BEET [1-3]. Flea beetles in SPRING OILSEED RAPE, WINTER OILSEED RAPE [1]. Insect pests in CARROTS *(off-label)*, PARSNIPS *(off-label)*, RED BEET *(off-label)*, SOYA BEANS *(off-label)* [1]. Leaf miners in SUGAR BEET [2, 3]. Pea and bean weevils in PEAS, SPRING FIELD BEANS, WINTER FIELD BEANS [1-3]. Pea aphid in PEAS [1]. Pea moth in PEAS [1-3]. Pear sucker in PEARS [1-3]. Pod midge in SPRING OILSEED RAPE, WINTER OILSEED RAPE [1, 2]. Pollen beetles in SPRING OILSEED RAPE, WINTER OILSEED RAPE [1-3]. Red spider mites in HOPS [1, 3]. Seed weevil in SPRING OILSEED RAPE, WINTER OILSEED RAPE [1, 2]. Silver y moth in CELERY *(off-label)*, DWARF BEANS *(off-label)*, NAVY BEANS *(off-label)*, RUNNER BEANS *(off-label)* [1]. Two-spotted spider mite in GOOSEBERRIES *(off-label)* [1]. Virus vectors in WINTER OILSEED RAPE [2, 3]. Yellow cereal fly in WINTER WHEAT [1].

Notes

Efficacy

- Best results normally obtained from treatment when pest attack first seen. See label for detailed recommendations on each crop
- Repeat applications recommended in some crops where prolonged attack occurs, up to maximum total dose. See label for details
- Where aphids in hops or pear suckers resistant to lambda-cyhalothrin occur control is unlikely to be satisfactory
- Addition of wetter recommended for control of certain pests in brassicas and oilseed rape [1, 2]
- Use of sufficient water volume to ensure thorough crop penetration recommended for optimum results
- Use of drop-legged sprayer gives improved results in crops such as Brussels sprouts

Crop Safety/Restrictions

- Maximum number of applications 1 per crop for winter wheat and barley; 2 per crop for winter oilseed rape, peas and field beans; 3 per crop for apples and pears; 4 per crop for brassicas [2, 3]; 6 per crop for hops [1-3]
- Maximum total dose 300 ml/crop for field beans, sugar beet, peas; 400 ml/crop for cereals; 450 ml/crop for oilseed rape; 540 ml/yr for pears; 800 ml/crop for brassicas [1, 2]

Special precautions/Environmental safety

- Harmful in contact with skin or if swallowed. Irritating to eyes and skin
- Flammable
- Keep in original container, tightly closed, in a safe place, under lock and key [3]
- Extremely dangerous to bees. Do not apply to crops in flower or to those in which bees are actively foraging, except as directed in peas and oilseed rape and hops. Do not apply when flowering weeds are present [2, 3]
- Do not apply to a cereal crop if any product containing a pyrethroid insecticide or dimethoate has been applied to the crop after the start of ear emergence (GS 51)
- Do not spray cereals in the spring/summer (ie after 1 Apr) within 6 m of edge of crop
- Extremely dangerous to fish or other aquatic life. Do not contaminate surface waters or ditches with chemical or used container

- Do not allow direct spray from vehicle mounted/drawn hydraulic sprayers to fall within 6 m, or from hand-held sprayers to within 2 m, of surface waters or ditches. Direct spray away from water
- Do not allow spray from air-assisted applications to hops to fall within 50 m of surface waters or ditches
- Do not allow spray from air-assisted applications to pears to fall within 38 m of surface waters or ditches

Protective clothing/Label precautions

- A, C, H, J, K, L, M
- M03, R03a, R03c, R04a, R04b, R07d, U02, U05a, U08, C03, E01, E13a, E16, E30a, E34 [1-3]; U20a, E31a [1, 2]; U20b, E17, E31b [3]; E12b, E26 [2, 3]; E17 (hops - 50 m; pears 38 m) [2]

Latest application/Harvest Interval(HI)

- Before late milk stage (GS 77) in cereals; before end flowering for oilseed rape
- HI 6 wk for spring oilseed rape; 8 wk for sugar beet [1]

Approval

- Off-label approval unlimited (OLA 1192/93) for use on potatoes for Colorado beetle control as required by Statutory Notice [1]; unlimited for use on carrots and parsnips (OLA 1737/96)[1]; to Sep 2000 for use on dwarf beans, navy beans and runner beans (OLA 1247/95)[1]; unlimited for use on red beet (OLA 1133/97)[1]; unlimited for use on outdoor celery (OLA 2066/97)[1]; to Oct 2002 for use on outdoor gooseberries (OLA 2340/97)[1]; to Mar 2002 for use on soya beans (OLA 0354/97)[1]

Maximum Residue Levels (mg residue/kg food)

- hops 10; cress, lamb's lettuce, lettuce, scarole, chervil, chives, parsley, celery leaves, tea 1; meat (except poultry) 0.5; apricots, peaches, grapes, head cabbages, beans (with pods), peas (with pods) 0.2; pome fruits, cherries, plums, currants, gooseberries,cucumbers, gherkins, courgettes 0.1; tree nuts, Brussels sprouts, barley, milk and dairy produce 0.05; blackberries, dewberries, loganberries, raspberries, bilberries, cranberries, wild berries, miscellaneous fruits, root and tuber vegetables, garlic, onions, shallots, sweet corn, watercress, beans (without pods), peas (without pods), asparagus, wild mushrooms, pulses, oilseed, potatoes, wheat, rye, oats, triticale, maize, rice, poultry meat, eggs 0.02

362 lambda-cyhalothrin + pirimicarb

An insecticide mixture combining translaminar, contact, fumigant and stomach activity

Products	Dovetail	Zeneca	5:100 g/l	EC	07973

Uses Aphids in BRASSICAS, CARROTS, CEREALS, LETTUCE, PEAS, POTATOES, SUGAR BEET. Beet virus yellows vectors in OILSEED RAPE. Cabbage stem flea beetle in OILSEED RAPE. Caterpillars in BRASSICAS. Cutworms in CARROTS, LETTUCE, POTATOES, SUGAR BEET. Flea beetles in OILSEED RAPE, SUGAR BEET. Leaf miners in SUGAR BEET. Mealy aphids in BRASSICAS, OILSEED RAPE. Pea and bean weevils in FIELD BEANS, PEAS. Pea midge in PEAS. Pea moth in PEAS. Pod midge in OILSEED RAPE. Pollen beetles in BRASSICAS, OILSEED RAPE. Seed weevil in OILSEED RAPE. Whitefly in BRASSICAS.

Notes

Efficacy

- Best results obtained from treatment when pest attack first seen or after warning issued
- Repeat applications recommended in some crops where prolonged attacks occur, up to maximum total dose
- Addition of Agral recommended for certain uses in brassicas and oilseed rape. See label
- Use of drop-leg sprayers improves efficacy in crops such as Brussels sprouts
- Control unlikely to be satisfactory if aphids resistant to lambda-cyhalothrin or pirimicarb present

Crop Safety/Restrictions

- Maximum total dose (product): peas, field beans, cereals, carrots, lettuce 3.0 l/ha; oilseed rape 4.5 l/ha; sugar beet, brassicas 6.0 l/ha; potatoes 12.0 l/ha

Special precautions/Environmental safety

- Harmful if swallowed. Irritating to skin. Risk of serious damage to eyes
- May cause sensitization by skin contact
- Flammable
- Extremely dangerous to bees. Do not apply to crops in flower or to those in which bees are actively foraging. Do not apply when flowering weeds are present
- Dangerous to fish or other aquatic life. Do not contaminate surface waters or ditches with chemical or used container
- Do not allow direct spray from vehicle mounted/drawn hydraulic sprayers to fall within 6 m, or from hand-held sprayers to within 2 m, of surface waters or ditches. Direct spray away from water
- Do not spray cereals after 1 Apr within 6 m of the edge of the crop
- Must not be applied to cereals if any product containing a pyrethroid insecticide or dimethoate has been sprayed after the start of ear emergence (GS 51)

Protective clothing/Label precautions

- A, C, H, J, K, M
- M03, R03c, R04b, R04d, R04e, R07d, U02, U05b, U08, U20a, C03, E01, E06b (7 d), E12b, E13b, E16, E30a, E31c, E34

Withholding period

- Harmful to livestock. Keep all livestock out of treated areas for at least 7 d following treatment

Latest application/Harvest Interval(HI)

- Before late milky ripe (GS 77) for cereals; before end of flowering for oilseed rape
- HI carrots 9 wk; sugar beet 8 wk; brassicas, lettuce, potatoes, peas, field beans 3 d;

Maximum Residue Levels (mg residue/kg food)

- see lambda-cyhalothrin entry

---

363 lenacil

A residual, soil-acting uracil herbicide for horticultural crops

See also chloridazon + lenacil

Products

1 Stefes Lenacil	Stefes	80% w/w	WP	07103
2 Venzar Flowable	DuPont	440 g/l	SC	06907

Uses

Annual dicotyledons in FARM WOODLAND *(off-label)* [2]. Annual dicotyledons in FODDER BEET, MANGELS, RED BEET, SUGAR BEET [1, 2]. Annual meadow grass in FARM

WOODLAND *(off-label)* [2]. Annual meadow grass in FODDER BEET, MANGELS, RED BEET, SUGAR BEET [1, 2].

Notes

Efficacy

- Weeds controlled as they germinate, not after emergence. Rain or irrigation necessary after spraying to activate chemical. Effectiveness may be reduced by dry conditions
- May be used in beet crops on more highly organic soils with incorporation [1]

Crop Safety/Restrictions

- Maximum number of treatments 1 per crop for red and fodder beet, mangels, spinach, spinach beet and strawberries; 1 per crop pre- and 3 per crop post-emergence for sugar beet
- Apply to beet crops pre-drilling incorporated [1], pre- or post-emergence [1, 2]
- Recommended in tank mixture with other beet herbicides. See label for details
- See label for soil type restrictions
- Heavy rain after application may cause damage to beet and maiden strawberries
- Do not apply other residual herbicides within 3 mth before or after spraying
- Do not treat crops under stress from drought, low temperatures, nutrient deficiency, pest or disease attack, or waterlogging
- Succeeding crops should not be planted or sown for at least 4 mth after treatment following ploughing to at least 150 mm

Special precautions/Environmental safety

- Irritating to eyes, skin and respiratory system
- Harmful to fish or other aquatic life. Do not contaminate surface waters or ditches with chemical or used container

Protective clothing/Label precautions

- A, C
- R04a, R04b, R04c, U05a, U08, U19, U20a, C03, E01, E13c, E30a, E31a, E32a

Latest application/Harvest Interval(HI)

- Pre-emergence for red beet, fodder beet, spinach, spinach beet, mangels; before leaves meet over rows for sugar beet

Approval

- Off-label approval unlimited for use in farm woodland (OLA 1282/97)[2]

364 lenacil + phenmedipham

A contact and residual herbicide for use in sugar beet

Products

1 DUK 880	DuPont	440:114 g/l	KL	04121
2 Stefes 880	Stefes	440:114	KL	08914

Uses

Annual dicotyledons in SUGAR BEET.

Notes

Efficacy

- Apply at any stage of crop when weeds at cotyledon stage

- Germinating weeds controlled by root uptake for several weeks
- Best results achieved under warm, moist conditions on a fine, firm seedbed
- Treatment may be repeated up to a maximum of 3 applications on later weed flushes
- Residual activity may be reduced on highly organic soils or under very dry conditions
- Rain falling 1 h after application does not reduce activity

Crop Safety/Restrictions
- Maximum number of treatments 3 per crop
- Do not apply when temperature above or likely to exceed 21°C on day of spraying or under conditions of high light intensity
- Do not spray any crop under stress from drought, waterlogging, cold, wind damage or any other cause
- Heavy rain after application may reduce stand of crop particularly in very hot weather. In severe cases yield may be reduced
- Do not sow or plant any crop within 4 mth of treatment. In case of crop failure only sow or plant beet crops, strawberries or other tolerant horticultural crop within 4 mth

Special precautions/Environmental safety
- Irritating to eyes, skin and respiratory system
- Harmful to fish or other aquatic life. Do not contaminate surface waters or ditches with chemical or used container

Protective clothing/Label precautions
- A, C
- R04a, R04b, R04c, U05a, U08, U19, U20a, C03, E01, E13c, E26, E30a, E31b

---

365 lindane

See gamma-HCH

---

366 linuron

A contact and residual urea herbicide for various field crops

See also 2,4-DB + linuron + MCPA
chlorpropham + linuron

Products

1 Afalon	AgrEvo	450 g/l	SC	08186
2 Alpha Linuron 50 SC	Makhteshim	500 g/l	SC	06967
3 Ashlade Linuron FL	Ashlade	480 g/l	SC	06221
4 MSS Linuron 500	Mirfield	480 g/l	SC	08893
5 PBI Linuron Flowable	PBI	450 g/l	SC	02965
6 Stefes Linuron	Stefes	450 g/l	SC	07902
7 UPL Linuron 45% Flowable	United Phosphorus	450 g/l	SC	07435

Uses

Annual dicotyledons in SPRING CEREALS [2-4, 7]. Annual dicotyledons in CELERY [1, 5, 6]. Annual dicotyledons in PARSNIPS [1, 2, 5-7]. Annual dicotyledons in CARROTS, POTATOES [1-7]. Annual dicotyledons in PARSLEY [1-4, 6, 7]. Annual dicotyledons in CELERIAC *(off-label)*, GARLIC *(off-label)*, LEEKS *(off-label)*, ONIONS *(off-label)* [1]. Annual grasses in CELERIAC *(off-label)*, GARLIC *(off-label)*, LEEKS *(off-label)*, ONIONS *(off-label)* [1]. Annual meadow grass in SPRING CEREALS [2-4, 7]. Annual meadow grass in CELERY [1, 5, 6]. Annual meadow grass in PARSNIPS [1, 2, 5-7]. Annual meadow grass in CARROTS, POTATOES [1-7]. Annual meadow grass in PARSLEY [1-4, 6, 7]. Annual weeds in CELERIAC *(off-label)*, GARLIC *(off-label)*, HERBS *(off-label)*, LEEKS *(off-label)*, ONIONS *(off-label)* [5]. Black bindweed in SPRING CEREALS [2-4, 7]. Black bindweed in CELERY [1, 5, 6]. Black bindweed in PARSNIPS [1, 2, 5-7]. Black bindweed in CARROTS, POTATOES [1-7]. Black bindweed in PARSLEY [1-4, 6, 7]. Chickweed in SPRING CEREALS [2-4, 7]. Chickweed in CELERY [1, 5, 6]. Chickweed in PARSNIPS [1, 2, 5-7]. Chickweed in CARROTS, POTATOES [1-7]. Chickweed in PARSLEY [1-4, 6, 7]. Corn marigold in CARROTS, PARSLEY, POTATOES, SPRING CEREALS [2-4]. Corn marigold in PARSNIPS [2]. Fat-hen in SPRING CEREALS [2-4, 7]. Fat-hen in CELERY [1, 5, 6]. Fat-hen in PARSNIPS [1, 2, 5-7]. Fat-hen in CARROTS, POTATOES [1-7]. Fat-hen in PARSLEY [1-4, 6, 7]. Redshank in SPRING CEREALS [2-4, 7]. Redshank in CELERY [1, 5, 6]. Redshank in PARSNIPS [1, 2, 5-7]. Redshank in CARROTS, POTATOES [1-7]. Redshank in PARSLEY [1-4, 6, 7].

Notes

Efficacy

- Many weeds controlled pre-emergence or post-emergence to 2-3 leaf stage, some (annual meadow grass, mayweed) only susceptible pre-emergence. See label for details
- Best results achieved by application to firm, moist soil of fine tilth
- Little residual effect on soil with more than 10% organic matter

Crop Safety/Restrictions

- Maximum number of treatments 1 per crop for celery, potatoes, cereals; 2 per crop for carrots, parsley, parsnips
- Apply only pre-crop emergence [7]
- Drill spring cereals at least 3 cm deep and apply pre-emergence of crop or weeds
- Do not use on undersown cereals or crops grown on Sands or Very Light soils or soils heavier than Sandy Clay Loam or with more than 10% organic matter
- Apply to potatoes well earthed up to a rounded ridge pre-crop emergence and do not cultivate after spraying. May be used in tank mix with glufosinate-ammonium [1]
- Apply within 4 d of drilling carrots or post-emergence after first rough leaf stage.
- Do not apply to emerged crops of carrots, parsnips or parsley under stress
- Recommendations for parsnips, parsley and celery vary. See label for details
- Potatoes, carrots and parsnips may be planted at any time after application. Lettuce should not be grown within 12 mth of treatment. Transplanted brassicas may be grown from 3 mth after treatment

Special precautions/Environmental safety

- Irritating to eyes and skin [3, 5, 7]
- Irritating to respiratory system [3, 5]
- Harmful to fish or other aquatic life. Do not contaminate surface waters or ditches with chemical or used container
- Do not allow direct spray from vehicle mounted/drawn hydraulic sprayers to fall within 6 m of surface waters or ditches. Direct spray away from water
- Do not apply by hand-held sprayers

Protective clothing/Label precautions

- A, C, H
- M03, U09a, E34 [7]; R04a, R04b [3-5, 7]; R04c [3-5]; U05a, U19, C03, E01 [2-5, 7]; U08 E16, E30a [1-7]; U20a [5, 6]; U20b [1-4, 7]; E07 (5 mth) [3, 4]; E13b [1-4, 6, 7]; E13c, E31a [5]; E26 [1, 6, 7]; E31b [1, 3, 4, 6, 7]; E32a [2]

Latest application/Harvest Interval(HI)

- Pre-emergence for cereals and linseed; 2 rough leaf stage for carrots, parsnips, parsley and celery; pre- or up to 40% emergence for potatoes (products differ, see label for details).
- HI onions, garlic, herbs 7 d (pre-emergence [5]), leeks 21 d

Approval

- Off-label approval to Jun 2001 for use on onions, leeks, garlic, celeriac (OLA 1660/98)[1]; unlimited for use on herbs (OLA 0209/93)[5]

367 linuron + trifluralin

A residual pre-emergence herbicide for use in winter cereals

Products

1 Linnet	PBI	106:192 g/l	EC	01555
2 Trifluron	United Phosphorus	250:480 g/l	KL	07854

Uses

Annual dicotyledons in WINTER BARLEY, WINTER WHEAT. Annual grasses in WINTER BARLEY, WINTER WHEAT. Annual meadow grass in WINTER BARLEY, WINTER WHEAT. Perennial ryegrass in WINTER BARLEY, WINTER WHEAT. Rough meadow grass in WINTER BARLEY, WINTER WHEAT.

Notes

Efficacy

- Effective against weeds germinating near soil surface. Best results achieved by application to fine, firm, moist seedbed free of clods, crop residues or established weeds
- Effectiveness reduced by long dry periods after application or on waterlogged soil
- Do not use on peaty soils or where organic matter exceeds 10%
- With autumn application residual effects normally last until spring but further herbicide treatment may be needed on thin or backward crops
- With loose seedbed on lighter soils results improved by rolling after drilling

Crop Safety/Restrictions

- Maximum number of treatments 1 per crop
- Apply without incorporation as soon as possible after drilling and before crop emergence, within 3 d on early drilled crops
- Do not treat durum wheat or undersown crops
- Crop seed must be well covered, minimum depth specified varies from 12 to 30 mm
- Do not use on soils classed as Sands. Do not harrow after treatment
- Only spring barley or spring wheat may be sown within 6 mth of application

Special precautions/Environmental safety

- Irritating to eyes, skin and respiratory system
- Flammable [2]
- Harmful [1], dangerous [2] to fish or other aquatic life. Do not contaminate surface waters or ditches with chemical or used container
- Do not allow direct spray from vehicle mounted/drawn hydraulic sprayers to fall within 6 m of surface waters or ditches. Direct spray away from water
- Do not apply by hand-held sprayers

Protective clothing/Label precautions

- A, C, H [1, 2]
- R04a, R04b, R04c, U05a, U08, U13, U19, U20b, C03, E01, E16, E26, E30a, E31b [1, 2]; R07d, E13b [2]; U02, U14, U15, E13c, E27 [1]

Latest application/Harvest Interval(HI)
- Pre-emergence of crop

368 magnesium phosphide

A phosphine generating compound used to control insect pests in stored commodities

Products	Degesch Plates	Rentokil	56% w/w	GE	07603

Uses Insect pests in STORED GRAIN, STORED PLANT MATERIAL.

Notes

Efficacy
- Product acts as fumigant by releasing poisonous hydrogen phosphide gas on contact with moisture in the air
- Place plates on the floor or wall of the building or on the surface of the commodity. Exposure time varies depending on temperature and pest. See label

Special precautions/Environmental safety
- Very toxic in contact with skin, by inhalation and if swallowed
- Highly flammable
- Do not contaminate surface waters or ditches with chemical or used container
- Only to be used by professional operators trained in the use of magnesium phosphide and familiar with the precautionary measures to be observed. See label for full precautions
- Keep in original container, tightly closed, in a safe place, under lock and key
- Do not allow plates or their spent residues to come into contact with food other than raw cereal grains
- Do not bulk spent plates and residues. Spontaneous ignition could result

Protective clothing/Label precautions
- A
- M04, R01a, R01b, R01c, R07c, U05b, U07, U13, U19, U20a, E02, E07, E15, E29, E30b, E32b

Withholding period
- Keep livestock out of treated areas

369 malathion

A broad-spectrum contact organophosphorus insecticide and acaricide

Products	Malathion 60	United Phosphorus	600 g/l	EC	08018

Uses Aphids in APPLES, APRICOTS, CHERRIES, CURRANTS, FLOWERS, GOOSEBERRIES, LETTUCE, NECTARINES, PEACHES, PEARS, PLUMS, POTATOES, ROSES, STRAWBERRIES, TOMATOES. Bryobia mites in APPLES, PEARS. Codling moth in APPLES,

PEARS. Flea beetles in WATERCRESS *(off-label)*. Leafhoppers in APPLES, FLOWERS, PEARS, ROSES, TOMATOES. Mealy bugs in FLOWERS, ROSES, TOMATOES. Midges in WATERCRESS *(off-label)*. Mustard beetle in WATERCRESS *(off-label)*. Red spider mites in APPLES, CHERRIES, CURRANTS, GOOSEBERRIES, PEARS, PLUMS. Sawflies in GOOSEBERRIES. Scale insects in ROSES. Suckers in APPLES, PEARS. Thrips in FLOWERS, ONIONS, ROSES, TOMATOES. Whitefly in FLOWERS, ROSES, TOMATOES. Woolly aphid in APPLES, PEARS.

Notes

Efficacy

- Spray when pest first seen and repeat as necessary, usually at 7-14 d intervals
- Number and timing of sprays vary with crop and pest. See label for details
- Repeat spray routinely for scale insect and whitefly control in glasshouses
- Where aphids, spider mites or pear suckers resistant to malathion occur control is unlikely to be satisfactory
- Addition of wetting agent recommended against certain pests eg woolly aphids and leaf-curling aphids

Crop Safety/Restrictions

- Maximum number of treatments 2 per season for apples, pears; not specified for other crops
- Do not use on antirrhinums, crassula, ferns, fuchsias, gerberas, petunias, pileas, sweet peas or zinnias

Special precautions/Environmental safety

- This product contains an anticholinesterase organophosphorus compound. Do not use if under medical advice not to work with such compounds
- Dangerous to bees. Do not apply to crops in flower or to those in which bees are actively foraging. Do not apply when flowering weeds are present
- Extremely dangerous to fish or other aquatic life. Do not contaminate surface waters or ditches with chemical or used container

Protective clothing/Label precautions

- A, H, M
- M01, U08, U19, U20b, E12d, E13a, E26, E30a, E31a, E34

Latest application/Harvest Interval(HI)

- HI 4 d; crops for processing 7 d

Approval

- Off-label approval unlimited for use on watercress beds (OLA 2719/97)[1]

Maximum Residue Levels (mg residue/kg food)

- cereals 8; garlic, onions, shallots, tomatoes, peppers, aubergines, cucumbers, gherkins, courgettes, cauliflowers, Brussels sprouts, head cabbages, lettuce, beans (with pods), peas (with pods), celery, leeks, rhubarb, cultivated mushrooms 3; citrus fruits 2; pome fruits, apricots, peaches, nectarines, plums, grapes, cane fruits, bilberries, cranberries, currants, gooseberries, bananas, carrots, horseradish, parsnips, parsley root, salsify, swedes, turnips, potatoes 0.5

370 maleic hydrazide

A pyridazinone plant growth regulator suppressing sprout and bud growth

See also dicamba + maleic hydrazide + MCPA

Products

1 Fazor	Dow	60% w/w	SG	05558
2 Mazide 25	Vitax	250 g/l	SL	02067
3 Regulox K	RP Amenity	250 g/l	SL	05405
4 Royal MH 180	Mirfield	180 g/l	SL	07840
5 Source II	Chiltern	80% w/w	WB	08314

Uses

Growth retardation in HEDGES [2]. Growth suppression in AMENITY GRASS [2-4]. Growth suppression in GOLF COURSES, ROAD VERGES [2, 4]. Growth suppression in GRASS NEAR WATER [3, 4]. Growth suppression in INDUSTRIAL SITES, ROADSIDE GRASS, WASTE GROUND [3]. Growth suppression in AREAS AROUND FARM BUILDINGS [4]. Sprout suppression in ONIONS [1, 2, 4, 5]. Sprout suppression in POTATOES [1]. Sucker inhibition in AMENITY TREES AND SHRUBS [2, 3]. Volunteer suppression in POTATOES [1]. Volunteer suppression in WARE POTATOES [5].

Notes

Efficacy

- Apply to grass at any time of yr when growth active, best when growth starting in Apr-May and repeated when growth recommences
- Uniform coverage and 8 h dry weather necessary for effective results
- Mow 2-3 d before and 5-10 d after spraying for best results. Need for mowing reduced for up to 6 wk
- Apply to onions at 10% necking and not later than 50% necking stage when the tops are still green. Rain or irrigation within 24 h may reduce effectiveness
- Apply to second early or maincrop potatoes at least 3 wk before haulm destruction. Accurate timing essential but rain or irrigation within 24 h may reduce effectiveness. See label for details [1]
- To control suckers wet trunks thoroughly, especially pruned and basal bud areas [2]
- Spray hawthorn hedges in full leaf, privet 7 d after cutting, in Apr-May [2]

Crop Safety/Restrictions

- Maximum number of treatments 2 per yr on amenity grass, land not intended for cropping and land adjacent to aquatic areas; 1 per crop on onions, potatoes and on or around tree trunks
- Do not apply in drought or when crops are suffering from pest, disease or herbicide damage. Do not treat fine turf or grass seeded less than 8 mth previously
- May be applied to grass along water courses but not to water surface [3, 4]
- Do not treat onions more than 2 wk before maturing. Treated onions may be stored until Mar but must then be removed to avoid browning
- Only use on potatoes of good keeping quality; not on seed, first earlies or crops grown under polythene
- Do not treat potatoes within 3 wk of applying a haulm desiccant or if temperatures above 26°C
- Avoid drift onto nearby vegetables, flowers or other garden plants

Special precautions/Environmental safety

- Irritant. Risk of serious damage to eyes [5]
- Only apply to grass not to be used for grazing
- Do not use treated water for irrigation purposes within 3 wk of treatment or until concentration in water falls below 0.02 ppm [3, 4]
- Maximum permitted concentration in water 2 ppm [3, 4]
- Do not contaminate surface waters or ditches with chemical or used container
- Do not dump surplus product in water or ditch bottoms [3, 4]
- Do not apply by knapsack sprayer [1]

Protective clothing/Label precautions

- A [1, 5]; C [5]; H [1]
- R04, R04d, U02, U05a, U13, U14, U15, U19, U20a, U22, C03, E01 [5]; U08, E32a [1]; U20b [1, 4]; U20c [2, 3]; E15, E30a [1-5]; E19, E21 [3, 4]; E26 [3, 5]; E31a [2, 4]; E31b [3]

Latest application/Harvest Interval(HI)

- 3 wk before haulm destruction for potatoes; before 50% necking for onions
- HI onions 4 d [2-5], 7 d [1]; potatoes 3 wk

Approval

- Approved for use on grass near water [3, 4]. See notes in Section 1 on use of herbicides in or near water

Maximum Residue Levels (mg residue/kg food)

- ware potatoes 50; garlic, onions, shallots 10; all other products 1

371 mancozeb

A protective dithiocarbamate fungicide for potatoes and other crops

See also benalaxyl + mancozeb
carbendazim + mancozeb
chlorothalonil + mancozeb
cymoxanil + mancozeb
cymoxanil + mancozeb + oxadixyl
dimethomorph + mancozeb

Products

1 Agrichem Mancozeb 80	Agrichem	80% w/w	WP	06354
2 Ashlade Mancozeb FL	Ashlade	412 g/l	SC	06226
3 Dithane 945	PBI	80% w/w	WP	04017
4 Dithane Dry Flowable	PBI	75% w/w	WG	04255
5 Headland Kor Flo	Headland	455 g/l	SC	08019
6 Headland Quell	Headland	455 g/l	SC	08317
7 Headland Zebra Flo	Headland	455 g/l	SC	07442
8 Headland Zebra WP	Headland	80% w/w	WP	07441
9 Helm 75 WG	PBI	75% w/w	WG	08309
10 Karamate Dry Flo	Rohm & Haas	75% w/w	WG	04250
11 Landgold Mancozeb 80 W	Landgold	80% w/w	WP	06507
12 Luxan Mancozeb Flowable	Luxan	455 g/l	SC	06812
13 Manex II	Agrichem	455 g/l	SC	07637
14 Manzate 200	DuPont	80% w/w	WP	01281
15 Manzate 200 PI	DuPont	80% w/w	WP	07209
16 Opie WP	PBI	80% w/w	WP	08301
17 Penncozeb WDG	Mirfield	75% w/w	WG	07833
18 Portman Mandate 80	Portman	80% w/w	WP	06320
19 Stefes Mancozeb DF	Stefes	75% w/w	WP	08010
20 Stefes Mancozeb WP	Stefes	80% w/w	WP	07655
21 Tariff 75 WG	PBI	75% w/w	WG	08308
22 Unicrop Mancozeb 80	Unicrop	80% w/w	WP	07451

Uses

Black spot in ROSES [10, 22]. Blight in POTATOES [1-9, 11-22]. Brown rust in SPRING WHEAT, WINTER WHEAT [3-9, 12, 16, 18, 21, 22]. Brown rust in SPRING BARLEY, WINTER BARLEY [18]. Brown rust in BARLEY, WHEAT [11, 17]. Currant leaf spot in BLACKCURRANTS, GOOSEBERRIES [10]. Downy mildew in LETTUCE [10, 22]. Downy mildew in PROTECTED LETTUCE [10]. Downy mildew in WINTER OILSEED RAPE [1, 3-9, 12, 16, 19-22]. Fire in TULIPS [10, 22]. Net blotch in SPRING BARLEY [18]. Net blotch in WINTER BARLEY [1, 5-9, 12, 16, 18, 20-22]. Net blotch in BARLEY [11, 17]. Ray blight in CHRYSANTHEMUMS [10, 22]. Rhynchosporium in WINTER BARLEY [1, 5-9, 12, 16, 20-22]. Rhynchosporium in BARLEY [17]. Rust in PELARGONIUMS, ROSES [10, 22]. Rust in PROTECTED CARNATIONS [10]. Rust in CARNATIONS [22]. Scab in APPLES [10, 17, 18]. Scab in PEARS [10]. Septoria diseases in SPRING WHEAT, WINTER WHEAT [11]. Septoria diseases in BARLEY, WHEAT [17]. Septoria leaf spot in SPRING WHEAT [3-9, 12, 16, 20-22]. Septoria leaf spot in WINTER WHEAT [1, 3-9, 12, 16, 20-22]. Septoria in SPRING WHEAT, WINTER WHEAT [18]. Sooty moulds in SPRING WHEAT, WINTER WHEAT [1, 3-9, 16, 19-22]. Sooty moulds in WINTER BARLEY [1, 5-9, 12, 16, 20-22]. Sooty moulds in WHEAT [11, 12, 17]. Sooty moulds in BARLEY [11, 17]. Yellow rust in SPRING WHEAT [18, 20]. Yellow rust in SPRING BARLEY, WINTER BARLEY [18]. Yellow rust in WINTER WHEAT [1, 3-9, 12, 16, 18, 20-22]. Yellow rust in BARLEY, WHEAT [11, 17].

FOR FULL CONDITIONS OF USE ALWAYS READ THE PRODUCT LABEL

Notes

Efficacy

- Mancozeb is a protectant fungicide and will give moderate control, suppression or reduction of the cereal diseases listed but in many cases mixture with carbendazim is essential to achieve satisfactory results. See labels for details
- Apply to cereals before any disease is well established. See label for recommended timing
- May be recommended for suppression or control of mildew in cereals depending on product and tank mix. See label for details
- Apply to potatoes before haulm meets across rows (usually mid-Jun) or at earlier blight warning, and repeat every 7-14 d depending on conditions and product used (see label)
- May be used on potatoes up to desiccation of haulm
- On oilseed rape apply as soon as disease develops between cotyledon and 5-leaf stage (GS 1,0-1,5)

Crop Safety/Restrictions

- Maximum number of treatments varies with crop and product used -check labels for details
- Check labels for minimum interval that must elapse between treatments
- Apply to cereals from 4-leaf stage to before early milk stage (GS 71). Recommendations vary, see labels for details
- Treat winter oilseed rape before 6 true leaf stage (GS 1,6) and before 31 Dec
- On protected lettuce only 2 post-planting applications of mancozeb or of any combination of products containing EBDC fungicide (mancozeb, maneb, thiram, zineb) either as a spray or a dust are permitted within 2 wk of planting out and none thereafter.
- Avoid treating wet cereal crops or those suffering from drought or other stress

Special precautions/Environmental safety

- Irritating to eyes and skin [2, 11, 14, 17, 18]
- Irritating to respiratory system [1-4, 8-11, 14-22]
- May cause sensitization by skin contact [1, 3-10, 12, 13, 15, 16, 19-22]
- Harmful to fish or other aquatic life. Do not contaminate surface waters or ditches with chemical or used container
- Keep away from fire and sparks [1, 3, 14, 16, 22]
- Use contents of opened container as soon as possible and do not store opened containers until following season. If product becomes wet, effectiveness may be reduced and inflammable vapour produced [1, 3, 14, 16, 22]

Protective clothing/Label precautions

- A [1-22]; C [11, 14, 18]; D [11, 15, 18, 22]; H [11, 18]
- R04 [5-7, 13]; R04a, R04b [2, 11, 14, 17, 18]; R04c [1-4, 8-11, 14-22]; R04e [1, 3-10, 12, 13, 15, 16, 19-22]; U05a, C03, E13c, E30a, E32a [1-22]; U08 [1-6, 9-14, 16-18, 20-22]; U13 [17, 18]; U19 [2, 14, 17, 18]; U20a [1-4, 7-9, 11, 14-16, 18, 20-22]; U20b [5, 6, 10, 12, 13, 17, 19]; E01 [1-4, 7-22]; E26 [2, 5, 6, 12-14, 19, 20, 22]; E29 [5, 6, 13, 22]; E34 [2, 5, 6, 12, 14]

Latest application/Harvest Interval(HI)

- Before early milk stage (GS 73) [1, 7, 8, 12] for cereals; before 6 true leaf stage and before 31 Dec for winter oilseed rape.
- HI potatoes 7 d; outdoor lettuce 14 d; protected lettuce 21 d; apples, pears, blackcurrants, gooseberries 4 wk

Approval

- Approved for aerial application on potatoes [1-4, 7, 8, 12, 14-17, 19, 22]; on cereals [2]. See notes in Section 1
- Approval expiry: 30 Sep 99 [14]

Maximum Residue Levels (mg residue/kg food)

- hops 25; lettuces, herbs 5; oranges, apricots, peaches, nectarines, grapes, strawberries, peppers, aubergines 2; cherries, plums 1; garlic, onions, shallots, cucumbers 0.5; celeriac, witloof 0.2; tree nuts, oilseeds (except rape seed), tea 0.1; bilberries, cranberries, wild berries, blackberries, loganberries, miscellaneous fruit, root and tuber vegetables, horseradish, Jerusalem artichokes, parsley root, sweet potatoes, swedes, turnips, yams, spinach, beet leaves, watercress, asparagus, cardoons, fennel, globe artichokes, rhubarb, mushrooms, potatoes, maize, rice, animal products 0.05

372 mancozeb + metalaxyl

A systemic and protectant fungicide for potatoes and some other crops

Products

1 Fubol 58 WP	Ciba Agric.	48:10% w/w	WP	00927
2 Fubol 58 WP	Novartis	48:10% w/w	WP	08534
3 Fubol 75 WP	Ciba Agric.	67.5:7.5% w/w	WP	03462
4 Fubol 75 WP	Novartis	67.5:7.5% w/w	WP	08409
5 Osprey 58 WP	Ciba Agric.	48:10% w/w	WP	05717
6 Osprey 58 WP	Novartis	48:10% w/w	WP	08428

Uses

Blight in POTATOES [3-6]. Cavity spot in CARROTS [1, 2]. Downy mildew in ONIONS *(off-label)* [1-4]. Downy mildew in SHALLOTS *(off-label)* [4]. Fungus diseases in FLOWERHEAD BRASSICAS *(off-label)*, LEAF BRASSICAS *(off-label)* [1, 2]. Fungus diseases in KOHLRABI *(off-label)*, SWEDES *(off-label)*, TURNIPS *(off-label)* [1]. Phytophthora fruit rot in APPLES *(off-label)* [1-4]. Root rot in RASPBERRIES *(off-label)* [1, 2]. White blister in BRASSICAS *(off-label)* [1-4]. White blister in RADISHES *(off-label)* [3, 4]. White tip in LEEKS *(off-label)*[1, 3, 4].

Notes

Efficacy

- Apply as protectant spray on potatoes immediately risk of blight in district or as crops begin to meet in (not across) rows and repeat every 10-21 d according to blight risk (every 10-14 d on irrigated crops)
- Use for first part of spray programme to mid-Aug. Later use a protectant fungicide (preferably tin-based) up to complete haulm destruction or harvest
- Do not treat potato crops showing active blight infection or a second potato crop in the same field in the same season
- Recommended for potato blight control in Northern Ireland but specific Northern Ireland label must be consulted [5, 6]
- Best results on carrots and parsnips achieved by application to damp soil. Do not use where these crops have been grown on same site during either of 2 previous years [1, 2]
- Do not apply on potatoes for 2-3 h before rainfall or when raining

Crop Safety/Restrictions

- Maximum number of treatments 5 per crop (including other phenylamide-based fungicides) for potatoes [3-6]; 3 per crop on Brussels sprouts, cabbages, cauliflowers, broccoli, calabrese, onions, leeks; 2 per crop on oilseed rape, field and broad beans, apple orchards; 1 per crop on carrots

Special precautions/Environmental safety

- Irritating to skin, eyes and respiratory system

- Harmful to fish or other aquatic life. Do not contaminate surface waters or ditches with chemical or used container

Protective clothing/Label precautions

- A
- R04a, R04b, R04c, U05a, U08, U20a, C02, C03, E01, E13c, E30a, E32a

Latest application/Harvest Interval(HI)

- Within 6 wk of drilling for carrots and parsnips; pre-planting for seedling brassicas.
- HI early and maincrop potatoes 1 d [3, 4]; early potatoes 7 d, maincrop potatoes 14 d [5, 6]; Brussels sprouts, onions 14 d; apples 4 wk; carrots 8 wk; raspberries 3 mth

Approval

- Approved for aerial application on potatoes [3-5]. See notes in Section 1
- Off-label approval unlimited for use as drench on raspberries (OLA 1189/96)[1], (OLA 1379/97)[2]; to Jul 2000 for use on outdoor radishes (OLA 0997/95)[3], (OLA 1707/97)[4]; unlimited for use on protected radishes (OLA 0996/95)[3], (OLA 1706/97)[4]; unlimited for use on apples (orchard floor) (OLA 0195/97)[3], (OLA 1704/97)[4], (OLA 0197/97)[1], (OLA 1701/97)[2]; to Feb 2001 for use on cauliflowers, broccoli, calabrese, cabbages, kale, onions, shallots, leeks, Brussels sprouts (OLA 0196/97)[3], (OLA 1705/97)[4], (OLA 0198/97)[1]; to Feb 2001 for use on a range of flowerhead and leaf brassicas, bulb onions, shallots (OLA 1702/97)[2]

Maximum Residue Levels (mg residue/kg food)

- see mancozeb and metalaxyl entries

---

373 mancozeb + metalaxyl-M

A systemic and protectant fungicide mixture

Products

Fubol Gold	Novartis	64:4% w/w	WB	08812

Uses Blight in POTATOES.

Notes

Efficacy

- Commence blight programme before risk of infection occurs as crops begin to meet along the rows and repeat every 10-14 d according to blight risk. Do not exceed a 14 d interval between sprays
- Monitor sprayed crops and stop using any phenylamide product if active blight is identified. Switch to a tin-based product
- Complete the blight programme using a protectant fungicide starting no later than 10 d after the last phenylamide spray
- After treating early potatoes destroy and remove any remaining haulm after harvest to minimise blight pressure on neighbouring maincrop potatoes

Crop Safety/Restrictions

- Maximum number of treatments 5 per crop (including all other phenylamide fungicides)

Special precautions/Environmental safety

- Dangerous to fish or other aquatic life. Do not contaminate surface waters or ditches with chemical or used container
- Do not harvest crops for human consumption for at least 7 d after final application

Protective clothing/Label precautions
- A
- U05a, U08, U20b, U22, C02 (7 d), E13b, E26, E30a, E32a, E34

Latest application/Harvest Interval(HI)
- HI 7 d

Maximum Residue Levels (mg residue/kg food)
- see mancozeb and metalaxyl entries

374 mancozeb + ofurace

A systemic and protectant fungicide mixture for potatoes

Products

Patafol	Mirfield	67:5.8% w/w	WP	07397

Uses

Blight in POTATOES, POTATOES GROWN FOR SEED *(off-label)*.

Notes

Efficacy
- Apply to potatoes at first blight warning or just before foliage meets in the row and repeat at 10-14 d intervals
- Do not apply after the end of Aug
- Complete blight programme up to haulm destruction with organotin products
- Always apply before the crop is infected
- Do not apply if rain likely within 2-3 h of treatment
- After rain or irrigation wait until foliage is dry before spraying

Crop Safety/Restrictions
- Maximum number of treatments 5 per crop

Special precautions/Environmental safety
- Irritating to eyes, skin and respiratory system
- Harmful to fish or other aquatic life. Do not contaminate surface waters or ditches with chemical or used container

Protective clothing/Label precautions
- A, C
- R04a, R04b, R04c, U05a, U08, U19, U20b, C02 (7 d), C03, E01, E13c, E29, E30a, E32a

Latest application/Harvest Interval(HI)
- HI 7 d

Approval
- Off-label approval unlimited for use on protected crops of potatoes for seed (OLA 1599/96)[1]

Maximum Residue Levels (mg residue/kg food)
- see mancozeb entry

375 mancozeb + oxadixyl

A systemic and contact protectant fungicide for potato blight control

Products

1 Recoil	Sandoz	56:10% w/w	WP	04039
2 Recoil	Novartis	56:10% w/w	WP	08483

Uses

Blight in POTATOES. Root rot in BLACKBERRIES *(off-label)*, CANE FRUIT *(off-label)*, DEWBERRIES *(off-label)*, LOGANBERRIES *(off-label)*, RASPBERRIES *(off-label)*.

Notes

Efficacy
- Commence treatment for potato blight early in season as soon as there is risk of infection
- In absence of a blight warning treatment should start before potatoes meet within row
- Repeat treatments every 10-14 d depending on degree of blight infection risk
- Use as a protectant treatment. Potatoes showing blight should not be treated
- Do not spray within 2-3 h of rainfall and only apply to dry foliage
- To complete blight spray programme after end of Aug make subsequent treatments up to haulm destruction with another fungicide, preferably fentin based

Crop Safety/Restrictions
- Maximum number of treatments 5 per crop (including other phenylamide-based fungicides) for potatoes; 2 per yr for raspberries
- Do not treat raspberries on light soils with less than 2% organic matter

Special precautions/Environmental safety
- Irritating to skin, eyes and respiratory system. May cause sensitization by skin contact
- Do not spray within 2 m of surface water courses (raspberries, blackberries, dewberries, loganberries and cane fruit only)

Protective clothing/Label precautions
- A, C
- R04a, R04b, R04c, R04e, U02, U05a, U11, U12, U19, U20b, C03, E01, E15, E30a, E32a

Latest application/Harvest Interval(HI)
- Before end of Aug.
- HI zero

Approval
- Approved for aerial application on potatoes (see notes in Section 1) [1, 2]. Consult firm for details
- Off-label approval unlimited for use as a soil drench on raspberries, blackberries, dewberries, loganberries, cane fruit (other than wild) (OLA 0750/95)[1], (OLA 1711/97)[2]

Maximum Residue Levels (mg residue/kg food)
- see mancozeb entry

376 mancozeb + propamocarb hydrochloride

A systemic and contact protectant fungicide for potato blight control

Products

Tattoo	AgrEvo	301.6:248 g/l	SC	07293

Uses Blight in POTATOES.

Notes **Efficacy**

- Commence treatment as soon as there is risk of blight infection
- In the absence of a warning treatment should start before the crop meets within the row
- Repeat treatments every 10-14 d depending on degree of infection risk
- Irrigated crops should be regarded as at high risk and treated every 10 d
- Do not use once blight has become readily visible
- Do not spray when rainfall is imminent and apply only to dry foliage
- To complete the spray programme after the end of Aug, make subsequent treatments up to haulm destruction with a protectant fungicide, preferably fentin based.

Crop Safety/Restrictions

- Maximum number of treatments 5 per crop

Special precautions/Environmental safety

- Irritant. May cause sensitization by skin contact
- Harmful to fish or other aquatic life. Do not contaminate surface waters or ditches with chemical or used container

Protective clothing/Label precautions

- A, H
- R04e, U05a, U08, U14, U19, U20a, C03, E01, E13c, E30a, E31b

Latest application/Harvest Interval(HI)

- Before end of Aug.
- HI 14 d

Maximum Residue Levels (mg residue/kg food)

- see mancozeb entry

---

377 maneb

A protectant dithiocarbamate fungicide

See also carbendazim + chlorothalonil + maneb
carbendazim + maneb
carbendazim + maneb + sulfur
carbendazim + maneb + tridemorph
copper oxychloride + maneb + sulfur
fentin acetate + maneb
ferbam + maneb + zineb

Products

1 Agrichem Maneb 80	Agrichem	80% w/w	WP	05474
2 Ashlade Maneb Flowable	Ashlade	480 g/l	SC	06477
3 Luxan Maneb 80	Luxan	80% w/w	WP	06570
4 Maneb 80	Rohm & Haas	80% w/w	WP	01276
5 Mazin	Unicrop	75% w/w	WP	06061
6 Unicrop Maneb 80	Unicrop	80% w/w	WP	06926
7 Unicrop Maneb FL	Unicrop	480 g/l	SC	08025
8 X-Spor SC	United Phosphorus	480 g/l	SC	08077

Uses

Blight in POTATOES [1-8]. Brown rust in SPRING BARLEY, WINTER BARLEY [1, 4, 6, 8]. Brown rust in WINTER WHEAT [1, 3, 4, 6, 8]. Brown rust in BARLEY [3]. Didymella in TOMATOES *(off-label)* [6]. Eyespot in SPRING BARLEY, WINTER BARLEY, WINTER WHEAT [8]. Net blotch in SPRING BARLEY, WINTER BARLEY [4]. Powdery mildew in WINTER WHEAT [1, 6, 8]. Powdery mildew in SPRING BARLEY, WINTER BARLEY [1, 6]. Powdery scab in POTATOES *(tuber borne)* [5]. Rhynchosporium in SPRING BARLEY, WINTER BARLEY [1, 6, 8]. Rhynchosporium in BARLEY [3]. Septoria diseases in WINTER WHEAT [8]. Septoria in WINTER WHEAT [1, 3, 4, 6]. Sooty moulds in WINTER WHEAT [1, 3, 6, 8]. Sooty moulds in SPRING BARLEY, WINTER BARLEY [1, 6]. Yellow rust in SPRING BARLEY [4, 6, 8]. Yellow rust in WINTER BARLEY [1, 4, 6, 8]. Yellow rust in WINTER WHEAT [1, 3, 4, 6, 8]. Yellow rust in BARLEY [3].

Notes

Efficacy

- Maneb is a protectant fungicide and will give moderate control, suppression or reduction of the cereal diseases listed but in many cases mixture with carbendazim is essential to achieve satisfactory results - see labels for details
- Apply to potatoes before blight infection occurs, at blight warning or before haulms meet in row and repeat every 10-14 d
- May also be used as dust application for reduction in tuber-borne powdery scab (not in situations of high disease pressure). Apply during planting using the "sandwich" technique in the planter hopper. See label for details [5]
- Most effective on cereals as a preventative treatment before disease established, an early application at about first node stage (GS 31), a late application after flag leaf emergence (GS 37) and during ear emergence before watery ripe stage (GS 71)
- Apply to tomatoes at first sign of disease and repeat every 7-14 d

Crop Safety/Restrictions

- Maximum number of treatments normally 2 per crop in cereals (check labels); 1 per crop as potato tuber dust [5]
- Do not apply if frost or rain expected, if crop wet or suffering from drought or physical or chemical stress
- A minimum of 10 d must elapse between applications to potatoes

Special precautions/Environmental safety

- Harmful if swallowed [3, 6-8]
- Irritating to respiratory system [1, 4], eyes and skin [2, 3, 5-8]
- May cause sensitization by skin contact [1, 4, 5]
- Dangerous (harmful [4]) to fish or other aquatic life. Do not contaminate surface waters or ditches with chemical or used container [1, 5]
- Keep away from fire and sparks [1, 3-5]
- Use contents of opened container as soon as possible and do not store opened containers until following season. If product becomes wet, effectiveness may be reduced and inflammable vapour produced [1, 3-5]
- Use on manual or semi-automatic potato planters not recommended because of the dusty nature of the exercise and the irritating nature of maneb [5]

Protective clothing/Label precautions

- A [1-8]; C [5, 8]; D [1, 6]; F [5]
- M03 [8]; R03c, U04a [3, 7, 8]; R04a, R04b [2, 3, 5, 7, 8]; R04c, C03, E01 [1-8]; R04e [1, 4-6]; U05a [1-7]; U08, E30a [1-4, 6-8]; U09a, U19 [5]; U20a [2-4, 7]; U20b [1, 5, 6, 8]; C02 (7 d), E13c, E34, [4]; C02 [3, 7]; E13b [1, 5, 6]; E15 [2, 3, 7, 8]; E26 [6, 8]; E29 [4, 6]; E31b [2, 8]; E32a [1, 3-7]; E34 [2]

Latest application/Harvest Interval(HI)

- Before grain milky-ripe (GS 71) for winter wheat [8]; before flag leaf sheath opening (GS 45) for barley [8]; full ear emergence (GS 59) for cereals [6]; before early milk stage (GS 71) for cereals [1]; at planting for potato tuber dust [5]
- HI potatoes, outdoor crops 7 d; protected crops 2 d

Approval

- Approved for aerial application on potatoes [1, 2, 4-6]. See notes in Section 1.
- Off-label approval unlimited for use on protected tomatoes (OLA 2811/96)[6]

Maximum Residue Levels (mg residue/kg food)

- hops 25; lettuces, herbs 5; oranges, apricots, peaches, nectarines, grapes, strawberries, peppers, aubergines 2; cherries, plums 1; garlic, onions, shallots, cucumbers 0.5; celeriac, witloof 0.2; tree nuts, oilseeds (except rape seed), tea 0.1; bilberries, cranberries, wild berries, blackberries, loganberries, miscellaneous fruit, root and tuber vegetables, horseradish, Jerusalem artichokes, parsley root, sweet potatoes, swedes, turnips, yams, spinach, beet leaves, watercress, asparagus, cardoons, fennel, globe artichokes, rhubarb, mushrooms, potatoes, maize, rice, animal products 0.05

378 MCPA

A translocated phenoxyacetic herbicide for cereals and grassland

See also 2,4-D + dichlorprop + MCPA + mecoprop
2,4-DB + linuron + MCPA
2,4-DB + MCPA
benazolin + 2,4-DB + MCPA
bentazone + MCPA + MCPB
clopyralid + 2,4-D + MCPA
clopyralid + diflufenican + MCPA
clopyralid + fluroxypyr + MCPA
dicamba + dichlorprop + ferrous sulfate + MCPA
dicamba + dichlorprop + MCPA
dicamba + maleic hydrazide + MCPA
dicamba + MCPA + mecoprop-P

Products

1 Agricorn 500	FCC	500 g/l	SL	00055
2 Agritox 50	Nufarm	500 g/l	SL	07400
3 Agroxone 50	Mirfield	485 g/l	SL	08345
4 Atlas MCPA	Atlas	480 g/l	SL	07717
5 Barclay Meadowman II	Barclay	500 g/l	SL	08525
6 BASF MCPA Amine 50	BASF	500 g/l	SL	00209
7 Campbell's MCPA 50	MTM Agrochem.	500 g/l	SL	00381
8 Cirsium	Mirfield	485 g/l	SL	08729
9 Headland Spear	Headland	500 g/l	SL	07115
10 HY-MCPA	Agrichem	500 g/l	SL	06293
11 Luxan MCPA 500	Luxan	500 g/l	SL	07470
12 MSS MCPA 50	Mirfield	500 g/l	SL	01412
13 Quad MCPA 50%	Quadrangle	500 g/l	SL	01669
14 Stefes Phenoxylene 50	Stefes	500 g/l	SL	07612

FOR FULL CONDITIONS OF USE ALWAYS READ THE PRODUCT LABE

Uses Annual dicotyledons in WINTER RYE [1, 6, 7, 9, 11, 13]. Annual dicotyledons in ESTABLISHED GRASSLAND [1-13]. Annual dicotyledons in BARLEY, OATS, WHEAT [1-9, 11-14]. Annual dicotyledons in UNDERSOWN BARLEY, UNDERSOWN OATS, UNDERSOWN WHEAT [6]. Annual dicotyledons in SPRING BARLEY, SPRING OATS, SPRING WHEAT, WINTER BARLEY, WINTER WHEAT [10]. Annual dicotyledons in GRASS SEED CROPS [2, 5, 11, 13, 14]. Annual dicotyledons in RYE [2, 5, 14]. Annual dicotyledons in LINSEED [2, 5]. Annual dicotyledons in NEWLY SOWN GRASS [11]. Annual dicotyledons in ASPARAGUS, PERMANENT PASTURE [14]. Annual dicotyledons in YOUNG LEYS [4, 10, 14]. Annual dicotyledons in ROAD VERGES, UNDERSOWN CEREALS [3, 8, 12]. Annual dicotyledons in LAND NOT INTENDED TO BEAR VEGETATION [3, 8]. Annual dicotyledons in MANAGED AMENITY TURF [3]. Annual dicotyledons in LAWNS, TURF [8, 12]. Buttercups in ESTABLISHED GRASSLAND [1, 3, 4, 6-10, 12, 13]. Buttercups in PERMANENT PASTURE [14]. Charlock in WINTER RYE [1, 6, 7, 9, 11, 13]. Charlock in BARLEY, OATS, WHEAT [1-9, 11-14]. Charlock in UNDERSOWN BARLEY, UNDERSOWN OATS, UNDERSOWN WHEAT [6]. Charlock in SPRING BARLEY, SPRING OATS, SPRING WHEAT, WINTER BARLEY, WINTER WHEAT [10]. Charlock in ESTABLISHED GRASSLAND [2, 5, 11]. Charlock in RYE [2, 5, 14]. Charlock in GRASS SEED CROPS, NEWLY SOWN GRASS [11]. Charlock in YOUNG LEYS [4, 10]. Charlock in UNDERSOWN CEREALS [3, 8]. Creeping thistle in ROAD VERGES [3, 8, 12]. Creeping thistle in LAND NOT INTENDED TO BEAR VEGETATION [3, 8]. Creeping thistle in MANAGED AMENITY TURF [3]. Creeping thistle in LAWNS, TURF [8, 12]. Daisies in ROAD VERGES [12]. Daisies in MANAGED AMENITY TURF [3]. Daisies in LAWNS, TURF [8, 12]. Dandelions in ESTABLISHED GRASSLAND [1, 3, 4, 6-8, 10, 12, 13]. Dandelions in PERMANENT PASTURE [14]. Dandelions in MANAGED AMENITY TURF [3]. Dandelions in LAWNS, TURF [8, 12]. Docks in ESTABLISHED GRASSLAND [1, 3, 4, 6-10, 12, 13]. Docks in PERMANENT PASTURE [14]. Docks in ROAD VERGES [3, 8, 12]. Docks in LAND NOT INTENDED TO BEAR VEGETATION [3, 8]. Fat-hen in WINTER RYE [1, 6, 7, 9, 11, 13]. Fat-hen in BARLEY, OATS, WHEAT [1-9, 11-14]. Fat-hen in UNDERSOWN BARLEY, UNDERSOWN OATS, UNDERSOWN WHEAT [6]. Fat-hen in SPRING BARLEY, SPRING OATS, SPRING WHEAT, WINTER BARLEY, WINTER WHEAT [10]. Fat-hen in ESTABLISHED GRASSLAND [2, 5, 11]. Fat-hen in RYE [2, 5, 14]. Fat-hen in GRASS SEED CROPS, NEWLY SOWN GRASS [11]. Fat-hen in YOUNG LEYS [4, 10]. Fat-hen in UNDERSOWN CEREALS [3, 8]. Hemp-nettle in BARLEY, OATS, WHEAT [1-9, 11-14]. Hemp-nettle in WINTER RYE [1, 3, 6-9, 11, 13]. Hemp-nettle in UNDERSOWN BARLEY, UNDERSOWN OATS, UNDERSOWN WHEAT [6]. Hemp-nettle in SPRING BARLEY, SPRING OATS, SPRING WHEAT, WINTER BARLEY, WINTER WHEAT [10]. Hemp-nettle in ESTABLISHED GRASSLAND [2, 5, 11]. Hemp-nettle in RYE [2, 5, 14]. Hemp-nettle in GRASS SEED CROPS, NEWLY SOWN GRASS [11]. Hemp-nettle in YOUNG LEYS [4, 10]. Hemp-nettle in UNDERSOWN CEREALS [3, 8]. Perennial dicotyledons in WINTER RYE [1, 6, 7, 9, 11, 13]. Perennial dicotyledons in ESTABLISHED GRASSLAND [1-13]. Perennial dicotyledons in BARLEY, WHEAT [1-9, 11-14]. Perennial dicotyledons in YOUNG LEYS [10, 14]. Perennial dicotyledons in SPRING BARLEY, SPRING OATS, SPRING WHEAT, WINTER BARLEY, WINTER WHEAT [10]. Perennial dicotyledons in OATS [2-9, 11-14]. Perennial dicotyledons in RYE [2, 5, 14]. Perennial dicotyledons in GRASS SEED CROPS [11, 13, 14]. Perennial dicotyledons in NEWLY SOWN GRASS [11]. Perennial dicotyledons in ASPARAGUS, PERMANENT PASTURE [14]. Perennial dicotyledons in ROAD VERGES [3, 8, 12]. Perennial dicotyledons in LAND NOT INTENDED TO BEAR VEGETATION [3, 8]. Perennial dicotyledons in MANAGED AMENITY TURF [3]. Perennial dicotyledons in LAWNS, TURF [8, 12]. Ragwort in ESTABLISHED GRASSLAND [9]. Rushes in ESTABLISHED GRASSLAND [9]. Thistles in PERMANENT PASTURE [14]. Thistles in ESTABLISHED GRASSLAND [3, 4, 6-10, 12, 13]. Thistles in LAND NOT INTENDED TO BEAR VEGETATION, ROAD VERGES [3, 8]. Wild radish in WINTER RYE [1, 6, 7, 9, 11, 13]. Wild radish in BARLEY, OATS, WHEAT [1-9, 11-14]. Wild radish in UNDERSOWN BARLEY, UNDERSOWN OATS, UNDERSOWN WHEAT [6]. Wild radish in SPRING BARLEY, SPRING OATS, SPRING WHEAT, WINTER BARLEY, WINTER WHEAT [10]. Wild radish in ESTABLISHED GRASSLAND [2, 5, 11]. Wild radish in RYE [2, 5, 14]. Wild radish in GRASS

SEED CROPS, NEWLY SOWN GRASS [11]. Wild radish in YOUNG LEYS [4, 10]. Wild radish in UNDERSOWN CEREALS [3, 8].

Notes

Efficacy

- Best results achieved by application to weeds in seedling to young plant stage under good growing conditions when crop growing actively
- Spray perennial weeds in grassland before flowering. Most susceptible growth stage varies between species. See label for details
- Do not spray during cold weather, drought, if rain or frost expected or if crop wet

Crop Safety/Restrictions

- Maximum number of treatments normally 1 per crop or yr except grass (2 per yr) for some products. See label
- Apply to winter cereals in spring from fully tillered, leaf sheath erect stage to before first node detectable (GS 31)
- Apply to spring barley and wheat from 5-leaves unfolded (GS 15), to oats from 1-leaf unfolded (GS 11) to before first node detectable (GS 31)
- Apply to cereals undersown with grass after grass has 2-3 leaves unfolded
- Do not use on cereals before undersowing
- Do not roll, harrow or graze for a few days before or after spraying; see label
- Recommendations for crops undersown with legumes vary. Red clover may withstand low doses after 2-trifoliate leaf stage, especially if shielded by taller weeds but white clover is more sensitive. See label for details
- Do not use on grassland where clovers are an important part of the sward
- Apply to grass seed crops from 2-3 leaf stage to 5 wk before head emergence
- Do not use on any crop suffering from stress or herbicide damage
- Do not direct drill brassicas or legumes within 6 wk of spraying grassland
- Avoid spray drift onto nearby susceptible crops

Special precautions/Environmental safety

- Harmful in contact with skin [3, 8, 10-12, 14]
- Harmful if swallowed [1-14]
- Irritating to skin. Risk of serious damage to eyes [6, 11]
- Harmful to fish or other aquatic life. Do not contaminate surface waters or ditches with chemical or used container [1, 6]

Protective clothing/Label precautions

- A, C [1-9, 11-14]
- M03, R03c, U05a, U08, C03, E01, E07 [1-14]; R03a [3, 8, 10-12, 14]; R04b, R04d [6, 11] U02 [11]; U20a [1, 4, 7, 9, 10, 13, 14]; U20b [2, 3, 5, 6, 8, 11, 12]; E13c [1, 6]; E15 [2-5, 7-14]; E26 [1-3, 5-8, 10-13]; E30a [1, 2, 4-7, 9-11, 13, 14]; E31a [1, 2, 7, 10, 13]; E31b [3-5, 8, 11, 12, 14]; E31c [6]; E34 [1-13]

Withholding period

- Keep livestock out of treated areas for at least 2 wk and until foliage of poisonous weeds such as ragwort has died and become unpalatable

Latest application/Harvest Interval(HI)

- Before 1st node detectable (GS 31) for cereals

379 MCPA + MCPB

A translocated herbicide for undersown cereals and grassland

Products

1 Bellmac Plus	United Phosphorus	38:262 g/l	SL	07521
2 MSS MCPB + MCPA	Mirfield	25:275 g/l	SL	01413
3 Trifolex-Tra	Cyanamid	34:216 g/l	SL	07147
4 Tropotox Plus	Unicrop	37.5:262.5 g/l	SL	06156

Uses

Annual dicotyledons in CEREALS [2, 4]. Annual dicotyledons in ESTABLISHED GRASSLAND [2, 3]. Annual dicotyledons in CLOVER SEED CROPS, DIRECT-SOWN SEEDLING CLOVERS, DREDGE CORN CONTAINING PEAS, ESTABLISHED LEYS, PERMANENT PASTURE [4]. Annual dicotyledons in PEAS [3]. Annual dicotyledons in SAINFOIN [1, 2, 4]. Annual dicotyledons in UNDERSOWN CEREALS [1-4]. Annual dicotyledons in LEYS [1-3]. Perennial dicotyledons in CEREALS [2, 4]. Perennial dicotyledons in ESTABLISHED GRASSLAND [2, 3]. Perennial dicotyledons in CLOVER SEED CROPS, DIRECT-SOWN SEEDLING CLOVERS, DREDGE CORN CONTAINING PEAS, ESTABLISHED LEYS, PERMANENT PASTURE [4]. Perennial dicotyledons in PEAS [3]. Perennial dicotyledons in SAINFOIN [1, 2, 4]. Perennial dicotyledons in UNDERSOWN CEREALS [1-4]. Perennial dicotyledons in LEYS [1-3].

Notes

Efficacy

- Best results achieved by application to weeds in seedling to young plant stage under good growing conditions when crop growing actively. Spray perennials when adequate leaf surface before flowering. Retreatment often needed in following year
- Spray leys and sainfoin before crop provides cover for weeds
- Rain, cold or drought may reduce effectiveness

Crop Safety/Restrictions

- Maximum number of treatments 1 per crop or yr
- Apply to cereals from 2-expanded leaf stage to before jointing (GS 12-30) and, where undersown, after 1-trifoliate leaf stage of clover
- Apply to direct-sown seedling clover after 1-trifoliate leaf stage
- Apply to mature white clover for fodder at any stage. Do not spray red clover after flower stalk has begun to form
- Do not spray clovers for seed [2, 3]
- Apply to sainfoin after first compound leaf stage [1], second compound leaf [2]
- Do not spray peas [1, 2]. Only spray peas in tank mix with Fortrol from 4-leaf stage to before bud visible (GS 104-201) [3]
- Do not roll or harrow for a few days before or after spraying [1-3]
- Avoid spray drift onto nearby sensitive crops

Special precautions/Environmental safety

- Harmful to fish or other aquatic life. Do not contaminate surface waters or ditches with chemical or used container

Protective clothing/Label precautions

- U05a, E01, E07 [2-4]; U08, E13c, E26, E30a, E31b [1-4]; U20a, C03 [2, 3]; U20b [1, 4]; E07 (2 wk), E34 [1]

Withholding period

- Keep livestock out of treated areas until foliage of any poisonous weeds such as ragwort has died and become unpalatable

Latest application/Harvest Interval(HI)
- Before first node detectable (GS 31) for cereals

380 MCPA + mecoprop-P

A translocated selective herbicide for amenity grass

Products

Greenmaster Extra	Scotts	0.49:0.29 % w/w	GR	07594

Uses

Annual dicotyledons in AMENITY GRASS. Perennial dicotyledons in AMENITY GRASS.

Notes

Efficacy
- Apply from Apr to Sep, when weeds growing actively and have large leaf area available for chemical absorption
- Granules contain NPK fertiliser to encourage grass growth
- Do not apply when heavy rain expected or during prolonged drought. Irrigate after 1-2 d unless rain has fallen
- Do not mow within 2-3 d of treatment

Crop Safety/Restrictions
- Maximum total dose 105 g/sq m per yr. The total amount of mecoprop-P applied in a single yr must not exceed the maximum total dose approved for any single product for use on turf
- Avoid contact with cultivated plants
- Do not use first 4 mowings as compost or mulch unless composted for 6 mth
- Do not treat newly sown or turfed areas for at least 6 mth
- Do not reseed bare patches for 8 wk after treatment

Special precautions/Environmental safety
- Harmful to fish or other aquatic life. Do not contaminate surface waters or ditches with chemical or used container

Protective clothing/Label precautions
- A, C, H, M
- U20a, E07 (2 wk), E13c, E15, E30a, E32a

Withholding period
- Keep livestock out of treated areas for at least 2 wk and until foliage of any poisonous weeds such as ragwort has died and become unpalatable

381 MCPB

A translocated phenoxybutyric herbicide

See also bentazone + MCPA + MCPB
bentazone + MCPB
MCPA + MCPB

Products

1 Bellmac Straight	United Phosphorus	400 g/l	SL	07522
2 Tropotox	Unicrop	400 g/l	SL	06179

Uses

Annual dicotyledons in BLACKCURRANTS, PEAS, WHITE CLOVER SEED CROPS [1, 2]. Annual dicotyledons in LEYS, UNDERSOWN CEREALS [1]. Perennial dicotyledons in BLACKCURRANTS, PEAS, WHITE CLOVER SEED CROPS [1, 2]. Perennial dicotyledons in LEYS, UNDERSOWN CEREALS [1].

Notes

Efficacy
- Best results achieved by spraying young seedling weeds in good growing conditions
- Best results on perennials by spraying before flowering
- Effectiveness may be reduced by rain within 12 h, by very cold or dry conditions

Crop Safety/Restrictions
- Maximum number of treatments 1 per crop
- Apply to undersown cereals from 2-leaves unfolded to first node detectable (GS 12-31), and after first trifoliate leaf stage of clover
- Red clover seedlings may be temporarily damaged but later growth is normal
- Apply to white clover seed crops in Mar to early Apr, not after mid-May, and allow 3 wk before cutting and closing up for seed
- Apply to peas from 3-6 leaf stage but before flower bud detectable (GS 103-201). Consult PGRO (see Appendix 2) for information on susceptibility of cultivars
- Do not use on beans, vetches or established red clover
- Apply to blackcurrants in Aug after harvest and after shoot growth ceased but while bindweed still growing actively; direct spray onto weeds as far as possible
- Do not roll or harrow for 7 d before or after treatment
- Avoid drift onto nearby sensitive crops

Special precautions/Environmental safety
- Harmful in contact with skin and if swallowed
- Harmful to fish or other aquatic life. Do not contaminate surface waters or ditches with chemical or used container

Protective clothing/Label precautions
- A, C [1, 2]
- M03, R03a, R03c, U05a, U08, U20b, C03, E01, E07, E13c, E26, E30a, E31b, E34 [1, 2]; U19 [2]

Withholding period
- Keep livestock out of treated areas until foliage of any poisonous weeds such as ragwort has died and become unpalatable

Latest application/Harvest Interval(HI)
- First node detectable stage (GS 31) for cereals; before flower buds appear in terminal leaf (GS 201) for peas

382 mecoprop

A translocted phenoxypropionic herbicide for cereals and grassland

See also 2,4-D + dicamba + mecoprop
2,4-D + dichlorprop + MCPA + mecoprop
2,4-D + mecoprop
bifenox + isoproturon + mecoprop
dicamba + MCPA + mecoprop
dicamba + mecoprop
diclorprop + mecoprop

Products

1 Campbell's CMPP	MTM Agrochem.	570 g/l	SL	02918
2 Clenecorn II	FCC	570 g/l	SL	08880
3 Clovotox	RP Amenity	300 g/l	SL	05354
4 Luxan CMPP	Luxan	560 g/l	SL	06037

Uses

Annual dicotyledons in BARLEY, OATS, WHEAT [1, 2, 4]. Annual dicotyledons in YOUNG LEYS [1, 2]. Annual dicotyledons in TURF [3]. Annual dicotyledons in ESTABLISHED GRASSLAND [4]. Chickweed in BARLEY, OATS, WHEAT [1, 2, 4]. Chickweed in YOUNG LEYS [1, 2]. Chickweed in TURF [3]. Chickweed in ESTABLISHED GRASSLAND [4]. Cleavers in BARLEY, OATS, WHEAT [1, 2, 4]. Cleavers in YOUNG LEYS [1, 2]. Clovers in TURF [3]. Perennial dicotyledons in BARLEY, OATS, WHEAT [1, 2, 4]. Perennial dicotyledons in YOUNG LEYS [1, 2]. Perennial dicotyledons in TURF [3]. Perennial dicotyledons in ESTABLISHED GRASSLAND [4].

Notes

Efficacy

- Best results achieved by application to weeds in seedling to young plant stage growing actively in a strongly competing crop
- Do not spray when rain imminent or when growth hard from cold or drought. Rainfall within 8-12 h may reduce effectiveness
- For dock control in grassland cut in mid-summer and spray regrowth in Aug-Sep
- Spray turf weeds in spring or early summer when in active growth. More resistant perennials may need repeat treatment after 4-6 wk

Crop Safety/Restrictions

- Maximum number of treatments 1 per crop for cereals; 2 per yr for grassland and turf; 1 per yr for newly sown grass
- Range of timings recommended for cereals varies with product. See label for details
- Do not spray rye, crops undersown with legumes or crops about to be undersown
- If hard frosts occur within 3-4 wk of application to barley of low vigour on light soils crop may be scorched or stunted
- Do not roll or harrow within a few days before or after treatment
- Apply to grasses after 3-fully expanded leaves and 1 tiller stage
- Clovers present in sprayed grass will be severely damaged
- Do not direct-drill brassicas or legumes within 6 wk after application
- Do not use first 4 mowings for mulching. Compost for at least 6 mth before use
- Avoid drift onto nearby fruit trees or other susceptible crops

Special precautions/Environmental safety

- Harmful in contact with skin and if swallowed [1, 3-5]
- Irritating to eyes and skin [1, 3, 4]
- Avoid drift onto all broad-leaved plants outside the target area

Protective clothing/Label precautions

- A, C [1-4]; H, M [4]
- M03, R03a, R03c [1, 2, 4]; R04a [1, 3, 4]; R04b [1, 4]; R05a [1]; U02, E13c [4]; U05a, C03, E01, E07, E30a, E31b [1-4]; U08, U20b, E26 [3, 4]; U09a, U20a [1, 2]; E15 [1-3]; E34 [2, 4]

Withholding period

- Keep livestock out of treated areas until foliage of any poisonous weeds such as ragwort has died and become unpalatable

Latest application/Harvest Interval(HI)

- Varies between before first node detectable (GS 31) and before third node detectable (GS 33) for cereals. See product label

---

383 mecoprop-P

A translocated phenoxypropionic herbicide for cereals and grassland

See also 2,4-D + mecoprop-P
bromoxynil + ioxynil + mecoprop-P
carfentrazone-ethyl + mecoprop-P
dicamba + MCPA + mecoprop-P
dicamba + mecoprop-P
fluroxypyr + mecoprop-P
MCPA + mecoprop-P

Products

1 Clovotox	RP Amenity	142.5 g/l	SL	05354
2 Compitox Plus	Nufarm	600 g/l	SL	07930
3 Duplosan	BASF	600 g/l	SL	05889
4 Landgold Mecoprop-P 600	Landgold	600 g/l	SL	06461
5 MSS Optica	Mirfield	600 g/l	SL	04973
6 Standon Mecoprop-P	Standon	600 g/l	SL	05651

Uses

Annual dicotyledons in TURF [1]. Annual dicotyledons in ESTABLISHED GRASSLAND [4, 6]. Annual dicotyledons in LEYS [4]. Annual dicotyledons in GRASS SEED CROPS, GRASSLAND, YOUNG LEYS [2, 3, 5]. Annual dicotyledons in BARLEY, OATS, WHEAT [2-6]. Chickweed in TURF [1]. Chickweed in ESTABLISHED GRASSLAND [4, 6]. Chickweed in LEYS [4]. Chickweed in GRASS SEED CROPS, GRASSLAND, YOUNG LEYS [2, 3, 5]. Chickweed in BARLEY, OATS, WHEAT [2-6]. Cleavers in ESTABLISHED GRASSLAND [4, 6]. Cleavers in LEYS [4]. Cleavers in GRASS SEED CROPS, GRASSLAND, YOUNG LEYS [2, 3, 5]. Cleavers in BARLEY, OATS, WHEAT [2-6]. Clovers in TURF [1]. Perennial dicotyledons in TURF [1]. Perennial dicotyledons in ESTABLISHED GRASSLAND [4, 6]. Perennial dicotyledons in LEYS [4]. Perennial dicotyledons in GRASS SEED CROPS, GRASSLAND, YOUNG LEYS [2, 3, 5]. Perennial dicotyledons in BARLEY, OATS, WHEAT [2-6].

Notes

Efficacy

- Best results achieved by application to seedling weeds which have not been frost hardened, when soil warm and moist and expected to remain so for several days

- Do not spray during cold weather, periods of drought, if rain or frost expected or if crop wet

Crop Safety/Restrictions
- Maximum number of treatments normally 1 per crop for spring cereals and newly sown grass; 2 per crop for winter cereals and established grass. Check labels for details
- The total amount of mecoprop-P applied in a single yr must not exceed the maximum total dose approved for any single product for the crop/situation
- Spray winter cereals from 1 leaf stage in autumn up to and including first node detectable in spring (GS 10-31) or up to second node detectable (GS 32) if necessary. Apply to spring cereals from first fully expanded leaf stage (GS 11) but before first node detectable (GS 31)
- Use maximum rate of 2.3 l/ha only after crop has reached GS 30
- Spray cereals undersown with grass after grass starts to tiller
- Do not spray cereals undersown with clovers or legumes or to be undersown with legumes or grasses
- Spray newly sown grass leys when grasses have at least 3 fully expanded leaves and have begun to tiller. Any clovers will be damaged
- Do not spray grass seed crops within 5 wk of seed head emergence
- Do not spray crops suffering from herbicide damage or physical stress
- Do not roll or harrow for 7 d before or after treatment
- Avoid drift onto beans, beet, brassicas, fruit and other sensitive crops

Special precautions/Environmental safety
- Harmful if swallowed [3, 5, 6] and in contact with skin [2, 4]
- Irritating to skin [4] and eyes [2]
- Irritating to eyes [1]
- Risk of serious damage to eyes [3, 5, 6]
- Harmful to fish or other aquatic life. Do not contaminate surface waters or ditches with chemical or used container

Protective clothing/Label precautions
- A, C [1-6]; H [1, 3]; M [3]
- M03, R03c, E34 [2-6]; R03a, E15 [2, 4]; R04a [1, 2, 4]; R04b [4]; R04d [3, 5, 6]; U05a, U08, C03, E01, E07, E26, E30a [1-6]; U13, E31a [2]; U20a [3-6]; U20b [1, 2]; E13c [1, 3, 5, 6]; E31b [1, 4-6]; E31c [3]

Withholding period
- Keep livestock out of treated areas for at least 2 wk and until foliage of any poisonous weed, such as ragwort, has died and become unpalatable

Latest application/Harvest Interval(HI)
- Before 1st node detectable (GS 31) for spring cereals; before 3rd node detectable (GS 33) for winter cereals [3, 6]; before second node detectable (GS 32) for cereals [5]; 5 wk before emergence of seed head for grass seed crops [3, 5]

384 mefluidide

A sulfonanilide plant growth regulator for suppression of grass

Products	Embark Lite	Intracrop	24 g/l	SL	08749

FOR FULL CONDITIONS OF USE ALWAYS READ THE PRODUCT LABEL

Uses Growth suppression in AMENITY GRASS, MANAGED AMENITY TURF.

Notes **Efficacy**

- Apply when grass growth active in Apr/May and repeat when growth re-commences
- Grass growth suppressed for up to 8 wk
- Mow (not below 2 cm) 2-3 d before or 5-10 d after spraying
- Only apply to dry foliage but do not use in drought
- A rainfree period of 8 h is needed after spraying
- May be tank-mixed with suitable 2,4-D or mecoprop product for control of dicotyledons
- Delay application for 3-4 wk after application of high-nitrogen fertilizer

Crop Safety/Restrictions

- Maximum number of treatments 2 per yr
- Do not use on fine turf areas or on turf established for less than 4 mth
- Do not use on grass suffering from herbicide damage
- Treated turf may appear less dense and be temporarily discoloured but normal colour returns in 3-4 wk
- Do not reseed within 2 wk after application

Special precautions/Environmental safety

- Do not contaminate surface waters or ditches with chemical or used container
- Do not allow animals to graze treated areas

Protective clothing/Label precautions

- U05a, U20c, E01, E15, E26, E30a, E31b

---

385 mepiquat chloride

A quaternary ammonium plant growth regulator available only in mixtures

See also 2-chloroethylphosphonic acid + mepiquat chloride
chlormequat + mepiquat chloride
chlormequat +2-chloroethylphosphonic acid + mepiquat chloride

---

386 mercuric oxide

An inorganic mercury fungicide, all approvals for which were revoked in 1992. Mercuric oxide is banned under the EC "Prohibition Directive"

---

387 mercurous chloride

A soil applied inorganic mercury fungicide, all approvals for which were revoked in 1992. Mercurous chloride is banned under the EC "Prohibition Directive"

---

388 metalaxyl

A phenylamide (acylalanine) fungicide available only in mixtures

See also carbendazim + metalaxyl
chlorothalonil + metalaxyl
copper oxychloride + metalaxyl
mancozeb + metalaxyl

389 metalaxyl + thiabendazole + thiram

A protective fungicide seed treatment for peas and beans

See also thiabendazole

Products

1 Apron Combi FS	Ciba Agric.	233:120:100 g/l	FS	07203
2 Apron Combi FS	Novartis	233:120:100	FS	08386

Uses

Ascochyta in BEANS *(seed treatment)*, PEAS - ALL TYPES *(seed treatment)*. Damping off in BEANS *(seed treatment)*, PEAS - ALL TYPES *(seed treatment)*. Downy mildew in BEANS *(seed treatment)*, PEAS - ALL TYPES *(seed treatment)*.

Notes

Efficacy

- Apply through continuous flow seed treaters which should be calibrated before use

Crop Safety/Restrictions

- Ensure moisture content of treated seed satisfactory and store in a dry place
- Check calibration of seed drill with treated seed before drilling

Special precautions/Environmental safety

- Harmful if swallowed. Irritating to eyes
- Harmful to fish or other aquatic life. Do not contaminate surface waters or ditches with chemical or used container
- Do not use treated seed as food or feed
- Treated seed harmful to game and wildlife

Protective clothing/Label precautions

- A, C, D, H, M
- M03, R03c, R04a, U05a, U09a, U13, U20a, C03, E01, E13c, E30a, E31a, S01, S02, S04b, S05, S06

Maximum Residue Levels (mg residue/kg food)

- see metalaxyl and thiabendazole entries

390 metalaxyl + thiram

A systemic and protectant fungicide for use in lettuce

Products

1 Favour 600 SC	Ciba Agric.	100:500 g/l	SC	05842
2 Favour 600 SC	Novartis	100:500 g/l	SC	08405

Uses Downy mildew in LETTUCE, PROTECTED LETTUCE.

Notes

Efficacy

- On lettuce apply from 100% emergence and repeat at 14 d intervals
- Resistant strains of lettuce downy mildew may develop against which product may be ineffective. See label for details

Crop Safety/Restrictions

- Maximum number of treatments; 2 post-planting applications of thiram or any combination of products containing thiram or other EBDC compound (mancozeb, maneb, zineb) either as a spray or a dust are permitted on protected lettuce within 2 wk of planting out and none thereafter. On winter lettuce, if only thiram based products are being used, 3 applications may be made within 3 wk of planting out and none thereafter
- Maximum number of treatments 5 per crop for outdoor lettuce; 3 per yr on outdoor parsley
- Do not use as a soil block treatment

Special precautions/Environmental safety

- Harmful if swallowed. Irritating to eyes, skin and respiratory system
- Harmful to fish or other aquatic life. Do not contaminate surface waters or ditches with chemical or used container

Protective clothing/Label precautions

- A, C
- M03, R03c, R04a, R04b, R04c, U05a, U08, U20b, C02 (2 wk outdoor; 3 wk protected), C03, E01, E13c, E30a, E31b, E34

Latest application/Harvest Interval(HI)

- HI protected lettuce 3 wk; outdoor lettuce 2 wk; outdoor parsley 4 wk

Maximum Residue Levels (mg residue/kg food)

- see metalaxyl entry

---

391 metalaxyl-M

A phenylamide systemic fungicide

See also mancozeb + metalaxyl-M

Products	SL 567A	Novartis	480 g/l	EC	08811

Uses Cavity spot in CARROTS.

Notes

Efficacy

- Best results achieved when applied to damp soil
- Efficacy may be reduced in prolonged dry weather. Irrigation immediately before or after application may be necessary
- Results may not be satisfactory on soils with high organic matter content
- Control of cavity spot on carrots overwintered in the ground or lifted in winter may be lower than expected

Crop Safety/Restrictions

- Maximum number of treatments 1 per crop
- Do not use where carrots have been grown on the same site for either of the previous two yrs

- Consult before use on crops intended for processing

Special precautions/Environmental safety
- Harmful if swallowed
- Irritating to eyes
- May cause sensitization by skin contact
- Harmful to fish or other aquatic life. Do not contaminate surface waters or ditches with chemical or used container

Protective clothing/Label precautions
- A, C, H
- M03, R03c, R04a, R04e, U02, U04a, U05a, U10, U20b, C03, E01, E13c, E26, E29, E30a, E31b, E34

Latest application/Harvest Interval(HI)
- 6 wk after drilling for carrots

Maximum Residue Levels (mg residue/kg food)
- see metalaxyl entry

392 metaldehyde

A molluscicide bait for controlling slugs and snails

Products

1 Devcol Morgan 6% Metaldehyde Slug Killer	Nehra	6% w/w	PT	07422
2 Doff Agricultural Slug Killer with Animal Repellent	Doff Portland	6% w/w	PT	06058
3 Doff Horticultural Slug Killer Blue Mini Pellets	Doff Portland	3% w/w	PT	05688
4 Doff Metaldehyde Slug Killer Mini Pellets	Doff Portland	6% w/w	PT	00741
5 Escar-Go 6	Chiltern	6% w/w	PT	06076
6 Hardy	Chiltern	6% w/w	PT	06948
7 Luxan 9363	Luxan	6% w/w	PT	07359
8 Luxan Metaldehyde	Luxan	6% w/w	PT	06564
9 Metarex Green	De Sangosse	6% w/w	PT	08131
10 Metarex RG	De Sangosse	6% w/w	PT	06754
11 Mifaslug	FCC	6% w/w	PT	01349
12 Molotov	Chiltern	3% w/w	PT	08295
13 Optimol	PBI	4% w/w	PT	09061
14 PBI Slug Pellets	PBI	6% w/w	PT	01558
15 Regel	De Sangosse	6% w/w	PT	08155
16 Superflor 6% Metaldehyde Slug Killer Mini Pellets	CMI	6% w/w	PT	05453
17 Tripart Mini Slug Pellets	Tripart	6% w/w	PT	02207

Uses Slugs in BRASSICAS *(seed admixture)* [4, 14]. Slugs in CEREALS *(seed admixture)* [4]. Slugs in WINTER WHEAT *(seed admixture)* [14]. Slugs in CEREALS [2, 5, 6, 8, 12]. Slugs in GRASSLAND [2, 8]. Slugs in FRUIT CROPS, VEGETABLES [1, 3-7, 9-17]. Slugs in

FOR FULL CONDITIONS OF USE ALWAYS READ THE PRODUCT LABEL

ORNAMENTALS, PROTECTED CROPS [1, 3, 4, 7, 9-11, 13-17]. Slugs in FIELD CROPS [1-17]. Snails in BRASSICAS *(seed admixture)*, CEREALS *(seed admixture)* [4]. Snails in CEREALS [2, 5, 6, 8, 12]. Snails in GRASSLAND [2, 8]. Snails in FRUIT CROPS, VEGETABLES [1, 3-7, 9-17]. Snails in ORNAMENTALS, PROTECTED CROPS [1, 3, 4, 7, 9-11, 13-17]. Snails in FIELD CROPS [1-17].

Notes

Efficacy

- Apply pellets by hand, fiddle drill, fertilizer distributor, by air (check label) or in admixture with seed. See labels for rates and timing.
- Best results achieved from an even spread of granules applied during mild, damp weather when slugs and snails most active. May be applied in standing crops
- Varieties of oilseed rape low in glucosinalates can be more acceptable to slugs than "single low" varieties and control may not be as good
- Do not apply when rain imminent or water glasshouse crops within 4 d of application
- To prevent slug build up apply at end of season to brassicas and other leafy crops
- To reduce tuber damage in potatoes apply twice in Jul and Aug

Special precautions/Environmental safety

- Dangerous to game, wild birds and animals
- Harmful to fish or other aquatic life. Do not contaminate surface waters or ditches with chemical or used container
- Product contains proprietary cat and dog deterrent [2, 3, 7, 9, 10, 13-15]
- Take care to avoid lodging of pellets in the foliage when making late applications to edible crops

Protective clothing/Label precautions

- A, H [7-10, 12, 13, 15]; J [8, 13]
- U20a, E31b, E34 [1, 2, 9-11, 13-15, 17]; U20b [5, 6, 16]; U20c [3, 4, 7, 8, 12]; E01, E30a [1, 2, 5-17]; E05a (7 d), E10a, E13c, E32a [1-17]; E29 [3-6, 12]

Withholding period

- Keep poultry out of treated areas for at least 7 d

Approval

- Approved for aerial application on all crops [2, 2-4, 10, 11, 13, 14, 16, 17]. See notes in Section 1

393 metamitron

A contact and residual triazinone herbicide for use in beet crops

See also ethofumesate + metamitron + phenmedipham

Products

1 Goltix 90	Bayer	90% w/w	WG	08654
2 Goltix Flowable	Bayer	700 g/l	SC	08986
3 Goltix WG	Bayer	70% w/w	WG	02430
4 Landgold Metamitron	Landgold	70% w/w	WG	06287
5 Standon Metamitron	Standon	70% w/w	WG	07885
6 Stefes Metamitron	Stefes	70% w/w	WG	05821
7 Tripart Accendo	Tripart	70% w/w	WG	06110

Uses

Annual dicotyledons in ASPARAGUS *(off-label)*, OUTDOOR LEAF HERBS *(off-label)* [3]. Annual dicotyledons in FODDER BEET, MANGELS, RED BEET, SUGAR BEET [1-7]. Annual grasses in FODDER BEET, MANGELS, RED BEET, SUGAR BEET [1-7]. Annual meadow grass in FODDER BEET, MANGELS, RED BEET, SUGAR BEET [1-7]. Fat-hen in FODDER BEET, MANGELS, RED BEET, SUGAR BEET [1-7].

Notes

Efficacy

- Low dose programme (LDP). Apply post-weed emergence as a fine spray, including adjuvant oil, timing treatment according to weed emergence and size. See label for details and for recommended tank mixes and sequential treatments. On mineral soils the LDP should be preceeded by pre-drilling or pre-emergence treatment
- Use up to 5 LDP sprays on mineral soils (up to 3 where also used pre-emergence) and up to 6 sprays on organic soils
- Traditional application. Apply either pre-drilling before final cultivation with incorporation to 8-10 cm, or pre-crop emergence at or soon after drilling into firm, moist seedbed to emerged weeds from cotyledon to first true leaf stage
- On emerged weeds at or beyond 2-leaf stage addition of adjuvant oil advised
- Up to 3 post-emergence sprays may be used on soils with over 10% organic matter
- For control of wild oats and certain other weeds, tank mixes with other herbicides or sequential treatments are recommended. See label for details

Crop Safety/Restrictions

- Maximum total dose equivalent to three full dose treatments
- Using traditional method post-crop emergence on mineral soils do not apply before first true leaves have reached 1 cm long
- Crop tolerance may be reduced by stress caused by growing conditions, effects of pests, disease or other pesticides, nutrient deficiency etc
- Only sugar beet, fodder beet or mangels may be drilled within 4 mth after treatment. Winter cereals may be sown in same season after ploughing, provided 16 wk passed since last treatment

Special precautions/Environmental safety

- Harmful: If swallowed [2]
- Do not contaminate surface waters or ditches with chemical or used container

Protective clothing/Label precautions

- A [2]
- M03, R03c, C03, E01, E26 [2]; U20a [1, 3, 4, 6, 7]; U20b [2, 5]; E15, E30a [1-7]; E32a [2, 4, 5]; E34 [1-3, 6, 7]

Approval

- Off-label approval unlimited for use on outdoor leafy herbs (OLA 0802/95)[3]; unlimited for use on outdoor asparagus (OLA 2415/96)[3]

394 metam-sodium

A methyl isothiocyanate producing sterilant for glasshouse, nursery and outdoor soils

Products

1 Metham Sodium 400	United Phosphorus	400 g/l	SL	08051
2 Sistan	Unicrop	380 g/l	SL	01957

Uses Dutch elm disease in ELM TREES *(off-label)* [2]. Nematodes in GLASSHOUSE SOILS, NURSERY SOILS, OUTDOOR SOILS, POTTING SOILS [2]. Nematodes in CHRYSANTHEMUMS, ORNAMENTALS, OUTDOOR TOMATOES *(Jersey)*, PROTECTED CARNATIONS, PROTECTED CROPS, TOMATOES [1]. Soil insects in CHRYSANTHEMUMS, ORNAMENTALS, OUTDOOR TOMATOES *(Jersey)*, PROTECTED CARNATIONS, PROTECTED CROPS, TOMATOES [1]. Soil pests in GLASSHOUSE SOILS, NURSERY SOILS, OUTDOOR SOILS, POTTING SOILS [2]. Soil-borne diseases in GLASSHOUSE SOILS, NURSERY SOILS, OUTDOOR SOILS, POTTING SOILS [2]. Soil-borne diseases in CHRYSANTHEMUMS, ORNAMENTALS, OUTDOOR TOMATOES *(Jersey)*, PROTECTED CARNATIONS, PROTECTED CROPS, TOMATOES [1]. Weed seeds in GLASSHOUSE SOILS, NURSERY SOILS, OUTDOOR SOILS, POTTING SOILS [2]. Weed seeds in CHRYSANTHEMUMS, ORNAMENTALS, OUTDOOR TOMATOES *(Jersey)*, PROTECTED CARNATIONS, PROTECTED CROPS, TOMATOES [1].

Notes

Efficacy

- Product acts by breaking down in contact with soil to release methyl isothiocyanate
- Apply to glasshouse soils as a drench, inject undiluted to 20 cm at 30 cm intervals and seal with water or apply to surface, rotavate and seal
- After treatment allow sufficient time (several weeks) for residues to dissipate and aerate soil by forking. Time varies with season. See label for details
- Apply to outdoor soils by drenching or sub-surface application
- Apply when soil temperatures exceed 7°C, preferably above 10°C, between 1 Apr and 31 Oct. Soil must be of fine tilth and adequate moisture content
- Product may also be used by mixing into potting soils
- Avoid using in equipment incorporating natural rubber parts
- Use in control programme against dutch elm disease to sever root grafts which can transmit disease from tree to tree. See OLA notice for details [2]

Crop Safety/Restrictions

- No plants must be present during treatment. Do not treat glasshouses within 2 m of growing crops. Do not plant until soil is entirely free of fumes
- Cress test advised to check for absence of residues. See label for details

Special precautions/Environmental safety

- Harmful in contact with skin and if swallowed [2]
- Irritating to eyes, skin and respiratory system

Protective clothing/Label precautions

- A, C [1, 2]; D [1]; H, M [1, 2]
- M03, R03a, R03c, U20b, E02 (24 h), E34 [2]; R04a, R04b, R04c, U02, U04a, U05a, U08, U19, C03, E01, E15, E30a, E31a [1, 2]; U20a [1]

Latest application/Harvest Interval(HI)

- Pre-planting of crop

Approval

- Off-label approval unlimited for use around elm trees infected with Dutch elm disease (OLA 0592/93)[2]

395 metazachlor

A residual anilide herbicide for use in brassicas, nurseries and forestry

Products

1 Barclay Metaza	Barclay	500 g/l	SC	07354
2 Butisan S	BASF	500 g/l	SC	00357
3 Landgold Metazachlor 50	Landgold	500 g/l	SC	08062
4 Standon Metazachlor 50	Standon	500 g/l	SC	05581

Uses

Annual dicotyledons in SPRING OILSEED RAPE, WINTER OILSEED RAPE [2-4]. Annual dicotyledons in FARM FORESTRY, FORESTRY, HARDY ORNAMENTAL NURSERY STOCK, HONESTY *(off-label)*, KOHLRABI *(off-label)*, NURSERY FRUIT TREES AND BUSHES, ORNAMENTALS [2]. Annual dicotyledons in BROCCOLI, BRUSSELS SPROUTS, CABBAGES, CALABRESE, CAULIFLOWERS, SWEDES, TURNIPS [1-4]. Annual dicotyledons in OILSEED RAPE [1, 2]. Annual grasses in FARM FORESTRY, FORESTRY, KOHLRABI *(off-label)* [2]. Annual meadow grass in SPRING OILSEED RAPE, WINTER OILSEED RAPE [2-4]. Annual meadow grass in HARDY ORNAMENTAL NURSERY STOCK, NURSERY FRUIT TREES AND BUSHES, ORNAMENTALS [2]. Annual meadow grass in BROCCOLI, BRUSSELS SPROUTS, CABBAGES, CALABRESE, CAULIFLOWERS, SWEDES, TURNIPS [1-4]. Annual meadow grass in OILSEED RAPE [1, 2]. Blackgrass in SPRING OILSEED RAPE, WINTER OILSEED RAPE [2-4]. Blackgrass in HARDY ORNAMENTAL NURSERY STOCK, NURSERY FRUIT TREES AND BUSHES, ORNAMENTALS [2]. Blackgrass in BROCCOLI, BRUSSELS SPROUTS, CABBAGES, CALABRESE, CAULIFLOWERS, SWEDES, TURNIPS [1-4]. Blackgrass in OILSEED RAPE [1, 2]. Triazine resistant groundsel in BROCCOLI, BRUSSELS SPROUTS, CABBAGES, CALABRESE, CAULIFLOWERS, HARDY ORNAMENTAL NURSERY STOCK, NURSERY FRUIT TREES AND BUSHES, ORNAMENTALS, SPRING OILSEED RAPE, SWEDES, TURNIPS, WINTER OILSEED RAPE [2].

Notes

Efficacy

- Activity is dependent on root uptake. For pre-emergence use apply to firm, moist, clod-free seedbed
- Some weeds (chickweed, mayweed, blackgrass etc) susceptible up to 2- or 4-leaf stage. Moderate control of cleavers achieved provided weeds not emerged and adequate soil moisture present
- Split pre- and post-emergence treatments recommended for certain weeds in winter oilseed rape on light and/or stony soils
- Effectiveness is reduced on soils with more than 10% organic matter
- Various tank-mixtures recommended. See label for details

Crop Safety/Restrictions

- Maximum number of treatments 1 per crop for spring oilseed rape, swedes, turnips and brassicas; 2 per crop for winter oilseed rape and honesty (split dose treatment); 3 per yr for ornamentals, nursery stock, nursery fruit trees, forestry and farm forestry
- On winter oilseed rape may be applied pre-emergence from drilling until seed chits, post-emergence after fully expanded cotyledon stage (GS 1,0) or by split dose technique depending on soil and weeds. See label for details
- On spring oilseed rape, swedes, turnips and transplanted brassicas recommended as a pre-emergence sequential treatment following trifluralin

FOR FULL CONDITIONS OF USE ALWAYS READ THE PRODUCT LABEL

- On spring oilseed rape may also be used pre-weed-emergence from cotyledon to 10-leaf stage of crop (GS 1,0-1,10)
- With pre-emergence treatment ensure seed covered by 15 mm of well consolidated soil. Harrow across slits of direct-drilled crops
- Ensure brassica transplants have roots well covered and are well established
- In ornamentals and hardy nursery stock apply after plants established and hardened off as a directed spray or, on some subjects, as an overall spray. See label for list of tolerant subjects. Do not treat plants in containers
- Do not use on sand, very light or poorly drained soils
- Do not treat protected crops or spray overall on ornamentals with soft foliage
- Do not spray crops suffering from wilting, pest or disease
- Do not spray broadcast crops or if a period of heavy rain forecast
- Any crop can follow normally harvested treated oilseed rape. See label for details of crops which may be planted in event of crop failure

Special precautions/Environmental safety

- Harmful if swallowed [1-4]. Irritating to eyes [4] and skin [1-3]
- Harmful to fish or other aquatic life. Do not contaminate surface waters or ditches with chemical or used container

Protective clothing/Label precautions

- A, C [1-4]; D [3]; H [1, 2, 4]; M [2-4]
- M03, R03c, R04b, U05a, U08, U19, C03, E01, E30a, E34 [1-4]; R04a, U20b, E07 (2 wk or 5 wk for swedes, urnips) [4]; U20a [1-3]; C02, E31c [2]; E07 (swedes/turnips 5 wk) [3]; E07 [1]; E13c, E31b [1, 3, 4]; E26 [3, 4]

Withholding period

- Keep livestock out of treated areas until foliage of any poisonous weeds such as ragwort has died and become unpalatable
- Keep livestock out of treated areas of swede and turnip for at least 5 wk following treatment

Latest application/Harvest Interval(HI)

- Pre-emergence for swedes and turnips; before 10 leaf stage for spring oilseed rape; before end of Jan for winter oilseed rape; before 6 pairs of true leaves for honesty.
- HI brassicas 6 wk. Do not use on nursery fruit trees within 1 yr of fruit production

Approval

- Off-label approval unlimited for honesty grown outdoors (OLA 0551/93)[2]; to Mar 2001 for use on outdoor kohlrabi (OLA 0498/96)[2]

---

396 metazachlor + quinmerac

A residual herbicide mixture for oilseed rape

Products Katamaran | BASF | 375:125 g/l | SC | 09049

Uses Annual dicotyledons in WINTER OILSEED RAPE. Annual meadow grass in WINTER OILSEED RAPE. Blackgrass in WINTER OILSEED RAPE. Cleavers in WINTER OILSEED RAPE. Poppies in WINTER OILSEED RAPE.

Notes

Efficacy

- Activity is dependant on root uptake. Pre-emergence treatments should be applied to firm moist seedbeds. Applications to dry soil do not become effective until after rain has fallen
- Maximum activity achieved from treatment before weed emergence
- Weed control may be reduced if excessive rain falls shortly after application especially on light soils

Crop Safety/Restrictions

- Maximum total dose equivalent to one full dose treatment
- May be used on all soil types except Sands, Very Light Soils, and soils containing more than 10% organic matter. Crop vigour and/or plant stand may be reduced on brashy and stony soils
- To ensure crop safety it is essential that crop seed is well covered with soil to 15 mm. Loose or puffy seedbeds must be consolidated before treatment. Do not use on broadcast crops
- Crop vigour and possibly plant stand may be reduced if excessive rain falls shortly after treatment especially on light soils. The effects may be increased where TCA has been used
- Damage may occur in waterlogged conditions. Do not use on poorly drained soils
- Do not treat stressed crops. In frosty conditions transient scorch may occur
- In the event of crop failure after use, wheat or barley may be sown in the autumn after ploughing to 15 cm. Spring cereals or brassicas may be planted after ploughing in the spring

Special precautions/Environmental safety

- Irritant. May cause sensitization by skin contact
- Dangerous to fish or other aquatic life. Do not contaminate surface waters or ditches with chemical or used container
- Do not allow direct spray from vehicle mounted/drawn hydraulic sprayers to fall within 6 m of surface waters or ditches. Direct spray away from water
- To reduce risk of movement to water do not apply to dry soil or if heavy rain is forecast. On clay soils create a fine consolidated seedbed

Protective clothing/Label precautions

- A
- R04, R04e, U05a, U08, U14, U19, U20b, C03, E01, E07, E13b, E16, E30a, E31c

Latest application/Harvest Interval(HI)

- End Jan in yr of harvest

397 methabenzthiazuron

A contact and residual urea herbicide for cereals

Products Tribunil | Bayer | 70% w/w | WP | 02169

Uses Annual dicotyledons in DURUM WHEAT, SPRING BARLEY, TRITICALE, WINTER BARLEY, WINTER OATS, WINTER RYE, WINTER WHEAT. Annual meadow grass in DURUM WHEAT, SPRING BARLEY, TRITICALE, WINTER BARLEY, WINTER OATS, WINTER RYE, WINTER WHEAT. Blackgrass in WINTER BARLEY, WINTER WHEAT

Rough meadow grass in DURUM WHEAT, SPRING BARLEY, TRITICALE, WINTER BARLEY, WINTER OATS, WINTER RYE, WINTER WHEAT.

Notes

Efficacy

- Apply pre-weed emergence on fine, firm seedbed or post-emergence to 1-2 true leaf stage of weeds, blackgrass (at higher rates) not beyond 2-leaf stage or after end Nov
- Do not use lower rates on soils with more than 10% organic matter
- Do not use higher rate for blackgrass control on soils with more than 4% organic matter or where seedbed preparation has not been preceded by ploughing

Crop Safety/Restrictions

- Maximum number of treatments 1 per crop
- Apply in winter wheat or autumn sown spring wheat pre- or post-emergence up to 6 wk after drilling, provided weeds in correct stage
- Apply pre-emergence in winter barley, but not in crops drilled after mid-Oct on sand in E. Anglia
- Only use pre-emergence in spring barley drilled to at least 25 mm and not on sand
- Do not use on crops to be undersown with clover
- Use lower rate on direct drilled crops and roll before spraying where harrowed or disced after drilling

Special precautions/Environmental safety

- Harmful to fish or other aquatic life. Do not contaminate surface waters or ditches with chemical or used container

Protective clothing/Label precautions

- U08, U19, U20a, E07, E13c, E30a, E32a

Withholding period

- Keep livestock out of treated areas for at least 14 d after application

Latest application/Harvest Interval(HI)

- Pre-emergence for spring barley, winter oats, winter rye, triticale and autumn sown spring wheat; post-emergence before 30 Oct for winter barley and durum wheat (pre-emergence before 30 Nov at higher rate); post-emergence before 1st node detectable (GS 31) for winter wheat (pre-emergence up to 6 wk after drilling but before 30 Nov at higher rate)

Approval

- Approved for aerial application on cereals [1]. See notes in Section 1

398 methiocarb

A stomach acting carbamate molluscicide and insecticide

Products

1 Decoy	Bayer	2% w/w	PT	06535
2 Draza	Bayer	4% w/w	PT	00765
3 Draza ST	Bayer	4% w/w	PT	05315
4 Exit	Bayer	4% w/w	PT	07632

Uses

Leatherjackets in CEREALS *(reduction)* [1, 2, 4]. Slugs in PROTECTED CROPS *(off-label)* [2, 4]. Slugs in CEREALS *(seed admixture)*[3]. Slugs in BRASSICAS, CEREALS, FIELD CROPS, FRUIT CROPS, GRASSLAND, OILSEED RAPE, ORNAMENTALS, POTATOES [1, 2, 4]. Snails in PROTECTED CROPS *(off-label)* [2, 4]. Snails in BRASSICAS, CEREALS, FIELD

CROPS, FRUIT CROPS, GRASSLAND, OILSEED RAPE, ORNAMENTALS, POTATOES [1, 2, 4]. Strawberry seed beetle in STRAWBERRIES [1, 2].

Notes

Efficacy
- Use as a surface, overall application when pests active (normally mild, damp weather), pre-drilling or post-emergence. May also be used on cereals, ryegrass, oilseed rape or other brassicas in admixture with seed at time of drilling [1, 2, 4]
- Also reduces populations of cutworms and millipedes [1, 2, 4]
- Best on potatoes in late Jul to Aug, on peas before start of pods formation [1, 2, 4]
- Apply to strawberries before strawing down, to blackcurrants at grape stage to prevent snails contaminating crop [1, 2, 4]
- See label for details of suitable application equipment

Crop Safety/Restrictions
- Do not allow pellets to lodge in edible crops [1, 2, 4]
- Product not suitable for broadcast surface application. For use only by seed merchants as a bait for admixture with cereal seed to be drilled [3]

Special precautions/Environmental safety
- Harmful if swallowed
- This product contains an anticholinesterase carbamate compound. Do not use if under medical advice not to work with such compounds
- Dangerous to game, wild birds and animals
- Dangerous to fish or other aquatic life. Do not contaminate surface waters or ditches with chemical or used container
- Keep poultry out of treated areas for at least 7 d
- Avoid surface application within 6 m of field boundary to reduce effects on non-target species

Protective clothing/Label precautions
- A, H, J
- M02, M03, R03c, U05a, U20b, C03, E01, E05a, E10a, E13b, E22b, E30a, E32a, E34 [1-4]; S01, S02, S04b, S05, S06, S07 [3]

Latest application/Harvest Interval(HI)
- HI 7 d

Approval
- Approved for aerial application on all crops [1-4]. See notes in Section 1
- Off-label Approval unlimited for use on all protected edible and non-edible crops (OLA 0186/92)[2]

399 methomyl

An oxime carbamate insecticide available only in mixtures

400 methomyl + (Z)-9-tricosene

An insecticide bait for fly control in animal houses

Products	Golden Malrin Fly Bait	Novartis A H	1:0.025% w/w	GR	08821

Uses Flies in LIVESTOCK HOUSES.

Notes

Efficacy
- Product contains a pheromone attractant and can be applied as a scatter bait.
- Alternatively sprinkle granules on to polystyrene ceiling tiles, old egg trays or similar coated with adhesive and hang in infested areas out of reach of animals
- Re-apply every 4-5 d, or when granules disappear, until fly population has been reduced to acceptable levels, then apply every 7 d

Crop Safety/Restrictions
- Prevent access to bait by children and livestock
- Do not contaminate animal feed or use in areas where bait might come into contact with food

Special precautions/Environmental safety
- Product contains an anticholinesterase carbamate compound. Do not use if under medical advice not to work with such compounds
- Dangerous to fish or other aquatic life. Do not contaminate surface waters or ditches with chemical or used container

Protective clothing/Label precautions
- A, H
- M02, U13, U20a, C03, E13b, E30a, E31a, V01a, V02

401 methoprene

A pheromone analogue insect growth regulator

Products	Apex 5E	Novartis A H	600 g/l	EC	05730

Uses Sciarid flies in MUSHROOM HOUSES.

Notes

Efficacy
- Use as surface spray either on the compost or the casing
- Use pre-casing treatment only when early pest infestation occurs
- Product controls sciarids at pupal stage. Treatment will not control the generation of sciarids present at application

Crop Safety/Restrictions
- Maximum number of treatments 1 per spawning on compost or casing
- Product prevents development of adults from eggs laid up to 14 d before or after treatment. Alternative treatments may be necessary if adult numbers increase subsequently
- Pre-casing treatment may not prevent development from larvae able to move into untreated casing prior to pupation

Special precautions/Environmental safety

- Dangerous to fish or other aquatic life. Do not contaminate surface waters or ditches with chemical or used container

Protective clothing/Label precautions

- A, C, H
- U05a, U19, U20c, C03, E01, E13b, E26, E30a, E32a

Latest application/Harvest Interval(HI)

- Before casing on the compost or immediately after casing

---

402 2-methoxyethyl mercury acetate

An organomercury fungicide seed treatment all approvals for which were revoked with effect from 31 March 1992

---

403 methyl bromide

A highly toxic alkyl halide fumigant for food and non-food commodities

Products					
	1 Methyl Bromide 100	Bromine & Chem.	99.7% w/w	GA	01336
	2 Sobrom BM 100	Brian Jones	100%	GA	04381

Uses

Food storage pests in FOOD STORAGE AREAS [1, 2]. Grain storage pests in GRAIN STORES [1]. Insect pests in STORED PLANT MATERIAL [1, 2].

Notes

Efficacy

- Chemical is released from containers as a highly toxic, colourless gas with only a slight odour
- Dose depends on temperature. At lower temperatures higher doses and longer exposures should be used
- See label for dosage and exposure required in different situations

Crop Safety/Restrictions

- Maximum number of treatments 1 per infestation [2]; 1 for flour, 2 for other food and non-food commodities [1]
- Air animal feed for at least 5 d following treatment
- Some types of produce, cut flowers and plants are susceptible to damage by methyl bromide fumigation. Check before treating
- Methyl bromide can cause taint to flours and foodstuffs containing reactive sulfur compounds. See label for guidance

Special precautions/Environmental safety

- Methyl bromide is subject to the Poisons Rules 1982 and the Poisons Act 1972. See notes in Section 1
- Not for retail sale
- Very toxic by inhalation. Fatal if swallowed.
- Wear suitable protective clothing (See HSE Guidance Note CS 12)

- Wear suitable respiratory equipment during fumigation and while removing sheets. Do *not* wear gloves or rubber boots
- Do not use application equipment incorporating natural rubber, PVC or aluminium, zinc, magnesium or their alloys
- Extinguish all naked flames, including pilot lights when applying fumigant
- Keep in original container, tightly closed in a safe place under lock and key. Do not re-use container and return to supplier when empty
- Keep unprotected persons out of treated area until it is shown by test to be clear of methyl bromide
- Harmful to fish or other aquatic life. Do not contaminate surface waters or ditches with chemical or used container [2]
- Special equipment is needed for application and the chemical may only be used by professional operators trained in its use and familiar with the precautionary measures to be observed
- All operations must be in accordance with the HSE guidance note CS 12. A minimum of 2 operators are required

Protective clothing/Label precautions
- See HSE Guidance Note CS 12 [1, 2]
- M04, R01a, E02 (until clear), E13c, E27 [2]; R01b, U04a, U05a, C03, E01, E30b, E34 [1, 2]; E02, E33 [1]

Withholding period
- Keep animals and livestock out of treated area until it is shown by test to be clear of methyl bromide

Latest application/Harvest Interval(HI)
- 3 d before loading grains; 7 d before use of all other commodities and spaces [1]

Maximum Residue Levels (mg residue/kg food)
- oilseeds, cereals 0.1; all other products (except tree nuts, stone fruits, grapes, figs, pulses, animal products) 0.05

---

404 methyl bromide with amyl acetate

A highly toxic alkyl halide fumigant for soil and food storage areas

Products Fumyl-O-Gas | Brian Jones | 99.7% | GA | 04833

Uses Food storage pests in FOOD STORAGE. Nematodes in FIELD CROPS, PROTECTED CROPS. Soil insects in FIELD CROPS, PROTECTED CROPS. Soil-borne diseases in FIELD CROPS, PROTECTED CROPS. Weed seeds in FIELD CROPS, PROTECTED CROPS.

Notes

Efficacy
- Chemical is released from containers as a highly volatile, highly toxic gas and must be released under gas-proof sheeting
- Amyl acetate is added as a warning odourant
- Soil must be clear of trash and cultivated to at least 40 cm before treatment
- Gas must be confined in soil for 48-96 h

Crop Safety/Restrictions
- Maximum number of treatments 1 per yr

- Crops may be planted 3-21 d after removal of gas-proof sheeting. For some crops a leaching irrigation is essential. See label for details

Special precautions/Environmental safety

- Methyl bromide is subject to the Poisons Rules 1982 and the Poisons Act 1972. See notes in Section 1
- Product not for retail sale
- Special equipment is needed for application and the chemical may only be used by professional operators trained in its use and familiar with the precautionary measures to be observed. Refer to HSE Guidance Note CS 12 before applying
- Very toxic by inhalation. Risk of serious damage to health by prolonged exposure
- Keep in original container, tightly closed, in a safe place, under lock and key
- Wear suitable respiratory equipment during fumigation and while removing sheeting
- Do not wear gloves or rubber boots
- Ventilate treated areas thoroughly when gas has cleared
- Keep unprotected persons out of treated areas for at least 24 h
- Dangerous to game, wild birds, animals and bees
- Harmful to fish or other aquatic life. Do not contaminate surface waters or ditches with chemical or used container

Protective clothing/Label precautions

- See HSE Guidance Note CS 12
- M04, R01a, R01b, U04a, U05a, C03, E01, E02 (until clear), E30b, E34

Withholding period

- Keep livestock out of treated areas for at least 24 h

Latest application/Harvest Interval(HI)

- Before sowing or planting

Maximum Residue Levels (mg residue/kg food)

- see methyl bromide entry

405 methyl bromide with chloropicrin

A highly toxic alkyl halide fumigant for treatment of soil and stored products

Products					
	1 Methyl Bromide 98	Bromine & Chem.	980:20 g/l	GA	01335
	2 Sobrom BM 98	Brian Jones	980:20 g/l	GA	04189

Uses Grain storage pests in GRAIN STORES [1]. Nematodes in BULBS/CORMS, FLOWERS NURSERY STOCK, STRAWBERRIES, TURF, VEGETABLES [1]. Nematodes in FIELD CROPS, ORCHARDS, PROTECTED CROPS [2]. Soil insects in BULBS/CORMS, FLOWERS NURSERY STOCK, STRAWBERRIES, TURF, VEGETABLES [1]. Soil insects in FIELD CROPS, ORCHARDS, PROTECTED CROPS [2]. Soil-borne diseases in BULBS/CORMS FLOWERS, NURSERY STOCK, STRAWBERRIES, VEGETABLES [1]. Soil-borne diseases in FIELD CROPS, ORCHARDS, PROTECTED CROPS [2]. Weed seeds in BULBS/CORMS FLOWERS, NURSERY STOCK, STRAWBERRIES, TURF, VEGETABLES [1]. Weed seeds in FIELD CROPS, ORCHARDS, PROTECTED CROPS [2].

Notes

Efficacy

- Chemical is released from containers as a highly volatile, highly toxic gas and for soil fumigation must be released under a plastic tarpaulin
- Chloropicrin is added as a warning odourant tear gas
- Soil must be clear of trash and cultivated to at least 40 cm before treatment
- Gas must be confined in soil for 48-96 h

Crop Safety/Restrictions

- Maximum number of treatments 1 per yr
- Crops may be planted 3-21 d after removal of tarpaulin. For some crops a leaching irrigation is essential. See labels for details

Special precautions/Environmental safety

- Methyl bromide and chloropicrin are subject to the Poisons Rules 1982 and the Poisons Act 1972. See notes in Section 1
- Product not for retail sale
- Special equipment is needed for application and the chemical may only be used by professional operators trained in its use and familiar with the precautionary measures to be observed. Refer to HSE Guidance Note CS 12 before applying
- Very toxic by inhalation, in contact with skin and if swallowed
- Irritating to eyes, respiratory system and skin
- Keep in original container, tightly closed, in a safe place, under lock and key
- Wear suitable respiratory equipment during fumigation and while removing sheeting
- Do *not* wear gloves or rubber boots
- Ventilate treated areas thoroughly when gas has cleared
- Keep unprotected persons out of treated areas for at least 24 h
- Harmful to fish or other aquatic life. Do not contaminate surface waters or ditches with chemical or used container [2]

Protective clothing/Label precautions

- See HSE Guidance Note CS 12 [1, 2]
- M04, E13c [2]; R01a, R01b, R01c, R04a, R04b, R04c, U04a, U05a, C03, E01, E02, E07, E30b, E34 [1, 2]; E33 [1]

Withholding period

- Keep livestock out of treated areas for at least 24 h

Latest application/Harvest Interval(HI)

- Before sowing or planting

Maximum Residue Levels (mg residue/kg food)

- see methyl bromide entry

406 metosulam

A sulfonanilide herbicide available only in mixture

See also fluroxypyr + metosulam

407 metoxuron

A contact and residual urea herbicide for cereals and carrots

Products

1 Dosaflo	Sandoz	500 g/l	SC	00754
2 Dosaflo	Novartis	500 g/l	SC	08473

Uses

Annual dicotyledons in CARROTS, DURUM WHEAT, TRITICALE, WINTER BARLEY, WINTER WHEAT. Annual grasses in CARROTS, DURUM WHEAT, TRITICALE, WINTER BARLEY, WINTER WHEAT. Barren brome in DURUM WHEAT, TRITICALE, WINTER BARLEY, WINTER WHEAT. Blackgrass in DURUM WHEAT, TRITICALE, WINTER BARLEY, WINTER WHEAT. Mayweeds in CARROTS.

Notes

Efficacy

- Best results achieved by application to weeds in seedling to young plant stage
- Best control of barren brome in autumn or winter from 1-3 leaf stage
- Effects on blackgrass reduced on soils of high organic matter or low pH

Crop Safety/Restrictions

- Maximum number of treatments 2 per crop for cereals with minimum of 6 wk between applications
- Apply to named winter cereal varieties from 2-leaf unfolded stage to before first node detectable (GS 12-31). See label for lists of resistant and susceptible varieties
- Apply to carrots after 2-true leaf stage. Do not spray when soil very dry or wet
- Do not spray cereals on Sands or Very Light soils, or carrots on soils with more than 80% sand or less than 1% organic matter
- Do not apply during prolonged frosty weather, when frost or heavy rain imminent or ground waterlogged
- If treatment repeated in spring at least 6 wk must elapse after autumn/winter spray
- Do not roll for 7 d before or after spraying
- Do not cultivate after spraying or harrow or cultivate within 14 d beforehand
- Do not spray crops checked by pests, wind, frost or waterlogging until recovered
- No crop should be sown within 6 wk of treatment

Protective clothing/Label precautions

- U08, U19, U20b, E15, E26, E30a, E31b, E34

Latest application/Harvest Interval(HI)

- Before 1st node detectable (GS 31) for cereals

408 metribuzin

A contact and residual triazinone herbicide for use in potatoes

Products

1 Citation	United Phosphorus	70% w/w	WG	08685
2 Lexone 70DF	DuPont	70% w/w	WG	04991
3 Sencorex WG	Bayer	70% w/w	WG	03755

Uses Annual dicotyledons in POTATOES [1-3]. Annual grasses in POTATOES [1-3]. Annual weeds in CARROTS *(off-label)*, PARSNIPS *(off-label)* [3]. Volunteer oilseed rape in POTATOES [1-3].

Notes

Efficacy

- May be applied pre- or post-emergence of crop. Best results achieved on weeds at cotyledon to 1-leaf stage
- Apply to moist soil with well-rounded ridges and few clods
- Activity reduced by dry conditions and on soils with high organic matter content
- Do not cultivate after treatment
- On fen and moss soils pre-planting incorporation to 10-15 cm gives increased activity
- With named maincrop and second early varieties on soils with more than 10% organic matter shallow pre- or post-planting incorporation may be used. See label for details
- Effective control using a programme of reduced doses is made possible by using a spray of smaller droplets, thus improving retention. See label for details

Crop Safety/Restrictions

- Maximum number of treatments 3 per crop subject to maximum permitted dose - see labels
- Apply pre-emergence only on named first earlies, pre- or post-emergence on named second earlies. On named maincrop varieties apply pre-emergence (except for certain varieties on Sands or Very Light soils) or post-emergence before longest shoots reach 15 cm. See label for details
- On stony or gravelly soils there is risk of crop damage, especially if heavy rain falls soon after application
- When days are hot and sunny delay spraying until evening
- Some varieties may be sensitive to post-emergence treatment if crop under stress
- Ryegrass, cereals or winter beans may be sown in same season provided at least 16 wk elapsed after treatment and ground ploughed to 15 cm and thoroughly cultivated soon after harvest
- Do not grow any vegetable brassicas on silt soils in Lincs, and lettuces or radishes anywhere in UK on land treated the previous yr. Other crops may be sown normally in spring of next yr

Special precautions/Environmental safety

- Do not contaminate surface waters or ditches with chemical or used container

Protective clothing/Label precautions

- U08, U13, U19, E15, E30a, E32a [1-3]; U20a [2, 3]; U20b, E29 [1]

Latest application/Harvest Interval(HI)

- Pre-emergence for earlies; before most advanced shoots reach 15 cm for maincrop

Approval

- Off-label approval to Aug 2001 for use on carrots, parsnips (OLA 1769/96)[3]

409 metsulfuron-methyl

A contact and residual sulfonylurea herbicide for use in cereals and setaside

See also carfentrazone-ethyl + metsulfuron-methyl
flupyrsulfuron-methyl + metsulfuron-methyl

Products

1 Ally	DuPont	20% w/w	WG	02977
2 Ally WSB	DuPont	20% w/w	WB	06588
3 Jubilee (WSB)	DuPont	20% w/w	WB	06082
4 Jubilee 20 DF	DuPont	20% w/w	WG	06136
5 Landgold Metsulfuron	Landgold	20% w/w	WG	06280
6 Lorate 20 DF	DuPont	20% w/w	WG	06135
7 Standon Metsulfuron	Standon	20% w/w	WG	05670

Uses

Annual dicotyledons in LINSEED, TRITICALE [1-4, 6]. Annual dicotyledons in BARLEY OATS, WHEAT [1-7]. Chickweed in LINSEED, TRITICALE [1-4, 6]. Chickweed in BARLEY OATS, WHEAT [1-7]. Green cover in LAND TEMPORARILY REMOVED FROM PRODUCTION [1]. Green cover in SETASIDE [4, 6]. Mayweeds in LINSEED, TRITICALE [1-4, 6]. Mayweeds in BARLEY, OATS, WHEAT [1-7].

Notes

Efficacy

- Best results achieved on small, actively growing weeds up to 6-true leaf stage. Good spray cover is important
- Commonly used in tank-mixture on wheat and barley with other cereal herbicides to improve control of resistant dicotyledons (cleavers, fumitory, ivy-leaved speedwell), larger weeds and grasses. See label for recommended mixtures
- When tank mixing, always add metsulfuron-methyl product to tank first
- May be used on setaside where the green cover is made up predominantly of grass, wheat barley, oats or triticale [1, 4, 6]

Crop Safety/Restrictions

- Maximum number of treatments 1 per crop or per yr on setaside
- Product must be applied only after 1 Feb
- Apply to wheat and oats from 2-leaf (GS 12), to barley, and triticale from 3-leaf stage (GS 13) until flag-leaf fully emerged (GS 39). Do not spray Igri barley before leaf sheath erect stage (GS 30)
- Apply to linseed from first pair of true leaves unfolded up to 30 cm high or before flower bud visible, whichever is the sooner
- Recommendations for oats, triticale and linseed apply to product alone
- Do not use on cereal crops undersown with grass or legumes
- Do not use on any crop suffering stress from drought, waterlogging, frost, deficiency, pest or disease attack or apply within 7 d of rolling
- Do not spray in tank mixture, or in sequence, with a product containing any other sulfonylurea
- Use recommended procedure to clean out spraying equipment

- Only cereals, oilseed rape, field beans or grass may be sown in same calendar year after treating cereals with the product alone. Other restrictions apply to tank mixtures. See label for details. In the event of crop failure sow only wheat within 3 mth after treatment. Only cereals should be planted within 16 mth of applying to a linseed crop

Special precautions/Environmental safety

- Extremely dangerous to aquatic higher plants. Do not contaminate surface waters or ditches with chemical or used container
- Do not allow direct spray from field mounted/drawn hydraulic sprayers to fall within 6 m, or from hand-held sprayers to within 2 m, of surface waters or ditches. Direct spray away from water
- Take extreme care to avoid damage by drift onto broad-leaved plants outside the target area, onto surface waters or ditches or onto land intended for cropping
- A range of broad leaved species will be fully or partially controlled when used in setaside, hence product may be suitable where wild flower borders or other forms of conservation headland are being developed [1, 4, 6]
- Before use on setaside as part of grant-aided scheme, ensure compliance with the management rules [1, 4, 6]
- Do not open the water soluble bag or touch it with wet hands or gloves [2, 3]

Protective clothing/Label precautions

- U05a, U20a [1, 3-7]; U08, U19, C03, E01, E14a, E16, E30a, E32a [1-7]; U22 [2, 3]

Latest application/Harvest Interval(HI)

- Before flag leaf sheath extending stage for cereals (GS 41); before flower bud visible or up to 30 cm tall, whichever is earlier, for linseed; before 1 Aug in yr of treatment for setaside

---

410 metsulfuron-methyl + thifensulfuron-methyl

A contact residual and translocated sulfonylurea herbicide mixture for use in cereals

Products

Harmony M	DuPont	7:68% w/w	WG	03990

Uses

Annual dicotyledons in SPRING BARLEY, SPRING WHEAT, WINTER WHEAT. Cleavers in SPRING BARLEY, SPRING WHEAT, WINTER WHEAT. Field pansy in SPRING BARLEY, SPRING WHEAT, WINTER WHEAT. Polygonums in SPRING BARLEY, SPRING WHEAT, WINTER WHEAT. Speedwells in SPRING BARLEY, SPRING WHEAT, WINTER WHEAT.

Notes

Efficacy

- Best results by application to small, actively growing weeds up to 6-true leaf stage
- Effectiveness may be reduced if heavy rain occurs within 4 h of application or if soil conditions very dry
- Tank-mixture with mecoprop, mecoprop-P or reduced rate of fluroxypyr improves control of cleavers and other problem weeds
- When tank mixing, always add metsulfuron-methyl/thifensulfuron-methyl product to tank first and fully disperse before adding second product

Crop Safety/Restrictions

- Maximum number of treatments 1 per crop
- Apply in spring to crops from 3-leaf stage to flag-leaf fully emerged (GS 13-39)
- Do not use on durum wheat or winter barley or on any crop suffering stress from drought, waterlogging, frost, deficiency, pest or disease attack

- Do not spray in tank mixture, or in sequence, with a product containing any other sulfonylurea herbicide
- Do not apply within 7 d of rolling
- Take extreme care to avoid drift onto broad-leaved crops or contamination of land planted or to be planted with any crops other than cereals
- Use recommended procedure to clean out spraying equipment
- Only cereals, oilseed rape, field beans or grass may be sown in same calendar year after treatment. In the event of crop failure sow only wheat within 3 mth after treatment

Special precautions/Environmental safety

- Extremely dangerous to aquatic higher plants. Do not contaminate surface waters or ditches with chemical or used container
- Do not allow direct spray from field mounted/drawn hydraulic sprayers to fall within 6 m, or from hand-held sprayers to within 2 m, of surface waters or ditches. Direct spray away from water

Protective clothing/Label precautions

- U08, U19, U20a, E14a, E16, E30a, E32a

Latest application/Harvest Interval(HI)

- Before flag leaf ligule/collar just visible (GS 39)

411 monolinuron

A residual urea herbicide for potatoes, leeks and French beans

Products				
Arresin	AgrEvo	200 g/l	EC	07303

Uses

Annual dicotyledons in FRENCH BEANS, LEEKS, POTATOES. Annual grasses in FRENCH BEANS, LEEKS, POTATOES. Annual meadow grass in FRENCH BEANS, LEEKS, POTATOES. Fat-hen in FRENCH BEANS, LEEKS, POTATOES. General weed control in HORSERADISH *(off-label)*. Polygonums in FRENCH BEANS, LEEKS, POTATOES.

Notes

Efficacy

- Best results achieved by application on moist, firm, clod-free seedbed pre-weed emergence or on seedlings up to 2-3 leaf stage
- Light rain after application can improve control
- In potatoes or French beans on fen or peat soils apply post-weed emergence but before emergence of crop

Crop Safety/Restrictions

- Maximum number of treatments 1 per crop
- May be used in tank-mixture with Challenge on potatoes up to 10% emergence on earlies, 40% emergence on maincrop. Do not apply alone post-emergence
- On early sown French beans apply at least 5 d pre-emergence
- Some bean cultivars may be sensitive on light soils, with application close to emergence with heavy rain after treatment or in unevenly drilled crop. See label for list of cultivars
- Apply to transplanted or direct-sown leeks after established and 180 mm high
- Only apply pre-emergence to crops grown under polythene

- Do not use on Very Light soils
- Do not sow or plant other crops within 2-3 mth, depending on rate applied. Lettuce must not be sown in same season

Special precautions/Environmental safety

- Irritating to skin, eyes and respiratory system
- Flammable
- Extremely dangerous to fish or other aquatic life. Do not contaminate surface waters or ditches with chemical or used container

Protective clothing/Label precautions

- A, C, H, J, M
- R04a, R04b, R04c, R07d, U04a, U05a, U08, U19, U20a, E01, E13a, E30a, E31b, E34

Latest application/Harvest Interval(HI)

- Pre-emergence for French beans; 3 wk after transplanting for leeks; pre-emergence or up to 10% emergence in seed or early potatoes, pre-emergence or up to 40% emergence (only in mixture with glufosinate-ammonium) in maincrop potatoes

Approval

- Approved for aerial application on potatoes and French beans [1]. See notes in Section 1
- Off-label approval to Apr 1999 for use on outdoor horseradish (OLA 0920/96)[1]

---

412 myclobutanil

A systemic, protectant and curative conazole fungicide

Products

1 Systhane 6 Flo	Promark	60 g/l	EW	07334
2 Systhane 6 W	PBI	6% w/w	WP	04571

Uses

American gooseberry mildew in BLACKCURRANTS, GOOSEBERRIES [1]. Black spot in ROSES [2]. Blossom wilt in CHERRIES *(off-label)*, MIRABELLES *(off-label)* [1]. Mildew in ROSES [2]. Powdery mildew in APPLES, HOPS *(off-label)*, STRAWBERRIES [1]. Rust in ROSES [2]. Scab in APPLES, PEARS [1].

Notes

Efficacy

- Best results achieved when used as part of routine preventive spray programme from bud burst to end of flowering in apples and pears and from just before the signs of mildew infection in blackcurrants and gooseberries [1]
- In strawberries commence spraying at, or just prior to, first flower [1]
- Spray at 7-14 d intervals depending on disease pressure and dose applied [1]
- For improved scab control in post-blossom period tank-mix with mancozeb or captan [1]
- Apply alone from mid-Jun for control of secondary mildew on apples and pears [1]
- On roses spray at first signs of disease and repeat every 2 wk. In high risk areas spray when leaves emerge in spring, repeat 1 wk later and then continue normal programme [2]

Crop Safety/Restrictions

- Total application for season in apples and pears should not exceed 15 l/ha (ie 10 sprays at highest recommended rate) and 9 l/ha in soft fruit [1]

Special precautions/Environmental safety

- Harmful to fish or other aquatic life. Do not contaminate surface waters or ditches with chemical or used container

- Do not harvest apples, pears, blackcurrants or gooseberries for at least 14 d, and strawberries for at least 3 d, after last application [1]

Protective clothing/Label precautions
- A, H, J [1]
- U08, C03, E01 [1]; U20a, E13c, E32a [1, 2]; E30a [2]

Latest application/Harvest Interval(HI)
- HI 14 d (apples, pears, blackcurrants, gooseberries); 3 d (strawberries)

Approval
- Off-label approval unlimited for use on hops (OLA 0849/95)[1]; unlimited for use on cherries, mirabelles (OLA 0991/98)[1]

413 2-(1-naphthyl)acetic acid

A plant growth regulator and herbicide for sucker control

See also 4-indol-3-yl-butyric acid + 2-(1-naphthyl)acetic acid with dichlorophen

Products

1 Rhizopon B Powder (0.1%)	Fargro	0.1% w/w	SP	07133
2 Rhizopon B Powder (0.2%)	Fargro	0.2% w/w	SP	07134
3 Rhizopon B Tablets	Fargro	25 mg a.i.	TB	07135
4 Tipoff	Unicrop	57.8 g/l	EC	05878

Uses

Control of water shoots in APPLES, PEARS [4]. Rooting of cuttings in ORNAMENTALS [1-3]. Sucker control in APPLES, CHERRIES, PEARS, PLUMS, RASPBERRIES [4].

Notes

Efficacy
- Dip moistened base of cuttings into powder immediately before planting [1, 2]
- Spray root suckers of top fruit trees when 10-15 cm high. Best results obtained on sucker regrowth cut back the previous winter. A second application can be made later in year (Jul) if there are any signs of regrowth [4]
- Spray water shoots of apple and pear when about 10 cm long [4]
- Spray raspberry suckers in alleys when 10-20 cm high and repeat if necessary [4]
- It is important that foliage of suckers is thoroughly wetted [4]
- See label for details of concentrations recommended for promotion of rooting in cuttings of different species [3]
- Dip prepared cuttings in solution for 4-24 h depending on species [3]

Crop Safety/Restrictions
- Maximum number of treatments 2 per yr [4]
- Avoid spray drift on to non-target areas of fruit trees [4]
- In raspberries only spray suckers in alleys, not replacement or fruiting canes or suckers in the crop rows [4]

Protective clothing/Label precautions
- U08, U19, U20a, E34 [1-3]; U20c [4]; E15, E30a, E32a [1-4]

414 (2-naphthyloxy)acetic acid

A plant growth regulator for setting tomato fruit

Products

Betapal Concentrate	Vitax	16 g/l	SL	00234

Uses

Increasing fruit set in TOMATOES.

Notes

Efficacy

- Apply with any fine sprayer or syringe when first half dozen flowers are open
- Spray actual trusses in flower and repeat every 2 wk

Crop Safety/Restrictions

- Do not spray growing head of tomato plants

415 napropamide

A soil applied amide herbicide for oilseed rape, fruit and woody ornamentals

Products

1 Devrinol	RP Agric.	450 g/l	SC	06195
2 Devrinol	United Phosphorus	450 g/l	SC	06653

Uses

Annual dicotyledons in BLACKCURRANTS, CONTAINER-GROWN WOODY ORNAMENTALS, ESTABLISHED WOODY ORNAMENTALS, GOOSEBERRIES, NEWLY PLANTED WOODY ORNAMENTALS, RASPBERRIES, STRAWBERRIES, WINTER OILSEED RAPE [1, 2]. Annual dicotyledons in FARM WOODLAND *(off-label)* [1]. Annual grasses in BLACKCURRANTS, CONTAINER-GROWN WOODY ORNAMENTALS, ESTABLISHED WOODY ORNAMENTALS, GOOSEBERRIES, NEWLY PLANTED WOODY ORNAMENTALS, RASPBERRIES, STRAWBERRIES, WINTER OILSEED RAPE [1, 2]. Annual grasses in FARM WOODLAND *(off-label)* [1]. Cleavers in BLACKCURRANTS, CONTAINER-GROWN WOODY ORNAMENTALS, ESTABLISHED WOODY ORNAMENTALS, GOOSEBERRIES, NEWLY PLANTED WOODY ORNAMENTALS, RASPBERRIES, STRAWBERRIES, WINTER OILSEED RAPE [1, 2]. Groundsel in BLACKCURRANTS, CONTAINER-GROWN WOODY ORNAMENTALS, ESTABLISHED WOODY ORNAMENTALS, GOOSEBERRIES, NEWLY PLANTED WOODY ORNAMENTALS, RASPBERRIES, STRAWBERRIES, WINTER OILSEED RAPE [1, 2].

Notes

Efficacy

- Best results obtained from treatment pre-emergence of weeds but product may be used in conjunction with contact herbicide such as paraquat for control of emerged weeds
- Product broken down by sunlight, so application during summer not recommended. Most crops recommended for treatment between Nov and end-Feb
- Apply to winter oilseed rape as pre-drilling treatment in tank-mixture with trifluralin and incorporated within 30 min. Post-emergence use of specific grass weedkilller recommended where volunteer cereals serious
- Where minimal cultivation used to establish oilseed rape, tank-mixture may be applied directly to stubble and mixed into top 25 mm as part of surface cultivations
- Increase water volume to ensure adequate dampening of compost when treating containerised nursery stock with dense leaf canopy

- Do not use on soils with more than 10% organic matter

Crop Safety/Restrictions

- Maximum number of treatments 1 per crop or yr
- Apply up to 14 d prior to drilling winter oilseed rape
- Do not use on Sands.
- Bush and cane fruit must be established for at least 10 mth before spraying and treated between 1 Nov and end Feb
- Apply to strawberries established for at least one season or to maiden crops as long as planted carefully and no roots exposed, between 1 Nov and end Feb. Do not treat runners of poor vigour or with shallow roots, or runner beds
- Nursery stock should be planted carefully with no roots exposed. Spray between 1 Nov and end Apr but do not treat stock of poor vigour or with shallow roots. Do not treat liners or containers less than 1 litre
- Some phytotoxicity seen on yellow and golden varieties of conifers and container grown alpines. On any ornamental variety treat only a small number of plants in first season
- After use in fruit or ornamentals no crop can be drilled within 7 mth of treatment. Leaf, flowerhead, root and fodder brassica crops may be drilled after 7 mth; potatoes, maize, peas or dwarf beans after 9 mth; autumn sown wheat or grass after 18 mth; any crop after 2 yr
- After use in oilseed rape only oilseed rape, swedes, fodder turnips, brassicas or potatoes should be sown within 12 mth of application
- Soil should be mould-board ploughed to a depth of at least 200 mm before drilling or planting any following crop

Special precautions/Environmental safety

- Irritating to skin and eyes
- Harmful to fish or other aquatic life. Do not contaminate surface waters or ditches with chemical or used container
- Do not apply by knapsack or hand-held sprayers

Protective clothing/Label precautions

- A, C, H
- U05a, U09a, U20b, C03, E01, E13c, E26, E30a, E31a, E34

Latest application/Harvest Interval(HI)

- Pre-sowing for winter oilseed rape; before end Feb for strawberries, bush and cane fruit; before end of Apr for field and container grown ornamental trees and shrubs

Approval

- Off-label approval unlimited for use on farm woodland (OLA 1280/97)[1]

416 nicotine

A general purpose, non-persistent, contact, alkaloid insecticide

Products

1 Nico Soap	United Phosphorus	75 g/l	LI	07517
2 Nicotine 40% Shreds	Dow	40% w/w	FU	0572[illegible]
3 No-Fid	Hortichem	75 g/l	LI	0795[illegible]
4 XL-All Insecticide	Vitax	70 g/l	LI	0236[illegible]
5 XL-All Nicotine 95%	Vitax	950 g/l	LI	0740[illegible]

Uses Aphids in CELERY, FLOWER BULBS, FLOWERS, NAVY BEANS, NURSERY STOCK, PARSLEY, PROTECTED ASPARAGUS, PROTECTED CELERY, PROTECTED CROPS, PROTECTED FLOWERS, PROTECTED MARROWS, PROTECTED MELONS, PROTECTED ROSES, SOFT FRUIT, TOP FRUIT, VEGETABLES [4, 5]. Aphids in PROTECTED COURGETTES, PROTECTED CUCUMBERS, PROTECTED LETTUCE, PROTECTED PEPPERS, PROTECTED TOMATOES [2, 4, 5]. Aphids in AUBERGINES, PROTECTED HERBS, PROTECTED POTATOES, PROTECTED SALAD ONIONS [2]. Aphids in PROTECTED ORNAMENTALS [1-5]. Aphids in BROAD BEANS, CARROTS, LEAF BRASSICAS, LETTUCE, ORNAMENTALS, PARSNIPS, PEAS, RED BEET, RUNNER BEANS, SPINACH, SWEDES, TURNIPS [1, 3-5]. Aphids in APPLES, ARTICHOKES, ASPARAGUS, CELERIAC, CHICORY, CHINESE CABBAGE, CHIVES, CHRYSANTHEMUMS, CUCUMBERS, CUCURBITS, DWARF BEANS, EDIBLE PODDED PEAS, ENDIVES, FENNEL, FIELD BEANS, FLOWERHEAD BRASSICAS, GARLIC, KALE, LEEKS, ONIONS, PEPPERS, POTATOES, RADICCHIO, RADISHES, SHALLOTS, STRAWBERRIES, SWEETCORN, TOMATOES [1, 3]. Capsids in FLOWER BULBS, FLOWERS, PROTECTED CROPS, SOFT FRUIT, TOP FRUIT, VEGETABLES [4, 5]. Capsids in APPLES [1, 3]. Caterpillars in APPLES, ARTICHOKES, ASPARAGUS, BROAD BEANS, BRUSSELS SPROUTS, CABBAGES, CARROTS, CELERIAC, CELERY, CHICORY, CHIVES, CUCURBITS, DWARF BEANS, FENNEL, FIELD BEANS, FLOWERHEAD BRASSICAS, FRENCH BEANS, GARLIC, GOOSEBERRIES, KALE, LEEKS, ONIONS, ORNAMENTALS, PARSNIPS, PEAS, POTATOES, PROTECTED ORNAMENTALS, RED BEET, RUNNER BEANS, SHALLOTS, SPINACH, STRAWBERRIES, SWEDES, SWEETCORN, TURNIPS [1, 3]. General insect control in FRENCH BEANS *(off-label)* [4, 5]. Glasshouse whitefly in AUBERGINES, PROTECTED COURGETTES, PROTECTED CUCUMBERS, PROTECTED HERBS, PROTECTED LETTUCE, PROTECTED ORNAMENTALS, PROTECTED PEPPERS, PROTECTED POTATOES, PROTECTED SALAD ONIONS, PROTECTED TOMATOES [2]. Green leafhopper in AUBERGINES, PROTECTED COURGETTES, PROTECTED CUCUMBERS, PROTECTED HERBS, PROTECTED LETTUCE, PROTECTED ORNAMENTALS, PROTECTED PEPPERS, PROTECTED POTATOES, PROTECTED SALAD ONIONS, PROTECTED TOMATOES [2]. Leaf miners in FLOWER BULBS, FLOWERS, LEAF BRASSICAS, PROTECTED CROPS [4, 5]. Leaf miners in CELERY, CHRYSANTHEMUMS [1, 3]. Leafhoppers in FLOWER BULBS, FLOWERS, PROTECTED CROPS, SOFT FRUIT, VEGETABLES [4, 5]. Potato virus vectors in CHITTING POTATOES [2]. Sawflies in FLOWER BULBS, FLOWERS, PROTECTED CROPS, SOFT FRUIT, TOP FRUIT, VEGETABLES [4, 5]. Sawflies in GOOSEBERRIES [1, 3]. Thrips in FLOWER BULBS, FLOWERS, PROTECTED CROPS, VEGETABLES [4, 5]. Thrips in AUBERGINES, PROTECTED COURGETTES, PROTECTED CUCUMBERS, PROTECTED HERBS, PROTECTED LETTUCE, PROTECTED ORNAMENTALS, PROTECTED PEPPERS, PROTECTED POTATOES, PROTECTED SALAD ONIONS, PROTECTED TOMATOES [2]. Woolly aphid in FLOWERS, FRUIT CROPS [4, 5].

Notes **Efficacy**

- Apply as foliar spray, taking care to cover undersides of leaves and repeat as necessary or dip young plants, cuttings or strawberry runners before planting out
- Best results achieved by spraying at air temperatures above 16°C but do not treat in bright sunlight or windy weather [2]
- Fumigate glasshouse crops at temperatures of at least 16°C. See label for details of recommended fumigation procedure [2]
- Potatoes may be fumigated in chitting houses to control virus-spreading aphids [2]. Field crops of potatoes may also be treated [1, 3]
- May be used in integrated control systems to give partial control of organophosphorus-, organochlorine- and pyrethroid-resistant whitefly

Crop Safety/Restrictions

- Maximum number of treatments normally not specified but 3 per crop for protected vegetables and hops [1]
- On plants of unknown sensitivity test first on a small scale

Special precautions/Environmental safety

- Nicotine is subject to the Poisons Rules 1982 and the Poisons Act 1972. See notes in Section 1 [2, 5]
- Toxic in contact with skin, by inhalation and if swallowed [5]
- Harmful in contact with skin, by inhalation and if swallowed [1, 2, 4, 5]
- Highly flammable [2]
- Keep unprotected persons out of treated glasshouses for at least 12 h
- Wear overall, hood, rubber gloves and respirator if entering glasshouse within 12 h of fumigating [2]
- Dangerous to livestock.
- Dangerous to game, wild birds and animals
- Harmful to bees. Do not apply to crops in flower or to those in which bees are actively foraging. Do not apply when flowering weeds are present
- Dangerous to fish or other aquatic life. Do not contaminate surface waters or ditches with chemical or used container
- Keep unprotected persons out of treated areas for at least 12 h following treatment and ensure adequate ventilation before re-entering

Protective clothing/Label precautions

- A [1-5]; C [1, 3, 5]; D [2]; H [2, 5]; J [2]; K, M [5]
- M03 [1-4]; M04, R02a, R02b, R02c [5]; R03a, R03b, R03c, U19, U20a, C03, E01, E10a, E12e, E13b [1-5]; R07c, U05b, C02 (24 h) [2]; U02, U04a, U05a, U13, E30a, E31b [1, 3-5]; U08, U10, U15, C02, E02, E06b, E31a [4, 5]; U09a [1, 3]; E02 (12 h), E06a (12 h) [1-3]; E30b [2, 5]; E32a, E34 [2, 4, 5]

Withholding period

- Keep all livestock out of treated areas for at least 12 h

Latest application/Harvest Interval(HI)

- HI 2 d for most crops - check labels

Approval

- Off-label Approval unlimited for use on French beans (OLA 0080/92)[4]

417 nuarimol

A protectant and eradicant pyrimidine fungicide available only in mixture

See also captan + nuarimol

418 octhilinone

An isothiazolone fungicide for treating canker and tree wounds

Products	Pancil T	Rohm & Haas	1% w/w	PA	01540

FOR FULL CONDITIONS OF USE ALWAYS READ THE PRODUCT LABEL

Uses Canker in APPLES, PEARS, WOODY ORNAMENTALS. Pruning wounds in APPLES, CHERRIES, PEARS, PLUMS, WOODY ORNAMENTALS. Silver leaf in APPLES, CHERRIES, PEARS, PLUMS, WOODY ORNAMENTALS.

Notes

Efficacy

- Treat canker by removing infected shoots, cutting back to healthy tissue and painting evenly over wound
- Apply to pruning wounds immediately after cutting. Cover all cracks in bark and ensure that coating extends beyond edges of wound
- Best results achieved by application in dry conditions

Crop Safety/Restrictions

- Only apply after harvest and before bud burst
- Do not apply to grafting cuts or bud. Do not apply under freezing conditions

Special precautions/Environmental safety

- Irritating to skin, risk of serious damage to eyes
- May cause sensitization by skin contact
- Dangerous to fish or other aquatic life. Do not contaminate surface waters or ditches with chemical or used container

Protective clothing/Label precautions

- A, C
- M03, R04b, R04d, R04e, U05a, U09a, U13, U19, U20a, C03, E01, E13b, E26, E30a, E31a, E34

419 ofurace

A phenylamide (acylalanine) fungicide available only in mixtures

See also mancozeb + ofurace

420 oxadiazon

A residual and contact oxadiazolone herbicide for fruit and ornamentals

See also glyphosate + oxadiazon

Products

1 Ronstar 2G	Hortichem	2% w/w	GR	06492
2 Ronstar Liquid	Hortichem	250 g/l	EC	06493

Uses Annual dicotyledons in CONTAINER-GROWN ORNAMENTALS [1]. Annual dicotyledons in APPLES, BLACKCURRANTS, GOOSEBERRIES, GRAPEVINES, HOPS, PEARS, RASPBERRIES, WOODY ORNAMENTALS [2]. Annual grasses in CONTAINER-GROWN ORNAMENTALS [1]. Annual grasses in APPLES, BLACKCURRANTS, GOOSEBERRIES, GRAPEVINES, HOPS, PEARS, RASPBERRIES, WOODY ORNAMENTALS [2]. Bindweeds in APPLES, BLACKCURRANTS, GOOSEBERRIES, GRAPEVINES, HOPS, PEARS, RASPBERRIES, WOODY ORNAMENTALS [2]. Cleavers in APPLES, BLACKCURRANTS, GOOSEBERRIES, GRAPEVINES, HOPS, PEARS, RASPBERRIES, WOODY ORNAMENTALS [2]. Knotgrass in APPLES, BLACKCURRANTS, GOOSEBERRIES, GRAPEVINES, HOPS, PEARS, RASPBERRIES, WOODY ORNAMENTALS [2].

Notes

Efficacy
- Rain or overhead watering is needed soon after application for effective results
- Pre-emergence activity reduced on soils with more than 10% organic matter and, in these conditions, post-emergence treatment is more effective [2]
- Best results on bindweed when first shoots are 10-15 cm long
- Do not cultivate after treatment [2]

Crop Safety/Restrictions
- See label for list of ornamental species which may be treated with granules. Treat small numbers of other species to check safety. Do not treat hydrangea, spiraea or genista
- Do not treat container stock under glass or use on plants rooted in media with high sand or non-organic content. Do not apply to plants with wet foliage
- Apply spray to apples and pears from Jan to Jul, avoiding young growth
- Treat bush fruit from Jan to bud-break, avoiding bushes, grapevines in Feb/Mar before start of new growth or in Jun/Jul avoiding foliage
- Treat hops cropped for at least 2 yr in Feb or in Jun/Jul after deleafing
- Treat woody ornamentals from Jan to Jun, avoiding young growth. Do not spray container stock overall
- Do not use more than 8 l/ha in any 12 mth period [2]

Special precautions/Environmental safety
- Irritating to eyes
- Flammable [2]
- Dangerous to fish or other aquatic life. Do not contaminate surface waters or ditches with chemical or used container

Protective clothing/Label precautions
- A, C [1, 2]
- R04a, U05a, U08, U14, U15, U20b, C03, E01, E13b, E30a, E32a, E34 [1, 2]; R07d [2]

Latest application/Harvest Interval(HI)
- Jul for apples, grapevines, hops, pears; Jun for raspberries, woody ornamentals

421 oxadixyl

A phenylamide (acylalanine) fungicide available only in mixtures

See also carbendazim + cymoxanil + oxadixyl + thiram
cymoxanil + mancozeb + oxadixyl
mancozeb + oxadixyl

422 oxamyl

A soil-applied, systemic oxime carbamate nematicide and insecticide

Products

Vydate 10G	DuPont	10% w/w	GR	02322

Uses

American serpentine leaf miner in AUBERGINES *(off-label)*, BROAD BEANS *(off-label* CUCURBITS *(off-label)*, GARLIC *(off-label)*, ORNAMENTALS *(off-label)*, PEPPERS *(of*

FOR FULL CONDITIONS OF USE ALWAYS READ THE PRODUCT LABEL

label), SHALLOTS *(off-label)*, SOYA BEAN *(off-label)*, TOMATOES *(off-label)*. Aphids in POTATOES, SUGAR BEET. Docking disorder vectors in SUGAR BEET. Free-living nematodes in POTATOES. Mangold fly in SUGAR BEET. Millipedes in SUGAR BEET. Non-indigenous leaf miners in AUBERGINES *(off-label)*, BROAD BEANS *(off-label)*, CUCURBITS *(off-label)*, GARLIC *(off-label)*, ORNAMENTALS *(off-label)*, PEPPERS *(off-label)*, SHALLOTS *(off-label)*, SOYA BEAN *(off-label)*, TOMATOES *(off-label)*. Potato cyst nematode in POTATOES. Pygmy beetle in SUGAR BEET. Spraing vectors in POTATOES. Stem nematodes in GARLIC *(off-label)*, PROTECTED GARLIC *(off-label)*, PROTECTED ONIONS *(off-label)*.

Notes

Efficacy
- Apply granules with suitable applicator before drilling or planting. See label for details of recommended machines
- In potatoes incorporate thoroughly to 10 cm and plant within 3-4 d
- In sugar beet apply in seed furrow at drilling

Crop Safety/Restrictions
- Maximum number of treatments 1 per crop or yr

Special precautions/Environmental safety
- This product contains an anticholinesterase carbamate compound. Do not use if under medical advice not to work with such compounds
- Harmful by inhalation or if swallowed
- Keep in original container, tightly closed, in a safe place, under lock and key
- Dangerous to fish or other aquatic life. Do not contaminate surface waters or ditches with chemical or used container
- Dangerous to game and wildlife. Cover granules completely. Bury spillages
- Wear protective gloves if handling treated compost or soil within 2 wk after treatment
- Allow at least 12 h, followed by at least 1 h ventilation, before entry of unprotected persons into treated glasshouses

Protective clothing/Label precautions
- A, B, C or D, H, K, M
- M02, M04, R03b, R03c, U02, U04a, U05a, U09a, U13, U19, U20a, C02, C03, E01, E10a, E13b, E30b, E32a, E34

Latest application/Harvest Interval(HI)
- At drilling/planting for vegetables; before drilling/planting for potatoes and peas.
- HI tomatoes, ornamentals 2 wk

Approval
- Off-label Approval unlimited for use on outdoor and protected ornamentals, protected tomatoes, aubergines, peppers, soya beans and broad beans (OLA 0020/93)[1]; unlimited for use on protected onions and garlic (OLA 0925/94)[1]; unlimited for use on outdoor garlic for stem nematode control (OLA 0163/92)[1]

423 oxycarboxin

A protectant, eradicant and systemic carboxamide fungicide

Products

1 Plantvax 20	Mirfield	200 g/l	EC	01600
2 Plantvax 75	Fargro	75% w/w	WP	01601
3 Ringmaster	RP Amenity	200 g/l	EC	05334

Uses Fairy rings in AMENITY GRASS, SPORTS TURF [3]. Rust in CARNATIONS, CHRYSANTHEMUMS, PELARGONIUMS, ROSES [2]. Yellow rust in WHEAT [1].

Notes

Efficacy

- Apply in wheat at first signs of infection. Where heavy infections occur a second application may be made after a minimum interval of 21 d [1]
- Apply as a routine to established carnation beds from Sep to Feb at 7-10 d intervals. If rust appears spray immediately and repeat several times [2]
- On stock plants or cuttings of carnations or geraniums apply after striking and repeat 3 times at 10-14 d intervals. Remove and burn any infected cuttings [2]
- Add wetter for first spray on carnations, geraniums and roses if rust already established, or on very rust susceptible rose varieties [2]
- Can be used as a drench on carnations grown in peat bags [2]
- Apply to fairy rings after spiking. On dry soil water before spraying or add wetter. Best results when fungus in active growth [3]

Crop Safety/Restrictions

- On chrysanthemums apply alone, not within 2 d of other sprays. Do not apply spraying oil for 14 d before or after. Can scorch chrysanthemum leaves when taken up by roots. Avoid spraying when blooms open [2]

Special precautions/Environmental safety

- Harmful if swallowed [1, 3]
- Irritating to eyes [2], skin and respiratory system [1, 3]
- Not to be used on food crops [3]
- Dangerous to fish or other aquatic life. Do not contaminate surface waters or ditches with chemical or used container

Protective clothing/Label precautions

- A, C
- M03, R03c, R04b, R04c, U08, U19, E34 [1, 3]; R04a, U05a, U20b, C03, E01, E13b, E30a [1-3]; U02 [1, 2]; U04a, E26, E31b [1]; U09a, E32a [2]; E31a [3]

424 paclobutrazol

A conazole plant growth regulator for ornamentals and fruit

See also dicamba + paclobutrazol

Products

1 Bonzi	Zeneca	4 g/l	SC	06640
2 Cultar	Zeneca	250 g/l	SC	06649

Uses Controlling vigour in APPLES, CHERRIES, PEARS, PLUMS [2]. Improving colour in POINSETTIAS [1]. Increasing flowering in AZALEAS, BEDDING PLANTS, BEGONIAS, KALANCHOES, LILIES, ROSES, TULIPS [1]. Increasing fruit set in APPLES, CHERRIES, PEARS, PLUMS [2]. Stem shortening in AZALEAS, BEDDING PLANTS, BEGONIAS, KALANCHOES, LILIES, ROSES, TULIPS [1].

Notes

Efficacy

- Chemical is active via both foliage and root uptake
- Apply as spray to produce compact pot plants and to improve bract colour of poinsettias [1]
- Apply as drench to reduce flower stem length of potted tulips [1]
- Timing is critical and varies with species. See label for details [1]
- Apply to apple and pear trees under good growing conditions as pre-blossom spray (apples only) and post-blossom at 7-14 d intervals [2]
- Timing and dose of orchard treatment vary with species and cultivar. See label [2]
- Apply to plums and cherries as a trench drench in the spring [2]

Crop Safety/Restrictions

- Maximum concentration 250 ml/10 l water [1]
- Maximum total dose 3.0 l/ha (pears), 4.0 l/ha (apples) [2]
- Maximum number of treatments 1 per yr for cherries, plums [2]
- Some varietal restrictions apply in top fruit (see label for details) [2]
- Do not use on trees of low vigour or under stress [2]
- Do not tank mix with or apply to apples and pears on same day as Thinsec [2]
- Do not use on trees from green cluster to 2 wk after full petal fall [2]
- Do not use in underplanted orchards. Consult label for guidance on following crops after grubbing [2]

Special precautions/Environmental safety

- Do not use on food crops [1]
- Harmful if swallowed [2]
- Harmful to fish or other aquatic life. Do not contaminate surface waters or ditches with chemical or used container

Protective clothing/Label precautions

- A, H, M [2]
- M03, R03c, U05a, C02 (2 wk), C03, E01, E07 [2]; U08, C01 [1]; U20a, E13c, E26, E30a, E31c, E34 [1, 2]

Withholding period

- Keep livestock out of treated areas [2]

Latest application/Harvest Interval(HI)

- Before end Apr for cherries; before beginning Apr for plums [2]
- HI apples, pears 14 d [2]

425 paraffin oil (commodity substance)

An agent for the control of birds by egg treatment

Products

paraffin oil	various	OL

Uses

Birds in MISCELLANEOUS SITUATIONS *(egg treatment)*.

Notes

Efficacy

- Egg treatment should be undertaken as soon as clutch is complete
- Eggs should be treated by complete immersion in liquid paraffin

Crop Safety/Restrictions

- Treat eggs once only
- Use to control eggs of birds covered by licences issued by the Agriculture and Environment Departments under Section 16(1) of the Wildlife and Countryside Act (1981)

Protective clothing/Label precautions

- A, C

Approval

- Only to be used where a licence has been approved in accordance with Section 16(1) of the Wildlife and Countryside Act (1981)

426 paraquat

A non-selective, non-residual contact bipyridilium herbicide

See also diquat + paraquat
diuron + paraquat

Products

1 Barclay Total	Barclay	200 g/l	SL	08822
2 Dextrone X	Nomix-Chipman	200 g/l	SL	00687
3 Gramoxone 100	Zeneca	200 g/l	SL	06674
4 Landgold Paraquat	Landgold	200 g/l	SL	06025
5 Speedway	Miracle	8% w/w	WG	07743
6 Standon Paraquat	Standon	200 g/l	SL	05621

Uses

Annual dicotyledons in WOODY ORNAMENTALS [2, 5]. Annual dicotyledons in FIELD CROPS *(conventional cultivation)*, FIELD CROPS *(direct drilling)*, GRASSLAND *(sward desiccation)* [4, 6]. Annual dicotyledons in GRAPEVINES, GRASSLAND *(sward destruction/direct drilling)*, HARDY ORNAMENTALS, HERBS *(off-label)*, LETTUCE *(off-label)*, LUCERNE *(off-label)*, MINT *(off-label)*, PROTECTED VEGETABLES *(off-label)*, TOP FRUIT [3]. Annual dicotyledons in NON-CROP AREAS [1-6]. Annual dicotyledons in FLOWER BULBS, FOREST NURSERY BEDS *(stale seedbed)*, FORESTRY, FORESTRY TRANSPLANT LINES [1-3]. Annual dicotyledons in FIELD CROPS *(minimum cultivation)*, POTATOES, ROW CROPS [1, 3, 4, 6]. Annual dicotyledons in BLACKCURRANTS, FIELD CROPS *(stubble treatment)*, GOOSEBERRIES, HOPS, RASPBERRIES, SUGAR BEET [1, 3]. Annual dicotyledons in ORCHARDS, REDCURRANTS, WHITECURRANTS [1]. Annual grasses in WOODY ORNAMENTALS [2, 5]. Annual grasses in FIELD CROPS *(conventional cultivation)*, FIELD CROPS *(direct drilling)*, GRASSLAND *(sward desiccation)* [4, 6]. Annual grasses in GRAPEVINES, HARDY ORNAMENTALS, HERBS *(off-label)*, LETTUCE *(off-label)*, LUCERNE *(off-label)*, MINT *(off-label)*, PROTECTED VEGETABLES *(off-label)*, TOP FRUIT [3]. Annual grasses in NON-CROP AREAS [1-6]. Annual grasses in FLOWER BULBS, FOREST NURSERY BEDS *(stale seedbed)*, FORESTRY, FORESTRY TRANSPLANT LINES [1-3]. Annual grasses in FIELD CROPS *(minimum cultivation)*, POTATOES, ROW CROPS [1, 3, 4, 6]. Annual grasses in BLACKCURRANTS, FIELD CROPS *(stubble treatment)*, GOOSEBERRIES, GRASSLAND *(sward destruction/direct drilling)*, HOPS, RASPBERRIES, SUGAR BEET [1, 3]. Annual grasses in ORCHARDS, REDCURRANTS, WHITECURRANTS [1]. Barren brome in FIELD CROPS *(conventional drilling)* [4, 6]. Barren brome in FIELD CROPS *(stubble treatment)* [1, 3]. Chemical stripping in HOPS [1, 3]. Creeping bent in WOODY ORNAMENTALS [2, 5]. Creeping bent in FIELD CROPS *(conventional drilling)*, FIELD CROPS *(direct drilling)*[4, 6]. Creeping bent in HARDY ORNAMENTALS, TOP FRUIT [3].

Creeping bent in NON-CROP AREAS [1-6]. Creeping bent in FLOWER BULBS, FORESTRY, FORESTRY TRANSPLANT LINES [1-3]. Creeping bent in FIELD CROPS *(minimum cultivation)* [1, 3, 4, 6]. Creeping bent in BLACKCURRANTS, FIELD CROPS *(stubble treatment)*, GOOSEBERRIES, GRASSLAND *(sward destruction/direct drilling)*, HOPS, RASPBERRIES [1, 3]. Creeping bent in ORCHARDS, REDCURRANTS, WHITECURRANTS [1]. Firebreak desiccation in FORESTRY [1-3]. Green cover in FIELD MARGINS [3]. Green cover in LAND TEMPORARILY REMOVED FROM PRODUCTION [1, 3]. Perennial ryegrass in FIELD CROPS *(direct drilling)* [4, 6]. Perennial ryegrass in TOP FRUIT [3]. Perennial ryegrass in NON-CROP AREAS [1-6]. Perennial ryegrass in BLACKCURRANTS, GOOSEBERRIES, GRASSLAND *(sward destruction/direct drilling)*, HOPS, RASPBERRIES [1, 3]. Perennial ryegrass in ORCHARDS, REDCURRANTS, WHITECURRANTS [1]. Rough meadow grass in FIELD CROPS *(direct drilling)* [4, 6]. Rough meadow grass in TOP FRUIT [3]. Rough meadow grass in NON-CROP AREAS [1-6]. Rough meadow grass in BLACKCURRANTS, GOOSEBERRIES, GRASSLAND *(sward destruction/direct drilling)*, HOPS, RASPBERRIES [1, 3]. Rough meadow grass in ORCHARDS, REDCURRANTS, WHITECURRANTS [1]. Runner desiccation in STRAWBERRIES [1, 3]. Volunteer cereals in FIELD CROPS *(conventional drilling)* [4, 6]. Volunteer cereals in FIELD CROPS *(minimum cultivation)*, POTATOES, ROW CROPS [1, 3, 4, 6]. Volunteer cereals in FIELD CROPS *(stubble treatment)*, SUGAR BEET [1, 3]. Wild oats in FIELD CROPS *(conventional drilling)* [4, 6]. Wild oats in FIELD CROPS *(stubble treatment)*[1, 3].

Notes

Efficacy

- Apply to green weeds preferably less than 15 cm high. Spray is rainfast after 10 min (15 min [6])
- Addition of wetter recommended with lower application rates [1, 3, 6]
- Only use clean water for mixing up spray
- Allow at least 4 h before cultivating, leave overnight if possible
- Spray in autumn to suppress couch when shoots have 2 leaves, repeat as necessary and plough after last treatment
- For direct-drilling land should be free of perennial weeds and protection against slugs should be provided. Allow 7-10 d before drilling into sprayed grass
- Best time for use in top fruit is Nov-Apr
- For forestry fire-break use apply in Jul-Aug and fire 7-10 d later

Crop Safety/Restrictions

- Maximum number of treatments 1 per yr for lucerne, outdoor lettuce, protected vegetables and ornamentals; 2 per yr for mint and leafy herbs
- Chemical is inactivated on contact with soil. Crops can be sown or planted soon after spraying on most soils 24 h [2], 4 h [3, 6], after 3 d on sandy or immature peat soils
- Do not use on straw or other artificial growing media
- Apply to lucerne in late Feb/early Mar when crop dormant
- Do not use on potatoes under hot, dry conditions or on hops under drought conditions
- For interrow use in row crops apply with guarded, no-drift sprayer
- Use as a carefully directed spray in blackcurrants, gooseberries, grapevines and other fruit crops, in raspberries only when dormant
- For stawberry runner control use guarded sprayer, not when flowers or fruit present
- Apply to bulbs pre-emergence (at least 3 d pre-emergence on sandy soils) or at end of season, provided no attached foliage and bulbs well covered (not on sandy soils)
- If using around glasshouses ensure vents and doors closed
- In forestry seedbed apply up to 3 d before seedling emergence

Special precautions/Environmental safety

- Paraquat is subject to the Poisons Rules 1982 and the Poisons Act 1972. See notes in Section 1 [1-4, 6]

- Toxic if swallowed. Paraquat can kill if swallowed [1-4, 6]
- Harmful if swallowed [5]
- Harmful in contact with skin [1-4, 6]
- Irritating to eyes and skin
- Keep in original container, tightly closed, in a safe place, under lock and key
- Do not put in a food or drinks container
- Products in this guide are for professional use only
- Harmful to animals
- Do not contaminate surface waters or ditches with chemical or used container
- Paraquat may be harmful to hares. Where possible spray stubbles early in the day [1-3, 6]

Protective clothing/Label precautions
- A, C
- M04, R04a, R04b, U05a, U19, C03, E01, E30b, E34 [1-6]; R02c, R03a, E26 [1-4, 6]; R03c, E07 [5]; U02 [1, 4, 5]; U04a [1, 4]; U04b [2, 3, 5, 6]; U08 [3, 6]; U09a [1, 2, 4, 5]; U20a [1, 4, 6]; U20b [2, 3, 5]; E06a [6]; E06b (24 h) [3]; E06b, E31b [1]; E11 (1 d) [4]; E11 (24 h), E13c [2]; E11 [1, 3, 6]; E15 [1, 3-6]; E31a [2-6]

Withholding period
- Keep all livestock out of treated areas for at least 24 h

Latest application/Harvest Interval(HI)
- Before shoots 15 cm high and before 10% of shoots emerged for early potatoes, before 40% of shoots emerged for maincrop potatoes [1, 3]; before 31 Mar for lucerne; see labels for other crops
- HI mint, leafy herbs 8 wk [3]

Approval
- Off-label Approval to Apr 2001 for use on leafy herbs (OLA 0832/96)[3]; unlimited for use on dormant lucerne (OLA 1188/96)[3]; unlimited for use on a wide range of protected and outdoor vegetables (see OLA notice for details)(OLA 1497/96)[3]

Maximum Residue Levels (mg residue/kg food)
- tea, hops 0.1; all other products (except cereals, animal products) 0.05

---

427 penconazole

A protectant conazole fungicide with antisporulant activity

See also captan + penconazole

Products

1 Topas	Novartis	100 g/l	EC	08458
2 Topas 100 EC	Ciba Agric.	100 g/l	EC	03231

Uses

Powdery mildew in APPLES, HOPS, ORNAMENTAL TREES. Rust in ROSES. Scab ir ORNAMENTAL TREES.

Notes

Efficacy
- Apply every 10-14 d (every 7-10 d in warm, humid weather) at first sign of infection or as a protective spray ensuring complete coverage. See label for details of timing

- Increase dose and volume with growth of hops but do not exceed 2000 l/ha
- Antisporulant activity reduces development of secondary mildew in apples

Crop Safety/Restrictions
- Maximum number of treatments 10 per yr for apples, 6 per yr for hops, 4 per yr for blackcurrants
- Check for varietal susceptibility in roses. Some defoliation may occur after repeat applications on Dearest

Special precautions/Environmental safety
- Irritating to eyes and skin
- Harmful to fish or other aquatic life. Do not contaminate surface waters or ditches with chemical or used container

Protective clothing/Label precautions
- A, C
- R04a, R04b, U02, U05a, U08, U20a, C02, C03, E01, E13c, E30a, E31b, E34

Latest application/Harvest Interval(HI)
- HI apples, hops 14 d

---

428 pencycuron

A non-systemic urea fungicide for use on seed potatoes

See also imazalil + pencycuron

Products

1 Monceren DS	Bayer	12.5% w/w	DS	04160
2 Monceren Flowable	Bayer	250 g/l	FS	04907
3 Standon Pencycuron DP	Standon	12.5% w/w	DS	08774

Uses Black scurf in POTATOES. Stem canker in POTATOES.

Notes

Efficacy
- Provides control of tuber-borne disease and gives some reduction of stem canker
- Apply to seed tubers in chitting trays, to bulk bins immediately before planting or in hopper at planting [1, 3]
- Apply to clean seed tubers at any time, into or out of store or in hopper at planting [2]
- If rain interrupts planting cover tubers in hopper

Crop Safety/Restrictions
- Maximum number of treatments 1 per batch of seed potatoes
- Treated tubers must be used only as seed and not for human or animal consumption

Special precautions/Environmental safety
- Use of suitable dust mask is mandatory when applying dust, filling the hopper or riding on planter
- Do not contaminate surface waters or ditches with chemical or used container
- Do not use treated seed as food or feed
- Treated seed harmful to game and wildlife

Protective clothing/Label precautions
- A [1-3]; F [1, 3]
- U20a [2]; U20b [1, 3]; E03, E15, E30a, E32a, S01, S02, S03, S04b, S05, S06 [1-3]

Latest application/Harvest Interval(HI)

- At planting

Approval

- May be applied by misting equipment mounted over roller table (see label for details) [2]

429 pendimethalin

A residual dinitroaniline herbicide for cereals and other crops

See also cyanazine + pendimethalin
isoproturon + pendimethalin

Products

1 Sovereign	Ciba Agric.	400 g/l	SC	08152
2 Sovereign	Novartis	400 g/l	SC	08533
3 Stomp 400 SC	Cyanamid	400 g/l	SC	04183

Uses

Annual dicotyledons in COMBINING PEAS, DURUM WHEAT, EVENING PRIMROSE *(off-label)*, FARM FORESTRY *(off-label)*, FODDER MAIZE, PARSLEY *(off-label)*, POTATOES, SAGE *(off-label)*, SPRING BARLEY, SUNFLOWERS, TRITICALE, WINTER BARLEY, WINTER RYE, WINTER WHEAT [3]. Annual dicotyledons in HERBS *(off-label)*, OUTDOOR LEAF HERBS *(off-label)* [1-3]. Annual dicotyledons in APPLES, BLACKBERRIES, BLACKCURRANTS, CARROTS, CHERRIES, GOOSEBERRIES, HOPS, LEEKS, LEEKS *(off-label)*, LOGANBERRIES, ONIONS, ONIONS *(off-label)*, PARSLEY, PARSNIPS, PEARS, PLUMS, RASPBERRIES, STRAWBERRIES, SWEETCORN *(off-label)*, TAYBERRIES, TRANSPLANTED BRASSICAS [1, 2]. Annual dicotyledons in PROTECTED LETTUCE *(off-label)*, RADICCHIO *(off-label)*, RUNNER BEANS *(off-label)* [2]. Annual grasses in COMBINING PEAS, DURUM WHEAT, EVENING PRIMROSE *(off-label)*, FARM FORESTRY *(off-label)*, PARSLEY *(off-label)*, POTATOES, SAGE *(off-label)*, SPRING BARLEY, SUNFLOWERS, TRITICALE, WINTER BARLEY, WINTER RYE, WINTER WHEAT [3]. Annual grasses in HERBS *(off-label)*, OUTDOOR LEAF HERBS *(off-label)* [1-3]. Annual grasses in APPLES, BLACKBERRIES, BLACKCURRANTS, CARROTS, CHERRIES, GOOSEBERRIES, HOPS, LEEKS, LEEKS *(off-label)*, LOGANBERRIES, ONIONS, ONIONS *(off-label)*, PARSLEY, PARSNIPS, PEARS, PLUMS, RASPBERRIES, STRAWBERRIES, SWEETCORN *(off-label)*, TAYBERRIES, TRANSPLANTED BRASSICAS [1, 2]. Annual grasses in PROTECTED LETTUCE *(off-label)*, RADICCHIO *(off-label)*, RUNNER BEANS *(off-label)* [2]. Annual meadow grass in COMBINING PEAS, DURUM WHEAT, FODDER MAIZE, POTATOES, SPRING BARLEY, SUNFLOWERS, TRITICALE, WINTER BARLEY, WINTER RYE, WINTER WHEAT [3]. Blackgrass in COMBINING PEAS, DURUM WHEAT, POTATOES, SPRING BARLEY, SUNFLOWERS, TRITICALE, WINTER BARLEY, WINTER RYE, WINTER WHEAT [3]. Cleavers in COMBINING PEAS, DURUM WHEAT, POTATOES, SPRING BARLEY, SUNFLOWERS, TRITICALE, WINTER BARLEY, WINTER RYE, WINTER WHEAT [3]. Speedwells in COMBINING PEAS, DURUM WHEAT, FODDER MAIZE, POTATOES, SPRING BARLEY, SUNFLOWERS, TRITICALE, WINTER BARLEY, WINTER RYE, WINTER WHEAT [3]. Wild oats in COMBINING PEAS, DURUM WHEAT POTATOES, SPRING BARLEY, SUNFLOWERS, TRITICALE, WINTER BARLEY, WINTER RYE, WINTER WHEAT [3].

Notes

Efficacy

- Apply as soon as possible after drilling. Weeds are controlled as they germinate and emerged weeds will not be controlled by use of the product alone
- For effective blackgrass control apply not more than 2 d after final cultivation and before weed seeds germinate
- Tank mixes with approved formulations of isoproturon or chlorotoluron recommended for improved pre- and post-emergence control of blackgrass, with isoxaben for additional broad-leaved weeds [3]
- Tank mixture with atrazine recommended for weed control in fodder maize [3]
- Best results by application to fine firm, moist, clod-free seedbeds when rain follows treatment. Effectiveness reduced by prolonged dry weather after treatment
- Do not use on spring barley after end Mar (mid-Apr in Scotland) because dry conditions likely. Do not apply to dry seedbeds in spring unless rain imminent [3]
- Effectiveness reduced on soils with more than 6% organic matter. Do not use where organic matter exceeds 10%
- Any trash, ash or straw should be incorporated evenly during seedbed preparation
- Do not disturb soil after treatment
- On peas drilled after end Mar (mid-Apr in Scotland) tank-mix with cyanazine [3]
- Apply to potatoes as soon as possible after planting and ridging in tank-mix with cyanazine or metribuzin [3]

Crop Safety/Restrictions

- Maximum number of treatments 1 per crop or yr
- May be applied pre-emergence of cereal crops sown before 30 Nov provided seed covered by at least 32 mm soil, or post-emergence to early tillering stage (GS 23) [3]
- Do not undersow treated crops
- Do not use on crops suffering stress due to disease, drought, waterlogging, poor seedbed conditions or chemical treatment or on soils where water may accumulate
- Apply to combining peas as soon as possible after sowing, not when plumule within 13 mm of surface [3]
- Apply to potatoes up to 7 d before first shoot emerges [3]
- In the event of crop failure specified crops may be sown after at least 2 mth following ploughing to 150 mm. See label for details
- After a dry season land must be ploughed to 150 mm before drilling ryegrass
- Apply to drilled crops as soon as possible after drilling but before crop and weed emergence
- Apply in top fruit, bush fruit and hops from autumn to early spring when crop dormant [1]
- In cane fruit apply to weed free soil from autumn to early spring, immediately after planting new crops and after cutting out canes in established crops [1]
- Apply in strawberries from autumn to early spring (not before Oct on newly planted bed). Do not apply pre-planting or during flower initiation period (post-harvest to mid-Sep) [1]
- Apply pre-emergence in drilled onions or leeks as tank-mixture with propachlor, not on Sands, Very Light, organic or peaty soils or when heavy rain forecast [1]
- Apply to brassicas after final plant-bed cultivation but before transplanting. Avoid unnecessary soil disturbance after application and take care not to introduce treated soil into the root zone when transplanting. Follow transplanting with specified post-planting treatments - see label [1]
- Do not use on protected crops or in greenhouses [1]

Special precautions/Environmental safety

- Dangerous to fish or other aquatic life. Do not contaminate surface waters or ditches with chemical or used container

Protective clothing/Label precautions

- U05a, U08, U13, U19, U20a, C03, E01, E13b, E30a, E31b [1-3]; E26 [1, 2]; E34 [3]

Latest application/Harvest Interval(HI)

- Pre-emergence for spring barley, carrots, lettuce, fodder maize, parsnips, parsley, sage, peas, runner beans, potatoes, onions and leeks; before transplanting for brassicas; before main shoot and 3 tillers stage (GS 23) for winter cereals; before bud burst for blackcurrants, gooseberries, cane fruit; before flower trusses emerge for strawberries.
- HI evening primrose, outdoor parsley and sage 5 mth

Approval

- Off-label Approval unlimited for use on outdoor parsley and sage (OLA 0175/93)[3]; unlimited for use in farm forestry (OLA 0226/94)[3]; unlimited for use on evening primrose (pre-emergence) (OLA 0319/92, 0660/92)[3]; unlimited for use on herbs and outdoor leaf herbs (OLA 1214/95)[1], (OLA 1727/97)[2], (OLA 1215/95)[3]; unlimited for use on outdoor leaf herbs grown under covers (OLA 1225/95)[1], (OLA 1726/97)[2]; unlimited for use on sweet corn under covers and mulches (OLA 0716/97)[1], (OLA 1109/97)[2]; unlimited for use on onions, leeks (OLA 0938/97)[1], (OLA 1808/98)[2]; unlimited for use on runner beans (OLA 1729/97)[2]

---

430 pendimethalin + simazine

A contact and residual herbicide for use in winter cereals

Products	Merit	Cyanamid	300:100 g/l	SC	04976

Uses

Annual dicotyledons in WINTER BARLEY, WINTER WHEAT. Meadow grasses in WINTER BARLEY, WINTER WHEAT.

Notes

Efficacy

- Apply from pre-emergence to early tillering stage of crop
- Best results achieved on fine, firm, moist seedbed
- Do not use on soils with more than 10% organic matter
- Any straw residues should be incorporated before spraying

Crop Safety/Restrictions

- Maximum number of treatments 1 per crop
- Do not use on durum wheat
- Ensure crop seed is evenly covered with at least 32 mm of settled soil
- Do not use on Sands, Very Light, stony or gravelly soils
- Do not apply to crops under stress or on waterlogged soils
- Avoid overlapping spray swathes

Special precautions/Environmental safety

- Irritating to eyes and skin
- Dangerous to fish or other aquatic life. Do not contaminate surface waters or ditches with chemical or used container

Protective clothing/Label precautions
- A, C
- R04a, R04b, U05a, U08, U13, U19, U20a, C03, E01, E13b, E16, E26, E30a, E31b, E34

Latest application/Harvest Interval(HI)
- Before main shoot and 2 tillers (GS 21)

---

431 pentanochlor

A contact anilide herbicide for various horticultural crops

See also chlorpropham + pentanochlor

Products

1 Atlas Solan 40	Atlas	400 g/l	EC	07726
2 Croptex Bronze	Hortichem	400 g/l	EC	04087

Uses

Annual dicotyledons in ANEMONES, ANNUAL FLOWERS, CARROTS, CHRYSANTHEMUMS, FREESIAS, NURSERY STOCK, PARSNIPS, SWEET PEAS [1, 2]. Annual dicotyledons in APPLES, CARNATIONS, CELERIAC, CELERY, CHERRIES, CONIFER SEEDLINGS, CURRANTS, FENNEL, FOXGLOVES, GOOSEBERRIES, LARKSPUR, PARSLEY, PEARS, PLUMS, ROSES, SWEET WILLIAMS, TOMATOES, UMBELLIFEROUS HERBS *(off-label)*, WALLFLOWERS [1]. Annual meadow grass in ANEMONES, ANNUAL FLOWERS, CARROTS, CHRYSANTHEMUMS, FREESIAS, NURSERY STOCK, PARSNIPS, SWEET PEAS [1, 2]. Annual meadow grass in APPLES, CARNATIONS, CELERIAC, CELERY, CHERRIES, CONIFER SEEDLINGS, CURRANTS, FENNEL, FOXGLOVES, GOOSEBERRIES, LARKSPUR, PARSLEY, PEARS, PLUMS, ROSES, SWEET WILLIAMS, TOMATOES, UMBELLIFEROUS HERBS *(off-label)*, WALLFLOWERS [1].

Notes

Efficacy
- Best results achieved on young weed seedlings under warm, moist conditions
- Weeds most susceptible in cotyledon to 2-leaf stage, up to 3 cm high, redshank, fat-hen, fumitory and some others also controlled at later stages
- Some residual effect in early spring when adequate soil moisture and growing conditions good. Effectiveness reduced by very cold weather or drought

Crop Safety/Restrictions
- Maximum number of treatments 1 or 2 per crop depending on dose applied. See label for details
- Apply pre-emergence in anemones, freesias, foxgloves, larkspur
- Not recommended for flower crops on light sandy soils
- Apply pre-emergence or after fully expanded cotyledon stage of carrots and related crops
- Apply as directed spray in fruit crops, nursery stock, perennial flowers and tomatoes
- Any crop may be planted after 4 wk following ploughing and cultivation

Special precautions/Environmental safety
- Irritating to skin and eyes
- Do not contaminate surface waters or ditches with chemical or used container

Protective clothing/Label precautions
- A, C
- R04a, R04b, U05a, U08, U19, U20a, C03, E01, E15, E30a, E31b

Latest application/Harvest Interval(HI)
- HI 28 d for parsnips, carrots

Approval
- Off-label Approval unlimited for use on chervil, coriander, dill, bronze and green fennel, wild celery leaf, lovage, sweet cicely (OLA 0465/92)[1]

432 permethrin

A broad spectrum, contact and ingested pyrethroid insecticide

See also fenitrothion + permethrin + resmethrin

Products

1 Coopex Maxi Smoke Generators	AgrEvo Environ.	13.5% w/w	FU	H5131
2 Coopex Mini Smoke Generators	AgrEvo Environ.	13.5% w/w	FU	H5130
3 Coopex WP	AgrEvo Environ.	25% w/w	WP	H5096
4 Darmycel Agarifume Smoke Generator	Sylvan	10% w/w	FU	07904
5 Fumite Permethrin Smoke	Hortichem	10% w/w	FU	00940
6 Permasect 10 EC	Mitchell Cotts	90 g/l	EC	03920
7 Permasect 25 EC	Mitchell Cotts	230 g/l	EC	01576

Uses

Aphids in AUBERGINES, CHILLIES, CUCUMBERS, TOMATOES [7]. Aphids in PROTECTED ORNAMENTALS [6, 7]. Aphids in PROTECTED CUCUMBERS, PROTECTED TOMATOES [6]. Black pine beetle in CONIFER SEEDLINGS *(off-label)*, CUT LOGS/TIMBER *(off-label)*, FORESTRY TRANSPLANTS *(off-label)*, FORESTRY *(off-label)* [7]. Caterpillars in PROTECTED ORNAMENTALS [7]. Flies in MANURE HEAPS [3]. Grain storage pests in GRAIN STORES [1-3]. Pine weevil in CONIFER SEEDLINGS *(off-label)*, CONIFERS *(off-label)*, CUT LOGS/TIMBER *(off-label)*, FORESTRY TRANSPLANTS *(off-label)*, FORESTRY *(off-label)* [7]. Sciarid flies in AUBERGINES, MUSHROOMS, PROTECTED CELERY, PROTECTED CUCUMBERS, PROTECTED LETTUCE, PROTECTED ORNAMENTALS, PROTECTED PEPPERS, PROTECTED TOMATOES [4]. Tomato fruitworm in AUBERGINES, CHILLIES, CUCUMBERS, TOMATOES [7]. Whitefly in AUBERGINES [5, 7]. Whitefly in PROTECTED ORNAMENTALS [5-7]. Whitefly in PROTECTED CUCUMBERS, PROTECTED TOMATOES [5, 6]. Whitefly in PROTECTED CELERY, PROTECTED LETTUCE, PROTECTED PEPPERS [5]. Whitefly in CHILLIES, CUCUMBERS, TOMATOES [7].

Notes

Efficacy
- Best results from glasshouse fumigation achieved by treating in late afternoon or evening [4, 5]
- Sprays give rapid knock-down effect with persistent protection on leaf surfaces [3, 5-7]
- Apply as soon as pest or damage appears, or as otherwise recommended and repeat as necessary usually at 14 d intervals [5-7]
- Number and timing of sprays vary with crop and pest. See label for details
- For whitefly or mushroom fly control use at least 3 fumigations at 5-7 d intervals [4, 5]
- Where glasshouse whitefly resistant to permethrin occurs control is unlikely to be satisfactory [4, 5]

- Apply in grain stores by fumigation or surface spraying. Foodstuffs should not come into contact with treated surfaces unless otherwise directed [1, 2]

Crop Safety/Restrictions
- No of treatments not restricted
- Cut any open blooms before fumigating flower crops [4, 5]
- Do not treat rare or unusual plants without first testing on a small scale [4, 5]
- Do not fumigate young seedlings or plants being hardened off [4, 5]
- Allow 7 d after fumigating before introducing *Encarsia* or *Phytoseiulus* [4, 5]

Special precautions/Environmental safety
- Irritating to eyes and skin [6, 7]; irritating to eyes and respiratory system [4, 5]
- Flammable [1, 2, 5, 6]
- Dangerous to bees. Do not apply to crops in flower or to those in which bees are actively foraging. Do not apply when flowering weeds are present
- Extremely dangerous to fish or other aquatic life. Do not contaminate surface waters or ditches with chemical or used container
- Domestic animals, birds and fish should be removed from the vicinity of buildings to be treated. Foodstuffs should be protected from smoke [4, 5]

Protective clothing/Label precautions
- A [1-3, 5-7]; C [3, 6, 7]; D [1-3, 5]; E [1, 2, 5]; H [1-3, 5]
- R04a [4-7]; R04b, U04a, U20b, E31a [6, 7]; R04c, U20c, E12d, E29 [4, 5]; R07d [1, 2, 6]; U05a, U19, C03, E13a, E30a [1, 2, 4-7]; U08 [1-3, 6, 7]; U20a, E31b, E34 [1, 2]; U22, S06 [3]; E01 [1-7]; E12c [1, 2, 6, 7]; E32a [1, 2, 4, 5]

Latest application/Harvest Interval(HI)
- HI zero

Approval
- Off-label Approval unlimited for use on forestry trees to control pine weevil (OLA 0135/92)[7] - see OLA notice for restrictions; unlimited for use on containerised forestry seedlings (OLA 0451/94)[7]; unlimited for use on forest saplings, felled trees, cut logs and timber (OLA 1210/97)[7]

Maximum Residue Levels (mg residue/kg food)
- lettuces, herbs, celery, rhubarb, cereals (except maize) 2; pome fruits, stone fruits, grapes, kiwi fruit, Chinese cabbage, kale, spinach, beet leaves 1; citrus fruits, tomatoes, peppers, aubergines, beans (with pods), leeks, meat 0.5; cotton seed, maize 0.2; almond, celeriac, radishes, cucumbers, gherkins, courgettes, melons, squashes, watermelons, sweetcorn, cauliflowers, peas (with pods), peanuts, rape seed, mustard seed, hops 0.1; tree nuts (except almond), cane fruits, bilberries, cranberries, blackberries, loganberries, raspberries, wild berries, currants, miscellaneous fruit (except kiwi fruit), root and tuber vegetables (except celeriac, radishes), garlic, broccoli, kohlrabi, watercress, witloof, beans (without pods), peas (without pods), asparagus, cardoons, celery, globe artichokes, mushrooms, pulses, linseed, poppy seed, sesame seed, sunflower seed, soya bean, potatoes, milk, eggs 0.05

433 permethrin + thiram

An insecticide/fungicide mixture for chrysanthemums and pot plants

Products	Combinex	Fargro	4:25 g/l	EC	00562

Uses Capsids in BEGONIAS, CHRYSANTHEMUMS, CYCLAMENS, HYDRANGEAS, PRIMULAS. Caterpillars in BEGONIAS, CHRYSANTHEMUMS, CYCLAMENS, HYDRANGEAS, PRIMULAS. Earwigs in BEGONIAS, CHRYSANTHEMUMS, CYCLAMENS, HYDRANGEAS, PRIMULAS. Leaf miners in BEGONIAS *(partial control)*, CHRYSANTHEMUMS *(partial control)*, CYCLAMENS *(partial control)*, HYDRANGEAS *(partial control)*, PRIMULAS *(partial control)*.

Notes

Efficacy

- Spray chrysanthemums as soon as plants established and repeat every 14 d, if powdery mildew infection particularly severe every 7 d
- Spray pot plants as soon as established in final pots and repeat as necessary

Crop Safety/Restrictions

- Spray up to stage immediately before new florets unfold. Later applications to unfolded petals may result in petal damage, particularly under adverse growing conditions
- Do not apply to carnations or vines
- Do not apply in bright sunshine or when plants are flagging
- Do not apply within 21 d before or after application of sulfur or dinocap
- Product not compatible with dinocap, magnesium sulfate, manganese sulfate or sulfur
- Do not mix with other chemicals when spraying at low volume

Special precautions/Environmental safety

- Irritating to skin, eyes and respiratory system
- Dangerous to bees. Do not apply to crops in flower or to those in which bees are actively foraging. Do not apply when flowering weeds are present
- Extremely dangerous to fish or other aquatic life. Do not contaminate surface waters or ditches with chemical or used container

Protective clothing/Label precautions

- A
- R04a, R04b, R04c, U02, U05a, U08, U19, U20a, C01, C03, E01, E12c, E13a, E30a, E31a

Latest application/Harvest Interval(HI)

- Immediately before the outer florets become unfolded for chrysanthemums

Maximum Residue Levels (mg residue/kg food)

- see permethrin entry

434 petroleum oil

An insecticidal and acaricidal hydrocarbon oil

Products

Hortichem Spraying Oil	Hortichem	710 g/l	EC

Uses Mealy bugs in PROTECTED CUCUMBERS, PROTECTED GRAPEVINES, PROTECTED POT PLANTS, PROTECTED TOMATOES. Red spider mites in PROTECTED CUCUMBERS, PROTECTED GRAPEVINES, PROTECTED POT PLANTS, PROTECTED TOMATOES. Scale insects in PROTECTED CUCUMBERS, PROTECTED GRAPEVINES, PROTECTED POT PLANTS, PROTECTED TOMATOES.

Notes

Efficacy
- Spray at 1% (0.5% on tender foliage) to wet plants thoroughly, particularly the underside of leaves, and repeat as necessary
- Apply under quick drying conditions but not in bright sun unless glass well shaded

Crop Safety/Restrictions
- Treat grapevines before flowering
- On plants of unknown sensitivity test first on a small scale
- Mixtures with certain pesticides may damage crop plants. If mixing, spray a few plants to test for tolerance before treating larger areas
- Do not mix with sulfur or use sulfur sprays within 28 d of treatment

Special precautions/Environmental safety
- Harmful if swallowed. Irritating to eyes, skin and respiratory system

Protective clothing/Label precautions
- A, C
- M03, R03c, R04a, R04b, R04c, U05a, U08, U19, C03, E01, E30a, E31a, E34

Latest application/Harvest Interval(HI)
- HI zero

435 phenmedipham

A contact carbamate herbicide for beet crops and strawberries

See also desmedipham + phenmedipham
ethofumesate + metamitron + phenmedipham
ethofumesate + phenmedipham
lenacil + phenmedipham

Products

1 Atlas Protrum K	Atlas	114 g/l	EC	07723
2 Barclay Punter XL	Barclay	114 g/l	EC	08047
3 Beetup	United Phosphorus	114 g/l	EC	07520
4 Betanal Flo	AgrEvo	160 g/l	SC	08898
5 Betosip	Sipcam	114 g/l	EC	06787
6 Herbasan	Mirfield	160 g/l	SC	07161
7 Hickson Phenmedipham	Hickson & Welch	118 g/l	EC	02825
8 Luxan Phenmedipham	Luxan	118 g/l	EC	06933
9 MSS Betaren Flow	Mirfield	160 g/l	SC	08022
10 MSS Protrum G	Mirfield	114 g/l	EC	08342
11 Stefes Forte 2	Stefes	114 g/l	EC	08204
12 Stefes Medipham 2	Stefes	114 g/l	EC	08203
13 Tripart Beta	Tripart	118 g/l	EC	03111

Uses

Annual dicotyledons in FODDER BEET, MANGELS, SUGAR BEET [1-13]. Annual dicotyledons in RED BEET [1-4, 6, 9-13]. Annual dicotyledons in STRAWBERRIES [2, 3, 11, 12]. Annual dicotyledons in SPINACH BEET *(off-label)*, SPINACH *(off-label)* [4].

Notes

Efficacy
- Best results achieved by application to young seedling weeds, preferably cotyledon stage, under good growing conditions when low doses are effective
- 2-3 repeat applications at 7-10 d intervals using a low dose are recommended on mineral soils, 3-5 applications may be needed on organic soils

- Do not spray wet foliage or if rain imminent
- Addition of adjuvant oil may improve effectiveness on some weeds
- Various tank-mixtures with other beet herbicides recommended. See label for details
- Use of certain pre-emergence herbicides is recommended in combination with post-emergence treatment. See label for details

Crop Safety/Restrictions

- Maximum number of treatments 1 per crop for spinach; see labels for other crops
- Apply to beet crops at any stage as low dose/low volume spray or from fully developed cotyledon stage with full rate. Apply to red beet after fully developed cotyledon stage
- At high temperatures (above 21°C) reduce rate and spray after 5 pm
- Do not apply immediately after frost or if frost expected
- Do not spray crops stressed by wind damage, nutrient deficiency, pest or disease attack etc. Do not roll or harrow for 7 d before or after treatment
- Apply to strawberries at any time when weeds in susceptible stage, except in period from start of flowering to picking
- Do not use on strawberries under cloches or polythene tunnels

Special precautions/Environmental safety

- Harmful if swallowed [1, 3, 7, 10, 13]
- Irritating to eyes, skin and respiratory system [1-3, 5, 7, 10-13]
- Harmful (dangerous [6]) to fish or other aquatic life. Do not contaminate surface waters or ditches with chemical or used container [1-3, 5, 7, 8, 10-13]

Protective clothing/Label precautions

- A, C [1-3, 5, 7, 8, 10-13]; H [7, 8]; P [13]
- M03, R03c [1, 3, 7, 8, 10, 13]; R04a, R04b, R04c, E13c [1-3, 5, 7, 8, 10-13]; U05a, U08, C03, E01 [1-3, 5-8, 10-13]; U09a [4]; U13 [3]; U19, E30a, E31b [1-8, 10-13]; U20a [1, 2, 5, 7]; U20b [3, 4, 6, 8, 10-13]; E13b [4, 6]; E26 [3, 5, 7, 8, 10]; E34 [1-3, 5, 7, 8, 10]

Latest application/Harvest Interval(HI)

- Before crop leaves meet between rows for beet crops; before flowering for strawberries; before 5 leaf stage for spinach
- HI 5 d [11]

Approval

- Off-label approval to Dec 1999 for use on outdoor spinach and spinach beet (OLA 1368/98)[4]

---

436 phenothrin

A non-systemic contact and ingested pyrethroid insecticide

Products	Sumithrin 10 Sec	Sumitomo	103 g/l	EC	H3762

Uses Flies in LIVESTOCK HOUSES.

FOR FULL CONDITIONS OF USE ALWAYS READ THE PRODUCT LABEL

Notes

Efficacy

- Spray walls and other structural surfaces of milking parlours, dairies, cow byres and other animal housing

Crop Safety/Restrictions

- Do not apply directly to livestock
- Remove exposed milk and collect eggs before application. Protect milk machinery and containers from contamination

Special precautions/Environmental safety

- Irritating to eyes
- Wear approved respiratory equipment and eye protection (goggles) when applying with Microgen equipment
- Not to be sold or supplied to amateur users
- Dangerous to fish or other aquatic life. Do not contaminate surface waters or ditches with chemical or used container

Protective clothing/Label precautions

- A, C, D, E, F, H
- R04a, U05a, U08, U20b, C03, C06, C07, C10, E01, E05, E13b, E30a, E32a

Approval

- May be applied with Microgen ULV equipment

437 phenothrin + tetramethrin

A pyrethroid insecticide mixture for control of flying insects

Products

1 Deosan Fly Spray	DiverseyLever	0.066:0.028% w/w	AE	H5246
2 Killgerm ULV 500	Killgerm	40:20 g/l	UL	H4647
3 Sorex Fly Spray RTU	Sorex	0.02:0.10% w/w	RH	H5718
4 Sorex Super Fly Spray	Sorex		AE	H4468

Uses

Flies in AGRICULTURAL PREMISES [1-4]. Mosquitoes in AGRICULTURAL PREMISES [1, 2, 4]. Wasps in AGRICULTURAL PREMISES [1, 2, 4].

Notes

Efficacy

- Close doors and windows and spray in all directions for 3-5 sec. Keep room closed for at least 10 min
- May be used in the presence of poultry and livestock

Crop Safety/Restrictions

- Do not use space sprays containing pyrethrins or pyrethroid more than once per week in intensive or controlled environment animal houses in order to avoid development of resistance. If necessary, use a different control method or product [2]

Special precautions/Environmental safety

- Do not spray directly on food, livestock or poultry
- Remove exposed milk and collect eggs before application. Protect milk machinery and containers from contamination
- Flammable [1, 2, 4]
- Dangerous (extremely dangerous [3]) to fish or other aquatic life. Do not contaminate surface waters or ditches with chemical or used container

Protective clothing/Label precautions
- A [2, 3]; C [2]; H [2, 3]
- R04a, R04b, U05a, U17, E26, E32a [2]; U02, C10, E13a [3]; U09a, U20b, C04, C06, C07, C08, C09, E05, E30a [2, 3]; U19 [1-4]; U20a [1, 4]; C03, E01, E13b [1, 2, 4]

438 phenylmercury acetate

An organomercury fungicide seed treatment approvals for which were revoked with effect from 31 March 1992

439 phorate

A systemic organophosphorus insecticide with vapour-phase activity

Products

Phorate 10G	United Phosphorus	10% w/w	GR	08007

Uses

Aphids in BROCCOLI, BRUSSELS SPROUTS, CABBAGES, CAULIFLOWERS, LETTUCE, POTATOES. Cabbage root fly in BROCCOLI, BRUSSELS SPROUTS, CABBAGES, CAULIFLOWERS. Capsids in POTATOES. Carrot fly in CELERY. Frit fly in MAIZE, SWEETCORN. Leafhoppers in POTATOES. Lettuce root aphid in LETTUCE. Wireworms in POTATOES.

Notes

Efficacy
- Must be applied as a soil treatment. Application method, rate and timing vary with crop, pest and soil type. See label for details
- Effectiveness of soil application reduced by hot, dry weather, excessive rainfall or on soils with more than 10% organic matter content

Crop Safety/Restrictions
- Maximum number of treatments 1 per crop
- When applied at time of sowing granules must not be in direct contact with seed

Special precautions/Environmental safety
- This product contains an anticholinesterase organophosphorus compound. Do not use if under medical advice not to work with such compounds
- Toxic in contact with skin, by inhalation or if swallowed
- Keep in original container, tightly closed in a safe place, under lock and key
- Harmful to livestock
- Dangerous to game, wild birds and animals
- Dangerous to fish or other aquatic life. Do not contaminate surface waters or ditches with chemical or used container

Protective clothing/Label precautions
- A, B, C, D, E, H, J, K, M
- M01, M04, R02b, R02c, R03a, R04a, R04b, R04e, U02, U04a, U05a, U10, U11, U13, U19 U20a, C02 (6 wk), C03, E01, E06a (6 wk), E10a, E13b, E26, E30b, E32a, E34

FOR FULL CONDITIONS OF USE ALWAYS READ THE PRODUCT LABEL

Withholding period
- Keep all livestock out of treated areas for at least 6 wk. Bury or remove spillages

Latest application/Harvest Interval(HI)
- At sowing or transplanting
- HI outdoor lettuce 6 wk

Approval
- Approval for use on carrots and parsnips suspended in Dec 1997

Maximum Residue Levels (mg residue/kg food)
- milk and dairy produce 0.2; peanuts, tea, hops 0.1; all other produce 0.01

---

440 phosalone

A contact and ingested organophosphorus insecticide and acaricide

Products

Zolone Liquid	Hortichem	350 g/l	EC	06206

Uses

Aphids in SPRING BARLEY, SPRING WHEAT, WINTER BARLEY, WINTER WHEAT. Brassica pod midge in CABBAGE SEED CROPS, FODDER RAPE SEED CROPS, KALE SEED CROPS, MUSTARD, OILSEED RAPE. Bryobia mites in APPLES, PEARS. Codling moth in APPLES, PEARS. Pollen beetles in CABBAGE SEED CROPS, FODDER RAPE SEED CROPS, KALE SEED CROPS, MUSTARD, OILSEED RAPE. Red spider mites in APPLES, PEARS, PLUMS. Seed weevil in CABBAGE SEED CROPS, FODDER RAPE SEED CROPS, KALE SEED CROPS, MUSTARD, OILSEED RAPE. Tortrix moths in APPLES, PEARS.

Notes

Efficacy
- Spray oilseed rape and brassica seed crops during flowering period. Number and timing of sprays vary with pest and crop. See label for details
- Spray apples and pears pre-blossom and up to 4 times post-blossom at intervals of about 14 d. Spray plums pre-blossom and up to twice post-blossom
- Chemical is most effective during warm weather but should not be applied at low volume during period of high light intensity or high temperature
- Do not mix with highly alkaline materials such as lime sulfur

Special precautions/Environmental safety
- This product contains an anticholinesterase organophosphorus compound. Do not use if under medical advice not to work with such compounds
- Harmful in contact with skin or if swallowed. Irritating to eyes
- Harmful to livestock. Keep all livestock out of treated areas for at least 4 wk
- Harmful to fish or other aquatic life. Do not contaminate surface waters or ditches with chemical or used container
- Toxicity hazard to bees and ladybirds is low at recommended rates. To reduce hazard further spray in late evening or early morning

Protective clothing/Label precautions
- A, C
- M01, M03, R03a, R03c, R04a, U02, U04a, U05a, U10, U11, U13, U19, U20a, C02, C03, E01, E06b, E13c, E30a, E31b, E34

Latest application/Harvest Interval(HI)
- HI 3 wk

Approval
- Approved for aerial application on brassica seed crops [1]. See notes in Section 1

Maximum Residue Levels (mg residue/kg food)
- pome fruits, apricots, peaches, nectarines 2; citrus fruits, plums, grapes, cane fruits, bilberries, cranberries, currants, gooseberries, bananas, garlic, onions, shallots, tomatoes, peppers, aubergines, cucumbers, gherkins, courgettes, cauliflowers, Brussels sprouts, head cabbages, lettuce, beans (with pod), peas (with pod), celery, leeks, rhubarb, cultivated mushrooms 1; carrots, horseradish, parsnips, parsley root, salsify, swedes, turnips, potatoes 0.1

441 picloram

A persistent, translocated pyridine carboxylic acid herbicide for non-crop areas

See also 2,4-D + picloram
bromacil + picloram

Products

Tordon 22K	Nomix-Chipman	240 g/l	SL	05790

Uses

Annual dicotyledons in NON-CROP AREAS, NON-CROP GRASS. Bracken in NON-CROP AREAS, NON-CROP GRASS. Japanese knotweed in NON-CROP AREAS, NON-CROP GRASS. Perennial dicotyledons in NON-CROP AREAS, NON-CROP GRASS. Woody weeds in NON-CROP AREAS, NON-CROP GRASS.

Notes

Efficacy
- May be applied at any time of year. Best results achieved by application as foliage spray in late winter to early spring
- For bracken control apply 2-4 wk before frond emergence
- Clovers are highly sensitive and eliminated at very low doses
- Persists in soil for up to 2 yr

Crop Safety/Restrictions
- Maximum number of treatments 1 per yr
- Do not apply around desirable trees or shrubs where roots may absorb chemical
- Do not apply on slopes where chemical may be leached onto areas of desirable plants

Special precautions/Environmental safety
- Irritating to eyes
- Harmful to fish or other aquatic life. Do not contaminate surface waters or ditches with chemical or used container

Protective clothing/Label precautions
- R04a, U05a, U08, U20b, C03, E01, E07, E13c, E30a, E31b

Withholding period
- Keep livestock out of treated areas until foliage of any poisonous weeds such as ragwort has died and become unpalatable

442 pirimicarb

A carbamate insecticide for aphid control

See also deltamethrin + pirimicarb
lambda-cyhalothrin + pirimicarb

Products

1 Aphox	Zeneca	50% w/w	WG	06633
2 Barclay Pirimisect	Barclay	50% w/w	WG	09057
3 Landgold Pirimicarb 50	Landgold	50% w/w	WG	09018
4 Phantom	Bayer	50% w/w	WG	04519
5 Pirimor	Zeneca	50% w/w	WP	06694
6 Portman Pirimicarb	Portman	50% w/w	WG	06922
7 Standon Pirimicarb 50	Standon	50% w/w	WG	08878
8 Standon Pirimicarb H	Standon	50% w/w	WP	[illegible]

Uses

Aphids in PROTECTED LETTUCE [8]. Aphids in CHICORY *(off-label)*, HORSERADISH *(off-label)*, MARROWS *(off-label)*, PARSLEY ROOT *(off-label)*, PROTECTED COURGETTES *(off-label)*, PROTECTED GHERKINS *(off-label)*, RADISHES *(off-label)*, SPINACH BEET *(off-label)*, SPINACH *(off-label)*, SWEETCORN *(off-label)* [1, 5]. Aphids in CARROTS, CELERY, DWARF BEANS, GRASSLAND, MAIZE, OILSEED RAPE, PARSNIPS, RUNNER BEANS, SWEETCORN [1, 2, 4, 6, 7]. Aphids in BLACKCURRANTS, CHINESE CABBAGE, COLLARDS, GOOSEBERRIES, RASPBERRIES, REDCURRANTS [1, 2, 4, 7]. Aphids in CHERRIES [1, 2, 4]. Aphids in STRAWBERRIES [1-8]. Aphids in BARLEY, BROAD BEANS, BROCCOLI, BRUSSELS SPROUTS, CABBAGES, CALABRESE, CAULIFLOWERS, FIELD BEANS, KALE, OATS, PEARS, PEAS, POTATOES, SUGAR BEET, SWEDES, TURNIPS, WHEAT [1-4, 6, 7]. Aphids in APPLES, DURUM WHEAT, RYE, TRITICALE [1-4, 7]. Aphids in CELERIAC *(off-label)*, ENDIVES *(off-label)*, FENNEL *(off-label)*, HONESTY *(off-label)*, KOHLRABI *(off-label)*, PARSLEY *(off-label)*, PLUMS *(off-label)*, RED BEET *(off-label)* [1]. Aphids in CUCUMBERS, LETTUCE, OUTDOOR, PROTECTED CARNATIONS, PROTECTED CHRYSANTHEMUMS, TOMATOES [5, 8]. Aphids in BARLEY *(off-label - research/breeding)*, BRUSSELS SPROUTS *(off-label - research/breeding)*, COURGETTES *(off-label)*, FOREST NURSERIES, GHERKINS *(off-label)*, ORNAMENTALS, PEAS *(off-label - research/breeding)*, PEPPERS, PROTECTED CINERARIAS, PROTECTED CYCLAMEN, PROTECTED ROSES, WHEAT *(off-label - research/breeding)* [5]. Blackfly in CHERRIES [7].

Notes

Efficacy

- Chemical has contact, fumigant and translaminar activity
- Best results achieved under warm, calm conditions when plants not wilting and spray does not dry too rapidly. Little vapour activity at temperatures below 15°C
- Apply as soon as aphids seen or warning issued and repeat as necessary
- Addition of non-ionic wetter recommended for use on brassicas
- Chemical has little effect on bees, ladybirds and other insects and is suitable for use in integrated control programmes on apples and pears
- On cucumbers and tomatoes a root drench is preferable to spraying when using predators in an integrated control programme
- Where aphids resistant to pirimicarb occur on protected crops control is unlikely to be satisfactory [8]

Crop Safety/Restrictions

- Maximum number of treatments normally 2 per crop but not specified in some cases

Special precautions/Environmental safety

- This product contains an anticholinesterase carbamate compound. Do not use if under medical advice not to work with such compounds urface waters or

ng position is within
assisted applications to

, E32a, E34 [1-8]; U20a [2-8];
3b [1-5, 7, 8]; E13c [6]

treated areas for at least 7 d. Bury or remove spillages

Harvest Interval(HI)

e, cereals, maize, sweetcorn, lettuce under glass 14 d; grassland 7 d;
tomatoes and peppers under glass 2 d; protected courgettes and gherkins 21 h;
ble crops 3 d; flowers and ornamentals zero

Approval

- Approved for aerial application on cereals [1, 2, 4, 6]. See notes in Section 1
- Off-label approval unlimited for use on honesty (OLA 0584/93)[1]; unlimited for use on a range of outdoor and protected vegetables - see OLA notices for details (OLA 1626/95)[1], (OLA 1634/95)[5]; unlimited for use on protected courgettes, gherkins (OLA 1302/96)[1]; to Sep 2001 for use on plums (OLA 2178/96)[1]; to Feb 2001 for use on red beet, celeriac, fennel, kohlrabi (OLA 0328/96)[1]; unlimited for use on protected courgettes, gherkins (OLA 1203/96)[5]; unlimited for use in research/breeding programmes in protected crops of wheat, barley, peas, Brussels sprouts (OLA 2448/96)[5]; unlimited for use on parsley and endives (OLA 1303/96)[1]; unlimited for use on chicory (OLA 1078/98)[1], (OLA 1079/98)[5]

443 pirimiphos-methyl

A contact, fumigant and translaminar organophosphorus insecticide

Products

1 Actellic 2% Dust	Zeneca	2% w/w	DP	06931
2 Actellic D	Zeneca	250 g/l	EC	06930
3 Actellic Smoke Generator No 20	Zeneca	20 g a.i.	FU	06627
4 Actellifog	Hortichem	100 g/l	HN	06628
5 Blex	Zeneca	500 g/l	EC	06639
6 Fumite Pirimiphos Methyl Smoke	Hortichem	22.5% w/w	FU	00941

Uses Ants in AUBERGINES, PROTECTED CUCUMBERS, PROTECTED ORNAMENTALS, PROTECTED PEPPERS, PROTECTED TOMATOES [6]. Aphids in AUBERGINES, PROTECTED CUCUMBERS, PROTECTED ORNAMENTALS, PROTECTED PEPPERS, PROTECTED TOMATOES [6]. Aphids in APPLES [5]. Capsids in AUBERGINES,

PROTECTED CUCUMBERS, PROTECTED ORNAMENTALS, PROTECTED PEPPERS, PROTECTED TOMATOES [6]. Carrot fly in CARROTS, PARSLEY *(off-label)* [5]. Caterpillars in APPLES, BROCCOLI, BRUSSELS SPROUTS, CABBAGES, CALABRESE, CAULIFLOWERS [5]. Earwigs in AUBERGINES, PROTECTED CUCUMBERS, PROTECTED ORNAMENTALS, PROTECTED PEPPERS, PROTECTED TOMATOES [6]. Flour beetles in STORED GRAIN [1, 2]. Flour moths in STORED GRAIN [1, 2]. French fly in CUCUMBERS [5]. Grain beetles in STORED GRAIN [1, 2]. Grain storage mites in STORED GRAIN [1, 2]. Grain storage pests in GRAIN STORES [3, 4]. Grain weevils in STORED GRAIN [1, 2]. Leaf miners in AUBERGINES, PROTECTED CUCUMBERS, PROTECTED ORNAMENTALS, PROTECTED PEPPERS, PROTECTED TOMATOES [6]. Leaf miners in SUGAR BEET [5]. Red spider mites in AUBERGINES, PROTECTED CUCUMBERS, PROTECTED ORNAMENTALS, PROTECTED PEPPERS, PROTECTED TOMATOES [6]. Rust mite in APPLES, PEARS [5]. Sawflies in AUBERGINES, PROTECTED CUCUMBERS, PROTECTED ORNAMENTALS, PROTECTED PEPPERS, PROTECTED TOMATOES [6]. Storage pests in STORED LINSEED, STORED OILSEED RAPE [1, 2]. Thrips in AUBERGINES, PROTECTED CUCUMBERS, PROTECTED ORNAMENTALS, PROTECTED PEPPERS, PROTECTED TOMATOES [6]. Warehouse moth in STORED GRAIN [1, 2]. Wheat bulb fly in WINTER WHEAT [5]. Whitefly in AUBERGINES, PROTECTED CUCUMBERS, PROTECTED ORNAMENTALS, PROTECTED PEPPERS, PROTECTED TOMATOES [6]. Whitefly in CUCUMBERS, ORNAMENTALS, PEPPERS, TOMATOES [4].

Notes

Efficacy

- Chemical acts rapidly and has short persistence in plants, though spray or dust persists for long period on inert surfaces
- Product to be used in conjunction with Actellic Dust [3]
- Apply spray when pest first seen or at time of egg hatch and repeat as necessary
- For wheat bulb fly control apply on receipt of ADAS warning
- Rate, number and timing of sprays vary with crop and pest. See label for details
- For use on Brussels sprouts use of drop-legged sprayer is recommended
- For protection of stored grain disinfect empty stores by spraying surfaces and/or fumigation and treat grain by full or surface admixture or spray bagged grain. See label for details of treatment and suitable application machinery
- Best results obtained when grain stored at 15% moisture or less. Dry and cool moist grain coming into store before treatment
- For control of whitefly and other glasshouse pests apply as smoke or by thermal fogging. See label for details of techniques and suitable machines
- Where glasshouse whitefly resistant to pirimiphos-methyl occur control is unlikely to be satisfactory
- Actellifog may be used to clean up houses before introducing whitefly parasites but do not use when *Encarsia* present

Crop Safety/Restrictions

- On carrots the maximum total dose applied per crop must not exceed the equivalent of 3 (on mineral soils) or 4 (on organic soils) full dose applications
- Grain treated by admixture as specified may be consumed by humans and livestock
- Do not fumigate glasshouses in bright sunshine or when foliage is wet or roots dry
- Do not fumigate young seedlings or plants being hardened off
- Do not fog open flowers of ornamentals without first consulting firm
- Do not fog mushrooms when wet as slight spotting may occur
- Consult processors before using on crops for processing [5]

Special precautions/Environmental safety

- This product contains an organophosphorus anticholinesterase compound. Do not use if under medical advice not to work with such compounds

- Irritating to eyes and respiratory system [3, 4, 6]; irritating to eyes and skin [2, 5]
- Highly flammable [3]; flammable [2, 4-6]
- Ventilate fumigated or fogged spaces thoroughly before re-entry
- Unprotected persons must be kept out of treated areas within 3 h of ignition [3]
- Dangerous to bees. Do not apply to crops in flower or to those in which bees are actively foraging. Do not apply when flowering weeds are present
- Harmful to fish or other aquatic life. Do not contaminate surface waters or ditches with chemical or used container

Protective clothing/Label precautions
- A [1, 2, 4, 5]; C [5]; D [3, 4]; H [1-3]; J, L, M [4]
- M01, U05a, U19, E13c, E30a [1-6]; M02, M03, U20a, C02 [4, 6]; R03c [5]; R04a, C03, E01 [2-6]; R04b [2, 5]; R04c, E32a [3, 4, 6]; R07c, E02 (3 h) [3]; R07d [2, 4-6]; U04a, U11, U16, U17, C09, E31a [2]; U08, E34 [1, 2, 4, 6]; U10 [2, 3, 5]; U20b [1-3, 5]; E12d, E31b [4-6]; S05, S07 [1, 2]

Latest application/Harvest Interval(HI)
- HI zero [1-4, 6]; brassicas 3 d [5]; cereals, sugar beet, carrots, apples, pears 7 d [5]; cucumbers 3 wk [5]

Approval
- Product formulated for application by thermal fogging (see label for details) [4]
- Off-label approval unlimited for use on outdoor parsley (OLA 1324/95)[5]

Maximum Residue Levels (mg residue/kg food)
- cereals 5; mandarins, kiwi fruit, Brussels sprouts, mushrooms 2; citrus fruits (except mandarins), carrots, broccoli, cauliflower 1; other crops 0.05

444 prochloraz

A broad-spectrum protectant and eradicant conazole fungicide

See also carbendazim + prochloraz
cyproconazole + prochloraz
fenpropidin + prochloraz
fenpropimorph + prochloraz

Products

1 Alpha Mirage 40 EC	Makhteshim	400 g/l	EC	06770
2 Barclay Eyetak	Barclay	450 g/l	EC	06813
3 Levington Octave	Levington	46% w/w	WP	07505
4 Octave	AgrEvo	46% w/w	WP	07267
5 Prelude 20LF	Agrichem	200 g/l	LS	04371
6 Scotts Octave	Scotts	46% w/w	WP	09275
7 Sporgon 50WP	Sylvan	46% w/w	WP	03829
8 Sportak 45 EW	AgrEvo	450 g/l	EW	07996
9 Sportak Sierra EW	AgrEvo	450 g/l	EW	08003
10 Stefes Poraz	Stefes	450 g/l	EC	07528

Uses Alternaria in OILSEED RAPE [2, 8-10]. Botrytis in ORNAMENTALS [3, 4, 6]. Cobweb in MUSHROOMS [7]. Dry bubble in MUSHROOMS [7]. Eyespot in WINTER BARLEY, WINTER WHEAT [1, 2, 8-10]. Eyespot in SPRING WHEAT [1]. Eyespot in WINTER RYE [2,

8-10]. Fungus diseases in CONTAINER-GROWN STOCK, HARDY ORNAMENTAL NURSERY STOCK [3, 4, 6]. Fungus diseases in HERBS *(off-label)*, LETTUCE *(off-label)* [3]. Fusarium bulb rot in FLOWER BULBS *(off-label)* [8, 9]. Glume blotch in WINTER WHEAT [2, 8-10]. Grey mould in OILSEED RAPE [8-10]. Leaf spot in WINTER RYE, WINTER WHEAT [2, 8-10]. Light leaf spot in OILSEED RAPE [2, 8-10]. Net blotch in SPRING BARLEY, WINTER BARLEY [2, 8-10]. Penicillium rot in FLOWER BULBS *(off-label)* [8, 9]. Phoma in OILSEED RAPE [2, 8-10]. Powdery mildew in SPRING BARLEY, WINTER BARLEY, WINTER RYE [2, 8-10]. Powdery mildew in SPRING WHEAT *(protection)* [2, 8, 9]. Powdery mildew in SPRING WHEAT, WINTER WHEAT [10]. Rhynchosporium in SPRING BARLEY, WINTER BARLEY, WINTER RYE [2, 8-10]. Sclerotinia stem rot in OILSEED RAPE [2, 8-10]. Seed-borne diseases in FLAX, LINSEED [5]. Wet bubble in MUSHROOMS [7]. White leaf spot in OILSEED RAPE [8-10].

Notes

Efficacy

- Spray cereals at first signs of disease. Protection of winter crops through season usually requires at least 2 treatments. See label for details of rates and timing. Treatment active against strains of eyespot resistant to benzimidazole fungicides
- Tank mixes with other fungicides recommended to improve control of rusts in wheat and barley. See label for details
- A period of at least 3 h without rain should follow spraying
- Can be used through most seed treatment machines if good even seed coverage is obtained. Check drill calibration before drilling treated seed [5]
- Apply as drench against soil diseases, as a spray against aerial diseases or as dip at propagation. Spray applications may be repeated at 10-14 d intervals. Under mist propagation use 7 d intervals [3, 4]
- Apply to mushrooms as casing treatment or spray between flushes. Timing determined by anticipated disease occurrence [7]

Crop Safety/Restrictions

- Maximum number of treatments 1 per batch for flax, linseed [5]; 2 at 120g or 3 at 60 g/100 sq m for mushrooms [7]; varies with dose and crop for other uses - see labels for details
- Do not treat linseed varieties Linda, Bolas, Karen, Laura, Mikael, Norlin, Moonraker, Abbey [5]

Special precautions/Environmental safety

- Harmful in contact with skin [2, 10]
- Irritating to eyes [1-5, 7, 10]; irritating to skin [1-4, 7, 10]
- May cause sensitization by skin contact [1]
- Flammable [1, 2]
- Dangerous to fish or other aquatic life. Do not contaminate surface waters or ditches with chemical or used container
- Do not use treated seed as food or feed [5]
- Treated seed harmful to game and wildlife [5]

Protective clothing/Label precautions

- A, C [1-10]; H, J [7]
- M03 [10]; R03a [2, 10]; R04a [1-3, 3, 4, 4-6, 6, 7, 10]; R04b [1-3, 3, 4, 4, 6, 6, 7, 10]; R04e [1]; R07d [1, 2]; U05a, E01, E13b, E30a [1-10]; U08 [1, 2, 5, 10]; U09a [7]; U11, E03, E31a, S01, S02, S03, S04b, S05, S06, S07 [5]; U20a [2, 8-10]; U20b [1, 3-7]; C03 [1-4, 6, 7, 10]; E26 [1, 2, 5, 8, 9]; E31b [1-4, 6-10]; E34 [2, 7-10]

Latest application/Harvest Interval(HI)

- Milky ripe stage (GS 77) for cereals; before drilling flax or linseed [5]
- HI 6 wk for oilseed rape and cereals; 2 d for mushrooms [7]

Approval

- Off-label Approval to Mar 2000 for use on leafy herbs and lettuce. See notice for list of species (OLA 0649/97)[3]

445 prochloraz + propiconazole

A broad spectrum fungicide mixture for wheat and barley

Products	Bumper P	Makhteshim	400:90 g/l	EC	08548

Uses Rhynchosporium in BARLEY. Septoria in WHEAT.

Notes

Efficacy

- Best results obtained from treatment when disease is active but not well established
- Treat wheat normally from flag leaf ligule just visible stage (GS 39) but earlier if there is a high risk of Septoria
- Treat barley from the first node detectable stage (GS 31)
- If disease pressure persists a second application may be necessary

Crop Safety/Restrictions

- Maximum number of treatments 2 per crop

Special precautions/Environmental safety

- Irritating to eyes
- Dangerous to fish or other aquatic life. Do not contaminate surface waters or ditches with chemical or used container

Protective clothing/Label precautions

- A, C
- R04a, U05a, U08, U20b, C03, E01, E13b, E26, E30a, E31b, E34

Latest application/Harvest Interval(HI)

- Before grain watery ripe (GS 71)
- HI 6 wk

Maximum Residue Levels (mg residue/kg food)

- see propiconazole entry

446 prochloraz + tebuconazole

A broad spectrum systemic fungicide mixture for cereals

Products	Agate	Bayer	267:133 g/l	EC	08826

Uses Brown rust in BARLEY, RYE, WHEAT. Eyespot in BARLEY, RYE, WHEAT. Glume blotch in WHEAT. Late ear diseases in WHEAT. Net blotch in BARLEY. Powdery mildew in BARLEY,

FOR FULL CONDITIONS OF USE ALWAYS READ THE PRODUCT LABEL

RYE, WHEAT. Rhynchosporium in BARLEY, RYE. Septoria leaf spot in WHEAT. Yellow rust in BARLEY, RYE, WHEAT.

Notes

Efficacy

- Best results achieved from applications at an early stage of disease development before infection spreads to new crop growth
- Adequate protection of winter cereals will usually require a programme of at least two treatments
- Optimum application timing is normally when disease first seen but varies with main target disease - see label
- To minimise possibility of development of resistance repeated applications should not be made against the same pathogen. Alternation with fungicides with a different mode of action is a preferred strategy

Crop Safety/Restrictions

- Maximum total dose equivalent to two full dose treatments
- Occasionally transient leaf speckling may occur after treating wheat. Yield responses should not be affected

Special precautions/Environmental safety

- Harmful if swallowed and in contact with skin
- Risk of serious damage to eyes
- May cause sensitization by skin contact
- Dangerous to fish or other aquatic life. Do not contaminate surface waters or ditches with chemical or used container

Protective clothing/Label precautions

- A, C, H
- M03, R03a, R03c, R04d, R04e, U05a, U09b, U20a, C03, E01, E13b, E26, E30a, E31c, E34

Latest application/Harvest Interval(HI)

- Before grain milky ripe (GS 73)

---

447 prometryn

A contact and residual triazine herbicide for various field crops

Products

1 Alpha Prometryne 50 WP	Makhteshim	50% w/w	WP	04871
2 Gesagard	Novartis	50% w/w	WP	08410
3 Gesagard 50 WP	Ciba Agric.	50% w/w	WP	00981

Uses

Annual dicotyledons in CARROTS, CELERY, PARSLEY [1-3]. Annual dicotyledons in PEAS, POTATOES, TRANSPLANTED CELERY, TRANSPLANTED LEEKS [1]. Annual dicotyledons in CELERIAC *(off-label)*, COMBINING PEAS, DRILLED LEEKS *(off-label)*, EARLY POTATOES, GARLIC *(off-label)*, KOHLRABI *(off-label)*, ONIONS *(off-label)*, OUTDOOR HERBS *(off-label)*, PROTECTED HERBS *(off-label)*, SWEDES *(off-label)*, TURNIPS *(off-label)*, VINING PEAS [2, 3]. Annual grasses in CARROTS, CELERY, PARSLEY [1-3]. Annual grasses in PEAS, POTATOES, TRANSPLANTED CELERY, TRANSPLANTED LEEKS [1]. Annual grasses in CELERIAC *(off-label)*, COMBINING PEAS, DRILLED LEEKS *(off-label)*, EARLY POTATOES, GARLIC *(off-label)*, KOHLRABI *(off-label)*, ONIONS *(off-label)*, OUTDOOR HERBS *(off-label)*, PROTECTED HERBS *(off-label)*, SWEDES *(off-label)*, TURNIPS *(off-label)*, VINING PEAS [2, 3].

Notes

Efficacy

- Best results achieved by application to young seedling weeds up to 5 cm high (cotyledon stage for knotgrass, mayweed and corn marigold) on fine, moist seedbed when rain falls afterwards. Do not use on very cloddy soils
- On organic soils only contact action effective and repeat application may be needed (certain crops only)

Crop Safety/Restrictions

- Maximum number of treatments 1 per crop for early potatoes, peas and transplanted leeks; 2 per crop for transplanted celery
- Apply to peas pre-emergence up to 3 d before crop expected to emerge
- Spring sown vining and drying peas may be treated. Damage may occur with Vedette or Printana, especially if emerging under adverse conditions. Do not treat forage peas
- Do not use on peas on Very Light soils, Sands, gravelly or stony soils
- Apply to early potatoes up to 10% emergence
- Apply to carrots, celery, parsley or coriander post-emergence after 2-rough leaf stage or after transplants established
- Apply to transplanted leeks or celery after transplants established. Do not use on drilled leeks
- Excessive rain after treatment may check crop
- In the event of crop failure only plant recommended crops within 8 wk

Special precautions/Environmental safety

- Harmful to fish or other aquatic life. Do not contaminate surface waters or ditches with chemical or used container [1]

Protective clothing/Label precautions

- M03, R04, C02, C03, E01, E15, E34 [2, 3]; U20b, E30a [1-3]; C02 (6 wk), E13c, E32a [1]

Latest application/Harvest Interval(HI)

- Pre-emergence for peas; 6 wk before harvest and before 10% emergence for early potatoes; before 10% crop emergence for swedes and turnips; 3 wk after planting out for onions and garlic.
- HI 6 wk

Approval

- Off-label approval unlimited for use on drilled leeks (OLA 0054/92)[3], (OLA 2196/97)[2], outdoor and protected herbs, swedes, turnips, kohlrabi, onions and garlic (OLA 0185/93)[3]; unlimited for use on outdoor and protected herbs, swedes, turnips, kohlrabi (OLA 2081/97)[2]; unlimited for use on outdoor onions (OLA 2080/97)[2]; unlimited for use on outdoor celeriac (OLA 0329/96)[3]; to Feb 2001 for use on outdoor celeriac (OLA 1382/97)[2]

448 propachlor

A pre-emergence chloroacetanilide herbicide for various horticultural crops

See also chloridazon + propachlor

Products

1 Alpha Propachlor 50 SC	Makhteshim	500 g/l	SC	04873
2 Brasson	Portman	480 g/l	SC	08159
3 Portman Propachlor 50 FL	Portman	500 g/l	SC	06892
4 Ramrod Flowable	Monsanto	480 g/l	SC	01688
5 Ramrod Granular	Monsanto	20% w/w	GR	01687
6 Tripart Sentinel	Tripart	500 g/l	SC	03250
7 Tripart Sentinel 2	Tripart	480 g/l	SC	05140

Uses

Annual dicotyledons in HERBACEOUS PERENNIALS, WOODY ORNAMENTALS [5]. Annual dicotyledons in BLACKCURRANTS *(off-label)*, BLUEBERRIES *(off-label)*, BROWN MUSTARD, CHINESE CABBAGE *(off-label)*, GOOSEBERRIES *(off-label)*, KOHLRABI *(off-label)*, LETTUCE *(off-label)*, OUTDOOR LEAF HERBS *(off-label)*, REDCURRANTS *(off-label)*, STRAWBERRIES *(off-label)*, WHITE MUSTARD, WHITECURRANTS *(off-label)* [4]. Annual dicotyledons in STRAWBERRIES [1, 6]. Annual dicotyledons in OILSEED RAPE [1, 3, 6, 7]. Annual dicotyledons in BROCCOLI, BRUSSELS SPROUTS, CABBAGES, CAULIFLOWERS, KALE, LEEKS, ONIONS, SWEDES, TURNIPS [1-7]. Annual dicotyledons in RAPE [7]. Annual dicotyledons in CALABRESE [2, 4, 5, 7]. Annual dicotyledons in FODDER RAPE [2, 4, 5]. Annual dicotyledons in SAGE, SPRING OILSEED RAPE [2, 4]. Annual dicotyledons in MUSTARD, ORNAMENTALS [2]. Annual grasses in HERBACEOUS PERENNIALS, WOODY ORNAMENTALS [5]. Annual grasses in BLACKCURRANTS *(off-label)*, BLUEBERRIES *(off-label)*, BROWN MUSTARD, CHINESE CABBAGE *(off-label)*, KOHLRABI *(off-label)*, LETTUCE *(off-label)*, OUTDOOR LEAF HERBS *(off-label)*, REDCURRANTS *(off-label)*, STRAWBERRIES *(off-label)*, WHITE MUSTARD, WHITECURRANTS *(off-label)* [4]. Annual grasses in STRAWBERRIES [1, 6]. Annual grasses in OILSEED RAPE [1, 3, 6, 7]. Annual grasses in BROCCOLI, BRUSSELS SPROUTS, CABBAGES, CAULIFLOWERS, KALE, LEEKS, ONIONS, SWEDES, TURNIPS [1-7]. Annual grasses in RAPE [7]. Annual grasses in ORNAMENTALS [2, 5]. Annual grasses in CALABRESE [2, 4, 5, 7]. Annual grasses in FODDER RAPE [2, 4, 5]. Annual grasses in SAGE, SPRING OILSEED RAPE [2, 4]. Annual grasses in MUSTARD [2].

Notes

Efficacy

- Controls germinating (not emerged) weeds for 6-8 wk
- Best results achieved by application to fine, firm, moist seedbed free of established weeds in spring, summer and early autumn
- Effective results with granules depend on rainfall soon after application [5]
- Use higher rate on soils with more than 10% organic matter
- Recommended as tank-mix with glyphosate or chlorthal-dimethyl on onions, leeks, swedes, turnips; with chlorpropham on leeks, onions [5]. See label for details

Crop Safety/Restrictions

- Maximum number of treatments 1 or 2 per crop, varies with product and crop - see label for details
- Apply in brassicas from drilling to time seed chits or after 3-4 true leaf stage but before weed emergence, in swedes and turnips pre-emergence only
- Apply in transplanted brassicas within 48 h of planting in warm weather. Plants must be hardened off and special care needed with block sown or modular propagated plants
- Apply in onions and leeks pre-emergence or from post-crook to young plant stage

- Apply to newly planted strawberries soon after transplanting, to weed-free soil in established crops in early spring before new weeds emerge
- Granules may be applied to onion, leek and brassica nurseries and to most flower crops after bedding out and hardening off
- Do not use pre-emergence of drilled wallflower seed
- Do not use on crops under glass or polythene
- Do not use under extremely wet, dry or other adverse growth conditions
- In the event of crop failure only replant recommended crops in treated soil

Special precautions/Environmental safety
- Harmful in contact with skin [1, 3]; harmful if swallowed [1, 2]
- Irritating to eyes [5] and skin [1-4, 6, 7]
- Irritating to respiratory system [4, 6, 7]
- May cause sensitization by skin contact [2-4, 6, 7]
- Do not contaminate surface waters or ditches with chemical or used container [2-7]

Protective clothing/Label precautions
- A [1-7]; C [1-4, 6, 7]
- M03 [2-4]; R03a [1, 3]; R03c [1, 2, 4]; R04a, U04a, U08, U13, U19, E31b, E34 [1-4, 6, 7]; R04b, U02, U05a, C03, E01, E15, E30a [1-7]; R04c [7]; R04e [2-4, 7]; U20a [2, 3, 5, 7]; U20b [1, 4, 6]; E26 [1, 3, 4, 6, 7]; E32a [5]

Approval
- Off-label approval unlimited for use on outdoor lettuce, lamb's lettuce, frise, radicchio, cress, scarole (all under covers) (OLA 0364/96)[4]; unlimited for use on outdoor leaf herbs (see OLA notice for details) (OLA 0652/95, 0957/95)[4]; to Feb 2001 for use on outdoor kohlrabi (OLA 0448/96), gooseberries, blueberries, other Ribes species (OLA 1047/96)[4]; unlimited for use on Chinese cabbage (OLA 1572/97)[4]; unlimited for use on strawberries (OLA 1520/98)[4]

449 propamocarb hydrochloride

A translocated protectant carbamate fungicide

See also chlorothalonil + propamocarb hydrochloride
mancozeb + propamocarb hydrochloride

Products					
	1 Filex	Scotts	722 g/l	SL	07631
	2 Proplant	Fargro	722 g/l	SL	08572

Uses Botrytis in LETTUCE *(off-label)* [1]. Damping off and foot rot in INERT SUBSTRATE CUCUMBERS *(off-label)*, INERT SUBSTRATE TOMATOES *(off-label)*, NFT TOMATOES *(off-label)*, PEPPERS *(off-label)* [1]. Damping off in AUBERGINES, BEDDING PLANTS, BRASSICAS, CUCUMBERS, FLOWERS, LEEKS, ONIONS, ORNAMENTALS, PEPPERS, POT PLANTS, ROCKWOOL AUBERGINES, ROCKWOOL CUCUMBERS, ROCKWOOL PEPPERS, ROCKWOOL TOMATOES, TOMATOES [1, 2]. Damping off in VEGETABLE SEEDLINGS [1]. Downy mildew in BRASSICAS [1, 2]. Downy mildew in LETTUCE *(off-label)*, RADISHES *(off-label)* [1]. Phytophthora in AUBERGINES, BEDDING PLANTS, CUCUMBERS, FLOWERS, LEEKS, NURSERY STOCK, ONIONS, ORNAMENTALS PEPPERS, POT PLANTS, TOMATOES, TULIPS [1, 2]. Phytophthora in CONTAINER-

GROWN STOCK, WATERCRESS *(off-label)* [1]. Phytophthora in CONTAINER-GROWN ORNAMENTALS, HARDY ORNAMENTAL NURSERY STOCK [2]. Pythium in HARDY ORNAMENTAL NURSERY STOCK, ORNAMENTALS, POT PLANTS, TULIPS [1, 2]. Pythium in WATERCRESS *(off-label)* [1]. Root rot in AUBERGINES, CUCUMBERS, LEEKS, ONIONS, PEPPERS, ROCKWOOL AUBERGINES, ROCKWOOL CUCUMBERS, ROCKWOOL PEPPERS, ROCKWOOL TOMATOES, TOMATOES [1, 2]. White blister in HORSERADISH *(off-label)* [1].

Notes

Efficacy

- Chemical is absorbed through roots and translocated throughout plant
- Incorporate in compost before use or drench moist compost or soil before sowing, pricking out, striking cuttings or potting up. Use treated compost within 2 wk
- Drench treatment can be repeated at 3-6 wk intervals
- Concentrated solution is corrosive to all metals other than stainless steel
- May also be applied in trickle irrigation systems or feed solution
- To prevent root rot in tulip bulbs apply as dip for 20 min or as a pre-planting drench

Crop Safety/Restrictions

- Maximum number of treatments 4 per crop for cucumbers, tomatoes, peppers, aubergines; 1 per crop for listed brassicas; 1 compost incorporation and/or 1 drench treatment for leeks, onions, flower bulbs, ornamentals
- When applied over established seedlings rinse off foliage with water and do not apply under hot, dry conditions
- Do not treat young seedlings with overhead drench
- On plants of unknown tolerance test first on a small scale

Special precautions/Environmental safety

- Do not contaminate surface waters or ditches with chemical or used container
- Store away from seeds and fertilisers

Protective clothing/Label precautions

- A, B, C, H, K, M [1, 2]
- U08, U20b, E15, E30a, E32a [1, 2]; E26, E34 [2]; E29 [1]

Latest application/Harvest Interval(HI)

- Before transplanting for brassicas
- HI 14 d for cucumbers, tomatoes, peppers, aubergines; 4 wk for calabrese, cauliflower, sprouting broccoli, Chinese cabbage; 10 wk for cabbage; 19 wk for leeks, onions

Approval

- Off-label unlimited for use on outdoor and protected radishes (OLA 1211/95)[1]; to Nov 2000 for use on outdoor and protected lettuces, scarole, cress, lamb's lettuce (OLA 1625/95[1]; unlimited for use on NFT tomatoes and peppers, and tomatoes, cucumbers, peppers on inert substrate (OLA 1036/96)[1]; to Jan 2002 for use on water cress during propagation (OLA 0177/97)[1]; unlimited for use on outdoor and protected horseradish (OLA 1469/97)[1]

450 propaquizafop

A phenoxy alkanoic acid foliar acting grass herbicide

Products

1 Barclay Rebel II	Barclay	100 g/l	EC	08897
2 Falcon	Cyanamid	100 g/l	EC	08288
3 Landgold PQF 100	Landgold	100 g/l	EC	08976
4 Standon Propaquizafop	Standon	100 g/l	EC	09120

Uses

Annual grasses in FARM FORESTRY [2-4]. Annual grasses in RED BEET *(off-label)*, WARE POTATOES [2]. Annual grasses in BULB ONIONS, CARROTS, COMBINING PEAS, FODDER BEET, LINSEED, OILSEED RAPE, PARSNIPS, SPRING FIELD BEANS, SUGAR BEET, SWEDES, TURNIPS, WINTER FIELD BEANS [1-4]. Annual grasses in POTATOES [1]. Annual grasses in MAINCROP POTATOES [3, 4]. Perennial grasses in FARM FORESTRY [2-4]. Perennial grasses in RED BEET *(off-label)*, WARE POTATOES [2]. Perennial grasses in BULB ONIONS, CARROTS, COMBINING PEAS, FODDER BEET, LINSEED, OILSEED RAPE, PARSNIPS, SPRING FIELD BEANS, SUGAR BEET, SWEDES, TURNIPS, WINTER FIELD BEANS [1-4]. Perennial grasses in MUSTARD, POTATOES [1]. Perennial grasses in MAINCROP POTATOES [3, 4].

Notes

Efficacy

- Apply to emerged weeds when they are growing actively with adequate soil moisture
- Activity is slower under cool conditions
- Broad-leaved weeds and any weeds germinating after treatment are not controlled
- Annual meadow grass up to 3 leaves checked at low doses and severely checked at highest dose
- Spray barley cover crops when risk of wind blow has passed and before there is serious competition with the crop
- Various tank mixtures and sequences recommended for broader spectrum weed control in oilseed rape, peas and sugar beet. See label for details
- Severe couch infestations may require a second application at reduced dose when regrowth has 3-4 leaves unfolded

Crop Safety/Restrictions

- Maximum total dose 2 l/ha per crop (or per yr for forestry)
- See label for earliest crop growth stages for treatment
- See label for list of tolerant tree species
- Application in high temperatures and/or low soil moisture content may cause chlorotic spotting especially on combining peas and field beans
- Overlaps at the highest dose can cause damage from early applications to carrots and parsnips
- Do not treat seed potatoes
- Products contain surfactants. Tank mixing with adjuvants not required or recommended
- An interval of 4 wk must elapse before redrilling a failed treated crop. Only broad-leaved crops may be redrilled

Special precautions/Environmental safety

- Irritating to eyes and skin

- Harmful to fish or other aquatic life. Do not contaminate surface waters or ditches with chemical or used container
- Do not allow direct spray from vehicle mounted/drawn sprayers to fall within 6 m, or from hand-held sprayers within 2 m, of surface waters or ditches. Direct spray away from water

Protective clothing/Label precautions

- A, C
- M05, R04a, R04b, U02, U05a, U08, U14, U15, U19, U20b, C03, E01, E13c, E22b, E26, E30a, E31b [1-4]; E16 [1]

Latest application/Harvest Interval(HI)

- Before crop flower buds visible for winter oilseed rape, linseed, field beans; before 8 fully expanded leaf stage for spring oilseed rape, mustard; before weeds are covered by the crop for potatoes, sugar beet, fodder beet; when flower buds visible for peas
- HI early potatoes, carrots, parsnips, bulb onions 4 wk; peas 7 wk; sugar beet, fodder beet, maincrop potatoes, swedes, turnips 8 wk; field beans 14 wk

Approval

- Off-label approval to May 2001 for use on red beet (OLA 1034/96)[2]

---

451 propham

A pre-sowing carbamate herbicide, all approvals for which were revoked in 1997

---

452 propiconazole

A systemic, curative and protectant conazole fungicide

See also carbendazim + propiconazole
chlorothalonil + propiconazole
fenbuconazole + propiconazole
fenpropidin + propiconazole
fenpropidin + propiconazole + tebuconazole
fenpropimorph + propiconazole
prochloraz + propiconazole

Products

1 Barclay Bolt	Barclay	250 g/l	EC	08341
2 Landgold Propiconazole	Landgold	250 g/l	EC	06291
3 Mantis	Novartis	250 g/l	EC	08423
4 Mantis 250 EC	Ciba Agric.	250 g/l	EC	06240
5 Radar	Zeneca	250 g/l	EC	06747
6 Standon Propiconazole	Standon	250 g/l	EC	07037
7 Tilt	Novartis	250 g/l	EC	08456
8 Tilt 250 EC	Ciba Agric.	250 g/l	EC	02138

Uses

Alternaria in OILSEED RAPE [3, 4, 7, 8]. Brown rust in BARLEY, WHEAT [1-8]. Brown rust in RYE [1, 3-8]. Cladosporium leaf blotch in LEEKS [7, 8]. Crown rust in GRASS FOR ENSILING, GRASS SEED CROPS [1, 3-8]. Drechslera leaf spot in GRASS FOR ENSILING, GRASS SEED CROPS [1, 3-8]. Eyespot in WINTER BARLEY *(with carbendazim)*, WINTER WHEAT *(with carbendazim)* [2]. Eyespot in WINTER BARLEY *(low levels only)* [1, 6]. Eyespot in WINTER WHEAT *(low levels only)* [1, 3-6]. Fungus diseases in HONESTY *(off-label)* [7, 8]. Leaf blotch in GARLIC *(off-label)*, ONIONS *(off-label)*, SHALLOTS *(off-label)* [8]. Light leaf spot in OILSEED RAPE [1, 3-8]. Mildew in GRASS FOR ENSILING, GRASS SEED CROPS, SUGAR

BEET [1, 3-8]. Net blotch in BARLEY [1-8]. Powdery mildew in BARLEY, WHEAT [1-8]. Powdery mildew in OATS, RYE [1, 3-8]. Ramularia leaf spots in SUGAR BEET [1, 3-8]. Rhynchosporium in BARLEY [1-8]. Rhynchosporium in GRASS FOR ENSILING, GRASS SEED CROPS, RYE [1, 3-8]. Rust in SUGAR BEET [1, 3-8]. Rust in LEEKS [7, 8]. Septoria in WHEAT [1-8]. Septoria in RYE [1, 3-8]. Snow rot in WINTER BARLEY [3, 4]. Sooty moulds in WHEAT [7, 8]. White rust in CHRYSANTHEMUMS *(off-label)* [5, 7, 8]. Yellow rust in BARLEY, WHEAT [1-8].

Notes

Efficacy

- Best results achieved by applying at early stage of disease. Recommended spray programmes vary with crop, disease, season, soil type and product. See label for details
- On leeks apply a programme of 3 treatments at 2-3 wk intervals [7, 8]

Crop Safety/Restrictions

- Maximum number of treatments 4 per crop for wheat (up to 3 in yr of harvest); 3 per crop for barley, oats, rye, triticale (up to 2 in yr of harvest), leeks; 2 per crop or yr for oilseed rape, sugar beet, grass seed crops, honesty; 1 per yr on grass for ensiling
- Maximum total dose 0.5 l/ha in yr of sowing + 1.5 l/ha in yr of harvest for wheat; 0.5 l/ha in yr of sowing + 1.0 l/ha in yr of harvest for other cereals; 1.0 l/ha for oilseed rape, sugar beet, grass seed; 0.5 l/ha for grass for ensiling [1]
- On oilseed rape do not apply during flowering
- A minimum interval of 14 d must elapse between treatments on leeks [7, 8]
- Avoid spraying crops under stress, e.g. during cold weather or periods of frost

Special precautions/Environmental safety

- Irritating to eyes and skin
- Dangerous to fish or other aquatic life. Do not contaminate surface waters or ditches with chemical or used container

Protective clothing/Label precautions

- A, C
- R04a, R04b, U02, U05a, U19, U20a, C03, E01, E13b, E30a [1-8]; U04a, U09b, E31b [1]; U08, U09a, C02 (35 d cereals/grass seed/leeks; 28 d rape/silage/sugar beet), E31c, E34 [2-8]

Latest application/Harvest Interval(HI)

- Before 4 pairs of true leaves for honesty.
- HI cereals, grass and seed crops 5 wk; oilseed rape, sugar beet and grass for ensiling 4 wk; leeks 35 d

Approval

- Approved for aerial application on wheat and barley [4, 5, 7, 8], on wheat, barley, rye, oats [6]. See notes in Section 1
- Off-label approval unlimited for use only on *confirmed outbreaks* of white rust on chrysanthemums (OLA 1211/96)[5], (OLA 0019/92)[8], (OLA 1388/97)[8345]; unlimited for use on outdoor crops of honesty undersown in cereals (OLA 0481/93)[8], (OLA 1703/97)[7]; to Aug 2000 for use on onions, garlic and shallots (OLA 2280/97)[8]

Maximum Residue Levels (mg residue/kg food)

- grapes 0.5; apricots, peaches 0.2; bananas, tea, hops, ruminant liver 0.1; citrus fruits, tree nuts, pome fruits, strawberries, blackberries, dewberries, loganberries, raspberries, bilberries, cranberries, currants, gooseberries, wild berries, miscellaneous fruits (except bananas), root and tuber vegetables, bulb vegetables, tomatoes, aubergines, sweet corn, brassica vegetables, leaf vegetables and herbs, legume vegetables, asparagus, cardoons, fennel, leeks, rhubarb, fungi, pulses, peanuts, poppy seed, sesame seed, sunflower seed, soya beans, mustard seed, cotton seed, potatoes, cereals, meat (except ruminant liver) 0.05; milk and dairy produce 0.01

453 propiconazole + tebuconazole

A broad spectrum systemic fungicide mixture for cereals

Products

1 Cogito	Ciba Agric.	250:250 g/l	EC	07384
2 Cogito	Novartis	250:250 g/l	EC	08397
3 Endeavour	Bayer	250:250 g/l	EC	07385

Uses

Brown rust in BARLEY, WHEAT. Glume blotch in WHEAT. Net blotch in BARLEY. Powdery mildew in BARLEY, WHEAT. Rhynchosporium in BARLEY. Septoria leaf spot in WHEAT. Yellow rust in BARLEY, WHEAT.

Notes

Efficacy

- Best disease control and yield benefit obtained when applied at early stage of disease development before infection spread to new growth
- To protect the flag leaf and ear from Septoria diseases apply from flag leaf emergence to ear fully emerged (GS 37-59)
- Applications once foliar symptoms of *Septoria tritici* are already present on upper leaves will be less effective

Crop Safety/Restrictions

- Maximum total dose 1 l/ha per crop
- Occasional slight temporary leaf speckling may occur on wheat but this has not been shown to reduce yield response or disease control
- Do not treat durum wheat

Special precautions/Environmental safety

- Irritating to eyes and skin. May cause sensitization by skin contact
- Dangerous to fish or other aquatic life. Do not contaminate surface waters or ditches with chemical or used container
- Do not allow direct spray from ground based/vehicle drawn sprayers to fall within 6 m, or from hand-held sprayers to within 2 m, of surface waters or ditches. Direct spray away from water

Protective clothing/Label precautions

- A, C, H
- R04a, R04b, R04e, U05a, U09b, U20a, C03, E01, E13b, E16, E30a, E31c

Latest application/Harvest Interval(HI)

- Before watery ripe stage (GS 71)

Maximum Residue Levels (mg residue/kg food)

- see propiconazole entry

454 propiconazole + tridemorph

A broad spectrum contact and systemic fungicide for cereals

Products Joust | Unicrop | 125:350 g/l | EC | 08122

Uses Brown rust in SPRING WHEAT, WINTER BARLEY, WINTER WHEAT. Net blotch in SPRING BARLEY, WINTER BARLEY. Powdery mildew in SPRING BARLEY, SPRING WHEAT, WINTER BARLEY, WINTER WHEAT. Rhynchosporium in SPRING BARLEY, WINTER BARLEY. Septoria diseases in SPRING WHEAT, WINTER WHEAT. Yellow rust in SPRING WHEAT, WINTER WHEAT.

Notes

Efficacy

- Best results obtained from treatment when disease first noticed
- Most effective time for treatment varies with disease, See label
- Yield response is reduced by spraying when disease established and crop under stress

Crop Safety/Restrictions

- Maximum number of treatments 2 per crop for spring barley, spring wheat; 3 per crop (only 2 after 1 Jan) for winter barley, winter wheat
- Treatment may increase any flag leaf tip scorch of wheat caused by high temperatures and soil moisture deficits

Special precautions/Environmental safety

- Irritating to eyes and skin
- Dangerous to fish or other aquatic life. Do not contaminate surface waters or ditches with chemical or used container

Protective clothing/Label precautions

- A, C
- R04a, R04b, U02, U04a, U05a, U09a, U19, U20b, C02 (35 d), C03, E01, E13b, E26, E29, E30a, E31c, E34

Latest application/Harvest Interval(HI)

- HI 35 d

Maximum Residue Levels (mg residue/kg food)

- see propiconazole entry

455 propoxur

A carbamate insecticide for glasshouse and general use

Products Fumite Propoxur Smoke | Hortichem | 50% w/w | FU | 0094

Uses Aphids in PROTECTED CARNATIONS, PROTECTED CHRYSANTHEMUMS, PROTECTE CUCUMBERS, PROTECTED TOMATOES. Whitefly in PROTECTED CARNATION PROTECTED CHRYSANTHEMUMS, PROTECTED CUCUMBERS, PROTECTE TOMATOES.

Notes

Efficacy

- Fumigate glasshouse at first sign of whitefly infestation and, where serious, repeat after 7 d. Provides partial control of aphids
- Make house smoke tight, water plants several hours previously so that leaves are dry and damp down paths. Fumigate in late afternoon or evening in calm weather when temperature is 16°C or over. Do not fumigate in bright sunshine
- Keep house closed overnight and ventilate well next morning
- For aphid control in chrysanthemums treat when flower colour first appears in bud

Crop Safety/Restrictions

- Do not treat tomatoes until after third truss has set
- Allow 7 d after fumigation before introducing *Encarsia* or *Phytoseiulus*
- Open flowers of carnations and chrysanthemums should be cut before fumigating

Special precautions/Environmental safety

- This product contains an anticholinesterase carbamate compound. Do not use if under medical advice not to work with such compounds
- Irritating to eyes and respiratory system
- Harmful to bees. Do not apply to crops in flower or to those in which bees are actively foraging. Do not apply when flowering weeds are present
- Harmful to fish or other aquatic life. Do not contaminate surface waters or ditches with chemical or used container

Protective clothing/Label precautions

- D
- U05a, U19, U20a, C03, E01, E12e, E13c, E30a, E34

Latest application/Harvest Interval(HI)

- HI 2 d

Maximum Residue Levels (mg residue/kg food)

- citrus fruit, pome fruit, stone fruit, grapes, strawberries, raspberries, olives, beetroot, celeriac, peppers, aubergines, gherkins, melons, squashes, watermelons, brassica vegetables, lettuce, spinach, spinach beet, herbs, beans (with pods), peas (with pods), cardoons, celery, fennel, globe artichokes 3; currants, gooseberries 0.2; tea, hops 0.1; all other produce (except tomatoes, courgettes) 0.05

456 propyzamide

A residual amide herbicide for use in a wide range of crops

See also clopyralid + propyzamide

Products

1 Barclay Piza 500	Barclay	50% w/w	WP	05283
2 Headland Judo	Headland	400 g/l	SC	08339
3 Headland Redeem Flo	Headland	400 g/l	SC	08340
4 Kerb 50 W	PBI	50% w/w	WP	02986
5 Kerb Flo	PBI	400 g/l	SC	04521
6 Kerb Granules	PBI	4% w/w	GR	08917
7 Kerb Pro Flo	PBI	400 g/l	SC	08679
8 Kerb Pro Granules	PBI	4% w/w	GR	08698
9 Landgold Propyzamide 50	Landgold	50% w/w	WP	05916
10 Precis	PBI	400 g/l	SC	08678
11 Rapier	MTM Agrochem.	450 g/l	SC	05314
12 Standon Propyzamide 50	Standon	50% w/w	WP	09054
13 Stefes Pride	Stefes	50% w/w	WP	05616

Uses

Annual dicotyledons in FODDER RAPE SEED CROPS, KALE SEED CROPS, LETTUCE, RHUBARB, TURNIP SEED CROPS [4, 5, 10, 12]. Annual dicotyledons in CHICORY *(off-label)*, EVENING PRIMROSE *(off-label)*, HONESTY *(off-label)*, PROTECTED HERBS *(off-label)* [4, 5]. Annual dicotyledons in FORESTRY, TREES AND SHRUBS [4, 6-8]. Annual dicotyledons in RASPBERRIES, ROSES [4, 12]. Annual dicotyledons in CHAMOMILE *(off-label)*, FARM WOODLAND, FENUGREEK *(off-label)*, GRAPEVINES *(off-label)*, HOPS *(off-label)*, PROTECTED LETTUCE *(off-label)*, RADICCHIO *(off-label)*, SAGE *(off-label)*, SEED BRASSICAS, TARRAGON *(off-label)*, VETCHES *(off-label)* [4]. Annual dicotyledons in HERBS *(off-label)* [5]. Annual dicotyledons in WINTER OILSEED RAPE [1-5, 9-13]. Annual dicotyledons in WINTER FIELD BEANS [1-5, 9, 10, 12, 13]. Annual dicotyledons in APPLES, BLACKBERRIES, BLACKCURRANTS, CLOVER SEED CROPS, GOOSEBERRIES, LOGANBERRIES, LUCERNE, PEARS, PLUMS, REDCURRANTS, SUGAR BEET SEED CROPS [2-5, 10, 12]. Annual dicotyledons in RASPBERRIES *(England only)*, STRAWBERRIES [2-5, 10]. Annual dicotyledons in WOODY ORNAMENTALS [2-4, 6-8]. Annual dicotyledons in BRASSICA SEED CROPS, OUTDOOR LETTUCE, RHUBARB *(outdoor)* [2, 3]. Annual dicotyledons in ORNAMENTALS [12]. Annual grasses in FODDER RAPE SEED CROPS, KALE SEED CROPS, LETTUCE, RHUBARB, TURNIP SEED CROPS [4, 5, 10, 12]. Annual grasses in CHICORY *(off-label)*, EVENING PRIMROSE *(off-label)*, HONESTY *(off-label)*, PROTECTED HERBS *(off-label)* [4, 5]. Annual grasses in FORESTRY, TREES AND SHRUBS [4, 6-8]. Annual grasses in RASPBERRIES, ROSES [4, 12]. Annual grasses in CHAMOMILE *(off-label)*, FARM WOODLAND, FENUGREEK *(off-label)*, GRAPEVINES *(off-label)*, HOPS *(off-label)*, PROTECTED LETTUCE *(off-label)*, RADICCHIO *(off-label)*, SAGE *(off-label)*, SEED BRASSICAS, TARRAGON *(off-label)*, VETCHES *(off-label)* [4]. Annual grasses in HERBS *(off-label)*[5]. Annual grasses in WINTER OILSEED RAPE [1-5, 9-13]. Annual grasses in WINTER FIELD BEANS [1-5, 9, 10, 12, 13]. Annual grasses in APPLES, BLACKBERRIES, BLACKCURRANTS, CLOVER SEED CROPS, GOOSEBERRIES, LOGANBERRIES, LUCERNE, PEARS, PLUMS, REDCURRANTS, SUGAR BEET SEED CROPS [2-5, 10, 12]. Annual grasses in RASPBERRIES *(England only)*, STRAWBERRIES [2-5, 10]. Annual grasses in WOODY ORNAMENTALS [2-4, 6-8]. Annual grasses in BRASSICA SEED CROPS, OUTDOOR LETTUCE, RHUBARB *(outdoor)* [2, 3]. Annual grasses in ORNAMENTALS [12]. Horsetails in FORESTRY [2, 3, 7]. Horsetails in WOODY ORNAMENTALS [7]. Perennial grasses in CLOVER SEED CROPS, FODDER RAPE SEED CROPS, KALE SEED CROPS,

LETTUCE, LUCERNE, RHUBARB, SUGAR BEET SEED CROPS, TURNIP SEED CROPS, WINTER FIELD BEANS, WINTER OILSEED RAPE [4, 5, 10, 12]. Perennial grasses in STRAWBERRIES [4, 5, 10]. Perennial grasses in CHICORY *(off-label)*, EVENING PRIMROSE *(off-label)*, HONESTY *(off-label)*, PROTECTED HERBS *(off-label)* [4, 5]. Perennial grasses in TREES AND SHRUBS [4, 6-8]. Perennial grasses in RASPBERRIES, ROSES [4, 12]. Perennial grasses in CHAMOMILE *(off-label)*, FARM WOODLAND, FENUGREEK *(off-label)*, GRAPEVINES *(off-label)*, HOPS *(off-label)*, PROTECTED LETTUCE *(off-label)*, RADICCHIO *(off-label)*, SAGE *(off-label)*, SEED BRASSICAS, TARRAGON *(off-label)*, VETCHES *(off-label)* [4]. Perennial grasses in HERBS *(off-label)* [5]. Perennial grasses in APPLES, BLACKBERRIES, BLACKCURRANTS, GOOSEBERRIES, LOGANBERRIES, PEARS, PLUMS, REDCURRANTS [2-5, 10, 12]. Perennial grasses in RASPBERRIES *(England only)* [2-5, 10]. Perennial grasses in FORESTRY, WOODY ORNAMENTALS [2-4, 6-8]. Perennial grasses in RHUBARB *(outdoor)* [2, 3]. Perennial grasses in ORNAMENTALS [12]. Sedges in FORESTRY [2, 3, 7]. Sedges in WOODY ORNAMENTALS [7]. Volunteer cereals in WINTER OILSEED RAPE [1-3, 9, 11, 13]. Volunteer cereals in WINTER FIELD BEANS [1-3, 9, 13]. Volunteer cereals in SUGAR BEET SEED CROPS [2, 3]. Wild oats in WINTER OILSEED RAPE [1-3, 9, 11, 13]. Wild oats in WINTER FIELD BEANS [1-3, 9, 13]. Wild oats in SUGAR BEET SEED CROPS [2, 3].

Notes

Efficacy

- Active via root uptake. Weeds controlled from germination to young seedling stage, some species (including many grasses) also when established
- Best results achieved by winter application to fine, firm, moist soil. Rain is required after application if soil dry
- Do not use on soils with more than 10% organic matter except in forestry
- Excessive organic debris or ploughed-up turf may reduce efficacy
- For heavy couch infestations a repeat application may be needed in following winter
- In lettuce lightly incorporate in top 25 mm pre-drilling or irrigate on dry soil

Crop Safety/Restrictions

- Maximum number of treatments 1 per crop or yr
- Apply to most crops from 1 Oct to 31 Jan, to lettuce at any time
- Apply as soon as possible after 3-true leaf stage of oilseed rape (GS 1,3) and seed brassicas, after 4-leaf stage of sugar beet for seed, within 7 d after sowing but before emergence for field beans, after perennial crops established for at least 1 season, strawberries after 1 yr
- Only apply to strawberries on heavy soils, to field beans on medium and heavy soils, to established lucerne not less than 7 d after last cut. Use reduced dose on matted row strawberries [4, 5, 10]
- Do not treat protected crops
- See label for lists of ornamental and forest species which may be treated
- Lettuce may be sown or planted immediately after treatment, with other crops period varies from 5 to 40 wk. See label for details
- Do not apply to the same land less than 9 mth after an earlier application

Special precautions/Environmental safety

- Do not contaminate surface waters or ditches with chemical or used container
- Do not harvest crops for human or animal consumption for at least 6 wk after last application

Protective clothing/Label precautions

- U20a [1, 6, 8, 13]; U20b [2, 3, 9, 12]; U20c [4, 5, 7, 10, 11]; C02 (6 wk) [4, 5, 9, 10, 12]; C02 [1, 2, 6, 8, 13]; E01 [1, 4-6, 8-10, 12, 13]; E15, E30a, E32a [1-13]; E26 [11]; E34 [1, 2, 4-6, 8, 10, 13]

Latest application/Harvest Interval(HI)

- Before 31 Dec for strawberries and winter field beans; before 31 Jan for oilseed rape,sugar beet seed crops, lucerne, seed crops of clover, fodder rape, kale and turnips, top, bush and cane fruit, rhubarb, woody ornamentals, hops, vines, evening primrose, tarragon, chamomile, strawberries, lettuce, radicchio, fenugreek; pre-emergence for chicory.
- HI strawberries, field beans, oilseed rape, sage 6 wk; protected herbs 3 mth; winter vetches 4 mth

Approval

- May be applied through CDA equipment [5], through CDA equipment in forestry [4]. See label for details
- Off-label approval unlimited for use on evening primrose (OLA 0489/92)[4]; tarragon, chamomile, radicchio, sage, fenugreek, protected herbs (OLA 0624/92)[4]; unlimited for use on honesty (OLA 1364/93)[4]; unlimited for use on outdoor chicory for forcing (OLA 1505/96)[4]

Maximum Residue Levels (mg residue/kg food)

- linseed, tea, meat 0.05; citrus fruit, tree nuts, pome fruit, stone fruit, berries and small fruit (except strawberries, currants, gooseberries), miscellaneous fruit, root and tuber vegetables, bulb vegetables, fruiting vegetables, brassica vegetables (except head cabbage), spinach, watercress, witloof, peas, stem vegetables, fungi, pulses, poppy seed, sesame seed, sunflower seed, soya bean, mustard seed, potatoes, cereals, eggs 0.02; milk and dairy produce 0.01

457 prosulfuron

A contact and residual sulfonyl urea herbicide available only in mixtures

See also bromoxynil + prosulfuron

458 pyrazophos

A systemic organophosphorus fungicide with insecticidal activity

Products	Afugan	Promark	300 g/l	EC	07301

Uses

Insect pests in BEETROOT *(off-label)*, CARROTS *(off-label)*, CELERIAC *(off-label)*, NON-EDIBLE ORNAMENTALS *(off-label)*, PARSNIPS *(off-label)*, SALSIFY *(off-label)*, VEGETABLES *(off-label)*. Powdery mildew in APPLES, HOPS, POT PLANTS, ROSES.

Notes

Efficacy

- Spray apples at pink bud and repeat every 10-14 d until extension growth ceased
- Spray hops at first sign of disease or when shoots 10-12 cm long and repeat every 10-14 d. Treatment suppresses damson-hop aphid
- Spray pot plants and roses at first sign of infection and repeat every 10-14 d
- Spray Brussels sprouts from young plant stage to 2 wk before harvest. Use of pendant lances recommended to give good cover
- Temperatures above 30°C may reduce efficacy

- Where aphids resistant to pyrazophos occur repeat treatments are likely to result in lower levels of control

Crop Safety/Restrictions

- Maximum number of treatments 4 per crop for beetroot, carrots, celeriac, parsnips, salsify; 7 per crop for protected marrows and melons; 10 per crop for cucumbers, other vegetable crops and ornamentals
- Do not tank-mix with sulfur, dinocap or other organophosphorus compounds
- Treatment may cause some yellowing of hop foliage but effect normally outgrown
- Do not spray roses under glass, aquilegia or scorzonera. Some outdoor roses may be slightly sensitive

Special precautions/Environmental safety

- This product contains an anticholinesterase organophosphorous compound. Do not use if under medical advice not to work with such compounds
- Harmful if swallowed. Irritating to eyes and skin
- Flammable
- Harmful to livestock
- Harmful to game, wild birds and animals
- Dangerous to bees. Do not apply to crops in flower or to those in which bees are actively foraging. Do not apply when flowering weeds are present
- Dangerous to fish or other aquatic life. Do not contaminate surface waters or ditches with chemical or used container

Protective clothing/Label precautions

- A, C
- M01, M03, R03c, R04a, R04b, R07d, U02, U04a, U05a, U10, U11, U13, U19, U20a, C02 (2 wk), C03, E01, E06b (2 wk), E10b, E12c, E13b, E30a, E31b, E34

Withholding period

- Keep all livestock out of treated areas for at least 2 wk

Latest application/Harvest Interval(HI)

- HI cucumbers 3 d; other vegetable crops 14 d

Approval

- Off-label approval unlimited for use on beetroot, carrots, salsify, celeriac, parsnips, (OLA1403/94)[1]; unlimited for use on non-edible ornamentals, Brussels sprouts, cucumbers, protected crops of courgettes, gherkins, pumpkins, squashes (OLA 0681/97)[1]

---

459 pyrethrins

A non-persistent, contact acting insecticide extracted from Pyrethrum

Products

1 Alfadex	Ciba Agric.	0.75 g/l	AL	00074
2 Alfadex	Novartis A H	0.75 g/l	AL	08591
3 Dairy Fly Spray	B H & B	0.75 g/l	AL	H5579
4 Killgerm ULV 400	Killgerm	30 g/l	UL	H4838
5 Multispray	AgrEvo Environ.	0.065% w/w	AL	H5165
6 Pybuthrin 33	AgrEvo Environ.	3 g/l	AL	H5106

Uses

Flies in DAIRIES, FARM BUILDINGS, LIVESTOCK HOUSES, POULTRY HOUSES [1-4]. Flies in REFUSE TIPS [5, 6].

Notes

Efficacy

- For fly control close doors and windows and spray or apply fog as appropriate
- For best results outdoors spray during early morning or late afternoon and evening when conditions are still

Crop Safety/Restrictions

- Do not allow spray to contact open food products or food preparing equipment or utensils
- Remove exposed milk and collect eggs before application
- Do not treat plants
- Do not use space sprays containing pyrethrins or pyrethroid more than once per week in intensive or controlled environment animal houses in order to avoid development of resistance. If necessary, use a different control method or product [4]

Special precautions/Environmental safety

- Irritating to eyes, skin [5, 6] and respiratory system [1, 3, 4]
- Do not apply directly to livestock
- Harmful to fish or other aquatic life. Do not contaminate surface waters or ditches with chemical or used container [1, 3, 4, 6]
- Extremely dangerous to fish or other aquatic life and reptiles. Do not contaminate water courses or ground.
- Exclude all persons and animals during treatment [3]

Protective clothing/Label precautions

- A [1-4]; B [3, 5, 6]; C [1, 2, 4]; D [1, 2, 4-6]; E [3, 5, 6]; H [1, 2, 4-6]
- M01, U10, U17, E13a [5]; R03c, U09a, U20b, C04, C06, C07, C08, C10, E04 [3]; R04a, R04b, E13c [1-4, 6]; R04c, E31a [1, 2, 4]; U02 [5, 6]; U05a, U19, C03, E01, E05, E30a [1-6]; U08, E02 [6]; U20a [1, 2, 4-6]; C09 [3, 5, 6]; E32a [3, 6]

Approval

- Product formulated for ULV application [3]. May be applied through fogging machine or sprayer [1, 3, 5, 6]. See label for details

460 pyrethrins + resmethrin

A contact insecticide for many glasshouse and horticultural crops

Products

Pynosect 30 Water Miscible	Mitchell Cotts	1.4:9.1% w/w	EC	01653

Uses

Aphids in PROTECTED CUCUMBERS, PROTECTED ORNAMENTALS, PROTECTED TOMATOES. Phorid flies in MUSHROOMS *(Off-label)*. Sciarid flies in MUSHROOMS *(Off-label)*. Whitefly in PROTECTED CUCUMBERS, PROTECTED ORNAMENTALS, PROTECTED TOMATOES.

Notes

Efficacy

- Apply when pest first seen and repeat as necessary, for whitefly every 3-4 d
- Apply as high volume spray. See label for details of suitable equipment
- Where glasshouse whitefly resistant to pyrethrins occur control is unlikely to be satisfactory

Crop Safety/Restrictions
- Do not spray when temperature above 24°C
- Care should be taken in spraying 'soft' cucumber plants grown during winter
- Do not treat crops suffering from drought or stress
- On ornamentals a small test spraying is recommended before treating whole crop
- Damage may occur to open flowers of ornamentals
- Do not apply to ferns

Special precautions/Environmental safety
- Extremely dangerous to bees. Do not apply to crops in flower or to those in which bees are actively foraging. Do not apply when flowering weeds are present
- Extremely dangerous to fish or other aquatic life. Do not contaminate surface waters or ditches with chemical or used container

Protective clothing/Label precautions
- U08, U19, U20b, E12b, E13a, E30a, E31b

Latest application/Harvest Interval(HI)
- HI zero

Approval
- Off-label approval unlimited for use on mushrooms for control of sciarid and phorid flies (OLA 0505/95)[1]

461 pyridate

A contact pyridazine herbicide for cereals, maize and brassicas

Products

1 Barclay Pirate	Barclay	45% w/w	WP	07104
2 Lentagran WP	Sandoz	45% w/w	WB	07556
3 Lentagran WP	Novartis	45% w/w	WB	08478

Uses

Annual dicotyledons in BARLEY, DURUM WHEAT, MAIZE, OATS, OILSEED RAPE, RYE, SWEETCORN, TRITICALE, WHEAT [1]. Annual dicotyledons in SPRING OILSEED RAPE *(off-label)*, WINTER OILSEED RAPE *(off-label)* [2]. Black nightshade in BRUSSELS SPROUTS, CABBAGES, MAIZE [2, 3]. Black nightshade in SWEETCORN [2]. Black nightshade in ONIONS [3]. Cleavers in MAIZE [1-3]. Cleavers in SWEETCORN [1, 2]. Cleavers in BARLEY, DURUM WHEAT, OATS, OILSEED RAPE, RYE, TRITICALE, WHEAT [1]. Cleavers in BRUSSELS SPROUTS, CABBAGES [2, 3]. Cleavers in ONIONS [3]. Dead nettle in BARLEY, DURUM WHEAT, MAIZE, OATS, OILSEED RAPE, RYE, SWEETCORN, TRITICALE, WHEAT [1]. Fat-hen in BRUSSELS SPROUTS, CABBAGES, MAIZE [2, 3]. Fat-hen in SWEETCORN [2]. Fat-hen in ONIONS [3]. Speedwells in BARLEY, DURUM WHEAT, MAIZE, OATS, OILSEED RAPE, RYE, SWEETCORN, TRITICALE, WHEAT [1].

Notes

Efficacy
- Best results achieved by application to actively growing weeds at 6-8 leaf stage when temperatures are above 8°C before crop foliage forms canopy

Crop Safety/Restrictions
- Maximum number of treatments 1 per crop
- Apply to oilseed rape in winter after 6-true leaf stage (GS 1,6) but before mid-Dec, in spring after crop growth started but before 15 cm of stem extension

- Do not apply in mixture with or within 14 d of any other product which may result in dewaxing of crop foliage
- Apply to winter or spring cereals from first-tiller stage (GS 21)
- Apply to maize and sweetcorn after first-leaf stage. Do not use on cv. Meritos, Sunrise or Tainon 236
- Apply to cabbages and Brussels sprouts after 4 fully expanded leaf stage. Allow 2 wk after transplanting before treating [2]
- Do not use on crops suffering stress from frost, drought, disease or pest attack
- Do not apply to oilseed rape or cereals when night temperature consistently below 2°C or to oilseed rape when daytime temperature exceeds 16°C
- Do not apply to spring sown oilseed rape [2]

Special precautions/Environmental safety

- Irritating to eyes and skin
- May cause sensitisation by skin contact [2]
- Harmful to fish or other aquatic life. Do not contaminate surface waters or ditches with chemical or used container

Protective clothing/Label precautions

- A, C
- R04a, R04b, U05a, U08, C03, E01, E13c, E30a [1-3]; R04e, U20b, U22, E26 [2, 3]; U20a, E32a [1]

Latest application/Harvest Interval(HI)

- Before flag leaf ligule just visible (GS 39) for cereals; before flower bud visible (GS 3,1) for oilseed rape; before 7 leaf stage for maize, sweetcorn
- HI maize, sweetcorn 2 mth [1]; cabbage, Brussels sprouts 1 mth [2]

Approval

- Off-label approval unlimited for use on winter and spring oilseed rape (OLA 0177/96)[2]

---

462 pyrifenox

A systemic pyridine fungicide for use on apples

Products

Dorado	Zeneca	200 g/l	EC	06657

Uses

Leaf spots in BLACKCURRANTS, GOOSEBERRIES. Powdery mildew in APPLES, BLACKCURRANTS, GOOSEBERRIES, HARDY ORNAMENTAL NURSERY STOCK, ORNAMENTALS, PEARS, STRAWBERRIES. Scab in APPLES, PEARS.

Notes

Efficacy

- Apply to apples and pears from bud-burst at 7-14 d intervals, depending on dose applied, until danger of scab infection ceases or extension growth completed
- Spray programmes on apples and pears which incorporate other suitable fungicides (such as Captan or Nimrod) will help to avoid development of tolerant strains
- Treat from early grape stage (blackcurrants and gooseberries) or just before blossom (strawberries) or at first signs of disease and repeat at 14 d intervals
- Treat ornamentals at first sign of disease and repeat at 7-14 d intervals

Crop Safety/Restrictions
- Maximum number of treatments 5 per yr (blackcurrants and gooseberries); 4 per yr (strawberries)
- Do not leave spray liquid in sprayer for long period

Special precautions/Environmental safety
- Harmful if swallowed. Irritating to eyes and skin
- Flammable
- Harmful to fish or other aquatic life. Do not contaminate surface waters or ditches with chemical or used container

Protective clothing/Label precautions
- A, C, H
- M03, R03c, R04a, R04b, R07d, U04b, U05a, U08, U14, U15, U19, U20a, C02 (14 d), C03, E01, E13c, E30a, E31c, E34

Latest application/Harvest Interval(HI)
- HI 14 d (apples, pears, blackcurrants, gooseberries), 3 d (strawberries)

463 pyrimethanil

An anilinopyrimidine fungicide for apples and strawberries

Products	Scala	Promark	400 g/l	SC	07806

Uses Botrytis in GRAPEVINES *(off-label)*. Grey mould in STRAWBERRIES. Scab in APPLES.

Notes

Efficacy
- On apples a programme of sprays will give early season control of scab. Season long control can be achieved by continuing programme with other approved fungicides
- In strawberries product should be used as part of a programme of disease control treatments which should alternate with other materials to prevent or limit development of less sensitive strains of grey mould

Crop Safety/Restrictions
- Maximum number of treatments 5 per yr for apples, 2 per yr for strawberries
- All varieties of apples and strawberries may be treated
- Treat apples from bud burst at 10-14 d intervals
- In strawberries start treatments at white bud to give maximum protection of flowers against grey mould and treat every 7-10 d. Product should not be used more than once in a 3 or 4 spray programme
- Product does not taint apples. Processors should be consulted before use on strawberries

Special precautions/Environmental safety
- Harmful to fish or other aquatic life. Do not contaminate surface waters or ditches with chemical or used container
- Product has negligible effect on hoverflies and lacewings. Limited evidence indicates some margin of safety to *Typhlodromus pyri*

Protective clothing/Label precautions
- U08, U20b, C03, E01, E13c, E26, E30a, E31b

Latest application/Harvest Interval(HI)
- Before end of flowering for apples

- HI 1 d for strawberries

Approval
- Off-label approval unlimited for use on outdoor grapevines (OLA 1203/98)[1]

464 quinalbarbitone-sodium

A stupefying agent for control of pest bird species

Products

Killgerm Seconal	Killgerm	100% w/w	SP	04715

Uses Birds in FIELD CROPS.

Notes

Efficacy
- To be used under licence for making up baits

Special precautions/Environmental safety
- Under the Misuse of Drugs Act 1971 and Misuse of Drugs Regulations 1985 a licence is required to produce, possess or supply this product. Licences are issued by the Home Office, Drugs Branch
- Only to be used by operators trained in the use of stupefying baits and familiar with precautions to be observed
- Toxic by inhalation and if swallowed
- Keep in original container, tightly closed, in a safe place, under lock and key
- Any non-target species affected must be allowed to recover and then released
- Affected pest birds must be searched for and humanely killed. The bodies must be disposed of by burning or burial
- All remains of bait must be removed after treatment and burned or buried

Protective clothing/Label precautions
- A, D
- M04, R02b, R02c, U02, U04a, U05a, U13, U20a, C03, E01, E30b, E32a, E34, V01a, V02, V03a

Approval
- Only to be used where a licence has been approved in accordance with section 16 (1) of the Wildlife and Countryside Act 1981

465 quinmerac

A residual herbicide available only in mixtures

See also metazachlor + quinmerac

466 quinoxyfen

A systemic protectant fungicide for cereals

See also cyproconazole + quinoxyfen
fenpropimorph + quinoxyfen

Products

1 Apres	Dow	500 g/l	SC	08881
2 Erysto	Dow	250 g/l	SC	08697
3 Fortress	Dow	500 g/l	SC	08279
4 Standon Quinoxyfen 500	Standon	500 g/l	SC	08924

Uses

Powdery mildew in DURUM WHEAT, RYE, SPRING OATS, TRITICALE, WINTER OATS [3]. Powdery mildew in SPRING BARLEY, SPRING WHEAT, WINTER BARLEY, WINTER WHEAT [1-4].

Notes

Efficacy

- For best results treat at early stage of disease development before infection spreads to new crop growth. Further treatment may be necessary if disease pressure remains high
- Product not curative and will not control latent or established disease infections
- For broad spectrum control use in tank mixtures - see label
- Product rainfast after 1 h
- Systemic activity may be reduced in severe drought

Crop Safety/Restrictions

- Maximum total dose equivalent to two full dose treatments
- Apply only in the spring from mid-tillering stage (GS 25)

Special precautions/Environmental safety

- Irritant. May cause sensitization by skin contact
- Dangerous to fish or other aquatic life. Do not contaminate surface waters or ditches with chemical or used container
- Do not allow direct spray from vehicle mounted/drawn hydraulic sprayers to fall within 6 m of surface waters or ditches. Direct spray away from water

Protective clothing/Label precautions

- A, C, H
- R04, R04e, U05a, U14, C03, E01, E13b, E16, E34 [1-4]; E26 [2]

Latest application/Harvest Interval(HI)

- First awns visible stage (GS 49)

467 quintozene

A protectant soil applied chlorophenyl fungicide for horticultural crops

Products

1 Quintozene WP	RP Amenity	50% w/w	WP	05404
2 Terraclor 20D	Hortichem	20% w/w	DP	06578
3 Terraclor Flo	Hortichem	497 g/l	SC	08666

Uses

Botrytis in BEDDING PLANTS, CHRYSANTHEMUMS, DAHLIAS, FUCHSIAS, LETTUCE, PELARGONIUMS, POT PLANTS, PROTECTED CARNATIONS, TOMATOES [2]. Botrytis in OUTDOOR LETTUCE, PROTECTED LETTUCE [3]. Dollar spot in TURF [1]. Fusarium patch in TURF [1]. Red thread in TURF [1]. Rhizoctonia in ANEMONES, BEDDING PLANTS, CHRYSANTHEMUMS, CUCUMBERS, DAHLIAS, FLOWER BULBS, FUCHSIAS, HYACINTHS, IRISES, LETTUCE, NARCISSI, PELARGONIUMS, POT PLANTS, PROTECTED CARNATIONS, TOMATOES, TULIPS [2]. Rhizoctonia in OUTDOOR LETTUCE, PROTECTED LETTUCE [3]. Sclerotinia in ANEMONES, BEDDING PLANTS, CHICORY *(forcing)*, CHRYSANTHEMUMS, CORMS, DAHLIAS, FLOWER BULBS, FUCHSIAS, HYACINTHS, IRISES, LETTUCE, NARCISSI, PELARGONIUMS, POT PLANTS, PROTECTED CARNATIONS, TOMATOES, TULIPS [2]. Sclerotinia in OUTDOOR LETTUCE, PROTECTED LETTUCE [3]. Wirestem in BRASSICAS [2, 3].

Notes

Efficacy

- Apply to soil or compost and incorporate before planting or sowing [2, 3]
- For use in turf apply drenching spray when fungal growth active [1]

Crop Safety/Restrictions

- Maximum number of treatments 1 per crop for ornamental bulbs, brassicas, chicory, cucumber, lettuce, tomato, bedding and pot plants [2, 3]; 6 per yr for turf [1]
- In general leave treated soil or compost for 2-3 wk before planting unrooted cuttings, 4 wk before sowing tomatoes or cucumbers and 2-3 d before planting other crops
- Do not allow soil to dry out leading to reduction of plant vigour [2, 3]

Special precautions/Environmental safety

- Irritating to skin, eyes and respiratory system [1, 2]
- Dangerous to fish or other aquatic life. Do not contaminate surface waters or ditches with chemical or used container [3]
- Do not contaminate surface waters or ditches with chemical or used container [1, 2]

Protective clothing/Label precautions

- A [2, 3]
- R04a, R04b, R04c, U05a, U08, U20a, C03, E01 [2]; U19, E15, E34 [1, 2]; U20b [1, 3]; E13b, E26 [3]; E30a, E32a [1-3]

Latest application/Harvest Interval(HI)

- Before sowing or planting for listed crops other than turf

Maximum Residue Levels (mg residue/kg food)

- lettuce 3; bananas 1; potatoes 0.2; tomatoes, peppers, aubergines 0.1; cauliflowers, head cabbages 0.02; beans (with pods) 0.01

468 quizalofop-P-ethyl

An aryl phenoxypropionic acid post-emergence grass herbicide

Products

1 CoPilot	AgrEvo	100 g/l	EC	0804[illegible]
2 Pilot D	AgrEvo	250 g/l	SC	0804[illegible]
3 Sceptre	AgrEvo	250 g/l	SC	0804[illegible]

Uses

Annual grasses in CARROTS, FODDER BEET, LINSEED, MANGELS, MUSTAR[D] PARSNIPS, RED BEET, SPRING OILSEED RAPE, SUGAR BEET, WINTER OILSEE[D] RAPE. Couch in CARROTS, FODDER BEET, LINSEED, MANGELS, MUSTAR[D]

PARSNIPS, RED BEET, SPRING OILSEED RAPE, SUGAR BEET, WINTER OILSEED RAPE. Perennial grasses in CARROTS, FODDER BEET, LINSEED, MANGELS, MUSTARD, PARSNIPS, RED BEET, SPRING OILSEED RAPE, SUGAR BEET, WINTER OILSEED RAPE. Volunteer cereals in CARROTS, FODDER BEET, LINSEED, MANGELS, MUSTARD, PARSNIPS, RED BEET, SPRING OILSEED RAPE, SUGAR BEET, WINTER OILSEED RAPE.

Notes

Efficacy

- Apply to emerged weeds. Weeds emerging after treatment are not controlled. Effective on annual grasses from 2-leaf stage to fully tillered, on perennials from 4-6 leaf stage to before jointing
- Best results achieved by application to weeds growing actively in warm conditions with adequate soil moisture. Use split treatment to extend period of control
- Must be used with a recommended adjuvant - see label for details [3]
- Annual meadow-grass is not controlled
- Various spray programmes and tank-mixtures recommended to control mixed dicotyledon/grass weed populations. See label for details
- For effective couch control do not hoe beet crops within 21 d after spraying
- At least 2 h rain free period required for effective results

Crop Safety/Restrictions

- Maximum number of treatments 2 as split dose on oilseed rape and mustard and 2 for couch control in all other crops. See labels for maximum total dose of individual products
- Apply to beet crops from 2 fully expanded leaves when weeds at appropriate stage and growing actively but before crop meets across rows
- Apply to oilseed rape and mustard from expanded cotyledon stage (GS 1,0), before crop covers larger weeds
- Do not spray crops under stress from any cause or in frosty weather
- Consult processor before using on carrot crops for processing
- An interval of at least 3 d must elapse between treatment and use of another herbicide on beet crops, 14 d on oilseed rape and mustard
- On all other recommended crops an interval of at least 3 wk must elapse after use of a broad-leaved herbicide
- In the event of crop failure broad-leaved crops may be resown at any time, cereals, onions, leeks or maize may be sown after 2-6 wk depending on dose used

Special precautions/Environmental safety

- Irritant. Risk of serious damage to eyes [1]
- Irritating to eyes and skin [2, 3]
- Flammable [1]
- Dangerous to fish or other aquatic life. Do not contaminate surface waters or ditches with chemical or used container

Protective clothing/Label precautions

- A, C
- R04, R04d, R07d, U05a, U11, U20b, C03, E01 [1]; R04a, R04b, U09a, U20a [2, 3]; E13b, E26, E30a, E31b [1-3]

Latest application/Harvest Interval(HI)

- HI 16 wk for beet crops; 11 wk for oilseed rape, mustard linseed; 10 wk for parsnips; 9 wk for carrots

469 resmethrin

A contact acting pyrethroid insecticide available only in mixtures

See also fenitrothion + permethrin + resmethrin
pyrethrins + resmethrin

470 resmethrin + tetramethrin

A contact acting pyrethroid insecticide mixture

Products

Sorex Wasp Nest Destroyer	Sorex	0.10:0.05% w/w	AE	H4410

Uses

Wasps in MISCELLANEOUS SITUATIONS.

Notes

Efficacy
- Product acts by lowering temperature in and around nest and producing a stupefying vapour in addition to its contact insecticidal effect. Spray only on surfaces
- Apply by directing jet at nest from up to 3 m away and approach nest gradually until jet can be directed into entrance hole. Continue spraying until nest saturated

Special precautions/Environmental safety
- For use only by trained operators and owners of commercial and agricultural premises
- Harmful by inhalation. Contains perchloroethylene
- Keep off skin. Ensure adequate ventilation
- Harmful to caged birds and pets
- Dangerous to bees. Do not apply to crops in flower or to those in which bees are actively foraging. Do not apply when flowering weeds are present
- Extremely dangerous to fish or other aquatic life. Remove fish tanks before spraying

Protective clothing/Label precautions
- A, C, H
- R03b, U14, U19, U20a, C03, E01, E10b, E12c, E13a, E15, E30a

471 rimsulfuron

A selective systemic sulfonylurea herbicide

Products

1 Landgold Rimsulfuron	Landgold	25% w/w	SG	0895
2 Titus	DuPont	25% w/w	SG	0790

Uses

Annual dicotyledons in FODDER MAIZE, POTATOES. Volunteer oilseed rape in FODDER MAIZE, POTATOES.

Notes

Efficacy
- Product should be used with a suitable adjuvant or a suitable herbicide tank-mix partner. See label for details

- Product acts by foliar action. Best results obtained from good spray cover of small actively growing weeds. Effectiveness is reduced in very dry conditions
- Split application provides control over a longer period and improves control of fat hen and polygonums
- Weed spectrum can be broadened by tank mixture with other herbicides. See label for details
- Susceptible weeds cease growth immediately and symptoms can be seen 10 d later

Crop Safety/Restrictions
- Maximum number of treatments 2 per crop
- All varieties of ware potatoes and fodder maize may be treated, but variety restrictions of any tank-mix partner must be observed
- Crops should be treated in spring on potatoes up to 25 cm high and on maize up to the 4-collar stage
- Do not treat maize treated with organophosphorus insecticides
- Avoid high light intensity (full sunlight) and high temperatures on the day of spraying
- Do not treat during periods of substantial diurnal temperature fluctuation or when frost anticipated
- Do not apply to any crop stressed by drought, water-logging, low temperatures, pest or disease attack, nutrient or lime deficiency
- Only winter wheat should follow a treated crop in the same calendar yr. Only barley, wheat or maize should be sown in the spring of the yr following treatment. In the second autumn after treatment any crop except brassicas or oilseed rape may be drilled

Special precautions/Environmental safety
- Extremely dangerous to fish or other aquatic life. Do not contaminate surface waters or ditches with chemical or used container
- Herbicide is very active. Take particular care to avoid drift onto plants outside the target area
- Immediate and thorough cleaning of spray equipment after use is vital to prevent subsequent damage to susceptible crops. Do not drain or flush sprayers on land to be used for crops other than potatoes or fodder maize

Protective clothing/Label precautions
- U08, U19, U20a, E13a, E30a, E32a

Latest application/Harvest Interval(HI)
- Before most advanced potato plants are 25 cm high; before 4-collar stage of fodder maize

---

72 rotenone

A natural, contact insecticide of low persistence

oducts

1 Devcol Liquid Derris	Nehra	50 g/l	EC	06063
2 FS Liquid Derris	Ford Smith	50 g/l	EC	01213

ses

Aphids in FLOWERS, ORNAMENTALS, PROTECTED CROPS, SOFT FRUIT, TOP FRUIT, VEGETABLES. Raspberry beetle in CANE FRUIT. Sawflies in GOOSEBERRIES. Slug sawflies in PEARS, ROSES.

tes

Efficacy
- Apply as high volume spray when pest first seen and repeat as necessary

- Spray raspberries at first pink fruit, loganberries when most of blossom over, blackberries as first blossoms open. Repeat 2 wk later if necessary
- Spray to obtain thorough coverage, especially on undersurfaces of leaves

Special precautions/Environmental safety
- Dangerous to fish or other aquatic life. Do not contaminate surface waters or ditches with chemical or used container
- Flammable

Protective clothing/Label precautions
- R07d, U20a, C02, E13b, E30a, E31a, E34

Latest application/Harvest Interval(HI)
- HI 1 d

473 sethoxydim

A post-emergence oxime grass weed herbicide

Products	Checkmate	Hortichem	193 g/l	EC	06129

Uses Annual grasses in BEDDING PLANTS *(off-label)*, COMBINING PEAS, FLAX, FLOWE BULBS *(off-label)*, HERBACEOUS PERENNIALS *(off-label)*, LINSEED, LUPIN MUSTARD, SUNFLOWERS *(off-label)*, TREES AND SHRUBS *(off-label)*, VINING PEA WINTER OILSEED RAPE, WOODY ORNAMENTALS *(off-label)*. Awned canary grass SPRING OILSEED RAPE, WINTER OILSEED RAPE. Barren brome in FIELD BEANS, FLA FODDER BEET, LINSEED, LUPINS, MANGELS, MUSTARD, RASPBERRIES, SPRIN OILSEED RAPE, SUGAR BEET. Black bent in FIELD BEANS, FODDER BEET, MANGEL POTATOES, RASPBERRIES, SUGAR BEET. Blackgrass in COMBINING PEAS, FIEL BEANS, FLAX, FODDER BEET, LINSEED, LUPINS, MANGELS, MUSTARD, POTATOE RASPBERRIES, SPRING OILSEED RAPE, SUGAR BEET, VINING PEAS, WINTE OILSEED RAPE. Couch in FIELD BEANS, FODDER BEET, MANGELS, POTATOE RASPBERRIES, SUGAR BEET. Creeping bent in FIELD BEANS, FODDER BEE MANGELS, POTATOES, RASPBERRIES, SUGAR BEET. Perennial grasses in BEDDIN PLANTS *(off-label)*, FLOWER BULBS *(off-label)*, HERBACEOUS PERENNIALS *(off-labe* SUNFLOWERS *(off-label)*, TREES AND SHRUBS *(off-label)*, WOODY ORNAMENTA *(off-label)*. Volunteer cereals in COMBINING PEAS, FIELD BEANS, FLAX, FODDER BEE LINSEED, LUPINS, MANGELS, MUSTARD, POTATOES, RASPBERRIES, SPRIN OILSEED RAPE, SUGAR BEET, VINING PEAS, WINTER OILSEED RAPE. Wild oats COMBINING PEAS, FIELD BEANS, FLAX, FODDER BEET, LINSEED, LUPIN MANGELS, MUSTARD, POTATOES, RASPBERRIES, SPRING OILSEED RAPE, SUG BEET, VINING PEAS.

Notes

Efficacy
- Apply to emerged weeds in combination with Adder or Actipron adjuvant oil (except on peas). Weeds emerging after treatment are not controlled
- Best results when weeds growing actively. Action is faster under warm conditions
- Apply to annual grasses from 2-leaf stage to end of tillering, to couch when largest shoo at least 30 cm long. For effective couch control do not cultivate for 2 wk

- Annual meadow-grass is not controlled
- Apply when rain not expected for at least 2 h
- Various tank mixes recommended for grass/dicotyledon weed populations. See label for details of mixtures and sequential treatments
- Leave at least 7 d before applying another herbicide

Crop Safety/Restrictions

- Maximum number of treatments 1 per crop or per yr
- See label for details of crop growth stage and timing
- Do not mix with adjuvant oil on peas. Check pea leaf-wax by Crystal Violet test and do not spray where wax insufficient or damaged
- Do not spray crops suffering damage from other herbicides
- In raspberries apply to inter-row or to base of canes only
- Brassicas, field beans and onions may be sown 3 d after treatment, grass and cereal crops after 4 d

Special precautions/Environmental safety

- Irritating to eyes and skin
- Harmful to fish or other aquatic life. Do not contaminate surface waters or ditches with chemical or used container

Protective clothing/Label precautions

- A, C
- R04a, R04b, U02, U05a, U08, U20b, C03, E01, E13c, E26, E30a, E31b

Latest application/Harvest Interval(HI)

- HI linseed, lupins, mustard, oilseed rape 12 wk; combining peas 11 wk; fodder beet, mangels 9 wk; maincrop potatoes, sugar beet 8 wk; seed potatoes 7 wk; early potatoes, vining peas, raspberries 4 wk

Approval

- Off-label approval unlimited for use on bedding plants, bulbs, herbaceous perennials, ornamental trees and shrubs (OLA 1187/94)[1]; unlimited for use on sunflowers (OLA 1447/96)[1]

---

74 simazine

A soil-acting triazine herbicide with restricted permitted uses

See also isoproturon + simazine
pendimethalin + simazine

oducts

1 Alpha Simazine 50 SC	Makhteshim	500 g/l	SC	04801
2 Ashlade Simazine 50 FL	Ashlade	500 g/l	SC	06482
3 Atlas Simazine	Atlas	500 g/l	SC	07725
4 Gesatop	Novartis	500 g/l	SC	08412
5 Gesatop 500 SC	Ciba Agric.	500 g/l	SC	05846
6 Sipcam Simazine Flowable	Sipcam	500 g/l	SC	07622
7 Unicrop Simazine FL	Unicrop	500 g/l	SC	08032

es

Annual dicotyledons in APPLES, BLACKCURRANTS, BROAD BEANS, CANE FRUIT, FIELD BEANS, GOOSEBERRIES, HOPS, PEARS, REDCURRANTS, STRAWBERRIES [1-7]. Annual dicotyledons in FORESTRY TRANSPLANT LINES [1, 2, 4-7]. Annual dicotyledons in FORESTRY PLANTATIONS [1]. Annual dicotyledons in SWEETCORN, WOODY ORNAMENTALS [2, 6]. Annual dicotyledons in ASPARAGUS, MAIZE [2, 3, 6]. Annual dicotyledons in ROSES [2-5]. Annual dicotyledons in RHUBARB [2, 3]. Annual dicotyledons in FOREST NURSERY BEDS [2, 4-6]. Annual dicotyledons in TREES AND SHRUBS [6]. Annual dicotyledons in WHITECURRANTS, WOODY NURSERY STOCK [3]. Annual dicotyledons in FORESTRY *(off-label)* [7]. Annual dicotyledons in ASPARAGUS *(off-label)*, HERBS *(off-label)*, MANGE-TOUT PEAS *(off-label)*, RHUBARB *(off-label)*, RUNNER BEANS *(off-label)* [4, 5]. Annual dicotyledons in GRAPEVINES *(off-label)* [4]. Annual grasses in APPLES, BLACKCURRANTS, BROAD BEANS, CANE FRUIT, FIELD BEANS, GOOSEBERRIES, HOPS, PEARS, REDCURRANTS, STRAWBERRIES [1-7]. Annual grasses in FORESTRY TRANSPLANT LINES [1, 2, 4-7]. Annual grasses in FORESTRY PLANTATIONS [1]. Annual grasses in SWEETCORN, WOODY ORNAMENTALS [2, 6]. Annual grasses in ASPARAGUS, MAIZE [2, 3, 6]. Annual grasses in ROSES [2-5]. Annual grasses in RHUBARB [2, 3]. Annual grasses in FOREST NURSERY BEDS [2, 4-6]. Annual grasses in TREES AND SHRUBS [6]. Annual grasses in WHITECURRANTS, WOODY NURSERY STOCK [3]. Annual grasses in FORESTRY *(off-label)* [7]. Annual grasses in ASPARAGUS *(off-label)*, HERBS *(off-label)*, MANGE-TOUT PEAS *(off-label)*, RHUBARB *(off-label)*, RUNNER BEANS *(off-label)* [4, 5]. Annual grasses in GRAPEVINES *(off-label)* [4].

Notes

Efficacy

- Active via root uptake. Best results achieved by application to fine, firm, moist soil, free of established weeds, when rain falls after treatment
- Do not use on highly organic soils (soils with more than 10% organic matter [7])
- Following repeated use of simazine or other triazine herbicides resistant strains of groundsel and some other annual weeds may develop

Crop Safety/Restrictions

- Maximum number of treatments 1 per yr for maize, sweetcorn, broad and field beans, asparagus, rhubarb, sage, strawberries, hops, bush and cane fruit, apples (lower rate), forest transplant lines; 1 every 2 yr for apples (higher rate); 2 per yr for mint
- See labels for maximum total dose of simazine and atrazine products for each situation
- May be applied to spring beans up to 2-true leaf stage but control may be poor if weeds already germinated [5]
- Do not spray beans on sandy or gravelly soils or cultivate after treatment. Do not treat varieties Beryl, Feligreen or Rowena
- Apply in top, bush and cane fruit and woody ornamentals established at least 12 mth in Feb-Mar. Roses may be sprayed immediately after planting. See label for lists of resistant and susceptible species
- Apply to strawberries in Jul-Dec, not in spring. Do not treat spring-planted crops established less than 6 mth, or winter-planted crops less than 9 mth. Do not spray varieties Huxley Giant, Madame Moutot or Regina
- Apply to hops overall in Feb-Apr before weeds emerge. Newly planted sets must be covered with minimum of 50 mm soil
- Apply to forest nursery seedbed in second yr or to transplant lines after plants 5 cm tall. Do not treat Norway spruce (Christmas trees)
- To reduce soil run-off on gradients especially in orchard and forest plantations, users are advised to plant grass strips or leave 6 m wide strips between treated areas and surface waters
- On Sands, stony or gravelly soils there is risk of crop damage, especially with heavy rain
- Allow at least 7 mth before drilling or planting other crops, longer if weather dry
- Do not sow oats in autumn following spring application in maize

Special precautions/Environmental safety

- Dangerous to fish or other aquatic life and aquatic higher plants. Do not contaminate surface waters or ditches with chemical or used container
- Do not allow direct spray from vehicle-mounted/drawn hydraulic sprayers to fall within 6 m, or from hand-held sprayers to within 2 m, of surface waters or ditches. Direct spray away from water
- Use must be restricted to one product containing atrazine or simazine, and either to a single application at the maximum approved rate or (subject to any existing maximum permitted number of treatments) to several applications at lower doses up to the maximum approved rate for a single application

Protective clothing/Label precautions

- A [1-7]; B [6]; C [1-7]; D [1, 6, 7]; H [1-7]; M [1, 6, 7]
- U02, U08, U19, U20a [4, 5]; U20b [1]; U20c [6, 7]; E13b, E16, E30a [1-7]; E26 [1-6]; E31a [2, 3, 6]; E31b [1-5]; E31c [7]; E32a, E34 [2, 3]

Latest application/Harvest Interval(HI)

- Before emergence of spears for asparagus; 7 d after drilling for maize and sweetcorn; before end of Feb (autumn planted), 10 d after drilling for spring planted field and broad beans; before end of Mar for top, bush and cane fruit; after harvest, before end of Nov for strawberries; after harvest, before 1 May for hops.
- HI mint and sage 4 mth

Approval

- Approvals for sale, supply and use of simazine on non-crop land by the approval holder or his agents were revoked with effect from 1 Sep 1992 and by persons other than the approval holder from 1 Sep 1993
- Off-label approval unlimited for use on edible-podded peas, runner beans (OLA 0297/94)[5], (OLA 1723/97)[4]; unlimited for use on asparagus, outdoor rhubarb (OLA 2506/96)[5], (OLA 1721/97)[4]; unlimited for use in forestry tree establishment (OLA 1267/97)[7]; unlimited for use on a range of outdoor herbs (see notices for details) (OLA 0162/97)[5], (OLA 1722/97)[4]
- Approval expiry: 31 Mar 99 [2]

---

475 simazine + trietazine

A pre-emergence residual herbicide for peas, field and broad beans

Products Remtal SC | AgrEvo | 57.5:402.5 g/l | SC | 07270

Uses Annual dicotyledons in BROAD BEANS, COMBINING PEAS, EDIBLE PODDED PEAS *(off-label)*, FIELD BEANS, VINING PEAS.

Notes

Efficacy

- Chemical acts mainly via roots but some contact effect on cotyledon stage weeds. Gives residual control of germinating weeds for season
- Effectiveness increased by rain after spraying
- Weeds of intermediate susceptibility, including annual meadow grass and blackgrass, may be controlled if adequate rainfall occurs soon after spraying
- Control of deep germinating weeds on heavier soils may be incomplete
- Do not use on soils with more than 10% organic matter
- With repeated use of simazine or other triazine herbicides resistant strains of groundsel and some other annual weeds may develop

Crop Safety/Restrictions
- Maximum number of treatments 1 per crop
- Apply to spring peas, winter or spring field and broad beans between drilling and 5% emergence
- On early-drilled peas apply when seed chitting, on later crops as soon as possible after drilling. Do not spray winter sown peas. Vedette peas may be checked by treatment
- Do not spray any forage pea varieties
- Crop seed should be covered by at least 25 mm of settled soil
- Do not use on sand or on gravelly or cloddy soils
- Other crops may be sown or planted 10 wk or more after spraying. With drilled brassicas allow a minimum of 14 wk
- In the event of crop failure only drill peas, field or broad beans without ploughing but do not respray
- To reduce soil run-off users are advised to plant grass strips or leave strips 6 m wide between treated areas and surface waters

Special precautions/Environmental safety
- Harmful if swallowed
- Dangerous to fish or other aquatic life and aquatic higher plants. Do not contaminate surface waters or ditches with chemical or used container
- Do not allow direct spray from vehicle-mounted/drawn hydraulic sprayers to fall within 6 m, or from hand-held sprayers to within 2 m, of surface waters or ditches. Direct spray away from water
- Use must be restricted to one product containing atrazine or simazine either to a single application at maximum approved rate or (subject to any existing maximum permitted number of treatments) to several applications at lower doses up to the maximum approved rate for single application

Protective clothing/Label precautions
- A, B, C, D, H, M
- M03, R03c, U05a, U20b, C03, E01, E13b, E16, E26, E30a, E31b, E34

Latest application/Harvest Interval(HI)
- Before 5% total crop emergence

Approval
- May be applied through CDA equipment (see label for details) [1]

476 sodium chlorate

A non-selective inorganic herbicide for total vegetation control

Products					
	1 Atlacide Soluble Powder	Nomix-Chipman	58.2% w/w	SP	00125
	2 Cooke's Weedclear	Nehra	50% w/w	SL	06512
	3 Deosan Chlorate Weedkiller (Fire Suppressed)	DiverseyLever	55% w/w	SP	08521
	4 Doff Sodium Chlorate Weedkiller	Doff Portland	53% w/w	SP	06049
	5 TWK Total Weedkiller	Yule Catto	89 g/l	SL	06393

Uses Total vegetation control in NON-CROP AREAS, PATHS AND DRIVES.

Notes

Efficacy
- Active through foliar and root uptake
- Apply as overall spray at any time during growing season. Best results obtained from application to moist soil in spring or early summer
- Do not apply before heavy rain

Crop Safety/Restrictions
- Maximum number of treatments 2 per yr for non-crop areas, paths and drives [4]
- Clothing, paper, plant debris etc become highly inflammable when dry if contaminated with sodium chlorate
- Fire risk has been reduced by inclusion of fire depressant but product should not be used in areas of exceptionally high fire risk e.g. oil installations, timber yards
- Treated ground must not be replanted for at least 6 mth after treatment [4]

Special precautions/Environmental safety
- Harmful if swallowed
- Oxidizing - contact with combustible material may cause fire
- Wash clothing thoroughly after use
- If clothes become contaminated do not stand near an open fire
- Must not be sold to persons under 18
- Do not contaminate surface waters or ditches with chemical or used container

Protective clothing/Label precautions
- A, H [2-5]; L [1]; M
- M03, R03c, E15, E30a, E34 [1-5]; R08 [1, 2, 4]; U02, U05a, C03, E01 [1-3, 5]; U20a [3, 5]; U20b [1, 4]; U20c, E31a [2]; E07 [1, 3, 5]; E32a [1, 3-5]

Withholding period
- Keep livestock out of treated areas until foliage of any poisonous weeds such as ragwort has died and become unpalatable

477 sodium chloride (commodity substance)

An inorganic salt for use as a sugar beet herbicide

Products

sodium chloride	various	SL

Uses Polygonums in SUGAR BEET. Volunteer potatoes in SUGAR BEET.

Notes

Efficacy
- Best results obtained from good spray cover treatment in hot, humid weather
- Use as a follow-up treatment where earlier sprays have failed to control large volunteer potatoes or polygonums
- Acts by contact scorch and has no direct effect on daughter potato tubers
- Saturated spray solution should contain 0.1% w/w non-ionic wetter and be applied at 1000 l/ha

Crop Safety/Restrictions
- Spray between three true leaf stage of crops and end Jul

- Crop scorch may occur after treatment

478 sodium cyanide

A poisonous gassing compound for control of rabbits and rats

Products	Cymag	Zeneca	40% w/w	GE	06651

Uses Rabbits in OUTDOOR SITUATIONS. Rats in OUTDOOR SITUATIONS.

Notes

Efficacy

- Hydrogen cyanide gas is produced when chemical is placed on moist earth. Use only in rabbit and rat holes out of doors and well away from farm or domestic buildings
- Place powder in burrows with a spoon or blow it in with a pump and seal openings with sod of grass. See label for details
- Do not use in wet or windy weather

Special precautions/Environmental safety

- Sodium cyanide is subject to the Poisons Rules 1982 and the Poisons Act 1972. See notes in Section 1
- For use only outdoors by persons trained in the use of sodium cyanide
- Very toxic in contact with skin, by inhalation or if swallowed
- Cymag and the gas it evolves are deadly poisons. Use only in the presence of another person aware of the symptoms and first aid treatment for hydrogen cyanide poisoning and provided with amyl nitrite for use in an emergency
- Maximum exposure limits apply to this chemical. See HSE Approved Code of Practice for the Control of Substances Hazardous to Health. See also under Occupational Exposure Limits in Section 1
- Keep in original container, tightly closed, in a safe place, under lock and key. See label for specific label disposal instructions
- Dangerous to people and livestock. Keep them out of treated areas during gassing operations
- Dangerous to fish or other aquatic life. Do not contaminate surface waters or ditches with chemical or used container

Protective clothing/Label precautions

- A, C, E, G, H, P
- M04, R01a, R01b, R01c, R04a, R04b, R04c, U01, U05a, U13, U14, U15, U19, U20a, C03, E01, E06a, E13a, E30b, E32b, E34

479 sodium hypochlorite (commodity substance)

An inorganic horticultural bactericide for use in mushrooms

Products	sodium hypochlorite	various	100% w/v

Uses Bacterial blotch in MUSHROOMS.

Notes

Crop Safety/Restrictions
- Maximum concentration 315 mg/litre of water

Special precautions/Environmental safety
- Mixing and loading must only take place in a ventilated area
- Must only be used by suitably trained and competent operators
- Harmful to fish or other aquatic life. Do not contaminate surface waters or ditches with chemical or used container

Protective clothing/Label precautions
- A, C, H
- E13c

Latest application/Harvest Interval(HI)
- HI 1 d

---

480 sodium monochloroacetate

A contact herbicide for various horticultural crops

Products

1 Atlas Somon	Atlas	96% w/w	SP	07727
2 Croptex Steel	Hortichem	95% w/w	SP	02418

Uses Annual dicotyledons in APPLES *(directed spray)*, BLACKCURRANTS *(directed spray)*, BROCCOLI *(off-label)*, CALABRESE *(off-label)*, CAULIFLOWERS *(off-label)*, GOOSEBERRIES *(directed spray)*, KOHLRABI *(off-label)*, PEARS *(directed spray)*, PLUMS *(directed spray)*, REDCURRANTS *(directed spray)* [2]. Annual dicotyledons in BRUSSELS SPROUTS, CABBAGES, KALE, LEEKS, ONIONS [1, 2]. Annual dicotyledons in APPLES, BLACKCURRANTS, GOOSEBERRIES, PEARS, PLUMS, REDCURRANTS [1]. Basal defoliation in HOPS *(off-label)* [2]. Basal defoliation in HOPS [1]. Sucker control in BLACKBERRIES *(off-label)*, HYBRID BERRIES *(off-label)*, LOGANBERRIES *(off-label)*, RASPBERRIES *(off-label)* [2].

Notes

Efficacy
- Best results achieved by application to emerged weed seedlings up to young plant stage under good growing conditions
- Effectiveness reduced by rain within 12 h
- May be used for hop defoliation in tank-mixture with tar-oil [1], with Wayfarer adjuvant [2]. Apply as a directed spray to the base of the plant when they are up to 20 cm high, after training when main shoots are at least 100 cm high

Crop Safety/Restrictions
- Maximum number of treatments 1 per crop or yr; 2 per yr for sucker control in cane fruit and hop defoliation
- Apply to brassicas from 2-4 leaf stage or after recovery from transplanting
- Do not spray cabbage that has begun to heart
- Apply to onions and leeks after crook stage but before 4-leaf stage
- Safety on brassicas, onions and leeks depends on presence of adequate leaf wax, check by crystal violet wax test. Do not add wetters, pesticides or nutrients to spray
- Apply in fruit crops established for at least 1 yr as a directed spray

- Do not spray if frost likely or if temperature likely to exceed 27°C
- Apply for sucker control in raspberries when canes 10-20 cm high with addition of Wayfarer adjuvant [2]

Special precautions/Environmental safety
- Harmful if swallowed, in contact with skin and by inhalation [1]
- Irritating to eyes, skin and respiratory system [2]
- Irritating to respiratory system. Risk of serious damage to eyes [1]
- Corrosive, causes burns [1]
- Do not apply directly to livestock/poultry [1, 2]
- Harmful to bees. Do not apply to crops in flower or to those in which bees are actively foraging. Do not apply when flowering weeds are present
- Do not contaminate surface waters or ditches with chemical or used container [2]

Protective clothing/Label precautions
- A, C, D, E [1, 2]
- R03a, R03b, R03c, R04d, R05, R05b, U10, U11, E07, E34 [1]; R04a, R04b, E15 [2]; R04c, U05a, U08, U19, U20a, C03, E01, E05, E05a, E12e, E30a, E31a [1, 2]

Withholding period
- Keep livestock, especially poultry, out of treated areas for at least 2 wk

Latest application/Harvest Interval(HI)
- Before 4 true leaf stage for onions and leeks.
- HI 21 d for cabbage, kale, Brussels sprouts; 28 d for cane fruit; 14 wk for hops

Approval
- Off-label approval unlimited for use on cane fruit (OLA 0115/93)[2], hops (OLA 0357/93)[2]; unlimited for use on hops to within 6 wk of harvest (OLA 0357/93)[2]; unlimited for use on cauliflowers, broccoli, calabrese (OLA 0961/95)[2]; to Mar 2001 for use on kohlrabi (OLA 0615/96)[2]

---

481 sodium silver thiosulfate

A plant-growth regulator used to extend life of flowers

Products	Argylene	Fargro	8% w/w	SP	03386

Uses Prolonging flower life in GLASSHOUSE CUT FLOWERS, POT PLANTS.

Notes

Efficacy
- Acts by inhibiting production of ethylene
- Spray flowering pot plants to run off 8-14 d before shipment from glasshouse or at 3 wk intervals. Best time for spraying is late afternoon
- Vase life of cut flowers extended by dip treatment immediately after cutting

Crop Safety/Restrictions
- Spraying pot plants just before shipping may cause damage

Special precautions/Environmental safety
- Not to be used on food crops

Protective clothing/Label precautions
- U08, U20a, C01, E15, E30a, E32a

482 spiroxamine

A spiroketal amine fungicide for cereals

Products

1 Neon	Bayer	500 g/l	EW	08337
2 Standon Spiroxamin 500	Standon	500 g/l	EC	08916
3 Torch	Bayer	500 g/l	EW	08336

Uses

Brown rust in BARLEY, RYE *(moderate control)*, WHEAT [1, 2]. Powdery mildew in BARLEY, RYE, WHEAT [1-3]. Rhynchosporium in BARLEY *(reduction)* [1, 2]. Yellow rust in BARLEY, RYE, WHEAT [1, 2].

Notes

Efficacy
- For best results treat at an early stage of disease development before infection spreads to new growth
- To reduce the risk of development of resistance avoid repeat treatments on diseases such as powdery mildew. If necessary tank mix or alternate with other non-morpholine fungicides

Crop Safety/Restrictions
- Maximum total dose 3.0 l product per ha

Special precautions/Environmental safety
- Harmful if swallowed
- Irritating to skin. May cause sensitization by skin contact
- Risk of serious damage to eyes
- Dangerous to fish or other aquatic life. Do not contaminate surface waters or ditches with chemical or used container

Protective clothing/Label precautions
- A, C, H
- M03, R03c, R04b, R04d, R04e, U05a, U11, U13, U14, U15, U20a, C03, E01, E13b, E26, E30a, E31b, E34

Latest application/Harvest Interval(HI)
- Before caryopsis watery ripe (GS 71) for spring wheat, winter rye, winter wheat; ear emergence complete (GS 59) for spring barley, winter barley

483 spiroxamine + tebuconazole

A broad spectrum systemic fungicide mixture for cereals

Products

Beam	Bayer	250:133 g/l	EW	08332

Uses

Brown rust in BARLEY, RYE, WHEAT. Fusarium ear blight in WHEAT. Glume blotch in WHEAT. Net blotch in BARLEY. Powdery mildew in BARLEY, RYE, WHEAT.

Rhynchosporium in BARLEY. Septoria leaf spot in WHEAT. Sooty moulds in WHEAT. Yellow rust in BARLEY, RYE, WHEAT.

Notes

Efficacy

- Best results achieved from treatment at early stage of disease development before infection spreads to new growth
- To protect flag leaf and ear from Septoria apply from flag leaf emergence to ear fully emerged (GS 37-59). Earlier treatment may be necessary where there is high disease risk
- Control of rusts, powdery mildew, leaf blotch and net blotch may require second treatment 2-3 wk later

Crop Safety/Restrictions

- Maximum total dose 3.0 l product per ha
- Do not use on durum wheat
- Some transient leaf speckling may occur on wheat but this has not been shown to reduce yield reponse or disease control

Special precautions/Environmental safety

- Harmful if swallowed
- Irritating to skin. May cause sensitization by skin contact
- Risk of serious damage to eyes
- Dangerous to fish or other aquatic life. Do not contaminate surface waters or ditches with chemical or used container

Protective clothing/Label precautions

- A, C, H
- M03, R03c, R04b, R04d, R04e, U05a, U11, U13, U14, U15, U20b, C03, E01, E13b, E26, E30a, E31b, E34

Latest application/Harvest Interval(HI)

- Before caryopsis watery ripe (GS 71) for rye, wheat; ear emergence complete (GS 59) for barley

484 strychnine hydrochloride (commodity substance)

A vertebrate control agent for destruction of moles underground

Products strychnine various

Uses Moles in GRASSLAND *(areas of restricted public access)*, MISCELLANEOUS SITUATIONS *(areas of restricted public access)*.

Notes

Efficacy

- For use as poison bait against moles on commercial agricultural/horticultural land where public access restricted, on grassland associated with aircraft landing strips, horse paddocks, race and golf courses and other areas specifically approved by Agriculture Departments

FOR FULL CONDITIONS OF USE ALWAYS READ THE PRODUCT LABEL

Special precautions/Environmental safety
- Strychnine is subject to the Poisons Rules 1982 and the Poisons Act 1972. See notes in Section 1
- Must only be supplied to holders of an Authority to Purchase issued by appropriate Agricultural Department. Authorities to Purchase may only be issued to persons who satisfy the appropriate authority that they are trained and competent in its use
- Only to be supplied in original sealed pack in units up to 2 g
- Store in original container under lock and key and only on premises under control of holder of Authority to Purchase or a named individual
- A written COSHH assessment must be made before using
- Must be prepared for application with great care so that there is no contamination of the ground surface
- Any prepared bait remaining at the end of the day must be buried
- Operators must be supplied with a Section 6 (HSW) Safety Data Sheet before commencing work
- Other restrictions apply, see *Pesticides 1998*, Annex D

Protective clothing/Label precautions
- A

485 sulfonated cod liver oil

An animal repellent

Products Scuttle | Fine | 800 g/l | EC | 06232

Uses Deer in BARLEY, CABBAGES, CAULIFLOWERS, FORESTRY, LETTUCE, OATS, OILSEED RAPE, WHEAT. Rabbits in BARLEY, CABBAGES, CAULIFLOWERS, FORESTRY, LETTUCE, OATS, OILSEED RAPE, WHEAT.

Notes

Efficacy
- Apply diluted as a spray before crop is being grazed or attacked or undiluted as a dip for forestry seedlings
- A rain-free period of about 5 h is required after application

Crop Safety/Restrictions
- Before using on vegetables users should consider risk of taint and consult proccssor before using on crops for processing

Special precautions/Environmental safety
- Do not apply directly to animals
- Dangerous to bees. Do not apply to crops in flower or to those in which bees are actively foraging. Do not apply when flowering weeds are present.
- Dangerous to fish or other aquatic life. Do not contaminate surface waters or ditches with chemical or used container

Protective clothing/Label precautions
- A, C
- U05a, U20a, C03, E01, E12c, E13b, E30a, E31b

486 sulfur

A broad-spectrum inorganic protectant fungicide, foliar feed and acaricide

See also carbendazim + maneb + sulfur
copper oxychloride + maneb + sulfur
copper sulfate + sulfur

Products

1 Ashlade Sulphur FL	Ashlade	720 g/l	SC	06478
2 Atlas Sulphur 80 FL	Atlas	800 g/l	SC	07729
3 Headland Sulphur	Headland	800 g/l	SC	03714
4 Kumulus DF	BASF	80% w/w	SG	04707
5 Luxan Micro-Sulphur	Luxan	80% w/w	SG	06565
6 Microsul Flowable Sulphur	Stoller	960 g/l	SC	03907
7 Microthiol Special	PBI	80% w/w	MG	06268
8 MSS Sulphur 80	Mirfield	80% w/w	WG	05752
9 Solfa	Atlas	80% w/w	MG	03529
10 Stoller Flowable Sulphur	Stoller	720 g/l	SC	03760
11 Sulphur Flowable	United Phosphorus	800 g/l	SC	07526
12 Thiovit	Sandoz	80% w/w	WG	05572
13 Thiovit	Novartis	80% w/w	WG	08493
14 Tripart Imber	Tripart	720 g/l	SC	04050

Uses

Foliar feed in OILSEED RAPE [4, 5, 7, 9, 12-14]. Foliar feed in WHEAT [4, 5, 7, 12-14]. Foliar feed in BARLEY [4, 5, 7, 8, 12-14]. Foliar feed in WINTER OILSEED RAPE, WINTER WHEAT [8]. Foliar feed in BRASSICAS, POTATOES, SOFT FRUIT, TOP FRUIT, TURF, TURNIPS, VEGETABLES [7, 9]. Foliar feed in OATS [7, 8, 14]. Foliar feed in SUGAR BEET [7]. Foliar feed in SWEDES [5, 7, 9]. Foliar feed in GRASSLAND [2, 5, 7-9, 12, 13]. Gall mite in BLACKCURRANTS [1, 4, 8, 14]. Powdery mildew in SWEDES [3, 7, 10-14]. Powdery mildew in PEARS [3, 11]. Powdery mildew in GRASSLAND [10]. Powdery mildew in OATS [6, 7, 9, 10, 14]. Powdery mildew in AUBERGINES *(off-label)*, HERB CROPS *(off-label)*, PARSNIPS *(off-label)*, PROTECTED PEPPERS *(off-label)*, PROTECTED TOMATOES *(off-label)* [12, 13]. Powdery mildew in TURNIPS [7, 14]. Powdery mildew in BARLEY, WHEAT [7]. Powdery mildew in APPLES, STRAWBERRIES [1-4, 6, 8, 10, 11, 14]. Powdery mildew in HOPS [1-4, 6, 8, 10-14]. Powdery mildew in SUGAR BEET [1-14]. Powdery mildew in GRAPEVINES [1, 2, 4, 6, 8, 10, 14]. Powdery mildew in GOOSEBERRIES [1, 2, 4, 8, 14]. Scab in PEARS [3, 6, 10, 11]. Scab in APPLES [2-4, 6, 8, 10, 11, 14].

Notes

Efficacy

- Apply when disease first appears and repeat 2-3 wk later. Details of application rates and timing vary with crop, disease and product. See label for information
- Sulfur acts as foliar feed as well as fungicide and with some crops product labels vary in whether treatment recommended for disease control or growth promotion
- In grassland best at least 2 wk before cutting for hay or silage, 3 wk before grazing
- Treatment unlikely to be effective if disease already established in crop

Crop Safety/Restrictions

- Maximum number of treatments 2 per crop for sugar beet (3 per yr [1, 14]), parsnips, swedes; 3 per yr for blackcurrants, gooseberries; variable on cereals - see label
- May be applied to cereals up to grain watery-ripe stage (GS 71) [1-3, 6, 8-10, 14], up to milky ripe stage (GS 75) [12, 13], to first node detectable (GS 31) [5], at any time [4]

- Do not use on sulfur-shy apples (Beauty of Bath, Belle de Boskoop, Cox's Orange Pippin, Lanes Prince Albert, Lord Derby, Newton Wonder, Rival, Stirling Castle) or pears (Doyenne du Comice)
- Do not use on gooseberry cultivars Careless, Early Sulphur, Golden Drop, Leveller, Lord Derby, Roaring Lion, or Yellow Rough
- Do not use on apples or gooseberries when young, under stress or if frost imminent
- Do not use on fruit for processing, on grapevines during flowering or near harvest on grapes for wine-making
- Do not use on hops at or after burr stage
- Do not spray top or soft fruit with oil or within 30 d of an oil-containing spray

Special precautions/Environmental safety

- Harmful to fish or other aquatic life. Do not contaminate surface waters or ditches with chemical or used container [2]
- Do not contaminate surface waters or ditches with chemical or used container [1, 3-14]

Protective clothing/Label precautions

- R07c [12, 13]; U20a [1-3]; U20b [4, 6-11, 14]; U20c [5, 12, 13]; E01 [2, 5, 7]; E13c [2]; E15 [1, 3-14]; E26 [11]; E30a [1-14]; E31a [1]; E31b [2, 3, 11]; E32a [4-10, 12-14]; E34 [5]

Latest application/Harvest Interval(HI)

- Late Jun for apples [1]; before 30 Sep (first wk of Aug [3]) in yr of harvest for sugar beet, swedes; up to and including fruit swell for gooseberries; up to burr stage for hops; not specified on other crops
- HI cutting grass for hay or silage 2 wk; grazing grassland 3 wk

Approval

- Approved for aerial application on wheat, barley, oilseed rape [3, 4]; sugar beet [12]. See notes in Section 1
- Off-label approval unlimited for use on protected tomatoes (OLA 0909/96)[12], (OLA 1717/97)[13]; unlimited for use on sorrel, summer savory, tarragon (OLA 0910/96)[12], (OLA 1716/97)[13]; unlimited for use on parsnips (OLA 0911/96)[12], (OLA 1715/97)[13]; unlimited for use on protected aubergines (OLA 1269/97)[13]; unlimited for use on protected peppers (OLA 1714/97)[13]; unlimited for use on protected cucumbers (OLA 1713/97)[13]

---

487 sulfuric acid (commodity substance)

A strong acid used as an agricultural desiccant

Products

sulfuric acid	various	77% w/w	SL

Uses

Haulm destruction in POTATOES. Pre-harvest desiccation in LINSEED, NARCISSI, ONIONS.

Notes

Efficacy

- Apply with suitable equipment between 1 Mar and 15 Nov for potatoes or narcissus bulbs, between 1 Aug and 15 Oct for linseed, between 1 Apr and 30 Aug for onions

Crop Safety/Restrictions

- Maximum number of treatments 3 per crop for potatoes; 2 per crop for linseed and onions, 1 per season for narcissus bulbs

Special precautions/Environmental safety

- Sulfuric acid is subject to the Poisons Rules 1982 and the Poisons Act 1972. See notes in Section 1
- Must only be used by suitably trained operators competent in use of equipment for applying sulfuric acid
- Not to be applied using hand-held or pedestrian controlled applicators
- A written COSHH assessment must be made before use. Operators should observe OES set out in HSE guidance note EH40/90 or subsequent issues
- Operators must have liquid suitable for eye irrigation immediately available at all times throughout spraying operation
- Spray must not be deposited within 1 m of public footpaths
- Written notice of any intended spraying must be given to owners of neighbouring land and warning notices posted. See PR 1990, Issue 12 for details
- Unprotected persons must be kept out of treated areas for at least 96 h after treatment

Protective clothing/Label precautions

- A, C, H, K, M
- R05, R05b

488 tar acids

A contact fungicide, soil sterilant and insecticide

Products

1 Armillatox	Armillatox	30% v/v (phenol)	EC	06234
2 Bray's Emulsion	Fargro	48% v/v	EC	00323

Uses

Algae in PATHS [2]. Ants in GLASSHOUSE STRUCTURES AND SURROUNDS [1, 2]. Canker in TREES AND SHRUBS [2]. Clubroot in BRASSICAS, WALLFLOWERS [1]. Crown gall in TREES AND SHRUBS [2]. Honey fungus in TREES AND SHRUBS [1, 2]. Lichens in PATHS [2]. Liverworts in PATHS [2]. Mosses in PATHS [2]. Mosses in TURF [1]. Overwintering pests in CANE FRUIT, CURRANTS, FRUIT TREES [2]. Rose replant disease in ROSES [1]. Slugs in GLASSHOUSE STRUCTURES AND SURROUNDS [1, 2]. Soil-borne diseases in PROTECTED CROPS [2]. Woodlice in GLASSHOUSE STRUCTURES AND SURROUNDS [1, 2].

Notes

Efficacy

- To control honey fungus drench around the collar region of woody subjects. Avoid treating waterlogged or frozen soil. Apply a foliar and/or root feed in following season
- Plants should be re-treated annually to prevent reinvasion
- For canker control prune back branch to healthy wood, cut out diseased tissues and paint wound and surrounding area [2]
- For crown gall control loosen soil round collar and saturate area for 30 cm around plant [2]
- To control soil-borne diseases apply as drench in autumn as soon as crop removed
- Apply as winter wash in Dec or Jan to cover entire plant, with particular attention to cracks and crevices in bark [2]

Crop Safety/Restrictions

- Maximum number of treatments 1 per crop for brassicas

- Do not apply honey fungus control treatment to new plantings until established at least 12 mth. Minimize contact with feeding roots
- Do not replant sterilized areas around treated trees for 6-8 mth
- Do not allow spray to contact any green tissues
- Soil sterilized against damping off and foot rot may be planted after 7 wk
- Protect plants or grass under trees treated with winter wash to avoid scorch

Special precautions/Environmental safety
- Irritating to eyes, skin and respiratory system
- Dangerous to fish or other aquatic life. Do not contaminate surface waters or ditches with chemical or used container

Protective clothing/Label precautions
- A
- R04a, R04b, R04c, U05a, U08, U19, U20a, C03, E01, E13b, E30a, E31a

Latest application/Harvest Interval(HI)
- 3 wk before planting for brassicas

---

489 tar oils

Hydrocarbon and phenolic oils used as insecticidal and fungicidal winter washes

See also anthracene oil

Products

1 Sterilite Tar Oil Winter Wash 60% Stock Emulsion	Coventry Chemicals	636 g/l	EC	05061
2 Sterilite Tar Oil Winter Wash 80% Miscible Quality	Coventry Chemicals	800 g/l	EC	05062

Uses

Aphids in APPLES, CANE FRUIT, CHERRIES, CURRANTS, GOOSEBERRIES, NECTARINES, PEACHES, PEARS, PLUMS. Scale insects in APPLES, CANE FRUIT, CHERRIES, CURRANTS, GOOSEBERRIES, GRAPEVINES, NECTARINES, PEACHES, PEARS, PLUMS. Suckers in APPLES. Winter moth in APPLES, CANE FRUIT, CHERRIES, CURRANTS, GOOSEBERRIES, NECTARINES, PEACHES, PEARS, PLUMS.

Notes

Efficacy
- Spray kills hibernating insects and eggs as well as moss and lichens on trunk
- Spray winter wash over dormant branches and twigs. See labels for detailed timings
- Ensure that all parts of trees or bushes are completely wetted, especially cracks and crevices. Use higher concentration for cleaning up neglected orchard
- Treatment also partially controls winter and raspberry moths and reduces Botrytis on currants and powdery mildew on currants and gooseberries
- Combine with spring insecticide treatment for control of caterpillars
- When used in cold weather always add product to water and stir thoroughly before use
- Apply hop defoliant by spraying when bines 1.5-3 m high and direct spray downward at 45°

Crop Safety/Restrictions
- Maximum number of treatments 2 per yr for hops
- Only spray fruit trees or bushes when fully dormant

- Do not spray hops if temperature is above 21°C or after cones have formed. Do not drench rootstocks
- Do not spray on windy, wet or frosty days

Special precautions/Environmental safety
- Harmful if swallowed. Irritating to eyes, skin and respiratory system
- Dangerous to fish or other aquatic life. Do not contaminate surface waters or ditches with chemical or used container

Protective clothing/Label precautions
- A, C, H
- M03, R03c, R04a, R04b, R04c, U05a, U08, U19, U20a, C03, E01, E13b, E26, E27, E30a, E31a, E34

Latest application/Harvest Interval(HI)
- Before any signs of bud swelling for fruit crops

490 tau-fluvalinate

A contact pyrethroid insecticide for cereals and oilseed rape

Products

Mavrik Aquaflow	Novartis	240 g/l	EW	08347

Uses

Aphids in OILSEED RAPE, SPRING BARLEY, SPRING WHEAT, WINTER BARLEY, WINTER WHEAT. Barley yellow dwarf virus vectors in WINTER BARLEY, WINTER WHEAT. Pollen beetles in OILSEED RAPE.

Notes

Efficacy
- For BYDV control on winter cereals follow local warnings or spray high risk crops in mid-Oct and make repeat application in late autumn/early winter if aphid activity persists
- For summer aphid control on cereals spray once when aphids present on two thirds of ears and increasing
- On oilseed rape treat peach potato aphids in autumn in response to local warning and repeat if necessary
- Best control of pollen beetle in oilseed rape obtained from treatment at green to yellow bud stage and repeat if necessary
- Good spray cover of target essential for best results

Crop Safety/Restrictions
- Maximum total dose varies with crop. See label for details
- A minimum of 14 d must elapse between applications to cereals

Special precautions/Environmental safety
- Irritating to eyes and skin
- Extremely dangerous to fish or other aquatic life. Do not contaminate surface waters or ditches with chemical or used container
- Do not allow direct spray from vehicle mounted/drawn hydraulic sprayers to fall within 6 m, or from hand-held sprayers to within 2 m, of surface waters or ditches. Direct spray away from water

- Presents low risk to many beneficial species when used as directed but some reduction in numbers may occur
- Avoid spraying oilseed rape within 6 m of field boundary to reduce effects on certain non-target species or other arthropods

Protective clothing/Label precautions
- A, C, H
- R04a, R04b, U05a, U10, U11, U19, U20a, C03, E01, E13a, E16, E22a, E26, E30a, E31b

Latest application/Harvest Interval(HI)
- Before caryopsis watery ripe (GS 71) for barley; before flowering for oilseed rape; before kernel medium milk (GS 75) for wheat

491 tebuconazole

A systemic conazole fungicide for cereals and other field crops

See also carbendazim + tebuconazole
fenpropidin + propiconazole + tebuconazole
fenpropidin + tebuconazole
imidacloprid + tebuconazole + triazoxide
prochloraz + tebuconazole
propiconazole + tebuconazole
spiroxamine + tebuconazole

Products

1 Folicur	Bayer	250 g/l	EW	08691
2 Halt	Bayer	250 g/l	EW	08693
3 Landgold Tebuconazole	Landgold	250 g/l	EW	08063
4 Standon Tebuconazole	Standon	250 g/l	EC	09056

Uses

Alternaria in OILSEED RAPE [1-4]. Alternaria in BRUSSELS SPROUTS, CABBAGES, CARROTS *(off-label)*, HORSERADISH *(off-label)*, KOHLRABI *(off-label)* [1]. Botrytis in LINSEED *(reduction)*[1, 2, 4]. Brown rust in SPRING BARLEY, SPRING WHEAT, WINTER BARLEY, WINTER RYE, WINTER WHEAT [3, 4]. Brown rust in BARLEY, RYE, WHEAT [1, 2]. Canker in PARSNIPS *(off-label)* [1]. Chocolate spot in FIELD BEANS [1, 4]. Crown rust in OATS *(reduction)* [1]. Fusarium ear blight in WHEAT [1, 2]. Glume blotch in WHEAT [1, 2]. Light leaf spot in OILSEED RAPE [1-4]. Light leaf spot in BRUSSELS SPROUTS, CABBAGES [1, 4]. Light leaf spot in KOHLRABI *(off-label)* [1]. Net blotch in SPRING BARLEY, WINTER BARLEY [3, 4]. Net blotch in BARLEY [1, 2]. Phoma leaf spot in OILSEED RAPE [1, 2]. Powdery mildew in SPRING BARLEY, SPRING WHEAT, WINTER BARLEY, WINTER RYE, WINTER WHEAT [3, 4]. Powdery mildew in BARLEY, LINSEED, RYE, WHEAT [1, 2]. Powdery mildew in BRUSSELS SPROUTS, CABBAGES, OATS, SWEDES, TURNIPS [1, 4]. Powdery mildew in HERBS *(off-label)* [1]. Rhynchosporium in SPRING BARLEY, WINTER BARLEY, WINTER RYE [3, 4]. Rhynchosporium in BARLEY, RYE [1, 2]. Ring spot in OILSEED RAPE *(reduction)* [1, 2]. Ring spot in BRUSSELS SPROUTS, CABBAGES [1, 4]. Ring spot in BROCCOLI *(off-label)*, CALABRESE *(off-label)*, CAULIFLOWERS *(off-label)*, CHINESE CABBAGE *(off-label)*, COLLARDS *(off-label)*, KALE *(off-label)*, KOHLRABI *(off-label)* [1]. Rust in FIELD BEANS [1, 4]. Rust in DWARF BEANS *(off-label)*, FRENCH BEANS *(off-label)*, HERBS *(off-label)*, LEEKS, RUNNER BEANS *(off-label)* [1]. Sclerotinia stem rot in OILSEED RAPE [1-4]. Septoria diseases in SPRING WHEAT, WINTER WHEAT [3, 4]. Septoria leaf spot in WHEAT [1, 2]. Sooty moulds in WHEAT [1, 2]. Stem canker in OILSEED RAPE [1, 2]. White rot in BULB ONIONS *(off-label)*, GARLIC *(off-label)*, SHALLOTS *(off-label)* [1]. Yellow rust in SPRING BARLEY, SPRING WHEAT, WINTER BARLEY, WINTER RYE, WINTER WHEAT [3, 4]. Yellow rust in BARLEY, RYE, WHEAT [1, 2].

Notes

Efficacy

- For best results apply at an early stage of disease development before infection spreads to new crop growth
- To protect flag leaf and ear from Septoria diseases apply from flag leaf emergence to ear fully emerged (GS 37-59). Earlier application may be necessary where there is a high risk of infection
- For light leaf spot control in oilseed rape apply in autumn/winter with a follow-up spray in spring/summer if required
- For control of most other diseases spray at first signs of infection with a follow-up spray 2-4 wk later if necessary. See label for details
- For disease control in Brussels sprouts and cabbages a 3-spray programme at 21-28 d intervals will give good control

Crop Safety/Restrictions

- Maximum total dose per crop 1.0 l/ha for linseed; 2 l/ha for wheat, barley, rye, field beans, swedes, turnips; 2.25 l/ha per crop for Brussels sprouts, cabbages; 2.5 l/ha for oilseed rape; 3.0 l/ha for leeks
- Some transient leaf speckling on wheat or leaf reddening/scorch on oats may occur but this has not been shown to reduce yield response to disease control
- Do not treat durum wheat
- Apply only to listed oat varieties (see label) [1]
- Do not apply before swedes and turnips have a root diameter of 2.5 cm, or before start of button formation in Brussels sprouts or heart formation in cabbages

Special precautions/Environmental safety

- Harmful if swallowed. Risk of serious damage to eyes
- Irritating to skin
- Harmful to fish or other aquatic life and aquatic plants. Do not contaminate surface waters or ditches with chemical or used container

Protective clothing/Label precautions

- A, C, H
- M03, R03c, R04b, R04d, U05a, U11, U20a, C03, E01, E13c, E26, E30a, E31b, E34

Latest application/Harvest Interval(HI)

- Before grain milky-ripe for cereals (GS 71); when most seed green-brown mottled for oilseed rape (GS 6,3); before brown capsule for linseed
- HI field beans, swedes, turnips 35 d; Brussels sprouts, cabbages 21 d; linseed 35 d; leeks 14 d

Approval

- Off-label approval to May 2001 for use on kohlrabi (OLA 2063/97)[1]; unlimited for use on cauliflower, calabrese, broccoli, Chinese cabbage (OLA 2048/97)[1]; to Nov 2001 for use on kale, collards (OLA 2050/97)[1]; to Apr 2002 for use on bulb onions, garlic, shallots (OLA 2062/97)[1]; to Jun 2002 for use on outdoor leafy herbs - see notice for list of species (OLA 2059/97)[1]; to May 2003 for use on runner beans, French beans, dwarf green beans (OLA 1220/98)[1]; to July 2002 for use on outdoor parsnips, carrots and horseradish (OLA 1588/98)[1]

FOR FULL CONDITIONS OF USE ALWAYS READ THE PRODUCT LABEL

492 tebuconazole + triadimenol

A broad spectrum systemic fungicide for cereals

Products

1 Garnet	Bayer	250:125 g/l	EC	06391
2 Silvacur	Bayer	250:125 g/l	EC	06387
3 Veto F	Bayer	225:75 g/l	EC	08057

Uses

Brown rust in BARLEY, RYE, WHEAT [1-3]. Crown rust in OATS [1, 2]. Fusarium ear blight in WHEAT [1-3]. Glume blotch in WHEAT [1-3]. Leaf spot in WHEAT [1-3]. Net blotch in BARLEY [1-3]. Powdery mildew in BARLEY, RYE, WHEAT [1-3]. Powdery mildew in OATS [1, 2]. Rhynchosporium in BARLEY [1-3]. Sooty moulds in WHEAT [1-3]. Yellow rust in BARLEY, RYE, WHEAT [1-3].

Notes

Efficacy

- For best results apply at an early stage of disease development before infection spreads to new crop growth
- To protect flag leaf and ear from Septoria diseases apply from flag leaf emergence to ear fully emerged (GS 37-59). Earlier application may be necessary where there is a high risk of infection
- For control of rust, powdery mildew, leaf and net blotch apply at first signs of disease with a second application 2-3 wk later if necessary

Crop Safety/Restrictions

- A maximum total dose of 2 l product/ha per crop must not be exceeded
- Do not use on durum wheat
- Use only on listed varieties of oats. See label [1, 2]
- Some transient leaf speckling may occur on wheat or leaf reddening/scorch on oats but this has not been shown to reduce yield response or disease control

Special precautions/Environmental safety

- Irritating to eyes
- Harmful to fish or other aquatic life. Do not contaminate surface waters or ditches with chemical or used container

Protective clothing/Label precautions

- A, C, H
- M03, R04a, U05a, U09b, U20a, C03, E01, E13c, E30a, E31c, E34

Latest application/Harvest Interval(HI)

- Before grain milky-ripe (GS 71)

493 tebuconazole + triazoxide

A triazole and benzotriazine fungicide seed treatment for use in barley

Products

Raxil S	Bayer	20:20 g/l	FS	06974

Uses

Leaf stripe in BARLEY. Loose smut in BARLEY. Net blotch in BARLEY.

Notes

Efficacy

- Best applied through recommended seed treatment machines
- Evenness of seed cover improved by simultaneous application of equal volumes of product and water or dilution of product with an equal volume of water
- Diluted product must be used immediately
- Drill treated seed in the same season

Crop Safety/Restrictions

- Maximum number of treatments 1 per batch of seed
- Slightly delayed and reduced emergence may occur but this is normally outgrown
- Any delay in field emergence, for whatever reason, may be accentuated by treatment
- Do not use on seed with more than 16% moisture content, or on sprouted, cracked or skinned seed

Special precautions/Environmental safety

- Harmful to fish or other aquatic life. Do not contaminate surface waters or ditches with chemical or used container
- Do not use treated seed as food or feed
- Treated seed harmful to game and wildlife
- Product also supplied in returnable containers. See label for guidance on handling, storage, protective clothing and precautions

Protective clothing/Label precautions

- A, H
- U20b, E03, E13c, E30a, E32a, S01, S02, S03, S04b, S05, S06, S07

Latest application/Harvest Interval(HI)

- Before drilling

494 tebuconazole + tridemorph

A broad spectrum fungicide mixture for cereals

Products				
Allicur	Bayer	125:165 g/l	EC	06468

Uses

Botrytis in WINTER WHEAT. Brown rust in BARLEY, WINTER WHEAT. Fusarium ear blight in WINTER WHEAT. Glume blotch in WINTER WHEAT. Net blotch in BARLEY. Powdery mildew in BARLEY, WINTER WHEAT. Rhynchosporium in BARLEY. Septoria leaf spot in WINTER WHEAT. Sooty moulds in WINTER WHEAT. Yellow rust in BARLEY, WINTER WHEAT.

Notes

Efficacy

- For best results apply at an early stage of disease development before infection spreads to new growth
- Repeat treatments may be necessary under high disease pressure or when reinfection occurs
- To protect flag leaf and ear from Septoria diseases apply from flag leaf emergence (GS 37) but before grain is milky ripe (GS 73)

Crop Safety/Restrictions
- Maximum total dose 4 l/ha per crop
- Occasional slight temporary leaf speckling may occur on wheat but this has not been shown to reduce yield response or disease control
- Do not treat durum wheat

Special precautions/Environmental safety
- Harmful if swallowed
- Irritating to eyes and skin
- Harmful to fish or other aquatic life. Do not contaminate surface waters or ditches with chemical or used container

Protective clothing/Label precautions
- A, C, H
- M03, R03c, R04a, R04b, U04a, U05a, U09a, U20a, C03, E01, E07 (14 d), E13c, E30a, E31b, E34

Withholding period
- Keep livestock out of treated areas for at least 14 d following treatment

Latest application/Harvest Interval(HI)
- Before grain is milky ripe (GS 73)

---

495 tebufenpyrad

A pyrazole mitochondrial electron transport inhibitor aphicide and acaricide

Products Masai | Cyanamid | 20% w/w | WB | 07452

Uses Damson-hop aphid in HOPS. Red spider mites in APPLES, PEARS. Two-spotted spider mite in HOPS, PROTECTED ROSES.

Notes

Efficacy
- Acts on eggs (except winter eggs) and all motile stages of spider mites up to adults
- Treat spider mites from 80% egg hatch but before mites become established
- Repeat treatment may be necessary after 2-4 wk
- Treat hops before they reach tops of the wires and not later than end of burr stage. For best results good spray coverage of the whole bine is necessary
- Product can be used in a programme to give season-long control of damson/hop aphids coupled with mite control
- Where aphids resistant to tebufenpyrad occur in hops control is unlikely to be satisfactory and repeat treatments may result in lower levels of control

Crop Safety/Restrictions
- Maximum total dose 1.0 kg/ha per yr for top fruit; 6.0 kg/ha per yr for protected roses, hops
- Other mitochondrial electron transport inhibitor (METI) acaricides should not be applied to the same crop in the same calendar yr either separately or in mixture
- Do not treat apples between pink bud and end of flowering
- Product has no effect on fruit quality or finish
- Small-scale testing of rose varieties to establish tolerance recommended before use

- Product not recommended for use in hand-held sprayers

Special precautions/Environmental safety

- Dangerous to fish or other aquatic life. Do not contaminate surface waters or ditches with chemical or used container
- Harmful to bees. Do not apply to crops in flower or to those in which bees are foraging, except as directed on hops. Do not apply when flowering weeds are present.
- Do not allow direct spray from ground based vehicle mounted/drawn sprayers to fall within 6 m of surface waters or ditches. Do not allow direct spray from air-assisted sprayers to fall within 18 m of surface waters or ditches. Direct spray away from water

Protective clothing/Label precautions

- A, C, H
- U02, U09a, U13, U14, U20a, U22, C02 (7 d), E12e, E13b, E16, E30a, E32a

Latest application/Harvest Interval(HI)

- End of burr stage for hops
- HI apples 7d

496 tebutam

A pre-emergence amide herbicide used in brassica crops

Products Comodor 600 | Agrichem | 600 g/l | EC | 06808

Uses Annual dicotyledons in BROCCOLI, BRUSSELS SPROUTS, CABBAGES, CALABRESE, CAULIFLOWERS, FODDER RAPE, KALE, SWEDES, TURNIPS, WINTER OILSEED RAPE. Annual grasses in BROCCOLI, BRUSSELS SPROUTS, CABBAGES, CALABRESE, CAULIFLOWERS, FODDER RAPE, KALE, SWEDES, TURNIPS, WINTER OILSEED RAPE. Volunteer cereals in BROCCOLI, BRUSSELS SPROUTS, CABBAGES, CALABRESE, CAULIFLOWERS, FODDER RAPE, KALE, SWEDES, TURNIPS, WINTER OILSEED RAPE.

Notes

Efficacy

- Acts via roots and inhibits germination of weeds. Emerged weeds are not controlled
- Best results are achieved on seedbeds which are weed free and have a good tilth without clods
- Effectiveness requires adequate soil moisture. Under very dry conditions light rain or irrigation may be needed but heavy rain on light soils may result in loss of control
- On some heavy soils there may be a reduction in weed control. Do not use on soils with more than 10% organic matter
- Recommended for use in tank-mixture with propachlor and as a tank-mix or in sequence with trifluralin. Apply trifluralin tank-mix as a pre-plant incorporated treatment
- Emerged weeds can be controlled pre-crop emergence by tank mixtures with paraquat products. See label for details

Crop Safety/Restrictions

- Maximum number of treatments 1 per crop
- Apply to brassica crops before drilling or planting in tank-mixture with trifluralin as a soil incorporated treatment. Do not use on brassica seedbeds

- Apply to oilseed rape or sown or transplanted brassicas shortly after drilling or planting but before weeds or crop germinate
- Transplants must be hardened off before treatment. Do not use on block or modular propagated brassicas
- Soil must be deep mouldboard ploughed before drilling or planting any following crop except brassicas
- In the event of crop failure only brassicas, oilseed rape, dwarf beans, maize, peas or potatoes may be grown in the same season

Special precautions/Environmental safety

- Irritating to eyes and skin
- Flammable
- Harmful to fish or other aquatic life. Do not contaminate surface waters or ditches with chemical or used container

Protective clothing/Label precautions

- A
- R04a, R04b, R07d, U05a, U08, U19, U20a, C03, E01, E13c, E30a, E31b

Latest application/Harvest Interval(HI)

- Pre-crop emergence for oilseed rape and other drilled named brassicas; 3 wk after transplanting for transplanted brassicas

497 tecnazene

A protectant chlorobenzene fungicide and potato sprout suppressant

See also carbendazim + tecnazene

Products

Product	Company	Content	Formulation	MAFF No.
1 Atlas Tecgran 100	Atlas	10% w/w	GR	07730
2 Atlas Tecnazene 6% Dust	Atlas	6% w/w	DP	07731
3 Bygran F	Wheatley	10% w/w	GR	00365
4 Bygran S	Wheatley	10% w/w	GR	00366
5 Fusarex Granules	Zeneca	10% w/w	GR	06668
6 Hickstor 10 Granules	Hickson & Welch	10% w/w	GR	03121
7 Hickstor 5 Granules	Hickson & Welch	5% w/w	GR	03180
8 Hortag Tecnazene 10% Granules	Hortag	10% w/w	GR	03966
9 Hortag Tecnazene Double Dust	Hortag	6% w/w	DP	01072
10 Hortag Tecnazene Dust	Hortag	3% w/w	DP	01074
11 Hortag Tecnazene Potato Granules	Hortag	5% w/w	GR	01075
12 Hystore 10	Agrichem	10% w/w	GR	03581
13 Hytec	Agrichem	3% w/w	DP	01099
14 Hytec 6	Agrichem	6% w/w	DP	03580
15 Nebulin	Wheatley	300 g/l	HN	01469
16 New Hickstor 6	Hickson & Welch	6% w/w	DS	04221
17 New Hystore	Agrichem	5% w/w	GR	01485
18 Tripart Arena 10G	Hickson & Welch	10% w/w	GR	05603
19 Tripart Arena 5G	Hickson & Welch	5% w/w	GR	05604
20 Tripart New Arena 6	Hickson & Welch	6% w/w	DP	05813

Uses Dry rot in SEED POTATOES, WARE POTATOES. Sprout suppression in WARE POTATOES.

Notes

Efficacy

- Apply dust or granules evenly to potatoes with appropriate equipment (see label for details) during loading. Works by vapour phase action so clamps or bins should be covered to avoid loss of vapour
- Best results achieved on dry, dirt-free potatoes. Fungicidal activity best at 12-14°C, sprouting control best at 9-12°C. For best sprouting control maintain temperature at 10-14°C for first 14 d, then reduce to 7°C [5]
- Under suitable storage conditions sprouting prevented for 4-6 mth
- Excessive ventilation shortens control period
- Sprouting not controlled in tubers which have already broken dormancy
- Treatment can be used in sequence with chlorpropham treatment [1, 4, 12]
- Treatment gives some control of skin spot, gangrene and silver scurf

Crop Safety/Restrictions

- Maximum number of treatments 1 per batch of potatoes
- Air tubers for 6 wk before planting; chitting should have commenced
- Treated tubers must not be removed for sale or processing, including washing, for at least 6 wk after application
- To prevent contamination of other crops remove all traces of treated potatoes and soil after the store has been emptied

Special precautions/Environmental safety

- Dangerous to fish or other aquatic life. Do not contaminate surface waters or ditches with chemical or used container [1]
- Do not remove from store for sale or processing (including washing) for at least 6 wk after application

Protective clothing/Label precautions

- A, C, H [15]
- U05a [5]; U08, U16 [15]; U19 [1-5, 7-10, 12-20]; U20a, C02 [5, 13, 14, 17]; U20b [1, 2, 12, 15]; U20c [3, 4, 6-11, 16, 18-20]; E13b [1, 3, 6, 7, 15, 18, 19]; E13c [4]; E15 [2, 5, 8-14, 16, 17, 20]; E30a [1-20]; E31a, E34 [5, 13-15, 17]; E32a [1-14, 16-20]

Latest application/Harvest Interval(HI)

- 6 wk before removal from store for sale or processing

Approval

- One product formulated for application by thermal fogging [15]. See label for details

Maximum Residue Levels (mg residue/kg food)

- potatoes 10 (ACP review - see PR 1995, Issue 7); lettuce 2

498 tecnazene + thiabendazole

A fungicide and potato sprout suppressant

Products

1 Hytec Super	Agrichem	6:1.8% w/w	DS	01100
2 Storite SS	Seedcote	300:100 g/l	FS	08702

FOR FULL CONDITIONS OF USE ALWAYS READ THE PRODUCT LABEL

Uses Dry rot in SEED POTATOES, WARE POTATOES. Gangrene in SEED POTATOES, WARE POTATOES. Silver scurf in SEED POTATOES, WARE POTATOES. Skin spot in SEED POTATOES, WARE POTATOES. Sprout suppression in WARE POTATOES.

Notes

Efficacy

- Apply dust evenly with suitable dusting machine (not by hand) as tubers enter store and cover as soon as possible [1]
- Apply liquid with suitable low-volume mist applicator or spinning disc [2]
- Treat tubers as soon as possible after lifting, no longer than 14 d
- Under suitable storage conditions sprouting prevented for 3-6 mth. Effectiveness may be decreased with excessive ventilation or inadequate covering
- Best results achieved on dry, dirt-free tubers
- Sprouting not controlled on tubers that have broken dormancy

Crop Safety/Restrictions

- Maximum number of treatments 1 per batch
- Treated tubers must not be removed for sale or processing, including washing, for at least 6 wk after application
- Air tubers for 6 wk before planting, chitting should have commenced
- Do not mix Storite SS with thiabendazole

Special precautions/Environmental safety

- Harmful to fish or other aquatic life. Do not contaminate surface waters or ditches with chemical or used container

Protective clothing/Label precautions

- A, C, D, H [1, 2]
- U08, U20c, E26, E31a, S01, S05 [2]; U19, E13c, E30a [1, 2]; U20a, E32a, E34 [1]

Latest application/Harvest Interval(HI)

- 6 wk before removal from store for sale or processing (including washing)

Maximum Residue Levels (mg residue/kg food)

- see tecnazene and thiabendazole entries

499 teflubenzuron

A benzoylurea insecticide for use on ornamentals

Products				
Nemolt	Fargro	150 g/l	SC	07012

Uses Browntail moth in ORNAMENTALS. Caterpillars in ORNAMENTALS. Whitefly in ORNAMENTALS.

Notes

Efficacy

- Product acts as larval stomach poison interfering with moulting process leading to cessation of feeding and larval death
- Apply as soon as first stage larvae seen. This will often coincide the peak of moth flight

Crop Safety/Restrictions

- Maximum number of treatments 3 per yr
- Test specific varieties before carrying out extensive treatments

Special precautions/Environmental safety

- Do not contaminate surface waters or ditches with chemical or used container
- Limited evidence suggests some margin of safety to *Encarsia formosa*. Effects on other parasites and predators not fully tested

Protective clothing/Label precautions

- A, C
- U20c, E15, E26, E30a, E32a

500 tefluthrin

A soil acting pyrethroid insecticide seed treatment

Products

1 Evict	Bayer	100 g/l	CS	08731
2 Force ST	Bayer	200 g/l	LS	06665

Uses

Bean seed flies in BULB ONIONS *(off-label)*, LEEKS *(off-label)*, SALAD ONIONS *(off-label)* [2]. Carrot fly in OUTDOOR CARROTS *(off-label)*, OUTDOOR PARSNIPS *(off-label)*[2]. Millipedes in FODDER BEET, SUGAR BEET [2]. Onion fly in BULB ONIONS *(off-label)*, LEEKS *(off-label)*, SALAD ONIONS *(off-label)* [2]. Pygmy beetle in FODDER BEET, SUGAR BEET [2]. Springtails in FODDER BEET, SUGAR BEET [2]. Symphylids in FODDER BEET, SUGAR BEET [2]. Wheat bulb fly in BARLEY *(seed treatment)*, WHEAT *(seed treatment)* [1]. Wireworms in BARLEY *(seed treatment)*, OATS *(seed treatment)*, WHEAT *(seed treatment)* [1].

Notes

Efficacy

- Apply during process of pelleting seed. Consult manufacturer for details of specialist equipment required [2]
- Apply to cereal seed through a suitable liquid seed treater calibrated to achieve even coverage [1]
- Micro-capsule formulation allows slow release to provide a protection zone around treated seed during establishment [1]
- Product is non-systemic and may not protect against wireworm attack at the soil surface [1]
- Where egg counts indicate a high wheat bulb fly attack, or where there is severe risk of wireworm attack, follow-up spray treatments may be needed [1]

Crop Safety/Restrictions

- Maximum number of treatments 1 per batch of seed
- Sow treated seed as soon as possible. Do not store treated seed from one drilling season to next
- Must be co-applied with suitable fungicide seed treatment to protect against seed and soil borne diseases [1]
- Do not use on seed that is sprouted, cracked, damaged or over 16% moisture content [1]
- Co-application with other seed treatments may block or damage application equipment

Special precautions/Environmental safety

- Irritant. May cause sensitization by skin contact
- Can cause a transient tingling or numbing sensation to exposed skin. Avoid skin contact with product and treated seed

- Extremely dangerous to fish or other aquatic life. Do not contaminate surface waters or ditches with chemical or used container
- Do not use treated seed as food or feed
- Treated seed harmful to game and wildlife

Protective clothing/Label precautions

- A [1]; B [2]; C, D, H [1, 2]
- R04, U07, U14, E26, E32a, E36, S03 [1]; R04e, U05a, C03, E01, E13a, S01, S02, S04b, S05, S06, S07 [1, 2]; U02, U04a, U08, U20a, E30a, E31a [2]

Withholding period

- Keep livestock out of areas drilled with treated seed for at least 80 d [2]

Latest application/Harvest Interval(HI)

- Before drilling seed

Approval

- Off-label approval unlimited for use on carrots and parsnips (OLA 0537/95)[2]; to Jun 2001 for use on outdoor leeks, bulb and salad onions (OLA 1406/96)[2]

---

501 terbacil

A residual uracil herbicide for use in asparagus and herbs

Products

Sinbar	DuPont	80% w/w	WP	01956

Uses

Annual dicotyledons in ASPARAGUS, OUTDOOR HERBS *(off-label)*, PROTECTED HERBS *(off-label)*. Annual grasses in ASPARAGUS, OUTDOOR HERBS *(off-label)*, PROTECTED HERBS *(off-label)*. Bent grasses in ASPARAGUS, OUTDOOR HERBS *(off-label)*, PROTECTED HERBS *(off-label)*. Couch in ASPARAGUS, OUTDOOR HERBS *(off-label)*, PROTECTED HERBS *(off-label)*.

Notes

Efficacy

- Has some foliar activity but uptake mainly through roots
- Adequate rainfall needed to ensure penetration to weed root zone
- Best results achieved by application in early spring, before mid-Apr

Crop Safety/Restrictions

- Maximum number of treatments 1 per yr
- Apply to asparagus established at least 2 yr. Do not apply after first spears emerge
- Do not use on Sands, Loamy Sand or gravels or soils with less than 1% organic matter
- Do not use for at least 2 yr prior to grubbing and replanting

Protective clothing/Label precautions

- U08, U19, U20a, E15, E30a, E31a

Latest application/Harvest Interval(HI)

- Before spears emerge for asparagus; pre-emergence of new growth in spring for herbs

Approval

- Off-label approval unlimited for use in range of herbs, see off-label approval notice for details (OLA 0082/93)[1]

---

502 terbuthylazine

A triazine herbicide available only in mixtures

See also cyanazine + terbuthylazine
diflufenican + terbuthylazine
isoxaben + terbuthylazine

503 terbuthylazine + terbutryn

A pre-emergence herbicide for peas, beans, lupins and potatoes

Products

1 Batallion	Makhteshim	150:350 g/l	SC	08305
2 Opogard	Novartis	150:350 g/l	SC	08427
3 Opogard 500 SC	Ciba Agric.	150:350 g/l	SC	05850

Uses

Annual dicotyledons in BROAD BEANS, COMBINING PEAS, FIELD BEANS, POTATOES, VINING PEAS [1-3]. Annual dicotyledons in LUPINS [2, 3]. Annual grasses in BROAD BEANS, COMBINING PEAS, FIELD BEANS, POTATOES, VINING PEAS [1-3]. Annual grasses in LUPINS [2, 3].

Notes

Efficacy
- Active via root uptake and with foliar activity on cotyledon stage weeds
- Best results achieved by application to fine, firm, moist seedbed, preferably at weed emergence, when rain falls after spraying
- Effectiveness may be reduced by excessive rain, drought or cold
- Residual control lasts for up to 8 wk on mineral soils. Effectiveness reduced on highly organic soils and subsequent use of post-emergence treatment recommended
- Do not cultivate after treatment
- Do not use on soils which are very cloddy or have more than 10% organic matter

Crop Safety/Restrictions
- Maximum number of treatments 1 per crop
- Apply to spring sown peas, beans and lupins as soon as possible after drilling
- Crop seed must be covered by at least 25 mm of settled soil
- Vedette and Printana peas may be damaged by treatment. Do not use on forage peas
- Heavy rain after application may cause damage to peas on light soils
- Do not treat peas, beans or lupins on soils lighter than loamy fine sand, potatoes on soils lighter than loamy sand or on very stony soils or lupins on silty clay soil
- Subsequent crops may be sown or planted after 12 wk (14 wk if prolonged drought)

Special precautions/Environmental safety
- Harmful if swallowed
- Harmful to fish or other aquatic life. Do not contaminate surface waters or ditches with chemical or used container

Protective clothing/Label precautions
- A, C [2, 3]

- M03, U02, U13, U14, U15 [1]; R03c, U05a, U08, U19, U20a, C03, E01, E13c, E30a, E31b, E34 [1-3]

Latest application/Harvest Interval(HI)

- 3 d before emergence for peas, beans and lupins; before 10% of plants emerged for potatoes

Approval

- May be applied through CDA equipment for use on peas, beans and lupins [3]. See label for details

504 terbutryn

A residual triazine herbicide for cereals and aquatic weed control

See also fomesafen + terbutryn
terbuthylazine + terbutryn

Products

1 Alpha Terbutryn 50 SC	Makhteshim	500 g/l	SC	04809
2 Clarosan	Novartis	1% w/w	GR	08396
3 Clarosan 1 FG	Ciba Agric.	1% w/w	GR	03859
4 Prebane	Novartis	490 g/l	SC	08432
5 Prebane SC	Ciba Agric.	490 g/l	SC	07634

Uses

Annual dicotyledons in WINTER BARLEY, WINTER OATS, WINTER WHEAT [1, 4, 5]. Annual grasses in WINTER OATS [4, 5]. Annual meadow grass in WINTER BARLEY, WINTER OATS, WINTER WHEAT [1, 4, 5]. Aquatic weeds in AREAS OF WATER [2, 3]. Blackgrass in WINTER BARLEY, WINTER WHEAT [1, 4, 5]. Perennial ryegrass in WINTER BARLEY, WINTER WHEAT [1, 4, 5]. Rough meadow grass in WINTER BARLEY, WINTER OATS, WINTER WHEAT [1, 4, 5].

Notes

Efficacy

- Apply in autumn after drilling but before crop emergence. Best results given on fine, firm seedbed in which any trash or ash dispersed. Do not use on soils with more than 10% organic matter [1, 4, 5]
- Effectiveness reduced by long, dry period after spraying or by waterlogging [1, 4, 5]
- Effectiveness reduced by excessive surface trash and an impermeable surface layer [1, 4, 5]
- Do not use on soils with more than 10% organic matter [1, 4, 5]
- Do not harrow after treatment [1, 4, 5]
- For aquatic weed control apply granules to water surface with suitable equipment, when growth active but before heavy infestation develops, usually Apr-May, sometimes to Aug [2, 3]
- Use only in static or sluggishly moving water. The flow in moving water should be stopped for at least 7 d after treatment otherwise weed control may be reduced [2, 3]
- Effectiveness reduced in water with peaty bottom [2, 3]

Crop Safety/Restrictions

- Maximum number of treatments 1 per crop
- Do not use on crops drilled after 30 Nov [1, 4, 5]
- Do not use on spring barley cultivars sown in autumn [1, 4, 5]
- Do not treat winter oats with dose higher than that recommeded otherwise serious crop damage may result. Winter oats in unconsolidated seedbeds may also be damaged [1, 4, 5]
- Do not treat winter barley on Sands or Very Light soils at recommended dose [1, 4, 5]

- Only use in programme with tri-allate in winter barley on Medium and heavier soils [1, 4, 5]
- Before spraying direct-drilled crops ensure seed well covered and drill slots closed [1, 4, 5]
- Risk of crop damage on stony or gravelly soils especially if heavy rain falls soon after treatment [1, 4, 5]

Special precautions/Environmental safety

- Do not use treated water for irrigation purposes within 7 d of treatment [2, 3]
- Concentration in water must not exceed 10 ppm [2, 3]
- Harmful to fish or other aquatic life. Do not contaminate surface waters or ditches with chemical or used container [1, 4, 5]
- If dense weed growth in watercourses to be controlled without de-oxygenation treat in sections of about 400 m at a time with intervals of at least 14 d [2, 3]

Protective clothing/Label precautions

- U20a, E30a [1-5]; E13c, E31b [1, 4, 5]; E19, E21, E32a [2, 3]; E26 [4, 5]

Latest application/Harvest Interval(HI)

- Aug for areas of water [2, 3]; before crop emergence for cereals [1, 4, 5]

Approval

- Approved for aquatic weed control [3]. See notes in Section 1 on use of herbicides in or near water
- Approved for aerial application on winter wheat and winter barley [5]. See notes in Section 1

---

505 terbutryn + trietazine

A residual triazine herbicide for peas, potatoes and field beans

Products	Senate	AgrEvo	250:250 g/l	SC	07279

Uses Annual dicotyledons in PEAS, POTATOES, SPRING FIELD BEANS. Annual meadow grass in PEAS, POTATOES, SPRING FIELD BEANS.

Notes

Efficacy

- Product acts mainly through roots of germinating seedlings but also has some contact effect on cotyledon stage weeds
- Weeds germinating from deeper than 2.5 cm may not be completely controlled
- Weed control is improved by light rain after spraying but reduced by excessive rain, drought or cold
- Persistence may be reduced on soils with high organic matter content

Crop Safety/Restrictions

- Maximum number of treatments 1 per crop
- Apply between drilling and 5% emergence (peas), pre-emergence (field beans) between sowing and 10% emergence (potatoes), provided weeds not beyond cotyledon stage
- May be used on spring sown vining, combining or forage peas and field beans covered by not less than 3 cm soil

- Do not spray any winter sown peas or beans sown before Jan
- May be used on early and maincrop potatoes, including seed crops
- On stony or gravelly soils there is risk of crop damage, especially if heavy rain falls soon after treatment. Frost after application may also check crop
- Do not use on soils classed as Sands
- Plough or cultivate to 15 cm before sowing or planting another crop. Any crop may be sown or planted after 12 wk (14 wk with drilled brassicas)
- If sprayed pea crop fails redrill without ploughing but do not respray

Special precautions/Environmental safety

- Harmful if swallowed
- Harmful to fish or other aquatic life. Do not contaminate surface waters or ditches with chemical or used container

Protective clothing/Label precautions

- A, C
- M03, R03c, U05a, U08, U20b, C03, E01, E13c, E26, E30a, E31b, E34

Latest application/Harvest Interval(HI)

- Pre-emergence of crop for field beans; 5% emergence for peas; 10% emergence for potatoes

---

506 terbutryn + trifluralin

A residual, pre-emergence herbicide for winter cereals

Products

1 Alpha Terbalin 35 SC	Makhteshim	150:200 g/l	SC	04792
2 Ashlade Summit	Ashlade	150:200 g/l	SC	06214

Uses

Annual dicotyledons in WINTER BARLEY, WINTER WHEAT. Annual grasses in WINTER BARLEY, WINTER WHEAT. Annual meadow grass in WINTER BARLEY, WINTER WHEAT. Blackgrass in WINTER BARLEY, WINTER WHEAT. Chickweed in WINTER BARLEY, WINTER WHEAT. Mayweeds in WINTER BARLEY, WINTER WHEAT. Speedwells in WINTER BARLEY, WINTER WHEAT.

Notes

Efficacy

- Apply as soon as possible after drilling, before weed or crop emergence when soil moist
- Best results achieved by application to fine, firm, moist seedbed when adequate rain falls after application. Do not use on crops drilled after 30 Nov
- Do not use on uncultivated stubbles or soils with trash or more than 10% organic matter. Do not harrow after treatment
- Effectiveness reduced in mild, wet winters, especially in south-west

Crop Safety/Restrictions

- Maximum number of treatments 1 per crop
- Crops should only be treated if drilled 25-35 mm deep
- Do not use on sandy soils (Coarse Sandy Loam - Coarse Sand) or on spring-drilled crops
- Do not treat undersown crops or crops to be undersown
- Crops suffering stress due to waterlogging, deficiency, poor seedbed preparation or pest problems may be damaged
- Any crop may be sown following harvest of treated crop. In the event of crop failure only resow with wheat or barley

Special precautions/Environmental safety
- Irritating to skin, eyes and respiratory system
- Harmful to fish or other aquatic life. Do not contaminate surface waters or ditches with chemical or used container

Protective clothing/Label precautions
- A, C [1, 2]
- R04a, R04b, U05a, U08, U13, U19, C03, E01, E13c, E30a, E31b [1, 2]; R04c, U20a [2]; U20b [1]

Latest application/Harvest Interval(HI)
- Pre-emergence of crop

507 tetradifon

A bridged-diphenyl acaricide for use in horticultural crops

See also dicofol + tetradifon

Products

Tedion V-18 EC	Hortichem	80 g/l	EC	03820

Uses

Red spider mites in APPLES, APRICOTS *(protected crops)*, BLACKBERRIES, BLACKCURRANTS, CHERRIES, GOOSEBERRIES, GRAPEVINES, HOPS, LOGANBERRIES, NECTARINES *(protected crops)*, PEACHES *(protected crops)*, PEARS, PLUMS, PROTECTED BEANS, PROTECTED CUCUMBERS, PROTECTED FLOWERS, PROTECTED GRAPEVINES, PROTECTED MELONS, PROTECTED ORNAMENTALS, PROTECTED PEPPERS, PROTECTED TOMATOES, RASPBERRIES, STRAWBERRIES.

Notes

Efficacy
- Chemical active against summer eggs and all stages of red spider larvae but does not kill adult mites
- Apply as soon as first mites seen and repeat as required. See label for details of timing
- May be used in conjunction with biological control
- Resistant strains of red spider mite have developed in some areas

Crop Safety/Restrictions
- Some rose cultivars are slightly susceptible. Do not treat cissus, dahlia, ficus, kalanchoe or primula

Special precautions/Environmental safety
- Harmful in contact with skin and if swallowed. Irritating to eyes and skin
- Flammable

Protective clothing/Label precautions
- A, C
- M03, R03a, R03c, R04a, R04b, U05a, U08, U20a, C03, E01, E15, E30a, E31a, E34

FOR FULL CONDITIONS OF USE ALWAYS READ THE PRODUCT LABEL

508 tetramethrin

A contact acting pyrethroid insecticide

See also phenothrin + tetramethin
phenothrin + tetramethrin
resmethrin + tetramethrin

Products

Killgerm Py-Kill W	Killgerm	10.2% w/v	EC	H4632

Uses

Flies in AGRICULTURAL PREMISES, LIVESTOCK HOUSES.

Notes

Efficacy
- Dilute in accordance with directions and apply as space or surface spray

Crop Safety/Restrictions
- Do not apply directly on food or livestock
- Remove exposed milk before application. Protect milk machinery and containers from contamination
- Do not use space sprays containing pyrethrins or pyrethroid more than once per wk in intensive or controlled environment animal houses in order to avoid development of resistance. If necessary, use a different control method or product

Special precautions/Environmental safety
- Harmful to fish or other aquatic life. Do not contaminate surface waters or ditches with chemical or used container

Protective clothing/Label precautions
- R07d, U08, U19, U20c, C06, C10, E05, E13c, E30a, E31a

509 thiabendazole

A systemic, curative and protectant benzimidazole (MBC) fungicide

See also ethirimol + flutriafol + thiabendazole
imazalil + thiabendazole
tecnazene + thiabendazole

Products

1 Hykeep	Agrichem	2% ww	DS	06744
2 Storite Clear Liquid	Seedcote	220 g/l	LS	08982
3 Storite Flowable	Seedcote	450 g/l	FS	08703
4 Tecto Flowable Turf Fungicide	Vitax	450 g/l	SC	06273

Uses

Dollar spot in AMENITY GRASS, TURF [4]. Dry rot in WARE POTATOES *(post-harvest)* [1-3]. Dry rot in SEED POTATOES *(post-harvest)* [2, 3]. Dutch elm disease in ELM TREES *(injection - off-label)* [2]. Fungus diseases in ASPARAGUS *(off-label)* [2]. Fusarium basal rot in NARCISSI [2]. Fusarium patch in AMENITY GRASS, TURF [4]. Gangrene in WARE POTATOES *(post-harvest)* [1-3]. Gangrene in SEED POTATOES *(post-harvest)* [2, 3]. Red thread in AMENITY GRASS, TURF [4]. Silver scurf in WARE POTATOES *(post-harvest)* [1-3]. Silver scurf in SEED POTATOES *(post-harvest)* [2, 3]. Skin spot in WARE POTATOES *(post-harvest)* [1-3]. Skin spot in SEED POTATOES *(post-harvest)* [2, 3].

Notes

Efficacy

- Apply post-harvest treatment to potatoes using suitable equipment within 24 h of lifting. See label for details
- Apply to turf as spray after mowing and do not mow for at least 48 h. Best results on red thread obtained in combination with fertilizer [4]
- Apply to bulbs as dip. Ensure bulbs are clean [2]

Crop Safety/Restrictions

- Maximum number of treatments 1 per batch for seed potato treatments; 2 per yr for bulbs [2]; 1 per batch as dip and module drench, or 10 per crop for drench in field grown asparagus [2]

Special precautions/Environmental safety

- Harmful to fish or other aquatic life. Do not contaminate surface waters or ditches with chemical or used container
- Treated seed potatoes must not be used for food or feed [2, 3]

Protective clothing/Label precautions

- A [1, 2]; B [2]; C, D, H [1, 2]; K [2]; M [1, 2]
- U19, E01 [1-3]; U20c, E13c, E30a [1-4]; E26 [2, 3]; E29, E32a [1]; E31a [2-4]; S01, S05 [2]

Latest application/Harvest Interval(HI)

- Before planting for seed potatoes; 21 d before removal from store for sale, processing or consumption for ware potatoes.
- HI asparagus 6 mth [2]

Approval

- Product formulated for application with ULV equipment (see label for details) [2]
- Off-label Approval unlimited for use on asparagus (OLA 0525/95)[2]; unlimited for use on elm trees (OLA 0990/95)[2]

Maximum Residue Levels (mg residue/kg food)

- citrus fruits 6; pome fruit, strawberries, broccoli, potatoes 5; bananas 3; tea, hops, meat, eggs 0.1; other crops 0.05

510 thiabendazole + thiram

A fungicide seed dressing mixture for field and vegetable crops

Products	Hy-TL	Agrichem	225:300 g/l	FS	06246

Uses

Ascochyta in BROAD BEANS *(seed treatment)*, FIELD BEANS *(seed treatment)*, PEAS *(seed treatment)*. Damping off in BROAD BEANS *(seed treatment)*, FIELD BEANS *(seed treatment)*, PEAS *(seed treatment)*.

Notes

Efficacy

- Dress seed as near to sowing as possible
- Dilution may be needed with particularly absorbent types of seed. If diluted material used, seed may require drying before storage

Crop Safety/Restrictions
- Maximum number of treatments 1 per batch

Special precautions/Environmental safety
- Harmful if swallowed and in contact with skin. Irritating to eyes, skin and respiratory system
- Dangerous to fish or other aquatic life. Do not contaminate surface waters or ditches with chemical or used container
- Do not use treated seed as food or feed
- Treated seed harmful to game and wildlife

Protective clothing/Label precautions
- A, C, D, H, M
- M03, R03a, R03c, R04a, R04b, R04c, U05a, U08, U20a, U20b, C03, E01, E03, E13b, E26, E30a, E31c, E34, S01, S02, S03, S04b, S05, S06, S07

Latest application/Harvest Interval(HI)
- Before drilling

Maximum Residue Levels (mg residue/kg food)
- see thiabendazole entry

---

511 thifensulfuron-methyl

A translocated sulfonylurea herbicide

See also metsulfuron-methyl + thifensulfuron-methyl

Products

Prospect	DuPont	75% w/w	WB	06541

Uses

Docks in ESTABLISHED GRASSLAND, LAND TEMPORARILY REMOVED FROM PRODUCTION *(in green cover).*

Notes

Efficacy
- Best results achieved from application to young green dock foliage when growing actively
- Only broad-leaved docks (*Rumex obtusifolius*) are controlled; curled docks (*Rumex crispus*) are resistant
- Apply 7-10 d before grazing and do not graze for 7 d afterwards
- Docks with developing or mature seed heads should bc topped and the regrowth treated later
- Established docks with large tap roots may require follow-up treatment
- High populations or poached grassland will require further treatment in following yr
- Ensure good spray coverage and apply to dry foliage

Crop Safety/Restrictions
- Maximum number of treatments 1 per calender yr. Must only be applied from 1 Feb in yr of harvest
- Do not treat new leys in year of sowing
- Do not treat where nutrient imbalances, drought, waterlogging, low temperatures, lime deficiency, pest or disease attack have reduced sward vigour
- Do not roll or harrow within 7 d of spraying
- Product may cause a check to both sward and clover which is usually outgrown

Special precautions/Environmental safety
- Extremely dangerous to aquatic higher plants. Do not contaminate surface waters or ditches with chemical or used container
- Do not allow direct spray from ground-based vehicle mounted/drawn sprayers to fall within 6 m, or from hand-held sprayers to within 2 m, of surface waters or ditches. Direct spray away from water
- Take extreme care to avoid drift onto broad-leaved plants outside the target area or onto surface waters or ditches, or land intended for cropping
- Use recommended procedure to clean out spraying equipment
- Only grass or cereals may be sown within 4 wk of treatment

Protective clothing/Label precautions
- U08, U19, U20b, E07 (7 d), E14a, E16, E30a, E32a

Withholding period
- Keep livestock out of treated areas for at least 7 d following treatment

Latest application/Harvest Interval(HI)
- Before 1 Aug

512 thifensulfuron-methyl + tribenuron-methyl

A mixture of two sulfonylurea herbicides for cereals

Products

1 Calibre	DuPont	50:25% w/w	WG	07795
2 DUK 110	DuPont	50:25% w/w	WG	06266

Uses Annual dicotyledons in BARLEY, WHEAT. Charlock in BARLEY, WHEAT. Chickweed in BARLEY, WHEAT. Mayweeds in BARLEY, WHEAT.

Notes

Efficacy
- Apply after 1 Feb when weeds are small and actively growing
- Ensure good spray cover of the weeds
- Ensure that weeds present are those that are susceptible. See label
- Follow label mixing instructions
- Effectiveness reduced by rain within 4 h of treatment

Crop Safety/Restrictions
- Maximum number of treatments 1 per crop
- May be sprayed on all varieties of wheat and barley from 3 leaf stage (GS 13) up to and including flag leaf fully emerged (GS 39)
- Igri winter barley may suffer damage during period of rapid growth and must not be treated before leaf sheath erect stage (GS 30)
- Do not apply to cereals undersown with grass, clover or other legumes
- Do not apply within 7 d of rolling
- Various tank mixtures recommended to broaden weed control spectrum. Other mixtures are specifically excluded. See label
- Do not apply in sequence or in tank mixture with a product containing any other sulfonyl urea

- Do not apply to any crop suffering from stress or not actively growing
- Take particular care to avoid damage by drift onto broad-leaved plants outside the target area or onto surface waters or ditches
- Only cereals, field beans or oilseed rape may be sown in the same calendar year as harvest of a treated crop. In the event of failure of a treated crop sow only a cereal crop within 3 mth of product application

Special precautions/Environmental safety

- Extremely dangerous to aquatic higher plants. Do not contaminate surface waters or ditches wirth chemical or used container
- Extremely dangerous to fish or other aquatic life. Do not contaminate surface waters or ditches with chemical or used container [1]
- Do not allow direct spray from ground-based vehicle mounted/drawn sprayers to fall within 6 m, or from hand-held sprayers to within 2 m, of surface waters or ditches. Direct spray away from water

Protective clothing/Label precautions

- U08, U19, U20a, E16, E30a, E32a [1, 2]; E13a [1]; E14a [2]

Latest application/Harvest Interval(HI)

- Before flag leaf ligule first visible (GS 39)

---

513 thiodicarb

A carbamate insecticide with molluscicide uses

Products

1 Genesis	RP Agric.	4% w/w	RB	06168
2 Genesis ST	RP Agric.	4% w/w	RB	08211
3 Judge	RP Agric.	4% w/w	RB	08163

Uses

Slugs in BARLEY, DURUM WHEAT, OATS, OILSEED RAPE, POTATOES, TRITICALE, WHEAT [1, 3]. Slugs in BARLEY *(seed admixture)*, DURUM WHEAT *(seed admixture)*, OATS *(seed admixture)*, TRITICALE *(seed admixture)*, WHEAT *(seed admixture)* [2].

Notes

Efficacy

- Apply bait as broadcast treatment [1, 3] or admixed with seed [1, 2]
- Additional broadcast treatments may be needed after drilling admixed seed

Crop Safety/Restrictions

- Maximum number of treatments 1 per crop (admixture), 3 per crop (broadcast)
- Do not treat grain with more than 16% moisture content or allow treated seed to rise above this level [1, 2]
- Sow admixed seed as soon as possible after treatment [1, 2]

Special precautions/Environmental safety

- Product contains an anticholinesterase carbamate compound. Do not use if under medical advice not to work with such compounds
- Harmful if swallowed. Irritating to eyes. May cause sensitization by skin contact
- Harmful to game, wild birds and animals
- Dangerous to fish or other aquatic life. Do not contaminate surface waters or ditches with chemical or used container
- Do not use treated seed as food or feed [1, 2]

Protective clothing/Label precautions

- A [1-3]; D [1, 2]; H [1-3]; J [1, 3]
- M02, M04, R03c, R04a, R04e, U05a, U13, U20b, C03, E01, E13b, E30a, E32a, E34 [1-3]; E10b [1, 3]; S01, S02, S03, S04b, S05, S06, S07 [1, 2]

Latest application/Harvest Interval(HI)

- Before drilling when admixed with barley, durum wheat, oats, triticale, wheat [1, 2]; before first node detectable (GS 31) for barley, durum wheat, oats, triticale, wheat [1, 3]; before stem extension (GS 2,0) for oilseed rape [1, 3]
- HI potatoes 3 wk [1, 3]

514 thiophanate-methyl

A carbendazim precursor fungicide with protectant and curative activity

See also gamma-HCH + thiophanate-methyl
iprodione + thiophanate methyl

Products

1 Mildothane Liquid	Hortichem	500 g/l	SC	06211
2 Mildothane Turf Liquid	RP Amenity	500 g/l	SC	05331

Uses

Botrytis in DWARF BEANS [1]. Canker in APPLES, PEARS [1]. Chocolate spot in FIELD BEANS [1]. Dollar spot in MANAGED AMENITY TURF [2]. Fusarium patch in MANAGED AMENITY TURF [2]. Powdery mildew in APPLES, CUCUMBERS, PEARS [1]. Red thread in MANAGED AMENITY TURF [2]. Scab in APPLES, PEARS [1]. Storage rots in APPLES, PEARS [1]. Wormcast formation in MANAGED AMENITY TURF [2].

Notes

Efficacy

- Number and timing of sprays varies with crop and disease, see label for details. On tree fruit sprays should be repeated at 14 d intervals
- Spray treatments can reduce wood canker on apples
- Apply a post-blossom spray or dip fruit to reduce storage rot diseases
- Apply spray or drench treatment on cucumbers. High volume sprays are not recommended where *Phytoseiulus* is used to control red spider mites
- Apply to turf during period of active growth and do not mow for 48 h [2]
- Do not mix with MCPB herbicides or copper. See label for compatible mixtures

Crop Safety/Restrictions

- Maximum number of treatments (including applications of products containing carbendazim or benomyl) 12 per crop as pre-harvest spray on apples and pears, 1 per batch as post-harvest dip; 1 per crop for dwarf beans; 2 per crop for field beans; 6 per crop for cucumbers
- Spraying apples from blossom to fruitlet stage may increase russeting on prone cultivars
- Do not use on fruit trees at petal fall or early fruitlet stage when preceded by a protectant such as captan

Special precautions/Environmental safety

- To dispose of tip empty solution into an approved soakaway, not into drains, open ditches or soakaways

Protective clothing/Label precautions

- A [1, 2]; B [1]; C, H [1, 2]; K [1]; M [1, 2]
- U20b, E26, E32a [1]; E15, E30a [1, 2]; E31a [2]

Latest application/Harvest Interval(HI)

- Before end of flowering for field beans; before pods fully formed for dwarf beans.
- HI apples, pears 7 d; cucumbers 2 d

515 thiram

A protectant dithiocarbamate fungicide

See also carbendazim + cymoxanil + oxadixyl + thiram
carboxin + gamma-HCH + thiram
carboxin + thiram
gamma-HCH + fenpropimorph + thiram
gamma-HCH + thiram
metalaxyl + thiabendazole + thiram
metalaxyl + thiram
permethrin + thiram
thiabendazole + thiram

Products

1 Agrichem Flowable Thiram	Agrichem	600 g/l	FS	06245
2 Unicrop Thianosan DG	Unicrop	80% w/w	WG	05454

Uses

Botrytis fruit rot in APPLES, PEARS [2]. Botrytis in CHRYSANTHEMUMS, FREESIAS, ORNAMENTALS *(except Hydrangea)*, OUTDOOR LETTUCE, PROTECTED LETTUCE, RASPBERRIES, STRAWBERRIES, TOMATOES [2]. Cane spot in RASPBERRIES [2]. Damping off in BROAD BEANS *(seed treatment)*, CABBAGES *(seed treatment)*, CARROTS *(seed treatment)*, CAULIFLOWERS *(seed treatment)*, DWARF BEANS *(seed treatment)*, FIELD BEANS *(seed treatment)*, GRASS SEED *(seed treatment)*, LEEKS *(seed treatment)*, LETTUCE *(seed treatment)*, MAIZE *(seed treatment)*, OILSEED RAPE *(seed treatment)*, ONIONS *(seed treatment)*, PEAS *(seed treatment)*, RADISHES *(seed treatment)*, RUNNER BEANS *(seed treatment)*, SOYA BEANS *(off-label seed treatment)*, TURNIPS *(seed treatment)*[1]. Downy mildew in PROTECTED LETTUCE [2]. Fire in TULIPS [2]. Gloeosporium in APPLES, PEARS [2]. Rust in BLACKCURRANTS, CARNATIONS, CHRYSANTHEMUMS [2]. Scab in APPLES, PEARS [2]. Seed-borne diseases in CARROTS *(seed soak)*, CELERY *(seed soak)*, FODDER BEET *(seed soak)*, MANGELS *(seed soak)*, PARSLEY *(seed soak)*, RED BEET *(seed soak)*, SUGAR BEET *(seed soak)* [1]. Spur blight in RASPBERRIES [2].

Notes

Efficacy

- Spray before onset of disease and repeat every 7-14 d. Spray interval varies with crop and disease. See label for details [2]
- Apply as seed treatment for protection against damping off [1]
- Do not spray when rain imminent [2]
- For use on tulips, chrysanthemums and carnations add non-ionic wetter [2]

Crop Safety/Restrictions

- Maximum number of treatments 3 per crop for protected winter lettuce (thiram based products only [2]; 2 per crop if sequence of thiram and other EBDC fungicides used) [2]; 2 per crop for protected summer lettuce [2]; 1 per batch of seed for seed treatments [1]
- Do not apply to hydrangeas [2]
- Notify processor before dusting or spraying crops for processing [2]

- Do not dip roots of forestry transplants [2]
- Do not treat seed of tomatoes, peppers or aubergines [1]

Special precautions/Environmental safety
- Harmful in contact with skin and if swallowed [1]
- Irritating to eyes, skin and respiratory system
- Dangerous to fish or other aquatic life. Do not contaminate surface waters or ditches with chemical or used container
- Do not use treated seed as food or feed [1]
- Treated seed harmful to game and wildlife [1]

Protective clothing/Label precautions
- A [1, 2]
- M03, R03a, R03c, U02, E03, E15, E31c, E34, S01, S02, S03, S04b, S05, S07 [1]; R04a, R04b, R04c, U05a, U08, U20b, C03, E01, E30a [1, 2]; U19, E13b, E32a [2]

Latest application/Harvest Interval(HI)
- 21 d after planting out or 21 d before harvest, whichever is earlier, for protected winter lettuce; 14 d after planting out or 21 d before harvest, whichever is earlier, for protected summer lettuce.
- HI protected lettuce 21 d; outdoor lettuce 14 d; apples, pears, blackcurrants, raspberries, strawberries, tomatoes 7 d

Approval
- Off-label approval to Mar 2002 for seed treatment of soya beans (OLA 0372/97)[1]

516 tolclofos-methyl

A protectant organophosphorus fungicide for soil-borne diseases

Products

1 Basilex	Scotts	50% w/w	WP	07494
2 Rizolex	AgrEvo	10% w/w	DS	07271
3 Rizolex Flowable	AgrEvo	500 g/l	FS	07273

Uses

Black scurf and stem canker in POTATOES [2, 3]. Bottom rot in LETTUCE, PROTECTED LETTUCE [1]. Damping off and wirestem in LEAF BRASSICAS [1]. Damping off in SEEDLINGS OF ORNAMENTALS [1]. Foot rot in ORNAMENTALS, SEEDLINGS OF ORNAMENTALS [1]. Rhizoctonia in SEED POTATOES *(off-label)* [3]. Rhizoctonia in PROTECTED CELERY *(off-label)*, PROTECTED RADISHES *(off-label)*[1]. Root rot in ORNAMENTALS, SEEDLINGS OF ORNAMENTALS [1].

Notes

Efficacy
- Dust seed potatoes during planting with automatic potato planter (see label for details of suitable applicator) or apply normally during hopper loading [2]
- Apply flowable formulation to clean tubers with suitable misting equipment over a roller table. Spray as potatoes taken into store (first earlies) or as taken out of store (second earlies, maincrop, crops for seed) pre-chitting [3]
- Do not mix flowable formulation with any other product [3]

- To control rhizoctonia in vegetables and ornamentals apply as drench before sowing, pricking out or planting or incorporate into compost [1]
- On established seedlings and pot plants apply as drench and rinse off foliage [1]

Crop Safety/Restrictions

- Maximum number of treatments 1 per batch for seed potatoes; 1 per crop for pre-transplanted lettuce and brassicas; 1 at each stage of growth (ie sowing, pricking out, potting) to a maximum of 3, for ornamentals
- Only to be used with automatic planters [2]
- Not recommended for use on seed potatoes where hot water treatment used or to be used [3]
- Do not apply as overhead drench to vegetables or ornamentals when hot and sunny [1]
- Do not use on heathers [1]

Special precautions/Environmental safety

- Tolclofos-methyl is an atypical organophosphorus compound which has weak anticholinesterase activity. Do not use if under medical advice not to work with such compounds
- Irritating to eyes, skin and respiratory system [2, 3]
- Dangerous (harmful [1]) to fish or other aquatic life. Do not contaminate surface waters or ditches with chemical or used container [2, 3]
- Do not allow direct spray from vehicle mounted/drawn hydraulic sprayers to fall within 6 m, or from hand-held sprayers to within 2 m, of surface waters or ditches. Direct spray away from water [1]
- Treated tubers to be used as seed only, not for food or feed [2, 3]

Protective clothing/Label precautions

- A [1-3]; C [1, 3]; D, E [2, 3]; F [1]; H [1-3]; J [2]; M [1, 2]
- M01, E16 [1]; R04a, R04b, R04c, U05a, C03, E01 [2, 3]; U19, E30a [1-3]; U20a, E13b, E31a [3]; U20b, E13c, E32a [1, 2]

Latest application/Harvest Interval(HI)

- At planting of seed potatoes [2, 3]; before transplanting for lettuce and brassicas [1]

Approval

- May be applied by misting equipment mounted over roller table. See label for details
- Off-label approval unlimited for use on protected celery (OLA 0209/97)[1]; unlimited for use on protected radishes pre-emergence (OLA 0767/97)[1]; to Jan 2001 for use on protected radishes post-emergence (OLA 0769/97)[1]

---

517 tralkoxydim

A foliar applied oxime herbicide for grass weed control in cereals.

Products

1 Grasp	Zeneca	250 g/l	SC	06675
2 Landgold Tralkoxydim	Landgold	250 g/l	SC	08604
3 Standon Tralkoxydim	Standon	250 g/l	SC	08326

Uses Blackgrass in DURUM WHEAT, SPRING BARLEY, SPRING WHEAT, TRITICALE, WINTER BARLEY, WINTER RYE, WINTER WHEAT. Wild oats in DURUM WHEAT, SPRING BARLEY, SPRING WHEAT, TRITICALE, WINTER BARLEY, WINTER RYE, WINTER WHEAT.

Notes

Efficacy

- Product leaf-absorbed and translocated rapidly to growing points. Best results achieved when weeds growing actively in competitive crops under warm humid conditions with adequate soil moisture
- Activity not dependent on soil type. Weeds germinating after application will not be controlled
- Best control of wild oats obtained from 2 leaf to 1st node detectable stage of weeds, and of blackgrass up to 3 tillers
- Authorised adjuvant must always be added. See label

Crop Safety/Restrictions

- Maximum number of treatments 1 per crop
- Apply to winter cereals 2 leaves unfolded up to and including flag leaf ligule just visible (GS 12-39). If necessary winter cereals may be sprayed twice: once in autumn and once in spring
- Apply to spring cereals from end of tillering up to and including flag leaf ligule just visible (GS 29-39)
- Do not spray undersown crops or crops to be undersown
- Do not spray when foliage wet or covered in ice or crop otherwise under stress
- Do not spray if a protracted period of cold weather forecast
- Do not spray crops under stress from chemical treatment, grazing, pest attack, mineral deficiency or low fertility
- Do not roll or harrow within 1 wk of spraying
- Do not tank-mix with phenoxy hormone or sulfonylurea herbicides

Special precautions/Environmental safety

- Irritating to eyes
- May cause sensitization by skin contact
- Do not contaminate surface waters or ditches with chemical or used container

Protective clothing/Label precautions

- A, C, H
- M03, E34 [1, 3]; R04a, R04e, U02, U04a, U05a, U08, U20b, C03, E01, E15, E26, E30a, E31b [1-3]

Latest application/Harvest Interval(HI)

- Before booting (GS 41)

518 triadimefon

A systemic conazole fungicide with curative and protectant action

Products

1 Bayleton	Bayer	25% w/w	WP	00221
2 Standon Triadimefon	Standon	25% w/w	WP	05673

Uses Brown rust in BARLEY, WHEAT [1, 2]. Crown rust in GRASSLAND [1]. Powdery mildew in BARLEY, OATS, RYE, WHEAT [1, 2]. Powdery mildew in APPLES, BARLEY *(off-label (research/breeding))*, BLACKBERRIES *(off-label)*, BLACKCURRANTS *(off-label)*, BRUSSELS SPROUTS, BRUSSELS SPROUTS *(off-label (research/breeding))*, CABBAGES,

FODDER BEET, GOOSEBERRIES *(off-label)*, GRAPEVINES *(off-label)*, GRASSLAND, HOPS, LOGANBERRIES *(off-label)*, PARSNIPS, PEAS *(off-label (research/breeding))*, RASPBERRIES *(off-label)*, STRAWBERRIES *(off-label)*, SUGAR BEET, SWEDES, TURNIPS, WHEAT *(off-label (research/breeding))* [1]. Rhynchosporium in GRASSLAND [1]. Rhynchosporium in BARLEY [2]. Rust in BARLEY *(off-label (research/breeding))*, BRUSSELS SPROUTS *(off-label (research/breeding))*, LEEKS, PEAS *(off-label (research/breeding))*, WHEAT *(off-label (research/breeding))* [1]. Snow rot in BARLEY [1]. Yellow rust in BARLEY, WHEAT [1, 2].

Notes

Efficacy

- Apply at first sign of disease and repeat as necessary, applications to established infections are less effective. Spray programme varies with crop and disease. See label
- Applications to control mildew will also reduce crown rust in oats, snow rot in winter barley and rust in beet [1]

Crop Safety/Restrictions

- Maximum number of treatments 2 per crop for wheat, oats, rye, spring barley, grassland, sugar beet, fodder beet, turnips, swedes, parsnips; 3 per crop for winter barley (1 in autumn), Brussels sprouts, cabbages, cane fruit and herbs; 4 per crop for leeks; 6 per yr for grapevines; 8 per yr for hops; 12 per yr for apples
- Continued use of fungicides from the same group in cereals may lead to reduced effectiveness against mildew

Special precautions/Environmental safety

- Harmful to fish or other aquatic life. Do not contaminate surface waters or ditches with chemical or used container

Protective clothing/Label precautions

- U20a, C02, E06b, E13c, E30a, E32a

Withholding period

- Keep all livestock out of treated areas for 21 d [1]

Latest application/Harvest Interval(HI)

- Before grain milky ripe (GS 71) for cereals.
- HI sugar beet, fodder beet, Brussels sprouts, cabbages, turnips, swedes, parsnips, leeks, apples, hops, cane fruit, strawberries, blackcurrants, gooseberries 14 d; grassland, herbs 21 d; grapes 6 wk

Approval

- Approved for aerial application on sugar beet, brassicas, swedes, turnips [1]; cereals [1, 2]. Scc notes in Section 1
- Off-label approval unlimited for use on blackcurrants, gooseberries, strawberries, outdoor grapes, raspberries, loganberries, blackberries, other Rubus hybrids (OLA 0024/95)[1]; unlimited for use on protected peas, wheat, barley, Brussels sprouts (for research/breeding) (OLA 0158/97)[1]

19 triadimenol

A systemic conazole fungicide for cereals, beet and brassicas

See also fuberidazole + triadimenol
tebuconazole + triadimenol

oducts	Bayfidan	Bayer	250 g/l	EC	02672

Uses Alternaria in BRUSSELS SPROUTS, CABBAGES. Fungus diseases in CARROTS *(off-label)*, HORSERADISH *(off-label)*, PARSLEY *(off-label)*, PARSNIPS *(off-label)*, SALSIFY *(off-label)*. Light leaf spot in BRUSSELS SPROUTS, CABBAGES. Powdery mildew in BARLEY, BRUSSELS SPROUTS, CABBAGES, FODDER BEET, OATS, RYE, SUGAR BEET, SWEDES, TURNIPS, WHEAT. Rhynchosporium in BARLEY, OATS, RYE, WHEAT. Ring spot in BRUSSELS SPROUTS, CABBAGES. Rust in BARLEY, OATS, RYE, WHEAT. Septoria in WHEAT. Snow rot in BARLEY, OATS, RYE, WHEAT.

Notes

Efficacy

- Apply sprays at first signs of disease and ensure good cover
- When applied for mildew and rust control in wheat product also gives good reduction of *Septoria tritici* but for specific protection against *S. tritici* and *S. nodorum* use tank-mix with chlorothalonil
- Treatment also reduces crown rust in oats and rust in beet
- Continued use of fungicides from same group can result in reduced effectiveness against powdery mildew. See label for recommended tank mixtures

Crop Safety/Restrictions

- Maximum number of treatments 2 per crop for spring wheat, oats, spring barley, rye, sugar beet, fodder beet, turnips, swedes, carrots, parsnips; 3 per crop (only 2 in spring) for winter barley, winter wheat, cabbages, Brussels sprouts

Special precautions/Environmental safety

- Harmful if swallowed. Irritating to eyes
- Harmful to fish or other aquatic life. Do not contaminate surface waters or ditches with chemical or used container

Protective clothing/Label precautions

- A, C
- M03, R03c, R04a, U05a, U08, U10, U11, U13, U20a, C03, E01, E13c, E30a, E31b, E34

Latest application/Harvest Interval(HI)

- Before grain milky ripe (GS 71) for cereals.
- HI beet crops, brassicas 14 d; carrots, parsnips 21 d

Approval

- Off-label approval unlimited for use on outdoor crops of carrots, parsnips, parsley root, salsify, horseradish (OLA0836/95)[1]

520 triadimenol + tridemorph

A systemic fungicide mixture with protectant and curative action

Products

Dorin	Bayer	125:375 g/l	EC	0836

Uses Brown rust in SPRING BARLEY, SPRING WHEAT, WINTER BARLEY, WINTER WHEA Crown rust in SPRING OATS, WINTER OATS. Powdery mildew in SPRING BARLE SPRING OATS, SPRING WHEAT, WINTER BARLEY, WINTER OATS, WINTER WHEA Rhynchosporium in SPRING BARLEY, WINTER BARLEY. Rust in SPRING WHEA WINTER WHEAT. Septoria leaf spot in SPRING WHEAT, WINTER WHEAT. Snow rot

WINTER BARLEY. Yellow rust in SPRING BARLEY, SPRING WHEAT, WINTER BARLEY, WINTER WHEAT.

Notes

Efficacy

- Apply at first signs of disease or as 2 or 3-spray protectant programme
- When applied for mildew and rust control in wheat product also gives good reduction of *Septoria tritici* but for specific protection against *S. tritici* and *S. nodorum* use tank-mix with chlorothalonil
- Treatment for mildew control will reduce crown rust infection on oats

Crop Safety/Restrictions

- Maximum number of treatments 3 per crop
- Crop scorch may occur if sprayed in frosty weather
- Scorch may occur on wheat if applied under very warm or drought conditions

Special precautions/Environmental safety

- Irritating to skin. Risk of serious damage to eyes
- Dangerous to fish or other aquatic life. Do not contaminate surface waters or ditches with chemical or used container

Protective clothing/Label precautions

- A, C
- M03, R03c, R04b, R04d, U04a, U05a, U13, U20b, C03, E01, E07, E13b, E30a, E31b, E34

Withholding period

- Grazing livestock must be kept out of treated areas for at least 14 d

Latest application/Harvest Interval(HI)

- Before grain milky ripe (GS 71)

521 tri-allate

A soil-acting thiocarbamate herbicide for grass weed control

Products

1 Avadex BW Granular	Monsanto	10% w/w	GR	00174
2 Avadex Excel 15G	Monsanto	15% w/w	GR	07117
3 Landgold Triallate 480	Landgold	480 g/l	EC	08505

Uses

Annual meadow grass in BEET CROPS, DURUM WHEAT, FIELD BEANS, FORAGE LEGUMES, PEAS, SPRING BARLEY, TRITICALE, WINTER BARLEY, WINTER RYE, WINTER WHEAT [1]. Blackgrass in DURUM WHEAT, TRITICALE, WINTER RYE, WINTER WHEAT [1-3]. Blackgrass in FIELD BEANS, FORAGE LEGUMES, PEAS [1, 2]. Blackgrass in SPRING BARLEY, WINTER BARLEY [1, 3]. Blackgrass in BEET CROPS [1]. Blackgrass in FODDER BEET, MANGELS, SUGAR BEET [2, 3]. Blackgrass in BARLEY, RED BEET [2]. Blackgrass in SPRING RYE [3]. Meadow grasses in DURUM WHEAT, FODDER BEET, MANGELS, SUGAR BEET, TRITICALE, WINTER RYE, WINTER WHEAT [2, 3]. Meadow grasses in BARLEY, FIELD BEANS, FORAGE LEGUMES, PEAS, RED BEET [2]. Meadow grasses in SPRING BARLEY, SPRING RYE, WINTER BARLEY [3]. Wild oats in DURUM WHEAT, TRITICALE, WINTER RYE, WINTER WHEAT [1-3]. Wild oats in FIELD BEANS, FORAGE LEGUMES, PEAS [1, 2]. Wild oats in SPRING BARLEY, WINTER BARLEY [1, 3]. Wild oats in BEET CROPS [1]. Wild oats in FODDER BEET, MANGELS, SUGAR BEET [2, 3]. Wild oats in BARLEY, RED BEET [2]. Wild oats in SPRING RYE [3].

Notes **Efficacy**

- Incorporate or apply to surface pre-emergence (post-emergence application possible in winter cereals up to 2-leaf stage of wild oats) [1, 2]
- Do not use on soils with more than 10% organic matter
- Wild oats controlled up to 2-leaf stage [1, 2]
- If applied to dry soil, rainfall needed for full effectiveness, especially with granules on surface. Do not use if top 5-8 cm bone dry
- Do not apply with spinning disc granule applicator; see label for suitable types [1, 2]
- Do not apply to cloddy seedbeds
- Use sequential treatments to improve control of barren brome and annual dicotyledons (see label for details)
- Clean start technique recommended to maximize control of sterile brome and blackgrass in winter cereals involves use of glyphosate to kill emerged weeds in stubble followed by Avadex BW Granular immediately after drilling

Crop Safety/Restrictions

- Apply Avadex BW Granular to spring wheat post-emergence only to crops drilled in winter
- Consolidate loose, puffy seedbeds before drilling to avoid chemical contact with seed
- Drill cereals well below treated layer of soil (see label for safe drilling depths)
- Do not use on direct-drilled crops or undersow grasses into treated crops
- Do not sow oats or grasses within 1 yr of treatment

Special precautions/Environmental safety

- Irritating to skin [3] and eyes [1, 2]
- Harmful to fish or other aquatic life. Do not contaminate surface waters or ditches with chemical or used container

Protective clothing/Label precautions

- A [1-3]; C, H [3]
- R04a, E32a [1, 2]; R04b, U02, U05a, U20a, C03, E01, E13c, E30a [1-3]; U04a, U09b, U19, E26, E31b [3]

Latest application/Harvest Interval(HI)

- Pre-drilling for beet crops; before crop emergence for field beans, spring barley, peas, forage legumes; before first node detectable stage (GS 31) for winter wheat, winter barley, durum wheat, triticale, winter rye

Approval

- Approved for aerial application on wheat, barley, winter beans, peas [1, 2]. See notes in Section 1

522 triasulfuron

A sulfonylurea herbicide for annual broad-leaved weed control in cereals

See also bromoxynil + ioxynil + triasulfuron

Products					
	1 Lo-Gran 20 WG	Ciba Agric.	20% w/w	WG	05993
	2 Lo-Gran 20 WG	Novartis	20% w/w	WG	08421

Uses Annual dicotyledons in BARLEY, DURUM WHEAT, RYE, TRITICALE, WHEAT, WINTER OATS. Charlock in BARLEY, DURUM WHEAT, RYE, TRITICALE, WHEAT, WINTER OATS. Chickweed in BARLEY, DURUM WHEAT, RYE, TRITICALE, WHEAT. Mayweeds in BARLEY, DURUM WHEAT, RYE, TRITICALE, WHEAT, WINTER OATS.

Notes **Efficacy**

- Best results achieved on small, actively growing weeds up to 6 leaf or 50 mm growth stage (up to flower bud for charlock, chickweed and mayweed)
- Some species, although not controlled remain stunted and uncompetitive with crop

Crop Safety/Restrictions

- Maximum number of treatments 1 per crop
- Product must be applied only after 1 Feb from 1 leaf stage of crop to before 3rd node detectable (GS 11-33)
- Do not apply to undersown crops or those due to be undersown
- Do not spray in windy weather. Avoid drift onto neighbouring crops
- Do not spray during period of frosty weather, when frost imminent or onto crops under stress from frost, waterlogging or drought
- Do not spray in tank mixture, or in sequence, with a product containing any other sulfonylurea
- Ensure spraying equipment is washed thoroughly according to specific instructions. Do not allow washings to drain onto land intended for cropping or growing crops
- See label for restrictions on succeeding crops

Special precautions/Environmental safety

- Extremely dangerous to aquatic higher plants. Do not contaminate surface waters or ditches with chemical or used container
- Do not allow direct spray from ground-based sprayer vehicles to fall within 6 m, or from hand-held sprayers to within 2 m, of surface waters or ditches. Direct spray away from water

Protective clothing/Label precautions

- A, H
- U20a, E16, E30a, E32a

Latest application/Harvest Interval(HI)

- Before 3rd node detectable (GS 33)

---

523 triazamate

A carbamoyl triazole insecticide for sugar beet

Products

Aztec	Cyanamid	140 g/l	EW	07817

Uses Aphids in SUGAR BEET *(including Myzus persicae)*.

Notes **Efficacy**

- Treat as soon as aphids seen in crop or immediately after official warnings are issued. Repeat as necessary
- Controls aphids resistant to other chemical groups
- To reduce risk of development of resistance consider use of products with alternative modes of action in intensive pest control programmes

Crop Safety/Restrictions
- Maximum number of treatments 3 per crop
- Label warning that damage may result unless all recommendations carefully followed
- Product must be used with Swirl adjuvant

Special precautions/Environmental safety
- Harmful by inhalation and if swallowed
- Extremely dangerous to fish or other aquatic life. Do not contaminate surface waters or ditches with chemical or used container
- Do not allow direct spray from vehicle mounted/drawn hydraulic sprayers to fall within 6 m, or from hand-held sprayers to within 2 m, of surface waters or ditches. Direct spray away from water
- Product not harmful to bees when used as directed but spraying in late evening/early morning or in dull weather recommended to avoid unnecessary stress on foraging bees

Protective clothing/Label precautions
- A, H
- M02, M03, R03b, R03c, U05a, U19, C03, E01, E13a, E16, E26, E34,

Latest application/Harvest Interval(HI)
- HI 28 d

524 triazoxide

A benzotriazine fungicide available only in mixtures

See also imidacloprid + tebuconazole + triazoxide
tebuconazole + triazoxide

525 tribenuron-methyl

A foliar acting sulfonylurea herbicide with some root activity for use in cereals

See also thifensulfuron-methyl + tribenuron-methyl

Products	Quantum	DuPont	50% w/w	TB	06270

Uses

Annual dicotyledons in DURUM WHEAT, OATS, SPRING BARLEY, SPRING WHEAT, TRITICALE, WINTER BARLEY, WINTER RYE, WINTER WHEAT. Charlock in DURUM WHEAT, OATS, SPRING BARLEY, SPRING WHEAT, TRITICALE, WINTER BARLEY, WINTER RYE, WINTER WHEAT. Chickweed in DURUM WHEAT, OATS, SPRING BARLEY, SPRING WHEAT, TRITICALE, WINTER BARLEY, WINTER RYE, WINTER WHEAT. Mayweeds in DURUM WHEAT, OATS, SPRING BARLEY, SPRING WHEAT, TRITICALE, WINTER BARLEY, WINTER RYE, WINTER WHEAT.

Notes

Efficacy
- Best control achieved when weeds small and actively growing
- Good spray cover must be achieved since larger weeds often become less susceptible

- Susceptible weeds cease growth almost immediately after treatment and symptoms can be seen in about 2 wk
- Product can be used on all soil types
- Weed control may be reduced when conditions very dry
- See label for details of technique to be used for dissolving tablets in spray tank
- In tank mixing ensure tablets fully dispersed before adding other products

Crop Safety/Restrictions
- Maximum number of treatments 1 per crop
- Do not spray in tank mixture, or in sequence, with a product containing any other sulfonylurea
- Apply in autumn or in spring (after 1 Feb) from 3 leaf stage of crop up to and including flag leaf fully emerged (GS 13-39)
- Do not apply to crops undersown with grass, clover or other broad-leaved crops
- Do not apply to any crop suffering stress from any cause or not actively growing
- Do not apply within 7 d of rolling
- Special care must be taken to avoid damage by drift onto nearby broad-leaved crops, surface waters or ditches
- Special care needed in spray tank cleaning. See label for details
- In the event of crop failure sow only a cereal within 3 mth of application. See label for other restrictions on subsequent cropping

Special precautions/Environmental safety
- Irritant. May cause sensitization by skin contact
- Do not contaminate surface waters or ditches with chemical or used container

Protective clothing/Label precautions
- A
- R04e, U05a, U08, U20a, C03, E01, E15, E30a, E32a

Latest application/Harvest Interval(HI)
- Up to and including flag leaf ligule/collar just visible (GS 39)

---

526 trichlorfon

A contact and ingested organophosphorus insecticide

Products	Dipterex 80	Bayer	80% w/w	SP	00711

Uses Browntail moth in HEDGES. Cabbage root fly in BRUSSELS SPROUTS, MOOLI *(off-label)*, RADISHES *(off-label)*. Caterpillars in BRASSICAS, ROSES. Cherry bark tortrix in APPLES. Cutworms in HERBS, VEGETABLES. Flea beetles in FODDER BEET, MANGELS, RED BEET, SPINACH, SUGAR BEET. Flies in MANURE HEAPS, REFUSE TIPS *(off-label)*, RUBBISH DUMPS *(off-label)*. Leaf miners in BRASSICAS. Liriomyza huidobrensis in AUBERGINES *(off-label)*, BRASSICAS *(off-label)*, LETTUCE *(off-label)*, MARROWS *(off-label)*, ORNAMENTALS *(off-label)*, PEPPERS *(off-label)*, PROTECTED CUCUMBERS *(off-label)*, PROTECTED MELONS *(off-label)*, PROTECTED ORNAMENTALS *(off-label)*, RADICCHIO *(off-label)*, SORREL *(off-label)*, SPINACH BEET *(off-label)*, TOMATOES *(off-label)*. Mangold fly in FODDER BEET, MANGELS, RED BEET, SPINACH, SUGAR BEET. Small ermine moth in HEDGES. Strawberry tortrix in STRAWBERRIES.

Notes

Efficacy

- Apply at appearance of caterpillars or larvae for most uses and repeat at 7-10 d intervals if necessary. See label for full details
- To control cabbage root fly in Brussels sprouts buttons spray 1 mth before expected harvest and repeat twice at 7 d intervals using pendant lances to obtain good cover
- Addition of non-ionic wetter recommended for use on brassicas
- For cutworm control apply as drenching spray
- For cherry bark tortrix control spray trunks and main branches in mid-May, taking care to avoid leaves

Crop Safety/Restrictions

- Maximum number of treatments 1 per crop for protected spinach, spinach beet and sorrel; 3 per crop for tomatoes; 8 per crop for radishes and mooli

Special precautions/Environmental safety

- This product contains an anticholinesterase organophosphorus compound. Do not use if under medical advice not to work with such compounds
- Harmful to fish or other aquatic life. Do not contaminate surface waters or ditches with chemical or used container

Protective clothing/Label precautions

- M01, U08, U19, U20a, C02, E13c, E30a, E32a

Latest application/Harvest Interval(HI)

- HI 2 d

Approval

- Off-label approval unlimited for use on aubergines, ornamentals, peppers, sorrel, spinach beet, tomatoes (OLA 0400/91, 0497/91)[1]; unlimited for use on waste tips and rubbish dumps (OLA 0690/91)[1]; unlimited for use on protected ornamentals (OLA 1079/92)[1]; unlimited for use on outdoor brassicas, outdoor and protected marrows, outdoor and protected lettuce and radicchio, protected peppers, cucumbers and melons (OLA 1542/93)[1]; unlimited for use on outdoor radsihes and mooli (OLA 1130/95)[1]

Maximum Residue Levels (mg residue/kg food)

- cereals 0.1

527 triclopyr

An aryloxyalkanoic acid herbicide for perennial and woody weed control

See also clopyralid + fluroxypyr + triclopyr
clopyralid + triclopyr
fluroxypyr + triclopyr

Products

1 Chipman Garlon 4	Nomix-Chipman	480 g/l	EC	06016
2 Garlon 2	Zeneca	240 g/l	EC	06616
3 Garlon 4	Dow	480 g/l	EC	05090
4 Timbrel	Dow	480 g/l	EC	05815

Uses

Brambles in FORESTRY, LAND NOT INTENDED TO BEAR VEGETATION [1, 3, 4]. Brambles in ESTABLISHED GRASSLAND, NON-CROP AREAS [2]. Broom in FORESTRY

FOR FULL CONDITIONS OF USE ALWAYS READ THE PRODUCT LABEL

LAND NOT INTENDED TO BEAR VEGETATION [1, 3, 4]. Broom in ESTABLISHED GRASSLAND, NON-CROP AREAS [2]. Brush clearance in INDUSTRIAL SITES [1, 3, 4]. Docks in FORESTRY, LAND NOT INTENDED TO BEAR VEGETATION [1, 3, 4]. Docks in ESTABLISHED GRASSLAND, NON-CROP AREAS [2]. Gorse in FORESTRY, LAND NOT INTENDED TO BEAR VEGETATION [1, 3, 4]. Gorse in ESTABLISHED GRASSLAND, NON-CROP AREAS [2]. Hard rush in ESTABLISHED GRASSLAND, NON-CROP AREAS [2]. Perennial dicotyledons in FORESTRY, INDUSTRIAL SITES, LAND NOT INTENDED TO BEAR VEGETATION [1, 3, 4]. Perennial dicotyledons in ESTABLISHED GRASSLAND, NON-CROP AREAS [2]. Rhododendrons in FORESTRY, LAND NOT INTENDED TO BEAR VEGETATION [1, 3, 4]. Scrub clearance in INDUSTRIAL SITES [1, 3, 4]. Scrub clearance in NON-CROP AREAS [2]. Stinging nettle in FORESTRY, LAND NOT INTENDED TO BEAR VEGETATION [1, 3, 4]. Stinging nettle in ESTABLISHED GRASSLAND, NON-CROP AREAS [2]. Woody weeds in FORESTRY, INDUSTRIAL SITES, LAND NOT INTENDED TO BEAR VEGETATION [1, 3, 4]. Woody weeds in ESTABLISHED GRASSLAND, NON-CROP AREAS [2].

Notes

Efficacy

- Apply in grassland as spot treatment or overall foliage spray when weeds in active growth in spring or summer. Details of dose and timing vary with species. See label [2, 3]
- Apply to woody weeds as summer foliage, winter shoot, basal bark, cut stump or tree injection treatment [1, 3, 4]
- Apply foliage spray in water when leaves fully expanded but not senescent
- Apply winter shoot, basal bark or cut stump sprays in paraffin or diesel oil. Dose and timing vary with species. See label for details
- Inject undiluted or 1:1 dilution into cuts spaced every 7.5 cm round trunk [1, 3, 4]
- Do not spray in drought, in very hot or cold conditions
- Control may be reduced if rain falls within 2 h of application
- Control of rhododendron can be variable. If higher than 1.8 m cut stump treatment recommended. A follow-up shoot treatment may be required

Crop Safety/Restrictions

- Maximum number of treatments 1 per yr on non-crop land (as directed spray) [2]; 2 per yr on established grassland (including land not intended for cropping) and forestry
- See label for maximum concentrations when applying in oil, water or via watering can [2]
- Clover will be killed or severely checked by application in grassland. Do not apply to grass leys less than 1 yr old [2]
- Do not allow spray to drift onto agricultural or horticultural crops, amenity plantings, gardens, ponds, lakes or water courses. Vapour drift may occur under hot conditions
- Do not drill kale, swedes, turnips, grass or mixtures containing clover within 6 wk of treatment. Allow at least 6 wk before planting trees

Special precautions/Environmental safety

- Harmful in contact with skin and if swallowed [1, 3, 4]
- Irritating to skin. May cause sensitization by skin contact
- Irritating to eyes [2]
- Flammable [2]
- Not to be used on food crops
- Dangerous to fish or other aquatic life. Do not contaminate surface waters or ditches with chemical or used container
- Do not apply through hand held rotary atomisers
- Do not allow direct spray from vehicle mounted/drawn hydraulic sprayers to fall within 6 m, or from hand-held sprayers to within 2 m, of surface waters or ditches. Direct spray away from water [1, 3, 4]
- Not to be applied in or near water

- Do not apply from tractor-mounted sprayer within 250 m of susceptible crops, ponds, lakes or watercourses

Protective clothing/Label precautions
- A, C, H, M
- R03a, R03c, E16 [1, 3, 4]; R04a, R07d, U19 [2]; R04b, R04e, U02, U05a, U08, U20b, C01, C03, E01, E07 (7 d), E13b, E23, E30a, E31b, E34 [1-4]

Withholding period
- Keep livestock out of treated areas for at least 7 d and until foliage of any poisonous weeds such as buttercups or ragwort has died and become unpalatable

Latest application/Harvest Interval(HI)
- 6 wk before replanting; 7 d before grazing

528 tridemorph

A systemic, eradicant and protectant morpholine fungicide

See also carbendazim + maneb + tridemorph
cyproconazole + tridemorph
fenbuconazole + tridemorph
fenpropimorph + flusilazole + tridemorph
fenpropimorph + tridemorph
flusilazole + tridemorph
propiconazole + tridemorph
tebuconazole + tridemorph
triadimenol + tridemorph

Products

1 Calixin	BASF	750 g/l	EC	00369
2 Landgold Tridemorph 750	Landgold	750 g/l	EC	08522
3 Standon Tridemorph 750	Standon	750 g/l	EC	05667

Uses Powdery mildew in BARLEY, OATS, SWEDES, TURNIPS, WINTER WHEAT.

Notes **Efficacy**
- Apply to cereals when mildew starts to build up, normally May-early Jun and repeat once if necessary. Treatment may also be made in autumn if required
- Reduced rate recommended in tank mix with triazole fungicides for control of established mildew in barley
- Tank mix with carbendazim recommended for Rhynchosporium control in barley, with other products for eyespot and yellow rust, see label for details
- Apply to swedes and turnips at first signs of disease, normally Jul-Aug, and repeat at 2 wk intervals if required
- Product is rainfast after 2 h

Crop Safety/Restrictions
- Maximum total dose 2.1 l/ha but 1.4 l/ha max in spring/summer for winter barley; 1.4 l/ha for other recommended crops
- Permitted growth stages vary with crop and level of infection. See label for details

- Do not spray winter wheat at high temperatures or in drought
- Do not apply if frost expected
- Do not mix with other chemicals on swedes or turnips

Special precautions/Environmental safety

- Harmful if swallowed
- Irritating to eyes and skin
- Harmful to fish or other aquatic life. Do not contaminate surface waters or ditches with chemical or used container

Protective clothing/Label precautions

- A, C
- M03, R03c, R04a, R04b, U04a, U05a, C03, E01, E07 (14 d), E13c, E30a, E34 [1-3]; U08, U20b, E26, E31b [2]; U09a, U20a, E31c [1, 3]

Withholding period

- Keep all livestock out of treated areas for at least 14 d

Latest application/Harvest Interval(HI)

- HI swedes, turnips 2 wk; barley, oats 4 wk; winter wheat 6 wk

Approval

- Approved for aerial application on barley, oats, winter wheat, swedes, turnips [1, 3]. See notes in Section 1

529 trietazine

A triazine herbicide available only in mixtures

See also simazine + trietazine
terbutryn + trietazine

530 trifluralin

A soil-incorporated dinitroaniline herbicide for use in various crops

See also clodinafop-propargyl + trifluralin
diflufenican + trifluralin
isoproturon + trifluralin
isoxaben + trifluralin
linuron + trifluralin
terbutryn + trifluralin

Products

1 Alpha Trifluralin 48 EC	Makhteshim	480 g/l	EC	07406
2 Ashlade Trifluralin	Ashlade	480 g/l	EC	08303
3 Ashlade Trimaran	Ashlade	480 g/l	EC	06228
4 Atlas Trifluralin	Atlas	480 g/l	EC	08498
5 MSS Trifluralin 48 EC	Mirfield	480 g/l	EC	07753
6 MTM Trifluralin	MTM Agrochem.	480 g/l	EC	05313
7 Portman Trifluralin	Portman	480 g/l	EC	05751
8 Treflan	Dow	480 g/l	EC	05817
9 Triflur	Nufarm	480 g/l	EC	08311
10 Trigard	FCC	480 g/l	EC	02178
11 Tripart Trifluralin 48 EC	Tripart	480 g/l	EC	02215
12 Tristar	PBI	480 g/l	EC	02219

Uses

Annual dicotyledons in LINSEED *(off-label)* [12]. Annual dicotyledons in CELERIAC *(off-label)*, COMBINING PEAS *(off-label)*, EVENING PRIMROSE *(off-label)*, ORNAMENTALS *(off-label)*, SOYA BEANS *(off-label)*, SUNFLOWERS *(off-label)* [8, 12]. Annual dicotyledons in GOLD-OF-PLEASURE *(off-label)*, KOHLRABI *(off-label)*, NURSERY FRUIT TREES AND BUSHES *(off-label)*, RADISH SEED CROPS *(off-label)*, SOFT FRUIT *(off-label)* [8]. Annual dicotyledons in SPRING OILSEED RAPE, WINTER OILSEED RAPE [3, 12]. Annual dicotyledons in SUGAR BEET [3, 4, 6-12]. Annual dicotyledons in PARSLEY [3, 4, 8, 10-12]. Annual dicotyledons in SPRING LINSEED, WINTER LINSEED [3, 4, 8, 11]. Annual dicotyledons in MUSTARD [1, 3, 5-12]. Annual dicotyledons in FIELD BEANS [1, 5-7, 10]. Annual dicotyledons in LINSEED [1, 5, 9]. Annual dicotyledons in BROAD BEANS, BROCCOLI, BRUSSELS SPROUTS, CABBAGES, CARROTS, CAULIFLOWERS, FRENCH BEANS, KALE, LETTUCE, PARSNIPS, RASPBERRIES, RUNNER BEANS, STRAWBERRIES, SWEDES, TURNIPS [1-12]. Annual dicotyledons in WINTER BARLEY, WINTER WHEAT [1-9, 11, 12]. Annual dicotyledons in OILSEED RAPE [1, 2, 4-11]. Annual dicotyledons in CALABRESE [2-4, 6-8, 10-12]. Annual dicotyledons in NAVY BEANS [2-4, 8, 11, 12]. Annual dicotyledons in OUTDOOR HERBS *(off-label)* [4, 6, 8, 12]. Annual dicotyledons in PROTECTED HERBS *(off-label)* [4, 6]. Annual grasses in LINSEED *(off-label)* [12]. Annual grasses in CELERIAC *(off-label)*, COMBINING PEAS *(off-label)*, EVENING PRIMROSE *(off-label)*, ORNAMENTALS *(off-label)*, SOYA BEANS *(off-label)*, SUNFLOWERS *(off-label)* [8, 12]. Annual grasses in GOLD-OF-PLEASURE *(off-label)*, KOHLRABI *(off-label)*, NURSERY FRUIT TREES AND BUSHES *(off-label)*, RADISH SEED CROPS *(off-label)*, SOFT FRUIT *(off-label)* [8]. Annual grasses in SPRING OILSEED RAPE, WINTER OILSEED RAPE [3, 12]. Annual grasses in SUGAR BEET [3, 4, 6-12]. Annual grasses in PARSLEY [3, 4, 8, 10-12]. Annual grasses in SPRING LINSEED, WINTER LINSEED [3, 4, 8, 11]. Annual grasses in MUSTARD [1, 3, 5-12]. Annual grasses in FIELD BEANS [1, 5-7, 10]. Annual grasses in LINSEED [1, 5, 9]. Annual grasses in BROAD BEANS, BROCCOLI, BRUSSELS SPROUTS, CABBAGES, CARROTS, CAULIFLOWERS, FRENCH BEANS, KALE, LETTUCE, PARSNIPS, RASPBERRIES, RUNNER BEANS, STRAWBERRIES, SWEDES, TURNIPS [1-12]. Annual grasses in WINTER BARLEY, WINTER WHEAT [1-9, 11, 12]. Annual grasses in OILSEED RAPE [1, 2, 4-11]. Annual grasses in CALABRESE [2-4, 6-8, 10-12]. Annual grasses in NAVY BEANS [2-4, 8, 11, 12]. Annual grasses in OUTDOOR HERBS *(off-label)* [4, 6, 8, 12] Annual grasses in PROTECTED HERBS *(off-label)* [4, 6].

Notes

Efficacy

- Acts on germinating weeds and requires soil incorporation to 5 cm (10 cm for crops to be grown on ridges) within 30 min of spraying (2 h [11], except in mixtures on cereals). See label for details of suitable application equipment
- Best results achieved by application to fine, firm seedbed, free of clods, crop residues and established weeds
- Do not use on sand, fen soil or soils with more than 10% organic matter
- In winter cereals normally applied as surface treatment without incorporation in tank-mixture with other herbicides to increase spectrum of control. See label for details
- Follow-up herbicide treatment recommended with some crops. See label for details

Crop Safety/Restrictions

- Maximum number of treatments 1 per crop
- Apply and incorporate at any time during 2 wk before sowing or planting
- Do not apply to brassica plant raising beds
- Transplants should be hardened off prior to transplanting
- Apply in sugar beet after plants 10 cm high with 4-8 leaves and harrow into soil
- Apply in cereals after drilling up to and including 3 leaf stage (GS 13)

- Minimum interval between application and drilling or planting may be up to 12 mth. See label for details

Special precautions/Environmental safety
- Irritating to eyes and skin [1, 2, 4-7, 11]
- Irritating to respiratory system [2, 4, 7]
- Flammable
- Harmful to fish or other aquatic life. Do not contaminate surface waters or ditches with chemical or used container

Protective clothing/Label precautions
- A [1, 2, 4-7, 9]; C [1, 2, 4, 5, 7, 9]
- R04a [1, 2, 4-7, 11]; R04b [1, 2, 4-7, 9, 11]; R04c [2, 4, 7]; R04d, U11, E34 [9]; R07d [1-8, 10-12]; U05a, C03 [1, 2, 4-7, 9-12]; U08, U13, E31b [1-12]; U20a [3-6, 10-12]; U20b [1, 2, 7-9]; E01, E30a [1-7, 9-12]; E13c [1-6, 8-12]; E26 [1, 2, 5-8, 10-12]; E27 [1, 5, 8, 9]; E28 E30a [8]; E29 [1, 5]

Latest application/Harvest Interval(HI)
- Before 4 leaves unfolded (GS 13) for cereals; up to 6 leaf stage for sugar beet [5, 8], (up to 10 leaf stage [3]); pre-sowing/planting for other crops
- Pre-sowing/pre-planting for all crops [1]

Approval
- Off-label approval unlimited for use on outdoor herbs (OLA 0737)[4]; unlimited for use on combining peas, sunflower, linseed, celeriac, ornamentals (OLA 1564/98)[12], evening primrose (OLA 1276/92)[12], outdoor herbs, soya bean (OLA 0077/93)[12]; unlimited for use on evening primrose (OLA 1080/92)[8], gold-of-pleasure, radish seed crops, sunflowers, soya beans, combining peas, nursery fruit trees and bushes, ornamentals, outdoor herbs (OLA 0074/93)[8]; to Feb 2000 for use on outdoor kohlrabi (OLA 0447/96)[8]; unlimited for use on combining peas, sunflower, linseed, celeriac, ornamentals (OLA 1564/98)[8],

531 triflusulfuron-methyl

A sulfonyl urea herbicide for sugar beet

Products

1 Debut	DuPont	50% w/w	WG	07804
2 Landgold TFS 50	Landgold	50% w/w	WG	08941

Uses

Annual dicotyledons in SUGAR BEET [1, 2]. Annual dicotyledons in FODDER BEET [1].

Notes

Efficacy
- Product should be used with a recommended adjuvant or a suitable herbicide tank-mix partner - see label for details
- Product acts by foliar action. Best results obtained from good spray cover of small actively growing weeds
- Susceptible weeds cease growth immediately and symptoms can be seen 5-10 d later
- Best results achieved from a programme of up to 4 treatments starting when first weeds have emerged with subsequent applications every 5-14 d when new weed flushes at cotyledon stage
- Weed spectrum can be broadened by tank mixture with other herbicides. See label for details
- Product may be applied overall or via band sprayer

Crop Safety/Restrictions
- Maximum number of treatments 4 per crop
- All varieties of sugar beet (and fodder beet [1]) may be treated from early cotyledon stage until the leaves begin to meet between the rows
- Do not apply to any crop stressed by drought, water-logging, low temperatures, pest or disease attack, nutrient or lime deficiency
- Only winter cereals should follow a treated crop in the same calendar yr. Any crop may be sown in the next calendar yr
- After failure of a treated crop, sow only spring barley, linseed or sugar beet within 4 mth of spraying unless prohibited by tank-mix partner

Special precautions/Environmental safety
- Irritant. May cause sensitisation by skin contact
- Extremely dangerous to fish or other aquatic life. Do not contaminate surface waters or ditches with chemical or used container
- Do not allow direct spray from ground-based vehicle mounted/drawn sprayers to fall within 6 m of surface waters or ditches
- Herbicide is very active. Take particular care to avoid drift onto plants outside the target area
- Immediate and thorough cleaning of spray equipment after use is vital to prevent subsequent damage to susceptible crops. Do not drain or flush sprayers on land to be used for crops other than sugar beet

Protective clothing/Label precautions
- A
- R04, R04e, U08, U19, U20a, E13a, E16, E26, E30a, E32a

Latest application/Harvest Interval(HI)
- Before crop leaves meet between rows

532 triforine

A locally systemic fungicide with protectant and curative activity

See also bupirimate + triforine

Products					
	Fairy Ring Destroyer	Vitax	190 g/l	EC	05541

Uses Fairy rings in TURF.

Notes

Efficacy
- For fairy ring control apply as high volume spray or drench as soon as infection noted and repeat twice at 14 d intervals

Crop Safety/Restrictions
- Maximum number of treatments 3 per yr for turf

Special precautions/Environmental safety
- Harmful in contact with skin, irritating to eyes

Protective clothing/Label precautions

- A, C
- M03, R03a, R04a, U02, U05a, U08, U20b, C03, E01, E13c, E26, E30a, E31a, E34

Maximum Residue Levels (mg residue/kg food)

- hops 30; pome fruit, cherries, currants, gooseberries 2; plums 1; cucumbers, gherkins, courgettes 0.5; tea, wheat, rye, barley, oats, triticale 0.1; citrus fruit, tree nuts, cane fruit, bilberries, cranberries, wild berries, miscellaneous fruit, root and tuber vegetables, sweet corn, lettuce, spinach beet, watercress, witloof, chervil, chives, celery leaves, cardoons, fennel, rhubarb, fungi, pulses, oilseeds, potatoes, sorghum, maize, buckwheat, millet, rice, meat, milk and dairy produce, eggs 0.05

533 trinexapac-ethyl

A novel cyclohexanecarboxylate plant growth regulator for cereals

Products

1 Moddus	Ciba Agric.	250 g/l	SL	07830
2 Moddus	Novartis	250 g/l	EC	08801
3 Shortcut	Scotts	25% w/w	WB	09254

Uses

Growth retardation in AMENITY GRASS, AMENITY TURF [3]. Lodging control in SPRING BARLEY, WINTER BARLEY, WINTER WHEAT [1, 2]. Lodging control in DURUM WHEAT, RYE, TRITICALE, WINTER OATS [2].

Notes

Efficacy

- On wheat apply as single treatment between leaf sheath erect stage (GS 30) and flag leaf fully emerged (GS 39)
- On barley apply as single treatment between leaf sheath erect stage (GS 30) and second node detectable (GS 32), or at higher dose between flag leaf just visible (GS 37) and flag leaf fully emerged (GS 39)
- Best results on turf achieved from application to actively growing weed free turf grass that is adequately fertilised and watered and is not under stress [3]
- Product rainfast after 12 h [3]

Crop Safety/Restrictions

- Maximum total dose equivalent to one full dose on cereals [1, 2]
- Maximum number of treatments on turf 5 per yr [3]
- Do not apply if rain or frost expected or if crop wet
- Only use on crops at risk of lodging [1, 2]
- Treatment may cause ears to remain erect through to harvest [1, 2]
- Turf under stress when treated may show signs of damage. Do not use on newly mown turf [3]
- Not recommended for closely mown fine turf [3]

Special precautions/Environmental safety

- Irritating to eyes and skin [1, 2]
- Flammable [1, 2]
- Dangerous (Harmful [3]) to fish or other aquatic life. Do not contaminate surface waters or ditches with chemical or used container [1, 2]
- Avoid drift outside target area
- Do not compost or mulch clippings [3]

Protective clothing/Label precautions
- A [1-3]; C [1, 2]
- R04a, R04b, R07d, U11, E12e, E13b [1]; R04e, U20c [2]; U05a, U15, C03, E01 [1, 2]; U20b [1, 3]; C01 [3]; E13c [2, 3]; E26, E30a, E31a [1-3]

Latest application/Harvest Interval(HI)
- Before flag leaf sheath extending stage (GS 41)

534 Verticillium lecanii

A fungal parasite of aphids and whitefly

Products

1 Mycotal	Koppert	16.1% w/w	WP	04782
2 Vertalec	Koppert	20% w/w	WP	04781

Uses

Aphids in AUBERGINES, CUTTINGS, PROTECTED BEANS, PROTECTED CHRYSANTHEMUMS, PROTECTED CUCUMBERS, PROTECTED PEPPERS, PROTECTED ROSES, PROTECTED TOMATOES [2]. Whitefly in AUBERGINES, CUCUMBERS, GLASSHOUSE CUT FLOWERS, PEPPERS, PROTECTED BEANS, PROTECTED LETTUCE, PROTECTED ORNAMENTALS, PROTECTED TOMATOES [1].

Notes

Efficacy
- Apply spore powder as spray as part of biological control programme
- Spray during late afternoon and early evening directing spray onto underside of leaves and to growing points
- Best results require minimum 80% relative humidity and 18°C within the crop canopy

Crop Safety/Restrictions
- Maximum number of treatments 3 per crop [1]; 2 per crop at high dose for chrysanthemums, 1 per crop for other crops, 12 per crop at low dose, low volume [2]
- Never use in tank mixture
- A fungicide may not be used within 3 d of treatment. Pesticides containing captan, chlorothalonil, dichlofluanid, triforine, maneb, thiram or tolylfluanid may not be used on the same crop
- Keep in a refrigerated store at 2-6°C

Special precautions/Environmental safety
- Product does not affect natural predators or parasites
- Do not contaminate surface waters or ditches with product or used container

Protective clothing/Label precautions
- U20b, E15, E26, E29, E30a, E32a

535 vinclozolin

A protectant dicarboximide fungicide

See also carbendazim + vinclozolin

Products

1 Barclay Flotilla	Barclay	500 g/l	SC	07905
2 Landgold Vinclozolin SC	Landgold	500 g/l	SC	06459
3 Ronilan FL	BASF	500 g/l	SC	02960
4 Standon Vinclozolin	Standon	500 g/l	SC	07836

Uses

Alternaria in OILSEED RAPE [2-4]. Alternaria in SPRING OILSEED RAPE, WINTER OILSEED RAPE [1]. Ascochyta in COMBINING PEAS, VINING PEAS [2-4]. Blossom wilt in APPLES [3]. Botrytis in OILSEED RAPE [2-4]. Botrytis in COMBINING PEAS, DWARF BEANS, NAVY BEANS, RUNNER BEANS, VINING PEAS [1-4]. Botrytis in SPRING OILSEED RAPE, WINTER OILSEED RAPE [1]. Chocolate spot in BROAD BEANS, FIELD BEANS [1-4]. Mycosphaerella in COMBINING PEAS, VINING PEAS [2-4]. Sclerotinia stem rot in OILSEED RAPE [2-4]. Sclerotinia stem rot in SPRING OILSEED RAPE, WINTER OILSEED RAPE [1].

Notes

Efficacy

- Timing of sprays varies with crop and disease. See label for details
- May be used at reduced rate on peas and field beans in tank-mix with chlorothalonil. Mixture with chlorothalonil essential on oilseed rape [3]
- Where dicarboximide resistant strains have developed product may not be effective
- Do not spray if crop wet or if rain or frost expected

Crop Safety/Restrictions

- Maximum number of treatments 2 per crop
- Do not treat mange-tout varieties of peas

Special precautions/Environmental safety

- Irritant. May cause sensitization by skin contact
- Irritating to skin [2]
- Product presents a minimal hazard to bees when used as directed but consider informing local bee-keepers if intending to spray crops in flower
- Harmful to fish and aquatic life. Do not contaminate surface waters or ditches with chemical or used container
- Operator must use a vehicle fitted with a cab and forced air filtration unit with a pesticide filter complying with HSE Guidance Note PM 74 or to an equally effective standard

Protective clothing/Label precautions

- A, C, H, K [1-4]; M [1-3]
- R04, U09a, E34 [4]; R04b [2]; R04e, U05a, U19, U20a, C03, E01, E13c, E30a [1-4]; U08 [1-3]; E26 [2, 4]; E31b [1, 2, 4]; E31c [3]

Latest application/Harvest Interval(HI)

- Before end of petal fall for apples
- HI beans, peas 2 wk; oilseed rape 7 wk

Maximum Residue Levels (mg residue/kg food)

- hops 40; grapes, cane fruits, lettuces, celery 5; tomatoes, peppers, aubergines 3; apricots, peaches, nectarines, Chinese cabbage, witloof, beans (with pods), peas (with pods) 2; pome fruits, bulb vegetables, cucumbers, gherkins, courgettes, melons, squashes, watermelons, rape seed 1; cherries 0.5; tea 0.1; citrus fruits, tree nuts, bilberries, cranberries, gooseberries wild berries, miscellaneous fruit (except kiwi fruit), beetroot, celeriac, Jerusalem artichokes parsnips, parsley root, salsify, sweet potatoes, turnips, yams, sweetcorn, broccoli, cauliflowers, Brussels sprouts, head cabbages, kale, kohlrabi, spinach, beet leaves, watercress, herbs, stem vegetables (except celery), mushrooms, oilseed (except rape seed), potatoes, cereals, animal products 0.05

536 warfarin

A hydroxycoumarin rodenticide

Products

1 Grey Squirrel Liquid Concentrate	Killgerm	0.5% w/w	CB	0645
2 Sakarat Ready-to-Use (cut wheat)	Killgerm	0.025% w/w	RB	0434
3 Sakarat Ready-to-Use (whole wheat)	Killgerm	0.025% w/w	RB	0185
4 Sakarat X Ready-to-Use Warfarin Rat Bait	Killgerm	0.05% w/w	RB	0185
5 Sewarin Extra	Killgerm	0.05% w/w	RB	0342
6 Sewarin P	Killgerm	0.025% w/w	RB	0193
7 Sewercide Cut Wheat Rat Bait	Killgerm	0.05 w/w	RB	0376
8 Sewercide Whole Wheat Rat Bait	Killgerm	0.05% w/w	RB	0375
9 Sorex Warfarin 250 ppm Rat Bait	Sorex	0.025% w/w	RB	0737
10 Sorex Warfarin 500 ppm Rat Bait	Sorex	0.05% w/w	RB	0737
11 Sorex Warfarin Sewer Bait	Sorex	0.05% w/w	RB	0737
12 Warfarin 0.5% Concentrate	B H & B	0.5%	CB	0232
13 Warfarin Ready Mixed Bait	B H & B	0.025%	RB	0233

Uses

Grey squirrels in AGRICULTURAL PREMISES, FORESTRY, INDUSTRIAL SITES [1]. Mi in AGRICULTURAL PREMISES [2-13]. Rats in AGRICULTURAL PREMISES [2-13].

Notes

Efficacy

- For rodent control place ready-to-use or prepared baits at many points wherever rats and mice active. Out of doors shelter bait from weather
- Inspect baits frequently and replace or top up as long as evidence of feeding. Do not underbait
- For grey squirrel control mix with whole wheat and leave to stand for 2-3 h before use
- Use bait in specially constructed hoppers and inspect every 2-3 d. Replace as necessary

Crop Safety/Restrictions

- For use only between 15 Mar and 15 Aug for tree protection [1]

Special precautions/Environmental safety

- For use only by local authorities, professional operators providing a pest control service and persons occupying industrial, agricultural or horticultural premises
- Prevent access to baits by children and animals, especially cats, dogs and pigs
- Rodent bodies must be searched for and burned or buried, not placed in refuse bins or rubbish tips. Remains of bait and containers must be removed after treatment and burned or buried
- Bait must not be used where food, feed or water could become contaminated
- The use of warfarin to control grey squirrels is illegal unless the provisions of the Grey Squirrels Order 1973 are observed. See label for list of counties in which bait may not be used [1]
- Must not used outdoors at all in Scotland, nor in areas of England or Wales where pine martens occur naturally [1]

Protective clothing/Label precautions

- M05 [9-11]; U13 [1, 9-13]; U20a, V05 [1]; U20b [9-13]; E30a, V04a [1-13]; E31a [12, 13]; E32a [1-11]; V01a, V03a [1-10, 12, 13]; V02 [1, 9, 10, 12, 13]

Approval

- Product not approved for use in N Ireland [1]

---

537 zeta-cypermethrin

A contact and stomach acting pyrethroid insecticide

Products

1 Fury 10 EW	PBI	100 g/l	EW	08153
2 Minuet EW	PBI	100 g/l	EW	08820

Uses

Aphids in PEAS, SPRING BARLEY, SPRING WHEAT, WINTER BARLEY, WINTER WHEAT. Barley yellow dwarf virus vectors in WINTER BARLEY, WINTER WHEAT. Cabbage stem flea beetle in OILSEED RAPE. Pea and bean weevils in FIELD BEANS, PEAS. Pea moth in PEAS. Pod midge in OILSEED RAPE. Pollen beetles in OILSEED RAPE. Seed weevil in OILSEED RAPE.

Notes

Efficacy

- On winter cereals spray when aphids first found in the autumn for BYDV control. A second spray may be required on late drilled crops or in mild conditions
- For summer aphids on cereals spray when treatment threshold exceeded
- For listed pests in other crops spray when feeding damage first seen or when treatment threshold exceeded
- Under high infestation pressure a second treatment may be necessary

Crop Safety/Restrictions

- Maximum number of treatments 2 per crop
- Consult processors before use on crops for processing

Special precautions/Environmental safety

- Harmful if swallowed. May cause sensitization by skin contact
- Extremely dangerous to fish or other aquatic life. Do not contaminate surface waters or ditches with chemical or used container

- Do not allow direct spray from vehicle mounted/drawn hydraulic sprayers to fall within 6 m, or from hand-held sprayers to within 2 m, of surface waters or ditches. Direct spray away from water

Protective clothing/Label precautions
- A, C, H
- M03, R03c, R04e, U05a, U08, U14, U15, U19, U20b, C03, E01, E13a, E16, E17, E22a, E22b, E30a, E31b

Latest application/Harvest Interval(HI)
- Before end of flowering for oilseed rape; before flowering completed (GS 69) for cereals
- HI 14 d for peas, field beans

538 zinc phosphide

A phosphine generating rodenticide

Products

1 Grovex Zinc Phosphide	Killgerm	100%	CB	06230
2 RCR Zinc Phosphide	Killgerm	100%	CB	06231
3 ZP Rodent Pellets	Antec	7.0% w/w	GB	07814

Uses

Mice in AGRICULTURAL PREMISES [1, 2]. Mice in FARM BUILDINGS [3]. Rats in AGRICULTURAL PREMISES [1, 2]. Rats in FARM BUILDINGS [3].

Notes

Efficacy
- Use to prepare baits either by dry or wet baiting as directed

Special precautions/Environmental safety
- Zinc phosphide is subject to the Poisons Rules 1982 and the Poisons Act 1972. See notes in Section 1
- For use only by local authorities, professional operators providing a pest control service and persons occupying industrial, agricultural or horticultural premises
- Toxic if swallowed
- Spontaneously inflammable in contact with acid
- Keep in original container, tightly closed, in a safe place, under lock and key
- Wear respirator if mixing baits in a confined space
- Prevent access to baits by children and domestic animals, especially cats, dogs and pigs
- Search for and burn or bury all rodent bodies. Do not place in refuse bins or on rubbish tips
- Do not prepare or use baits where food or water could be contaminated
- Remove all remains of baits and bait containers after treatment and burn or bury
- Wash out all mixing equipment thoroughly at the end of every operation

Protective clothing/Label precautions
- A [1-3]; D, H [1, 2]
- M03, E15, V03b [3]; R02c, U02, U04a, U05a, C03, E01, E34, V03a [1, 2]; U13, U20a, E30b, E32a, V01a, V02, V04a [1-3]

539 zineb

A protectant dithiocarbamate fungicide for many horticultural crops

See also ferbam + maneb + zineb

Products

Unicrop Zineb	Unicrop	70 % w/w	WP	02279

Uses

Blight in POTATOES, TOMATOES. Botrytis in ANEMONES. Celery leaf spot in CELERY. Currant leaf spot in BLACKCURRANTS. Downy mildew in HOPS, OUTDOOR LETTUCE, PROTECTED LETTUCE. Fire in TULIPS. Leaf mould in TOMATOES. Root rot in TOMATOES. Rust in CARNATIONS.

Notes

Efficacy

- Best results obtained by treatment in settled weather conditions. Do not spray if rain imminent
- Follow local blight warnings to ensure accuracy of first treatment on potatoes
- Recommended timing and spray volume varies with crop and disease. See label for details
- For root rot control in tomatoes apply as a trench watering at about the time of set of 5th truss
- Addition of wetting agent recommended for tulips and anemones

Crop Safety/Restrictions

- Maximum number of treatments on protected lettuce (including zineb, maneb, mancozeb, other EBDC fungicides or thiram) 2 per crop post-planting up to 2 wk later, and none thereafter. If thiram-based products are used post-planting on crops that will mature from Nov to Mar, 3 treatments are permitted within 3 wk of planting out
- Do not apply pre-picking to blackcurrants intended for canning

Special precautions/Environmental safety

- Irritating to eyes, skin and respiratory system

Protective clothing/Label precautions

- A
- R04a, R04b, R04c, U05a, U08, U19, U20a, C03, E01, E15, E30a, E32a

Latest application/Harvest Interval(HI)

- HI blackcurrants 4 wk; protected lettuce 3 wk; outdoor lettuce 2 wk; other edible outdoor crops 1 wk; other edible glasshouse crops 2 d

Approval

- Approved for aerial application on potatoes [1]. See notes in Section 1

Maximum Residue Levels (mg residue/kg food)

- hops 25; lettuces, herbs 5; oranges, apricots, peaches, nectarines, grapes, strawberries, peppers, aubergines 2; cherries, plums 1; garlic, onions, shallots, cucumbers 0.5; celeriac, witloof 0.2; tree nuts, oilseeds (except rape seed), tea 0.1; bilberries, cranberries, wild berries, blackberries, loganberries, miscellaneous fruit, root and tuber vegetables, horseradish, Jerusalem artichokes, parsley root, sweet potatoes, swedes, turnips, yams, spinach, beet leaves, watercress, asparagus, cardoons, fennel, globe artichokes, rhubarb, mushrooms, potatoes, maize, rice, animal products 0.05

540 zineb-ethylene thiuram disulphide adduct

A protectant dithiocarbamate fungicide for potatoes

Products Polyram DF | BASF | 70% w/w | WG | 08234

Uses Blight in POTATOES.

Notes

Efficacy
- Apply to potatoes before blight infection begins, at blight warning, or before haulms meet in row and repeat every 10-14 d
- If blight infection periods occur spray interval should be reduced
- Increase water volume later in season when foliage is denser

Crop Safety/Restrictions
- Do not apply if crop wet or if rain or frost expected
- Product prolongs haulm growth. Haulms must be burned off or destroyed before late blight attacks

Special precautions/Environmental safety
- Harmful to game, wild birds and animals
- Do not contaminate surface waters or ditches with chemical or used container

Protective clothing/Label precautions
- A
- U05a, U08, U19, U20b, C03, E01, E15, E30a, E32a, S04b

Latest application/Harvest Interval(HI)
- HI 7 d

Approval
- Approved for aerial application on potatoes [1]

Maximum Residue Levels (mg residue/kg food)
- see zineb entry

541 ziram

A dithiocarbamate bird and animal repellent

Products AAprotect | Unicrop | 32% w/w | PA | 03784

Uses Birds in FIELD CROPS, FORESTRY, ORNAMENTALS, TOP FRUIT. Deer in FIELD CROPS, FORESTRY, ORNAMENTALS, TOP FRUIT. Hares in FIELD CROPS, FORESTRY, ORNAMENTALS, TOP FRUIT. Rabbits in FIELD CROPS, FORESTRY, ORNAMENTALS, TOP FRUIT.

Notes

Efficacy

- Apply undiluted to main stems up to knee height to protect against browsing animals at any time of yr or spray 1:1 dilution on stems and branches in dormant season
- Use dilute spray on fully dormant fruit buds to protect against bullfinches
- Only apply to dry stems, branches or buds
- Use of diluted spray can give limited protection to field crops in areas of high risk during establishment period

Crop Safety/Restrictions

- Do not spray elongating shoots or buds about to open
- Do not apply concentrated spray to foliage, fruit buds or field crops

Special precautions/Environmental safety

- Irritating to eyes, skin and respiratory system

Protective clothing/Label precautions

- A, C
- R04a, R04b, R04c, U05a, U08, U19, U20b, C02 (8 wk), C03, E01, E13c, E30a, E31c

Latest application/Harvest Interval(HI)

- HI edible crops 8 wk

SECTION 5
APPENDICES

Appendix 1
SUPPLIERS OF PESTICIDES AND ADJUVANTS

AgrEvo: AgrEvo UK Ltd
East Winch Hall
East Winch
King's Lynn
Norfolk PE32 1HN
Tel: (01553) 841581
Fax: (01553) 841090
Email: qgrevo.assist@agrevo.com
Web: www.agrevo.com

AgrEvo Environ.: AgrEvo Environmental Health
Hauxton
Cambridge CB2 5HU
Tel: (01223) 870312
Fax: (01223) 872142

Agrichem: Agrichem (International) Ltd
Industrial Estate
Station Road
Whittlesey
Cambs PE7 2EY
Tel: (01733) 204019
Fax: (01733) 204162

Aitken: R. Aitken Ltd
123 Harmony Row
Govan
Glasgow G51 3NB
Tel: (0141) 440 0033
Fax: (0141) 440 2744

Allen: Allen Power Equipment Ltd
The Broadway
Didcot
Oxon. OX11 8ES
Tel: (01235) 813936
Fax: (01235) 811491

Allied Colloids: Allied Colloids Limited
See Ciba Specialty Chemicals

Antec: Antec International
Windham Road
Chilton Industrial Estate
Sudbury
Suffolk CO10 6XD
Tel: (01787) 377305
Fax: (01787) 310846
Email: antec-international@compuserve.com
Web: www.antecint.com

Aquaspersions: Aquaspersions Ltd
Charlestown Works
Charlestown
Hebden Bridge
W. Yorks. HX7 6PL
Tel: (01422) 843715
Fax: (01422) 845067

Armillatox: Armillatox Ltd
121 Main Road
Morton
Alfreton
Derbyshire DE55 6HL
Tel: (01773) 590566
Fax: (01773) 590681

Ashlade: Ashlade Formulations Ltd
Denaby Lane Industrial Estate
Denaby Lane
Old Denaby
Doncaster
S. Yorks. DN12 4LQ
Tel: (01709) 772200
Fax: (01709) 772201

Atlas: Atlas Crop Protection Ltd
Denaby Lane Industrial Estate
Denaby Lane
Old Denaby
Doncaster
S. Yorks. DN12 4LQ
Tel: (01709) 772200
Fax: (01709) 772201

B H & B: Battle Hayward & Bower Ltd
Victoria Chemical Works
Crofton Drive
Allenby Road Industrial Esate
Lincoln LN3 4NP
Tel: (01522) 529206/541241
Fax: (01522) 538960

Barclay: Barclay Chemicals (UK) Ltd
Barclay House
Lilmar Industrial Estate
Santry
Dublin 9
Ireland
Tel: (00353) 1 842 5755
Fax: (00353) 1 842 5381

Barrettine: Barrettine Environmental Health
Barrettine Works
St. Ivel Way
Warmley
Bristol BS15 5TY
Tel: (0117) 967 2222
Fax: (0117) 961 4122

BASF: BASF plc.
Agricultural Divison
PO Box 4
Earl Road, Cheadle Hulme
Cheadle
Cheshire SK8 6QG
Tel: (0161) 485 6222
Fax: (0161) 485 2229

Batsons: Joseph Batsons Ltd
Dudley Road
Tipton
W. Midlands DY4 8EH
Tel: (0121) 557 2284
Fax: (0121) 557 8068

Bayer: Bayer plc.
Crop Protection Business Group
Eastern Way
Bury St Edmunds
Suffolk IP32 7AH
Tel: (01284) 763200
Fax: (01284) 702810
Email: crop.protection@bayer.co.uk

Brian Jones: Brian Jones and Associates Ltd
Fluorocarbon Building
Caxton Hill
Hertford
Herts. SG13 7NH
Tel: (01992) 553065
Fax: (01992) 551873

Bromine & Chem.: Bromine and Chemicals Ltd
201 Haverstock Hill
Hampstead
London NW3 4QG
Tel: (0171) 431 7707
Fax: (0171) 431 7797

Cheminova: Cheminova Agro (UK) Ltd
Bishop House
Bath Road
Taplow
Maidenhead
Berks. SL6 0NX
Tel: (01628) 664038
Fax: (01628) 663994

Chiltern: Chiltern Farm Chemicals Ltd
11 High Street
Thornborough
Buckingham MK18 2DF
Tel: (01280) 822400
Fax: (01280) 822082

Ciba Agric.: Ciba Agriculture
See Novartis Crop Protection UK Ltd

Ciba Specialty: Ciba Specialty Chemicals
Water Treatment Division
P O Box 38
Low Moor
Bradford
W. Yorks. BD12 0JZ
Tel: (01274) 417000
Fax: (01274) 606499
Email: heidi.waterworth@cibasc.com
Web: www.cibasc.com

CMI: CMI Ltd (incorporating Collingham Marketing)
United House
113 High Street
Collingham
Newark
Notts. NG23 7NG
Tel: (01636) 892078
Fax: (01636) 893037

Coalite: Coalite Chemicals
PO Box 152
Buttermilk Lane
Bolsover
Chesterfield
Derbyshire S44 6AZ
Tel: (01246) 826816
Fax: (01246) 240309

Coventry Chemicals: Coventry Chemicals Ltd
Woodhams Road
Siskin Drive
Coventry CV3 4FX
Tel: (01203) 639739
Fax: (01203) 639717

Cyanamid: Cyanamid Agriculture Ltd
Cyanamid House
Fareham Road
Gosport
Hants. PO13 0AS
Tel: (01329) 224000
Fax: (01329) 224335
Web: www.agricentre.co.uk

Dax: Dax Products Ltd
P O Box 119
76 Cyprus Road
Nottingham NG3 5NA
Tel: (0115) 926 9996
Fax: (0115) 966 1173

De Sangosse: De Sangosse (UK) SA
PO Box 135
Market Weighton
York YO4 3YY
Tel: (01430) 872525
Fax: (01430) 873123

Deosan: Deosan Ltd
See DiverseyLever

Dewco-Lloyd: Dewco-Lloyd Ltd
Cyder House
Ixworth
Suffolk IP31 2HT
Tel: (01359) 230555
Fax: (01359) 232553

DiverseyLever: DiverseyLever Ltd
Weston Favell Centre
Northampton NN3 8PD
Tel: (01604) 783505
Fax: (01604) 783506

Doff Portland: Doff Portland Ltd
Benneworth Close
Hucknall
Nottingham NG15 6EL
Tel: (0115) 963 2842
Fax: (0115) 963 8657
Email: sales@doff.co.uk
Web: www.doff.co.uk

Dow: Dow AgroSciences
Latchmore Court
Brand Street
Hitchin
Herts. SG5 1NH
Tel: (01462) 457272
Fax: (01462) 426605

DuPont: DuPont (UK) Ltd
Agricultural Products Department
Wedgwood Way
Stevenage
Herts. SG1 4QN
Tel: (01438) 734000
Fax: (01438) 734452

Elliott: Thomas Elliott Ltd
143A High Street
Edenbridge
Kent TN8 5AX
Tel: (01732) 866566
Fax: (01732) 864709

English Woodland: English Woodlands Biocontrol
Hoyle Depot
Graffham
Petworth
W. Sussex GU28 0LR
Tel: (01798) 867574
Fax: (01798) 867574

Euroagkem: Euroagkem Ltd
1 Cambray Mews
Wellington Street
Cheltenham
Glos GL50 1XZ
Tel: (0345) 697849

Fargro: Fargro Ltd
Toddington Lane
Littlehampton
Sussex BN17 7PP
Tel: (01903) 721591
Fax: (01903) 730737
Email: promos-fargro@btinternet.com
Web: www.btinternet.com/website-fargro

FCC: Farmers Crop Chemicals Ltd
Thorn Farm
Evesham Road
Inkberrow
Worcs. WR7 4LJ
Tel: (01386) 793401
Fax: (01386) 793184
Email: fcc.ltd@btinternet.com

Fine: Fine Agrochemicals Ltd
Hill End House
Whittington
Worcester WR5 2RL
Tel: (01905) 361800
Fax: (01905) 361810
Email: enquire@fine-agrochemicals.com

Ford Smith: Ford Smith & Co. Ltd
Lyndean Industrial Estate
Felixstowe Road
Abbey Wood
London SE2 9SG
Tel: (0181) 310 8127
Fax: (0181) 310 9563

Graincare: Graincare (Colchester) Ltd
17 Woodlands
Colchester
Essex CO4 3JA
Tel: (01206) 862436
Fax: (01206) 862436

Grampian: Grampian Pharmaceuticals Ltd
Marathon Place
Moss Side Industrial Estate
Leyland
Lancs. PR5 3QN
Tel: (01772) 452421
Fax: (01772) 456820

Growing Success: Growing Success Organics Ltd
Wessex House
1-3 Hilltop Business Park
Devizes Road
Salisbury
Wilts. SP3 4UF
Tel: (01722) 337744
Fax: (01722) 333177
Email:
growing@wessexhortgrp.demon.co.uk

Headland: Headland Agrochemicals Ltd
Norfolk House
Gt. Chesterford Court
Gt. Chesterford
Saffron Walden
Essex CB10 1PF
Tel: (01799) 530146
Fax: (01799) 530229

Heatherington: J.V. Heatherington Farm & Garden Supplies
29 Main Street
Glenary
Crumlin
Co. Antrim BT29 4LN
Tel: (018494) 22227

Helm: Helm Great Britain Chemicals Ltd
Wimbledon Bridge House
1 Hartfield Road
London SW19 3RU
Tel: (0181) 544 9000
Fax: (0181) 544 1011
Web: www.helmag.com

Hickson & Welch: Hickson & Welch Ltd
Wheldon Road
Castleford
W. Yorks. WF10 2JT
Tel: (01977) 556565
Fax: (01977) 550910

Hortag: Hortag Chemicals Ltd
Salisbury Road
Downton
Wilts. SP5 3JJ
Tel: (01725) 512822
Fax: (01725) 512840

Hortichem: Hortichem Ltd
1b, Mills Way
Boscombe Down Business Park
Amesbury
Wilts. SP4 7RX
Tel: (01980) 676500
Fax: (01980) 626555
Email: hortichem@hortichem.co.uk

I T Agro: I T Agro Ltd
805 Salisbury House
31 Finsbury Circus
London EC2M 5SQ
Tel: (0171) 628 2040

Interagro: Interagro (UK) Ltd
2 Ducketts Wharf
South Street
Bishop's Stortford
Herts. CM23 3AR
Tel: (01279) 501995
Fax: (01279) 501996
Email: info@interagro.co.uk
Web: www.interagro.co.uk/adjuvants

Intracrop: Intracrop
Brian Lewis Agriculture Ltd
Byemoor Farm
Melmerby
Leyburn
N. Yorks. DL8 4TW
Tel: (01969) 640655
Fax: (01969) 640633

Irish Drugs: Irish Drugs Ltd
Burnfoot
Lifford
Co. Donegal
Ireland
Tel: (00353) 77 68103/4
Fax: (00353) 77 68311

Isagro: Isagro Sp.A
See Sipcam UK Ltd

ISK Biosciences: ISK Biosciences Ltd
Kestrel House
Falconry Court
Bakers Lane
Epping
Essex CM16 5DQ
Tel: (01992) 571555
Fax: (01992) 571666

Killgerm: Killgerm Chemicals Ltd
115 Wakefield Road
Flushdyke
Ossett
W. Yorks. WF5 9BW
Tel: (01924) 265090
Fax: (01924) 265033

Koppert: Koppert (UK) Ltd
1 Wadhurst Business Park
Faircrouch Lane
Wadhurst
E. Sussex TN5 6PT
Tel: (01892) 784411
Fax: (01892) 782469

Lambson: Lambson Agricultural Services
Cinder Lane
Castleford
W. Yorks. WF10 1LU
Tel: (01977) 510511
Fax: (01977) 603049

Landgold: Landgold & Co. Ltd
PO Box 829
Charles House
Charles Street
St. Helier
Jersey JE4 9NZ
Tel: (01534) 68446
Fax: (01534) 32843

Lever Industrial: Lever Industrial Ltd
PO Box 100
Runcorn
Cheshire WA7 3JZ
Tel: (01928) 719000
Fax: (01928) 714628

Levington: Levington Horticulture Ltd
See The Scotts Company (UK) Limited

Luxan: Luxan (UK) Ltd
Sysonby Lodge
Nottingham Road
Melton Mowbray
Leics. LE13 0NU
Tel: (01664) 66372
Fax: (01664) 480137

Makhteshim: Makhteshim-Agan (UK) Ltd
Sir William Atkins House
2 Ashley Avenue
Epsom
Surrey KT18 5WA
Tel: (01372) 747372
Fax: (01732) 747270

Mandops: Mandops (UK) Ltd
36 Leigh Road
Eastleigh
Hants. SO50 9DT
Tel: (01703) 641826
Fax: (01703) 629106
Email: mandops@interalpha.co.uk
Web: www.mandops.co.uk

Maxicrop: Maxicrop International Ltd
Weldon Road
Corby
Northants. NN17 5US
Tel: (01536) 402182
Fax: (01536) 204254
Email: info@maxicrop.co.uk

Microcide: Microcide Ltd
Shepherds Grove
Stanton
Bury St. Edmunds
Suffolk IP31 2AR
Tel: (01359) 251077
Fax: (01359) 251545

Miracle: Miracle Professional
See The Scotts Company (UK) Limited

Mirfield: Mirfield Sales Services Ltd
Denaby Lane Industrial Estate
Denaby Lane
Old Denaby
Doncaster
S. Yorks. DN12 4LQ
Tel: (01709) 772200
Fax: (01709) 772201

Mitchell Cotts: Mitchell Cotts Chemicals Ltd
PO Box 6
Steanard Lane
Mirfield
W. Yorks. WF14 8QB
Tel: (01924) 493861
Fax: (01924) 490972

Monsanto: Monsanto plc
Agricultural Sector
PO Box 53
Lane End Road
High Wycombe
Bucks HP12 4HL
Tel: (01494) 474918
Fax: (01494) 474920
Web: www.monsanto.com

MSD AGVET: MSD AGVET
See Novartis Crop Protection UK Ltd

MTM Agrochem.: MTM Agrochemicals Ltd
See United Phosphorus Ltd

Nehra: Nehra Cookes Chemicals Ltd
16 Chiltern Close
Warren Wood
Arnold
Nottingham NG5 9PX
Tel: (0115) 973 5999
Fax: (0115) 973 6700

Newman: Newman Agrochemicals Ltd
Swaffham Bulbeck
Cambridge CB5 0LU
Tel: (01223) 811215
Fax: (01223) 812725

Nickerson Seeds: Nickerson Seeds Ltd
JNRC
Rothwell
Lincs. LN7 6DT
Tel: (01472) 371661
Email: Nickerson.Seeds@farmline.com

omix-Chipman: Nomix-Chipman Ltd
Portland Building
Portland Street
Staple Hill
Bristol BS16 4PS
Tel: (0117) 957 4574
Fax: (0117) 956 3461

ovartis: Novartis Crop Protection UK Ltd
Whittlesford
Cambridge CB2 4QT
Tel: (01223) 833621
Fax: (01223) 835211
Web: www.novartiscrop.co.uk

ovartis A H: Novartis Animal Health
K Ltd
Whittlesford
Cambridge CB2 4XW
Tel: (01223) 833634
Fax: (01223) 836526

ovartis BCM: Novartis BCM Ltd
Aldham Business Centre
New Road
Aldham
Colchester
Essex CO6 3PN
Tel: (01206) 243200
Fax: (01206) 243209

ufarm: Nufarm UK Limited
Crabtree Manorway North
Belvedere
Kent DA17 6BQ
Tel: (0181) 319 7246
Fax: (O181) 319 7200
Email:
dominic.scicchitano@uk.nufarm.com
Web: www.nufarm.com

BI: pbi Agrochemicals Ltd
Britannica Works
Sewardstone Road
Waltham Abbey
Essex EN9 1NP
Tel: (01992) 712579
Fax: (01992) 659800
Email: David.Roberts@pbi.co.uk
Web: www.pbi.co.uk

Portman: Portman Agrochemicals Ltd
Apex House
Grand Arcade
Tally-Ho Corner
North Finchley
London N12 0EH
Tel: (0181) 446 8383
Fax: (0181) 445 6045

Promark: Promark Ltd
See AgrEvo UK Crop Protection Ltd

Quadrangle: Quadrangle Agrochemicals
Crook Farm
North Deighton
Wetherby
Yorks LS22 5HW
Tel: (01937) 584228
Fax: (01937) 580937

Rentokil: Rentokil Initial plc
Felcourt
East Grinstead
W. Sussex RH19 2JY
Tel: (01342) 833022
Fax: (01342) 326229

Rigby Taylor: Rigby Taylor Ltd
Rigby Taylor House
Garside Street
Bolton
Lancs. BL1 4AE
Tel: (01204) 394888
Fax: (01204) 385276

Roebuck Eyot: Roebuck Eyot Ltd
7a Hatfield Way
South Church Enterprise Park
Bishop Auckland
Co. Durham DL14 6XF
Tel: (01388) 772233
Fax: (01388) 775233
Email: roebuck.eyot@onyxnet.co.uk
Web: www.roebuck-eyot.co.uk

Rohm & Haas: Rohm & Haas (UK) Ltd
Lennig House
2 Masons Avenue
Croydon
Surrey CR9 3NB
Tel: (0181) 774 5300
Fax: (0181) 774 5301

RP Agric.: Rhone-Poulenc Agriculture Ltd
Fyfield Road
Ongar
Essex CM5 0HW
Tel: (01277) 301301
Fax: (01277) 362610
Email: ian.cockram@rhone-poulenc.com
Web: www.rhone-poulenc.com/ag-uk

RP Amenity: Rhone-Poulenc Amenity
See above address

Sandoz: Sandoz Agro Ltd
See Novartis Crop Protection UK Ltd

Scotts: The Scotts Company (UK) Ltd
Paper Mill Lane
Bramford
Ipswich
Suffolk IP8 4BZ
Tel: (01473) 830492
Fax: (01473) 830386

Seedcote: Seedcote Systems Ltd
Telford Way
Thetford
Norfolk IP24 1HU
Tel: (01842) 766261
Fax: (01842) 66263

Service Chemicals: Service Chemicals Ltd
Lanchester Way
Royal Oak Industrial Estate
Daventry
Northants. NN11 5PH
Tel: (01327) 704444
Fax: (01327) 71154

Sinclair: William Sinclair Horticulture Ltd
Firth Road
Lincoln LN6 7AH
Tel: (01522) 537561
Fax: (01522) 513609

Sipcam: Sipcam UK Ltd
Sheraton House
Castle Park
Cambridge CB3 0AX
Tel: (01223) 370030
Fax: (01223) 354026

Sorex: Sorex Ltd
St Michael's Industrial Estate
Hale Road
Widnes
Cheshire WA8 8TJ
Tel: (0151) 420 7151
Fax: (0151) 495 1163

Sphere: Sphere Laboratories (London) Ltd
The Yews
Main Street
Chilton
Oxon. OX11 0RZ
Tel: (01235) 831802
Fax: (01235) 833896

Standon: Standon Chemicals Ltd
48 Grosvenor Square
London W1X 9LA
Tel: (0171) 493 8648
Fax: (0171) 493 4219

Stefes: Stefes UK Ltd
Standard House
Chord Business Park
London Road
Godmanchester
Cambs. PE18 8LN
Tel: (01480) 358358
Fax: (01480) 358444
Email: general.info@stefes.com

Stoller: Stoller Chemical Ltd
53 Bradley Hall Trading Estate
Bradley Lane
Standish
Lancs. WN6 0XQ
Tel: (01257) 427722
Fax: (01257) 427888

Sumitomo: Sumitomo Corporation (UK) PLC
Vintners' Place
68 Upper Thames Street
London EC4V 3BJ
Tel: (0171) 246 3796
Fax: (0171) 246 3933

Sylvan: Sylvan Spawn Limited
Broadway
Yaxley
Peterborough PE7 3EJ
Tel: (01733) 240412
Fax: (01733) 245020

Tech Nova: Tech Nova Products Ltd
The Arable Centre
Winterbourne Monkton
Swindon
Wilts. SN4 9NW
Tel: (01672) 539591
Fax: (01672) 539406

Techsol: Techsol Contract Manufacturing
Agricultural Division
116 High Street
Solihull
W. Midlands B91 3SX
Tel: (0121) 711 7698
Fax: (0121) 704 4024

Tripart: Tripart Farm Chemicals Ltd
The Grove
Cambridge Road
Godmanchester
Huntingdon
Cambs. PE18 8BW
Tel: (01480) 435102
Fax: (01480) 445705

Truchem: Truchem Ltd
Brook House
30 Larwood Grove
Sherwood
Nottingham NG5 3JD
Tel: (0115) 926 0762
Fax: (0115) 967 1153

Unicrop: Universal Crop Protection Ltd
Park House
Maidenhead Road
Cookham
Berks. SL6 9DS
Tel: (01628) 526083
Fax: (01628) 810457

Uniroyal: Uniroyal Chemical Ltd
Kennet House
4 Langley Quay
Slough
Berks. SL3 6EH
Tel: (01753) 603000
Fax: (01753) 603077

United Phosphorus: United Phosphorus Ltd
18 Liverpool Road
Great Sankey
Warrington
Cheshire WA5 1QR
Tel: (01925) 633232
Fax: (01925) 652679

Vass: L.W. Vass (Agricultural) Ltd
Springfield Farm
Silsoe Road
Maulden
Bedford MK45 2AX
Tel: (01525) 403041
Fax: (01525) 402282

Vitax: Vitax Ltd
Owen Street
Coalville
Leicester LE67 3DE
Tel: (01530) 510060
Fax: (01530) 510299

Wheatley: Wheatley Chemical Co.
Denaby Lane Industrial Estate
Denaby Lane
Old Denaby
Doncaster
S. Yorks. DN12 4LQ
Tel: (01709) 772200
Fax: (01709) 772201

Whyte: Whyte Chemicals Ltd
Marlborough House
298 Regents Park Road
Finchley
London N3 2UA
Tel: (0181) 346 5946
Fax: (0181) 349 4589

Yule Catto: Yule Catto Consumer Chemicals Ltd
Stanhope Road
Swadlincote
Burton-on-Trent
Staffs. DE11 9BE
Tel: (01283) 221044
Fax: (01283) 225731

Zeneca: Zeneca Crop Protection
Fernhurst
Haslemere
Surrey GU27 3JE
Tel: (01428) 656564
Fax: (01428) 657385
Web: www.zeneca-crop.co.uk

Zeneca Prof.: Zeneca Professional Products
See The Scotts Company (UK) Ltd

Appendix 2
USEFUL CONTACTS

BASIS Limited
Bank Chambers
34 St John Street
Ashbourne
Derbyshire DE6 1GH
Tel: (01335) 343945/346138
Fax: (01335) 346488

British Crop Protection Council (BCPC)
49 Downing Street
Farnham
Surrey GU9 7PH
Tel: (01252) 733072
Fax: (01252) 727194

BCPC Publications Sales
Bear Farm
Binfield
Bracknell
Berks. RG12 5QE
Tel: (0118) 934 2727
Fax: (0188) 934 1998

British Agrochemicals Association Ltd
4 Lincoln Court
Lincoln Road
Peterborough
Cambs. PE1 2RP
Tel: (01733) 349225
Fax: (01733) 562523

British Beekeepers' Association
National Agricultural Centre
Stoneleigh
Kenilworth
Warwickshire CV8 2LZ
Tel: (01203) 696679
Fax: (01203) 690682

British Pest Control Association
3 St James Court
Friar Gate
Derby DE1 1ZU
Tel: (01332) 294288
Fax: (01332) 295904

Department of Agriculture Northern Ireland
Pesticides Section
Dundonald House
Upper Newtownards Road
Belfast BT4 3SB
Tel: (01232) 524704
Fax: (01232) 524266

Department of the Environment
Water Directorate
Romney House
43 Marsham Street
London SW1P 3PY
Tel: (0171) 276 8220
Fax: (0171) 276 8639

Department of Health
Food Safety and Public Health Branch
Skipton House
80 London Road
London SE1 6LW
Tel: (0171) 972 5033
Fax: (0171) 972 5137

Farmers' Union of Wales
Llys Amaeth
Queen's Square
Aberystwyth
Dyfed SY23 2EA
Tel: (01970) 612755
Fax: (01970) 624369

Forestry Commission
231 Corstorphine Road
Edinburgh EH12 7AT
Tel: (0131) 334 0303
Fax: (0131) 334 3047

Global Crop Protection Federation
Avenue Louise 143
B-1050 Brussels
Belgium
Tel: (+32) 2 542 0410
Fax: (+32) 2 542 0419

Health and Safety Executive
Information Centre
Broad Lane
Sheffield
Yorkshire S3 7HQ
Tel: (0114) 289 2345/6
Fax: (0114) 289 2333

Health and Safety Executive
Pesticides Registration Section
Magdalen House
Bootle
Merseyside L20 3QZ
Tel: (0151) 951 4000

Health and Safety Executive - Books
PO Box 1999
Sudbury
Suffolk CO10 6FS
Tel: (01787) 881165
Fax: (01787) 313995

Lantra National Training Organisation Ltd (previously ATB-LandBase)
National Agricultural Centre
Stoneleigh
Kenilworth
Warwickshire CV8 2UG
Tel: (01203) 696996
Fax: (01203) 696732

Ministry of Agriculture Fisheries and Food
Food and Veterinary Science Division
Nobel House
17 Smith Square
London SW1P 3JR
Tel: (0171) 238 5526
Fax: (0171) 238 6129

National Association of Agricultural Contractors (NAAC)
Huts Corner
Tilford Road
Hindhead
Surrey GU26 6SF
Tel: (01428) 605360
Fax: (01428) 606531

National Farmers' Union
Agriculture House
164 Shaftesbury Avenue
London WC2H 8HL
Tel: (0171) 331 7200
Fax: (0171) 331 7313

National Poisons Information Service
Guys' and St Thomas' Hospital Trust
London SE14 5ER
Tel: (0171) 635 9191

The Royal Hospitals
Belfast BT12 6BA
Tel: (01232) 240503

City Hospital NHS Trust
Birmingham B18 7QH
Tel: (0121) 507 5588/5589

Llandough Hospital
Penarth CF64 2XX
Tel: (01222) 709901

Royal Infirmary
Edinburgh
Tel: (0131) 536 2300

The General Infirmary
Leeds LS1 3EX
Tel: (0113) 243 0715

Royal Victoria Hospital
Newcastle NE1 4LP
Tel: (0191) 232 5131

National Proficiency Tests Council (NPTC)
National Agricultural Centre
Stoneleigh
Kenilworth
Warwickshire CV8 2LG
Tel: (01203) 696553
Fax: (01203) 696128

National Turfgrass Council
Hunter's Lodge
Dr Brown's Road
Minchinhampton
Glos. GL6 9BT
Tel: (01453) 883588
Fax: (01453) 731449

Pesticides Safety Directorate
Mallard House
King's Pool
3 Peasholme Green
York YO1 2PX
Tel: (01904) 640500
Fax: (01904) 455733

Processors and Growers Research Organisation
The Research Station
Great North Road
Thornhaugh
Peterborough
Cambs. PE8 6HJ
Tel: (01780) 782585
Fax: (01780) 783993

Scottish Beekeepers' Association
North Trinity House
114 Trinity Road
Edinburgh EH5 3JZ
Tel: (0131) 552 5341

The Environment Agency (previously The National Rivers Authority)
Rio House
Waterside Drive
Aztec West
Almondsbury
Bristol BS12 4UD
Tel: (01454) 624400
Fax: (01454) 624409

The European Crop Protection Association (ECPA)
Avenue E van Nieuwenhuyse 6
B-1160
Brussels
Belgium
Tel: (+32) 2 663 1550
Fax: (+32) 2 663 1560

The Stationery Office (previously HMSO)
Publications Centre
PO Box 276
London SW8 5DT
Tel: (0171) 873 9090 (orders)
Tel: (0171) 873 0011 (enquiries)
Fax: (0171) 873 8200

UK Agricultural Supply Trade Association (UKASTA)
3 Whitehall Court
London SW1A 2EQ
Tel: (0171) 930 3611
Fax: (0170) 930 3952

Ulster Beekeepers' Association
57 Liberty Road
Carrickfergus
Co. Antrim BT38 9DJ
Tel: (01960) 362998

Welsh Beekeepers' Association
Trem y Clawdd
Fron Isaf
Chirk
Wrexham
Clwyd LL14 5AH
Tel: (01691) 773300
Fax: (01691) 773300

Appendix 3
KEYS TO CROP AND WEED GROWTH STAGES

Decimal Code for the Growth Stages of Cereals

Illustrations of these growth stages can be found in the reference indicated below and in some company product manuals.

0 Germination
- 00 Dryseed
- 03 Imbibition complete
- 05 Radicle emerged from caryopsis
- 07 Coleoptile emerged from caryopsis
- 09 Leaf at coleoptile tip

1 Seedling growth
- 10 First leaf through coleoptile
- 11 First leaf unfolded
- 12 2 leaves unfolded
- 13 3 leaves unfolded
- 14 4 leaves unfolded
- 15 5 leaves unfolded
- 16 6 leaves unfolded
- 17 7 leaves unfolded
- 18 8 leaves unfolded
- 19 9 or more leaves unfolded

2 Tillering
- 20 Main shoot only
- 21 Main shoot and 1 tiller
- 22 Main shoot and 2 tillers
- 23 Main shoot and 3 tillers
- 24 Main shoot and 4 tillers
- 25 Main shoot and 5 tillers
- 26 Main shoot and 6 tillers
- 27 Main shoot and 7 tillers
- 28 Main shoot and 8 tillers
- 29 Main shoot and 9 or more tillers

3 Stem elongation
- 30 ear at 1 cm
- 31 1st node detectable
- 32 2nd node detectable
- 33 3rd node detectable
- 34 4th node detectable
- 35 5th node detectable
- 36 6th node detectable
- 37 Flag leaf just visible
- 39 Flag leaf ligule/collar just visible

4 Booting
- 41 Flag leaf sheath extending
- 43 Boots just visibly swollen
- 45 Boots swollen
- 47 Flag leaf sheath opening
- 49 First awns visible

5 Inflorescence
- 51 First spikelet of inflorescence just visible
- 52 ¼ of inflorescence emerged
- 55 ½ of inflorescence emerged
- 57 ¾ of inflorescence emerged
- 59 Emergence of inflorescence completed

6 Anthesis
- 60, 61 } Beginning of anthesis
- 64, 65 } Anthesis half way
- 68, 69 } Anthesis complete

7 Milk development
- 71 Caryopsis watery ripe
- 73 Early milk
- 75 Medium milk
- 77 Late milk

8 Dough development
- 83 Early dough
- 85 Soft dough
- 87 Hard dough

9 Ripening
- 91 Caryopsis hard (difficult to divide by thumb-nail)
- 92 Caryopsis hard (can no longer be dented by thumb-nail)
- 93 Caryopsis loosening in daytime

(From Tottman, 1987. *Annals of Applied Biology*, **110**, 441–454)

Stages in Development of Oilseed Rape

Illustrations of these growth stages can be found in the reference indicated below and in some company product manuals.

0 Germination and emergence

1 Leaf production
- 1,0 Both cotyledons unfolded and green
- 1,1 First true leaf
- 1,2 Second true leaf
- 1,3 Third true leaf
- 1,4 Fourth true leaf
- 1,5 Fifth true leaf
- 1,10 About tenth true leaf
- 1,15 About fifteenth true leaf

2 Stem extension
- 2,0 No internodes ('rosette')
- 2,5 About five internodes

3 Flower bud development
- 3,0 Only leaf buds present
- 3,1 Flower buds present but enclosed by leaves
- 3,3 Flower buds visible from above ('green bud')
- 3,5 Flower buds raised above leaves
- 3,6 First flower stalks extending
- 3,7 First flower buds yellow ('yellow bud')

4 Flowering
- 4,0 First flower opened
- 4,1 10% all buds opened
- 4,3 30% all buds opened
- 4,5 50% all buds opened

5 Pod development
- 5,3 30% potential pods
- 5,5 50% potential pods
- 5,7 70% potential pods
- 5,9 All potential pods

6 Seed development
- 6,1 Seeds expanding
- 6,2 Most seeds translucent but full size
- 6,3 Most seeds green
- 6,4 Most seeds green-brown mottled
- 6,5 Most seeds brown
- 6,6 Most seeds dark brown
- 6,7 Most seeds black but soft
- 6,8 Most seeds black and hard
- 6,9 All seeds black and hard

7 leaf senescence

8 Stem senescence
- 8,1 Most stem green
- 8,5 Half stem green
- 8,9 Little stem green

9 Pod senescence
- 9,1 Most pods green
- 9,5 Half pods green
- 9.9 Few pods green

(From Sylvester-Bradley, 1985. *Aspects of Applied Biology*, **10**, 395–400)

Stages in Development of Peas

Illustrations of these growth stages can be found in the reference indicated below and in some company product manuals.

0 Germination and emergence
- 000 Dry seed
- 001 Imbibed seed
- 002 Radicle apparent
- 003 Plumule and radicle apparent
- 004 Emergence

1 Vegetative stage
- 101 First node (leaf with one pair leaflets, no tendril)
- 102 Second node (leaf with one pair leaflets, simple tendril)
- 103 Third node (leaf with one pair leaflets, complex tendril)
- •
- •
- 10x X nodes (leaf with more than one pair leaflets, complex tendril)
- •
- •
- 10n Last recorded node

2 Reproductive stage (main stem)
- 201 Enclosed buds
- 202 Visible buds
- 203 First open flower
- 204 Pod set (small immature pod)
- 205 Flat pod
- 206 Pod swell (seeds small, immature)
- 207 Podfill
- 208 Pod green, wrinkled
- 209 Pod yellow, wrinkled (seeds rubbery)
- 210 Dry seed

3 Senescence stage
- 301 Desiccant application stage. Lower pods dry and brown, middle yellow, upper green. Overall moisture content of seed less than 45%
- 302 Pre-harvest stage. Lower and middle pods dry and brown, upper yellow. Overall moisture content of seed less than 30%
- 303 Dry harvest stage. All pods dry and brown, seed dry

(From Knott, 1987. *Annals of Applied Biology*, **111**, 233–244)

Stages in Development of Faba Beans

Illustrations of these growth stages can be found in the reference indicated below and in some company product manuals.

0 Germination and emergence
000 Dry seed
001 Imbibed seed
002 Radicle apparent
003 Plumule and radicle apparent
004 Emergence
005 First leaf unfolding
006 First leaf unfolded

1 Vegetative stage
101 First node
102 Second node
103 Third node
•
•
10x X nodes
•
•
10n N, last reoorded node

2 Reproductive stage (main stem)
201 Flower buds visible
203 First open flowers
204 First pod set
205 Pods fully formed, green
207 Pod fill, pods green
209 Seed rubbcry, pods pliable, turning black
210 Seed dry and hard, pods dry and black

3 Pod senescence
301 10% pods dry and black
•
•
305 50% pods dry and black
•
•
308 80% pods dry and black, some upper pods green
309 90% pods dry and black, most seed dry. Desiccation stage.
310 All pods dry and black, seed hard. Pre-harvest (glyphosate application stage)

4 Stem senescence
401 10% stem brown/black
•
•
405 50% stem brown/black
•
•
409 90% stem brown/black
410 All stems brown/black. All pods dry and black, seed hard.

(From Knott, 1990. *Annals of Applied Biology*, **116**, 391–404)

Stages in Development of Potato

Illustrations of these growth stages can be found in the reference indicated below and in some company product manuals.

0 Seed germination and seedling emergence
- 000 Dry seed
- 001 Imbibed seed
- 002 Radicle apparent
- 003 Elongation of hypocotyl
- 004 Seedling emergence
- 005 Cotyledons unfolded

1 Tuber dormancy
- 100 Innate dormancy (no sprout development under favourable conditions)
- 150 Enforced dormancy (sprout development inhibited by environmental conditions)

2 Tuber sprouting
- 200 Dormancy break, sprout development visible
- 21x Sprout with 1 node
- 22x Sprout with 2 nodes
- •
- •
- 29x Sprout with 9 nodes
- 21x(2) Second generation sprout with 1 node
- 22x(2) Second generation sprout with 2 nodes
- •
- •
- 29x(2) Second generation sprout with 9 nodes

Where x = 1, sprout <2 mm;
2, 2–5 mm; 3, 5–20 mm;
4, 20–30 mm; 5, 50–100 mm;
6, 100–150 mm long

3 Emergence and shoot expansion
- 300 Main stem emergence
- 301 Node 1
- 302 Node 2
- •
- •
- 319 Node 19

Second order branch
- 321 Node 1
- •
- •

Nth order branch
- 3N1 Node 1
- •
- •
- 3N9 Node 9

4 Flowering

Primary flower
- 400 No flowers
- 410 Appearance of flower bud
- 420 Flower unopen
- 430 Flower open
- 440 Flower closed
- 450 Berry swelling
- 460 Mature berry

Second order flowers
- 410(2) Appearance of flower bud
- 420(2) Flower unopen
- 430(2) Flower open
- 440(2) Flower closed
- 450(2) Berry swelling
- 460(2) Mature berry

5 Tuber development
- 500 No stolons
- 510 Stolon initials
- 520 Stolon elongation
- 530 Tuber initiation
- 540 Tuber bulking (> 10 mm diam)
- 550 Skin set
- 560 Stolon development

6 Senescence
- 600 Onset of yellowing
- 650 Half leaves yellow
- 670 Yellowing of stems
- 690 Completely dead

(From Jefferies & Lawson, 1991. *Annals of Applied Biology*, **119**, 387–389)

Stages in Development of Linseed

Illustrations of these growth stages can be found in the reference indicated below and in some company product manuals.

0 Seed germination and seedling emergence
- 00 Dry seed
- 01 Imbibed seed
- 02 Radicle apparent
- 04 Hypocotyl extending
- 05 Emergence
- 07 Cotyledon unfolding from seed case
- 09 Cotyledons unfolded and fully expanded

1 Vegetative stage (of main stem)
- 10 True leaves visible
- 12 First pair of true leaves fully expanded
- 13 Third pair of true leaves fully expanded
- 1n n leaf fully expanded

2 Basal branching
- 21 One branch
- 22 Two branches
- 23 Three branches
- 2n n branches

3 Flower bud development (on main stem)
- 31 Enclosed bud visible in leaf axils
- 33 Bud extending from axil
- 35 Corymb formed
- 37 Buds enclosed but petals visible
- 39 First flower open

4 Flowering (whole plant)
- 41 10% of flowers open
- 43 30% of flowers open
- 45 50% of flowers open
- 49 End of flowering

5 Capsule formation (whole plant)
- 51 10% of capsules formed
- 53 30% of capsules formed
- 55 50% of capsules formed
- 59 End of capsule formation

6 Capsule senescence (on most advanced plant)
- 61 Capsules expanding
- 63 Capsules green and full size
- 65 Capsules turning yellow
- 67 Capsules all yellow brown but soft
- 69 Capsules brown, dry and senesced

7 Stem senescence (whole plant)
- 71 Stems mostly green below panicle
- 73 Most stems 30% brown
- 75 Most stems 50% brown
- 77 Stems 75% brown
- 79 Stems completely brown

8 Stems rotting (retting)
- 81 Other tissue rotting
- 85 Vascular tissue easily removed
- 89 Stems completely collapsed

9 Seed development (whole plant)
- 91 Seeds expanding
- 92 Seeds white but full size
- 93 Most seeds turning ivory yellow
- 94 Most seeds turning brown
- 95 All seeds brown and hard
- 98 Some seeds shed from capsule
- 99 Most seeds shed from capsule

(From Freer, 1991. *Aspects of Applied Biology*, **28**, 33–40)

Stages in Development of Annual Grass Weeds

Illustrations of these growth stages can be found in the reference indicated below and in some company product manuals.

0 Germination and emergence
- 00 Dry seed
- 01 Start of imbibition
- 03 Imbibition complete
- 05 Radicle emerged from caryopsis
- 07 Coleoptile emerged from caryopsis
- 09 Leaf just at coleoptile tip

1 Seedling growth
- 10 First leaf through coleoptile
- 11 First leaf unfolded
- 12 2 leaves unfolded
- 13 3 leaves unfolded
- 14 4 leaves unfolded
- 15 5 leaves unfolded
- 16 6 leaves unfolded
- 17 7 leaves unfolded
- 18 8 leaves unfolded
- 19 9 or more leaves unfolded

2 Tillering
- 20 Main shoot only
- 21 Main shoot and 1 tiller
- 22 Main shoot and 2 tillers
- 23 Main shoot and 3 tillers
- 24 Main shoot and 4 tillers
- 25 Main shoot and 5 tillers
- 26 Main shoot and 6 tillers
- 27 Main shoot and 7 tillers
- 28 Main shoot and 8 tillers
- 29 Main shoot and 9 or more tillers

3 Stem elongation
- 31 First node detectable
- 32 2nd node detectable
- 33 3rd node detectable
- 34 4th node detectable
- 35 5th node detectable
- 36 6th node detectable
- 37 Flag leaf just visible
- 39 Flag leaf ligule just visible

4 Booting
- 41 Flag leaf sheath extending
- 43 Boots just visibly swollen
- 45 Boots swollen
- 47 Flag leaf sheath opening
- 49 First awns visible

5 Inflorescence emergence
- 51 First spikelet of inflorescence just visible
- 53 ¼ of inflorescence emerged
- 55 ½ of inflorescence emerged
- 57 ¾ of inflorescence emerged
- 59 Emergence of inflorescence completed

6 Anthesis
- 61 Beginning of anthesis
- 65 Anthesis half-way
- 69 Anthesis complete

(From Lawson & Read, 1992. *Annals of Applied Biology*, **12**, 211–214)

Growth Stages of Annual Broad-leaved Weeds
Preferred Descriptive Phrases

Illustrations of these growth stages can be found in the reference indicated below and in some company product manuals.

Pre-emergence
Early cotyledons
Expanded cotyledons
One expanded true leaf
Two expanded true leaves
Four expanded true leaves
Six expanded true leaves
Plants up to 25 mm across/high
Plants up to 50 mm across/high
Plants up to 100 mm across/high
Plants up to 150 mm across/high
Plants up to 250 mm across/high
Flower buds visible
Plant flowering
Plant senescent

(From Lutman & Tucker, 1987. *Annals of Applied Biology*, **110**, 683–687)

Appendix 4
A. KEY TO LABEL REQUIREMENTS FOR PROTECTIVE CLOTHING

COSHH requires that exposure of anyone, including members of the public, who may be affected by work involving substances hazardous to health be either prevented or, where this is not reasonably practicable, adequately controlled. Engineering or other control measures should be used in preference to personal protective equipment (PPE). However, in the case of pesticides, spray operators are clearly most at risk from exposure and PPE will usually be needed in addition to engineering controls to achieve adequate control of exposure. Even if COSHH does not apply (because the product in use is not classed as hazardous), employers may still have responsibilities under the Protective Equipment at Work Regulations 1992.

European Council Directive 86/686/EEC defines the requirements for the use of PPE and these are enacted in UK under the Personal Protective Equipment (EC Directive) Regulations 1992. The Regulations require that any PPE made or imported after 1 July 1995 must meet the requirements set out in the Directive and be CE-marked. Older PPE may continue to be used provided it gives adequate protection and is properly maintained.

Where a product label specifies the use of PPE, the requirements are listed in the profiles under the heading **Protective clothing/label precautions**, using letter codes to denote the protective items, according to the list below. Often PPE requirements are different for specified operations, e.g. handling the concentrate, cleaning equipment etc., but it is not possible to list them separately. The lists of PPE are therefore an indication of what the user may require to have available to use the product in different ways. **When making a COSHH assessment it is therefore essential that the product label is consulted for information on the particular use that is being assessed.**

- A Suitable protective gloves (the product label should be consulted for any specific requirements about the material of which the gloves should be made)
- B Rubber gauntlet gloves
- C Face-shield
- D Approved respiratory protective equipment
- E Goggles
- F Dust mask
- G Full face-piece respirator
- H Coverall
- J Hood
- K Rubber apron
- L Waterproof coat
- M Rubber boots
- N Waterproof jacket and trousers
- P Suitable protective clothing

Appendix 4
B. KEY TO NUMBERED LABEL PRECAUTIONS

The series of numbers listed in the profile refer to the numbered precautions below. Where the generalised wording includes a phrase such as "... for xx days" the specific requirement for each pesticide is given in the **Special precautions/Environmental safety** section of the profile.

Medical Advice

M01 This product contains an anticholinesterase organophosphorus compound. DO NOT USE if under medical advice NOT to work with such compounds
M02 This product contains an anticholinesterase carbamate compound. DO NOT USE if under medical advice NOT to work with such compounds
M03 If you feel unwell, seek medical advice (show the label where possible)
M04 In case of accident or if you feel unwell, seek medical advice immediately (show the label where possible)
M05 If swallowed, seek medical advice immediately and show this container or label

Risk Phrases

R01 Very toxic
R01a Very toxic: In contact with skin
R01b Very toxic: By inhalation
R01c Very toxic: If swallowed
R02 Toxic
R02a Toxic: In contact with skin
R02b Toxic: By inhalation
R02c Toxic: If swallowed
R03 Harmful
R03a Harmful: In contact with skin
R03b Harmful: By inhalation
R03c Harmful: If swallowed
R04 Irritant
R04a Irritating to eyes
R04b Irritating to skin
R04c Irritating to respiratory system
R04d Risk of serious damage to eyes
R04e May cause sensitization by skin contact
R04f May cause sensitization by inhalation
R05 Corrosive
R05a Causes severe burns
R05b Causes burns

VERY TOXIC

TOXIC

HARMFUL

IRRITANT

CORROSIVE

R06 Danger of serious damage to health by prolonged exposure
R07a Extremely flammable liquefied gas
R07b Extremely flammable
R07c Highly flammable
R07d Flammable
R08 Oxidising agent. Contact with combustible material may cause fire
R09 Explosive when mixed with oxidising substances

User Safety Precautions

U01 To be used only by operators instructed or trained in the use of chemical/product/type of produce and familiar with the precautionary measures to be observed
U02 Wash all protective clothing thoroughly after use, especially the inside of gloves/Avoid excessive contamination of coveralls and launder regularly
U03 Wash splashes off gloves immediately
U04a Take off immediately all contaminated clothing
U04b Take off immediately all contaminated clothing and wash underlying skin. Wash clothes before re-use
U05a When using do not eat, drink or smoke
U05b When using do not eat, drink , smoke or use naked lights
U06 Handle with care and mix only in a closed container
U07 Open container only as directed
U08 Wash concentrate/dust from skin or eyes immediately
U09a Wash any contamination/splashes/dust/powder/concentrate from skin or eyes immediately
U09b Wash any contamination/splashes/dust/powder/concentrate from eyes immediately
U10 After contact with skin or eyes wash immediately with plenty of water
U11 In case of contact with eyes rinse immediately with plenty of water and seek medical advice
U12 In case of contact with skin rinse immediately with plenty of water and seek medical advice
U13 Avoid all contact by mouth
U14 Avoid all contact with skin
U15 Avoid all contact with eyes
U16 Ensure adequate ventilation in confined spaces
U17 Ventilate treated areas thoroughly when smoke has cleared/Ventilate treated rooms thoroughly before occupying
U18 Extinguish all naked flames, including pilot lights, when applying the fumigant/dust/liquid/product
U19 Do not breathe dust/fog/fumes/gas/smoke/spray mist/vapour. Avoid working in spray mist
U20a Wash hands and exposed skin before eating, drinking or smoking and after work
U20b Wash hands and exposed skin before meals and after work
U20c Wash hands before meals and after work
U21 Before entering treated crops, cover exposed skin areas, particularly arms and legs
U22 Do not touch sachet with wet hands or gloves/Do not touch water soluble bag directly

Consumer Safety Precautions

C01 Not to be used on food crops
C02 Do not harvest for human or animal consumption for at least xx days/weeks after last application
C03 Keep away from food, drink and animal feeding-stuffs
C04 Do not apply to surfaces on which food/feed is stored, prepared or eaten
C05 Remove/cover all foodstuffs before application
C06 Remove exposed milk before application
C07 Collect eggs before application
C08 Protect food preparing equipment and eating utensils from contamination during application
C09 Cover water storage tanks before application
C10 Protect exposed water/feed/milk machinery/milk containers from contamination

Environmental Safety Precautions (Public, Livestock, Wildlife etc.)

E01 Keep out of reach of children
E02 Keep unprotected persons out of treated areas for at least xx hours/days
E03 Label treated seed with the appropriate precautions, using the printed sacks, labels or bag tags supplied
E04 Remove all pets/livestock/fish tanks before treatment/spraying
E05 Do not apply directly to livestock/poultry
E05a Keep poultry out of treated areas for at least xx days/weeks
E06a Dangerous to livestock. Keep all livestock out of treated areas/away from treated water for at least xx days/weeks. Bury or remove spillages
E06b Harmful to livestock. Keep all livestock out of treated areas/away from treated water for at least xx days/weeks. Bury or remove spillages
E07 Keep livestock out of treated areas/Keep livestock out of treated areas for at least xx weeks and until foliage of any poisonous weeds such as ragwort has died and become unpalatable
E08 Do not feed treated straw or haulm to livestock within x days of spraying
E09 Do not use on crops if the straw is to be used as animal feed/bedding
E10a Dangerous to game, wild birds and animals
E10b Harmful to game, wild birds and animals
E11 Harmful to animals. Paraquat may be harmful to hares, where possible spray stubbles early in the day
E12a High risk to bees. Do not apply to crops in flower or to those in which bees are actively foraging. Do not apply when flowering weeds are present
E12b Extremely dangerous to bees. Do not apply at flowering stage. Keep down flowering weeds/Do not apply to crops in flower or to those in which bees are actively foraging. Do not apply when flowering weeds are present
E12c Dangerous to bees. Do not apply at flowering stage. Keep down flowering weeds/Do not apply to crops in flower or to those in which bees are actively foraging. Do not apply when flowering weeds are present
E12d Dangerous to bees. Do not apply to crops in flower or to those in which bees are actively foraging except as directed on [crop]. Do not apply when flowering weeds are present
E12e Harmful to bees. Do not apply at flowering stage. Keep down flowering weeds/ Do not apply to crops in flower or to those in which bees are actively foraging. Do not apply when flowering weeds are present
E13a Extremely dangerous to fish. Do not contaminate ponds, waterways or ditches with chemical or used container/Extremely dangerous to fish or other aquatic life. Do not contaminate surface waters or ditches with chemical or used container
E13b Dangerous to fish. Do not contaminate ponds, waterways or ditches with chemical or used container/Dangerous to fish or other aquatic life. Do not contaminate surface waters or ditches with chemical or used container
E13c Harmful to fish. Do not contaminate ponds, waterways or ditches with chemical or used container/Harmful to fish or other aquatic life. Do not contaminate surface waters or ditches with chemical or used container
E14a Extremely dangerous to aquatic higher plants. Do not contaminate surface waters or ditches with chemical or used container
E14b Dangerous to aquatic higher plants. Do not contaminate surface waters or ditches with chemical or used container
E15 Do not contaminate ponds, waterways or ditches with chemical or used container/Do not contaminate surface waters or ditches with chemical or used container
E16 Do not allow direct spray from vehicle mounted/drawn hydraulic sprayers to fall within 6m of surface waters or ditches/Do not allow direct spray from hand-held sprayers to fall within 2m of surface waters or ditches. Direct spray away from water

E17 Do not allow direct spray from broadcast air-assisted sprayers to fall within 18 m of surface waters or ditches. Direct spray away from water
E18 Do not spray from the air within 250 m horizontal distance of surface waters or ditches
E19 Do not dump surplus herbicide in water or ditch bottoms
E20 Prevent any surface run-off from entering storm drains
E21 Do not use treated water for irrigation purposes within xx days/weeks of treatment
E22a High risk to non-target insects or other arthropods. Do not spray within 6m of the field boundary
E22b Risk to non-target insects or other arthropods. See directions for use
E23 Avoid damage by drift onto susceptible crops or water courses
E24 Store away from seeds, fertilizers, fungicides and insecticides
E25 Store well away from corms, bulbs, tubers and seeds
E26 Store away from frost
E27 Store away from heat
E28 Flammable. Do not store near heat or open flame
E29 Store under cool, dry conditions
E30a Store/keep in original container, tightly closed, in a safe place
E30b Store/keep in original container, tightly closed, in a safe place, under lock and key
E31a Wash out container thoroughly and dispose of safely
E31b Wash out container thoroughly, empty washings into spray tank and dispose of safely
E31c Rinse container thoroughly by using an integrated pressure rinsing device or manually rinsing three times. Add washings to sprayer at time of filling and dispose of container safely
E32a Empty container completely and dispose of safely
E32b Empty container completely and dispose of it in the specified manner
E33 Return empty container as instructed by supplier
E34 Do not re-use this container for any purpose
E35 Do not burn this container
E36 Do not rinse out the container

Sack Label Precautions for Treated Seed

S01 Do not handle seed unnecessarily
S02 Do not use treated seed as food or feed
S03 Keep treated seed secure from people, domestic stock/pets and wildlife at all times during storage and use
S04a Bury or remove spillages
S04b Harmful to game and wild life. Bury spillages/Bury or remove spillages
S05 Do not re-use sack for food or feed/Do not reuse sacks or containers that have been used for treated seed for food or feed
S06 Wash hands and exposed skin before meals and after work
S07 Do not apply treated seed from the air

Vertebrate/Rodent Control Product Precautions

V01a Prevent access to baits/powder by children, birds and other animals, particularly cats, dogs, pigs and poultry
V01b Prevent access to bait/gel/dust by children, birds and non-target animals, particularly dogs, cats, pigs, poultry
V02 Do not prepare/use/lay baits/dust/spray where food/feed/water could become contaminated
V03a Remove all remains of bait, tracking powder or bait containers after use and burn or bury

V03b Remove all remains of bait and bait containers/exposed dust/after treatment (except where used in sewers) and dispose of safely (e.g. burn/bury). Do not dispose of in refuse sacks or on open rubbish tips.

V04a Search for and burn or bury all rodent bodies. Do not place in refuse bins or on rubbish tips

V04b Search for rodent bodies (except where used in sewers) and dispose of safely (e.g. burn/bury). Do not dispose of in refuse sacks or on open rubbish tips

V04c Dispose of safely any rodent bodies and remains of bait and bait containers that are recovered after treatment (e.g. burn/bury). Do not dispose of in refuse sacks or on open rubbish tips

V05 Use bait containers clearly marked POISON at all surface baiting points

Appendix 5
KEY TO ABBREVIATIONS AND ACRONYMS

The abbreviations of formulation types in the following list are used in Section 4 (Pesticide Profiles) and are derived from the Catalogue of Pesticide Formulation Types (GCPF Technical Monograph 2, 1989).

1 Formulation Types

AE	Aerosol generator
AL	Other liquids to be applied undiluted
BB	Block bait
CB	Bait concentrate
CG	Encapsulated granule (controlled release)
CR	Crystals
CS	Capsule suspension
DP	Dustable powder
DS	Powder for dry seed treatment
EC	Emulsifiable concentrate
ES	Emulsion for seed treatment
EW	Oil in water emulsion
FG	Fine granules
FP	Smoke cartridge
FS	Flowable concentrate for seed treatment
FT	Smoke tablet
FU	Smoke generator
FW	Smoke pellets
GA	Gas
GB	Granular bait
GE	Gas-generating product
GG	Macrogranules
GP	Flo-dust
GR	Granules
GS	Grease
HN	Hot fogging concentrate
KK	Combi-pack (solid/liquid)
KL	Combi-pack (liquid/liquid)
KN	Cold-fogging concentrate
KP	Combi-pack (solid/solid)
LA	Lacquer
LI	Liquid, unspecified
LS	Solution for seed treatment
ME	Microemulsion
MG	Microgranules
OL	Oil miscible liquid
PA	Paste
PC	Gel or paste concentrate
PS	Seed coated with a pesticide
PT	Pellet
RB	Ready-to-use bait
RH	Ready-to-use spray in hand-operated sprayer

SA Sand
SC Suspension concentrate (= flowable)
SE Suspo-emulsion
SG Water soluble granules
SL Soluble concentrate
SP Water soluble powder
SS Water soluble powder for seed treatment
SU Ultra low-volume suspension
TB Tablets
TC Technical material
TP Tracking powder
UL Ultra low-volume liquid
VP Vapour releasing product
WB Water soluble bags
WG Water dispersible granules
WP Wettable powder
WS Water dispersible powder for slurry treatment of s
XX Other formulations

2 Other Abbreviations and Acronyms

ACP Advisory Committee on Pesticides
ACTS Advisory Committee on Toxic Substances
ADAS Agricultural Development and Advisory Service
a.i. active ingredient
BAA British Agrochemicals Association
CDA controlled droplet application
cm centimetre
COPR Control of Pesticides Regulations 1986
COSHH Control of Substances Hazardous to Health Regulations
d day(s)
EA Environment Agency
EBDC ethylene-bis-dithiocarbamate fungicide
FEPA Food and Environment Protection Act 1985
g gram(s)
GCPF Global Crop Protection Federation
GS growth stage (but also "grease" as a formulation)
h hour(s)
ha hectare(s)
HBN hydroxybenzonitrile herbicide
HI harvest interval
HMIP HM Inspectorate of Pollution
HSE Health and Safety Executive
ICM integrated crop management
IPM integrated pest management
kg kilogram(s)
litre(s)
m metre(s)
MAFF Ministry of Agriculture, Fisheries and Food
MBC methyl benzimidazole carbamate fungicide
MEL Maximum Exposure Limit
min minute(s)
mm millimetre(s)
MRL Maximum Residue Level

mth	month(s)
NA	notice of approval
NFU	National Farmers Union
OES	Occupational Exposure Standard
OLA	off-label approval
PPE	personal protective equipment
PPPR	Plant Protection Products Regulations 1995
PR	The Pesticides Register
PSD	The Pesticides Safety Directorate
SOLA	specific off-label approval
ULV	ultra-low volume
wk	week(s)
w/v	weight/volume
w/w	weight/weight
yr	year(s)

INDEX OF PROPRIETARY NAMES OF PESTICIDES

The references are to entry numbers, not pages. Adjuvant names are referred to as *Adj* and are listed separately in Section 2.

The references are to entry numbers, not pages.

The references are to entry numbers, not pages.

The references are to entry numbers, not pages.

The references are to entry numbers, not pages.

The references are to entry numbers, not pages.

The references are to entry numbers, not pages.

The references are to entry numbers, not pages.

The references are to entry numbers, not pages.

The references are to entry numbers, not pages.

The references are to entry numbers, not pages.

The references are to entry numbers, not pages.

THE UK PESTICIDE GUIDE 1999
RE-ORDER FORM

Surname ________________________________Initials __________ Dr/Mr/Mrs/Ms __________

Address ____

____

Postcode ____________________ Country ____________________________________

Date ____________________ Signed ____

Please send me ________ more copies of the *The UK Pesticide Guide 1999* at £22.50 each

Please send me ________ copies of the *The Electronic UK Pesticide Guide 1999* at £60.00 each (£70.50 incl VAT), standalone version*

Please send me ________ copies of the package set of CD-ROM and book at £70.00 each (£78.31 incl VAT charged on book only)

* *The Electronic UK Pesticide Guide* can be networked. Please call for pricing information.

❒ Payment enclosed, cheques made payable to BCPE Ltd.

❒ Please invoice (please attach your official company order)

❒ Please debit my Access/Mastercard/Eurocard/Visa Card/American Express Account

by the sum of ______________ Name of cardholder ______________________________

Card no ____Expiry date ______________

Address of cardholder (if different from above) ______________________________

____

OR TELEPHONE YOUR ORDER NOW AND ASK FOR THE SALES DEPARTMENT

Discount for bulk sales of book

No. of copies	Discount
100+	30%
50 – 99	25%
10 – 49	15%
1 – 9	No Discount

2000 EDITION

❒ Please send me advance price details for the 2000 edition of *The UK Pesticide Guide* as soon as they are available.

Please detach and return to:

BCPC Publications Sales
Bear Farm, Binfield, Bracknell, Berkshire RG42 5QE
Telephone: 0118 934 2727
Fax: 0118 934 1998